Springer Proceedings in Complexity

Springer Proceedings in Complexity publishes proceedings from scholarly meetings on all topics relating to the interdisciplinary studies of complex systems science. Springer welcomes book ideas from authors. The series is indexed in Scopus.

Proposals must include the following:

- name, place and date of the scientific meeting
- a link to the committees (local organization, international advisors etc.)
- scientific description of the meeting
- list of invited/plenary speakers
- an estimate of the planned proceedings book parameters (number of pages/articles, requested number of bulk copies, submission deadline)

Submit your proposals to: Hisako.Niko@springer.com

Santo Banerjee · Asit Saha
Editors

Nonlinear Dynamics and Applications

Proceedings of the ICNDA 2022

Volume 1

Editors
Santo Banerjee
Department of Mathematics
Politecnico di Torino
Torino, Italy

Asit Saha
Sikkim Manipal Institute of Technology
Sikkim Manipal University
East-Sikkim, India

ISSN 2213-8684 ISSN 2213-8692 (electronic)
Springer Proceedings in Complexity
ISBN 978-3-030-99794-6 ISBN 978-3-030-99792-2 (eBook)
https://doi.org/10.1007/978-3-030-99792-2

This Springer imprint is published by the registered company Springer Nature Switzerland AG
The registered company address is: Gewerbestrasse 11, 6330 Cham, Switzerland

Contents

Graphs, Networks and Communication

Fractional System and Applications

Artificial Intelligence, Internet of Things and Smart Learning

Contributors

Abdikian A. Department of Physics, Malayer University, Malayer, 65719-95863, Iran

Ajith K. M. National Institute of Technology Karnataka, Mangaluru, India

Akleylek Sedat Ondokuz Mayıs University, Samsun, Turkey

Alam Shariful Indian Institute of Engineering Science and Technology, Shibpur, Howrah, West Bengal, India

Ali Demir Mehmet Computer Engineering Department, Kocaeli University, Kocaeli, Turkey

Amrutha R. Department of Physics, KCG College of Technology, Chennai, India

Amrutha S. V. Department of Physics, National Institute of Technology Karnataka, Surathkal, Mangalore, Karnataka, India

Ananthan Rohit Mechanical Engineering, Indian Institute of Information Technology, Design and Manufacturing, Kancheepuram, Chennai, India

Anh Mai The Vinh University, Vinh city, Nghean, Vietnam

Anshu Department of Mathematics, BITS Pilani, Pilani, Rajasthan, India

Ashwani School of Basic Sciences, Indian Institute of Technology Mandi, Mandi, Himachal Pradesh, India

Aslam Humaira Department of Mathematics, Adamas University, Kolkata, India

Atteya A. Department of Physics, Faculty of Science, Alexandria University, Alexandria, Egypt

Bandyopadhyay Anup Department of Mathematics, Jadavpur University, Kolkata, India

Banerjee Saayan Centre for Engineering Analysis and Design, Combat Vehicles R&D Establishment, DRDO, New Delhi, India

Barman Dipesh Indian Institute of Engineering Science and Technology, Shibpur, Howrah, West Bengal, India

Baroi Juhi School of Sciences, Christ (Deemed to Be University) Delhi NCR, Ghaziabad, India

Bhardwaj S. B. Department of Physics, SUS Govt. College Matak-Majri, Karnal, India

Bishi Binaya K. Department of Mathematics, Lovely Professional University, Jalandhar, Phagwara, Panjab, India;
Department of Mathematical Sciences, University of Zululand, Kwa-Dlangezwa, South Africa

Biswas Santanu Department of Mathematics, Adamas University, Kolkata, India;
Department of Mathematics, Jadavpur University, Kolkata, India

Borah Joydeep Department of Mathematics, D. D. R. College, Chabua, AS, India

Borah Rupjyoti Department of Mathematics, Dibrugarh University, Dibrugarh, AS, India

Boral Rony Department of Physics, Indian Institute of Technology Guwahati, Guwahati, Assam, India

Büyüksaraçoğlu Sakallı Fatma Trakya University, Edirne, Turkey

Chakrabarty Siddhartha P. Indian Institute of Technology Guwahati, Guwahati, Assam, India

Chand Fakir Department of Physics, Kurukshetra University, Kurukshetra, India

Chanda Koustav L&T Infotech, Ranaghat, India

Chandra Swarniv Jadavpur University, Kolkata, India;
Department of Physics, Govt. General Degree College at Kushmandi, Dakshin Dinajpur, India;
Institute of Natural Sciences and Applied Technology, Kolkata, India

Chandrashekhar G. Department of Mathematics, Osmania University, Hyderabad, TG, India

Chatterjee Prasanta Department of Mathematics, Siksha Bhavana, Visva-Bharati, Santiniketan, Santiniketan, West Bengal, India

Chaube Mithilesh Kumar Dr. SPM IIIT Naya Raipur, C.G., Raipur, India

Chaudhuri Kripasindhu Department of Mathematics, Jadavpur University, Kolkata, India

Chukwuma Victor Iheanacho Institute of Radioelectronics, Riga Technical University, Riga, Latvia

Chuong Le Van Vinh University, Vinh city, Nghean, Vietnam

Cuong Ngo Tri Nam Systemtec JSC, Hanoi, Vietnam

Das Ankush Department of Electrical Engineering, JIS College of Engineering, Kalyani, West Bengal, India

Das Apurba Narayan Department of Mathematics, Alipurduar University, Alipurduar, West Bengal, India

Das Chinmay Department of Mathematics, Govt. General Degree College at Kushmandi, Dakshin Dinajpur, India;
Institute of Natural Sciences and Applied Technology, Kolkata, India

Das Debabrata Cooch Behar Panchanan Barma University, Cooch Behar, India

Das Rituparna Department of Mathematics, Pandu College, Guwahati, India

Das Satyajit Department of Mathematics, Adamas University, Kolkata, India

Das Susmita Cooch Behar Panchanan Barma University, Cooch Behar, India

Deb Biswajit Department of Mathematics, Sikkim Manipal Institute of Technology, Sikkim Manipal University, East-Sikkim, India

Dehingia Hirak Jyoti Department of Mathematics, Dibrugarh University, Dibrugarh, Assam, India

Deka P. N. Department of Mathematics, Dibrugarh University, Dibrugarh, Assam, India

Devesh M. School of Mechanical Engineering, Vellore Institute of Technology, Chennai, India

Dey Mondal Sukarna Department of Mathematics, Dr. B.C. Roy Engineering College, MAKAUT, Kolkata, West Bengal, India

Dey Debasish Department of Mathematics, Dibrugarh University, Dibrugarh, AS, India

Dey Dipankar Depatment of Computer Science and Technology, Global Institute of Science and Technology, Haldia, WB, India

Dhar Subham Cooch Behar Panchanan Barma University, Cooch Behar, WB, India

Donnarumma Davide Università di Torino, Torino, Italy

Doss M. Arockia Department of Chemistry, St. Joseph University, Dimapur, India

Dubey Balram Department of Mathematics, BITS Pilani, Pilani Campus, Pilani, Rajasthan, India

Dutta Dharitree Department of Physics, Anandaram Dhekial Phookan College, Nagaon, Assam, India

Fataf Nur Aisyah Binti Abdul Cyber Security Centre, National Defence University of Malaysia (NDUM), Kuala Lumpur, Malaysia

Félix-Beltrán O. Faculty of Electronics Sciences, Benemérita Universidad Autónoma de Puebla, Puebla, Mexico

Gandhi Rishu Department of Mathematics, Birla Institute of Technology and Science, Pilani, Rajasthan, India

Gangele Rajkumar Department of Mathematics and Statistics, Dr. Harisingh Gour Vishwavidyalaya, Sagar, Madhya Pradesh, India

Gangwar Sanu Kumar Department of Physics, Indian Institute of Technology, Guwahati, Assam, India

Gangwar Sonali Department of Physics, Indian Institute of Technology, Guwahati, Assam, India

García López Álvaro Universidad Rey Juan Carlos, Madrid, Spain

Gautam Anjan Department of Mathematics, Sikkim Manipal Institute of Technology, Sikkim Manipal University, East-Sikkim, India

Gautam Manjeet Singh Department of Physics, Government College, Jind, India

Ghosh Aditya Department of Mathematics, Adamas University, Kolkata, India

Ghosh Basudev Jadavpur University, Kolkata, India

Ghosh S. S. Indian Institute of Geomagnetism, Mumbai, India

Ghosh Sabyasachi Indian Institute of Technology Bhilai, Sejbahar, Raipur, Chhattisgarh, India

Gomathy G. Department of Mathematics, School of Advanced Sciences, Vellore Institute of Technology, Vellore, Tamilnadu, India

Goswami Damodar Prasad Asutosh Mookerjee Memorial Institute, Sivotosh Mookerjee Science Centre, Kolkata, India

Goswami Jyotirmoy Jadavpur University, Kolkata, India

Goswami K. S. Centre of Plasma Physics—Institute for Plasma Research, Assam, India

Goswami Pinkimani Department of Mathematics, University of Science and Technology Meghalaya, Ri-Bhoi, ML, India

Gowrisankar A. Department of Mathematics, School of Advanced Sciences, Vellore Institute of Technology, Vellore, Tamil Nadu, India

Gupta Ashvini Department of Mathematics, BITS Pilani, Pilani Campus, Pilani, Rajasthan, India

Gupta Nishi National Institute of Technology Silchar, Silchar, Assam, India

Hassan M. K. Department of Physics, University of Dhaka, Dhaka, Bangladesh

Ipatovs Aleksandrs Institute of Radioelectronics, Riga Technical University, Riga, Latvia

Jana Chandan Department of Applied Mathematics with Oceanology and Computer Programming, Vidyasagar University, Midnapore, West Bengal, India

Jana Soovoojeet Department of Mathematics, Ramsaday College, Amta, Howrah, India

Jeppu Yogananda Honeywell Technology Solutions, Hyderabad, Telangana, India

Joseph Subin P. Government Engineering College, Wayanad, Thalapuzha, Kerala, India

Joshi Sudhir University of Petroleum & Energy Studies, Dehradun, Uttarakhand, India

Kairi Rishi Raj Department of Mathematics, Cooch Behar Panchanan Barma University, Cooch Behar, India

Kalita Pankaj School of Energy Science and Engineering, Indian Institute of Technology Guwahati, Guwahati, India

Kar T. K. Department of Mathematics, Indian Institute of Engineering Science and Technology, Shibpur, Howrah, India

Karakoc Seydi Battal Gazi Department of Mathematics, Faculty of Science and Art, Nevsehir Haci Bektas Veli University, Nevsehir, Turkey

Karmakar Mithun Department of CSE, CIT Kokrajhar, Kokrajhar, Assam, India

Kaur Manveet Department of Physics, Guru Nanak Dev University, Amritsar, India

Kaur Rajneet Department of Physics, Guru Nanak Dev University, Amritsar, India

Kavitha L. Department of Physics, School of Basic and Applied Sciences, Central University of Tamil Nadu, Thiruvarur, Tamil Nadu, India;
The Abdus Salam International Centre for Theoretical Physics, Trieste, Italy

Khan Aftab Department of Electrical Engineering, JIS College of Engineering, Kalyani, West Bengal, India

Khandale K. Y. Department of Physics, Shivaji University, Kolhapur, India

Khanduri Umesh Department of Mathematics, Birla Institute of Technology and Science, Pilani, Rajasthan, India

Khare Shivani Department of Mathematics and Statistics, Dr. Harisingh Gour Vishwavidyalaya, Sagar, Madhya Pradesh, India

Khasim Ahamed National Institute of Technology Karnataka, Mangaluru, India

Khatua Anupam Department of Mathematics, Indian Institute of Engineering Science and Technology, Shibpur, Howrah, India

Kodi Raghunath Department of Humanities and Sciences (Mathematics), Bheema Institute of Technology and Science, Adoni, AP, India

Kostić Srđan Geology Department, Jaroslav Černi Water Institute, Belgrade, Serbia

Kulkarni Vinayak School of Energy Science and Engineering, Indian Institute of Technology Guwahati, Guwahati, India

Kumar Bose Goutam Department of Mechanical Engineering, Haldia Institute of Technology, Haldia, West Bengal, India

Kumar Dey Pabitra Department of Computer Applications, Dr. B.C. Roy Engineering College, MAKAUT, Durgapur, West Bengal, India

Kumar Jha Saket Department of Electronics and Communication Engineering, Sikkim Manipal Institute of Technology, Sikkim Manipal University, Majitar, Sikkim, India

Kumar Maiti Dilip Department of Applied Mathematics with Oceanology and Computer Programming, Vidyasagar University, Midnapore, West Bengal, India

Kumar Mishra Mohit Department of Electronics and Communication Engineering, Manipal University, Jaipur, India

Kumar Sarkar Ashis Department of Mathematics, Jadavpur University, Kolkata, India

Kumar Abhishek Department of Electrical Engineering, Manipal University Jaipur, Jaipur, Rajasthan, India;
Department of Aerospace Engineering, University of Petroleum and Energy Studies, Dehradun, India

Kumar Ajay Department of Mathematics and Humanities, Sardar Vallabhbhai National Institute of Technology, Surat, Gujarat, India

Kumar Anand Department of Physics, Chaudhary Ranbir Singh University, Jind, India

Kumar Ankit Department of Mathematics, BITS Pilani, Pilani, Rajasthan, India

Kumar Ayush Department of Electronics and Communication Engineering, Sikkim Manipal Institute of Technology, Sikkim Manipal University, Majitar, Sikkim, India

Kumar B. Rushi Department of Mathematics, SAS, Vellore Institution of Technology, Vellore, Tamil Nadu, India

Kumar Bhagwan Department of Mathematics and Statistics, Dr. Harisingh Gour Vishwavidyalaya, Sagar, Madhya Pradesh, India

Kumar Hitender Department of Physics, Government College for Women, Gharaunda, India

Kumar Santosh Dr. SPM IIIT Naya Raipur, C.G., Raipur, India

Kumar Vijil Department of Mathematics and Computing, Indian Institute of Technology (ISM), Dhanbad, Jharkhand, India

Kumar Vinod Department of Physics, Chaudhary Devi Lal University, Sirsa, India

Kumari Agin Department of Mathematics, Chaudhary Bansi Lal University, Bhiwani, India

Kurt Pehlivanoğlu Meltem Kocaeli University, Kocaeli, Turkey

Kushvah Badam Singh Department of Mathematics and Computing, Indian Institute of Technology (ISM), Dhanbad, Jharkhand, India

Lavanya C. Department of Physics, Periyar University, Salem, Tamil Nadu, India

Lepse Pratik V. Department of Mathematics, Lovely Professional University, Jalandhar, Phagwara, Panjab, India

Mahata Animesh Mahadevnagar High School, Maheshtala, Kolkata, West Bengal, India

Majee Suvankar Department of Mathematics, Indian Institute of Engineering Science and Technology, Shibpur, Howrah, India

Makrariya Akshara School of Advanced Science-Mathematics, VIT Bhopal University, Bhopal, Madhya Pradesh, India

Malakar Bidhan Department of Electrical Engineering, JIS College of Engineering, Kalyani, West Bengal, India

Malgol Amit National Institute of Technology Calicut, Kattangal, Kerala, India

Mali Prakash Chandra Department of Mathematics, Jadavpur University, Kolkata, India

Malik Anand Department of Physics, Chaudhary Ranbir Singh University, Jind, India

Mamun Miah M. Department of Mathematics, Khulna University of Engineering & Technology, Khulna, Bangladesh

Mani Rajan M. S. Department of Physics, University College of Engineering, Anna University, Ramanathapuram, India

Mani Awadesh Condensed Matter Physics Division, Indira Gandhi Centre for Atomic Research, Kalpakkam, Tamil Nadu, India

Manigandan R. School of Mechanical Engineering, Vellore Institute of Technology, Chennai, India

Maqbul Md. National Institute of Technology Silchar, Silchar, Assam, India

Mathur Kunwer Singh Department of Mathematics and Statistics, Dr. Harisingh Gour Vishwavidyalaya, Sagar, Madhya Pradesh, India

Medellín-Neri N. Faculty of Electronics Sciences, Benemérita Universidad Autónoma de Puebla, Puebla, Mexico

Meher Ramakanta Department of Mathematics and Humanities, Sardar Vallabhbhai National Institute of Technology, Surat, Gujarat, India

Meiyazhagan J. Department of Nonlinear Dynamics, Bharathidasan University, Tiruchirappalli, Tamil Nadu, India

Micheal David Raj Division of Mathematics, School of Advanced Sciences, Vellore Institute of Technology, Chennai, Tamil Nadu, India

Mishra Pankaj K. Department of Physics, Indian Institute of Technology Guwahati, Guwahati, Assam, India

Mishra Subodha Department of Physics, Siksha 'O' Anusandhan, Deemed to be University, Bhubaneswar, 751030, India

Mondal Biswajit Raja N.L. Khan Women's College (Autonomous), Midnapore, West Bengal, India

Mondal Dipankar Indian Institute of Technology Guwahati, Guwahati, Assam, India

Mondal Kajal Kumar Department of Mathematics, Cooch Behar Panchanan Barma University, Cooch Behar, West Bengal, India

Mondal Supriya MMI College of Nursing, Pachpedi Naka, Raipur, Chhattisgarh, India;
VY Hospital, Adjacent to Kamal Vihar (Sector 12), Raipur, Chhattisgarh, India

Mortuja Md. Golam Dr. SPM IIIT Naya Raipur, C.G., Raipur, India

Mukherjee Supriya Department of Mathematics, Gurudas College, Kolkata, West Bengal, India

Munoz-Pacheco J. M. Faculty of Electronics Sciences, Benemérita Universidad Autónoma de Puebla, Puebla, Mexico

Murmu Tanushree Department of Mathematics, Jadavpur University, Kolkata, India

Murthy B. S. N. Department of Mathematics, Aditya College of Engineering and Technology, Surampalem, Andhra Pradesh, India

Murugesan Regan Vel Tech Rangarajan Dr. Sagunthala R & D Institute of Science and Technology, Chennai, Tamil Nadu, India

Nag Amitava Department of CSE, CIT Kokrajhar, Kokrajhar, Assam, India

Nagaram Nagadevi Bala Vel Tech Rangarajan Dr. Sagunthala R & D Institute of Science and Technology, Chennai, Tamil Nadu, India

Natarajan C. Department of Mathematics, Srinivasa Ramanujan Centre, SASTRA Deemed University, Kumbakonam, India

Nath Ghosh Dipendra Controller of Examinations, Kazi Nazrul University, Asansol, West Bengal, India

Nath Surojit Department of Electrical Engineering, JIS College of Engineering, Kalyani, West Bengal, India

Nayak Sima Department of Mechanical Engineering, Indian Institute of Technology Guwahati, Guwahati, India

Nikam P. P. Department of Physics, Devchand College, Arjunnagar, Kolhapur, India

Oligo Jubert B. College of Teacher Education, Nueva Vizcaya State University, Bayombong, Philippines

Ozah Jintu Dibrugarh University, Dibrugarh, Assam, India

Pain Pritam Department of Mechanical Engineering, Haldia Institute of Technology, Haldia, West Bengal, India

Pal Chahal Rishi Department of Physics, Chaudhary Bansi Lal University, Bhiwani, India

Pal Charli Chinmayee Department of Physics, Siksha 'O' Anusandhan, Deemed to be University, Bhubaneswar, 751030, India

Pal Debjit Department of Electrical Engineering, JIS College of Engineering, Kalyani, West Bengal, India

Pal Nikhil Department of Mathematics, Siksha-Bhavana, Visva-Bharati University, Santiniketan, India

Palit Aniruddha Department of Mathematics, Surya Sen Mahavidyalaya, Siliguri, India

Panigoro Hasan S. Department of Mathematics, State University of Gorontalo, Bone, Bolango, Indonesia

Patidar Neelam School of Advanced Science-Mathematics, VIT Bhopal University, Bhopal, Madhya Pradesh, India

Patil S. D. Department of Physics, Devchand College, Arjunnagar, Kolhapur, Maharashtra, India

Patil S. S. Department of Physics, Vivekanand College, Kolhapur, India

Patra Maiti Atasi Directorate of Distance Education, Vidyasagar University, Midnapore, West Bengal, India

Pattanayak Solanki Department of Computer Science, Haldia Institute of Management, Haldia, WB, India

Paul Anindya Department of Mathematics, Cooch Behar Panchanan Barma University, Cooch Behar, West Bengal, India

Paul Debraj Department of Electrical Engineering, JIS College of Engineering, Kalyani, West Bengal, India

Paul Kamakhya North Eastern Hill University, Shillong, ML, India

Paul Niranjan Department of Mathematics, Siksha Bhavana, Visva-Bharati, Santiniketan, Santiniketan, West Bengal, India

Paul Subrata Department of Mathematics, Arambagh Govt. Polytechnic, Arambagh, West Bengal, India

Pavithra T. Department of Physics, Central University of Tamil Nadu, Thiruvarur, Tamil Nadu, India

Pawar V. S. Department of Physics, Rajer Ramrao Mahavidyalay, Jath, Maharashtra, India

Pikulins Dmitrijs Institute of Radioelectronics, Riga Technical University, Riga, Latvia

Pio Ferreira Claudia São Paulo State University (UNESP), Botucatu, Brazil

Poddar Nanda Cooch Behar Panchanan Barma University, Cooch Behar, WB, India

Poonam Department of Mathematics, BITS Pilani, Pilani, Rajasthan, India

Prabhu A. Department of Physics, Periyar University, Salem, India

Prabhu Department of Chemistry, Periyar University, Salem, Tamil Nadu, India

Pradhan Barsha Department of Mathematics, Sikkim Manipal Institute of Technology, Sikkim Manipal University, Majitar, Rangpo, East-Sikkim, India

Prakash Om Department of Aerospace Engineering, University of Petroleum and Energy Studies, Dehradun, Uttarakhand, India

Pramanik Sukanta Department of Statistics, Siliguri College, North Bengal University, Siliguri, West Bengal, India

Prasad Punam Kumari Department of Mathematics, Sikkim Manipal Institute of Technology, Sikkim Manipal University, East-Sikkim, India

Priyanka T. M. C. Department of Mathematics, School of Advanced Sciences, Vellore Institute of Technology, Vellore, Tamil Nadu, India

Punacha Shreyas Department of Physics, National Institute of Technology Karnataka, Surathkal, Mangalore, Karnataka, India

Raghavi K. Department of Physics, School of Basic and Applied Sciences, Central University of Tamil Nadu, Thiruvarur, India

Rahmi Emli Department of Mathematics, State University of Gorontalo, Bone, Bolango, Indonesia

Rajan M. S. Mani Anna University, University College of Engineering, Ramanathapuram, India

Rao Shiva Dibrugarh University, Dibrugarh, Assam, India

Rasappan Suresh University of Technology and Applied Sciences- Ibri, Sultanate of Oman, Ibri, Oman

Raut Santanu Department of Mathematics, Mathabhanga College, Coochbehar, India

Ravisankar R. Department of Physics, Indian Institute of Technology, Guwahati, Assam, India

Ravuri Mohana Ramana Department of Basic Science and Humanities (Mathematics), Narasaraopeta Engineering College, Narasaraopeta, AP, India

Ray Rajendra K. School of Basic Sciences, Indian Institute of Technology Mandi, Mandi, Himachal Pradesh, India

Rehman Attiq ul Department of Mathematical Sciences, BGSB University, Rajouri, India

Remi T. Department of Physics, Farook College University of Calicut, Kozhikode, Kerala, India

Rino Nelson N. Mechanical Engineering, Indian Institute of Information Technology, Design and Manufacturing, Kancheepuram, Chennai, India

Rout Anil Kumar School of Energy Science and Engineering, Indian Institute of Technology Guwahati, Guwahati, India

Roy Antara Asansol Institute of Engineering and Management-Polytechnic, Asansol, India

Roy Ashim Department of Mathematics, Alipurduar Univeristy, Alipurduar, India

Roy Banamali Department of Mathematics, Bangabasi Evening College, Kolkata, West Bengal, India

Roy Jyotirmoy Indian Institute of Engineering Science and Technology, Shibpur, Howrah, West Bengal, India

Roy Subrata Department of Mathematics, Cooch Behar Panchanan Barma University, Cooch Behar, India

Rushi Kumar B. Department of Mathematics, School of Advanced Sciences, Vellore Institute of Technology, Vellore, Tamilnadu, India

Saha Abhijit Department of Mechanical Engineering, Haldia Institute of Technology, Haldia, West Bengal, India

Saha Ashesh National Institute of Technology Calicut, Kattangal, Kerala, India

Saha Dipan Advanced Centre for Nonlinear and Complex Phenomena, Kolkata, India

Saha Gourab Cooch Behar Panchanan Barma University, Cooch Behar, India

Saha Sandip Department of Mathematics, Madanapalle Institute of Technology & Science, Madanapalle, Andhra Pradesh, India;
School of Advance Sciences (SAS) Mathematics, Vellore Institute of Technology Chennai, Chennai, Tamilnadu, India

Sahoo Niranjan Department of Mechanical Engineering, School of Energy Science and Engineering, Indian Institute of Technology Guwahati, Guwahati, India

Sahu P. K. Department of Mathematics, Government Shyama Prasad Mukharjee College, Sitapur, Chhattisgarh, India

Saikia Banashree Department of Mathematics, Dibrugarh University, Dibrugarh, Assam, India

Saini Nareshpal Singh Department of Physics, Guru Nanak Dev University, Amritsar, India

Sajan Department of Mathematics, BITS Pilani, Pilani, Rajasthan, India

Samanta Sabyasachi Department of Information Technology, Haldia Institute of Technology, Haldia, WB, India

Sapkota Niraj Department of Mathematics, Sikkim Manipal Institute of Technology, Sikkim Manipal University, Sikkim, India

Sardar Sankirtan Department of Mathematics, Guru Ghasidas Vishwavidyalaya, Bilaspur, India

Sarkar Jit Jadavpur University, Kolkata, India

Sarkar Kshirod Raja N.L. Khan Women's College (Autonomous), Midnapore, West Bengal, India

Sarkar Sudipta Department of Mathematics, Heritage Institute of Technology, Anandapur, Kolkata, West Bengal, India

Sarkar Swarup Department of Physics, Indian Institute of Technology Guwahati, Guwahati, Assam, India

Sarkar Tanay Department of Mathematics, Jadavpur University, Kolkata, India

Sarma Ridip Department of Mathematics, Assam Don Bosco University, Tapesia, Sonapur, Assam, India

Sasmal Sourav Kumar Department of Mathematics, BITS Pilani, Pilani, Rajasthan, India

Savapondit Santonu Department of Mathematics, Sikkim Manipal Institute of Technology, Sikkim Manipal University, Sikkim, India

Sebastian Anupama Department of Physics, National Institute of Technology Karnataka, Surathkal, Mangalore, Karnataka, India

Sekhar Roy Deb Department of Electrical Engineering, JIS College of Engineering, Kalyani, West Bengal, India

Sengupta Debajyoti Department of Electrical Engineering, JIS College of Engineering, Kalyani, West Bengal, India

Senthilvelan M. Department of Nonlinear Dynamics, Bharathidasan University, Tiruchirappalli, Tamil Nadu, India

Shajahan T. K. Department of Physics, National Institute of Technology Karnataka, Surathkal, Mangalore, Karnataka, India

Shanmugavelan S. Department of Mathematics, Srinivasa Ramanujan Centre, SASTRA Deemed University, Kumbakonam, India

Sharma Bhupendra K. Department of Mathematics, Birla Institute of Technology and Science, Pilani, Rajasthan, India

Sharry Physics Department, Guru Nanak Dev University, Amritsar, India

Shibin Raj C. P. National Institute of Technology Calicut, Kattangal, Kerala, India

Shilpi Physics Department, Guru Nanak Dev University, Amritsar, India

Shukla Manoj Kumar Lovely Professional University, Phagwara, Punjab, India

Sibeesh Puthiyapurayil Department of Physics, National Institute of Technology Karnataka Surathkal, Mangalore, India

Singh Chandraketu School of Sciences, Christ (Deemed to Be University) Delhi NCR, Ghaziabad, India

Singh Tamang Jitendra Department of Electronics and Communication Engineering, Sikkim Manipal Institute of Technology, Sikkim Manipal University, Majitar, Sikkim, India

Singhal Abhinav School of Sciences, Christ (Deemed to Be University) Delhi NCR, Ghaziabad, India

Singh Kuldeep Department of Physics, Guru Nanak Dev University, Amritsar, India;
Department of Mathematics, Khalifa University of Science and Technology, Abu Dhabi, UAE

Singh Madan Mohan Department of Basic Sciences & Social Sciences, North Eastern Hill University, Shillong, ML, India

Singh Ram Department of Mathematical Sciences, BGSB University, Rajouri, India

Singh Ram Mehar Department of Physics, Chaudhary Devi Lal University, Sirsa, India

Singh Ritesh Manipal University Jaipur, Jaipur, Rajasthan, India;
University of Petroleum & Energy Studies, Dehradun, Uttarakhand, India

Singla Sunidhi Department of Physics, Guru Nanak Dev University, Amritsar, India

Sinha Kaushik Department of Electrical Engineering, JIS College of Engineering, Kalyani, West Bengal, India

Sinha Sudipta Burdwan Raj College, University of Burdwan, Burdwan, India

Slathia Geetika Department of Physics, Guru Nanak Dev University, Amritsar, India

Solo Peter Department of Chemistry, St. Joseph University, Dimapur, India;
Department of Chemistry, St. Joseph's College (Autonomous), Jakhama, India

Sonal Deepa Department of Computer Science, V.K.S. University, Arrah, India

Sowmiya C. Department of Mathematics, SAS, Vellore Institution of Technology, Vellore, Tamil Nadu, India

Srinivas M A S Department of Mathematics, Jawaharlal Nehru Technological University, Hyderabad, Telangana, India

Srinivas M N Department of Mathematics, School of Advanced Sciences, Vellore Institute of Technology, Vellore, Tamilnadu, India

Subha P. A. Department of Physics, Farook College University of Calicut, Kozhikode, Kerala, India

Subramanian K. Department of Physics, SRM Institute of Science and Technology, Ramapuram Campus, Chennai, India

Subramanian Venkatesh National Institute of Technology Karnataka, Mangaluru, India

Sucu Derya Yildirim Faculty of Science and Art, Department of Mathematics, Nevsehir Haci Bektas Veli University, Nevsehir, Turkey

Sunny Geo Department of Physics, School of Basic and Applied Sciences, Central University of Tamil Nadu, Thiruvarur, India

Takale M. V. Department of Physics, Shivaji University, Kolhapur, Maharashtra, India

Takale P. T. Department of Physics, Shivaji University, Kolhapur, India

Talukdar Pranabendra Department of Mathematics, Alipurduar University, Alipurduar, West Bengal, India

Tamang Jharna Department of Mathematics, Sikkim Manipal Institute of Technology, Sikkim Manipal University, Majitar, East-Sikkim, India;
Department of Mathematics, Sikkim Alpine University, Kamrang, Namchi, South-Sikkim, India

Thilakavathy J. Department of Science and Humanities, Jerusalem College of Engineering, Chennai, India

Tiwari Rakhi Department of Mathematics, Babasaheb Bhimrao Ambedkar Bihar University, Muzaffarpur, India

Tiwari Rakshit Indian Institute of Technology Guwahati, Guwahati-781039, Assam, India

Tjukovs Sergejs Institute of Radioelectronics, Riga Technical University, Riga, Latvia

Tolga Sakallı Muharrem Trakya University, Edirne, Turkey

Urunkar T. U. Department of Physics, Shivaji University, Kolhapur, India

Valderama Julius S. College of Arts and Sciences, Nueva Vizcaya State University, Bayombong, Philippines

Varghese Steffy Sara Space and Planetrary Science Center, Khalifa University, Abu Dhabi, UAE

Varunkumar M. Department of Basic Sciences and Humanities, GMR Institute of Technology, Rajam, Andhra Pradesh, India;
Department of Mathematics, School of Advanced Sciences, VIT-AP University, Amaravati, Andhra Pradesh, India

Vasović Nebojša Department of Applied Mathematics and Informatics, Faculty of Mining and Geology, University of Belgrade, Belgrade, Serbia

Veni Saravana Department of Physics, Amirta College of Engineering and Technology, Erachakulam Campus, Nagercoil, India

Venkatalaxmi A. Department of Mathematics, Osmania University, Hyderabad, TG, India

Venturino Ezio Università di Torino, Torino, Italy

Zambrano-Serrano E. Facultad de Ingeniería Mecánica y Eléctrica, Universidad Autónoma de Nuevo León, San Nicolás de los Garza, Nuevo León, Mexico

Nonlinear Waves and Plasma Dynamics

Offset Bipolar Pulses in Magnetospheric Plasma Systems

Steffy Sara Varghese and S. S. Ghosh

Abstract The satellite borne electric field instruments have been observed the signatures of Electrostatic Solitary Waves (ESWs) throughout the Earth's magnetospheric boundary layers and have been recorded as localized monopolar, bipolar or tripolar pulses in the electric field (E data). There are also reports of various kinds of electric field structures by the satellites which are different from conventional bipolar pulses. There was no generic theory for them so far. In this work, we are intended to introduce offset bipolar pulses, a possible kinds of non-conventional pulses which can be supported by different kinds of space plasma system. In this regard, we have analyzed a warm multi-ions and two electron temperature plasma model which is quite signifying in magnetospheric studies by adopting the Sagdeev pseudopotential technique. Through rigorous numerical analysis, we have identified and delineated the conditions for the existence of flat top solitary wave, and correlate them with the non-conventional electrostatic solitary wave structures in space plasma observations. It is expected that it will provide a new way of understanding the non-conventional localized pulses in the E-field data recorded during satellite expeditions which are known to be important in determining the microphysics of the Earth's magnetospheric plasma system.

Keywords Flat top solitary waves · Electrostatic solitary waves · Offset bipolar pulses

1 Introduction

The theory of coherent and organized nonlinear dynamical structures are often found to be relevant in interpreting Electrostatic Solitary Waves (ESWs) which are observed ubiquitously in the Earth's magnetosphere. The large gradient in particle properties

S. S. Varghese (✉)
Space and Planetrary Science Center, Khalifa University, Abu Dhabi 127788, UAE
e-mail: steffystephan28@gmail.com

S. S. Ghosh
Indian Institute of Geomagnetism, Mumbai, India

S. Banerjee and A. Saha (eds.), *Nonlinear Dynamics and Applications*,
Springer Proceedings in Complexity,
https://doi.org/10.1007/978-3-030-99792-2_1

at the magnetospheric boundary layers initiates the perturbations which lead to the generation of stable localized solitary structures. The satellite borne electrical instruments recorded the signatures of localized bipolar and monopolar pulses in the Electric Field (E) data, moving parallel to the background magnetic field. The fast moving positive amplitude bipolar structures are often interpreted as BGK phase space holes for electrons, or, alternatively, as an electron acoustic solitary wave while the low frequency, slowly moving E-field bipolar pulses in the auroral regions are generally interpreted as ion acoustic solitary waves. The monopolar structures, on the other hand, are interpreted as double layers, or shocks, governed by either electron or ion dynamics, and they influence the particle acceleration processes along the field lines.

Advent of high resolution satellite borne instruments, however, identified more kinds of composite and complex organized structures which yet to have a proper theoretical understanding. One such relatively lesser reported structures are the offset bipolar pulses (ofbp) and monopole pairs (mpp) which are the main focus for the present work. The first observational evidence of ESWs was reported in the auroral region by Temerin et al. [1] using S3-3 satellite data. Apart from the conventional bipolar and monopolar structures, they also bore the signatures of ofbp and mpp [2]. For an ofbp, the distance between the bipolar peaks is relatively large compared to the characteristic width of the each peak. Because of their "stretched" look, an ofbp has also been termed as a "stretched bipolar" or a "dispersed bipolar" in the literature. Witt and Lotko [3] theoretically modelled them as paired ion acoustic shocks and showed that they are associated with a square shaped potential well which is different from the conventional bell shaped solitary structures. Their unique theoretical finding, however, didn't seem to have any subsequent follow ups. Later, fast moving offset bipolar pulses were observed in the day side Polar Cap Boundary Layer (PCBL) [4], the downward current regions of the auroral zone [5, 6], the diffusion region of reconnection [7], and in the magnetosheath [8] by POLAR, FAST, GEOTAIL, and CLUSTER satellites. It was Tsurutani et al. [4] who identified them as unique kinds of coherent localized structures. Following their idea, and incorporating a Bernstein, Greene and Kruskal (BGK) model [5] and Particle In Cell (PIC) simulation [6], ofbp were interpreted as flat shaped electron phase space holes, emerging due to trapping of electrons. Such a theory, however, remain inadequate to explain slow moving ion mode ofbp.

During our theoretical analysis, we have indeed obtained such Flat Top Solitary Wave (FTSW) solutions with square shaped potential profiles [9]. The nomenclature not only describes the unique morphology of the potential profile but also rightly connects it to a more analytical and ideal set of solutions called Flat Top Solitons (FTS). Analogous to a solitary wave vis á vis a soliton solution, an FTSW is the realistic counter part of the FTS which is obtained by using Sagdeev pseudopotential technique. The technique is widely used to predict the existence and characteristics of any nonlinear coherent structure by studying the trajectory of the pseudoparticle in the pseudopotential well without the rigorous solution of any particular NLPDE, as is essential for a soliton, or FTS solution.

Here, for the first time, we have extended the same idea to space plasma for interpreting the ofbp. Rather than being event based, the theory proposes a more generic approach which correlates the observed ofbp with a more fundamental non-linear dynamical structure called FTS through FTSW where the latter retains all the physical characteristics and the boundary conditions of FTS.

The article is organized as follows. Assuming a simple plasma model, in Sect. 2 we have found an ion acoustic ofbp which we have further validated with the corresponding satellite observations. In Sect. 3 we have explored the generic characteristics of the associated FTSW which is the steady state generalization of an FTS. The concluding remarks are given in Sect. 4

2 Analytical Concept

Slow moving ESWs, moving with ion acoustic speed, have been observed in the low altitude auroral region by several spacecraft expeditions. S3-3 recorded slow moving ofbp at an altitude between 6000 and 8000 km [2]. At this altitude, the plasma has been found to have a significant contribution of O^+ ions along with its usual proton (H^+) population. Moreover, there is an admixing of hot magnetospheric electrons with a cooler component originated from the ionosphere. Satellite observations have recorded cooler electron temperature $T_{ec} \approx 0.5{-}5\,\text{eV}$. The recorded ambient plasma density of this region is of the order of $n_0 \approx 5{-}10\,\text{cm}^3$. The tenuous plasma condition and the absence of any physical boundary allow us to assume the plasma to be collisionless, homogeneous, and infinite. Following Temerin et al. [1] we have further assumed that the wave is moving along the ambient magnetic field, making the plasma isotropic and unmagnetized. It is observed that the speed of the ESW in this region is of the order of the ion acoustic speed of the medium, with the wave speed $V \approx 50\,\text{km/s}$, which indicates that they are governed by the ion dynamics. To sustain any such wave, the electron temperature should be higher than the corresponding ion temperatures. Besides, they have negligible inertia compared to ions. Theoretically it is well known that a secondary component of electrons is necessary to sustain an ion acoustic DL. The same condition was found to hold true for a Super Solitary Wave (SSW) [10], or an FTSW [9] as well. Hence we have assumed that the plasma has two electron temperatures, both obeying Boltzmann distributions and are separately in thermal equilibrium. The overall plasma is a four component one with warm multi-ion fluids comprising H^+ and O^+ ions so that the corresponding mass ratio $Q = 1/16$.

Following Temerin et al. [1], we have chosen an ambient plasma density $n_0 = 10\,\text{cm}^3$, giving rise to an overall proton plasma frequency for n_0 as $\omega_{pi} = 4.163\,\text{kHz}$. We have assumed a very low concentration of cooler electrons (viz., 0.12% of n_0) and sufficiently small presence of O^+ ions (10% of n_0), leading to the normalized ambient densities $\mu = 0.0012$ for cooler electrons and $\alpha_h = 0.1$ for O^+ ions, respectively. All the number densities were normalized by the ambient plasma density n_0. For our convenience, we have chosen $T_{ec} = 0.5\,\text{eV}$, and the electron temperature ratio $\tau = 0.0485$ so that $T_{ew} = 10.3\,\text{eV}$. Since both the electrons are taking part in the Debye shielding, we have estimated the effective temperature $T_{eff} = \frac{T_{ec}T_{ew}}{\mu T_{ew}+(1-\mu)T_{ec}} =$

10.0722 eV which further gives us the estimated effective Debye length $\lambda_{d_{eff}} = 7.4606$ m and the effective ion acoustic speed for protons $c_{isl} = 31.06$ km/s. These three parameters, together with ω_{pi}, determine the overall scale of our plasma system. This is consistent with the observations of Temerin et al. [1] who have reported a Debye length of $\lambda_d \approx 5$ m for the said region. For our theoretical analysis, we have normalized all the space variables by $\lambda_{d_{eff}}$, time by ω_{pi}^{-1}, temperatures by T_{eff}, and the electrostatic potential ϕ by T_{eff}/e. All the speeds, along with the wave Mach number M, are normalized by c_{isl}.

In the literature, there was a mention of the hotter ion temperature of the order of KeV. Such a high ion temperature would not sustain an ion acoustic wave and it may not be appropriate for the current region. Since ions are necessarily cooler than electrons, we have chosen the lighter ion temperature $T_{il} = 0.33$ eV which is cooler than the cooler electron temperature. Theoretically we have found that the ion temperature plays a marginal role in determining an FTSW solution compared to the corresponding electronic parameters, such as μ and τ. Besides, the effect of O^+ ions is expected to be even smaller because of its apparently low concentration. In the absence of any clear mention of the type the ion species for the hotter one, and for our analytical convenience, we have assumed that both the H^+ and O^+ ions have equal temperatures, giving rise to an overall normalized ion temperature $\sigma = 0.033$.

To ensure a steady state condition, or a wave frame, we have further assumed η to be the generalized coordinate where

$$\eta = x - Mt \tag{1}$$

The corresponding Sagdeev pseudopotential $\Psi(\Phi)$ for the chosen plasma model is [10].

$$\begin{aligned}\psi(\Phi) = -\Bigg[\Big\{\mu + (1-\mu)\tau\Big\}\Big\{\mu\Big(exp\frac{\Phi}{\mu+(1-\mu)\tau} - 1\Big) + \frac{1-\mu}{\tau}\Big(exp\frac{\tau\Phi}{\mu+(1-\mu)\tau} - 1\Big)\Big\} \\ + \frac{\alpha_l}{6\sqrt{3\sigma_l}}\Big\{[(M+\sqrt{3\sigma_l})^2 - 2\Phi]^{\frac{3}{2}} - (M+\sqrt{3\sigma_l})^3 \\ - [(M-\sqrt{3\sigma_l})^2 - 2\Phi]^{\frac{3}{2}} + (M-\sqrt{3\sigma_l})^3\Big\} \\ + \frac{\alpha_h}{6\sqrt{3\sigma_h}}\Big\{[(\frac{M}{\sqrt{Q}} + \sqrt{3\sigma_h})^2 - 2\Phi]^{\frac{3}{2}} - (\frac{M}{\sqrt{Q}} + \sqrt{3\sigma_h})^3 \\ - [(\frac{M}{\sqrt{Q}} - \sqrt{3\sigma_h})^2 - 2\Phi]^{\frac{3}{2}} + (\frac{M}{\sqrt{Q}} - \sqrt{3\sigma_h})^3\Big\}\Bigg]\end{aligned} \tag{2}$$

which satisfies the following 'energy equation'

$$\frac{1}{2}\left(\frac{d\Phi}{d\eta}\right)^2 + \Psi(\Phi) = 0 \tag{3}$$

Equation (3) is a modified form of the Poisson's equation where the slope of the pseudopotential is defined as the associated charge separation, i.e., $\frac{\partial\Psi(\phi)}{\partial\Phi} = \Delta n(\phi)$

for any ϕ, Δn being the charge separation. A Ψ versus Φ curve determines the existence of the localized nonlinear coherent structures, like SWs, SSW, or FTSW, provided following boundary conditions are satisfied

$$\Psi(\Phi = 0) = \frac{\partial \Psi}{\partial \Phi}\bigg|_0 = 0; \quad \frac{\partial^2 \Psi(0)}{\partial \Phi^2} < 0 \tag{4a}$$

$$\Psi(\Phi_0) = 0; \quad \frac{\partial \Psi(\Phi_0)}{\partial \Phi} \neq 0; \quad \Psi(\Phi) < 0 \text{ for } 0 \geq \Phi \geq \Phi_0. \tag{4b}$$

where Φ_0 is the amplitude of the wave structure.

The last condition in (4a, 4b) ensures the recurrence of the initial state for a soliton, or SW. For a DL, this condition modifies as

$$\frac{\partial \Psi(\Phi_0)}{\partial \Phi} = \Delta n_d = 0 \tag{5}$$

where Δn_d is the charge separation at the maximum amplitude for DL. Studying the trajectory of the "pseudoparticle" in the said pseudopotential well, and implementing the above mentioned 'boundary conditions' for the pseudoparticle, viz., (4a, 4b), the Sagdeev pseudopotential technique enables one to predict the presence of the corresponding steady state nonlinear structure.

For $M = 1.05663685$, and for the chosen set of parameters as mentioned above, in Fig. 1 we have plotted the corresponding Sagdeev pseudopotential profile. Since it satisfies (4a, 4b), it represents a SW which is a steady state analog of a soliton. For a FTSW, similar to a regular solitary wave profile the pseudopotential curve shows two extrema and two roots even though, near amplitude the pseudopotential curve meets the zero axis with almost a "grazing incidence". In order to under the significance of such solitary wave structures, we have analysed the corresponding electric field profile and density profile as mentioned below.

To find the corresponding electric field profile, we have deduced the generalized electric field $E = -\frac{d\Phi}{d\eta}$ from (3). The time duration Δt (non-normalized) is estimated

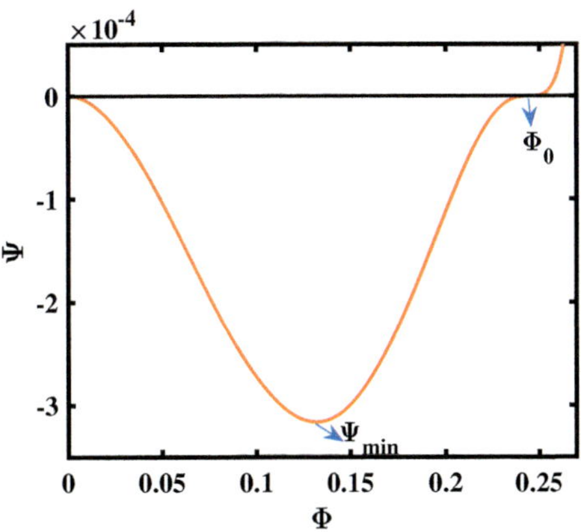

Fig. 1 Sagdeev pseudopotential profile corresponding to an FTSW

from from the half-width W assuming $x = 0$ in (1). The half-width W is defined from (3) as

$$W = 2\eta_{1/2} = 2\int_{\Phi_0}^{\frac{\Phi_0}{2}} \left(\frac{1}{\sqrt{(-2\Psi(\Phi))}}\right) d\Phi\,; \quad \Delta t = \left(\frac{W}{M}\right)' \text{ for } x = 0 \tag{6}$$

The normalized half-width $W = 422.55$ m and the prime ($\prime$) in (6) denotes the corresponding non-normalized parameters. Figure 2 shows the associated electric field profile in non-normalized parameters. In X-axis we have plotted the time and Y-axis the non-normalized electric field. Comparing the electric field profile with that of observed non-conventional bipolar pulses clearly revels an offset bipolar pulses (ofbp) [4]. We have further estimated the average $\left(E_{avg} = \left(\frac{\Phi_0}{W}\right)'\right)$ and the maximum $\left(E_{max} = -\left(\frac{d\Phi}{d\eta}\right)'\Big|_{\Psi=\Psi_{min}}\right)$ electric fields analytically, where $\Phi_0 = 0.2448$ is the normalized potential amplitude and Ψ_{min} is the minimum value of ψ for $0 \le \phi \le \phi_0$, both of them are marked by the respective arrows in Fig. 1. The estimated average electric field across the structure $E_{avg} = 5.8387$ mV/m, and the estimated time duration $\Delta t = 12.8746$ ms (Fig. 2). The S3-3 satellite observations have revealed an $E \le 15$ mV/m and $\Delta t = 2–20$ ms which are a close match to our analytical estimations. The analytically estimated peak to peak E-field amplitude $E_{p-p} = 2\,|E_{max}| \approx 37.02$ mV/m, have higher amplitude than the observed E-field. This still remain consistent with the qualitative agreement since a fluid approximation is known to overestimate the amplitude. We have further estimated the speed of the wave structure $V = 32.83$ km/s from our chosen Mach number $M(= 1.05663685)$ which is in accordance with the satellite observations, i.e.,$V \approx 50$ km/s, as mentioned earlier. Table 1 compares shape, size, and speed of an ESW in the auroral region with those estimated analytically for a possible ofbp. The latter shows a com-

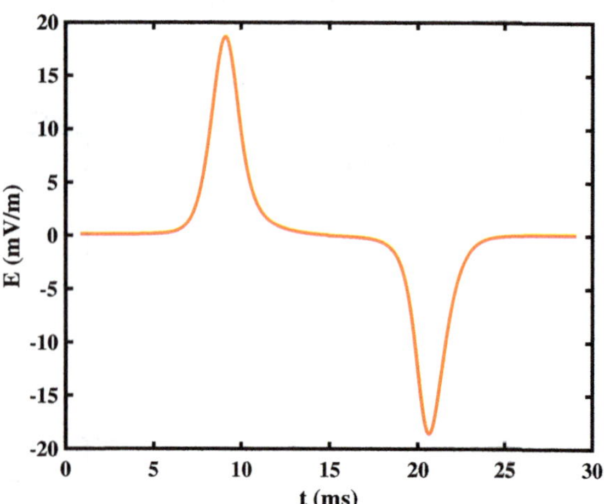

Fig. 2 Non-normalized electric field (E) profile corresponding to an FTSW

Table 1 Wave parameters

Theoretical	Observational
$E_{avg} = 5.8387$ mV/m	$E \leq 15$ mV/m
$\Delta t = 12.8746$ ms	$\Delta t = 2$–20 ms
$V = 32.82$ km/s	$V \approx 50$ km/s

paratively wider Δt, as expected from its stretched structure. Our results indicates that the ofbp obtained analytically here using Sagdeev pseudopotential is a feasible candidate to interpret the corresponding satellite observations.

3 Physical Properties of FTSW and ofbp

To explore the physical characteristics of an ofbp, we have plotted the corresponding potential (solid line) and charge separation (dotted line) profiles in in Fig. 3. As shown in Fig. 3 the potential profile corresponds to an FTSW confirms a flat top profile. Conceptually, an FTSW can be visualized as an amalgamation of a SW and a DL. It is well evident form the charge separation profile (dotted line). Its charge separation near the maximum amplitude turns vanishingly small (dotted line), approaching the condition of a DL (5), although it always remain finite and non-zero, eventually satisfying (4a, 4b) and the solution bounces back to its initial state like any conventional soliton. The modified condition for an FTSW can thus be written as

$$\Psi(\Phi_0) = 0; \quad \left.\frac{\partial \Psi}{\partial \Phi}\right|_{\Phi_0} = \epsilon, \quad \left.\frac{\partial^2 \Psi}{\partial \Phi^2}\right|_{\Phi_0} = \delta; \quad \epsilon, \delta \neq 0; \tag{7}$$

where ϵ, δ are two arbitrarily small, but finite numbers. The grazing incidence (i.e., low slope) of the curve to the Φ axis at Φ_0 satisfies (6) for an FTSW while its finite slope at Φ_0 satisfies (4a, 4b) as well. This implies that the 'pseudoparticle', associated with the solution, leaves $\Phi = 0$ at rest and reaches its reflection point at Φ_0 after a prolonged time, as ascertained by the grazing incidence of Ψ, and then it oscillates back to $\Phi = 0$ giving rise to an wider, but well localized, coherent structure similar to a conventional SW. The morphology of the structure, however, is different from that of a conventional one as is evident from the associated electric field and potential profiles.

Previously Roth et al. [6] achieved an ofbp by assuming flat top potential profile for their PIC simulation. In the present case, the FTSW and the associated ofbp have been obtained analytically from a simple plasma model without any prior assumption of the potential profile. It was previously indicated that the trapping of electrons in BGK phase space hole may cause an ofbp. Such an assumption will not be valid for an ion mode, positive amplitude ofbp. However, in spite of the differences in the

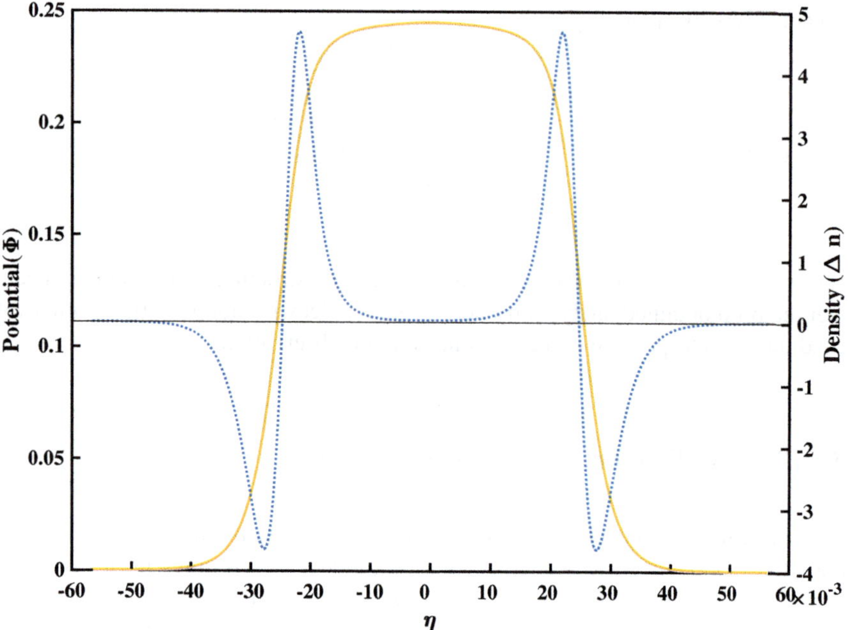

Fig. 3 Potential profile (solid line) and charge separation profile (dotted line) corresponding to an FTSW

respective plasma models and associated techniques, the charge separation profiles remain the same for both the cases. We have previously observed that, when a Regular Solitary Wave (RSW) transits to a DL in the parameter space, its charge separation at its maximum amplitude drops, becoming ideally zero for a DL. We can here visualize an FTSW as an 'incomplete DL' where the solution goes very close to a DL solution but stops short of it and bounces back retaining the characteristics of a SW along with a strong imprint of a DL-like solution as well. This is manifested in the long sustentation of the vanishingly small charge separation which eventually causes the flat top profile. It is the same characteristic which causes the stretching in the localized E-field making it offset bipolar.

It is now evident that the ofbp, and the FTSW, both indicate the same localized coherent structure. The concept now can be extended further to the more ideal solutions called FTS which, as we have discussed earlier, is the mathematical counterpart of an FTSW, in the same way as a soliton is correlated with a more general class of solutions we call as solitary waves. It is well known that the integrable NLPDEs, like Korteweg-de Vries (KdV), govern soliton solutions. Analogously, a modified, or extended form of KdV (eKdV), popularly often known as the Gardner equation, governs FTS solutions depending on its specific boundary conditions [11]. Like eKdV, there are other such modified NLPDEs which also have FTS as one of their possible solutions. As per previous literature, an ofbp has often been described as a

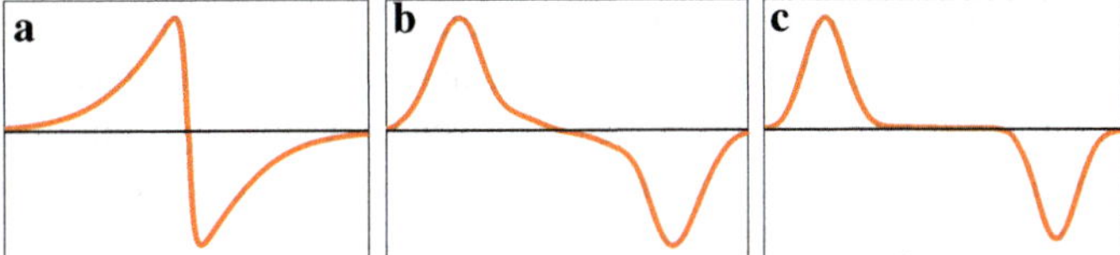

Fig. 4 Schematic diagram of **a** bipolar E-field pulse **b** offset bipolar E-field pulse and **c** paired monopolar E-field pulse

'stretched' or 'dispersed' bipolar pulse where it was conjectured that the stretching has been happened due to an extra dispersion in the medium. The solution obtained from an e-KdV or Gardner equation describes the significance of an extra cubic nonlinearity term which results in a secondary insurgence in the nonlinearity" that balances the excessive dispersion maintaining the solitary structure.

To complement our understanding on ofbp, here we recall a schematic diagram in Fig. 4 which explains the characteristics of an ofbp (Fig. 4b) vis á vis a conventional bipolar and and a monopole pair (mpp) in Fig. 4a and c respectively. For both an ofbp and an mpp, the distances between the two peaks are relatively large compared to the characteristic width of the each peak. The fine difference between an ofbp and mpp lies in the finite slope of the E-field connecting the two lobes for the former which categorizes it as a solitary wave in general. An mpp, on the other hand, is a pair of two simple monopoles with opposite polarities (Fig. 4c) while the slope of the electric field connecting the two poles goes ideally to zero. The significance of the finite slope in E reflects the low, but non zero slope of the Sagdeev pseudopotential at its maximum amplitude which in turn defines the respective boundary conditions of different nonlinear structures, viz., SW (4a, 4b), DL (5), and FTSW (7). Besides providing a physical explanation for the morphology of an ofbp, it further enhances the candidature of the FTSW to interpret the observed structures.

Recently Qureshi et al. [2] generalized the concept of ofbp beyond the acoustic mode as they predicted both ion acoustic and ion cyclotron ofbp for their theoretical model. Though their results are yet to be validated by the observational data, it indicates that the structure is more generic than it was thought so far. According to the present understanding, the offset bipolars are appearing like a sporadic "deformation" of the conventional bipolar electric field, resulting due to certain arbitrary local conditions at the spot. The proposed theory of FTSW, on the other hand, generalizes an ofbp beyond its local conditions correlating it to an FTS which is, like soliton, is known to exist across different physical situations, even beyond the realm of the plasma physics. Apart from its unique morphology, one common characteristic of an FTS is that they often define a boundary between two phases or nonlinear dynamical processes. Similar characteristic has also been reported for an FTSW as well. During our theoretical analysis, we have found that the FTSW is occurring at the boundary between two types of SSWs, viz. Type I and Type II, where an SSW is characterized by the extra wiggles in their otherwise bipolar electric field [9]. A Type I SSW associates it with a preceding DL/monopole while a Type II emerges due to a continuous deformation of the bipolar electric field. We here conjecture that, analogous to an FTSW or FTS, an ofbp, too, may define a boundary between two phases, or nonlinear dynamical processes in the space. A more rigorous mathematical derivation of an ofbp from the preliminary FTS solution is beyond the scope of the current paper and may be presented elsewhere.

4 Conclusion

Using a simple plasma model, we have analytically estimated the shape, size, and speed of an FTSW which was found to be consistent with the slow moving ofbp observed in the Earth's auroral region. It manifests that ofbps are eventually FTSWs, or even may be FTS where the latter is the more mathematical and ideal counterpart of FTSW. Following the theory of the FTSW and FTS, we have interpreted the ofbp as an amalgamation of SW and DL which determines the boundary of two distinct phases of nonlinear dynamical processes. The proposed theory not only explains the unique morphology of the E-field data but also provide a more generic interpretation for the ofbp which is eventually correlating the mathematical description of a coherent nonlinear dynamical structure with the satellite observations. It is expected to provide a new way of understanding the non conventional localized pulses in the E-field data recorded during satellite expeditions which are known to be important in determining the microphysics of the Earth's magnetospheric boundary layers.

Acknowledgements The first author gratefully acknowledges financial support from Khalifa University's Sapce and Planetary Science Center under grant No. KU-SPSC-8474000336.

References

1. Temerin, M., Cerny, K., Lotko, W., Mozer, F.Z.: Observations of double layers and solitary waves in the auroral plasma. Phys. Rev. Lett **48**, 1175–1179 (1982)
2. Qureshi, M.N.S., Shi, J., Torkar, K., Liu, Z.: Theoretical properties of offset bipolar electric field solitary structures in space plasmas. Adv. Space Res. **45**, 1219–1223 (2010)
3. Lotko, W., Kennel, C.F.: Spiky ion acoustic waves in collisionless auroral plasma. J. Geophys. Res. **88**, 381–394 (1983)
4. Tsurutani, B.T., Arballo, J.K., Lakhina, G.S., Ho, C.M., Buti, B., Pickett, J.S., Gurnett, D.A.: Plasma waves in the dayside polar cap boundary layer: bipolar and monopolar electric pulses and whistler mode waves. Geophys. Res. Lett. **25**(22), 4117–4120 (1998)
5. Muschietti, L., Roth, I., Carlson, C.W., Berthomier, M.: Modeling stretched solitary waves along magnetic field lines. Nonlinear Process. Geophys. **9**, 101–109 (2002)
6. Roth, I., Muschietti, L., Carlson, C.W., Mozer, F.S., Ergun, R.E.: Stability and interaction of fast auroral solitary structures in three dimensional plasma. J. Geophys. Res.: Space Phys. **107**, 13–14 (2002)
7. Deng, X., Tang, R., Matsumoto, H., Pickett, J., Fazakerley, A., Kojima, H., Baumjohann, W., Coates, A., Nakamura, R., Gurnett, D., Liu, Z.: Observations of electrostatic solitary waves associated with reconnection by geotail and cluster. Adv. Space Res. **37**, 1373–1378 (2005)

8. Pickett, J., Chen, L.-J., Mutel, R., Christopher, I., Santolk, O., Lakhina, G., Singh, S., Reddy, R., Gurnett, D., Tsurutani, B., Lucek, E., Lavraud, B.: Furthering our understanding of electrostatic solitary waves through cluster multispacecraft observations and theory. Adv. Space Res. **41**, 1666–1676 (2008)
9. Steffy, S.V., Ghosh, S.S.: Phase portrait analysis of super solitary waves and flat top solutions. Phys. Plasmas **25**, 302 (2018)
10. Steffy, S.V., Ghosh, S.S.: Transitional properties in a two electron temperature warm multi-ion plasma. Phys. Plasmas **23**, 082304 (2016)
11. Grimshaw, R., Pelinovsky, D., Pelinovsky, E., Slunyaev, A.: Generation of large-amplitude solitons in the extended kortewegde vries equation. Chaos: Interdiscip. J. Nonlinear Sci. **12**, 1070–1076 (2002)

Forced KdV Equation in Degenerate Relativistic Quantum Plasma

Geetika Slathia, Rajneet Kaur, Kuldeep Singh, and Nareshpal Singh Saini

Abstract The study of heavy ion acoustic solitary waves (HIASWs) in a relativistic degenerate dense plasma (RDDP) having relativistic degenerate lighter ions as well as electrons and inertial heavy ions fluid has been illustrated. By adopting the reductive perturbation method, the forced Korteweg-de Vries (fKdV) equation is obtained to examine the HIASWs. The solution for HIASWs is determined analytically in the presence of external periodic force. It has been observed that the impact of various plasma parameters viz., speed of the wave, frequency and strength of the periodic force significantly alter the basic characteristics of different HIASWs. The findings of this work may be convenient to understand the behaviour of HIASWs in white dwarfs.

Keywords Heavy ions acoustic waves · Quantum plasma · Forced KdV equation

1 Introduction

Quantum plasma has incited great deal of interest and has attracted many physicists for the comprehensive study to explore different nonlinear waves in white dwarfs, neutron stars and black holes [1–3] owing to its low temperature and extremely large density in dense astrophysical region [4–6]. The various onboard satellite observations [1] have reported that the average particles density per unit volume of the white dwarfs is 10^6 to 10^8 gcm^{-3} with average interparticle distance of 10^{-10} to 10^{-13} cm [7]. The prestellar is mainly having heavy nuclei which was compressed

G. Slathia (✉) · R. Kaur · K. Singh · N. S. Saini
Department of Physics, Guru Nanak Dev University, Amritsar 143005, India
e-mail: gslathia93@gmail.com

K. Singh
Department of Mathematics, Khalifa University of Science and Technology, Abu Dhabi, UAE

S. Banerjee and A. Saha (eds.), *Nonlinear Dynamics and Applications*,
Springer Proceedings in Complexity,
https://doi.org/10.1007/978-3-030-99792-2_2

into ultra high densities. It is remarkable that the white dwarfs are having number density of heavy nuclei is 10^{29} cm^{-3} while the average distance between particle is 10^{-10} cm [1]. The estimation of degeneracy of different species is followed by Heisenberg's uncertainty principle where the uncertainty in momenta is extremely large and fast moving degenerate elements yields high pressure, (i.e., degenerate pressure). In quantum regime, as the interspecies distance is comparable to the de Broglie wavelength of species then relativistic effects become important. The characteristics and propagation of nonlinear HIAWs have been explored in different the unmagnetized RDDP models [8–14]. Islam et al. [11] reported the investigation of extremely HIA shock waves in a DRQP system by obtaining Burgers' equation. Singh et al. [12] have examined the characteristics of HIA oscillatory as well as monotonic shocks in dense magnetoplasma. Saini et al. [13] have investigated the dynamics of cnoidal and solitons in a three component magnetised plasma. By incorporating Sagdeev pseudopotential technique, the energy balance equation has been obtained. Kaur et al. [14] have studied the heavy and lighter ions acoustic higher order shock waves in quantum plasma. They have examined that the inclusion of higher order corrections yield dressed shocks.

With the induction of the external periodic force, different kinds of nonlinear waves excite in the plasma system and they characterise completely different which has been investigated by numerous researchers [15–18]. The characteristics of HNA solitary structures in an unmagnetized dense plasma containing of degenerate electrons and lighter ions in a heavy ions fluid along with the source term obtained from various experiments and simulations have been studied [15–18]. A lot of different investigations have been reported which describe about the underlying phenomena of various nonlinear excitations in astrophysical environments [19–21]. Sen et al. [19] illustrated the characteristics of nonlinear waves in the Lower Orbital region of the Earth by derving fKdV equation under influence of external periodic perturbation. Ali et al. [20] explored the analytical solution and the effect of various plasma parameters on the electron acoustic solitons by obtaining the fKdV equation. Mir et al. [21] illustrated the nonlinear wave mixing obtained from the exact analytical solution of the fKdV equation in the dusty plasma.

In this investigation, the main aim is to explore the dynamics of HIASWs in RDDP comprising of inertial heavy ions with degenerate lighter ions and electrons. To the best of our knowledge, the study of various kinds of nonlinear coherent structures in a RDDP obtained from the fKdV equation in white dwarfs have not been reported so far. The layout of the manuscript is presented as follows: The basic fluid model is provided in Sect. 2. The derivation and analytical solution of fKdV is discussed in Sect. 3. Different kinds of nonlinear structures have been examined and their parametric analysis is given in Sect. 4. The last Sect. 5 presents the conclusions.

2 Basic Fluid Equations

We assume unmagnetised RDDP containing relativistic degenerate electrons and lighter ions and inertial heavy ions as fluid. Thus, we have $N_{e0} = Z_h N_{h0} + Z_l N_{l0}$, where N_{e0}, N_{l0}, N_{h0} are the unperturbed number density of electrons, lighter and heavy ions, respectively. The dynamics of HIAWs is described by the following normalized equations as [12]:

$$\frac{\partial N_h}{\partial t} + \frac{\partial (N_h U_h)}{\partial x} = 0, \tag{1}$$

$$\frac{\partial U_h}{\partial t} + U_h \frac{\partial U_h}{\partial x} = -\frac{\partial \phi}{\partial x}, \tag{2}$$

$$\frac{\partial^2 \phi}{\partial x^2} = (1 + \beta\alpha) N_e - \beta\alpha N_l - N_h + \Theta(x, t), \tag{3}$$

where N_i(for $i = h, l, e$) is normalised by its equilibrium density N_{i0}. Here, source term $\Theta(x, t)$ is an external periodic perturbation which appears in different laboratory experiments. The solitary wave structures or solitons experience external forces such as the resistive wall modes of the plasma when subjected to external magnetic force which may be constant or periodic. The evolution of HIASWs, whose key attributes change with time when subjected to an external perturbations are termed as forced KdV solitary wave structures [20, 22]. The ϕ is normalised by $\phi_0 = m_e c^2/e$ and U_h by $C_0 = (Z_h m_e c^2/m_h)^{1/2}$. The time and space coordinates are normalised by $\omega_{ph} = (4\pi N_{h0} Z_h^2 e^2/m_h)^{1/2}$ and $\lambda_s = (m_e c^2/4\pi Z_h N_{h0} e^2)^{1/2}$, respectively. Where, $\eta_1 = \eta_l N_{l0}^{\rho_l - 1}/Z_l m_e c^2$ and $\eta_2 = \eta_e N_{e0}^{\rho_e - 1}/Z_e m_e c^2$, $\alpha = Z_l/Z_h$ and $\beta = N_{l0}/N_{h0}$. The expression for lighter ions and electrons in RDDP is given as [12]:

$$N_l = \left(1 - \frac{\rho_l - 1}{\rho_l \eta_1}\phi\right)^{\frac{1}{\rho_l - 1}} = 1 - a_1\phi + a_2\phi^2 + ... \tag{4}$$

$$N_e = \left(1 + \frac{\rho_e - 1}{\rho_e \eta_2}\phi\right)^{\frac{1}{\rho_e - 1}} = 1 + b_1\phi + b_2\phi^2 + ... \tag{5}$$

Now, by putting the (4)–(5) into (3), we get;

$$\frac{\partial^2 \phi}{\partial X^2} = 1 - N_h + c_1\phi + c_2\phi^2 + \Theta(x, t), \tag{6}$$

where $c_1 = b_1 + a_1$ and $c_2 = b_2 - a_2$. Here, $a_1 = \frac{\beta\alpha}{\rho_l \eta_1}$, $b_1 = \frac{1+\beta\alpha}{\rho_e \eta_2}$, $a_2 = \frac{\beta\alpha(2-\rho_l)}{2\rho_l^2 \eta_1^2}$ and $b_2 = \frac{(1+\beta\alpha)(2\rho_e)}{2\rho_e^2 \eta_2^2}$.

3 Forced KdV Equation and Its Solution

The stretching of variables is given as:

$$X = \epsilon^{\frac{1}{2}}(x - \lambda t) \tag{7}$$

$$T = \epsilon^{\frac{3}{2}} t \tag{8}$$

The expansion of dependent variables is illustrated as

$$N_h = 1 + \epsilon N_{h1} + \epsilon^2 N_{h2} + ... \tag{9}$$

$$U = \epsilon U_1 + \epsilon^2 U_2 + ... \tag{10}$$

$$\phi = \epsilon \phi_1 + \epsilon^2 \phi_2 + ... \tag{11}$$

$$\Theta(x, t) = \epsilon^2 \Theta_2(x, t) + ... \tag{12}$$

Substituting (7)–(12) into (1)–(2) and (6) and equating the coefficients of smaller powers of ϵ, we get the following equations:

$$N_{h1} = \frac{1}{\lambda^2}\phi_1 \quad \text{and} \quad \mathrm{U}_1 = \frac{1}{\lambda}\phi_1, \tag{13}$$

$$\lambda = \frac{1}{\sqrt{c_1}} \tag{14}$$

Equation (14) represents the dispersion relation of HIASWs. In limiting case, the phase speed matches with Singh et al. [12]. From the next higher order, we get:

$$\frac{\partial N_{h1}}{\partial T} - \lambda \frac{\partial N_{h2}}{\partial X} + \frac{\partial U_2}{\partial X} + \frac{\partial N_{h1} U_1}{\partial X} = 0, \tag{15}$$

$$\frac{\partial U_1}{\partial T} - \lambda \frac{\partial U_2}{\partial X} + U_1 \frac{\partial U_1}{\partial X} + \frac{\partial \phi_2}{\partial X} = 0, \tag{16}$$

$$\frac{\partial^2 \phi_1}{\partial X^2} = c_1 \phi_2 + c_2 \phi_1^2 - N_{h2} + \Theta(X, T) \tag{17}$$

Equating second order coefficients of ϵ and after some algebraic manipulations, we get the following forced KdV (fKdV) equation,

$$\frac{\partial \phi_1}{\partial T} + A\phi_1 \frac{\partial \phi_1}{\partial X} + B \frac{\partial^3 \phi_1}{\partial X^3} = B \frac{\partial \Theta_2(X, T)}{\partial X} \tag{18}$$

where nonlinear coefficient $A = (\frac{3}{2\lambda} - \frac{\lambda c_2}{c_1})$ and dispersion coefficient $B = \frac{\lambda^3}{2}$.

Jun-Xiao and Bo-Ling [23] determined the solutions of the fKdV equation by adopting Hirota bilinear technique. Let us consider $\Theta_2 = \frac{f_0}{B} X \cos(\omega T)$, where f_0 represents the strength and ω is the frequency of the source. Therefore, (18) becomes;

$$\frac{\partial \phi_1}{\partial T} + A\phi_1 \frac{\partial \phi_1}{\partial X} + B\frac{\partial^3 \phi_1}{\partial X^3} = f_0 \cos(\omega T). \tag{19}$$

This type of source term is deduced from the experimental results. Equation (19) is called fKdV equation. If $f_0 = 0$, then (19) reduces the usual KdV equation;

$$\phi_1 = \phi_m \sec h^2 \left(\frac{X - \Lambda T}{W} \right), \tag{20}$$

where $\phi_m = \frac{3\Lambda}{A}$ is the maximum amplitude and $W = 2\sqrt{\frac{B}{\Lambda}}$ is width and Λ is the velocity of the HIASWs. The influence of periodic term $f_0 \cos(\omega T)$ on the HIASWs we have also determined the momentum conservation law. For small forcing term $f_0 \cos(\omega T)$, we obtain;

$$I = \int_{-\infty}^{\infty} \phi_1^2 dX, \tag{21}$$

$$I = \frac{24\sqrt{B}}{A^2} \Lambda^{\frac{3}{2}}(T), \tag{22}$$

and

$$\int_{-\infty}^{\infty} \phi_1 dX = \frac{12\sqrt{B\Lambda(T)}}{A}, \tag{23}$$

where $\Lambda(T)$ is the speed of the HIASWs of the fKdV equation. Now differentiate (21) w.r.t T, we get;

$$\frac{dI}{dT} = 2f_0 \cos(\omega T) \tag{24}$$

Putting (22) and (23) into (24) along with $\Lambda(0) = \Lambda$, we get;

$$\Lambda(T) = \Lambda + \frac{2Af_0}{3\omega} \sin(\omega T). \tag{25}$$

Then, the solution of fKdV equation can be written as;

$$\phi_2 = \phi_m(T) \sec h^2 \left(\frac{X - \Lambda(T)T}{W(T)} \right), \tag{26}$$

where the maximum amplitude and width of HIASWs are time dependent i.e., $\phi_m = \frac{3\Lambda(T)}{A}$ and $W(T) = 2\sqrt{\frac{B}{\Lambda(T)}}$, respectively.

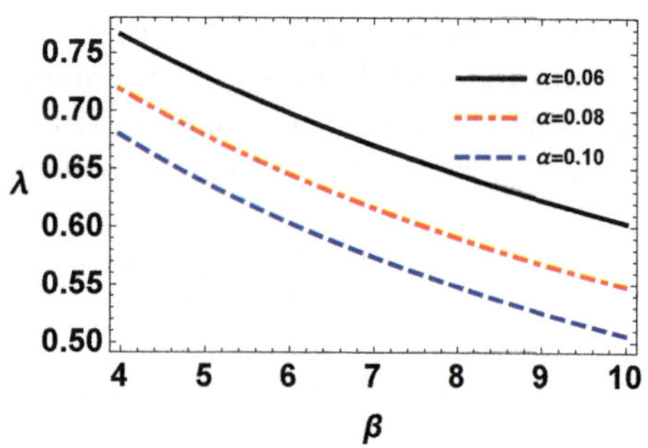

Fig. 1 The phase speed (λ) of HIASWs versus $\beta(=N_{l0}/N_{h0})$ for different values of $\alpha(=Z_l/Z_h)$

4 Parametric Analysis

Here, the parametric analysis has been performed to examine the characteristics of HIASWs in RDDP. We have assumed three species RDDP having relativistic degenerate lighter ions/electrons (viz., ${}^{1}_{1}H$ or ${}^{4}_{2}He$ or ${}^{12}_{6}C$ or ${}^{16}_{8}O$) and heavy ions (viz., ${}^{56}_{26}Fe$ or ${}^{87}_{37}Rd$ or ${}^{96}_{42}Mo$), and the data is taken from the region of white dwarfs [10, 12, 24].

Figure 1 shows the plot for phase speed (λ) of HIASWs versus β $(=N_{l0}/N_{h0})$ for different values of α $(=Z_l/Z_h)$. The phase speed of HIASWs enervates with rise in β and α which means that phase speed gets reduced as the charge and number densities of heavy ions are flourished.

Figure 2 illustrates the plot of nonlinear coefficient A of HIASWs versus β for different values of α. The A is increased with an increment in β and α. It is remarked that only positive potential (compressive) HIASWs are evolved for $A > 0$.

Figure 3 illustrates the plot of the compressive HIASWs profile for different values of α and β. The amplitude of the HIASWs decreases with the rise in α and β. It is emphasized that this shrink in the HIASWs amplitude is because of rise in the value of A.

In Fig. 4, profile of compressive HIASWs for various values of strength (f_0) and frequency ω of periodic force. It is found that as f_0 increases, the amplitude of the HIASWs is increased and is decreased with increment in ω. In Fig. 5, the 3D plot of compressive HIASWs profile of the fKdV equation for different values of α is given. It is found that as the values of α is increased, the amplitude of the HIASWs is decreased. Similarly, Fig. 6 shows the 3D plot of compressive HIASWs profile of the fKdV equation versus β. It is noticed that as the values of β is increased, the maximum amplitude of the HIASWs is shrinked.

In Fig. 7, the plot of compressive HIASWs profile versus f_0 for different α and β is shown. It is noticed that the maximum amplitude of HIASWs decreases with rise in the values of α and β but increases as the values of f_0 increases.

In Fig. 8, the plot of compressive HIASWs profile us ω for different values of α and β is shown. It is found that the width of HIASWs decreased with rise in the values of α and β but enhanced with rise in the values of ω.

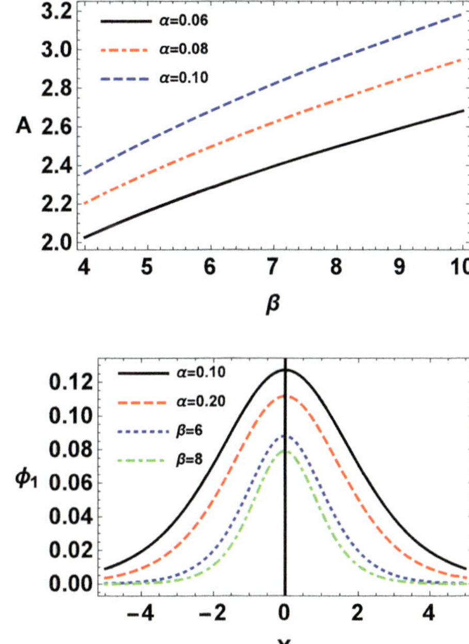

Fig. 2 The nonlinear coefficient (A) of HIASWs versus $\beta(=N_{l0}/N_{h0})$ for different values of $\alpha(=Z_l/Z_h)$

Fig. 3 The compressive HIASWs profile for different values of $\alpha = Z_l/Z_h$ and $\beta = N_{l0}/N_{h0}$

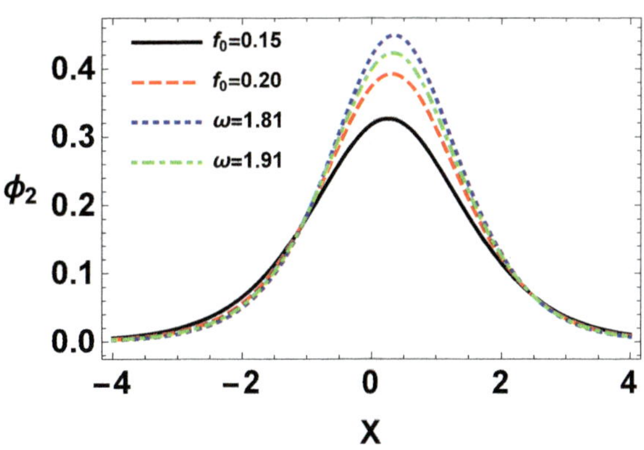

Fig. 4 The HIASWs profile of the fKdV equation for different values of f_0 and frequency ω with fixed values of $\Lambda = 0.1$, $\alpha = 0.08$, and $\beta = 5$

It is stressed that the various plasma parameters have significantly influenced the nonlinear and dispersion effects to modify the characteristics of HIASWs in the RDDP environment.

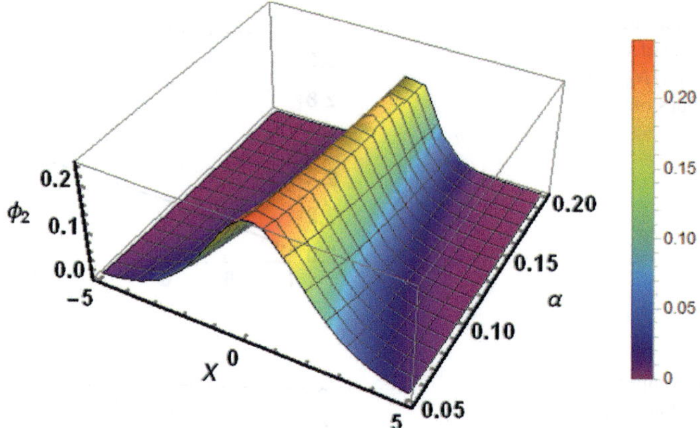

Fig. 5 The 3D plot of HIASWs profile of the fKdV equation vs α for $\Lambda = 0.1$, $f_0 = 0.1$, $\omega = 1.5$, $T = 1.5$

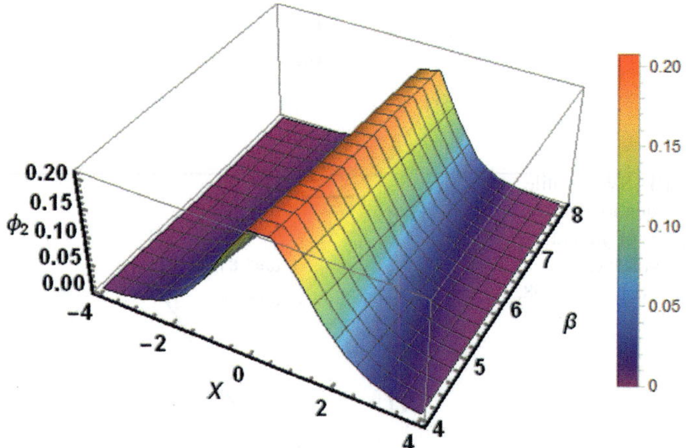

Fig. 6 The 3D plot of HIASWs profile versus β with $\Lambda = 0.1$, $f_0 = 0.1$, $\omega = 1.5$, and $T = 1.5$

Fig. 7 The maximum amplitude of HIASWs versus f_0 for different α and β, other parameters $\Lambda = 0.1$, $\omega = 1.2$, $T = 1$ are fixed

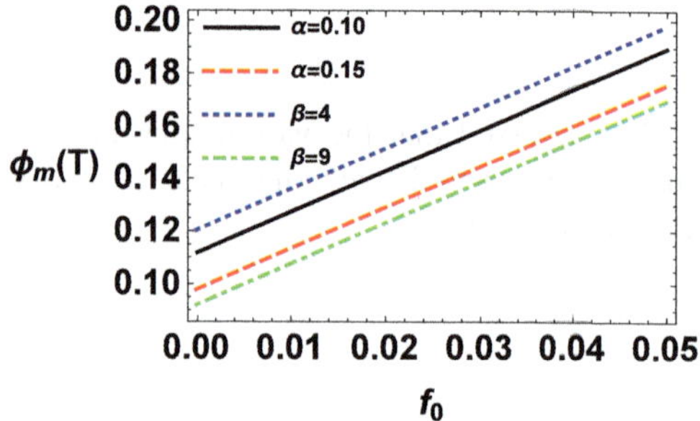

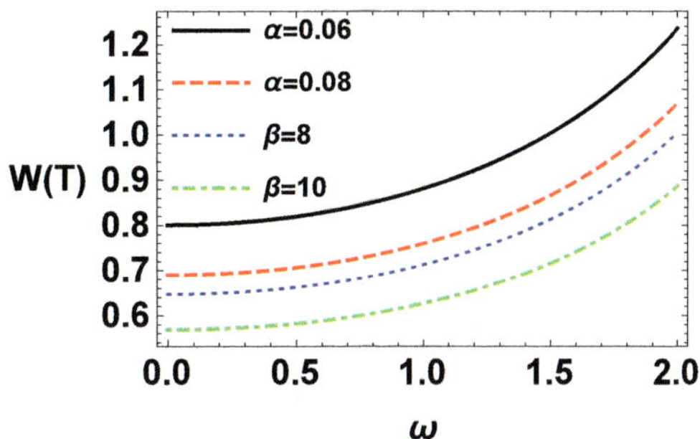

Fig. 8 The width of HIASWs wrt ω of the fKdV equation for different values of α and β, where other parameters $\Lambda = 0.1$, $f_0 = 0.7$, $T = 1.1$ are fixed

5 Conclusions

In this investigation, we have examined the salient features of HIASWs in RDDP having degenerate lighter ions/electrons and inertial heavy ion fluid. By adopting the reductive perturbation method, the forced Korteweg-de Vries (fKdV) equation is obtained for HIASWs. The solution of fKdV equation is determined to explore the behaviour of HIASWs under the influence of periodic force. Only compressive HIASWs are observed. It is shown that the impact of various plasma parameters like speed and strength of the periodic force significantly modify the basic properties of different HIASWs. The amplitude of HIASWs enervates with the rise in the density ratio while the amplitude of the HIASWs excels with rise in the value of periodic force. The findings of this investigation may highlight the basic features of HIASWs in RDDP in white dwarfs [10, 12, 24].

References

1. Shapiro, S.-L., Teukolsky, S.-A.: Black Holes, White Dwarfs, and Neutron Stars: The Physics of Compact Objects. Wiley-VCH Verlag, Weinheim (2004)
2. Koester, D., Chanmugam, G.: Physics of white dwarf stars. Rep. Prog. Phys. **53**(7), 837–915 (1990)
3. Koester, D.: White dwarfs: recent developments. Astron. Astrophys. Rev. **11**(1), 33–66 (2002)
4. Drake, R.-P.: High-energy-density physics. Phys. Today **63**(6), 28–33 (2010)
5. Killian, T.-C.: Plasma physics: cool vibes. Newly observed electron-density waves could become useful probes of how electrons behave in this exotic regime. Nature **441**(5), 297–298 (2006)
6. Glenzer, S.-H., Redmer, R.: X-ray Thomson scattering in high energy density plasmas. Rev. Mod. Phys. **81**(4), 1625–1663 (2009)
7. Azam, M., Sami, M.: Many-body treatment of white dwarf and neutron stars on the brane. Phys. Rev. D **72**(7), 024024(1–11) (2005)
8. Mamun, A.-A., Amina, M., Schlickeiser, R.: Nucleus-acoustic shock structures in a strongly coupled self-gravitating degenerate quantum plasma. Phys. Plasmas **23**(9), 094503(1–4) (2016)
9. Mamun, A.-A., Amina, M., Schlickeiser, R.: Heavy nucleus-acoustic spherical solitons in self-gravitating super-dense plasmas. Phys. Plasmas **24**(4), 042307(1–7) (2017)

10. Sultana, S., Islam, S., Mamun, A.-A., Schlickeiser, R.: Modulated heavy nucleus-acoustic waves and associated rogue waves in a degenerate relativistic quantum plasma system. Phys. Plasmas **25**(1), 012113(1–9) (2018)
11. Islam, S., Sultana, S., Mamun, A.-A.: Ultra-low frequency shock dynamics in degenerate relativistic plasmas. Phys. Plasmas **24**(9), 092308(1–5) (2017)
12. Singh, K., Sethi, P., Saini, N.-S.: Nonlinear excitations in a degenerate relativistic magneto-rotating quantum plasma. Phys. Plasmas **26**(9), 092104(1–10) (2019)
13. Saini, N.-S., Kaur, M., Singh, K.: Heavy nucleus acoustic periodic waves in a degenerate relativistic magneto-rotating quantum plasma. Waves Random Complex Media **30**, 1–12 (2020)
14. Kaur, R., Singh, K., Saini, N.-S.: Heavy-and light-nuclei acoustic dressed shock waves in white dwarfs. Chin. J. Phys. **72**(8), 286–298 (2021)
15. Saha, A., Pal, N., Chatterjee, P.: Dynamic behavior of ion acoustic waves in electron-positron-ion magnetoplasmas with superthermal electrons and positrons. Phys. Plasmas **21**(10), 102101(1–10) (2014)
16. Saha, A., Pal, N., Chatterjee, P.: Bifurcation and quasiperiodic behaviors of ion acoustic waves in magnetoplasmas with nonthermal electrons featuring Tsallis distribution. Braz. J. Phys. **45**(4), 325–333 (2015)
17. Zhen, H., Tian, B., Wang, Y., Sun, W., Liu, L.: Soliton solutions and chaotic motion of the extended Zakharov-Kuznetsov equations in a magnetized two-ion-temperature dusty plasma. Phys. Plasmas **21**(7), 073709(1–7) (2014)
18. Sadiq, S., Mahmood, S., Haque, Q., Ali, M.-Z.: Ion acoustic solitons in dense magnetised plasmas with nonrelativistic and ultrarelativistic degenerate electrons and positrons. Astrophys. J. **793**(9), 27(1–12) (2014)
19. Sen, A., Tiwari, S., Mishra, S., Kaw, P.: Nonlinear wave excitations by orbiting charged space debris objects. Adv. Space Res. **56**(3), 429–435 (2015)
20. Ali, R., Saha, A., Chatterjee, P.: Analytical electron acoustic solitary wave solution for the forced KdV equation in superthermal plasmas. Phys. Plasmas **24**(12), 122106(1–9) (2017)
21. Mir, A.-A., Tiwari, S.-K., Goree, J., Sen, A., Crabtree, C., Ganguli, G.: A forced Korteweg-de Vries model for nonlinear mixing of oscillations in a dusty plasma. Phys. Plasmas **27**(11), 113701(1–6) (2020)
22. Chandra, S.: Analytical and simulation studies of forced KdV solitary structures in a two-component plasma. J. Korean Phys. Soc. **76**(6), 469–478 (2020)
23. Xiao, Z.-J., Ling, G.-B.: Commun. Theor. Phys. **52**(2), 279–283 (2009)
24. Tout, C.-A., Wickramasinghe, D.-T., Ferrario, L.: Magnetic fields in white dwarfs and stellar evolution. Mon. Not. R. Astron. Soc. **355**(3), L13–L16 (2004)

Heliospheric Two Stream Instability with Degenerate Electron Plasma

Jit Sarkar, Swarniv Chandra, Jyotirmoy Goswami, and Basudev Ghosh

Abstract In this paper, we have studied the effect of relativistic degeneracy, electron spin exchange potential and quantum diffraction effect on the electron two-stream instability in a dense stellar body like our Sun. The effect of streaming motion has been studied along with other parameters. We have observed regions of stability and instability in this case considering two oppositely flowing electron jets. Under constrained environments, our findings adjust to traditional results. The results of other researchers in this field also correspond to our findings. We have made use of the Quantum Hydrodynamic Model and also incorporated the one-dimensional thermal anisotropy that slightly modifies the model. Our findings will be helpful in the study of various instabilities and associated phenomena in solar flares, solar storms and similar other indicates in the stellar and nebular environments.

Keywords Two stream instability · Electron spin exchange potential · Quantum hydrodynamic model

1 Introduction

Over the past few years quantum plasma has attracted much attention in the plasma community. With the growth of study in dense plasma and the ability of experimenting with high power laser beams, the study of quantum tunneling in plasma phenomena has gained utmost importance. Initially pioneering works were carried out by Haas [17], Manfredi [21], Shukla [29], Eliasson [13], Sarkar [26–28], Goswami [15, 16], Brodin [5], Chandra [6–10] on quantum plasmas. Later there were more works with applicability of quantum phenomenon in such dense plasmas. Various types of insta-

J. Sarkar · S. Chandra (✉) · J. Goswami · B. Ghosh
Jadavpur University, Kolkata, India
e-mail: swarniv147@gmail.com

S. Chandra
Government General Degree College at Kushmandi, Kushmandi, West Bengal, India

S. Banerjee and A. Saha (eds.), *Nonlinear Dynamics and Applications*,
Springer Proceedings in Complexity,
https://doi.org/10.1007/978-3-030-99792-2_3

bilities and other non-linear effects were observed in them. Quantum hydrodynamic model (QHD) model has been often used to address problem like this.

In plasma physics, the two-stream instability is a very common type of phenomenon which can be induced by an energetic stream of particles injected into the plasma. It can be also obtained by setting a current along with the plasma so that different species (electron & ions; positron and negative ions; etc.) can have different drift velocities depending on their inertia as well as response to the field. Such instability appears immediate when the beams comprise cold particle species with none of the particles being resonant with the wave. Conversely, they can be found from two hot beams in which particles from one or both the streams are resonant with the wave. This kind of instability under various limiting cases evolve into beam plasma instability, beam instability or bump on tail instability. The dispersion relation of the wave showing two-stream instability has two parts, a real and other imaginary. With totally real roots there can be no type of damping. On the other hand with the non-zero imaginary segment of the frequency, there may be either damping or instability(growing). Electron two stream Instability (ETSI) which is driven by oppositely streaming beams was first represented by Bohm and Gross [3] since then various space plasma and experimental instances have reported the existence of ETSI. Such instability produces strong electron heating and coherent plasma emissions at ω_{pe} (electron plasma frequency) or its harmonics. As the plasma density varies along the path of the electron beams, the emission frequency drifts. Solar emissions being in the radio band emit 'U' and 'J' type of radio bursts which correspond to various origin of the electrons and contain information about them.

In this paper, we will try to investigate the two-stream instability, its nature and evaluation in a plasma containing streams of electrons moving in the opposite direction in dense quantum plasma. We consider an anisotropy in the direction of propagation of the wave [14] which is more meaningful due to such temperature. The ions are homogenously distributed to form a neutralising background.

We organised the article in the following way. In Sect. 2 we start with the dynamical equations governing our model plasma. In Sect. 3 we derive the linear dispersion relation. In the next section, we analyzed the result and finally conclude with some remarks concerning recent findings and possible application.

2 Finite Temperature Quantum Hydrodynamic Model

Our mathematical model is based on the three dimensional equilibrium Fermi-Dirac distribution for electrons. Due to adiabatic compression, temperature anisotropy occurs in distribution with the motion of a longitudinal electrostatic wave propagating in collision-less plasma. The Fermi-Dirac equilibrium state for Fermions is given by Bransden and Joachain [4] in the following form

$$n_0 = \frac{1}{2\pi^2}\left(\frac{2m}{\hbar}\right)^{3/2}\int_0^\infty \frac{E^{1/2}dE}{e^{\beta(E-\mu)}+1}$$
$$= -\frac{1}{2\pi^2\beta^{3/2}}\left(\frac{2m}{\hbar}\right)^{3/2}\Gamma\left(\frac{3}{2}\right)Li_{3/2}(-e^{\beta\mu}) \tag{1}$$

where m is the electron mass, $\hbar$ is the reduced Plank's constant, n_0 is the equilibrium number density, $\beta = 1/k_B T_{e0}$, T_{0E} is the background temperature of electron, μ is the chemical potential and $Li_\nu(x)$ is the poly-logarithmic function in x of order (ν). When $\beta \to \infty$ i.e, cold temperature of electron, we have $\mu \to \varepsilon_F$, where ε_F is the fermi energy, and Eq. (1) becomes

$$n_0 = \frac{1}{3\pi^2}\left(\frac{2m}{\hbar^2}\right)^{3/2}\varepsilon_F^{3/2} \tag{2}$$

or in other way the Fermi energy is

$$\varepsilon_F = (3\pi^2 n_0)^{2/3}\frac{\hbar^2}{2m} \tag{3}$$

Here, we can obtain the Vlasov equations with $df/dt = 0$ as the fluid is incompressible in phase space. But the incompressibility is compromised by quantum tunneling according to Wigner equation [14], so a non-equilibrium particle distribution function can be represented as

$$f(x,u,t) = \frac{\alpha}{exp[(\beta m/2)\{(u_x-u_{ex})^2\eta + u_y^2 + u_x^2\} - \beta\mu] + 1} \tag{4}$$

where $u_{ex}(x,t)$ is mean velocity of the particles and η is the temperature anisotropy given by $\eta(x,t) = [n_0/n_e(x,t)]^2$ and $\alpha = -\frac{n_0}{Li_{3/2}(-e^{\beta\mu})}\left(\frac{\beta m}{2\pi}\right)^{(3/2)}$ (normalization constant). With a constant chemical potential and normalizing f over velocity space equals n_0; when $\eta = 1$ and $u_{ex} = 0$, we get $f_{max} = \alpha/[exp(-\beta\mu)+1]$.

In the present problem, we consider the dynamics of electrons in positive (+) and negative (−) x axis. We also considered that the streaming of particles attain such values so as to incorporate relativistic factor (γ) in the dynamical equations [20]. Now, using Eq. 4, we calculate the zeroth, first and second moments of the distribution function (f). So from the zeroth and first moments of the Fermi-Dirac distribution function, we obtain the continuity and momentum equation in the following form:

$$\frac{\partial\gamma_\pm n_\pm}{\partial t} + \vec{\nabla}\cdot(\gamma_\pm n_\pm \vec{u}_\pm) = 0 \tag{5}$$

$$\frac{\partial\gamma_\pm\vec{u}_\pm}{\partial t} + (\vec{u}_\pm\cdot\vec{\nabla})\gamma_\pm\vec{u}_\pm = \frac{e}{m_{e0}}\vec{\nabla}\phi_E + \frac{1}{m_e n_\pm}\vec{\nabla}\Pi + \frac{\hbar^2}{2m_e^2\gamma_\pm}\left[\vec{\nabla}\left(\frac{\nabla^2\sqrt{n_\pm}}{\sqrt{n_\pm}}\right)\right] \tag{6}$$

Here, Π is the total pressure due to relativistic degeneracy, spin exchange interaction and finite temperature effects ($\Pi = P_G + P_{ex} + P_{degeneracy}$), where $\gamma_{\pm} = 1/\sqrt{1 - u_0^2/c^2}$, $n_{\pm}$ and $u_{\pm}$ denote the relativistic factor, electron number density and speeds along $\pm x$ directions respectively. And ϕ_E, n_{i0} refer to the electric potential and ion equilibrium number density. The last term in the momentum equation (6) corresponds to the Bohm potential.

The Relativistic degeneracy potential ϕ_{ch} originates from the degeneracy pressure in dense stellar bodies [11] is given by

$$P_{degeneracy} = \frac{\pi m_e^4 c^5}{3h^3}[R(2R^2 - 3)\sqrt{1 + R^2} + 3\sinh^{-1} R] \tag{7}$$

Here, $R = (n/n_0)^{1/3}$ and n_0 is the equilibrium number density of electrons in the plasma. We obtained effective potentials corresponding to relativistic degeneracy as $(\phi_{ch})[ch \rightarrow$ Chandrasekhar pressure [11]. The effective potential contribution due to relativistic degeneracy can be simplified as [1]

$$\left.\begin{aligned} \phi_{ch} &= \sqrt{1 + R_0^2 n^{2/3}} \quad \textit{Classical case} \\ \phi_{ch} &= \sqrt{1 + R_0^6 \tfrac{n^2}{\beta}} \quad \textit{Quantum case} \end{aligned}\right\} \tag{8}$$

Here, $\beta = \frac{B_0}{B_d}$ with B_0 is the axial magnetic field strength and $B_d \approx 4.4 \times 10^9$ Tesla is the critical magnetic field intensity of typical strongly magnetised stars. The condition for quantum regime is $R^2 << 2\beta$ whereas $R^2 >> 2\beta$ corresponds to classical case.

$$R = (\frac{n_e^{cl}}{n_0})\frac{R_0^3}{\beta} \tag{9}$$

corresponds to classical case and

$$R = (\frac{n_e^Q}{n_0})^{1/3} R_0 \tag{10}$$

correspond to quantum regime.

Such a correlation can be incorporated by eliminating the exchange potential term (ϕ_{ex}) and replacing Γ by $R_0/\sqrt{\beta^2 + \beta R_0^6}$. This transformation was also checked by Akbari-Moghanjoughi and Ghorbanalilu [2].

The electron spin exchange interaction [ϕ_{ex}], though small can have significant effects on the electron wave at such high densities. The electron spin exchange potential [25] in generalised terms is given as

$$\left.\begin{array}{l}\phi_{ex} = -\phi_0 \left[A(\eta) - B(\eta)\right] \\ A(\eta) = \left[\frac{1}{32}\left(\eta^4 + \eta^{-4}\right) + \frac{1}{4}\left(\eta^2 + \eta^{-2}\right) - \frac{3}{4}\left(\eta^2 - \eta^{-2}\right)\ln\eta - \frac{9}{16} + \frac{3}{2}(\ln\eta)^2\right] \\ B(\eta) = \left[-\frac{\eta}{3}\left(\frac{\eta^2-1}{\eta^2+1}\right)\left\{\frac{1}{8}\left(\eta^3 - \eta^{-5}\right) - \frac{1}{4}\left(\eta - \eta^{-3}\right) - \frac{3}{2}\left(\eta + \eta^{-3}\right)\ln\eta + \frac{3\ln\eta}{\eta}\right\}\right]\end{array}\right\} \tag{11}$$

where $\eta = R + \sqrt{1 + R^2}$, $\phi_0 = \frac{2\alpha m_e^4 c^5}{h^3}$ and $\alpha = e^2/\hbar c$ is the fine structure constant.

This system is bounded by the Poisson's equation

$$\nabla^2 \phi_E = 4\pi e\left(\frac{n_+}{2} + \frac{n_-}{2} - n_{i0}\right) \tag{12}$$

The value of degeneracy parameters for different kinds of plasma is given below [6]:

Degeneracy parameter for different kinds of plasma			
Types of plasma	Density (m^{-3})	Temperature (K)	G
Tokamak	10^{20}	10^{18}	1
Inertial confinement fusion	10^{32}	10^{8}	1
Metal and metal clusters	10^{28}	10^{4}	1.4
Jupiter	10^{32}	10^{4}	1.4
White dwarf	10^{35}	10^{8}	4

3 Linear Dispersion Relation and Instability Criteria

In order to investigate the linear and non-linear behavior of electron acoustic wave in this three component electron-ion plasma we make the following perturbation expansion for the field quantities $n_\pm$, $u_\pm$ and ϕ_E about their equilibrium values:

$$\begin{bmatrix} n_\pm \\ u_\pm \\ \phi_E \end{bmatrix} = \begin{bmatrix} 1 \\ u_{(0)} \\ 0 \end{bmatrix} + \varepsilon \begin{bmatrix} n_\pm^{(1)} \\ u_\pm^{(1)} \\ \phi_E^{(1)} \end{bmatrix} + \epsilon^2 \begin{bmatrix} n_\pm^{(2)} \\ u_\pm^{(2)} \\ \phi_E^{(2)} \end{bmatrix} + \cdots \tag{13}$$

and normalization of the relativity parameter is carried out as $R \to \bar{R}R_0$ with $R_0 = (n_\pm/n_0)^{1/3}$. Now we express $A(\eta)$ and$B(\eta)$ in Eq. 11 in terms of n_0 & $n_\pm$ in the following manner

$$\eta = R + \sqrt{1 + R^2} = \left(\frac{n}{n_0}\right)^{1/3} + \left[1 + \left(\frac{1}{n_0}\right)^{1/3}\left(1 + \varepsilon n^{(1)} + \varepsilon^2 n^{(2)}\right)^{1/3}\right]^{1/2} \tag{14}$$

The different orders of η (omitting $\pm$ sign for brevity) can be obtained as,

$$\left.\begin{array}{l}\eta^1 = P^1\left[1+\varepsilon\frac{Q}{P}n^{(1)}\right];\quad \eta^2 = P^2\left[1+2\varepsilon\frac{Q}{P}n^{(1)}\right];\\ \eta^3 = P^3\left[1+3\varepsilon\frac{Q}{P}n^{(1)}\right];\quad \eta^4 = P^4\left[1+4\varepsilon\frac{Q}{P}n^{(1)}\right];\\ \eta^5 = P^5\left[1+5\varepsilon\frac{Q}{P}n^{(1)}\right];\quad \eta^{-1} = P^{-1}\left[1-\varepsilon\frac{Q}{P}n^{(1)}\right];\\ \eta^{-2} = P^{-2}\left[1-2\varepsilon\frac{Q}{P}n^{(1)}\right];\quad \eta^{-3} = P^{-3}\left[1-3\varepsilon\frac{Q}{P}n^{(1)}\right];\\ \eta^{-4} = P^{-4}\left[1-4\varepsilon\frac{Q}{P}n^{(1)}\right];\quad \eta^{-5} = P^{-5}\left[1-5\varepsilon\frac{Q}{P}n^{(1)}\right];\\ (\ln\eta)^2 = -2\left[1-4P\left(1+\varepsilon\frac{Q}{P}n^{(1)}\right)+2P^2\left(1+2\varepsilon\frac{Q}{P}n^{(1)}\right)\right]\end{array}\right\} \tag{15}$$

where

$$\left.\begin{array}{l}P = \left\{\left(\frac{1}{n_0}\right)^{(1/3)}+\frac{1}{2}\left(\frac{1}{n_0}\right)^{(2/3)}\right\}\\ Q = \left\{\frac{1}{3}\left(\frac{1}{n_0}\right)^{(2/3)}+\frac{1}{2}\left(\frac{1}{n_0}\right)^{(1/3)}n^{(1)}\right\}\end{array}\right\} \tag{16}$$

Putting all values from Eq. 15 omitting $\pm$ sign corresponding to streaming directions in Eq. 11, we get

$$\left.\begin{array}{l}A(\eta) = R_1+\varepsilon S_1 n^{(1)}\\ B(\eta) = V_1+\varepsilon W_1 n^{(1)}\end{array}\right\} \tag{17}$$

where $R_1 = \frac{1}{32}\left(P^4+P^{-4}\right)+\frac{1}{4}\left(P^2+P^{-2}\right)-\frac{3}{4}\Lambda-\frac{9}{16}+6\left(1-4P+P^2\right)$;
$\Lambda = 2\left(P^{-2}+P^2\right)(1-P)^2$;

$S_1 = \left\{\frac{Q}{8}\left(P^3-P^{-5}\right)+\frac{Q}{2}\left(P-P^{-3}\right)+\frac{3}{4}\beta-24Q+12QP\right\}$;

$$V_1 = \left\{\begin{array}{l}\frac{1}{24}P^4(P-1)^2-\frac{1}{24}P^{-4}(P-1)^2-\frac{P^2}{12}(P-1)^2\left(P-P^{-3}\right)\\ +P\left(P+P^{-3}\right)(P-1)^4-2P(P-1)^2\left(P-2+\frac{1}{P}\right)\end{array}\right\}$$

$$W_1 = \left\{\begin{array}{l}\frac{1}{8}P^3(P-1)^2Q+\frac{5}{24}P^{-5}(P-1)^2Q-\frac{1}{12}P(P-1)^2Q\\ +P\left(1-3P^{-4}\right)(P-1)^4Q-2P(P-1)^2\left(P+P^{-3}\right)\left(1-p^2\right)Q\\ -\frac{1}{3}P(P-1)^2\left(\frac{2}{P}-1+\frac{1}{P^2}\right)Q+\frac{P^3}{8}T_1-\frac{P^{-5}}{8}T_1-\frac{P}{4}(P-P^{-3})T_1\\ +3T_1(P+P^{-3})(P-1)^2-6T_1\left(P^{-1}+2+P\right)\end{array}\right\}$$

$T_1 = \left\{\frac{1}{3}(P-1)^2Q+\frac{4}{3}(P^4-P^2)Q\right\}$;

$\beta = \left\{2P^2\left(4QP-6Q+2\frac{Q}{P}\right)-2P^{-2}\left(4QP+2Q-2\frac{Q}{P}\right)\right\}$

Using linear perturbations of field quantities like $n_\pm = 1+n_\pm^{(1)}exp(ikx-\omega t)$; $u_\pm = u_0+u_\pm^{(1)}exp(ikx-\omega t)$ and $\phi_E = \phi_E^{(1)}exp(ikx-\omega t)$, the first order terms are obtained as

$$\left.\begin{array}{l}n_\pm^{(1)} = \frac{-k^2\phi_E^{(1)}}{\left[\left(\omega\mp ku_+^{(0)}\right)^2\gamma^3-k^2\Gamma-H^2k^4\right]}\\ u_\pm^{(1)} = \frac{-(\omega\mp ku_0)k\phi_E^{(1)}}{\left[(\omega\mp ku_0)^2\gamma^3-k^2\Gamma-H^2k^4\right]}\end{array}\right\} \tag{18}$$

where

$$\Gamma = \left\{ \frac{1}{12}\left(\frac{1}{n_0}\right)^{2/3} + \phi_{ex} S_1 - \phi_{ex} W_1 + 3G\beta^2 \right\} \tag{19}$$

Now the linear dielectric function can be written as $F(\omega, k) = 1 + \chi_{e+} + \chi_{e-}$ where $\chi_{\pm}$ are the dielectric susceptibilities of counter streaming electron fluids. The dispersion relation is obtained as

$$\frac{2}{R_0^3} + \frac{1}{[k^2\Omega - \gamma^3\omega(\omega + 2kv)]} + \frac{1}{[k^2\Omega - \gamma^3\omega(\omega - 2kv)]} = 0 \tag{20}$$

where $\Omega = (H^2k^2 - u_0^2\gamma^3 + \Gamma)$

For non-relativistic or weakly relativistic case ($R_0 \ll 1$) with non-relativistic streaming ($\gamma \to 1$) the Eq. 20 boils down into

$$\omega^2 = R_0^3 + \Gamma k^2 \tag{21}$$

Here we have ignored the finite temperature statistical pressure term. It is justified that if in the classical limit when $\beta\mu \to (-\infty)$, so that $G \to 0$, the dispersion relation (20) corresponds to the Bohm-Gross [3] dispersion relation for hot plasma i.e.;

$$\omega^2 = \omega_{pc}^2 + 3V_{Te}^2k^2 + \frac{\hbar^2k^4}{4m^2} \tag{22}$$

Now in the ultra cold limit we have to consider the finite temperature effect as $\beta\mu \to \infty$ i.e.; $G \to \frac{2}{5}\beta\mu$ & $\mu \to \varepsilon_F$, [$\equiv \frac{mV_{Fe}^2}{2}$ is the electron Fermi energy]. In this case the dispersion relation takes the form

$$\omega^2 = \omega_{pe}^2 + \frac{3}{5}V_{Fe}^2k^2 + \frac{\hbar k^4}{4m^2} - \frac{e^2V_{Fe}k^2}{5\pi\hbar} \tag{23}$$

where the second term of on the right hand side of Eq. 23 corresponds to the finite temperature contribution due to anisotropy, the third term corresponds to the quantum diffraction effect and the forth one is related to the electron exchange pressure. The result may be comparable to high frequency Langmuir oscillations [12]. These results correspond to the previous findings of von Roos and Zmuidzinas [24], Nozieres and Pines [23], Karazawa et al. [19] as well as recent works by Akbari-Moghanjoughi and Ghorbanalilu [2]. The terms in our dispersion relation have the same nature as previous researchers except with certain additional multiplicative terms on the RHS. The expansion of Lindhard dielectric function [18] for phase velocities ($V_{ph} \leq V_{Fe}$) has similar term like the second term in Eq. 23. The quantum diffraction term third on RHS of (23) correspond with the finding of Eliasson and Shukla [14]. The forth term too is relatable with Ekman and others [12] only with $\frac{3}{5\pi}$ co-factor distinct from Mohammadnejad and Akbari-Moghanjoughi [22]

Before Solving the linear dispersion relation (20), from the survey of the available literature we came to know that there is no well constructed model accounting for relativistic quantum hydrodynamics of finite temperature Fermi plasma. To some extent, we have knowledge on relativistic extension of (QHD) equations incorporating Bohm term from Wigner-Poisson's formulation. However, exchange pressure (Salpeter's [25]) gives a correct density dependence in this relativistic case. Considering all the factors the dispersion relation (20) is solvable and its solution gives four modes (or branches) as,

$$\omega = \pm\sqrt{\frac{[R_0^3 + (4u_0^2\gamma^3 + 2\Omega^2)k^2] \pm \sqrt{[(R_0^2 + 4u_0^2\gamma^3 k)^2 + 16u_0^2\gamma^3\Omega k^4]}}{2\gamma^3}} \tag{24}$$

or

$$\omega = \pm\sqrt{\frac{J(k) \pm L(k)}{2\gamma^3}}$$

where $J(k)=[R_0^3 + (4u_0^2\gamma^3 + 2\Omega^2)k^2]$ & $L(k)=\sqrt{[(R_0^2 + 4u_0^2\gamma^3 k)^2 + 16u_0^2\gamma^3\Omega k^4]}$

The term written within the inner square root can amount to imaginary frequency and hence may be set to zero to find possible real roots. From this above consideration we get a sixth order algebraic equation as

$$16u_0^2\gamma^3 H^2 k^6 + 4\Gamma k^4 + 8R_0^3 u_0^2\gamma^3 k^2 + R_0^6 = 0 \tag{25}$$

This can be rewritten as

$$A_1 k^6 + B_1 k^4 + C_1 k^2 + D_1 = 0 \tag{26}$$

Equation 25 has no real positive root. Under these algebraic considerations the total term under the square root in Eq. 24 is obviously real for all possible wave-numbers. This does not mean it will be non-negative. Such a negative value of the expression [i.e;$J(k) \pm L(k) < 0$] will correspond to imaginary values of frequency ω and determine the range of stable modes.

In order to obtain the wavenumber domain at which the frequency is imaginary, we set the term $[J(k) \pm L(k)]$ equal to zero and further constraining ourselves within the positive sign of ω in Eq. 24, we get the upper mode frequency $\omega = \sqrt{\frac{R_0^3}{\gamma^3}}$ to be always stable. The instability aries when we consider the negative sign in Eq. 24 i.e.; for $J(k) \pm L(k)$ quantity.

With these considerations and imposing stability criteria for possible frequency ranges we obtain from Eq. 20 after expanding the squared terms and with some algebraic simplifications.

$$\Omega k^2[R_0^3 + 8u_0^2\gamma_0^3 k^2 + 1] = 0 \tag{27}$$

which on further simplification provides

$$k^2(H^2k^2 - u_0^2\gamma^3 + \Gamma)(R_0^3 + 8u_0^2\gamma_0^3k^2 + 1) = 0 \tag{28}$$

This can be rewritten as,

$$A_2k^6 + B_2k^4 + C_2k^2 = 0 \tag{29}$$

$A_2 = H^28u_0^2\gamma^3$, $B_2 = H^2(R_0^3 + 1) + (\Gamma - u_0^2\gamma^3)8u_0^2\gamma^3$, $C_2 = (\Gamma - u_0^2\gamma^3)$

$$A_2k^6 + B_2k^4 + C_2k^2 = 0 \tag{30}$$

$$or \quad k^2\left(A_2k^4 + B_2k^2 + C_2\right) = 0 \tag{31}$$

$k = 0$ is a root
another roots are obtained by taking $k^2 = \kappa$

$$or \quad A_2\kappa^2 + B_2\kappa + C_2 = 0 \tag{32}$$

$$or \quad \kappa = \frac{-B_2 \pm \sqrt{B_2^2 - 4A_2C - 2}}{2A_2} \tag{33}$$

$$or \quad k = \pm\sqrt{\frac{\sqrt{B_2^2 - 4A_2C_2} - B_2}{2A_2}} \tag{34}$$

We have taken positive values of the square root of the discriminant because otherwise we would have obtained imaginary values. Now using standard techniques, the roots are obtained as κ_2 and κ_3 where $(\kappa_1 = 0) < k < \sqrt{\kappa_2}$ and $\sqrt{\kappa_3} < k$ are the range for unstable states whereas $\sqrt{\kappa_2} < k < \sqrt{\kappa_3}$ is the stable region in k-space. A classical correspondence can be drawn for the region of instability for wave number values

$$0 < k < \sqrt{-\frac{\Gamma\left(R_0^3 + 1\right)}{\left(\Gamma - u_0^2\gamma_0^3\right)8u_0^2\gamma_0^3}} \tag{35}$$

4 Analytic Results and It's Physical Interpretations

In Fig. 1, we plot the ω versus k corresponding to Eq. 20 in which we get four different modes in the classical case, in Fig. 2 we plot the dispersion relation corresponding to the quantum range. In both cases, we find a loop kind of structure along the

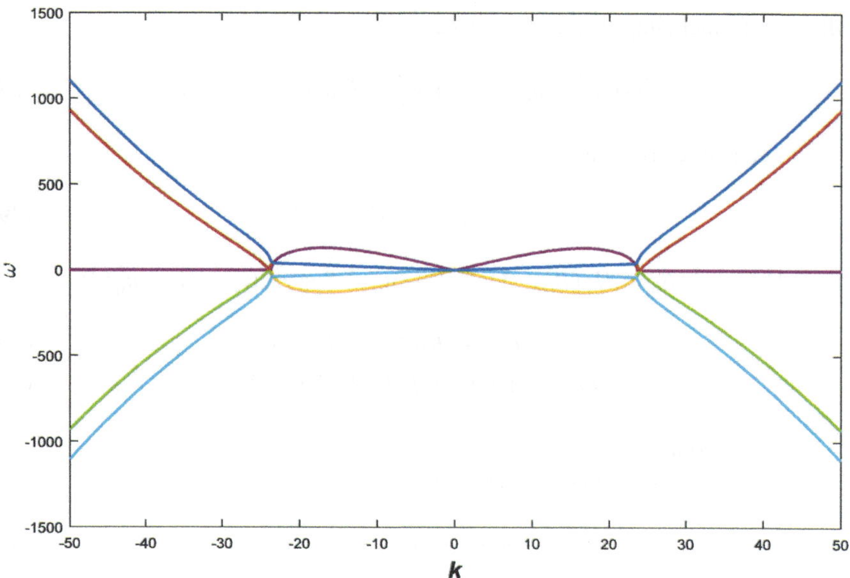

Fig. 1 Linear dispersion relation for two stream instability in classical regime

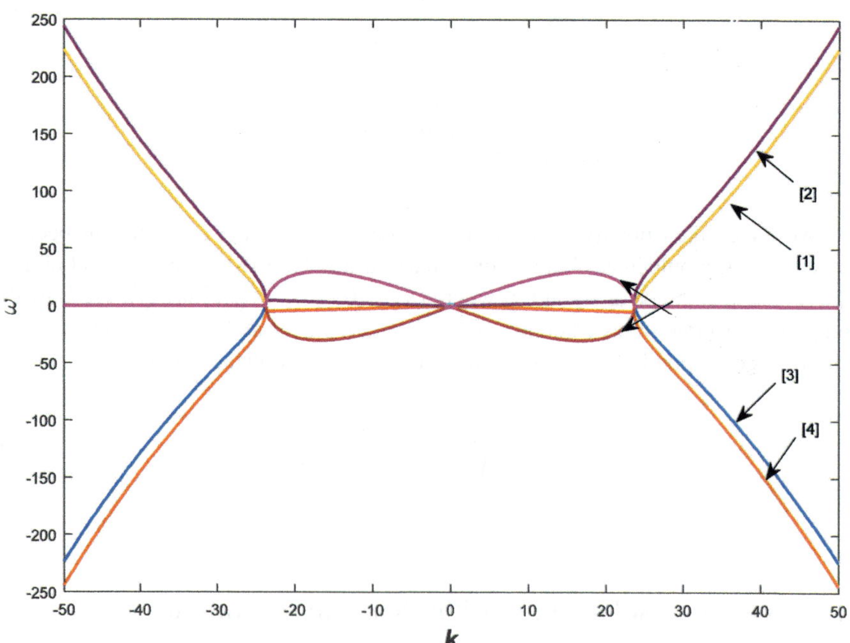

Fig. 2 Linear dispersion relation for two stream instability in quantum regime

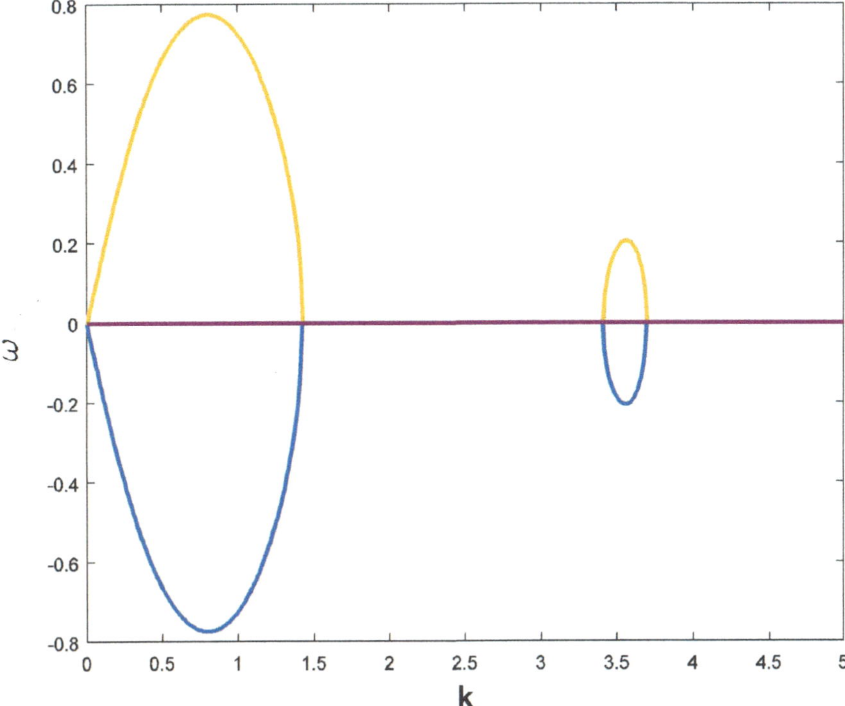

Fig. 3 Unstable modes with growth rate variation in classical regime

wavenumber axis which corresponds to unstable mode. Just where the unstable mode ceases to exist, we get corresponding curves for the stable modes. This reflects the fact that coupling is exact. In the quantum range, we also get similar behaviours but here the fast modes and the slow modes are not widely separated which means their phase velocity is close by. An electromagnetic wave in such a dispersive medium in which there is a beam of particles with streaming velocity (u_0), there can be an exchange of energy from the beam to the wave and vice-versa. When the phase velocity of the wave is slightly smaller than the velocity of the beam, there can be an energy transfer from the beam to the wave resulting in instability of the wave. The fast and slow mode (Figs. 1 and 2) in both cases immediately take up the energy of the beam at higher wavenumber values which means energy is easily dispersed preventing any unwanted instability. Here the perturbation is considered infinitesimally small. However, for a larger value of perturbation bounded solution may be absent. The application of the hydrodynamic model is mere meaningful in such a case where quasi-neutrality, spatial periodicity and stationary stable states can be studied extensively. In such a quantum plasma the instability originates from the free energy source and associated mode coupling resulting in stable and unstable regimes in the wavenumber. It deviates from the classical picture since the quantum effect causes changes in the dielectric

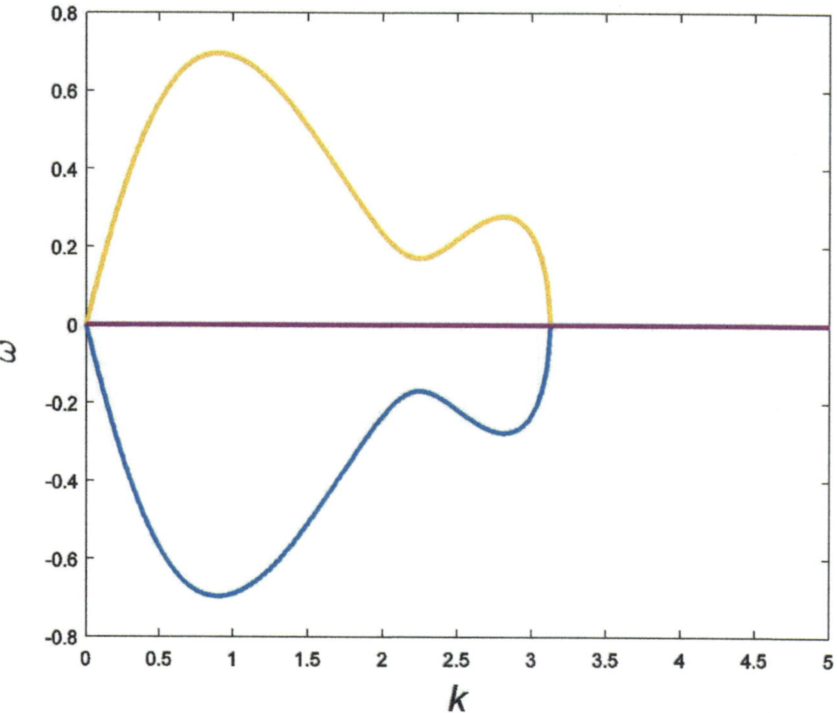

Fig. 4 Two interacting unstable mode in the quantum regime

function of the plasma. The positive energy modes correspond to the energy being given from the wave to the beam, whereas the negative energy modes are related to the fact the energy is given by the beam to wave. In Fig. 3, two separated loops have been shown corresponding to unstable modes with higher growth of instability at smaller wavenumbers. A small zone of instability appears isolatedly existent. This corresponds to the quick dissipation of energy acquired from the beam to the wave.

Figure 4, shows a semi-classical situation in which two unstable modes are coupled in such a way to exchange between them and give rise to a hybrid model of varying instability growth. Next in Figs. 5 and 6 we show that the quantum regime of all the four modes is stable. Two of them are fast whereas the other two are slow. Here, the wave phase speed is compared with the beam speed. In Figs. 7 and 8, we plot the imaginary roots of dispersion relation i.e; the unstable mode corresponding to parametric variation of quantum diffraction (H), relativistic degeneracy factor (R_0) and streaming motion (u_0). From Fig. 7, it is clear that an increase in the value of R_0 (represented by outward moving curve) increases the instability growth rate in such case the instability ceases to die out with increasing wavenumber. Quantum diffraction negatively affects instability. From the fourth plot of Fig. 7 , we can say that 'H' and 'R_0' have ceratin correlation at higher density in which it fails to contain the

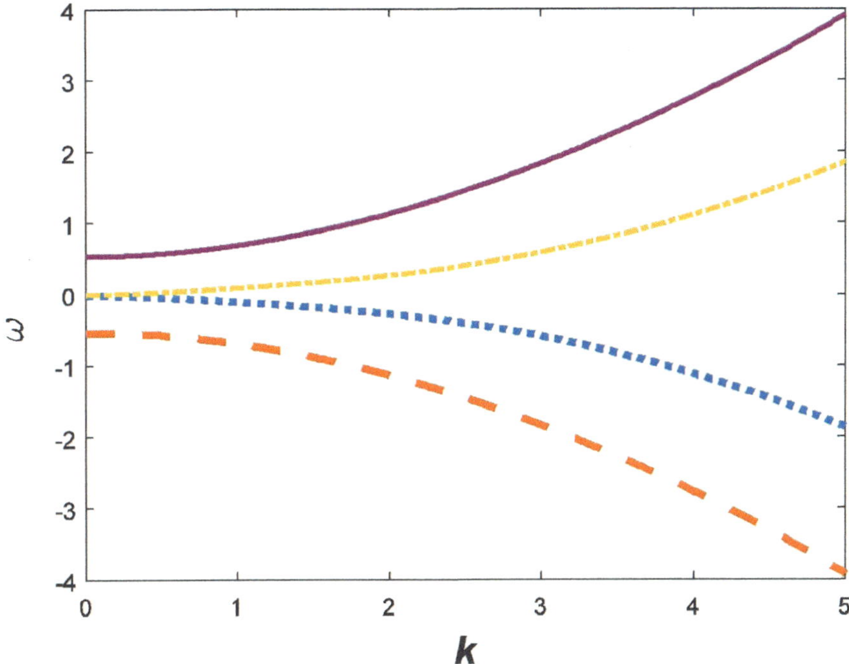

Fig. 5 Dispersion curve when value of relativistic degeneracy parameter (R_0) = 0.8 and quantum diffraction parameter (H) = 2

instability at higher wavenumber. From Fig. 8, we conclude, by increasing relativistic streaming velocity, the instability is enhanced but the wave number corresponding to unstable modes are independent of the streaming motion. In Fig. 9, we plot the classical and quantum variation of contour plot in the (k − R_0) plane. The zone of instability is less in the classical case and is extended in the quantum regime. In Fig. 10, we plot the contour variations in (k − R_0) plane corresponding to different beam velocities. The curves are almost similar about the wavenumber axis and it is clear from the figure that with the increase in beam velocity the instability zone shrinks. In Fig. 11 we plot the R_0 dependence in the (k − u_0) plane and conclude that with increasing value of 'R_0' the zone stability slightly get decreased. All this instability is dominant in the lower wavenumber values which is relatable with Figs. 1, 2 and 3. Our results will find application in interpreting various electrostatic two-stream instability phenomena and associated wave modes that are found to exist in stellar and solar flares. The resulting close correlation with the findings of Akbari, Haas and Eliasson.

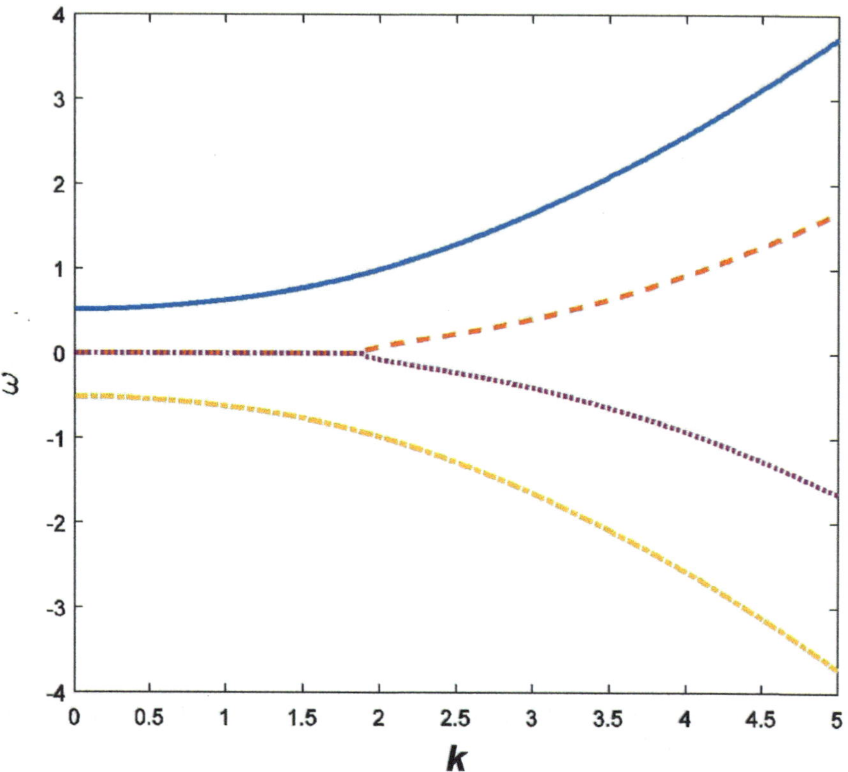

Fig. 6 Unstable modes with growth rate variation in classical regime

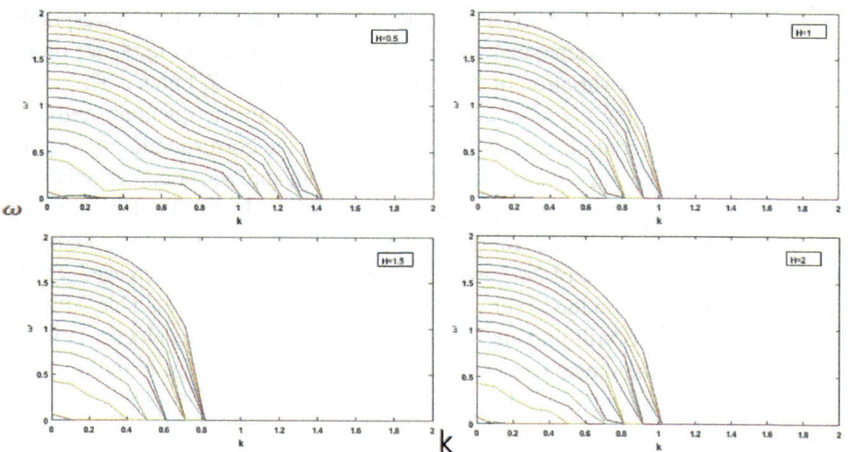

Fig. 7 Two interacting unstable mode in the quantum regime

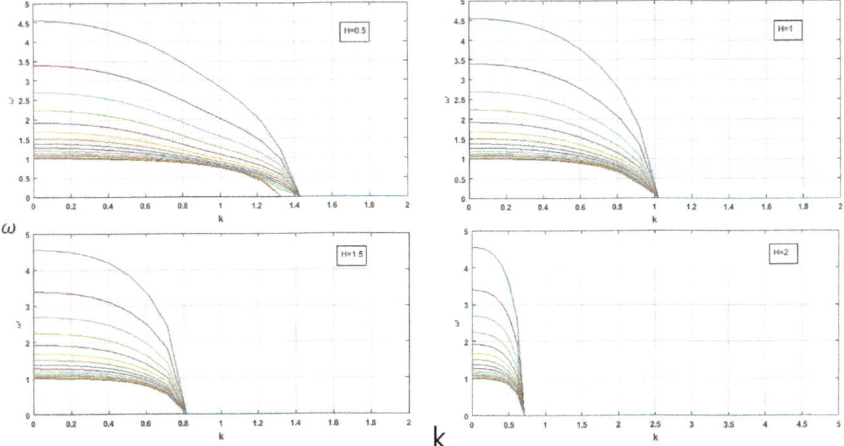

Fig. 8 Dispersion curve when value of relativistic degeneracy parameter (R_0) = 0.8 and quantum diffraction parameter (H) = 2

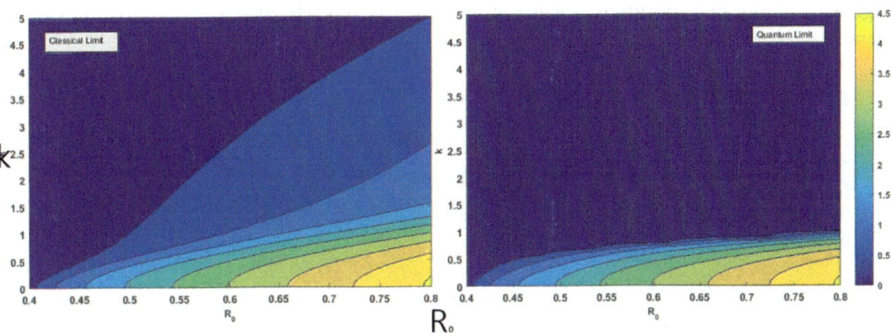

Fig. 9 Contour plot ($k - R_0$) plane in classical and quantum limit

5 Conclusion

In this study, we investigated analytically the properties encountered in counter-propagating plasma streams that contribute to two-stream instability. We designed the governing equations in a certain way that both classical and quantum regimes may be examined. For this reason, we have employed quantum hydrodynamic model and obtained dispersion characteristics. The physical situation of the problem is related to solar wind and heliospheric plasma. The effects of degeneracy factor, quantum diffraction and streaming motion have been studied with great detail. The stability criterion was encountered to be dependent on the wavenumber's range along with the parameter values. Interacting modes and their domain of interaction have been identified. In a nutshell, this work gives us a better view of the counter streaming instability in a dense plasma.

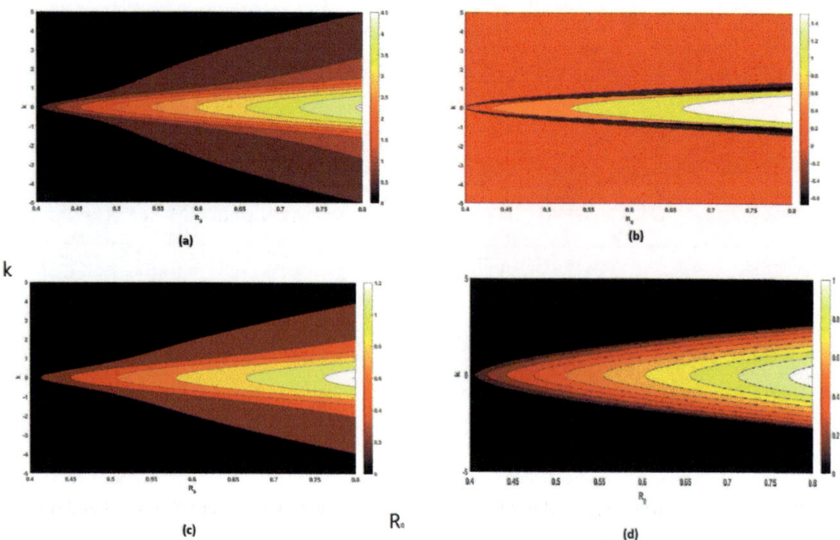

Fig. 10 Contour plot for different value of u_0 [0.8c (**a**), 0.6c (**b**), 0.4c (**c**), 0.2c (**d**)], in $(k - R_0)$ plane

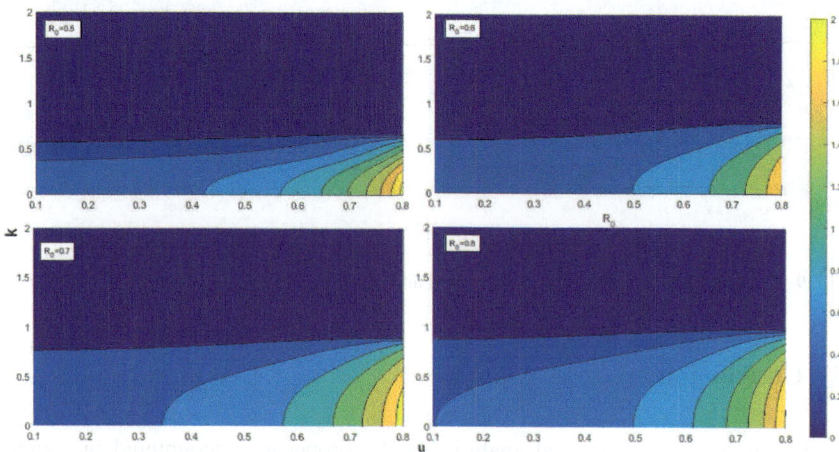

Fig. 11 Contour plot for different value of R_0 in $(k - u_0)$ plane

Acknowledgements We would like to thank Prof. A. Roy Chowdhuri and Mr. Chinmay Das for their inspiration and support. Authors would like to thank Physics departments of Jadavpur University and Government General Degree College at Kushmandi for providing facilities to carry this work.

References

1. Akbari-Moghanjoughi, M.: Field-induced degeneracy regimes in quantum plasmas. Phys. Plasmas **19**(3), 032703 (2012)
2. Akbari-Moghanjoughi, M., Ghorbanalilu, M.: Energy exchange in strongly coupled plasmas with electron drift. Phys. Plasmas **22**(11), 112111 (2015)
3. Bohm, D., Gross, E.P.: Theory of plasma oscillations. b. Excitation and damping of oscillations. Phys. Rev. **75**, 1864–1876 (Jun 1949). https://doi.org/10.1103/PhysRev.75.1864
4. Bransden, B.H., Joachain, C.J.: Quantum mechanics, 2nd edn. (Harlow 2000)
5. Brodin, G., Marklund, M., Manfredi, G.: Quantum plasma effects in the classical regime. Phys. Rev. Lett. **100**, 175001 (2008)
6. Chandra, S., Ghosh, B.: Modulational instability of electron-acoustic waves in relativistically degenerate quantum plasma. Astrophys. Space Sci. **342**(2), 417–424 (2012)
7. Chandra, S., Ghosh, B.: Non-linear propagation of electrostatic waves in relativistic fermi plasma with arbitrary temperature (2013)
8. Chandra, S., Goswami, J., Sarkar, J., Das, C.: Analytical and simulation studies of forced kdv solitary structures in a two-component plasma. J. Korean Phys. Soc. **76**(6), 469–478 (2020)
9. Chandra, S., Goswami, J., Sarkar, J., Das, C., Nandi, D., Ghosh, B.: Formation of electron acoustic shock wave in inner magnetospheric plasma. Indian J. Phys. 1–15 (2022)
10. Chandra, S., Sarkar, J., Das, C., Ghosh, B.: Self-interacting stationary formations in plasmas under externally controlled fields. Plasma Phys. Rep. **47**(3), 306–317 (2021)
11. Chandrasekhar, S., Chandrasekhar, S.: An Introduction to the Study of Stellar Structure, vol. 2. Courier Corporation (1957)
12. Ekman, R., Zamanian, J., Brodin, G.: Exchange corrections in a low-temperature plasma. Phys. Rev. E **92**, 013104 (2015)
13. Eliasson, B., Kogelschatz, U.: Modeling and applications of silent discharge plasmas. IEEE Trans. Plasma Sci. **19**(2), 309–323 (1991)
14. Eliasson, B., Shukla, P.K.: Nonlinear quantum fluid equations for a finite temperature fermi plasma. Physica Scripta **78**(2), 025503 (2008)
15. Goswami, J., Sarkar, J.: Kbm approach to electron acoustic envelope soliton in viscous astrophysical plasma. Physica Scripta **96**(8), 085601 (2021)
16. Goswami, J., Sarkar, J., Chandra, S., Ghosh, B.: Amplitude-modulated electron-acoustic waves with bipolar ions and kappa-distributed positrons and warm electrons. Pramana **95**(2), 1–10 (2021)
17. Haas, F., Bret, A., Shukla, P.K.: Physical interpretation of the quantum two-stream instability. Phys. Rev. E **80**, 066407 (2009)
18. Hu, G., O'connell, R.: Generalization of the Lindhard dielectric function to include fluctuation effects. Phys. Rev. B, Condensed Matter **40**, 3600–3604 (09 1989). https://doi.org/10.1103/PhysRevB.40.3600
19. Kanazawa, H., Misawa, S., Fujita, E.: Green function method for electron gas. ii: Dispersion relation of plasmons. Prog. Theor. Phys. **23**(3), 426–432 (1960)
20. Lee, N.C., Choi, C.R.: Ion-acoustic solitary waves in a relativistic plasma. Phys. Plasmas **14**(2), 022307 (2007)
21. Manfredi, G.: How to Model Quantum Plasmas (2005)
22. Mohammadnejad, M., Akbari-Moghanjoughi, M.: Two stream ion acoustic wave instability in warm dense plasmas. Astrophys. Space Sci. **364**(2), 23 (2019)

23. Nozieres, P., Pines, D.: Correlation energy of a free electron gas. Phys. Rev. **111**(2), 442 (1958)
24. von Roos, O., Zmuidzinas, J.S.: Effect of electron exchange on the dispersion relation of plasmons. Phys. Rev. **121**(4), 941 (1961)
25. Salpeter, E.E.: Energy and pressure of a zero-temperature plasma. Astrophys. J. **134**, 669 (1961)
26. Sarkar, J., Chandra, S., Ghosh, B.: Resonant interactions between the fundamental and higher harmonic of positron acoustic waves in quantum plasma. Zeitschrift für Naturforschung A **75**(10), 819–824 (2020)
27. Sarkar, J., Chandra, S., Goswami, J., Das, C., Ghosh, B.: Growth of rt instability at the accreting magnetospheric boundary of neutron stars. In: AIP Conference Proceedings, vol. 2319, p. 030006. AIP Publishing LLC (2021)
28. Sarkar, J., Chandra, S., Goswami, J., Ghosh, B.: Formation of solitary structures and envelope solitons in electron acoustic wave in inner magnetosphere plasma with suprathermal ions. Contrib. Plasma Phys. **60**(7), e201900202 (2020)
29. Shukla, P.K., Rao, N., Yu, M., Tsintsadze, N.: Relativistic nonlinear effects in plasmas. Phys. Rep. **138**(1–2), 1–149 (1986)

Bifurcation of Nucleus-Acoustic Superperiodic and Supersolitary Waves in a Quantum Plasma

Barsha Pradhan, Nikhil Pal, and David Raj Micheal

Abstract The main motivation of this work is to investigate the existence of arbitrary-amplitude nucleus-acoustic (NA) supernonlinear waves in a degenerate plasma system consisting of light and heavy nuclei and non-relativistic degenerate electrons. To achieve this objective, we plot phase portraits of the dynamical system and their corresponding Sagdeev's pseudopotential curves for different values of Mach number. Distinct topology of phase portraits along with two minima separated by a maxima in Sagdeev's pseudopotential curve ensure existence of NA supernonlinear waves. Further, we discuss influence of Mach number on periodic, superperiodic, solitary and supersolitary wave solutions.

Keywords Supersoliton · Supernonlinear periodic trajectory · Supernonlinear homoclinic trajectory · Phase plane analysis

Supported by Sikkim Manipal Institute of Technology.

B. Pradhan (✉)
Department of Mathematics, Sikkim Manipal Institute of Technology, Sikkim Manipal University, Majitar, Rangpo, East-Sikkim 737136, India
e-mail: barshapradhan09@gmail.com

N. Pal
Department of Mathematics, Siksha-Bhavana, Visva-Bharati University, Santiniketan 731235, India

D. R. Micheal
Division of Mathematics, School of Advanced Sciences, Vellore Institute of Technology, Chennai, Tamil Nadu, India
e-mail: davidraj.micheal@vit.ac.in

S. Banerjee and A. Saha (eds.), *Nonlinear Dynamics and Applications*,
Springer Proceedings in Complexity,
https://doi.org/10.1007/978-3-030-99792-2_4

1 Introduction

Compact astrophysical bodies have high density (of order $10^{36}\,\text{cm}^{-3}$ in neutron stars and $10^{29}\,\text{cm}^{-3}$ in white dwarf) such that the deBroglie wavelength of particles is comparable to the inter-particle distance [1, 2]. Therefore, roles of quantum effects and degeneracy pressure in such objects become vital. White dwarf mainly contains degenerate electrons, immobile heavy nuclei (e.g., ${}^{56}_{26}$Fe and/or ${}^{85}_{37}$Rd and/or ${}^{96}_{42}$Mo, etc.,) [3–5] and light nuclei (e.g., ${}^{1}_{1}$H and/or ${}^{12}_{6}$C and/or ${}^{16}_{8}$O, etc.) [1, 2, 5]. In his work, Chandrasekhar [6, 7] reported that the outward pressure in compact astrophysical objects produced by degenerate electrons balances the inward pull due to gravity. The degenerate electrons follow the equation of state $P_e = K_e n_e^{\gamma}$ with $\gamma = \frac{5}{3}$, $K_e = \frac{3}{5}\frac{\pi \hbar^2}{m_e}$ for non-relativistic (NR) limit and $\gamma = \frac{4}{3}$, $K_e = \frac{3}{4}\hbar c$ for ultra-relativistic (UR) limit [5–8]. Hence, we can employ either of the two limits for degenerate electrons.

Mammun et al. [9] studied NA waves in which inertia is supplied by nucleus mass density and restoring force is supplied by inertialess degenerate electron pressure which depends solely on electron number density. It is remarkable to note that NA waves exist in cold plasma limit unlike other modes (ion or electron or positron acoustic) which do no exist in cold plasma limits. Nonlinear features such as NA shocks [5, 9, 10] and NA solitons [8, 11] were extensively investigated in the recent years due to its potential implications in dense astrophysical objects. Jannat and Mammun [5] have briefly discussed the application of their results in white dwarfs.

Investigation of nonlinear waves in quantum plasmas by employing the concept of planar dynamical system (PDS) is gaining immense popularity [12–15]. In plasma, Dubinov and Kolotkov [16] developed a new category of nonlinear waves, publicized as supernonlinear waves (SNWs) that are marked by their distinct nonlinear topology of phase plots. Dubinov and Kolotkov [17] initiated the term "supersolitons" in plasmas by taking a model consisting of five components. Afterwards, Verheest et al. [18] verified that three-component in a plasma were sufficient to support supersolitons. Saha and Tamang [19] discussed SNWs using bifurcation theory of planar dynamical system. Since then, arbitrary amplitude SNWs [20, 21] as well as small-amplitude SNWs [15] were investigated in different plasmas using bifurcation theory. Very recently, Saha et al. [22] showed the existence of SNWs in a two-component Maxwellian plasma. But, there is no study of arbitrary amplitude nucleus-acoustic supernonlinear waves in quantum plasmas to the best of our knowledge. In this work, we have employed bifurcation theory to investigate arbitrary amplitude nucleus-acoustic superperiodic and supersolitary waves in a degenerate plasma system consisting of light and heavy nuclei and non-relativistic degenerate electrons.

The layout of this article is as follows: Normalised basic equations are presented in Sect. 2. Formation of planar dynamical system is briefly discussed in Sect. 3. Phase plots of the system are shown in Sect. 4. Section 5 gives the conclusion.

2 Normalised Basic Equations

A three-component quantum plasma is considered comprising of stationary heavy nuclei (mass = m_h and charge = $Z_h e$), non-degenerate mobile, cold, light inertial nuclei (mass = m_i and charge = $Z_i e$) together with NR degenerate electrons. The mass of stationary heavy nuclei that conserves background neutrality is much greater than the mass of inertial light nuclei that supplies inertia. Charge of light nuclei is greater or equal to unity. On the other hand, non-relativistic inertialess degenerate electrons supplies degenerate pressure that act as a restoring force. The normalized basic equations [8] that describe the dynamics of light nuclei are as follows:

$$\frac{\partial n_i}{\partial t} + \frac{\partial (n_i u_i)}{\partial x} = 0, \tag{1}$$

$$\frac{\partial u_i}{\partial t} + u_i \frac{\partial u_i}{\partial x} + \frac{\partial \phi}{\partial x} = 0, \tag{2}$$

$$K \frac{\partial n_e^{\gamma}}{\partial x} - n_e \frac{\partial \phi}{\partial x} = 0, \tag{3}$$

$$\frac{\partial^2 \phi}{\partial x^2} - \mu_e (n_e - 1) + n_i - 1 = 0. \tag{4}$$

Here number density n_s (where indices $s = h, i, e$ designate heavy nuclei, light nuclei and electron), electrostatic potential ϕ and velocity u_i are respectively normalized by equilibrium value n_{s0}, $(m_e c^2/e)$, where e is magnitude of protonic charge and $C_0 = (Z_i m_e c^2/m_i)^2$. Light nuclei plasma period $\omega_{pi}^{-1} = (4\pi e^2 Z_i^2 n_{i0}/m_i)^{-\frac{1}{2}}$ and modified Debye length $\lambda_m = (m_e c^2/4\pi e^2 Z_i n_{i0})^{\frac{1}{2}}$ are used to scale time and space variables respectively. Here $\mu_e = (n_{e0}/Z_i n_{i0})$ and $K = (K_e n_{e0}^{\gamma-1}/m_e c^2)$, where K_e obeys the equation of state $P_e = K_e n_e^{\gamma}$ with $\gamma = \frac{5}{3}$, $K_e = \frac{3}{5}\frac{\pi \hbar^2}{m_e}$ for NR limit.

Here, we have set $\frac{Z_h n_{h0}}{Z_i n_{i0}} = \mu_e - 1$ following the quasi-neutrality condition $Z_h n_{h0} + Z_i n_{i0} = n_{e0}$ in (4). Subsequently, the model equations (1)–(4) absorb the charge state of heavy nuclei Z_h as a result of normalization.

3 Planar Dynamical System (PDS)

Normalized basic equations (1)–(4) are transformed to a PDS by employing travelling wave transformation

$$\xi = x - Mt, \tag{5}$$

where travelling wave velocity M is taken in positive direction of x-axis. Applying this transformation to (1)–(3) with boundary conditions $\phi \to 0, n_i \to 1,\ u_i \to 0$ and $n_e \to 1$ as $\xi \to \pm\infty$, we acquire

$$n_i = \left(1 - \frac{2\phi}{M^2}\right)^{-\frac{1}{2}}, \tag{6}$$

$$n_e = \left[1 + \left(\frac{\gamma - 1}{k\gamma}\right)\phi\right]^{\frac{1}{\gamma-1}}. \tag{7}$$

Applying transformation (5) to (4) with boundary conditions and substituting (6) and (7), we get

$$\frac{d^2\phi}{d\xi^2} = a\phi + b\phi^2 + c\phi^3 + d\phi^4, \tag{8}$$

with $a = -\frac{1}{M^2} + \frac{\mu_e}{\gamma K}$, $b = -\frac{3}{2M^4} + \frac{\mu_e(2-\gamma)}{2\gamma^2 K^2}$, $c = -\frac{5}{2M^6} + \frac{\mu_e(2-\gamma)(3-2\gamma)}{6\gamma^3 K^3}$ and $d = -\frac{35}{8M^8} + \frac{\mu_e(2-\gamma)(3-2\gamma)(4-3\gamma)}{24\gamma^4 K^4}$.

The system (8) can be expressed as following PDS:

$$\begin{cases} \frac{d\phi}{d\xi} = z, \\ \frac{dz}{d\xi} = a\phi + b\phi^2 + c\phi^3 + d\phi^4. \end{cases} \tag{9}$$

The corresponding Hamiltonian function is:

$$H(\phi, z) = \frac{z^2}{2} - \left(\frac{a}{2}\phi^2 + \frac{b}{3}\phi^3 + \frac{c}{4}\phi^4 + \frac{d}{5}\phi^5\right) = h, \tag{10}$$

say. We consider Sagdeev's equation corresponding to the PDS (9) as

$$\frac{d^2\phi}{d\xi^2} = -\frac{d\psi}{d\phi}, \tag{11}$$

where ψ is the Sagdeev's pseudopotential. Therefore, from (8) and (11) we get

$$\psi(\phi) = -\left(\frac{a}{2}\phi^2 + \frac{b}{3}\phi^3 + \frac{c}{4}\phi^4 + \frac{d}{5}\phi^5\right). \tag{12}$$

4 Phase Plots

To find all equilibrium points the PDS (9), we have $\frac{d\phi}{d\xi} = 0$ and $\frac{dz}{d\xi} = 0$, which give

$$z = 0,\ \phi(d\phi^3 + c\phi^2 + b\phi + a) = 0, \tag{13}$$

$$\Rightarrow z = 0,\ \phi(\phi^3 + p\phi^2 + q\phi + r) = 0, \tag{14}$$

where $p = \frac{c}{d}$, $q = \frac{b}{d}$ and $r = \frac{a}{d}$.

(i) If $\frac{h^2}{4}+\frac{g^3}{27}>0$, then there are two equilibrium points of the PDS (9) at $P_0(\phi_0,0)$ and $P_1(\phi_1,0)$, where $\phi_0=0$, $\phi_1=A+B$ with $A=\sqrt[3]{-\frac{h}{2}+\sqrt{\frac{h^2}{4}+\frac{g^3}{27}}}$, $B=\sqrt[3]{-\frac{h}{2}-\sqrt{\frac{h^2}{4}+\frac{g^3}{27}}}$, $g=\frac{1}{3}(3q-p^2)$ and $h=\frac{1}{27}(2p^3-9pq+27r)$.

(ii) If $\frac{h^2}{4}+\frac{g^3}{27}<0$, then there are four equilibrium points of the PDS (9) at $P_0(\phi_0,0)$, $P_1(\phi_1,0)$, $P_2(\phi_2,0)$ and $P_3(\phi_3,0)$, where $\phi_0=0$, $\phi_{2,3,4}=-\frac{p}{3}+2\sqrt{-\frac{g}{3}}cos(\frac{\psi}{3}+\frac{2k\pi}{3})$, $k=0,1,2$, with

$cos\psi=-\sqrt{\frac{h^2/4}{-g^3/27}}$, if $h>0$;

$\sqrt{\frac{h^2/4}{-g^3/27}}$, if $h<0$.

We denote Jacobian matrix of the system (9) corresponding to the critical point $P_i(\phi_i,0)$ by $J(\phi_i,0)$. The critical point $P_i(\phi_i,0)$ is a saddle point if $|J(\phi_i,0)|<0$ and a center when $|J(\phi_i,0)|>0$ [23].

Phase plots of a system can differ relying on the number of surrounding separatrix layers [17] and the total number of equilibrium points. For a particular dynamical system, a wave solution can be obtained for a trajectory contained in the phase plot. In this article, we denote nonlinear periodic trajectory by $NPT_{m,n}$, nonlinear homoclinic trajectory by $NHT_{m,n}$, supernonlinear periodic trajectory by $SNPT_{m,n}$ and supernonlinear homoclinic trajectory by $SNHT_{m,n}$, where m designates the total number of stable critical points (centres) and n designates the total number of enveloping separatrix layers. Corresponding to $NPT_{m,n}$ and $NHT_{m,n}$, we can obtain NA periodic wave solution and NA solitary wave solution, respectively, while corresponding to $SNPT_{m,n}$ and $SNHT_{m,n}$, we can obtain NA superperiodic wave solution and NA supersolitary wave solution, respectively.

For our computation we will use the values of dense plasmas [1, 2, 4, 5] consisting of degenerate NR electrons, non-degenerate light nuclei ($^{12}_{6}$C) and heavy nuclei ($^{85}_{37}$Rd): $n_{e0}\sim10^{35}\,\mathrm{m}^{-3}$, $n_{i0}\sim1.666\times10^{34}\,\mathrm{m}^{-3}$, $\hbar=1.054\times10^{-34}\,\mathrm{Js}$, $m_e=9.10938356\times10^{-31}\,\mathrm{kg}$, $Z_i=6$, $c=3\times10^8\,\mathrm{m/s}$ and $\gamma=\frac{5}{3}$ (NR limit). The calculated values of μ_e and K are 1.0004 and 0.06040810205, respectively. We will use these values of γ, K and μ_e in this present work.

Figure 1a exhibits phase plot of PDS (9) for $M=0.51$. Here, the critical points P_0 and P_1 are saddle and centre, respectively. A family of periodic trajectories surround P_1 while there is a homoclinic trajectory at P_0. The plasma system has a family of periodic wave profile given by $H(\phi,z)=h$, $h\in(0,h_1)$ and a compressive solitary wave profile given by $H(\phi,z)=0$. In Fig. 1b, we present effect of Mach number M on NA periodic wave solution. One can see from Fig. 1b that only amplitude of NA periodic wave elevates on enhancing M. The Sagdeev's pseudopotential curve corresponding to Fig. 1a is displayed in Fig. 1c. One can clearly observe a minima at P_1 and a maxima at P_0.

In Fig. 2a, we present a phase plot of the PDS (9) for $M=0.7$. There exist four equilibrium points P_0, P_1, P_2 and P_3 of the PDS (9), of which P_2 and P_3 are centres and P_0 and P_1 are saddles. There are two families of periodic trajectories enclosing

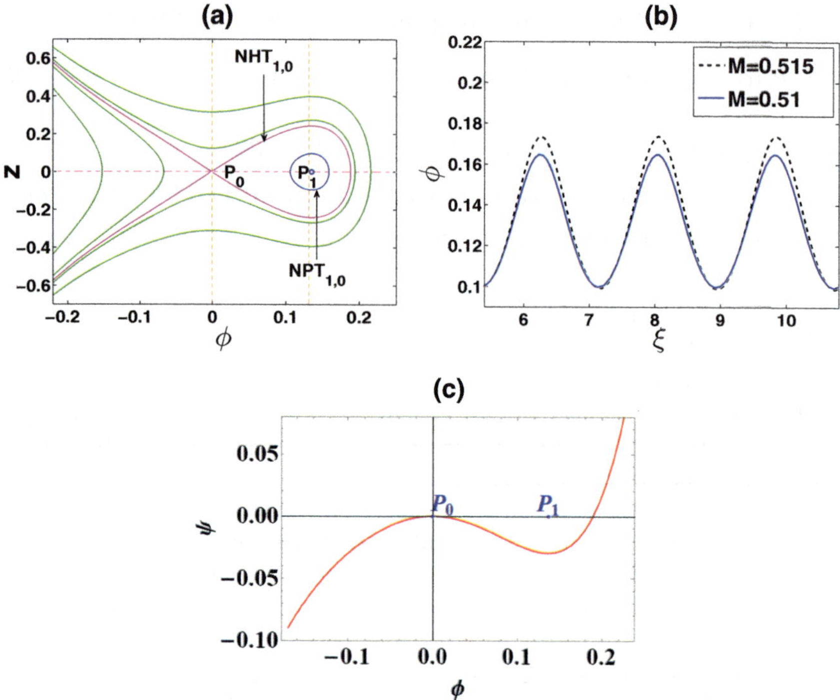

Fig. 1 **a** Phase portrait of the PDS (9) for $M = 0.51$, **b** effect of M on NA periodic wave solution and **c** Sagdeev pseudopotential ψ versus ϕ for $M = 0.51$

P_2 and P_3, respectively, a pair of homoclinic trajectories at P_0, a family of superperiodic trajectories enclosing these two homoclinic trajectories and a supernonlinear homoclinic trajectory at P_1. Effect of Mach number M on NA periodic, superperiodic, compressive solitary, rarefactive solitary and compressive supersolitary wave profiles are shown in Fig. 2b–f, respectively. Heights of NA periodic and rarefactive solitary wave profiles decrease while heights of NA superperiodic, compressive solitary and compressive supersolitary wave profiles increase on enhancing M. On the other hand, widths of superperiodic and compressive supersolitary wave profiles diminish on enhancing M. It has been numerically observed that such type of phase plot (Fig. 2a) is possible for this plasma system if M varies from $M = 0.666$ to $M = 0.747$ (approximately). The Sagdeev's pseudopotential curve corresponding to Fig. 2a is depicted in Fig. 2g. Clearly, one can observe two minima at P_2 and P_3 separated by a maxima at P_0, which is the criteria for SNWs [22].

We show a phase plot of the PDS (9) for $M = 0.77$ in Fig. 3a that contains four critical points P_0, P_1, P_2 and P_3, out of which P_0 and P_2 are saddles while P_1 and P_3 are centres. There are two families of periodic trajectories about P_1 and P_2, respectively, a pair of homoclinic trajectories at P_0, a family of superperiodic trajectories about these two homoclinic trajectories and a rarefactive supersolitary trajectory at P_3. We

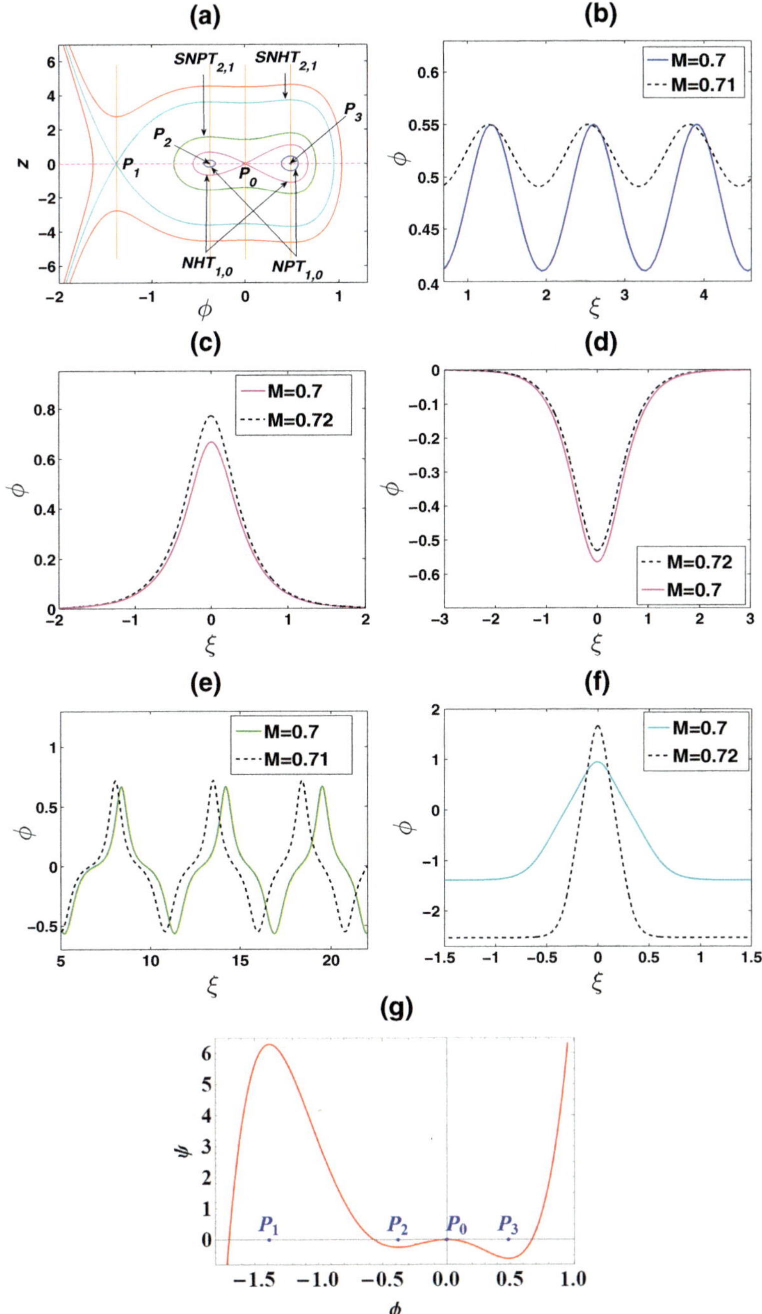

Fig. 2 **a** Phase portrait of the PDS (9) for $M = 0.7$. Influence of M on NA **b** periodic, **c** compressive solitary, **d** rarefactive solitary, **e** superperiodic and **f** compressive supersolitary waves. **g** Sagdeev pseudopotential ψ versus ϕ for $M = 0.7$

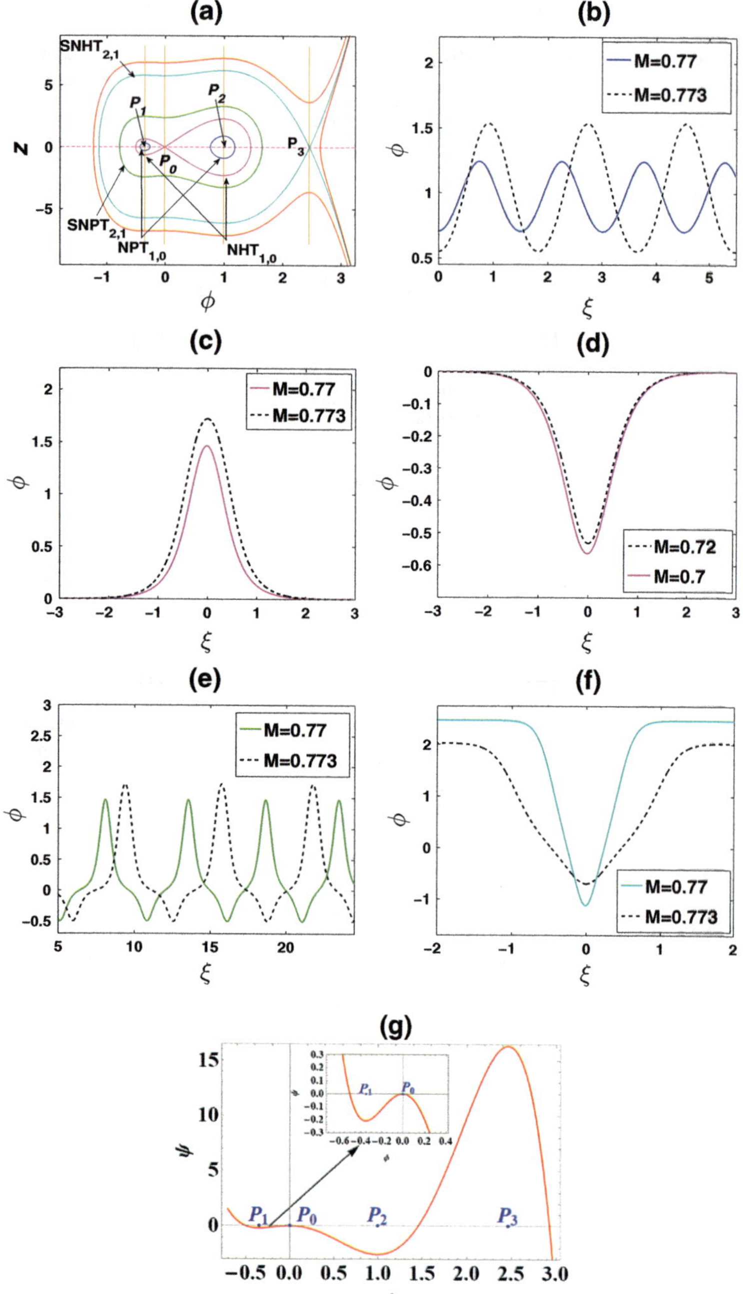

Fig. 3 **a** Phase portrait of the PDS (9) for $M = 0.77$. Influence of M on NA **b** periodic, **c** compressive solitary, **d** rarefactive solitary **e** superperiodic and **f** rarefactive supersolitary waves. **g** Sagdeev pseudopotential ψ versus ϕ for $M = 0.77$

show influence of Mach number M on NA periodic, superperiodic, compressive solitary, rarefactive solitary and compressive supersolitary wave structures in Fig. 3b–f, respectively. Clearly, one can see that enhancement of M results in increase in heights of NA periodic, superperiodic and compressive solitary wave structures and decrease in heights of NA rarefactive solitary and rarefactive supersolitary wave structures. Simultaneously, enhancement of M brings about expansion in widths of periodic, superperiodic, compressive solitary and rarefactive supersolitary wave structures and a slight decrease in width of rarefactive solitary wave. In fact, compressive solitary wave is seen to flourish on enhancing M. It has been numerically observed that such type of phase plot (Fig. 3a) is possible if M varies from $M = 0.748$ to $M = 0.776$ (approximately). Corresponding Sagdeev's pseudopotential is given in Fig. 3b, which presents two minima at P_1 and P_2 with a maxima at P_0 in between.

Figure 4a displays phase portrait of PDS (9) for the sonic case, i.e., $M = 1$. Clearly, there exist two critical points P_0 and P_1. The origin P_0 is a saddle while P_1 is a center. A family of periodic trajectories surround P_1 while a homoclinic trajectory begins and terminates at P_0. This plasma system has a family of periodic wave structures given by $H(\phi, z) = h$, $h \in (h_2, 0)$ and a rarefactive solitary wave structure given by $H(\phi, z) = 0$. Impact of M on NA periodic wave solution is shown Fig. 4b. Vividly,

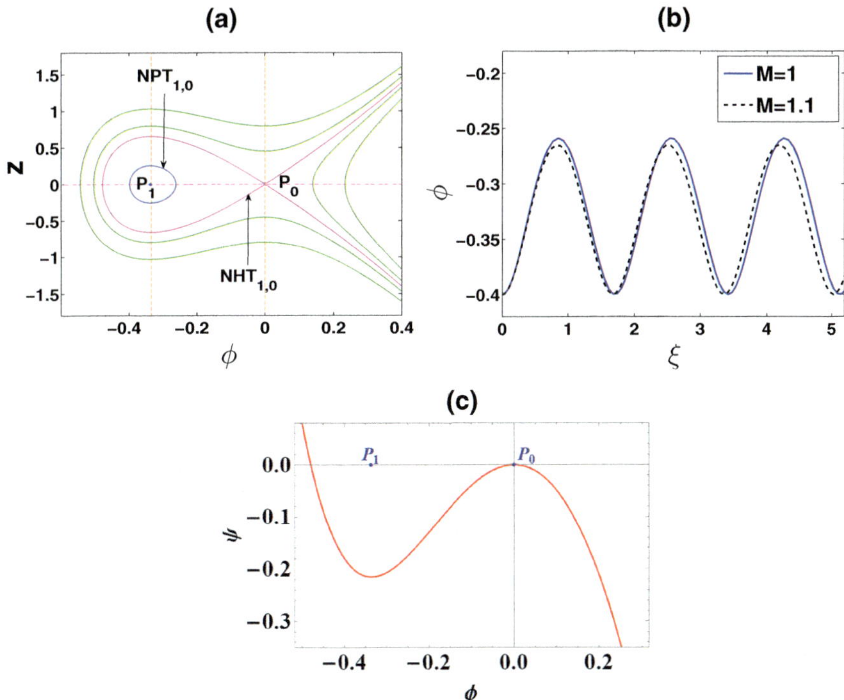

Fig. 4 **a** Phase portrait of the PDS (9) for $M = 1$, **b** impact of M on NA periodic wave solution and **c** Sagdeev pseudopotential ψ versus ϕ for $M = 1$

one can notice that amplitude as well as width of NA periodic wave decrease on enhancing the value of M. Corresponding Sagdeev's pseudopotential is depicted in Fig. 4c. Similar phase portrait can be observed for the supersonic case, i.e., $M > 1$.

5 Conclusion

We have shown the existence of arbitrary amplitude nucleus-acoustic superperiodic and supersolitary waves in a three component degenerate plasma system. Four qualitatively different phase plots of the system have been obtained for four distinct values of Mach number. Thus, we can say that bifurcation occurred and Mach number acted as a bifurcation parameter. For two different ranges of Mach number, two qualitatively different phase profiles distinguished by compressive and rarefactive type supernonlinear homoclinic trajectories have been obtained. It has been numerically observed that NA compressive supersoliton was possible if M varied from 0.666 to 0.747 (approximately) while NA rarefactive supersoliton was possible if M varied from 0.748 to 0.776 (approximately). In both the cases, pseudopotential curve formed a double well with two minima separated by a maxima satisfying the necessary condition for the existence of supernonlinear wave. Mach number M has significant influence on superperiodic and supersolitary wave solutions. An increase in M increased amplitudes of superperiodic and compressive supersolitary wave solutions and decreased amplitude of rarefactive supersolitary wave solution. The parameter values are taken in such a way that the results of this work find its application in examining NA supernonlinear waves in dense astrophysical objects.

Acknowledgements Barsha Pradhan is grateful to Sikkim Manipal University for providing TMA Pai Research Grant (Ref. Nos. 118/SMU/REG/UOO/104/2019).

References

1. Shapiro, S.L., Teukolsky, S.A.: Black Holes, White Dwarfs, and Neutron Stars: The Physics of Compact Objects. Wiley, New York (1983)
2. Koester, D., Chanmugam, G.: Physics of white dwarf stars. Rep. Prog. Phys. **53**, 837 (1990)
3. Van Horn, H.M.: Dense astrophysical plasmas. Science **252**, 384 (1991)
4. Vanderburg, A., Johnson, J.A., Rappaport, S., Bieryla, A., Irwin, J., Lewis, J.A., Kipping, D., Brown, W.R., Dufour, P., Ciardi, D.R., Angus, R., Schaefer, L., Latham, D.W., Charbonneau, D., Beichman, C., Eastman, J., McCrady, N., Wittenmyer, R.A., Wright, J.T.: A disintegrating minor planet transiting a white dwarf. Nature **526**, 546 (2015)
5. Jannat, S., Mammun, A.A.: Nucleus-acoustic shock waves in white dwarfs. Pramana J. Phys. **90**, 51 (2018)
6. Chandrasekhar, S.: The density of white dwarf stars. Philos. Mag. Ser. **7**(11), 592 (1931)
7. Chandrasekhar, S.: The maximum mass of ideal white dwarfs. Astrophys. J. **74**, 81 (1931)
8. Sultana, S., Schlickeiser, R.: Arbitrary amplitude nucleus-acoustic solitons in multi-ion quantum plasmas with relativistically degenerate electrons. Phys. Plasmas **25**, 022110 (2018)

9. Mamun, A.A., Amina, M., Schlickeiser, R.: Nucleus-acoustic shock structures in a strongly coupled self-gravitating degenerate quantum plasma. Phys. Plasmas **23**, 094503 (2016)
10. Zaman, D.M.S., Amina, M., Dip, P.R., Mamun, A.A.: Planar and non-planar nucleus-acoustic shock structures in self-gravitating degenerate quantum plasma systems. Eur. Phys. J. Plus **132**, 457 (2017)
11. Zaman, D.M.S., Amina, M., Dip, P.R., Mamun, A.A.: Nucleus-acoustic solitary waves in self-gravitating degenerate quantum plasmas. Chin. Phys. B **27**, 040402 (2018)
12. Samanta, U.K., Saha, A., Chatterjee, P.: Bifurcations of dust-ion-acoustic travelling waves in a magnetized quantum dusty plasma. Astrophys. Space Sci. **347**, 293 (2013)
13. Sahu, B., Poria, S., Roychoudhury, R.: Solitonic, quasi-periodic and periodic pattern of electron acoustic waves in quantum plasma. Astrophys. Space Sci. **341**, 567 (2012)
14. El-Labany, S.K., El-Taibany, W.F., Atteya, A.: Bifurcation analysis for ion acoustic waves in a strongly coupled plasma including trapped electrons. Phys. Lett. A **382**, 412 (2018)
15. Saha, A., Pradhan, B., Banerjee, S.: Multistability and dynamical properties of ion-acoustic wave for the nonlinear Schrödinger equation in an electron-ion quantum plasma. Phys. Scr. **95**, 055602 (2020)
16. Dubinov, A.E., Kolotkov, D.Y.: Ion-acoustic supersolitons in plasma. Plasma Phys. Rep. **38**, 909 (2012)
17. Dubinov, A.E., Kolotkov, D.Y.: Ion-acoustic super solitary waves in dusty multispecies plasmas. IEEE Trans. Plasma Sci. **40**, 1429 (2012)
18. Verheest, F., Hellberg, M.A., Kourakis, I.: Dust-ion-acoustic supersolitons in dusty plasmas with nonthermal electrons. Phys. Rev. E **87**, 043107 (2013)
19. Saha, A., Tamang, J.: Effect of q-nonextensive hot electrons on bifurcations of nonlinear and supernonlinear ion-acoustic periodic waves. Adv. Space Res. **63**, 1596 (2019)
20. Tamang, J., Saha, A.: Dynamical behavior of supernonlinear positron-acoustic periodic waves and chaos in nonextensive electron-positron-ion plasmas. Zeitschrift Für Naturforschung A **74**, 6 (2019)
21. Taha, R.M., El-Taibany, W.F.: Bifurcation analysis of nonlinear and supernonlinear dust-acoustic waves in a dusty plasma using the generalized (r, q) distribution function for ions and electrons. Contrib. Plasma Phys. **2020**, e202000022 (2020)
22. Saha, A., Chatterjee, P., Banerjee, S.: An open problem on supernonlinear waves in a two-component Maxwellian plasma. Eur. Phys. J. Plus **135**, 801 (2020)
23. Guckenheimer, J., Holmes, P.J.: Nonlinear Oscillations, Dynamical Systems and Bifurcations of Vector Fields. Springer, New York (1983)

Effect of q Parameter and Critical Beam Radius on Propagation Dynamics of q Gaussian Beam in Cold Quantum Plasma

P. T. Takale, K. Y. Khandale, V. S. Pawar, S. S. Patil, P. P. Nikam, T. U. Urunkar, S. D. Patil, and M. V. Takale

Abstract The q Gaussian intensity distribution is very interesting as in the limit $q \to \infty$, it reduces to the Gaussian intensity profile. Naturally, the freedom of exploring the q exponent enables us to study a wide range of propagation dynamics. The quantum plasma offers wide possibilities of its existence right from astrophysical situations to laboratory plasmas. Keeping in mind the wide applicability domain of cold quantum plasma, we have theoretically investigated the propagation behavior of q Gaussian laser beam in cold quantum plasma. The ordinary nonlinear differential equation is set up by following Akhmanov's parabolic equation approach under WKB and paraxial approximations. The effect of the q parameter on the critical curve is explored graphically. The variation in the beam width parameter f over normalized distance ζ due to variation in the q-parameter is graphically depicted and discussed at the end. It is observed that the supercritical region and self focusing length are affected by the q parameter significantly.

Keywords q-Gaussian · Cold quantum plasma · Critical beam radius · Self focusing

---

Supported by DST-SERB, New Delhi, the Special Assistance Program (SAP), Department of Physics, Shivaji University, Kolhapur.

P. T. Takale (✉) · K. Y. Khandale · T. U. Urunkar · M. V. Takale
Department of Physics, Shivaji University, Kolhapur 416004, India
e-mail: mansingtakale@gmail.com

V. S. Pawar
Department of Physics, Rajer Ramrao Mahavidyalay, Jath 416404, India

P. P. Nikam · S. D. Patil
Department of Physics, Devchand College, Arjunnagar, Kolhapur 591237, India

S. S. Patil
Department of Physics, Vivekanand College, Tarabai Park, Kolhapur 416003, India

S. Banerjee and A. Saha (eds.), *Nonlinear Dynamics and Applications*,
Springer Proceedings in Complexity,
https://doi.org/10.1007/978-3-030-99792-2_5

1 Introduction

Interaction of intense laser beam with plasma has received great attention due to its various application such as laser induced fusion [1–3], particle accelerators [4–7], high harmonic generations [8], compact X-ray sources [9–11]. It necessitates studying propagation behaviour of the laser beams through plasma. In all the above applications, the requirement of propagation of laser beam over several Rayleigh lengths is very essential. However, it is impossible to self trap the given laser beam due to diffraction. Due to the nonlinear interaction of the laser and plasma, plasma behaves as a converging lens. Laser beam alters dielectric behaviour of plasma. Intensity pattern of beam along wavefront causes change in refractive index. Eventually beam gets focused.

Quantum plasma has promising applications in cosmological and astrophysical situations, fusion science, laser solid interaction, and nanotechnology. Various studies have been done on cold quantum plasma. Patil et al. [12] have reported the propagation behaviour of beam in quantal-medium. They have found that quantum effects plays important role in laser plasma interaction. Habibi et al. [13] have investigated time independent self focusing in cold quantum plasma using ramp density profile. They have studied the effect on behavior of oscillatory beam width parameter due to quantum effects and inhomogeneity of plasma such as upward ramp density profile. Apart from these few studies, dynamical aspects of propagation of elliptical laser beams, Hermite Cosine Gaussian laser beams, cosh Gaussian laser beams in cold quantum plasma are also studied [14–16].

The study of Valcan Petawatt laser proposes that the intensity distribution deviates from the Gaussian distribution [17]. Nakatsutsumi et al. [18] on further study suggests that intensity distribution is in q-Gaussian which in function form can be written as

$$I(r) = I_0(r)\left(1 + \frac{r^2}{q^2 r_0^2}\right)^{-q} \tag{1}$$

Here real parameters q and r_0 are based on experimental data. The q parameter represents the departure from the Gaussian beam's field distribution. Variation from the Gaussian beam may be due to unexpected inclusion in the gain medium or small obstacles. The investigation of this intensity distribution could be more practical. Sharma and Kourakis [19] have studied relativistic propagation of q Gaussian laser beam analytically and numerically. Using higher order corrections, Kaur et al. [20] have explored the propagation in relativistic anisotropic medium for our plasma. Using the variational technique, Wang et al. [21] have reported the propagation behavior of q-Gaussian laser beam in a systematized plasma channel. Valkunde et al. [22] have studied self focusing in inhomogeneous plasma of exponential profile for varied q values. Vhanmore et al. [23] have investigated the influence of q parameter on the propagation of beam considering relativistic nonlinearity. Kashyap et al. [24] have studied self focusing of q Gaussian laser beam in relativistic plasma considering the effect of light absorption. Gupta and Kumar [25] and Gupta et al. [26] have also studied q Gaussian laser beam. We have explored the effect q and critical radius on propagation characteristics of laser beam in quantum medium by using parabolic equation. The dielectric function of cold quantum plasma and the intensity distribution of a q Gaussian laser beam are reported in Sect. 2. Second order nonlinear differential equation is obtained. Section 3 is dedicated to the result and discussion. Brief some important conclusions are given in Sect. 4.

2 Basic Formulation

Consider q-Gaussian laser beam propagating through relativistic cold quantum plasma along $\hat{z}$ direction. The electric field of beam is given as

$$E(r,z) = A(r,z)Exp[i(kz - \omega t)] \tag{2}$$

where $A(r,z)$ is complex amplitude of electric field. $k = \frac{\omega\sqrt{\epsilon_0}}{c}$ is propagation constant and ω is frequency of laser beam. A following z dependent intensity of q-Gaussian laser beam can be given as [18]

$$A^2 = \frac{E_0^2}{f}\left(1 + \frac{r^2}{qr_0^2 f^2}\right)^{-q} \tag{3}$$

where E_0 is the amplitude at $r = z = 0$, r_0 is the initial spot size and q parameter alters beam intensity profile from that of Gaussian profile. Intensity distribution which becomes Gaussian when $q \to \infty$ i.e

$$\lim_{q\to\infty} EE^* = \frac{E_0^2}{f} exp\left(-\frac{r^2}{r_0^2 f^2}\right) \tag{4}$$

Propagation of laser in plasma having effecting dielectric constant ϵ is governed by following equation

$$\nabla^2 E + \frac{\omega^2}{c^2}\epsilon E = 0 \tag{5}$$

The effective dielectric constant in general is given as [12]

$$\epsilon = \epsilon_0 + \phi(EE^*) \tag{6}$$

where ϵ_0 and $\phi(EE^*)$ are linear and nonlinear terms of dielectric constant. Here $\epsilon_0 = 1 - \left(\frac{\omega_p}{\omega}\right)^2$ and $\omega_p = \sqrt{\frac{4\pi n_0 e^2}{m}}$ where e, m, n_0 are electrons charge, mass of electron and electron density respectively. In case of cold quantum plasma, intensity dependent dielectric constant is given as [12]

$$\phi(EE^*) = \left(\frac{\omega_p}{\omega}\right)^2 \left[1 - \frac{1}{\gamma}\left(1 - \frac{\delta}{\gamma}\right)^{-1}\right] \tag{7}$$

where $\gamma = \sqrt{1 + \alpha EE^*}$, $\delta = \frac{4\pi^4 h^2}{m_0^2 \omega^2 \lambda^4}$, h is Planck's constant, λ is wavelength of laser source.

Following Akhmanov's parabolic approach et al. [27] and its extension by Sodha et al. [28], differential equation of beam width parameter (f) for q Gaussian beam in cold quantum plasma can be obtained as

$$\frac{\partial^2 f}{\partial \zeta^2} = \frac{4+q}{qf^3} - \frac{\alpha E_0^2 \sqrt{1 + \frac{\alpha E_0^2}{f}} \left(\frac{r_0 \omega_p}{c}\right)^2}{2 * \left(\delta - \sqrt{1 + \frac{\alpha E_0^2}{f}}\right)^2 f^2 \left(\alpha E_0^2 + f\right)} \tag{8}$$

where $\zeta = \frac{z}{kr_0^2}$ is dimensionless distance of propagation. Using boundary conditions $f = 1, \frac{\partial f}{\partial z} = 0$ corresponding to initial wave front at $\zeta = 0$, (8) is solved numerically.

3 Result and Discussion

Equation (8) is a non integrable that governs the propagation behavior of laser beam in plasma. It is observed that as $q \to \infty$ (8) reduces to similar differential equation which is obtained earlier [12]. On right hand side in (8), the first term is responsible for diffraction divergence and the second term is responsible for the convergence of beam. Equation (8) is solved numerically using the following numerical parameters $\omega = 1.77800 \times 10^{20}\ \frac{\text{rad}}{\text{s}}$, $m_0 = 9.10938 \times 10^{-28}$ g, $h = 6.62618 \times 10^{-27}$ erg.sec, $c = 2.99792 \times 10^{10}$ cm/sec. Under critical conditions, $\frac{\partial^2 f}{\partial z^2} = 0, \frac{\partial f}{\partial z} = 0, f = 1, \alpha E_0^2 = p_0, \frac{r_0 \omega_p}{c} = \rho_0$, (8) reduces to

$$\rho_0 = \sqrt{\frac{(4+q)\, 2 \left(\delta - \sqrt{1+p_0}\right)^2 \left(\sqrt{1+p_0}\right)}{q p_0}} \tag{9}$$

It is noted that as $q \to \infty$, one can obtain the critical curve equation for Gaussian beam in quantum-medium. Figure 1 gives the critical curve for varied values of q. It is a graphical representation of the relation between critical beam radius (ρ_0) and initial intensity parameter (p_0). Critical curve divides the graph into self focusing and defocusing regions. In Fig. 1, arbitrary choice of point (p_0, ρ_0) in a region above the critical curve (supercritical region) promises focusing of beam and point (p_0, ρ_0) below the critical curve (subcritical region) leads to defocusing of beam. Any point on the critical curve leads to self trapping of the laser beam. In Fig. 1, it is observed that, initially ρ_0 decreases rapidly within the short range of p_0 and attains minimum value. Then ρ_0 increases slowly with a further increase in p_0. It is observed that the

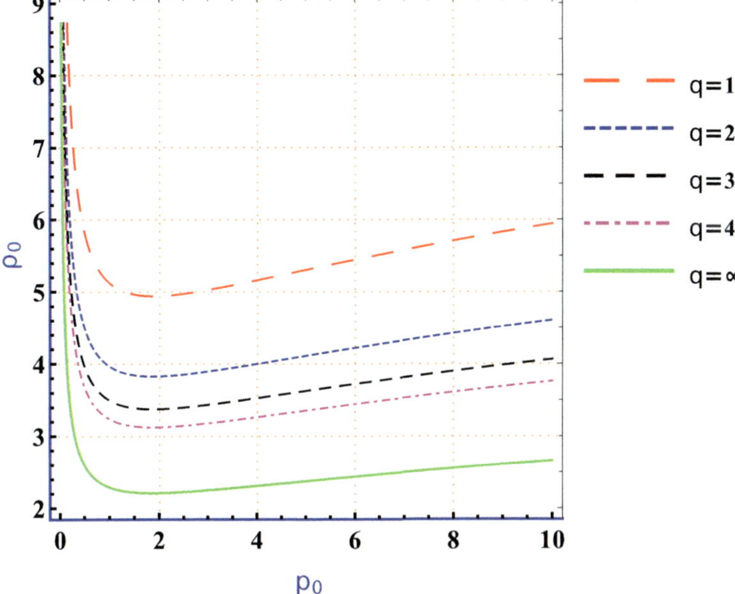

Fig. 1 Variations of ρ_0 versus p_0 (critical curves) for varied values of q i.e. ($q = 1, 2, 3, 4, \infty$)

critical curve shifts downward as q increases. For the Gaussian beam, the supercritical region is relatively the largest.

Figure 2, reveals propagation behavior of q Gaussian laser beam i.e. variation of f against ζ. Self focusing of beam in oscillatory mode is observed for $q = 1, 2, 3, 4, \infty$ (for $p_0 = 2$, $\rho_0 = 6$ i.e. in supercritical region). Also in Fig. 2, it is observed that as the value of parameter q increases, the beam width parameter and self focusing length decrease which indicates enhanced self focusing. At $q \to \infty$ i.e. Gaussian beam, self-focusing length is minimum. In Fig. 2, steady defocusing is observed (for $p_0 = 2$, $\rho_0 = 1.5$ i.e. in subcritical region). Also, it is observed that the rate of defocusing of f decreases as q increases.

4 Conclusion

In present paper, the study of propagation behavior of q-Gaussian beam in cold quantum plasma has been done. Using Akhmanov's approach et al. [27], nonlinear ordinary differential equation of beam width parameter has been obtained. Propaga-

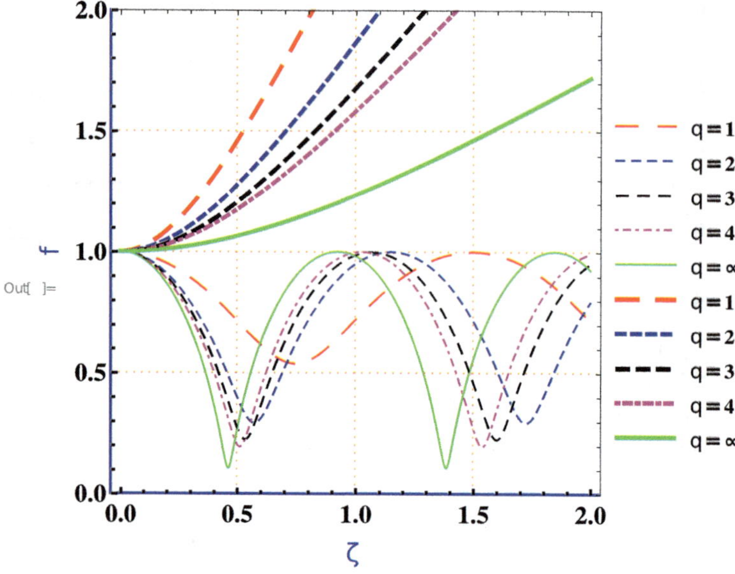

Fig. 2 Variation of f along ζ with $p_0 = 2$, $\rho_0 = 6$ (thin lines) and $p_0 = 2$, $\rho_0 = 1.5$ (thick lines) (Color Online)

tion behavior of q Gaussian laser beam can be effectively explored by varying the limits of q from extremely low value to infinity. The significant change in f and self focusing length takes place in the limit $q \to \infty$ as compared to lower values of q.

References

1. Deutsch, C., Furukawa, H., Mima, K., Murakami, M., Nishihara, K.: Interaction physics of the fast ignitor concept. Phys. Rev. Lett. **77**, 2483 (1996). https://doi.org/10.1103/PhysRevLett.77.2483
2. Tabak, M., Hammer, J., Glinsky, M.E., Kruer, W.L., Wilks, S.C., Woodworth, J., Campbell, E.M., Perry, M.D., Mason, R.J.: Ignition and high gain with ultrapowerful lasers. Phys. Plasmas **1**, 1626 (1994). https://doi.org/10.1063/1.870664
3. Regan, S.P., Bradley, D.K., Chirokikh, A.V., Craxton, R.S., Meyerhofer, D.D., Seka, W., Short, R.W., Simon, A., Town, R.P.J., Yaakobi, B.: Laser-plasma interactions in long-scale-length plasmas under direct-drive National Ignition Facility conditions. Phys. Plasmas **6**, 2072 (1999). https://doi.org/10.1063/1.873716
4. Malka, V.: Laser plasma accelerators. Phys. Plasmas **19**, 055501 (2012). https://doi.org/10.1063/1.3695389
5. Wiggins, S.M., Issac, R.C., Welsh, G.H., Brunetti, E., Shanks, R.P., Anania, M.P., Cipiccia, S., Manahan, G.G., Aniculaesei, C., Ersfeld, B., Islam, M.R., Burgess, R.T.L., Vieux, G., Gillespie, W.A., MacLeod, A.M., van der Geer, S.B., de Loos, M.J., Jaroszynski, D.A.: High quality electron beams from a laser wakefield accelerator. Plasma Phys. Control. Fus. **52**, 124032 (2010). https://doi.org/10.1088/0741-3335/52/12/124032

6. Fiuza, F., Stockem, A., Boella, E., Fonseca, R.A., Silva, L.O., Haberberger, D., Tochitsky, S., Gong, C., Mori, W.B., Joshi, C.: Laser-driven shock acceleration of monoenergetic ion beams. Phys. Rev. Lett. **109**, 215001 (2012). https://doi.org/10.1103/PhysRevLett.109.215001
7. Hooker, S.M.: Developments in laser-driven plasma accelerators. Nat. Photonics **7**, 775 (2013). https://doi.org/10.1038/nphoton.2013.234
8. Sprangle, P., Esarey, E., Ting, A.: Nonlinear theory of intense laser-plasma interactions. Phys. Rev. Lett. **64**, 2011 (1990). https://doi.org/10.1103/PhysRevLett.64.2011
9. Ferrari, H.E., Lifschitz, A.F., Maynard, G., Cros B.: Electron acceleration by laser wakefield and X-ray emission at moderate intensity and density in long plasmas. Phys. Plasmas **18**, 083108 (2011). https://doi.org/10.1063/1.3624771
10. Liu, Y., Dong, Q., Peng, X., Jin, Z., Zhang, J.: Soft X-ray emission, angular distribution of hot electrons, and absorption studies of argon clusters in intense laser pulses. Phys. Plasmas **16**, 043301 (2009). https://doi.org/10.1063/1.3125308
11. Bagchi, S., Kiran, P.P., Yang, K., Rao, A.M., Bhuyan, M.K., Krishnamurthy, M., Kumar, G.R.: Bright, low debris, ultrashort hard x-ray table top source using carbon nanotubes. Phys. Plasmas **18**, 014502 (2011). https://doi.org/10.1063/1.3531685
12. Patil, S.D., Takale, M.V., Navare, S.T., Dongare, M.B., Fulari, V.J.: Self-focusing of Gaussian laser beam in relativistic cold quantum plasma. Optik **124**, 180–183(2013). https://doi.org/10.1016/j.ijleo.2011.11.061
13. Habibi, M., Ghamari, F.: Stationary self-focusing of intense laser beam in cold quantum plasma using ramp density profile. Phys. Plasmas **19**, 103110 (2012). https://doi.org/10.1063/1.4762848
14. Walia, K., Tripathi, D.: Self-focusing of elliptical laser beam in cold quantum plasma. Optik **186**, 46–51(2019). https://doi.org/10.1016/j.ijleo.2019.04.081
15. Thakur, V., Kant, N.: Combined effect of chirp and exponential density ramp on relativistic self-focusing of Hermite-Cosine-Gaussian laser in collisionless cold quantum plasma. Braz. J. Phys. **49**, 113–118 (2019). https://doi.org/10.1007/s13538-018-00624-7
16. Habibi, M., Ghamari, F.: Improved focusing of a Cosh-Gaussian laser beam in quantum plasma: higher order paraxial theory. IEEE Trans. Plasma Sci. **43**, 2160–2165 (2015). https://doi.org/10.1109/TPS.2015.2440319
17. Patel, P.K., Key, M.H., MacKinnon, A.J., Berry, R., Borghesi, M., Chambers, D.M., Chen, H., Clarke, R., Damian, C., Eagleton, R., Freeman, R., Glenzer, S., Gregori, G., Heathcote, R., Hey, D., Izumi, N., Kar, S., King, J., Nikroo, A., Niles, A., Park, H.S., Pasley, J., Patel, N., Shepherd, R., Snavely, R.A., Steinman, D., Stoeckl, C., Storm, M., Theobald, W., Town, R., Van Maren, R., Wilks, S.C., Zhang, B.: Integrated laser-target interaction experiments on the RAL petawatt laser. Plasma Phys. Control Fus. **47**, B833–B840 (2005). https://doi.org/10.1088/0741-3335/47/12B/S65
18. Nakatsutsumi, M., Davies, J.R., Kodama, R.: Space and time resolved measurements of the heating of solids to ten million kelvin by a petawatt laser. New J. Phys. **10**, 043046 (2008). https://doi.org/10.1088/1367-2630/10/4/043046
19. Sharma, A., Kourakis, I.: Spatial evolution of a q-Gaussian laser beam in relativistic plasma. Laser Part. Beams **28**, 479–489 (2010). https://doi.org/10.1017/S0263034610000479
20. Kaur, R., Gill, T.S.: Relativistic effects on evolution of a q-Gaussian laser beam in magneto-plasma: application of higher order corrections. Phys. Plasmas **24**, 053105 (2017). https://doi.org/10.1063/1.4983309
21. Wang, L., Hong, X.-R., Sun, J.-A., Tang, R.-A., Yang, Y., Zhou, W.-J., Tian, J.-M., Duan, W.-S.: Effects of relativistic and channel focusing on q-Gaussian laser beam propagating in a preformed parabolic plasma channel. Phys. Lett. A, **381**, 2065–2071 (2017). https://doi.org/10.1016/j.physleta.2017.04.033
22. Valkunde, A.T., Vhanmore, B.D., Urunkar, T.U., Gavade, K.M., Patil, S.D., Takale, M.V.: Effect of exponential density transition on self-focusing of q-Gaussian laser beam in collisionless plasma. AIP Conf. Proc. **1953**, 140088 (2018). https://doi.org/10.1063/1.5033263
23. Vhanmore, B.D., Patil, S.D., Valkunde, A.T., Urunkar, T.U., Gavade, K.M., Takale, M.V., Gupta, D.N.: Effect of q-parameter on relativistic self focusing of q Gaussian laser beam in plasma. Optik **158**, 574–579 (2018). https://doi.org/10.1016/j.ijleo.2017.12.182

24. Kashyp, R., Aggrawal, M., Gill, T.S., Arora, N.S., Kumar, H., Moudhagill, D.: Self-focusing of q-Gaussian laser beam in relativistic plasma under the effect of light absorption. Optik Int. J. Light Electron. Opt. **182**, 1030–1038 (2019). https://doi.org/10.1016/j.ijleo.2018.12.105
25. Gupta, N., Kumar, S.: Generation of second harmonics of relativistically self-focused q-Gaussian laser beams in underdense plasma with axial density ramp. Opt. Quant. Electron. **53**, 193 (2021). https://doi.org/10.1007/s11082-021-02827-w
26. Gupta, N.: Second harmonic generation of q-Gaussian laser beam in plasma channel created by ignitor heater technique. Laser Part. Beams **37**, 184–196 (2019). https://doi.org/10.1017/S0263034619000193
27. Akhmanov, S.A., Sukhorukov, A.P., Khokhlov, R.V.: Self-focusing and diffraction of light in a nonlinear medium. Sov. Phys. Uspekhi **10**(5), 609–636 (1968). https://doi.org/10.1070/PU1968v010n05ABEH005849
28. Sodha, M.S., Ghatak, A.K., Tripathi, V.K.: Self focusing of laser beams in plasmas and semiconductors. Prog. Opt. **13**, 169–265 (1976). https://doi.org/10.1016/S0079-6638(08)70021-0

Study of Quantum-Electron Acoustic Solitary Structures in Fermi Plasma with Two Temperature Electrons

Shilpi, Sharry, Chinmay Das, and Swarniv Chandra

Abstract One dimensional Quantum Hydrodynamic (QHD) model is used to study linear and non-linear effects of Quantum Electron Acoustic waves. We have derived the KdV equations using Reductive Perturbation Technique and obtained solitary wave solutions for Quantum Electron acoustic waves. The mutual interaction of such stationary formations and the breakdown mechanism are studied. We have used a newly designed code to study the time evolution of wave-wave interaction and the breakdown mechanism. In order to discover the stability regime and chaotic scenario in quantum electron-acoustic waves, bifurcation analysis and dynamical system research are used. A dynamical system is investigated further to determine the stability factors and the parametric range for such stability. The chaotic behaviour of the system is studied through Largest Lyapunov exponent. This work will find applications in polar capes, magnetospheres and stellar media.

Keywords Electron-acoustics wave · Linear dispersion relation · Quantum hydrodynamic model · Two-temperature electron

Shilpi · Sharry
Physics Department, Guru Nanak Dev University, Amritsar 143001, India

C. Das
Department of Mathematics, Govt. General Degree College at Kushmandi, Dakshin Dinajpur 733121, India

S. Chandra (✉)
Department of Physics, Govt. General Degree College at Kushmandi, Dakshin Dinajpur 733121, India
e-mail: swarniv147@gmail.com

C. Das · S. Chandra
Institute of Natural Sciences and Applied Technology, Kolkata 700032, India

S. Banerjee and A. Saha (eds.), *Nonlinear Dynamics and Applications*, Springer Proceedings in Complexity,
https://doi.org/10.1007/978-3-030-99792-2_6

1 Introduction

The fourth state of matter is Plasma which is a hot ionizing gas created from roughly equal numbers of charged ions and charged electrons. The charged particles in plasma travel at random. They interact with each other via their own electromotive force and also respond to exterior disturbances [1]. The perturbations forces can be pressure force, microwave, or laser injection. Due to the coherent motion of plasma particles various collective wave phenomena arise. Since plasma is strongly influenced by electric and magnetic fields hence a possibility of a wide range of longitudinal and transverse waves. plasma acoustic wave is a longitudinal wave that propagates due to compression and rarefaction of plasma particles.

Plasma waves can be in electron acoustic, ion-acoustic, or dust acoustic mode. The present paper will stick to the Electrostatic modes in plasma. Electrostatic modes rely on the mass of the electrons whereas ions are supposed to be infinitely heavy i.e. stationary. Electron acoustic waves (EAWs) is one of the longitudinal waves in plasma. EAWs are created by two groups of electrons with differing temperatures, called cold and hot electrons [2–11]. The distinction between two electron groups comes from their energy. The thermal energy of warm electrons is higher than cold electrons. The corresponding pressure perturbations from fast and slow particles are hugely different; cool electrons, as they are immobile, are in-charge of providing the inertial effects [12–19] whereas extremely heated electrons can move freely and their pressure provides restoring force. EAWs are high frequency (as compared to ion plasma frequency) modes of electrostatic waves. For frequencies greater than ion plasma frequencies as in the case of EAWs, ions do not take part in dynamics, they play the role of maintaining the charge neutrality. Plasma with two temperature electrons exist in space plasma [20–23] and also occur in laboratory plasma [24–34]. EAWs are thrust areas of research as various phenomenons are explained by them like wave emission in different regions of Earth's magnetosphere (region of space around the earth in which charged particles are affected by Earth's magnetic field), hiss in the polar cusp region (where the particles from the sun have direct access to Earth's atmosphere) and the source of broadband electrostatic noise. The research on the nonlinear evolution of EAWs has increased in recent years [12, 35–44]. Initially, the research was carried out using classical non-relativistic less dense plasma in which thermal de-Broglie wavelength was much less than inter-Fermion distance. But for high-density plasma, as in space plasma, where thermal de-Broglie wavelength becomes comparable to or more than the inter-Fermion distance, quantum effects become important [45–48]. So an additional quantum diffraction term or Bohm potential term is added in the momentum equations of hot and cold electrons. Since ions do not take part in dynamics, no such equations are defined for them. Theoretical research was done on linear and nonlinear propagation of various electrostatics modes in quantum plasmas using QHD [49–63]. The Quantum hydrodynamic QHD model is the most widely used formula for describing the behavior of plasma particles at quantum sizes. Studies show that quantum effects can considerably alter the linear and non-linear properties of plasma.

We have studied various mechanism for soliton breakdown [64, 65] and soliton-soliton interaction [66, 67]. Solitons are linear flaws that bend randomly at a finite temperature. This causes solitons to collide, changing the nature of the soliton-soliton interaction [68–72]. The solitons may be decelerated/accelerated through the interactions of soliton with background waves which may be utilized to study tsunami waves and fiber soliton communications [73]. Dynamical systems theory was used to report bifurcations of nonlinear travelling wave solutions in such plasmas [74]. Various solitonic and quasiperiodic wave characteristics for ion-acoustic waves have been investigated by reducing the nonlinear equation to a Hamiltonian system with electrostatic potential and applying the bifurcation theory of dynamical systems [75–79]. The paper is organized in the following way. In the Sect. 2, the basic equations and it's normalized forms are presented and in Sect. 3, Linear Dispersion Relation is derived. Further, the Derivation of KdV Equation is there in Sect. 4, in which soliton-soliton interaction and breakdown of solitons is discussed under Sects. 5.1 and 5.2. Then, the Dynamical study of Electron Acoustic KdV Solitons is carried out in Sect. 6.

2 Basic Equations

We consider the plasma consisting of of two separate groups of electrons namely cold electron fluid, hot electron fluid [13, 18, 80–91] and ions forming uniform neutralizing background, this allows the generation of quantum electron acoustic modes in plasma. To study the linear and non-linear behavior we obtain normalized (dimensionless) fluid continuity, momentum, and Poisson's equations through some appropriate transformations. The unnormalized equations are as following.

2.1 Continuity Equations for Hot and Cold Electrons

$$\frac{\partial n_h}{\partial t} + \frac{\partial (n_h u_h)}{\partial x} = 0 \tag{1}$$

where n_h is hot electron density in plasma and u_h is the velocity of hot electrons.

$$\frac{\partial n_c}{\partial t} + \frac{\partial (n_c u_c)}{\partial x} = 0 \tag{2}$$

where n_c is cold electron density in plasma and u_c is the velocity of cold electrons.

2.2 Momentum Equations

$$0 = e\frac{\partial \phi}{\partial x} - \frac{1}{n_h}\frac{\partial P_h}{\partial x} + \frac{\hbar^2}{2m_e}\frac{\partial}{\partial x}\left\{\frac{1}{\sqrt{n_h}}\frac{\partial^2\sqrt{n_h}}{\partial x^2}\right\} \tag{3}$$

This is the momentum equation of hot electrons where ϕ is electrostatic wave potential, e is the electric charge, and p_h is the pressure of hot electrons which provides restoring force to the EAW's. The left-hand side of this equation i.e the inertial term is zero because of the very high mobility of hot electrons.

$$\left(\frac{\partial}{\partial t} + u_c\frac{\partial}{\partial x}\right)u_c = \frac{1}{m_e}\left[e\frac{\partial \phi}{\partial x} + \frac{\hbar^2}{2m_e}\frac{\partial}{\partial x}\left\{\frac{1}{\sqrt{n_c}}\frac{\partial^2\sqrt{n_c}}{\partial x^2}\right\}\right] \tag{4}$$

This is the momentum equation for cold electrons. Here inertial term is non-zero because of the very low mobility of cold electrons. cool electrons are providing restoring force for electron acoustic oscillations.

2.3 Poisson's Equation

The Poisson's equation closes the system of equations as

$$\frac{\partial^2 \phi}{\partial x^2} = 4\pi e\left(n_c + n_h - Z_i n_i\right) \tag{5}$$

where $\hbar$ is the reduced Planck's constant $(h/2\pi)$ and $Z_i e$ is the charge on ion.

2.4 The Pressure Law

Here we assume that plasma particles behave as 1-D Fermi gas and therefore taking the Fermi pressure P_j as [92–95]:

$$P_j = \frac{m_j c_{sj}^2}{3n_{j0}^2} n_j^3 \tag{6}$$

Normalized Fermi pressure term for hot electrons is given as-:

$$P_h = \frac{m_e c_{sh}^2}{3n_{h0}^2} n_h^3 n_{h0}^3 = \frac{m_e c_{sh}^2}{3} n_h^3 n_{h0} \tag{7}$$

Using the normalization scheme as $x \rightarrow x\omega_j/c_{sj}$, $t \rightarrow t\omega_j$, $\phi \rightarrow e\phi/2k_BT_{Fj}$, $u_j \rightarrow u_j/c_{sj}$, $n_j \rightarrow n_j/n_{j0}$, $\eta_c \rightarrow \eta_c\omega_j/m_ec_{sj}^2$, Eqs. (1–5) can be written as:

$$\frac{\partial n_h}{\partial t} + \frac{\partial (n_h u_h)}{\partial x} = 0 \tag{8}$$

$$\frac{\partial n_c}{\partial t} + \frac{\partial (n_c u_c)}{\partial x} = 0 \tag{9}$$

$$0 = \frac{\partial \phi}{\partial x} - n_h \frac{\partial n_h}{\partial t} + \frac{H^2}{2}\frac{\partial}{\partial x}\left[\frac{1}{\sqrt{n_h}}\frac{\partial^2 \sqrt{n_h}}{\partial x^2}\right] \tag{10}$$

$$\left(\frac{\partial}{\partial t} + u_c\frac{\partial}{\partial x}\right)u_c = \frac{\partial \phi}{\partial x} + \frac{H^2}{2}\frac{\partial}{\partial x}\left[\frac{1}{\sqrt{n_h}}\frac{\partial^2 \sqrt{n_h}}{\partial x^2}\right] \tag{11}$$

$$\frac{\partial^2 \phi}{\partial x^2} = n_c + \frac{n_h}{\delta} - \frac{\delta_i}{\delta} n_i \tag{12}$$

where $H = \hbar\omega_j/2k_BT_{Fj}$ is a non-dimensional quantum diffraction parameter, $\delta = n_{c0}/n_{h0}$ and $\delta_i = Zn_{i0}/n_{h0}$.

3 Linear Dispersion Relation

To investigate the nonlinear behavior of electron acoustic waves, we consider the following perturbation expansion for the field quantities:

$$\begin{bmatrix} n_h \\ n_c \\ u_h \\ u_c \\ \phi \end{bmatrix} = \begin{bmatrix} 1 \\ 1 \\ u_0 \\ u_0 \\ \phi_0 \end{bmatrix} + \varepsilon \begin{bmatrix} n_h^{(1)} \\ n_c^{(1)} \\ u_h^{(1)} \\ u_c^{(1)} \\ \phi^{(1)} \end{bmatrix} + \varepsilon^2 \begin{bmatrix} n_h^{(2)} \\ n_c^{(2)} \\ u_h^{(2)} \\ u_c^{(2)} \\ \phi^{(2)} \end{bmatrix} + \cdots \tag{13}$$

where $j = h, c$ is the subscript which is for both hot and cold electrons, n_j is the number density of hot and cold electrons, u_j is the velocity of hot and cold electrons in plasma, ϕ is electro-static wave potential and ϵ is the smallness parameter, powers of ϵ represents the order of perturbation. By assuming that all field quantities vary as $e^{i(kx-\omega t)}$ such that eigenvalues of $\frac{\partial}{\partial t}$ and $\frac{\partial}{\partial x}$ are $-i\omega$ and ik respectively. Now in order to obtain linear dispersion relation substituting perturbation expansion of field quantities in governing equations and keeping only first-order terms in ϵ we get the following linear dispersion relation for normalized wave frequency (ω) and wavenumber (k):

$$1 = \frac{1}{(\omega - ku_o)^2 - \frac{H^2k^4}{4}} - \frac{1}{\delta k^2\left\{1 + \frac{H^2k^2}{4}\right\}} \tag{14}$$

Equation (14) represents the linear dispersion relation for EAW's in Fermi plasma, the form of wave frequency ω is given as follows:

$$\omega = \left[\frac{\delta k^2(1 + \frac{H^2k^2}{4})}{1 + \delta k^2(1 + \frac{H^2k^2}{4})} + \frac{H^2k^4}{4}\right]^{1/2} + ku_0 \tag{15}$$

4 Derivation of KDV Equation

To study the nonlinear behavior of electron acoustic waves we consider inertialess warm electrons, inertial cool electrons and stationary ions. The pressure effect is only due to hot electrons which provide restoring force. We explore the collision of two solitary waves using an expanded PLK perturbation approach. The following are the perturbed quantities:

$$Y = Y_0 + \sum_{r=1}^{\infty} \epsilon^{r+1} Y_r, \quad \Gamma = \sum_{r=1}^{\infty} \epsilon^{r+2} \Gamma_r \tag{16}$$

where $Y = (n_{\mathrm{d}}, u_{\mathrm{dz}}, \psi)$, $Y_0 = (1, 0, 0)$, and $\Gamma = \left(u_{\mathrm{dx}}, u_{\mathrm{dy}}\right)$ The scaling variables x and t are stretched by the new coordinate system ξ, η and τ in many scale variables:

$$\xi = \epsilon\left(l_x x + l_y y + l_z z - V_0 t\right) + \epsilon^2 M_0(\eta, \tau) + \cdots \tag{17}$$

$$\eta = \epsilon\left(l_x x + l_y y + l_z z + V_0 t\right) + \epsilon^2 N_0(\xi, \tau) + \cdots \tag{18}$$

$$\tau = \epsilon^3 t \tag{19}$$

where ξ and η denote the opposite side trajectories of solitary waves and ϵ is a parameter that determines the strength of nonlinearity. l_x, l_y, and l_z are direction cosines along the x, y, and z directions, and hence $l_x^2 + l_y^2 + l_z^2 = 1$. Using (17)–(19) into (8)–(12) and comparing lowest order ϵ and taking direction cosines as (1, 0, 0) as wave propagation is along x-axes only, we get the normalized phase velocity for magnetized plasma as:

$$V_0 = \frac{\omega}{k} = \left[\frac{\delta(1 + \frac{H^2k^2}{4})}{1 + \delta k^2(1 + \frac{H^2k^2}{4})} + \frac{H^2k^2}{4}\right]^{1/2} + u_0 \tag{20}$$

Next taking higher order of ϵ, we get the following equations:

$$\frac{\partial \phi^{(1)}}{\partial \tau} + A_1 \phi^{(1)} \frac{\partial \phi^{(1)}}{\partial \xi} + B_1 \frac{\partial^3 \phi^{(1)}}{\partial \xi^3} = 0 \tag{21}$$

$$\frac{\partial \phi^{(1)}}{\partial \tau} - A_2 \phi^{(1)} \frac{\partial \phi^{(1)}}{\partial \eta} - B_2 \frac{\partial^3 \phi^{(1)}}{\partial \eta^3} = 0 \tag{22}$$

where

$$A_1 = \frac{\frac{p_2^2 (V_0 - u_0)^2}{\delta} - p_1^2 - 2 p_1 p_2}{p_1 + (V_0 - u_0)\, p_2} \tag{23}$$

$$A_2 = \frac{\frac{p_2^2 (V_0 - u_0)^2}{\delta} - p_1^2 - 2 p_1 p_2}{p_1 + (u_0 - V_0)\, p_2} \tag{24}$$

$$B_1 = \frac{(V_o - u_o)^2 - \frac{H^2}{4} p_2 (V_o - u_o)^2}{p_1 + (V_o - u_o)\, p_2} \tag{25}$$

$$B_2 = \frac{(V_o - u_o)^2 - \frac{H^2}{4} p_2 (V_o - u_o)^2}{p_1 + (u_o - V_o)\, p_2} \tag{26}$$

Equations (21) and (22) are the KdV equations of two solitary waves move towards one another in the reference frame of ξ and η.

5 Simulation Results

Soliton-soliton collision and decomposition of stationary structures into the secondary formation are the two mechanisms which explains the result of this analytic study. We start with the KdV equation and use our choice of initial and boundary conditions which show many intermittent phenomena that are crucial in understanding of these mechanisms. The simulation carried out here had been designed by Chinmay Das and Swarniv Chandra in the summer of 2021 at the computational faculty of Institute of Natural Sciences and Applied Technology, Kolkata. It explains Fourier transform coupled with Runge Kutta technique. The code has been named as INSAT-FORK code and is referred to in literature [41].

5.1 Interaction of Solitary Structures

Here we would study the time evolution of the stationary structures when two of them collide with each other. The interacting solitons can either be of the same or opposite structures. Density structures can be either solitary wavelets or of opposite polarity (expressed as a hyperbolic secant function) depending upon whether the solitons have similar structures or different ones at the beginning. These structures will give us different mechanisms if we let them propagate in space. The mechanisms we get from these are that the propagations are dependent on the scaling scheme i.e the order of scheme which we want to study. Here the scale of interaction is normalization. From figures, we get the information of soliton-soliton collision and their evolution with time. Figure 1 shows that initially two Gaussian pulses are separated by some distance and they are moving towards each other. The perturbation caused by their interaction stays a bit longer even though the interaction is over. Perturbation stays after the interaction of two hyperbolic secant squared waves too, but in this case, perturbation caused is more frequent and of sharp and high altitudes (Fig. 2). The cause behind this can be the higher peaks and sharpness of both waves. The interaction of two sinusoidal waves in Fig. 3 causes equal and opposite disturbance in opposite polarity regions of the wave at time $\tau = 3$, which immediately disappears.

Figure 4 shows the interaction of two Gaussian pulses caused by an external force. Even in this interaction too, the disturbance remains even after the interaction has finished. Next, in Fig. 5, the interaction of secant hyperbolic squared pulses after the external force is applied. As can be seen, the peaks of perturbation caused at time τ=4 and 10 are higher than when no external force was applied. Even in the interaction of two sinusoidal waves (Fig. 6, the external force caused to have a little disturbance after the interaction, which was not there without external force.

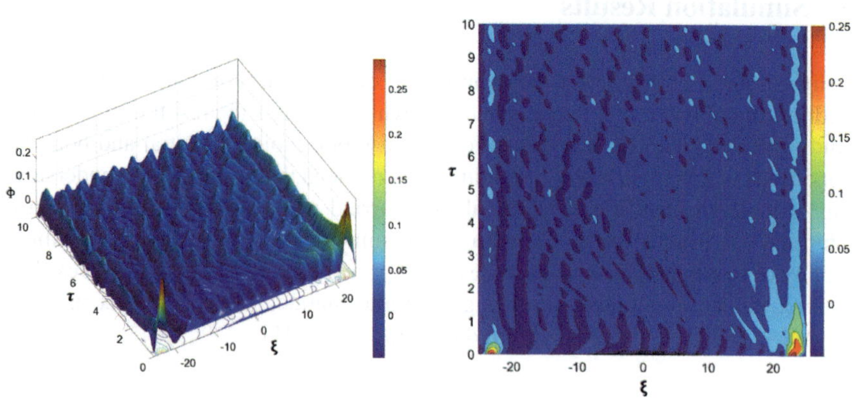

Fig. 1 (i) Variation in soliton-soliton collisions for two Gaussian pulses. (ii) Counter plots of variation in soliton-soliton collisions for two Gaussian pulses

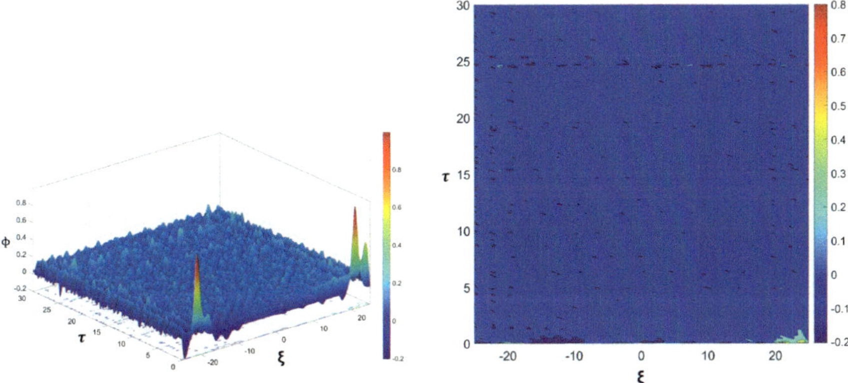

Fig. 2 (i) Variation in soliton-soliton collisions for two hyperbolic secant squared type pulses. (ii) Counter plots of variation in soliton-soliton collisions for two hyperbolic secant squared type pulses

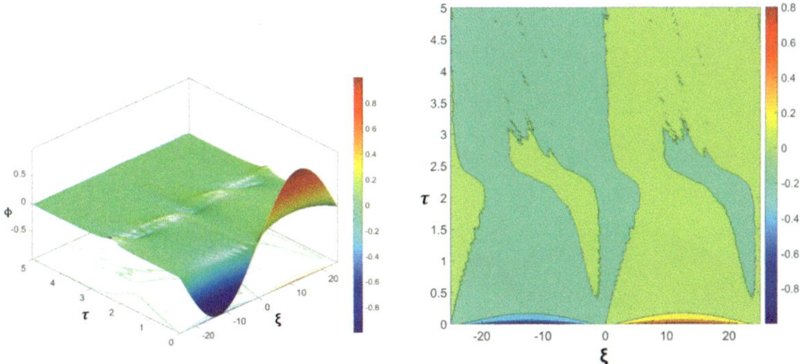

Fig. 3 (i) Variation in soliton-soliton collisions for two sinusoidal propagating pulses. (ii) Counter plots of variation in soliton-soliton collisions for two sinusoidal propagating pulses

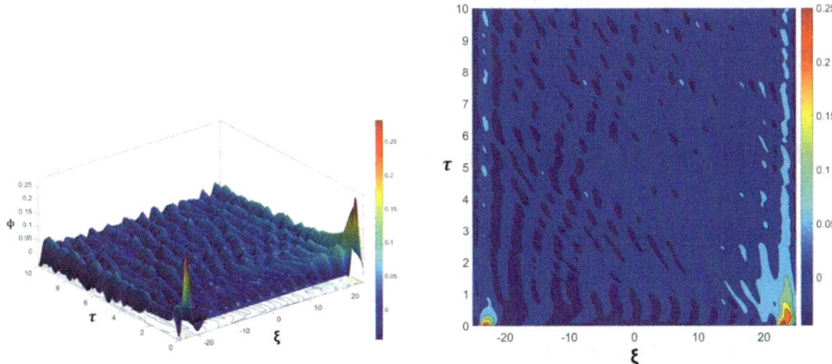

Fig. 4 (i) Interaction of two Gaussian pulses from both extremities towards a single point under the action of an external force. (ii) Counter plots of two Gaussian pulses from both extremities towards a single point under the action of an external force

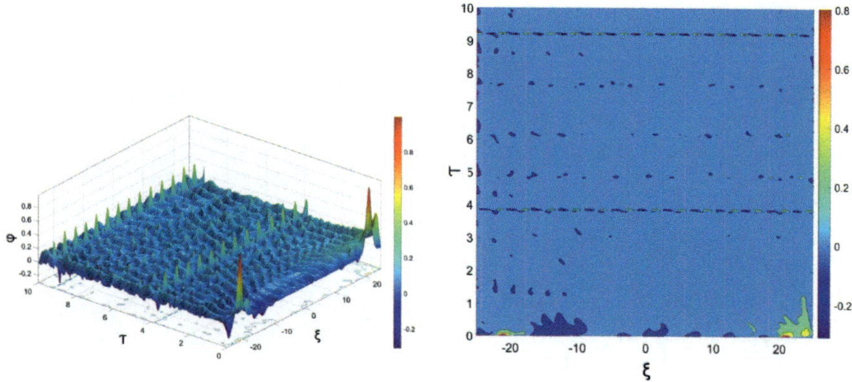

Fig. 5 (i) Time evolution of the potential profile formed by the collision of two secant hyperbolic squared type pulses under the action of an external force. (ii) Counter plots of the time evolution of the potential profile formed by the collision of two secant hyperbolic squared type pulses under the action of an external force

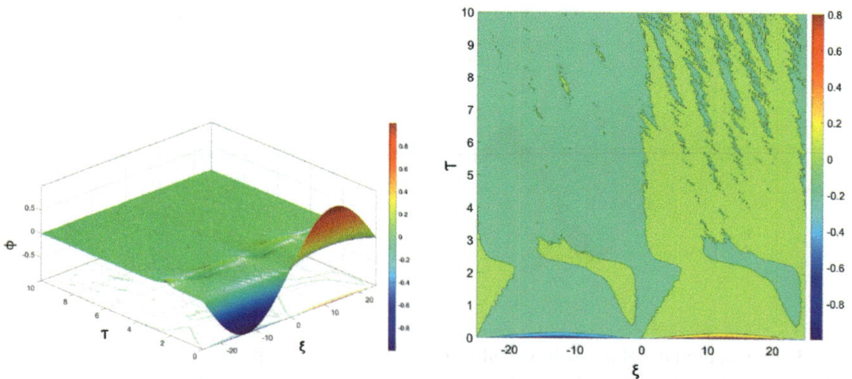

Fig. 6 (i) Time evolution of the potential profile formed by the collision of two sinusoidal pulses under the action of an external force. (ii) Counter plots of the time evolution of the potential profile formed by the collision of two sinusoidal pulses under the action of an external force

5.2 *Breakdown of Stationary Structures*

Here, we would study the decomposition mechanism of stationary structures for different initial conditions. An initial Gaussian function is seen to break into multiple fluctuations of potential (Fig. 7), which eventually take on distinct values at the boundary. This mechanism can be explained as follows: dispersion forces tear down the Gaussian profile, whereas nonlinear forces seek to strengthen it. Similar is the case of hyperbolic secant squared pulse in Fig. 8 but the fluctuations, in this case, are of lower amplitude. The effect of a sinusoidal waveform is spatially extended but

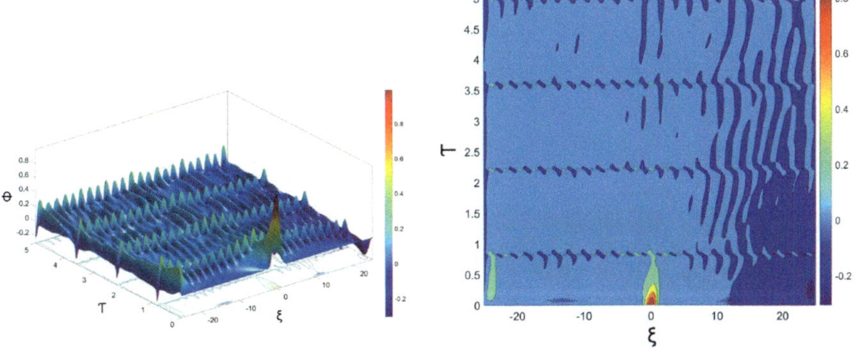

Fig. 7 (i) Breakdown of initial Gaussian pulse. (ii) Counter plots of breakdown of Gaussian pulses

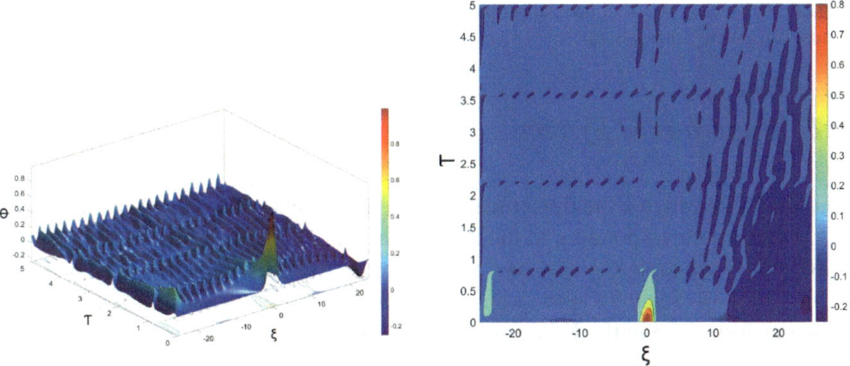

Fig. 8 (i) Breakdown of Initial hyperbolic secant squared hyperbolic function. (ii) Counter plots of Breakdown of secant hyperbolic squared function

temporally localized, implying that when the energy of the disturbance dissipates, the sinusoidal profile is effective at exciting the entire medium into small scale short-lived perturbations that occupy otherwise homogenous media (Fig. 9).

6 Dynamical Study of Electron Acoustic KdV Solitons

In this part, we would determine how EAWs propagates and evolves, how it behaves when the small perturbation is given to the system, its stability and its chaotic behavior i.e. when very small changes are made in the initial conditions how the system would respond to them. Dynamics of both perturbed and unperturbed would be discussed.

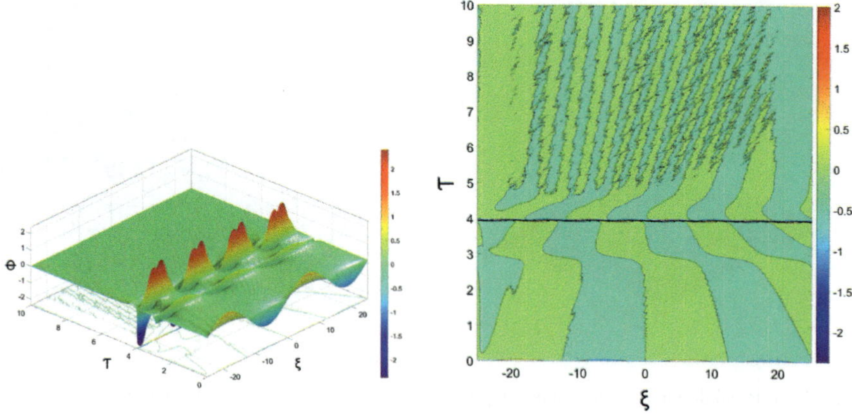

Fig. 9 (i) Breakdown of sinusoidal propagating pulses. (ii) Counter plots of breakdown of initial sinusoidal propagating pulses

6.1 Unperturbed System

In order to investigate the dynamical systems (DS) behavior of the system which is not perturbed corresponding to the KdV-equation (21), a one-dimensional traveling wave transformation $\eta = \xi - M\tau$ has been considered with boundary conditions $\psi \rightarrow 0$, $\frac{\partial \psi}{\partial \xi} \rightarrow 0$ as $\eta \rightarrow \pm\infty$, where M is the velocity of the wave frame. By using this transformation in the KdV equation (21) and after integrating w.r.t η, we get

$$\frac{\partial^2 \psi}{\partial \xi^2} = R\psi - T\psi^2 \tag{27}$$

where $R = \frac{M}{Bl^3}$, $T = \frac{A}{2Bl^2}$

By assuming

$$\frac{\partial \psi}{\partial \eta} = z_1 \tag{28}$$

The equation (27) transformed into the dynamical system of the form

$$\frac{\partial z}{\partial \eta} = R\psi - T\psi^2 \tag{29}$$

The two nonlinear differential equations (27) and (28) represent the dynamics of the unperturbed KdV equation (21). The system will be conservative if the divergence of field is zero ($\nabla \cdot \mathbf{F} = 0$) i.e the field is Solenoidal, where $\mathbf{F}(z, R\Psi^{(1)} - T\psi^{(1)^3})$. Hence the Hamiltonian of the planar system is given by:

$$H(\psi, z) = \frac{z^2}{4} - \frac{R\psi^2}{2} + \frac{T\psi^4}{4} \tag{30}$$

The term Bifurcation is used for the mathematical investigation of dynamical systems. Bifurcation is the study of unforeseen changes in the qualitative or topological structure of a given family, such as the integral curves of a family of vector fields, and the solutions of a family of differential equations when a small smooth change is made to the bifurcation parameters [96–100]. Hence we also analyze the Bifurcation as the phase plots describe all equations of EAWs with different initial conditions.

6.2 Perturbed System

The system is said to be perturbed when there is some external source of disturbance and the effects of these small external disturbances/forces/perturbations on the system are of great importance. Different types of forces affect the system differently. For our case, the provided external force is $f_0 cos(\omega\eta)$. Under such an external force the modeled equation will result in the forced KdV. Hence, we obtain

$$\frac{\partial \Psi}{\partial \eta} = z \tag{31}$$

$$\frac{\partial z}{\partial \eta} = R\psi^{(1)} - T\psi^2 + f_0 cos(\omega\eta) \tag{32}$$

These two DS equations represent the dynamics of forced KdV. Here η is the external perturbation, f_0 is the strength of the perturbation and ω is the compound frequency.

Figure 10 shows super nonlinear wave solution with variation of ψ for different values of δ, H and u_0. There is an increase in the amplitude and wavelength with every greater value of δ and so is the case of increase in value of H and u_0 but

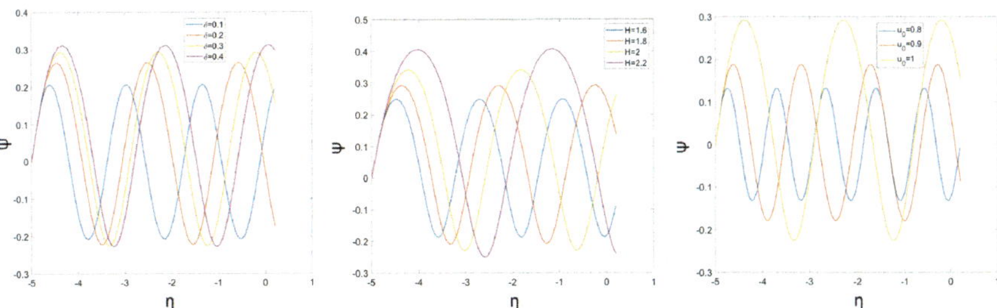

Fig. 10 Supernonlinear wave solution with variation of ψ corresponding to dynamical system for forced KdV equation with different values of (i) $\delta = 0.1$ (Sky Blue), 0.2 (yellow), 0.3 (Brown), 0.4 (purple), (ii) H = 1.6 (Sky Blue), 1.8 (yellow), 2 (Brown) 2.2 (Purple), (iii) $u_0 = 0.8$ (Sky blue), 0.9 (Brown), 1 (Yellow)

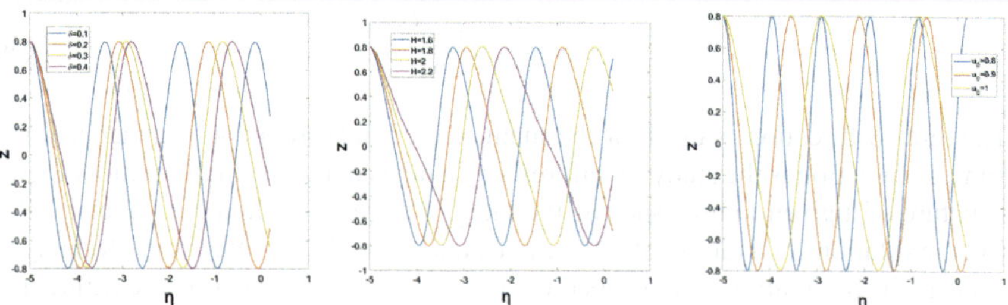

Fig. 11 Super nonlinear wave solution with variation of z corresponding to dynamical system for forced KdV equation with different values of (i) $\delta = 0.1$ (Sky Blue), 0.2 (yellow), 0.3 (Brown), 0.4 (purple), (ii) H = 1.6 (Sky Blue), 1.8 (yellow), 2 (Brown) 2.2 (Purple), (iii) $u_0 = 0.8$ (Sky blue), 0.9 (Brown), 1 (Yellow)

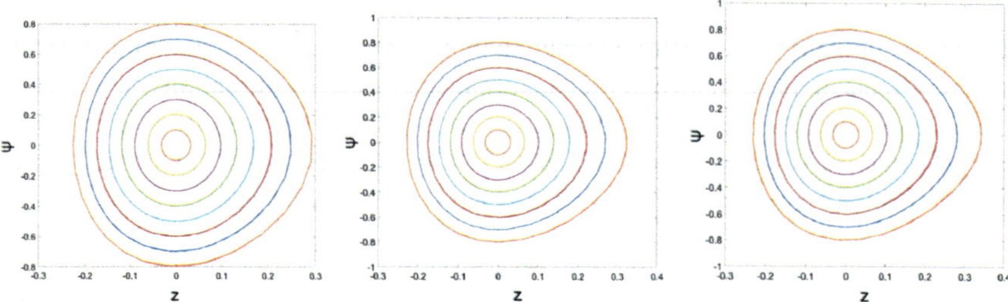

Fig. 12 Phase portrait of forced KdV for different values of (i) $\delta = 0.3$, (ii) $\delta = 0.5$, (iii) $\delta = 0.7$

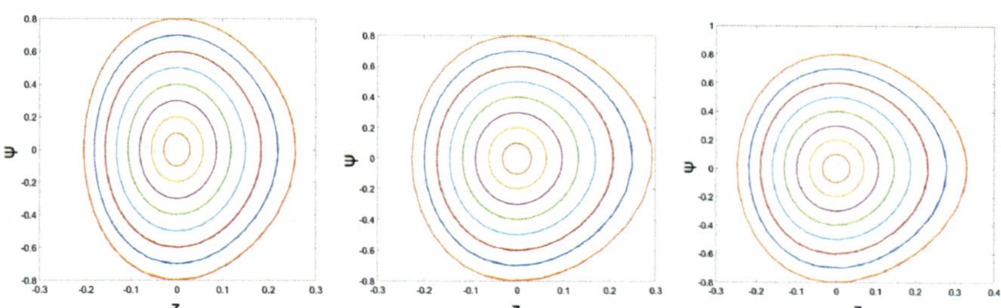

Fig. 13 Phase portrait of forced KdV for different values of (i) H = 1.8, (ii) H = 2, (iii) H = 2.2

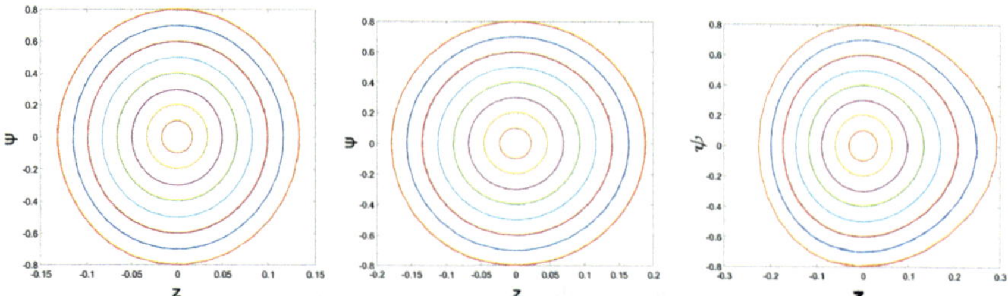

Fig. 14 Phase portrait of forced KdV for different values of (i) $u_0 = 0.8$, (ii) $u_0 = 0.9$, (iii) $u_0 = 1$

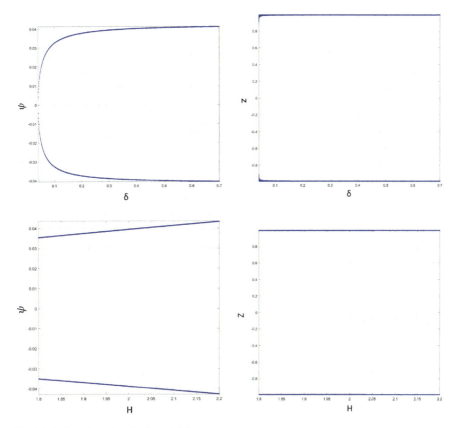

Fig. 15 Bifurcation plots for forced Kdv

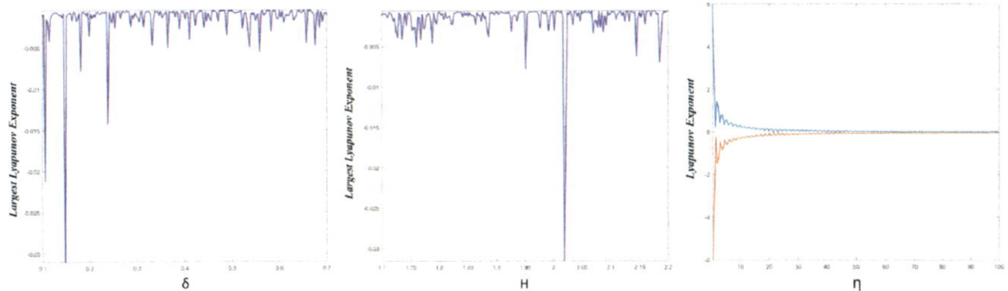

Fig. 16 Largest Lyapunov exponent for Forced KdV for variation of (i) δ, (ii) H and (iii) η

the increase in wavelength is much more dominating in case of higher value of H whereas in case of u_0 it is the increase in amplitude that dominates.

In Fig. 11, super nonlinear wave solution with variation of z for different values of δ, H and u_0 are shown. In all the three cases here, it is only the wavelength that increases with rise in any of the three parameters. The phase portrait plot, different initial conditions have closed loops. In Figs. 12 and 13, the loop is unsymmetrical along z and tend to elongate itself even more in the direction of z with rise in δ. In Fig. 14, the loop is initially symmetrical but with rise in u_0 becomes elongated and unsymmetrical.

Figure 15, the bifurcation plots for forced KdV are presented for different parameters. With variation in δ, the position varies exponentially whereas there is no change in momentum. And further, as H varies, the position varies linearly but momentum has no effect of H also. As Lyapunov exponent represents chaos, Chaotic behaviour for different parameters is shown in Fig. 16. For δ, the chaotic peaks lie between 0.1 and 0.3 and for H lies near 2. The chaotic behaviour is kind of perturbes and exponential in case of η.

7 Conclusion

We have studied two temperature plasma with Quantum-Hydrodynamic (QHD) Model, deriving its dispersion relation and further studied the decomposition of soliton and interaction of two solitons. The solution to the Forced-KdV equation demonstrates how the behaviour of these perturbed solitons is affected by the parameters used. The unperturbed plasma system is investigated first, followed by how the perturbation impacts the system.

The dynamical system is also investigated. The significance of researching Dynamical systems is demonstrated through phase pictures. The qualitative shift in this system's dynamics with changes in parameters is explained by bifurcation, which uses the Lyapunov exponent to characterise the rate of separation of infinitesimally close paths. The system's chaotic behaviour is investigated using the Largest Lyapunov exponent. This research can be used to a variety of astrophysical phenomena, such as stellar media, magnetospheres, and polar caps.

References

1. Akter, T., Deeba, F., Kamal-Al-Hassan, Md.: Electron-acoustic solitary waves in a two-temperature plasma having electrons with kappa distribution. IEEE Trans. Plasma Sci. **44**(8), 1449–1459 (2016)
2. Sahu, B.: Electron acoustic solitary waves and double layers with superthermal hot electrons. Phys. Plasmas **17**(12), 122305 (2010)

3. Saha, A., Chatterjee, P.: Electron acoustic blow up solitary waves and periodic waves in an unmagnetized plasma with kappa distributed hot electrons. Astrophys. Space Sci. **353**(1), 163–168 (2014)
4. Chandra, S., Ghosh, B.: Modulational instability of electron-acoustic waves in relativistically degenerate quantum plasma. Astrophys. Space Sci. **342**(2), 417–424 (2012)
5. Chatterjee, M., Dasgupta, M., Das, S., Halder, M., Chandra, S.: Study of dynamical properties in shock & solitary structures and its evolutionary stages in a degenerate plasma. African Rev. Phys. **15**, 75 (2021)
6. Chandra, S., Goswami, J., Sarkar, J., Das, C.: Analytical and simulation studies of forced kdv solitary structures in a two-component plasma. J. Korean Phys. Soc. **76**(6), 469–478 (2020)
7. Chandra, S., Das, C., Sarkar, J.: Evolution of nonlinear stationary formations in a quantum plasma at finite temperature. Zeitschrift für Naturforschung A **76**(4), 329–347 (2021)
8. Ballav, S., Das, A., Pramanick, S., Chandra, S.: Plasma shock wave in gamma-ray bursts: nonlinear phenomena and radiative process. IEEE Trans. Plasma Sci. (2021)
9. Chandra, S., Goswami, J., Sarkar, J., Das, C., Ghosh, B., Nandi, D.: Formation of electron acoustic shock wave in inner magnetospheric plasma. Indian J. Phys. (2021). https://doi.org/10.1007/s12648-021-02276-x
10. Ballav, S., Kundu, S., Das, A., Chandra, S.: Non-linear behaviour of dust acoustic wave mode in a dynamic dusty plasma containing negative dust particles and positrons. African Rev. Phys. **15**, 54 (2021)
11. Chandra, S., Sarkar, J., Das, C., Ghosh, B.: Self-interacting stationary formations in plasmas under externally controlled fields. Plasma Phys. Rep. **47**, 306–317 (2021)
12. Sultana, S., Kourakis, I.: Electrostatic solitary waves in the presence of excess superthermal electrons: modulational instability and envelope soliton modes. Plasma Phys. Control. Fusion **53**(4), 045003 (2011)
13. Watanabe, K., Taniuti, T.: Electron-acoustic mode in a plasma of two-temperature electrons. J. Phys. Soc. Jpn. **43**(5), 1819–1820 (1977)
14. Yu, M.Y., Shukla, P.K.: Linear and nonlinear modified electron-acoustic waves. J. Plasma Phys. **29**(3), 409–413 (1983)
15. Tokar, R.L., Gary, S.P.: Electrostatic hiss and the beam driven electron acoustic instability in the dayside polar cusp. Geophys. Res. Lett. **11**(12), 1180–1183 (1984)
16. Mace, R.L., Hellberg, M.A.: Higher-order electron modes in a two-electron-temperature plasma. J. Plasma Phys. **43**(2), 239–255 (1990)
17. Stix, H.: Waves in plasmas American Institute of Physics, New York (1992). Borg, G.G., Harris, J.H., Miljak, D.G., et al.: Plasma columns as radiating elements. IEEE Trans. Plasma Sci. submitted, 29
18. Treumann, R.A., Baumjohann, W.: Advanced Space Plasma Physics, vol. 30. Imperial College Press, London (1997)
19. Gary, S.P., Tokar, R.L.: The electron-acoustic mode. Phys. Fluids **28**(8), 2439–2441 (1985)
20. Feldman, W.C., Asbridge, J.R., Bame, S.J., Montgomery, M.D., Gary, S.P.: Solar wind electrons. J. Geophys. Res. **80**(31), 4181–4196 (1975)
21. Feldman, W.C., Anderson, R.C., Bame, S.J., Gary, S.P., Gosling, J.T., McComas, D.J., Thomsen, M.F., Paschmann, G., Hoppe, M.M.: Electron velocity distributions near the earth's bow shock. J. Geophys. Res. Space Phys. **88**(A1), 96–110 (1983)
22. Feldman, W.C., Anderson, R.C., Bame, S.J., Gosling, J.T., Zwickl, R.D., Smith, E.J.: Electron velocity distributions near interplantary shocks. J. Geophys. Res. Space Phys. **88**(A12), 9949–9958 (1983)
23. Barnes, W.L., Dereux, A., Ebbesen, T.W.: Surface plasmon subwavelength optics. Nature **424**, 824 (2003)
24. Ditmire, T., Springate, E., Tisch, J.W.G., Shao, Y.L., Mason, M.B., Hay, N., Marangos, J.P., Hutchinson, M.H.R.: Explosion of atomic clusters heated by high-intensity femtosecond laser pulses. Phys. Rev. A **57**(1), 369 (1998)
25. Gudmundsson, J.T., Alami, J., Helmersson, U.: Spatial and temporal behavior of the plasma parameters in a pulsed magnetron discharge. Surface Coat. Technol. **161**(2–3), 249–256 (2002)

26. Kadomtsev, B.B., Pogutse, O.P.: Trapped particles in toroidal magnetic systems. Nucl. Fus. **11**(1), 67 (1971)
27. Henry, D., Trguier, J.P.: Propagation of electronic longitudinal modes in a non-maxwellian plasma. J. Plasma Phys. **8**(3), 311–319 (1972)
28. Saberian, E., Esfandyari-Kalejahi, A.: Kinetic theory of acoustic-like modes in nonextensive pair plasmas. Astrophys. Space Sci. **349**(2), 799–811 (2014)
29. Das, C., Chandra, S., Ghosh, B.: Amplitude modulation and soliton formation of an intense laser beam interacting with dense quantum plasma: symbolic simulation analysis. Contrib. Plasma Phys. 10–1002 (2020)
30. Das, C., Chandra, S., Ghosh, B.: Nonlinear interaction of intense laser beam with dense plasma. Plasma Phys. Control. Fus. **63**, 015011 (2020)
31. Das, C., Chandra, S., Ghosh, B.: Effects of exchange symmetry and quantum diffraction on amplitude modulated electrostatic waves in quantum magnetoplasma. Pramana-J. Phys. 95(2) (2021)
32. Das, A., Ghosh, P., Chandra, S., Raj, V.: Electron acoustic peregrine breathers in a quantum plasma with 1-d temperature anisotropy. IEEE Trans. Plasma Sci. 1–12 (2021)
33. Dey, S., Maity, D., Ghosh, A., Samanta, P., De, A., Chandra, S.: Chaotic excitations of rogue waves in stable parametric region for highly-energetic pair plasmas. African Rev. Phys. **15**, 33 (2021)
34. Dey, A., Chandra, S., Das, C., Mandal, S., Das, T.: Rogue wave generation through non-linear self interaction of electrostatic waves in dense plasma. IEEE Trans. Plasma Sci. 10–1109 (2022)
35. Bains, A.S., Tribeche, M., Gill, T.S.: Modulational instability of ion-acoustic waves in a plasma with aq-nonextensive electron velocity distribution. Phys. Plasmas **18**(2), 022108 (2011)
36. Kourakis, I., Shukla, P.K.: Electron-acoustic plasma waves: oblique modulation and envelope solitons. Phys. Rev. E **69**(3), 036411 (2004)
37. Singh, S.V., Lakhina, G.S.: Generation of electron-acoustic waves in the magnetosphere. Planet. Space Sci. **49**(1), 107–114 (2001)
38. Ghosh, B., Chandra, S., Paul, S.N.: Relativistic effects on the modulational instability of electron plasma waves in quantum plasma. Pramana **78**(5), 779–790 (2012)
39. Ghosh, T., Pramanick, S., Sarkar, S., Dey, A., Chandra, S.: Chaotic scenario in three-component fermi plasma. African Rev. Phys. **15**, 45 (2021)
40. Ghosh, S., Saha, S., Chakraborty, T., Sadhukhan, K., Bhanja, R., Chandra, S.: Linear and nonlinear properties of electron acoustic waves in a viscous plasma. African Rev. Phys. **15**, 90 (2021)
41. Ghosh, A., Goswami, J., Chandra, S., Das, C., Arya, Y., Chhibber, H.: Resonant interactions and chaotic excitation in nonlinear surface waves in dense plasma. IEEE Trans. Plasma Sci. (2021)
42. Goswami, J., Chandra, S., Ghosh, B.: Study of small amplitude ion-acoustic solitary wave structures and amplitude modulation in e-p-i plasma with streaming ions. Laser Part. Beams **36**(1), 136–143 (2018)
43. Goswami, J., Chandra, S., Ghosh, B.: Shock waves and the formation of solitary structures in electron acoustic wave in inner magnetosphere plasma with relativistically degenerate particles. Astrophys. Space Sci. **364**(4), 1–7 (2019)
44. Goswami, J., Chandra, S., Sarkar, J., Chaudhuri, S., Ghosh, B.: Collision-less shocks and solitons in dense laser-produced fermi plasma. Laser Part. Beams **38**(1), 25–38 (2020)
45. Sah, O.P., Manta, J.: Nonlinear electron-acoustic waves in quantum plasma. Phys. Plasmas **16**(3), 032304 (2009)
46. Manfredi, G., Haas, F.: Self-consistent fluid model for a quantum electron gas. Phys. Rev. B **64**(7), 075316 (2001)
47. Manfredi, G.: Fields inst. commun. In: Proceedings of the Workshop on Kinetic Theory (The Fields Institute, Toronto, Canada, 29 March 2 April 2004), vol. 46, p. 263 (2005)
48. Santamato, E.: Geometric quantum mechanics. In: Fundamental Aspects of Quantum Theory, pp. 443–444. Springer (1986)

49. Gardner, C.L., Ringhofer, C.: Smooth quantum potential for the hydrodynamic model. Phys. Rev. E **53**(1), 157 (1996)
50. Shukla, P.K., Eliasson, B.: Formation and dynamics of dark solitons and vortices in quantum electron plasmas. Phys. Rev. Lett. **96**(24), 245001 (2006)
51. Sahu, B., Roychoudhury, R.: Electron acoustic solitons in a relativistic plasma with nonthermal electrons. Phys. Plasmas **13**(7), 072302 (2006)
52. Shukla, P.K., Ali, S.: Dust acoustic waves in quantum plasmas. Phys. Plasmas **12**(11), 114502 (2005)
53. Ali, S., Shukla, P.K.: Dust acoustic solitary waves in a quantum plasma. Phys. Plasmas **13**(2), 022313 (2006)
54. Ergun, R.E., Carlson, C.W., McFadden, J.P., Mozer, F.S., Muschietti, L., Roth, I., Strangeway, R.J.: Debye-scale plasma structures associated with magnetic-field-aligned electric fields. Phys. Rev. Lett. **81**(4), 826 (1998)
55. Ergun, R.E., Carlson, C.W., McFadden, J.P., Mozer, F.S., Delory, G.T., Peria, W., Chaston, C.C., Temerin, M., Roth, I., Muschietti, L., et al.: Fast satellite observations of large-amplitude solitary structures. Geophys. Res. Lett. **25**(12), 2041–2044 (1998)
56. Manfredi, G.: How to model quantum plasmas. Fields Inst. Commun. **46**, 263–287 (2005)
57. Haas, F., Garcia, L.G., Goedert, J., Manfredi, G.: Quantum ion-acoustic waves. Phys. Plasmas **10**(10), 3858–3866 (2003)
58. Mamun, A.A., Shukla, P.K.: Solitary waves in an ultrarelativistic degenerate dense plasma. Phys. Plasmas **17**(10), 104504 (2010)
59. Goswami, J., Chandra, S., Sarkar, J., Ghosh, B.: Electron acoustic solitary structures and shocks in dense inner magnetosphere finite temperature plasma. Radiat. Effects Defects Solids **175**(9–10), 961–973 (2020)
60. Goswami, J., Chandra, S., Sarkar, J., Ghosh, B.: Amplitude modulated electron acoustic waves with bipolar ions and kappa distributed positrons and warm electrons. Pramana-J. Phys. **95**, 54 (2021)
61. Goswami, J., Chandra, S., Das, C., Sarkar, J.: Nonlinear wave-wave interaction in semiconductor junction diode. IEEE Trans. Plasma Sci. (2021). https://doi.org/10.1109/TPS.2021.3124454
62. Goswami, J., Chandra, S., Sarkar, J., Ghosh, B.: Quantum two stream instability in a relativistically degenerate magnetised plasma. In: AIP Conference Proceedings, vol. 2319, pp. 030005. AIP Publishing LLC (2021)
63. Kapoor, S., Dutta, D., Ghosh, M., Chandra, S.: Magnetosonic shocks and solitons in fermi plasma with quasiperiodic perturbation. IEEE Trans. Plasma Sci. (2022)
64. Kockaert, P., Haelterman, M.: Stability and symmetry breaking of soliton bound states. JOSA B **16**(5), 732–740 (1999)
65. Ermakov, A., Stepanyants, Y.: Soliton interaction with external forcing within the korteweg-de vries equation. Chaos Interdisc. J. Nonlinear Sci. **29**(1), 013117 (2019)
66. Han, J.-N., Luo, J.-H., Li, S.-C., Liu, S.-W., Yang, Y., Duan, W.-S., Han, J.-F., Li, J.-X.: Composite nonlinear structure within the magnetosonic soliton interactions in a spin-1/2 degenerate quantum plasma. Phys. Plasmas **22**(6), 062101 (2015)
67. Howes, G.G.: Laboratory space physics: investigating the physics of space plasmas in the laboratory. Phys. Plasmas **25**(5), 055501 (2018)
68. Naumovets, A.G.: Two-Dimensional Crystals. Elsevier (2012)
69. Maiti, A., Chowdhury, S., Singha, P., Ray, S., Dasgupta, R., Chandra, S.: Study of small amplitude ion-acoustic bunched solitary waves in a plasma with streaming ions and thermal electrons. African Rev. Phys. **15**, 97 (2021)
70. Majumdar, A., Sen, A., Panda, B., Ghosal, R., Mallick, S., Chandra, S.: Study of shock fronts and solitary profile in a weakly relativistic plasma and its evolution into an amplitude modulated envelop soliton. African Rev. Phys. **15**, 18 (2021)
71. Mukhopadhyay, A., Bagui, D., Chandra, S.: Electrostatic shock fronts in two-component plasma and its evolution into rogue wave type solitary structures. African Rev. Phys. **15**, 25 (2021)

72. Paul, I., Chandra, S., Chattopadhyay, S., Paul, S.N.: W-type ion-acoustic solitary waves in plasma consisting of cold ions and nonthermal electrons. Indian J. Phys. **90**(10), 1195–1205 (2016)
73. Cheng, X.-P., Lou, S.Y., Chen, C.-L., Tang, X.-Y.: Interactions between solitons and other nonlinear schrödinger waves. Phys. Rev. E **89**(4), 043202 (2014)
74. Prasad, P.K., Gowrisankar, A., Saha, A., Banerjee, S.: Dynamical properties and fractal patterns of nonlinear waves in solar wind plasma. Physica Scripta. **95**(6), 065603 (2020)
75. Abdikian, A., Saha, A., Alimirzaei, S.: Bifurcation analysis of ion-acoustic waves in an adiabatic trapped electron and warm ion plasma. J. Taibah Univ. Sci. **14**(1), 1051–1058 (2020)
76. Roychowdhury, A., Banerjee, S., Chandra, S.: Stationary formation of dust-ion acoustic waves in degenerate dusty plasma at critical regime. African Rev. Phys. **15**, 102 (2021)
77. Sahoo, H., Chandra, S., Ghosh, B.: Dust acoustic solitary waves in magnetized dusty plasma with trapped ions and q-non-extensive electrons. African Rev. Phys. **10**(32), 235 (2015)
78. Sahoo, H., Das, C., Chandra, S., Ghosh, B., Mondal, K.K.: Quantum and relativistic effects on the kdv and envelope solitons in ion-plasma waves. IEEE Trans. Plasma Sci. (2021). https://doi.org/10.1109/TPS.2021.3120077
79. Samanta, P., De, A., Dey, S., Maity, D., Ghosh, A., Chandra, S.: Nonlinear excitations in dust-ion acoustic waves and the formation of rogue waves in stable parametric region in a 3-component degenerate plasma. African Rev. Phys. **15**, 10 (2021)
80. Mahmood, S., Masood, W.: Electron acoustic solitary waves in unmagnetized two electron population dense plasmas. Phys. Plasmas **15**(12), 122302 (2008)
81. ur Rehman, S. Linear and nonlinear quantum ion acoustic waves in a plasma with positive, negative ions and fermi electron gas. Phys. Plasmas **17**(6), 062303 (2010)
82. Lakhina, G.S., Kakad, A.P., Singh, S.V., Verheest, F.: Ion-and electron-acoustic solitons in two-electron temperature space plasmas. Phys. Plasmas **15**(6), 062903 (2008)
83. Jones, W.D., Lee, A., Gleman, S.M., Doucet, H.J.: Propagation of ion-acoustic waves in a two-electron-temperature plasma. Phys. Rev. Lett. **35**(20), 1349 (1975)
84. Ghosh, B., Chandra, S., Paul, S.N.: Amplitude modulation of electron plasma waves in a quantum plasma. Phys. Plasmas **18**(1), 012106 (2011)
85. Chandra, S., Paul, S.N., Ghosh, B.: Electron-acoustic solitary waves in a relativistically degenerate quantum plasma with two-temperature electrons. Astrophys. Space Sci. **343**(1), 213–219 (2013)
86. Sarkar, J., Goswami, J., Chandra, S., Ghosh, B.: Study of ion-acoustic solitary wave structures in multi-component plasma containing positive and negative ions and q-exponential distributed electron beam. Laser Part. Beams **35**(4), 641–647 (2017)
87. Sarkar, J., Chandra, S., Goswami, J., Ghosh, B.: Formation of solitary structures and envelope solitons in electron acoustic wave in inner magnetosphere plasma with suprathermal ions. Contrib. Plasma Phys. **60**(7), e201900202 (2020)
88. Sarkar, J., Chandra, S., Ghosh, B.: Resonant interactions between the fundamental and higher harmonic of positron acoustic waves in quantum plasma. Zeitschrift für Naturforschung A **75**(10), 819–824 (2020)
89. Sarkar, J., Chandra, S., Dey, A., Das, C., Marick, A., Chatterjee, P.: Forced kdv and envelope soliton in magnetoplasma with kappa distributed ions. IEEE Trans. Plasma Sci. 10–1109 (2021)
90. Sarkar, J., Chandra, S., Goswami, J., Das, C., Ghosh, B.: Growth of rt instability at the accreting magnetospheric boundary of neutron stars. In: AIP Conference Proceedings, vol. 2319, p. 030006. AIP Publishing LLC (2021)
91. Sarkar, S., Dey, A., Pramanick, S., Ghosh, T., Das, C., Chandra, S.: Homotopy study of spherical ion-acoustic waves in relativistic degenerate galactic plasma. IEEE Trans. Plasma Sci. 10–1109 (2022)
92. Ghosh, M., Sharry, K., Dutta, D., Chandra, S.: Propagation of rogue waves and cnoidal waves formations through low frequency plasma oscillations. African Rev. Phys. **15** (2021)
93. Singh, A.K., Chandra, S.: Electron acceleration by ponderomotive force in magnetized quantum plasma. Laser Part. Beams **35**(2), 252–258 (2017)

94. Singh, A.K., Chandra, S.: Second harmonic generation in high density plasma. African Rev. Phys. **12**(11), 84 (2018)
95. Thakur, S., Das, C., Chandra, S.: Stationary structures in a four component dense magnetoplasma with lateral perturbations. IEEE Trans. Plasma Sci. (2021)
96. Papavaritis, P., Pardo, W.B.: Numerical mode and boundary analysis of bifurcation phenomena in plasmas with stationary striations. In: IEEE 1989 International Conference on Plasma Science, p. 154. IEEE (1989)
97. Devaney, R.L.: An Introduction to Chaotic Dynamical Systems. CRC Press (2018)
98. Robinson, C.: Dynamical systems: stability, symbolic dynamics, and chaos. CRC Press (1998)
99. Arnold, V.I., Afrajmovic, V.S., Il'yasenko, U.S., Shil'nikov, L.P.: Bifurcation Theory and Catastrophe Theory. Springer (1999)
100. Saha, A., Chatterjee, P.: Bifurcations of electron acoustic traveling waves in an unmagnetized quantum plasma with cold and hot electrons. Astrophys. Space Sci. **349**(1), 239–244 (2014)

Motion of Adiabatic or Isothermal Flow Headed by a Magnetogasdynamic Cylindrical Shock Through Rotating Dusty Gas

P. K. Sahu

Abstract Expansion of cylindrical shocks pushed out through a dynamic piston via rotating perfect dust-pervade gas in the presence of spatially diminishing magnetic field is inquired. The velocity and magnetic field are presumed to comply with power rules. The gas should be conducting electrically. The shock wave proceeds by mutable velocity as well as the total energy being non-stationary. Numerical calculations are accomplished to access the flow variable's profiles. It is also assessed as to how the magnetic field affects the behaviour of the flow parameters. Further, it's far exciting to word that in attendance of an azimuthal magnetic field the density and pressure evanesce at expansive region and therefore void is constituted at the symmetry's axis, that's in great accordance with laboratory situations to generate shock-wave.

Keywords Magnetic field · Perfect dust-pervade gas · Rotating medium · Adiabatic and isothermal flows · Mechanics of fluids

1 Introduction

"When the energy of the electric field is much smaller than that of the magnetic field, then all the electromagnetic quantities can be expressed in terms of the magnetic field, then only the interaction between gas-dynamic field and magnetic field can be considered. Such analysis is known as Magnetogasdynamics." Prime persuasive for considering magnetogasdynamics is it has several implementations in the area of astrophysics, aerodynamics, as well as atmospheric sciences. The comprehensive investigation of magnetogasdynamics shock is offered by several researchers (see Hartmann [1], Balick and Frank [2], Nath [3], Nath and Sahu [4], Nath et al. [5, 6], Sahu [7–9] and the references cited therein). For evaluation of improvements

P. K. Sahu (✉)
Department of Mathematics, Government Shyama Prasad Mukharjee College, Sitapur 497111, Chhattisgarh, India
e-mail: praveensahu173@gmail.com

S. Banerjee and A. Saha (eds.), *Nonlinear Dynamics and Applications*,
Springer Proceedings in Complexity,
https://doi.org/10.1007/978-3-030-99792-2_7

of the statistical approach in MHD (Magnetohydrodynamics) turbulence, refer to Verma [10].

The topic of magneto-gas-dynamic shock wave expansion in a rotating interstellar environment is of special relevance in the investigation of astronomical events. The external atmospheres of the planets spin as a result of the planets' spinning, according to experimental research and astrophysical observations. In an interstellar environment with spin, macroscopic motion at supersonic speeds occurs, and shock waves are created. Furthermore, because the intergalactic magnetic field is linked to the rotating sun, a wide-scale magnetic field may arise in quickly rotating stars. Thus, the spin of planets or stars has a substantial impact on the processes occurring in their outermost parts. As a result, topics concerning detonation in spinning gas atmospheres are of great astronomical importance. Many academics have researched these issues in recent years (see, Levin and Skopina [11], Nath [12], Nath et al. [13], Nath and Sahu [14–16], Sahu [17–19]).

Recently, the perusal of fluid flow in a dust-pervade gas is a topic of high involvement because it has extensive implementations in environmental as well as industrial fields like lunar-ash flowing, nozzle flowing, volcanic explosions, under-ground explosion, cosmic explosion, the formation of polluted crystals, formation of the stars, supersonic flights and various others real-life problems of engineering and science. Several research articles based on the screening of the shock waves expansion in a dust-pervade gas (see Refs. Pai et al. [20], Higashino [21], Miura and Glass [22], Popel and Gisko [23], Pai [24], Nath and Sahu [25, 26], Sahu [27, 28] as well as the sources listed throughout).

Notwithstanding, in a dust-pervade gas, similarity approaches have been mentioned very drastically within the literature, but they have hardly been investigated taking magnetic field into account. In this article, a system of non-linear PDEs that describes the cylindrically symmetric flow is considered through a perfect dust-pervade gas in the existence of a magnetic field for both isothermal and adiabatic flows. When radiative transfer effects are included, the isothermal flow hypothesis is physically plausible. This supposition about the flow's nature correlates to the start of a highly powerful explosion (for instance, subterranean, volcano, and cosmological blasts; or coal-mine bursts) when the temperature of the gas is exceptionally high (Sedov [29], Laumbach and Probstein [30]). The extant work is the enhancement of Vishwakarma and Pandey's [31] work by recognising the consequences of the rotating medium as well magnetic fields in cylindrical coordinate. The extant work is also the enhancement of Nath's [3, 12] works by recognising the consequences of dust particles or magnetic fields respectively in perfect dust-pervade gas.

2 Governing Equations—Adiabatic Flow

The executive equations describing 1D adiabatic, non-stationary, cylindrical, perfect dust-pervade rotating gas flow together with an azimuthal or axial magnetic field can be compiled like (c.f. Nath [3, 12], Sahu [7, 17], Levin and Skopina [11], Pai et al. [20])

$$\frac{D\rho}{Dt} + \rho\left(\frac{\partial u}{\partial r} + \frac{u}{r}\right) = 0, \tag{1}$$

$$\rho\frac{Du}{Dt} + \frac{\partial p}{\partial r} + \frac{\partial h}{\partial r} + \frac{2ih}{r} - \frac{\rho v^2}{r} = 0, \tag{2}$$

$$\frac{Dv}{Dt} + \frac{uv}{r} = 0, \tag{3}$$

$$\frac{Dw}{Dt} = 0, \tag{4}$$

$$\frac{Dh}{Dt} + 2h\left(\frac{\partial u}{\partial r} + \frac{u(1-i)}{r}\right) = 0, \tag{5}$$

$$\frac{DE_m}{Dt} + p\frac{D}{Dt}\left(\frac{1}{\rho}\right) = 0. \tag{6}$$

in which r and t are unattached space as well as time coordinates; u, v, w designate dust-pervade gas's velocity components; p, ρ, and E_m designate dust-pervade gas's pressure, density, and internal energy; $h = \frac{\mu H^2}{2}$ designates magnetic pressure, H designates the intensity of the magnetic field, it might be alternatively axial ($i = 0$) or azimuthal ($i = 1$); μ designates magnetic permeability. $\frac{D}{Dt} = \left(\frac{\partial}{\partial t}\right) + u\left(\frac{\partial}{\partial r}\right)$ being material derivative.

The governing Eqs. (1–6) ought to be enclosed by an equation of state (see, Sahu [18], Nath and Sahu [25], Vishwakarma and Pandey [31]).

$$p = \frac{(1-\chi_p)}{(1-Z)}\rho R^* T, \qquad E_m = \frac{p(1-Z)}{(\Gamma-1)\rho}. \tag{7}$$

We assumed that the medium rotated around a symmetry axis. For details of that readers are referred to see, Levin and Skopina [11], Sahu [17–19]. Flow factors forthwith preceding the shock are considered as (see, Sahu [7, 17])

$$H = H_1 = H_0 r_s^{-\delta}, \rho = \rho_1 = constant,$$

$$u = u_1 = 0, v = v_1 = v_0 r_s^{\alpha}, w = w_1 = w_0 r_s^{\lambda},$$

$$p = p_1 = \left(\frac{i-\delta}{2\delta}\right)\mu H_0^2 r_s^{-2\delta} + \frac{\rho_1 v_0^2}{2\alpha} r_s^{2\alpha}, \delta \neq 0, \alpha \neq 0 \tag{8}$$

in which v_0, w_0, H_0, α, λ and δ designate dimensional constants; r_s designates shock-radius and relator 1 designates circumstances forthwith preceding shock.

The Rankine-Hugoniot stipulations are (c.f. Sahu [18], Vishwakarma and Pandey [31]) namely,

$$\rho_2 = \rho_{r=r_s} = \frac{\rho_1}{\beta}, \quad u_2 = u_{r=r_s} = (1-\beta)U_s, \quad v_2 = v_{r=r_s} = v_1$$

$$p_2 = p_{r=r_s} = \left[(1-\beta) + C_a\left(1 - \frac{1}{\beta^2}\right) + \frac{1}{\gamma M^2}\right]\rho_1 U_s^2, \tag{9}$$

$$w_2 = w_{r=r_s} = w_1, \quad h_2 = h_{r=r_s} = \frac{h_1}{\beta^2}, \quad Z_2 = \frac{Z_1}{\beta}$$

Relator 2 designates situations forthwith back of the shock, $U_s \left(= \frac{dr_s}{dt}\right)$ designates shock front velocity, $M\left(= \frac{\rho_1 U_s^2}{\gamma p_1}\right)^{\frac{1}{2}}$ and $C_a = \left(\frac{h_1}{\rho_1 U_s^2}\right)$ designate shock-Mach and Cowling number. Following interrelation is used to specify the density ratio β throughout the shock (see, Sahu [17, 18])

$$\beta^3(\Gamma+1) - \beta^2\left[\frac{2\Gamma}{\gamma M^2} + \Gamma(1+2C_a) + 2Z_1 - 1\right] + 2\beta C_a(Z_1 + \Gamma - 2) + 2C_a Z_1 = 0. \tag{10}$$

3 Self-similarity Transformations

Behind the shock, the interior extent of flow is avowed to be an expansive region. Following Sedov [29], in the formation of self-similarity, the expansive region's velocity is putative to adhere to a power-law that reads (see, Sahu [9, 28], Steiner and Hirschler [32], Zel'Dovich and Raizer [33])

$$u_p = \frac{dr_p}{dt} = U_0\left(\frac{t}{t_0}\right)^n, \tag{11}$$

where r_p designates expansive region's radius, t_0 designates reference-time, n designates constant and U_0 designates expansive region velocity at reference-time.

In terms of extent stipulation, similarity solution asserts that the shock velocity is proportional to expansive region velocity, as shown below

$$U_s = \frac{dr_s}{dt} = CU_0 \left(\frac{t}{t_0}\right)^n, \tag{12}$$

in which C designates dimensionless constant. Also,

$$\varpi(self - similarity\ variable) = \frac{r}{r_s} = \left[\frac{(n+1)t_0^n}{CU_0}\right]\left(\frac{r}{t^{n+1}}\right). \tag{13}$$

Certainly, onto the shock $\varpi = 1$ and $\varpi = \varpi_p \left(= \frac{r_p}{r_s}\right)$ upon the expansive region.

The unknown variables are note downed in the under-mentioned form to obtain the similarity solutions, (Sahu [7, 9, 28], Steiner and Hirschler [32])

$$u = \frac{r}{t} V(\varpi), v = \frac{r}{t} \phi(\varpi), w = \frac{r}{t} \psi(\varpi), p = \rho_1 \frac{r^2}{t^2} P(\varpi),$$

$$\rho = \rho_1 D(\varpi), h = \rho_1 \frac{r^2}{t^2} B(\varpi), Z = Z_1 D(\varpi), \tag{14}$$

where V, ϕ, ψ, P, D and B are functions of ϖ only.

Furthermore, the total energy of the shock is

$$E = 2\pi \int_{r_p}^{r_s} \rho \left[E_m + \frac{1}{2}\left(u^2 + v^2 + w^2\right) + \frac{h}{\rho}\right] r dr. \tag{15}$$

Using (7) and (14), Eq. (15) becomes

$$E = 2\pi\rho_1 \left[\frac{CU_0}{(n+1)t_0^n}\right]^{\frac{2}{n+1}} r_s^{\frac{4n+2}{n+1}} \int_{\varpi_p}^{1} \left[\frac{P(1 - Z_1 D)}{(\Gamma - 1)} + \frac{D}{2}\left(U^2 + \psi^2 + \phi^2\right) + B\right] \varpi^3 d\varpi. \tag{16}$$

Therefore, the total energy of the shock wave is non-stationary and varies as $r_s^{\frac{4n+2}{n+1}}$ (see, Freeman [34], Sahu [35–37]).

Given the similarity of the results, entities M and C_a are intended to be unchanging in this case

$$\alpha = -\delta = \frac{n}{n+1}. \tag{17}$$

We now obtain $\alpha < 1$ (see, Levin and Skopina [11], Sahu [17–19]). As a result of Eq. (17), we obtain $\frac{n}{n+1} < 1$. Thus

$$M^2 = \frac{\delta}{\gamma C_a \left[(i - \delta) - \frac{\rho_1 v_0^2}{\mu H_0^2}\right]}. \tag{18}$$

Equation (18) demonstrates that the solutions to the identified issue may be limited to the situation when the ambient media is non-rotating (i.e. $v_0 = 0$ and $w_0 = 0$).

Employing the Eq. (14), the governing apparatus of Eqs. (1)–(6) metamorphose into the consecutive apparatus of ODEs:

$$\frac{dD}{d\varpi} = \frac{1}{L}\left[\{V-(n+1)\}\,2\,V\,D - D\,V\,(V-1) - 2P - 2B(i+1) + \frac{2\,P\,(V-1) - 2B - (i-1)\,2\,B\,V}{\{V-(n+1)\}} + D\,\phi^2\right], \tag{19}$$

$$\frac{dV}{d\varpi} = -\frac{[V-(n+1)]}{D}\frac{dD}{d\varpi} - \frac{2\,V}{\varpi}, \tag{20}$$

$$\frac{dP}{d\varpi} = \frac{\Gamma\,P}{D\,(1-Z_1\,D)}\frac{dD}{d\varpi} - \frac{2\,P\,(V-1)}{\varpi\,[V-(n+1)]}, \tag{21}$$

$$\frac{dB}{d\varpi} = \frac{2\,B + (i-1)\,2\,B\,V}{\varpi\,[V-(n+1)]} + \frac{2\,B}{D}\frac{dD}{d\varpi}, \tag{22}$$

$$\frac{d\phi}{d\varpi} = \frac{(1-2\,V)\,\phi}{\varpi\,[V-(n+1)]}, \tag{23}$$

$$\frac{d\psi}{d\varpi} = \frac{(1-V)\,\psi}{\varpi\,[V-(n+1)]}, \tag{24}$$

where

$$L = L(\varpi) = \frac{\varpi}{D\,(1-Z_1\,D)}\left[\Gamma\,P + 2\,B\,(1-Z_1\,D) - (1-Z_1\,D)\,D\,\{V-(n+1)\}^2\right]. \tag{25}$$

Employing Eq. (14), the transformed shock circumstances be

$$B(1) = \frac{C_a}{\beta^2}\,(n+1)^2 D(1) = \frac{1}{\beta},\ V(1) = (1-\beta)\,(n+1),$$

$$P(1) = \left[(1-\beta) + C_a\left(1 - \frac{1}{\beta^2}\right) + \frac{1}{\gamma M^2}\right](n+1)^2. \tag{26}$$

$$\phi(1) = v_0\left[\frac{C\,U_0}{t_0^n\,(n+1)}\right]^{\frac{-1}{(n+1)}},\ \psi(1) = w_0\left[\frac{C\,U_0}{t_0^n\,(n+1)}\right]^{\frac{-1}{(n+1)}},$$

where $\alpha = \lambda = -\delta$ was required to get the similarity solutions.

4 Isothermal Flow

In this part, we show the solution in respect of isothermal, cylindrically symmetric, perfect dust-pervade rotating gas flow having a magnetic field that might be alternatively axial or azimuthal. The issue's boundary circumstances are the shock circumstances (9), i.e. same as the scenario of adiabatic flow.

In respect of isothermal flow, Eq. (6) is rewritten as (see, Sahu [7, 17])

$$\frac{\partial T}{\partial r} = 0. \tag{27}$$

Equation (27), when combined using Eq. (14), yields

$$\frac{p}{p_2} = \frac{\rho\,(1 - Z_2)}{\rho_2\,(1 - Z)}. \tag{28}$$

Equation (27), when combined using Eqs. (14) and (26), presents a kind of a relationship connecting P and D

$$P(\varpi) = \left[(1-\beta) + \frac{C_a(\beta^2 - 1)}{\beta^2} + \frac{1}{\gamma M^2}\right] \frac{(\beta - Z_1)\,D\,(n+1)^2}{\varpi^2\,[1 - Z_1)\,D]}. \tag{29}$$

Employing the Eq. (14), the governing apparatus of equations and (27) metamorphose into the consecutive apparatus of ODE :

$$\frac{dD}{d\varpi} = \frac{1}{N}\left[\{V - (n+1)\}\,2\,VD - DV\,(V-1) - 2B(i+1) - \frac{2B + (i-1)\,2\,BV}{\{V - (n+1)\}} + D\,\phi^2\right], \tag{30}$$

where

$$N = N(\varpi) = \frac{\varpi}{D\,(1 - Z_1\,D)}\left[P + 2\,B\,(1 - Z_1\,D) - (1 - Z_1\,D)\,D\,\{V - (n+1)\}^2\right]. \tag{31}$$

For obtaining numerical results, it's suitable to compose variables u, v, w, p, ρ and h in the under-mentioned form (non-dimensional)

$$\frac{u}{u_2} = \varpi\,\frac{V(\varpi)}{V(1)},\ \frac{v}{v_2} = \varpi\,\frac{\phi(\varpi)}{\phi(1)},\ \frac{w}{w_2} = \varpi\,\frac{\psi(\varpi)}{\psi(1)},$$

$$\frac{p}{p_2} = \varpi^2\,\frac{P(\varpi)}{P(1)},\ \frac{\rho}{\rho_2} = \frac{Z}{Z_2} = \frac{D(\varpi)}{D(1)},\ \frac{h}{h_2} = \varpi^2\,\frac{B(\varpi)}{B(1)}. \tag{32}$$

Also, we have

$$V(\varpi_p) = n + 1. \tag{33}$$

5 Results and Discussion

The flow parameters distribution between $\varpi = 1$ and $\varpi = \varpi_p$ are decided by means of numerically integrating Eqs. (19)– (24) in respect of adiabatic flow as well as

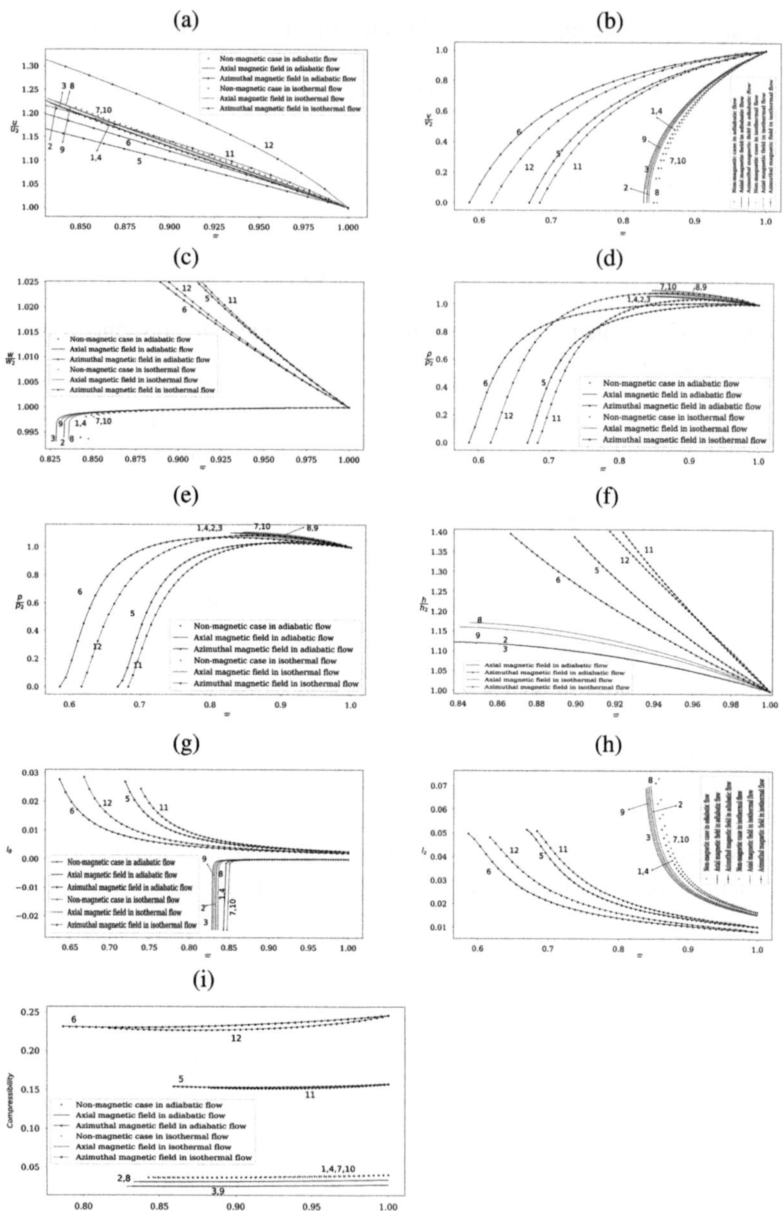

Fig. 1 Variability of dust-pervade gas parameters at the back of the shock: **a** radial component of fluid velocity $\frac{u}{u_2}$, **b** azimuthal component of fluid velocity $\frac{v}{v_2}$, **c** axial component of fluid velocity $\frac{w}{w_2}$, **d** density $\frac{\rho}{\rho_2}$, **e** pressure $\frac{p}{p_2}$, **f** magnetic field $\frac{h}{h_2}$, **g** azimuthal component of vorticity vector l_θ, **h** axial component of vorticity vector l_z, **i** compressibility : adiabatic (C_{adi}) p_1, and isothermal (τ_{iso}) p_1 (Reference Table 1 for further information on input variables)

Table 1 Taking into account different aspects of i and C_a with $\gamma = \frac{5}{3}$; $\beta' = 1$; $\chi_p = 0.1$; $G_1 = 10$; $n = -0.1$

i	C_a	β	ϖ_p	
			Adiabatic flow (Case No. in Fig. 1)	Isothermal flow (Case No. in Fig. 1)
0	0.0	0.300699	0.842824 (1)	0.847369 (7)
	0.01	0.309486	0.837828 (–)	0.841335 (–)
	0.02	0.317855	0.833051 (2)	0.835803 (8)
	0.03	0.325860	0.828462 (3)	0.830629 (9)
1	0.0	0.300699	0.842824 (4)	0.847369 (10)
	0.01	0.357026	0.755073 (–)	0.761960 (–)
	0.02	0.471998	0.669336 (5)	0.684021 (11)
	0.03	0.585440	0.586502 (6)	0.617276 (12)

by means of Eqs. (20), (22)–(24), (30) in respect of isothermal flow with (26, 33) using fourth-order Runge-Kutta method. The extant work is also the enhancement of Nath's [3, 12] works recognising the consequences of dust particles or magnetic fields respectively in perfect dust-pervade gas. Figure 1 shows that the procured solution is in nice concurrence with the established solutions of Nath [3, 12].

5.1 *Efficacy of the Existence of the Magnetic Field Is as Adhere*

By enhancing the value of C_a; the extent between expansive region and shock as well as β enhances (see Table 1). Physically it approaches that the gas at the back of the shock contracted lower, i.e. the shock strength reduces in the existence of both azimuthal or axial magnetic fields. $\frac{u}{u_2}$ enhances while moving inside towards expansive region through shock, though, it reduces in the existence of axial magnetic field for an isothermal flow. $\frac{v}{v_2}$ enhances, but to reduce $\frac{h}{h_2}$ and l_{z*} while moving inside towards expansive region through shock. $\frac{w}{w_2}$ and l_θ enhance, but to reduce $\frac{\rho}{\rho_2}$, $\frac{p}{p_2}$, (C_{adi}) p_1, and (τ_{iso}) p_1 in the existence of axial magnetic field; though, inverted conduct is executed in the existence of azimuthal magnetic field (see Fig. 1).

5.2 *The Differences of Adiabatic and Isothermal Motions*

It is obvious through Table 1 that ϖ_p in isothermal flow is bigger than ϖ_p in adiabatic flow. Physically, it indicates that the gas is condensed further in isothermal flow than in adiabatic flow. As a result, the shock strength is greater for isothermal flow than in the adiabatic flow.

6 Conclusions

On the ground of the aforementioned work, the succeeding conclusions could be sketched:

(i) Investigation of shock expansion in the existence of the magnetic field in an ideal dust-pervade gas through adiabatic and isothermal flows for cylindrical shock, has not been made previously.
(ii) The supposition of a constant temperature difference eliminates the discontinuities that occur in the scenario of adiabatic flow.
(iii) The azimuthal magnetic field contemplation produces outstanding distinction in the flow variables distribution.
(iv) In the attendance of magnetic field the density and pressure evanesce at the expansive region and therefore void is constituted at the symmetry's centre.

Acknowledgements The author is thankful to **Prof. M. K. Verma**, Department of Physics, Indian Institute of Technology Kanpur, Kanpur–208016, India for fruitful discussions. This work was supported by the research grant no. **TAR/2018/000150** under **Teachers Associateship for Research Excellence (TARE)** scheme from the **Science and Engineering Research Board (SERB)**, India. The author gracefully acknowledges financial support from **SERB**.

References

1. Hartmann, L.: Accretion Processes in Star Formation. Cambridge University Press, Cambridge (1998)
2. Balick, B., Frank, A.: Shapes and shaping of planetary nebulae. Ann. Rev. Astron. Astrophys. **40**(1), 439–486 (2002)
3. Nath, G.: Magnetogasdynamic shock wave generated by a moving piston in a rotational axisymmetric isothermal flow of perfect gas with variable density. Adv. Space Res. **47**(9), 1463–1471 (2011)
4. Nath, G., Sahu, P.K.: Flow behind an exponential shock wave in a rotational axisymmetric perfect gas with magnetic field and variable density. SpringerPlus **5**(1), 1–18 (2016)
5. Nath, G., Sahu, P.K., Chaurasia, S.: Modelling. Measur. Control B **87**(4), 236–243 (2018)
6. Nath, G., Sahu, P.K., Chaurasia, S.: Self-similar solution for the flow behind an exponential shock wave in a rotational axisymmetric non-ideal gas with magnetic field. Chin. J. Phys. **58**, 280–293 (2019)
7. Sahu, P.K.: Shock wave driven out by a piston in a mixture of a non-ideal gas and small solid particles under the influence of azimuthal or axial magnetic field. Braz. J. Phys. **50**(5), 548–565 (2020)
8. Sahu, P.K.: Magnetogasdynamic exponential shock wave in a self-gravitating, rotational axisymmetric non-ideal gas under the influence of heat-conduction and radiation heat-flux. Ricerche di Matematica 1–37 (2021)

9. Sahu, P.K.: The influence of magnetic and gravitational fields in a non-ideal dusty gas with heat conduction and radiation heat flux. Indian J. Phys. 1–15 (2022)
10. Verma, M.K.: Statistical theory of magnetohydrodynamic turbulence: recent results. Phys. Rep. **401**(5–6), 229–380 (2004)
11. Levin, V.A., Skopina, G.A.: Detonation wave propagation in rotational gas flows. J. Appl. Mech. Tech. Phys. **45**(4), 457–460 (2004)
12. Nath, G.: Self-similar solution of cylindrical shock wave propagation in a rotational axisymmetric mixture of a non-ideal gas and small solid particles. Meccanica **47**(7), 1797–1814 (2012)
13. Nath, G., Sahu, P.K., Dutta, M.: Magnetohydrodynamic cylindrical shock in a rotational axisymmetric non-ideal gas under the action of monochromatic radiation. Proc. Eng. **127**, 1126–1133 (2015)
14. Nath, G., Sahu, P.K.: Unsteady adiabatic flow behind a cylindrical shock in a rotational axisymmetric non-ideal gas under the action of monochromatic radiation. Proc. Eng. **144**, 1226–1233 (2016)
15. Nath, G., Sahu, P.K.: Flow behind an exponential shock wave in a rotational axisymmetric non-ideal gas with conduction and radiation heat flux. Int. J. Appl. Comput. Math. **3**(4), 2785–2801 (2017)
16. Nath, G., Sahu, P.K.: Similarity solution for the flow behind a cylindrical shock wave in a rotational axisymmetric gas with magnetic field and monochromatic radiation. Ain Shams Eng. J. **9**(4), 1151–1159 (2018)
17. Sahu, P.K.: Propagation of an exponential shock wave in a rotational axisymmetric isothermal or adiabatic flow of a self-gravitating non-ideal gas under the influence of axial or azimuthal magnetic field. Chaos Solitons Fractals **135**, 109739 (2020)
18. Sahu, P.K.: Shock wave propagation in perfectly conducting rotational axisymmetric two-phase medium with increasing energy under the action of heat conduction and radiation heat flux. Chin. J. Phys. **72**, 176–190 (2021)
19. Sahu, P.K.: Flow behind the magnetogasdynamical cylindrical shock wave in rotating non-ideal dusty gas with monochromatic radiation. Plasma Res. Exp. **3**(4), 045004 (2021)
20. Pai, S.I., Menon, S., Fan, Z.Q.: Similarity solutions of a strong shock wave propagation in a mixture of a gas and dusty particles. Int. J. Eng. Sci. **18**(12), 1365–1373 (1980)
21. Higashino, F., Suzuki, T.: The effect of particles on blast waves in a dusty gas. Zeitschrift für Naturforschung A **35**(12), 1330–1336 (1980)
22. Miura, H., Glass, I.I.: Proc. R. Soc. Lond. A. Math. Phys. Sci. **397**, 295–309 (1985)
23. Popel, S.I., Gisko, A.A.: Nonlinear Process. Geophys. **13**, 223–229 (2006)
24. Pai, S.I.: Two-Phase Flows, vol. 3. Springer (2013)
25. Nath, G., Sahu, P.K.: Self-similar solution of a cylindrical shock wave under the action of monochromatic radiation in a rotational axisymmetric dusty gas. Commun. Theor. Phys. **67**(3), 327 (2017)
26. Nath, G., Sahu, P.K.: Propagation of a cylindrical shock wave in a mixture of a non-ideal gas and small solid particles under the action of monochromatic radiation. Combust. Explos. Shock Waves **53**(3), 298–308 (2017)
27. Sahu, P.K.: Self-similar solution of spherical shock wave propagation in a mixture of a gas and small solid particles with increasing energy under the influence of gravitational field and monochromatic radiation. Commun. Theor. Phys. **70**(2), 197 (2018)
28. Sahu, P.K.: Analysis of magnetogasdynamic spherical shock wave in dusty real gas with gravitational field and monochromatic radiation. Eur. Phys. J. Plus **136**(4), 1–19 (2021)
29. Sedov, L.I.: Similarity and dimensional methods in mechanics. Academic Press, New York (1959)
30. Laumbach, D.D., Probstein, R.F.: Self-similar strong shocks with radiation in a decreasing exponential atmosphere. Phys. Fluids **13**(5), 1178–1183 (1970)
31. Vishwakarma, J.P., Pandey, S.N.: Propagation of strong spherical shock waves in a dusty gas. Phys. Scripta **68**(4), 259 (2003)
32. Steiner, H., Hirschler, T.: A self-similar solution of a shock propagation in a dusty gas. Eur. J. Mech. B Fluids **21**(3), 371–380 (2002)

33. Zel'Dovich, Y.B., Raizer, Y.P.: Physics of shock waves and high-temperature hydrodynamic phenomena. Courier Corporation (2002)
34. Freeman, R.A., Craggs, J.D.: Shock waves from spark discharges. J. Phys. D Appl. Phys. **2**(3), 421 (1969)
35. Sahu, P.K.: Similarity solution for a spherical shock wave in a non-ideal gas under the influence of gravitational field and monochromatic radiation with increasing energy. Math. Methods Appl. Sci. **42**(14), 4734–4746 (2019)
36. Sahu, P.K.: Similarity solution for the flow behind an exponential shock wave in a rotational axisymmetric non-ideal gas under the influence of gravitational field with conductive and radiative heat fluxes. In: International Conference on Innovation in Modern Science and Technology, pp. 1060-1070. Springer, Cham (2019)
37. Sahu, P.K.: Unsteady flow behind an MHD exponential shock wave in a rotational axisymmetric non-ideal gas with conductive and radiative heat fluxes. In: International Conference on Innovation in Modern Science and Technology, pp. 1049–1059. Springer, Cham (2019)

Structural Variations of Ion-Acoustic Solitons

Hirak Jyoti Dehingia and P. N. Deka

Abstract In this paper, we have presented our investigation on the variation in the structure of ion-acoustic solitons due to variation of densities in the presence of isothermal electrons in the plasma. We have considered the standard ion density profile and its variations and studied the effects of density variation on solitons. We have considered the governing fluid equations of plasma and derived the modified Kadomtsev–Petviashvili (KP) equation using the reductive perturbation technique (RPT). The solution of the KP equation indicates the variation in soliton structures.

Keywords Ion-acoustic solitons · Inhomogeneous plasma · Isothermal electrons

1 Introduction

Solitons or solitary waves are special wave packets or self-reinforcing waves that maintain their shape during their propagation at a constant speed. Solitons are formed by canceling the nonlinear and dispersive effects in the respective medium. Washimi and Taniuti [1] started extensively exploring the Korteweg de-Vries equation (KdV) for describing ion-acoustic solitons. Then, Nishikawa and Kaw [2] considered the inhomogeneous plasmas for studying ion-acoustic soliton propagation for the first time. Kuehl [3] has investigated the reflection of ion-acoustic solitons theoretically. He concluded that there were some changes in the amplitudes of both reflected and incident solitons. Later, Kuehl and Imen [4] studied soliton propagations in inhomogeneous plasmas considering the set of fluid equations. Nejoh [5] has investigated the various effects of ion temperature on the characteristics of soliton propagation in collisionless, inhomogeneous relativistic plasmas. Singh and Dahiya [6] extended the theory of the KdV equation by using the Reductive Perturbation Technique (RPT) in inhomogeneous plasmas. A Large member of the plasma physics community has studied soliton and their other characteristics in various physical situations. Some of

H. J. Dehingia (✉) · P. N. Deka
Department of Mathematics, Dibrugarh University, Dibrugarh 786004, Assam, India
e-mail: hirakjyotidehingia11@gmail.com

S. Banerjee and A. Saha (eds.), *Nonlinear Dynamics and Applications*,
Springer Proceedings in Complexity,
https://doi.org/10.1007/978-3-030-99792-2_8

the authors have studied soliton propagation in inhomogeneous plasma in the presence of finite ion temperature [7, 8], negative ions [9–12], dust effect [13], etc. Singh and Malik investigated the energy of solitons at both the critical and noncritical densities for negative ions under the influence of magnetized warm plasmas [14]. Hellberg and Verheest studied ion-acoustic solitary waves in inhomogeneous plasmas in the presence of two-temperature ions [15]. Baluku et al. studied dust ion-acoustic solitary waves in inhomogeneous plasmas under the influence of κ-distributed electrons [16]. Later, Kakad et al. [17] investigated experimentally the chain formation and validation of nonlinear fluid theory on ion-acoustic solitary waves in magnetized electron–ion plasmas. Gogoi and Deka [18] have studied the solitary waves for weakly inhomogeneous plasmas in the presence of nonthermal electrons. Mukherjee et al. [19] studied soliton bending for weak and slowly varying inhomogeneous unmagnetized plasmas. Again, Gogoi and Deka [20] studied the dust acoustic solitary wave's propagation in inhomogeneous plasmas in the presence of dust charge fluctuations. Zhou and Hutchinson investigated the motion of slow electron holes associated with one-dimensional ion-acoustic solitary waves [21]. Wang et al. studied the numerical simulation of dark envelope solitary waves in electron–ion plasmas. They concluded that the plasma waves are not described in the linear superposition of modes. They are described in the mode of nonlinear dynamical waves [22]. Later, Shi et al. studied the nonlocal Kundu-nonlinear Schrödinger equation (Kundu-NLS) for investigating the dynamics of solitary wave solutions which is obtained from coupled Kundu-NLS system [23]. Song et al. studied the various recent progress of solitons or solitary waves in optical fiber lasers [24]. Rani and Yadav studied the nonlinear propagation of electron acoustic solitary waves in the presence of dense magnetized plasma under the effect of degenerate quantum electrons [25]. Chen et al. investigated the dynamics of solitary waves in quantum plasmas with nonlinear effects and higher-order dispersion [26]. Prayitno and Budi studied the energy solution of solitary waves numerically in the KdV equation [27]. Lu and Liu investigated some ion-acoustic solitary waves having small amplitude with regularized κ-distributed electrons [28]. Then by using the inverse scattering method, Wu studied Kadomtsev–Petviashvili (KP) perturbed multi-line solitons [29]. Recently, the dust ion-acoustic solitary waves and double layers were studied [30] in the presence of adiabatic positive dust grains, ion species, and Cairns-distributed electrons. Thus, understanding the present importance of soliton study in different astrophysical situations, we have considered our investigation on the structural variations of ion-acoustic solitons due to variation of densities in the presence of isothermal electrons plasmas.

2 Governing Equations and Derivations

Let us consider two-dimensional, unmagnetized, collisionless inhomogeneous plasma consisting of cold and hot isothermal electrons. The dimensionless ion continuity equation and momentum equation, along with Poisson's equation and the equation for electron Boltzmann distribution, is taken as follows:

$$\left.\begin{array}{c}\frac{\partial n}{\partial t}+\frac{\partial(nu_x)}{\partial x}+\frac{\partial(nu_y)}{\partial y}=0\\ \frac{\partial u_x}{\partial t}+u_x\frac{\partial u_x}{\partial x}+u_y\frac{\partial u_x}{\partial y}+\frac{\partial\varnothing}{\partial x}=0\\ \frac{\partial u_y}{\partial t}+u_x\frac{\partial u_y}{\partial x}+u_y\frac{\partial u_y}{\partial y}+\frac{\partial\varnothing}{\partial y}=0\\ \frac{\partial^2\varnothing}{\partial x^2}=n_e-n\\ n_e=\mathrm{e}^{\varnothing}\end{array}\right\} \tag{1}$$

To study ion-acoustic solitary waves and their propagation in inhomogeneous plasmas, we have considered the following two dimensional stretched coordinates [19]:

$$\eta=\epsilon^{3/2}x,\,\lambda=\epsilon y,\,\xi=\epsilon^{\frac{1}{2}}(x-Vt) \tag{2}$$

where V is a constant or the phase velocity of the normalized wave, normalized by ion-acoustic speed C_s, ϵ is an expansion parameter.

We consider the plasma model [19] in which unperturbed number density for ions of the form $\tilde{n}_0(\eta)=1+\delta f_0(\eta)$, where δ is a very small parameter. To use the Reductive Perturbation Technique (RPT) method for the small expansion parameter ϵ we use some expanded variables are follows

$$\left.\begin{array}{c}n=1+\epsilon\{f_0(\eta)+n_1(\eta,\gamma,\xi)\}+\epsilon^2 n_2(\eta,\gamma,\xi)+\ldots\\ \phi=\epsilon\left\{f_0(\eta)2{c_1}^2+\phi_1(\eta,\gamma,\xi)\right\}+\epsilon^2\phi_2(\eta,\gamma,\xi)+\ldots\\ u_x=c_1+\epsilon\{u_1(\eta,\gamma,\xi)-c_1 f_0(\eta)\}+\epsilon^2 u_2(\eta,\gamma,\xi)+\ldots\\ u_y=\epsilon^{\frac{3}{2}}v_1(\eta,\gamma,\xi)+\epsilon^{\frac{5}{2}}v_2(\eta,\gamma,\xi)+\ldots\end{array}\right\} \tag{3}$$

After using the set of stretched coordinates referred by Eq. (2) and expanded variables referred by Eq. (3) in the governing Eq. (1) we get some sets of equations. Then the coefficients of various powers of ϵ are compared. After evaluating and combining the new set of equations, will generate an equation is of the form

$$\frac{\partial}{\partial\xi}\left[A\frac{\partial n_1}{\partial\eta}+2n_1\frac{\partial n_1}{\partial\xi}+\frac{\partial^3 n_1}{\partial\xi^3}\right]+\frac{\partial^2 n_1}{\partial\lambda^2}-2c_1 f_0\frac{\partial^2 n_1}{\partial\xi^2}=0 \tag{4}$$

Here A is a constant, $c_1=\frac{1}{\sqrt{2}}$ and Eq. (4) is a modified KP equation. But the last extra term arises in Eq. (4) due to inhomogeneity.

Now, let us consider a new frame of reference $X=\xi+p(\eta), Y=\lambda, T=\eta$ where $p(\eta)=\frac{1}{\sqrt{2+1}}\int f_0(\eta)d\eta$. Using this new frame of reference Eq. to (4) we get the modified form of KP equation is

$$\frac{\partial}{\partial X}\left[\frac{\partial S}{\partial\tau}+6S+\frac{\partial S}{\partial X}+\frac{\partial^3 S}{\partial X^3}\right]+\frac{\partial^2 S}{\partial Y^2}=0 \tag{5}$$

The solution of the above modified KP equation i.e., Eq. (5) will give the soliton solutions. In the new co-ordinate X, the structural variations of the soliton solutions can be seen due to the presence of $p(\eta)$ which is related to $f_0(\eta)$.

3 Results

The one soliton solution of (5) is given by [31]

$$S = \frac{{k_1}^2}{2} sech^2\left[\frac{1}{2}\left(k_1 X + m_1 Y - \frac{{k_1}^4 + {m_1}^2}{k_1}\right)\right]$$

$$\Rightarrow n_1 = \frac{3{k_1}^2}{2} sech^2\left[\frac{1}{2}\left(k_1\xi + k_1 p(\eta) + m_1\lambda - \frac{{k_1}^4 + {m_1}^2}{k_1\sqrt{2}\left(\sqrt{2}+1\right)}\eta\right)\right] \tag{6}$$

where k_1 and m_1 are arbitrary constants.

Similarly, two soliton solution of Eq. (4) is given by [31]

$$S = 2\frac{\partial^2}{\partial\xi^2}(lnC)$$

$$\Rightarrow n_1 = 6\frac{\partial^2}{\partial\xi^2}(lnC) \tag{7}$$

where

$$C = 1 + e^{\eta_1} + e^{\eta_1} + Be^{\eta_1+\eta_2},\ B = \frac{(K_1 - K_2)^2 - (M_1 - M_2)^2}{(K_1 + K_2)^2 - (M_1 - M_2)^2}$$

$$\eta_1 = K_1\left[\xi + p(\eta) + \sqrt{3}M_1\lambda - \frac{\left({K_1}^2 + 3{M_1}^2\right)}{\left(2+\sqrt{2}\right)}\eta\right],$$

$$\eta_2 = K_2\left[\xi + p(\eta) + \sqrt{3}M_2\lambda - \frac{\left({K_2}^2 + 3{M_2}^2\right)}{\left(2+\sqrt{2}\right)}\eta\right]$$

where K_1, K_2, M_1, and M_2 are arbitrary constants. The structural variations of a soliton will take place due to the presence of inhomogeneous ion number density and the term $p(\eta)$, which is related to $f_0(\eta)$. The inhomogeneities will be different based on the various choices of $f_0(\eta)$. So, for the trivial choice of f_0, i.e., $f_0 = 0$, the inhomogeneous ion number density profile will reproduce the homogenous line

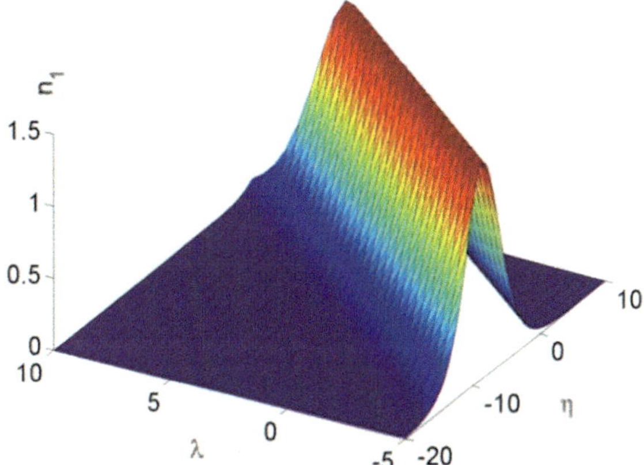

Fig. 1 Picture of one soliton solution given by Eq. (6) at $f_0 = 0$, representing a homogeneous line soliton solution reproduced from the chosen Inhomogeneous plasmas or ion number density

soliton, shown in Fig. 1. Similarly, the structural variations of one soliton solution will also be seen based on the various choices of $f_0(\eta)$, which are shown in Fig. 2. The different structural variations for two soliton solutions can also be observed by considering $f_0 = 0$ and based on the various choices of $f_0(\eta)$. Here, the graphical interpretations of the two soliton solutions are not presented in this work.

Now to investigate how many structural variations of a soliton will occur or what condition the structural variations will be seen.

In the case of one soliton solution given by Eq. (6), for the static case, i.e., $\xi = 0$, the locus of maximum amplitude is of from

$$k_1 p(\eta) + m_1\lambda - \frac{{k_1}^4 + {m_1}^2}{k_1\sqrt{2}\left(\sqrt{2}+1\right)}\eta = 0$$

Then for some chosen k_1, m_1, we have

$$\frac{du}{d\eta} = \frac{df_0}{d\eta} \tag{8}$$

where $u = \frac{d\lambda}{d\eta}$ is the slope for the locus of the maximum amplitude. The above Eq. (8) implies the larger variation of soliton structures in the larger rate of slope variation. Hence for the larger variation of soliton structures, RHS of Eq. (8) must be larger, i.e., the first-order derivative of f_0 w.r.t. η also must be immense. So we can choose some functional forms of f_0 to show the variations of various soliton structures.

In Fig. 2a, we have seen that if we increase the amplitude of f_0, then the structure of the soliton will be more deformed and will cause larger bending. Again, if we decrease the amplitude of f_0, then the structural variations of the solitons will be seen relatively less. Similarly, we can also observe similar things in Fig. 2b, where the sum of sine and cosine functions goes more rapidly. Also, based on the increasing/decreasing wave vector of f_0, the soliton will cause mores/less deformed. Thus based on both

Fig. 2 For some chosen k_1, m_1, K_1, K_2, and $\xi = 0$, various functional forms of $f_0(\eta)$ are considered above to show the structural variations of solitons by changing the phase of the solitary waves

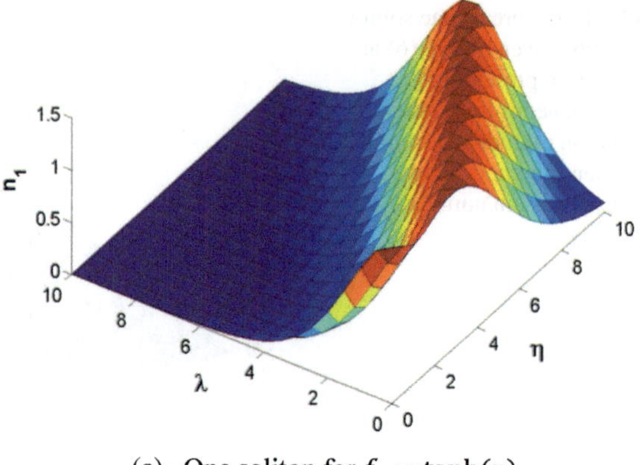

(a) One soliton for $f_0 = \tanh(\eta)$

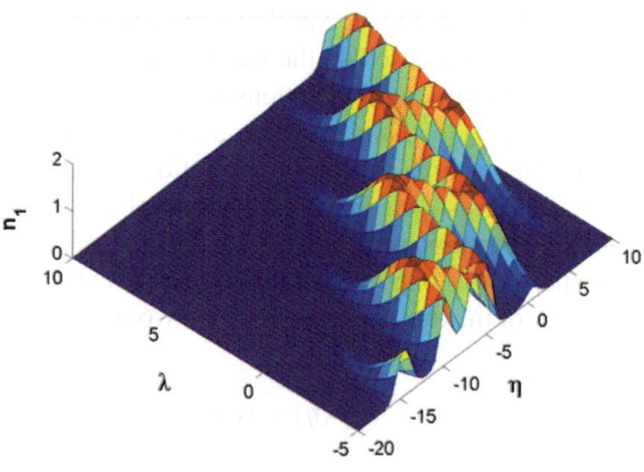

(b) One soliton for $f_0 = \sin\eta + cos\eta$

amplitude and wave vector of f_0, the first-order derivative of f_0 will increase/decrease and will case the structural variations of soliton in the more/less amount. Hence, some similar conclusions can also be given for various choices of $f_0(\eta)$.

Thus in this work, the structural variations of solitons are presented analytically. The structural variations are shown based on the dependency of f_0, is related to the ion number density for inhomogeneous plasmas. For the above investigations, here exactly the KP Eq. (5) is solved for the propagation of ion-acoustic soliton in two-dimensional inhomogeneous plasmas. As we have transformed the derived Eq. (4) into a standard, modified KP Eq. (6) with constant coefficient, we have seen that the phase modifications for each and every solution was controlled by the function f_0, which causes the structural modifications of the soliton solutions in the two-dimensional plane. Here the amplitude of the soliton solution is unchanged.

4 Conclusion

We have studied the ion-acoustic solitary waves in inhomogeneous unmagnetized plasmas in the presence of cold and hot isothermal electrons. During these investigations, we have solved the modified KP equation to understand the structural variations of soliton in the presence of the above-considered model. In solving the modified KP equation, we have seen the phase of the solitary wave solution gets modified by the function $f_0(\eta)$. Under the above considerations and constant amplitude, some structural variations of the soliton are seen accordingly. The study on the structural variation of soliton inhomogeneous plasmas is a very important feature like the other important features propagation, transmission, reflection, etc. In our future work, we wish to study the structural variations of soliton in inhomogeneous plasmas in the presence of magnetic field and dust effect.

References

1. Washimi, H., Taniuti, T.: Propagation of ion-acoustic solitary waves of small amplitude. Phys. Rev. Lett. **17**, 996–998 (1966)
2. Nishikawa, N., Kaw, K.: Propagation of solitary ion-acoustic waves in inhomogeneous plasmas. Phys. Lett. **50**(A), 455–456 (1975)
3. Kuehl, H.H.: Reflection of an ion-acoustic soliton by plasma inhomogeneities. Phys. Fluids **26**(6), 1577–1583 (1983)
4. Kuehl, H.H., Imen, K.: Finite amplitude ion-acoustic solitons in weakly inhomogeneous plasmas. Phys. Fluids **28**, 2375–2381 (1985)
5. Nejoh, Y.: The effect of the ion temperature on the ion-acoustic solitary waves in a collisionless relativistic plasma. J. Plasma Phys. **37**(3), 487–495 (1987)
6. Singh, S., Dahiya, R.P.: Effect of ion temperature and plasma density on an ion-acoustic soliton in a collisionless relativistic plasma: an application to radiation belts. Phys. Fluids B **2**(5), 901–906 (1990)
7. Singh, S., Dahiya, R.P.: Effect of zeroth-order density inhomogeneity on ion-acoustic soliton reflection in a finite ion temperature plasma. Phys. Fluids B **3**(1), 255–258 (1991)
8. Singh, S., Dahiya, R.P.: Propagation characteristics and reflection of an ion-acoustic soliton in an inhomogeneous plasma having warm ions. J. Plasma Phys. **41**(1), 185–197 (1989)
9. Malik, H.K., Dahiya, R.P.: Ion acoustic solitons in finite ion temperature inhomogeneous plasmas having negative ions. Phys. Plasmas **1**(9), 2872–2875 (1994)
10. Singh, D.K., Malik, H.K.: Soliton reflection in a negative ion containing plasma: Effect of magnetic field and ion temperature. Phys. Plasmas **13**(8), 082104(1–10) (2006)
11. Chauhan, S.S., Malik, H.K., Dahiya, R.P.: Reflection of ion acoustic solitons in a plasma having negative ions. Phys. Plasmas **3**(11), 3932–3938 (1996)
12. Singh, D.K., Malik, H.K.: Modified Korteweg-de Vries soliton evolution at critical density of negative ions in an inhomogeneous magnetized cold plasma. Phys. Plasmas **14**(6), 062113 (2007)
13. Xiao, D., Ma, J. X., Li, Y., Xia, Y., Yu, M. Y.: Evolution of nonlinear dust-ion-acoustic waves in an inhomogeneous plasma. Phys. Plasmas **13**(5), 052308(1–7) (2006)
14. Singh, D.K., Malik, H.K.: Modified Korteweg-de Vries soliton evolution at critical density of negative ions in an inhomogeneous magnetized cold plasma. Phys. Plasmas **14**(6), 112103(1–8) (2007)

15. Hellberg, M.A., Verheest, F.: Dust acoustic solitons in plasmas with kappa-distributed electrons and/or ions. Phys. Plasmas **15**(12), 062307(1–11) (2008)
16. Baluku, T.K., Hellberg, M.A., Kourakis, I., Saini, N.S.: Dust ion acoustic solitons in a plasma with kappa-distributed electrons. Phys. Plasmas **17**(5), 053702(1–11) (2010)
17. Kakad, A., Omura, K., Kakad, B.: Phys. Plasmas **20**(6), 062103(1–13) (2013)
18. Gogoi, L.B., Deka, P.N.: Solitary waves in weakly inhomogeneous plasma with nonthermal electrons. Int. J. Appl. Eng. Res. **93**(1), 51–64 (2015)
19. Mukherjee, A., Janaki, M.S., Kundu, A.: Bending of solitons in weak and slowly varying inhomogeneous plasma. Phys. Plasmas **22**(12), 122114(1–7) (2015)
20. Gogoi, L.B., Deka, P.N.: Propagation of dust acoustic solitary waves in inhomogeneous plasma with dust charge fluctuations. Phys. Plasmas **24**(3), 033708(1–6) (2017)
21. Zhou, C., Hutchinson, I.H.: Dynamics of a slow electron hole coupled to an ion-acoustic soliton. Phys. Plasmas **25**(3), 0823039(1–13) (2018)
22. Wang, F.P., Zhang, J.F., Gao, D.N., Li, Z.Z., Duan, W. S., Zhang, H.: Numerical simulation of dark envelope soliton in plasma. Phys. Plasmas **25**(3), 032121(1–6) (2018)
23. Shi, X., Li, J., Wu, C.: Dynamics of soliton solutions of the nonlocal Kundu-nonlinear Schrödinger equation. Chaos **29**, 023120(1–12) (2019)
24. Song, Y., Shi, X., Wu, C., Zhang, H.: Recent progress of study on optical solitons in fiber lasers. Appl. Phys. Rev. **6**, 021313(1–20) (2019)
25. Rani, N., Yadav, M.: Propagation of nonlinear electron acoustic solitons in magnetized dense plasma with quantum effects of degenerate electrons. In: AIP Conference Proceedings, vol. 2352, pp. 030008(1–8). AIP Publishing (2020)
26. Chen, C., Pan, Y., Guo, J., Wang, Y., Gao, G., Wang, W.: Soliton dynamics for quantum systems with higher-order dispersion and nonlinear interaction. AIP Adv. **10**, 065313(1–4) (2020)
27. Prayitno, T.B., Budi, E.: Numerical calculation on energy of static soliton solution for KdV equation. In: AIP Conference Proceedings 2021, 9th National Physics Seminar, vol. 2320, pp. 050014(1–3). AIP Publishing, Jakarta (2021)
28. Lu, F.F., Liu, S.Q.: Small amplitude ion-acoustic solitons with regularized κ-distributed electrons. AIP Adv. **11**, 085223(1–5) (2021)
29. Wu, D.: The direct scattering problem for perturbed Kadomtsev–Petviashvili multi line solitons. J. Math. Phys. **62**, 091513(1–19) (2021)
30. Mushinzimana, X., Nsengiyumva, F., Yadav, L.L., Baluku, T.K.: Dust ion acoustic solitons and double layers in a dusty plasma with adiabatic positive dust, adiabatic positive ion species, and Cairns-distributed electrons. AIP Adv. **12**, 015208(1–10) (2022)
31. Wazwaz, A.M.: Multiple-soliton solutions for the KP equation by Hirota's bilinear method and by the tanh–coth method. Appl. Math. Comput. **190**, 633 (2007)

Effect of Kappa Parameters on the Modulational Instability in a Polarized Dusty Plasma

A. Abdikian

Abstract In this paper, we have studied the propagation of dust-acoustic modulated waves in the polarized dusty plasmas. The distributions of electrons and ions are Boltzmann and Kappa, respectively. We have used the reductive perturbation method (RPM) to find out the nonlinear amplitude modulation of dust-acoustic waves in an unmagnetized collisionless polarized dusty plasma and a modified nonlinear Schrodinger equation governing the evolution of the dust-acoustic envelope waves has been derived. The effects of the Kappa parameter on the modulational instability (MI) was discussed by using the numerical values. It is found that increasing the value of κ causes to increase the value of the dispersion relation and the group velocity. It was shown that the MI maximum growth rate firstly increases and then decreases as the mentioned plasma parameter increases.

Keywords Polarized dusty plasmas · Modulational instability · Nonlinear Schrodinger equation

1 Introduction

Over the last decade, the study of dusty plasma was an important issue among researchers because of its applications in variety environments from interstellar space to laboratory plasma systems [1]. A dusty plasma defined as a common plasma with massive dust grains in it. The presence of these additional charged particles can remarkably change the characteristics and behaviors of a plasma [2, 3] and new low frequency dust mode so-called "dust-acoustic wave" (DAW) give rise. DAW would be due to the restoring force provided by the plasma thermal pressure while the inertia is due to the dust mass [4].

The polarization force is one of the important forces that change the linear and nonlinear dust dynamics and studies on it are increasing [5]. Physically, the polarization

A. Abdikian (✉)
Department of Physics, Malayer University, Malayer, 65719-95863, Iran
e-mail: abdykian@gmail.com

S. Banerjee and A. Saha (eds.), *Nonlinear Dynamics and Applications*,
Springer Proceedings in Complexity,
https://doi.org/10.1007/978-3-030-99792-2_9

force comes from the deformation of the spherical Debye screening of the dusty grains and it is defined as [6, 7]

$$\mathbf{F}_p = -\frac{q_d^2}{2\lambda_D^2}\nabla\lambda_D, \tag{1}$$

where q_d is the grain charge and $\lambda_D = \lambda_{Di}/\sqrt{1 + n_e T_i/n_i T_e}$ is the linearized Debye radius with the ion Debye radius $\lambda_{Di} = \epsilon_0 k_B T_i/(e^2 n_i)$, and $T_{i(e)}$ is the ion (electron) temperature and $n_{i(e)}$ is the ion (electron) number density. It has been shown that the linear dispersions and nonlinear peculiarities of the DAWs have been modified by polarization force [8, 9]. Khrapak et al. [8] found that the wave phase velocity would be decreased in the presence of the polarization effect. Recently, considering a system of dusty plasma including the negatively charged dust, Maxwellian and superthermally distributed for electrons and ions respectively, [10], Singh et al. have reported the impact of polarization force on dust-acoustic cnoidal waves. They found that an increase in superthermality index of ions leads to a decrease in polarization parameter. Very recently, the importance of the consideration of nonthermally polarization force on modulational instability (MI) of dust acoustic waves was investigated theoretically by Singh and Saini [11]. By supposing a dusty plasma system with negatively charged dust, Maxwellian and Cairns' nonthermal distributed for electrons and ions respectively, they have studied the evolution of DA breathers, namely rogue waves. The standard reductive perturbation method (RPM) is generally applied to obtain a nonlinear Schrodinger (NLS) equation. the Peregrine soliton is the analytical solution of the NLS equation [13]. The Peregrine solitary wave is a significant collective behavior in plasma, since it can be described as an rogue waves [14, 15]. The modulational instability (MI) of several plasma modes is amount of interest in studying because the wave propagation stability depend on its relevance [12, 16, 17]. The kappa distribution function, which represents superthermal particles, was first proposed by Vasyliunas [18]. This distribution function with spectral index κ, which determines the particle velocity distribution, is suitable for the behavior of particles in space and interplanetary environment and has attracted a lot of interest. This distribution function in the limit $\kappa \to \infty$ includes the Maxwellian distribution function [19]. Since electrons have less inertia than ions their thermal energy is higher than ions. Hence, in this paper, we have considered the Maxwellian distribution for electrons and the kappa distribution has been suggested for the ion particles. Also, since the ratio masses of the electron and ion to the dust particles are so small, these particles can be considered as non- inertial particles [20]. The aims of the present paper using of the hydrodynamic (HD) model are to peruse the effects Kappa parameters on the dust-acoustic modulational instability and the propagation of pulse waves in a polarized dusty plasma. The paper is organized as follows. In Sect. 2, the model equations that include the effects of polarization force and the superthermal ions are presented, and then using the standard reductive perturbation method, the NLS equation is derived. In Sect. 3, using the plasma parameter values, the numerical discussion is presented and the effects of the Kappa parameters on the modulational instability are also searched. Finally, in Sect. 4 the conclusion is reported.

2 Mathematical Formulation

Considering a dusty plasma media including electrons, ions and charged dust particles. The normalized dynamic of the DAWs is governed by the following equations including the polarization force [5, 21]

$$\frac{\partial n}{\partial t}+\frac{\partial (nu)}{\partial x}=0, \tag{2}$$

$$\frac{\partial u}{\partial t}+u\frac{\partial u}{\partial x}=\chi\frac{\partial \Phi}{\partial x}, \tag{3}$$

$$\frac{\partial^2 \Phi}{\partial x^2}=\mu_e n_e-\mu_i n_i+n, \tag{4}$$

where Φ is the electrostatic potential normalized by $(k_B T_i/e)$, n_j is densities of jth species normalized by its unperturbed densities (n_{j0}), u is dust fluid velocity normalized by DA speed $C_d=\sqrt{Z_d k_B T_i/m_d}$, x normalized by the dust Debye length $\lambda_{D0}=\epsilon_0 k_B T_i/Z_d^2 n_{d0}$, t normalized by $\omega_{pd}^{-1}=\epsilon_0 m_d/e^2 Z_d^2 nd0$, $\mu_e=n_{e0}/Z_d n_{d0}$, $\mu_i=n_{i0}/Z_d n_{d0}$, $\chi=1-R$ and $R\approx q_d e/16\pi\,\epsilon_0\lambda_D k_B T_i$ [5]. Here, we assume that the ions obey the Kappa distribution function given by

$$n_i=\left(1+\frac{\Phi}{\kappa-3/2}\right)^{-\kappa+1/2}, \tag{5}$$

while the electrons satisfy the Boltzmann relation routinely in the DAWs slow time regime,

$$n_e=exp\left(\sigma_i \Phi\right), \tag{6}$$

where $\sigma_i=T_i/T_e$.
we have applied the standard reductive perturbation method (RPM) to obtain a nonlinear Schrodinger (NLS) equation and then to study the modulation of the DAWs in the mentioned media [4, 22, 23]. We employ the following stretching of independent variables

$$\xi=\varepsilon(x-v_g t),\ \tau=\varepsilon^2 t, \tag{7}$$

in which ε is the nonlinearity strength and also is a small value ($0 < \varepsilon \ll 1$) and v_g is the group velocity of the wave propagating.

$$n = 1 + \sum_{m=1}^{\infty} \varepsilon^m \sum_{\ell=-m}^{m} n_\ell^{(m)}(\xi, \tau) exp[il(kx - \omega t)], \tag{8}$$

$$u = \sum_{m=1}^{\infty} \varepsilon^m \sum_{\ell=-m}^{m} u_\ell^{(m)}(\xi, \tau) exp[il(kx - \omega t)], \tag{9}$$

$$\Phi = \sum_{m=1}^{\infty} \varepsilon^m \sum_{\ell=-m}^{m} \Phi_\ell^{(m)}(\xi, \tau) exp[il(kx - \omega t)], \tag{10}$$

One can substitute the Expression (7) into the Eqs. (2)–(4) and then collect terms in the different powers of ε, by this method there are some mth-order equations. One can obtain the following equation for the first-order $m = 1$ and $\ell = 1$

$$\begin{pmatrix} n_1^{(1)} \\ u_1^{(1)} \end{pmatrix} = \begin{pmatrix} C_1 - k^2 \\ -k\, \chi/\omega \end{pmatrix} \Phi_1^{(1)}, \tag{11}$$

and the dispersion relation

$$\omega = \frac{k\,\sqrt{\chi}}{\sqrt{k^2 - C_1}}, \tag{12}$$

and group velocity

$$v_g = \frac{C_1 \omega^2}{\omega^2 - \chi} \equiv \frac{\partial \omega}{\partial k}, \tag{13}$$

And then one can get the following relations between second harmonic modes and terms of $\Phi_1^{(1)}$. For $m = 2$ and $\ell = 0$

$$\begin{pmatrix} n_0^{(2)} \\ u_0^{(2)} \\ \Phi_0^{(2)} \end{pmatrix} = \begin{pmatrix} -\frac{2\chi\left(C_1^2 k v_g \omega + C_2 \omega^2 - C_1 k^2 \left(\chi + k v_g \omega\right)\right)}{\left(C_1 v_g^2 - \chi\right)\omega^2} \\ \frac{2\chi\left(C_1 k \chi\left(k v_g - \omega\right) + \omega\left(k^3 \chi - C_2 v_g \omega\right)\right)}{\left(C_1 v_g^2 - \chi\right)\omega^2} \\ \frac{2\,\left(-k^2 \chi^2 + C_1 k v_g \chi \omega - k^3 v_g \chi\, \omega + C_2 v_g^2 \omega^2\right)}{(C_1 v_g^2 - \chi)\omega^2} \end{pmatrix} \left|\Phi_1^{(1)}\right|^2, \tag{14}$$

for $m = 2$, $\ell = 1$, we have

$$\begin{pmatrix} n_1^{(2)} \\ u_1^{(2)} \end{pmatrix} = -\begin{pmatrix} C_1 + k^2 \\ (k^2 - C_1)k/\omega \end{pmatrix} \Phi_1^{(2)} + 2\,i(k^2 - C_1) \begin{pmatrix} 1/k\omega \\ 1/k^2 \end{pmatrix} (k v_g - \omega) \frac{\partial \Phi_1^{(1)}}{\partial \xi}, \tag{15}$$

and for $m = 2$ and $\ell = 2$:

$$\begin{pmatrix} n_2^{(2)} \\ u_2^{(2)} \\ \Phi_2^{(2)} \end{pmatrix} = \begin{pmatrix} -\frac{k^2\chi(C_1k^2\chi+4k^4\chi-2C_1^2\omega^2-2C_2\omega^2-6C_1k^2\omega^2+8k^4\omega^2)}{2\omega^2(k^2\chi-C_1\omega^2-4k^2\omega^2)} \\ -\frac{k\chi(-C_1k^2\chi+6k^4\chi-2C_2\omega^2)}{2\omega(k^2\chi-C_1\omega^2-4k^2\omega^2)} \\ \frac{-k^4\chi^2+2C_1k^2\chi\omega^2-2k^4\chi\omega^2+2C_2\omega^4}{2\omega^2(-k^2\chi+C_1\omega^2+4k^2\omega^2)} \end{pmatrix} \left(\Phi_1^{(1)}\right)^2, \tag{16}$$

By continuing the same method for the third order and simultaneously considering the above equations, one can derive the NLS equation as follow

$$i\frac{\partial\Phi}{\partial\tau} + \frac{P}{2}\frac{\partial^2\Phi}{\partial\xi^2} + Q\,|\Phi|^2\,\Phi = 0, \tag{17}$$

where $\Phi = \Phi_1^{(1)}$. The dispersion coefficient relation (P) is

$$P = \frac{3C_1\omega^5}{k^4\chi^2}, \tag{18}$$

while the nonlinear coefficient relation (Q) defines as

$$\begin{aligned} Q = -\frac{\omega}{4\left(C_1-k^2\right)\left(2C_1+3k^2\right)\left(2C_1^3-3C_1^2k^2+3C_1k^4-k^6\right)}&\Big[18C_1^7+23C_1^6k^2 \\ +C_1^5\left(80C_2-274k^4\right)+2k^6\left(2C_2^2+9C_3k^2+12C_2k^4\right)& \\ -3C_1^4\left(8C_3+64C_2k^2-177k^6\right)+C_1^3\left(-24C_2^2+84C_2k^4-449k^8\right)& \\ -3C_1k^4\left(4C_2^2+14C_3k^2+36C_2k^4+9k^8\right)& \\ +2C_1^2k^2\left(-6C_2^2+9C_3k^2+56C_2k^4+89k^8\right)\Big]&, \end{aligned} \tag{19}$$

3 Results and Discussion

In this section, we want to study the stability/instability of the modulated envelope waves on the basis of the NLS Eq. (14) of the DAWs in polarized dusty plasma. For this purpose, we have chosen the parameters in dusty plasmas in which the grain radius $r_d = 1\,\mu\text{m}$, $n_{i0} = 1.65 \times 10^8 cm^{-3}$, $T_e = 3\,ev$ and $\sigma_i = 0.01$ are applied. It is proved that [24] the unstable of the modulated wave packets (bright solitary wave) occurs when the product PQ is positive and the stable of the modulated wave packets (dark envelope soliton) exists when the product $PQ < 0$ in the modulation wave number region $k^2 > 2Q\,|\Phi_0|^2\,/P$, where Φ_0 is the amplitude of the carrier waves. Besides, one can choose the wavenumber as $k = |\Phi_0|\,\sqrt{Q/P}$ and then attain the maximum growth rate as $Q\,|\Phi_0|^2$. It is proposed the following breather solution in the unstable regions for the NLS Eq. (14) [15, 25]

$$\Phi = \sqrt{\frac{P}{Q}}\left[-1 + \frac{4(1+2\,i\,P\tau)}{1+4\xi^2+4P^2\tau^2}\right] exp(i\,P\tau), \tag{20}$$

We numerically investigate the effect of the polarization force on the nature and behavior of the solution by identification the sign of the product PQ. The critical wave number is defined as the value of $k = k_c = |\Phi_0|\,\sqrt{Q/P}$ for the onset of the modulational instability. It should be noted that choosing Q=0 leads to $P/Q \to \pm\infty$. Figure 1 depicts the dispersion relation (ω) and the group velocity (v_g) versus wave number (k) for the different values of the Kappa parameters (κ). It is seen from Fig. 1a that although ω boosts as the value of k enhances, it shifts towards lower values as κ increases. This means that the wave energy would be decrease when the value of the κ parameter increases. The schematic of the group velocity of the solitary wave for three values of the Kappa parameters (κ) is illustrated in Figure 1b. For plotting, the Kappa parameters (κ) are chosen as $\kappa = 3,\ 5$ and 10 for blue, orange and green lines, respectively. The group velocity (v_g) would be decreased when the values of the Kappa parameters (κ) increase.

Figure 2 show the MI region as a blue area for relevant physical parameters. The stable modulational pulses are depicted by the white regions (where $PQ < 0$) and the dark solitary waves can propagate. While the blue area corresponds to the unstable modulational pulses where the bright solitary waves can exist. It should be notified that the solid blue line in this figure represents the critical wave number k_c. The behavior of the value k_c is depending to the Kappa parameter and increasing the value of κ causes to a shift of k_c towards higher values.

Figure 3a, b demonstrate the absolute of the DA rogue pulse $|\Phi|$ against ξ for several plasma values such as (a) the Kappa parameters (κ), (b) σ_i, (c) μ_e and R.

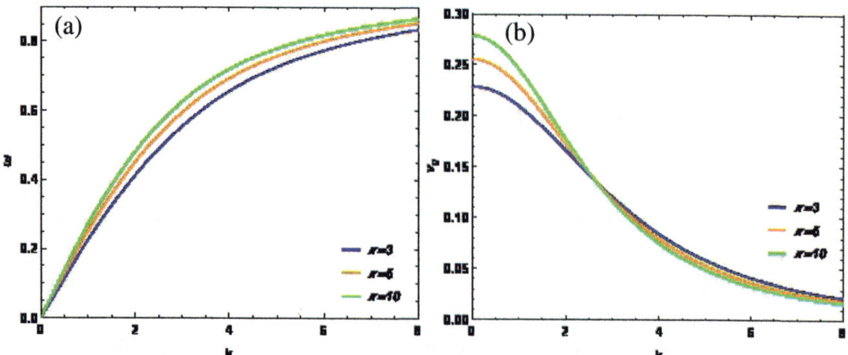

Fig. 1 The carrier frequency ω and the group velocity v_g are plotted against the wave number k for different values of κ. Here, $\kappa = 3$ (blue line), $\kappa = 5$ (orange line) and $\kappa = 10$ (green line)

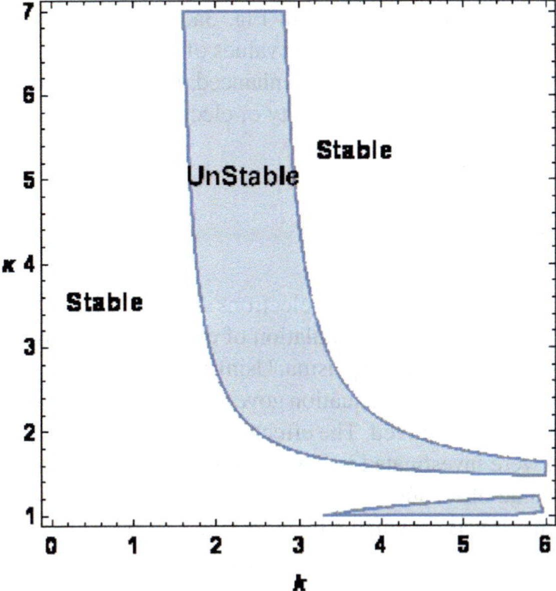

Fig. 2 The contour of PQ is plotted versus the wave number k and the κ parameter

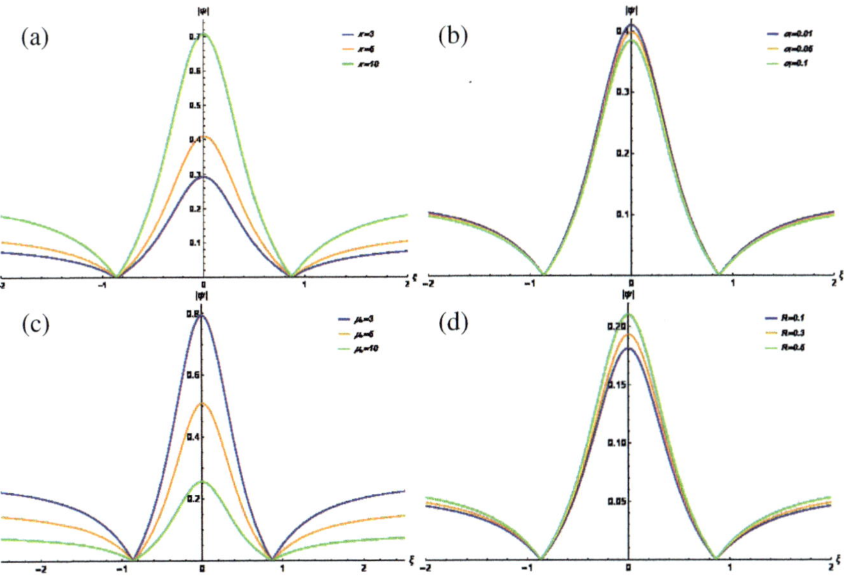

Fig. 3 **a** The absolute of the rogue wave against ξ for different plasma values **a** κ **b** σ_i, **c** μ_e and R

Although, as can be observed that from Fig. 3a, d that the amplitude of the DA rogue pulse would be increased when the values of the superthermal ions (κ) and the magnitude of the polarization force (R) enhanced, its amplitude decreased when the ratio of ion temperature σ_i and the density of electrons μ_e increased.

4 Conclusion

In the present work, using a Boltzmann electrons and Kappa ion distributions, we have studied the nonlinear amplitude modulation of dust-acoustic waves in an unmagnetized collisionless dusty polarized plasma. Using the reductive perturbation method, a modified nonlinear Schrodinger equation governing the evolution of the dust-acoustic envelope waves has been derived. The effects of the κ parameter on the modulational instability (MI) were investigated by changing the numerical values. It is found that increasing the value of κ causes to increase the value of the dispersion relation and the group velocity. It was shown that by increasing the superthermal ions (κ) and the magnitude of the polarization force (R), the amplitude of DA rogue waves decreased while by enhancing the ratio of ion temperature σ_i and the density of electrons μ_e caused to increased its amplitude.

References

1. Shukla, P.K., Mamun, A.: Introduction to Dusty Plasma Physics. CRC Press (2015)
2. Abdikian, A.: Dust-ion-acoustic solitary waves in a magnetized dusty pair-ion plasma with Cairns-Gurevich electrons and opposite polarity dust particles. Contrib. Plasma Phys. **59**, 20 (2019)
3. El Awady, E., El Tantawy, S., Abdikian, A.: Dissipative cylindrical magnetosonic solitary waves in a magnetized quantum dusty plasma. Rom. Rep. Phys. **71**, 105 (2019)
4. Bouzit, O., Tribeche, M.: Dust-acoustic waves modulational instability and rogue waves in a polarized dusty plasma. Phys. Plasmas **22**, 103703 (2015)
5. Chen, H., Zhou, S., Luo, R., Liu, S.: Nonlinear dust acoustic waves with polarization force effects in Kappa distribution plasma. Jpn. J. Appl. Phys. **56**, 016101 (2016)
6. Hamaguchi, S., Farouki, R.: Phys. Rev. E Polarization force on a charged particulate in a nonuniform plasma. Phys. Rev. E **49**, 4430 (1994)
7. Hamaguchi, S., Farouki, R.: Plasma-particulate interactions in nonuniform plasmas with finite flows. Phys. Plasmas **1**, 2110 (1994)
8. Khrapak, S., Ivlev, A., Yaroshenko, V., Morfill, G.: Influence of a polarization force on dust acoustic waves. Phys. Rev. Lett. **102**, 245004 (2009)
9. Mamun, A., Ashrafi, K., Shukla, P.: Effects of polarization force and effective dust temperature on dust-acoustic solitary and shock waves in a strongly coupled dusty plasma. Phys. Rev. E **82**, 026405 (2010)
10. Singh, K., Ghai, Y., Kaur, N., Saini, N.S.: Effect of polarization force on dust-acoustic cnoidal waves in dusty plasma. Eur. Phys. J. D **72**, 1 (2018)
11. Singh, K., Saini, N.S.: Breather structures and peregrine solitons in a polarized space dusty plasma. Front. Phys. **8**, 511 (2020)

12. Demiray, H., Abdikian, A.: Modulational instability of acoustic waves in a dusty plasma with nonthermal electrons and trapped ions. Chaos Soliton Fract. **121**, 50 (2019)
13. Peregrine, D.: Water waves, nonlinear Schrodinger equations and their solutions. J. Aust. Math. Soc. B Appl. Math. **25**, 16 (1983)
14. Draper, L.: "Freak" ocean waves. Weather **21**, 2 (1966)
15. Abdikian, A., Ismaeel, S.: Ion-acoustic rogue waves and breathers in relativistically degenerate electron-positron plasmas. Eur. Phys. J. Plus **132**, 368 (2017)
16. Bhowmik, C., Misra, A., Shukla, P.: Oblique modulation of electron-acoustic waves in a Fermi electron-ion plasma. Phys. Plasmas **14**, 122107 (2007)
17. Bains, A.S., Tribeche, M., Gill, T.S.: Modulational instability of electron-acoustic waves in a plasma with a q-nonextensive electron velocity distribution. Phys. Lett. A **375**, 2059 (2011)
18. Vasyliunas, V.M.: A survey of low-energy electrons in the evening sector of the magnetosphere with OGO 1 and OGO 3. J. Geophys. Res. **73**, 2839 (1968)
19. Hau, L.-N., Fu, W.-Z.: Mathematical and physical aspects of Kappa velocity distribution. Phys. Plasmas **14**, 110702 (2007)
20. Abdikian, A., Sultana, S.: Dust-acoustic solitary and cnoidal waves in a dense magnetized dusty plasma with temperature degenerate trapped electrons and nonthermal ions. Phys. Scr. **96**, 095602 (2021)
21. Mayout, S., Bentabet, K., Tribeche, M.: Effect of the polarization force on the dust-acoustic soliton energy. Contrib. Plasma Phys. **56**, 99 (2016)
22. Abdikian, A., Tamang, J., Saha, A.: Electron-acoustic supernonlinear waves and their multi-stability in the framework of the nonlinear Schrödinger equation. Commun. Theor. Phys. **72**, 075502 (2020)
23. Abdikian, A.: Modulational instability of ion-acoustic waves in magnetoplasma with pressure of relativistic electrons. Phys. Plasmas **24**, 052123 (2017)
24. Amin, M., Morfill, G., Shukla, P.: Modulational instability of dust-acoustic and dust-ion-acoustic waves. Phys. Rev. E **58**, 6517 (1998)
25. Pathak, P., Sharma, S., Nakamura, Y., Bailung, H.: Observation of second order ion acoustic Peregrine breather in multicomponent plasma with negative ions. Phys. Plasmas **23**, 022107 (2016)

Nonlinear Wave Structures in Six-Component Cometary Ion-Pair Dusty Plasma

Punam Kumari Prasad, Jharna Tamang, and Nur Aisyah Binti Abdul Fataf

Abstract The dust-ion-acoustic waves (DIAWs) in a six-component plasma constituting of ion-pair, negatively charged fluid dusts, superthermal light hydrogen ion, solar electrons and cometary tail electrons are studied. Employing the technique of reductive perturbation, the Korteweg-de Vries (KdV) equation is formulated and the corresponding phase plane is analyzed. The analytical wave solutions and electric field under the effect of plasma parameters are examined. The findings of the present work can be useful to understand nonlinear wave features in the region of the cometary tail where plasma particles are superthermal in nature.

Keywords Superthermal plasma · Electric field · Solitary wave

1 Introduction

A study of nonlinear wave features in dusty plasma enables us to understand the behavior and physical processes that occur in various space and astrophysical surroundings like interplanetary space, interstellar medium, planetary rings, cometary tails, asteroid zones, astroclouds, etc. The embedded charged dust particles in an electron-ion plasma modify the collective behavior of a plasma and also lead to the excitation of wave modes such as dust-acoustic waves (DAWs) [1], dust ion-acoustic waves (DIAWs) [2], and dust lattice waves [3]. Very recently, authors [4],

P. K. Prasad (✉) · J. Tamang
Department of Mathematics, Sikkim Manipal Institute of Technology, Sikkim Manipal University, Majitar, Rangpo, East-Sikkim 737136, India
e-mail: psikkim@ymail.com

J. Tamang
Department of Mathematics, Sikkim Alpine University, Kamrang, Namchi, South Sikkim 737126, India

N. A. B. A. Fataf
Cyber Security Centre, National Defence University of Malaysia (NDUM), Kuala Lumpur, Malaysia

S. Banerjee and A. Saha (eds.), *Nonlinear Dynamics and Applications*,
Springer Proceedings in Complexity,
https://doi.org/10.1007/978-3-030-99792-2_10

[5] investigated the existence of nonlinear arbitrary amplitude solitary waves in a six-component cometary dusty plasmas. Venugopal and Neethu [6] investigated the influence of variable dusts on dust acoustic shock waves in a six-component cometary dusty plasma with pair of oppositely charged heavy ions.

A vaporized gas of cloud at the tail region of comets formed due to the solar wind-comet interaction. In addition to the dust species, the plasma environment of a comet is essentially dominated by the newly generated multi-ion species that include water group ions (OH^+, H_2O^+ and H_3O^+) with solar wind protons and electrons [7]. The dissociation of molecules of water group leads to the formation of positively charged oxygen (O^+) and hydrogen (H^+) ions with photo-electrons [8]. Along with these positive ions, the Giotto spacecraft also identified negative ions in the plasma environment of the comet with energies ranging from 0.03 to 3.0 keV. Among all ionic species, negatively charged ions (O^-) are clearly identified [9].

The Rosetta spacecraft at comet 67P/Churyumov-Gerasimenko, has observed the signature of highly energetic superthermal electrons that deviate significantly from the Maxwellian distribution [10]. The source for the formation of energetic cometary plasma particles are the production of photo-electrons by photo-ionization of the cometary coma and inward movement of solar wind electrons into the coma region [11]. Thus, light hydrogen ions, solar and cometary electrons are characterized by superthermal $\kappa-$distributions having distinct temperatures and spectral indices (κ). Low values of κ signify the distributions with relatively large components of superthermal particles than thermal particles [12]. Moreover, superthermal $\kappa-$distribution tends to Maxwellian distribution for higher values of κ. Recently, authors [13, 14] investigated the influence of κ-distributed particles on small-amplitude nonlinear waves in multi-component plasma system.

The phase plane analysis is an effective concept to investigate the dynamical features for any nonlinear system. Implementing this concept, Samanta et al. [15] studied bifurcation behavior of the nonlinear DIAWs in a magneto-dusty plasma. Selim et al. [16] explored bifurcations behavior of nonlinear waves in a magnetized multi-component plasma with superthermal electrons. Recently, Monier and Atteya [17] explored the properties of the KdVB equation in a dusty plasma. Rahim et al. [18] investigated the dynamical behaviors of DA solitary structures in Thomas Fermi dusty plasma. Tamang et al. [19] explored the existence of localized structures which are featured by the nontrivial topology of their phase portraits under the framework of higher order Korteweg-deVries (KdV) equations. Very recently, the method of phase plane analysis is adopted to analyze small-amplitude nonlinear waves in various three-component plasma systems [20, 21]. This paper investigates the propagation of small-amplitude nonlinear DIAWs in six-component plasma constituting of ion-pair, negatively charged fluid dusts, superthermal light hydrogen ion, solar electrons and cometary tail electrons.

The paper is organized as: in Sect. 2, the basic equations which describe the dynamics of DIAWs in cometary plasma system are considered. Section 3 deals with the derivation of the nonlinear KdV equation and phase plane analysis with the analytical wave solution corresponding to the KdV equation are done in Sect. 4. The concluding remarks are presented in Sect. 5.

2 Basic Equations

One can consider a six-component dusty plasma system consisting of ion pair (negatively and positively charged oxygen ions indicated, respectively, by subscripts '$-$' and '$+$'), negatively charged dust particles (indicated by subscript 'd'), kappa distributed light hydrogen ions (indicated by subscript 'H'), hot solar electrons and cold cometary tail electrons (indicated respectively by subscripts 'se and 'ce'). The dynamics of DIAWs can be studied by considering the normalized fluid equations given as [4]:

$$\frac{\partial n_-}{\partial t} + \frac{\partial}{\partial x}(n_- v_-) = 0, \tag{1}$$

$$\frac{\partial v_-}{\partial t} + v_- \frac{\partial v_-}{\partial x} = \frac{z_- \alpha_-}{z_d} \frac{\partial \phi}{\partial x} - \frac{3\alpha_- \sigma_-}{z_d} n_- \frac{\partial n_-}{\partial x}, \tag{2}$$

$$\frac{\partial n_+}{\partial t} + \frac{\partial}{\partial x}(n_+ v_+) = 0, \tag{3}$$

$$\frac{\partial v_+}{\partial t} + v_+ \frac{\partial v_+}{\partial x} = -\frac{z_+ \alpha_+}{z_d} \frac{\partial \phi}{\partial x} - \frac{3\alpha_+ \sigma_+}{z_d} n_+ \frac{\partial n_+}{\partial x}, \tag{4}$$

$$\frac{\partial n_d}{\partial t} + \frac{\partial}{\partial x}(n_d v_d) = 0, \tag{5}$$

$$\frac{\partial v_d}{\partial t} + v_d \frac{\partial v_d}{\partial x} = \frac{\partial \phi}{\partial x}, \tag{6}$$

$$\frac{\partial^2 \phi}{\partial x^2} = n_d + \mu_- n_- - \mu_+ n_+ - \mu_H n_H + \mu_{se} n_{se} + \mu_{ce} n_{ce}, \tag{7}$$

where, $\mu_- = \frac{z_- n_{-0}}{z_d n_{d0}}$, $\mu_+ = \frac{z_+ n_{+0}}{z_d n_{d0}}$, $\mu_H = \frac{n_{H0}}{z_d n_{d0}}$, $\mu_{se} = \frac{n_{se0}}{z_d n_{d0}}$, $\mu_{ce} = \frac{n_{ce0}}{z_d n_{d0}}$, $\sigma_- = \frac{T_-}{T_d}$, $\sigma_+ = \frac{T_+}{T_d}$, $\alpha_- = \frac{m_d}{m_-}$ and $\alpha_+ = \frac{m_d}{m_+}$. Here, n_j is the number densities of jth plasma particles, where $j = -, +, H, d, se$, and ce, with '0' denoting the unperturbed state of plasma particles. ϕ denotes the electrostatic wave potential. Furthermore, $v_{-,+,d}$, $m_{-,+,d}$ and $z_{-,+,d}$ represent velocity, mass and charge number of negative oxygen ions, positive oxygen ions, and negative dust particles, respectively. The normalized variables are as follows: $v_j \to v_j/C_d, n_j \to n_j/n_{j0}$ (where, $j = -, +, H, d, se$ and ce), $\phi \to \phi e/k_B T_d$, $t \to \omega_{pd} t$ and $x \to x/\lambda_{Dd}$, where the DA wave speed $C_d = (z_{d0} k_B T_d/m_d)^{1/2}$, dust frequency $\omega_{pd}^{-1} = (m_d/4\pi z_{d0}^2 e^2 n_{d0})^{1/2}$ and Debye length for the dust plasma $\lambda_{Dd} = (k_B T_d/4\pi z_{d0}^2 n_{d0} e^2)^{1/2}$. Here k_B represents the Boltzmann constant, e denotes the electronic charge. The equation of state is $p_{-,+,d} = c n_{-,+,d}^{\nu}$, where $\nu = (f+2)/f$ with f as the degree of freedom. For one-dimensional adiabatic state, $f = 1$ and hence $\nu = 3$.

The normalized number densities of light hydrogen ions (n_H), hot solar electrons (n_{se}) and cold cometary tail electrons (n_{ce}) are given as follows:

$$n_H = \left(1 + \frac{\phi}{\sigma_H(\kappa_H - 3/2)}\right)^{-\kappa_H + 1/2}, \tag{8}$$

$$n_{se,ce} = \left(1 - \frac{\phi}{\sigma_{se,ce}(\kappa_{se,ce} - 3/2)}\right)^{-\kappa_{se,ce} + \frac{1}{2}}, \tag{9}$$

where, $\sigma_H = \frac{T_H}{T_d}$, $\sigma_{se} = \frac{T_{se}}{T_d}$, and $\sigma_{ce} = \frac{T_{ce}}{T_d}$. The spectral indices $\kappa_{H,se,ce}$ represents the superthermality of hydrogen ions, solar and cometary tail electrons, respectively. Evidently, one requires $\kappa_{H,se,ce} > 3/2$ and if $\kappa_{H,se,ce} \rightarrow \infty$ then distributions (8)–(9) reduce to Maxwell–Boltzmann distribution.

3 Formulation of Korteweg-DeVries (KdV) Equation

To study DIAWs in the considered cometary plasma system, one can employ the technique of reductive perturbation to formulate the KdV equation. For this purpose, dependent variables are expanded as:

$$\begin{aligned} n_j &= 1 + \varepsilon n_j^{(1)} + \varepsilon^2 n_j^{(2)} + \cdots \\ v_j &= 0 + \varepsilon v_j^{(1)} + \varepsilon^2 v_j^{(2)} + \cdots \\ \phi &= 0 + \varepsilon \phi^{(1)} + \varepsilon^2 \phi^{(2)} + \cdots \end{aligned} \tag{10}$$

where, $j = -, +, d$ and ε is an indicative of the magnitude of perturbation. Furthermore, stretching of the independent variables are introduced as:

$$\tau = \varepsilon^{3/2} t \quad \xi = \varepsilon^{1/2}(x - v_0 t), \tag{11}$$

where, the v_0 denotes phase velocity of DIAWs.

Substituting expressions (10) and (11) in system of normalized Eqs. (1)–(7) and equating the coefficients of various powers of ε to zero, one can get

$$\begin{cases} n_-^{(1)} = \dfrac{1}{v_0} v_-^{(1)} = \left(\dfrac{z_-\alpha_-}{3\alpha_-\sigma_- - v_0^2 z_d}\right)\phi^{(1)}, \\ n_+^{(1)} = \dfrac{1}{v_0} v_+^{(1)} = \left(\dfrac{z_+\alpha_+}{v_0^2 z_d - 3\alpha_+\sigma_+}\right)\phi^{(1)}, \\ n_d^{(1)} = \dfrac{1}{v_0} v_d^{(1)} = -\dfrac{1}{v_0^2}\phi^{(1)}, \\ \dfrac{z_-\alpha_-\mu_-}{3\alpha_-\sigma_- - v_0^2 z_d} + \dfrac{z_+\alpha_+\mu_+}{3\alpha_+\sigma_+ - v_0^2 z_d} - \dfrac{1}{v_0^2} + T_1 = 0, \end{cases} \tag{12}$$

$$\frac{\partial n_-^{(1)}}{\partial \tau} + \frac{\partial v_-^{(2)}}{\partial \xi} - v_0 \frac{\partial n_-^{(2)}}{\partial \xi} + \frac{\partial}{\partial \xi}(n_-^{(1)} v_-^{(1)}) = 0, \tag{13}$$

$$\frac{\partial n_+^{(1)}}{\partial \tau} + \frac{\partial v_+^{(2)}}{\partial \xi} - v_0 \frac{\partial n_+^{(2)}}{\partial \xi} + \frac{\partial}{\partial \xi}(n_+^{(1)} v_+^{(1)}) = 0, \tag{14}$$

$$\frac{\partial n_d^{(1)}}{\partial \tau} + \frac{\partial v_d^{(2)}}{\partial \xi} - v_0 \frac{\partial n_d^{(2)}}{\partial \xi} + \frac{\partial}{\partial \xi}(n_d^{(1)} v_d^{(1)}) = 0, \tag{15}$$

$$\frac{\partial v_-^{(1)}}{\partial \tau} - v_0 \frac{\partial v_-^{(2)}}{\partial \xi} + v_-^{(1)} \frac{\partial v_-^{(1)}}{\partial \xi} - \alpha_- \frac{z_-}{z_d} \frac{\partial \phi^{(2)}}{\partial \xi} + \frac{3\alpha_- \sigma_-}{z_d} n_-^{(1)} \frac{\partial n_-^{(1)}}{\partial \xi} + \frac{3\alpha_- \sigma_-}{z_d} \frac{\partial n_-^{(2)}}{\partial \xi} = 0, \tag{16}$$

$$\frac{\partial v_+^{(1)}}{\partial \tau} - v_0 \frac{\partial v_+^{(2)}}{\partial \xi} + v_+^{(1)} \frac{\partial v_+^{(1)}}{\partial \xi} + \alpha_+ \frac{z_+}{z_d} \frac{\partial \phi^{(2)}}{\partial \xi} + \frac{3\alpha_+ \sigma_+}{z_d} n_+^{(1)} \frac{\partial n_+^{(1)}}{\partial \xi} + \frac{3\alpha_+ \sigma_+}{z_d} \frac{\partial n_+^{(2)}}{\partial \xi} = 0, \tag{17}$$

$$\frac{\partial v_d^{(1)}}{\partial \tau} - v_0 \frac{\partial v_d^{(2)}}{\partial \xi} + v_d^{(1)} \frac{\partial v_d^{(1)}}{\partial \xi} - \frac{\partial \phi^{(2)}}{\partial \xi} = 0, \tag{18}$$

$$\frac{\partial^2 \phi^{(1)}}{\partial \xi^2} = n_d^{(2)} + \mu_- n_-^{(2)} - \mu_+ n_+^{(2)} + T_1 \phi^{(2)} + T_2 \left(\phi^{(1)}\right)^2, \tag{19}$$

where, $T_1 = \mu_{ce} \frac{1}{\sigma_{ce}} \frac{2\kappa_{ce} - 1}{2\kappa_{ce} - 3} + \mu_{se} \frac{1}{\sigma_{se}} \frac{2\kappa_{se} - 1}{2\kappa_{se} - 3} + \mu_H \frac{1}{\sigma_H} \frac{2\kappa_H - 1}{2\kappa_H - 3}$, and
$T_2 = \frac{\mu_{ce}}{2\sigma_{ce}^2} \frac{4\kappa_{ce}^2 - 1}{(2\kappa_{ce} - 3)^2} + \frac{\mu_{se}}{2\sigma_{se}^2} \frac{4\kappa_{se}^2 - 1}{(2\kappa_{se} - 3)^2} - \frac{\mu_H}{2\sigma_H^2} \frac{4\kappa_H^2 - 1}{(2\kappa_H - 3)^2}$.

Using above set of relations (12)–(19), one can formulate the following KdV equation

$$\frac{\partial \phi^{(1)}}{\partial \tau} + A \phi^{(1)} \frac{\partial \phi^{(1)}}{\partial \xi} + B \frac{\partial^3 \phi^{(1)}}{\partial \xi^3} = 0, \tag{20}$$

where, $A = -N/D$, and $B = 1/D$ with $N = \frac{3}{v_0^4} - \frac{3\mu_- \alpha_-^2 z_-^2 (v_0^2 z_d + \alpha_- \sigma_-)}{(3\alpha_- \sigma_- - v_0^2 z_d)^3} + \frac{3\mu_+ \alpha_+^2 z_+^2 (v_0^2 z_d + \alpha_+ \sigma_+)}{(3\alpha_+ \sigma_+ - v_0^2 z_d)^3} - 2T_2$, and $D = \frac{2}{v_0^3} + \frac{2 v_0 z_d \mu_- \alpha_- z_-}{(3\alpha_- \sigma_- - v_0^2 z_d)^2} + \frac{2 v_0 z_d \mu_+ \alpha_+ z_+}{(3\alpha_+ \sigma_+ - v_0^2 z_d)^2}$.

4 Phase Plane Analysis

To analyze the phase plane of dynamical function corresponding to the KdV Eq. (20), one can take a traveling wave transformation $\chi = \xi - V\tau$, where V is DIAW velocity. Thus, the following dynamical systems is obtained on using transformation χ into the Eq. (20):

$$\begin{cases} \frac{d\phi^{(1)}}{d\chi} = y, \\ \frac{dy}{d\chi} = \frac{V}{B}\phi^{(1)} - \frac{A}{2B}(\phi^{(1)})^2, \end{cases} \tag{21}$$

The above dynamical system depends on superthermal parameters ($\kappa_{H,se,ce}$), charge numbers ($z_{-,+,d}$), number densities ($n_{-,+,H,d,se,ce}$), temperatures ($n_{-,+,H,d,se,ce}$), mass ratios (α_-, α_+), and DIAW velocity (V). To investigate the influence of equilibrium plasma parameters associated with cometary plasma system, one can consider the typical range of parameters observed in the surroundings of various comets. In this work, the following values of parameters relevant to comet Halley [5, 8, 22, 23] are considered: hydrogen ions with temperature $T_H = 8 \times 10^4$ K and number density $n_H = 4.95$ cm^{-3}, solar electron with temperature $T_{se} = 2 \times 10^5$ K and number density same as that of hydrogen ions. Cometary electron with temperature $T_{ce} = 2 \times 10^4$ K and the negatively charged dust particles with temperature $T_d = 2 \times 10^3$ K, number density $n_{d0} = 10^{-5}$ cm^{-3}, and charge number $Z_{d0} = 10^3$. The temperatures of the negative and positive Oxygen ions are taken as $T_- = T_+ = 1.16 \times 10^4$ K with number densities $n_{-0} = 0.05$ cm^{-3}, $n_{+0} = 0.5$ cm^{-3} and charges $z_- = 1$, and $z_+ = 2$, respectively. The masses of dust particles, ion-pair of Oxygen are considered such that $\alpha_- = \alpha_+ = 10^3$.

The phase plane of any dynamical system determines the instantaneous behavior of its trajectories for a set of initial conditions. The stability or instability of the equilibrium points predicts the nature of nearby trajectories. Using the concept of phase plane analysis [24, 25], one can investigate the phase portraits of any nonlinear dynamical system which may vary consequentially on numbers of fixed points and separatrix layers [26]. To distinguish trajectories in the phase portrait, following representations are used: $NPT_{m,n}$, and $NHT_{m,n}$ for nonlinear periodic, and homoclinic trajectories, respectively, where m indicates number of stable center points and n denotes number of separatrix layers. Every trajectory in phase portrait corresponds to a traveling wave solution for the considered plasma system. For example, periodic, and homoclinic trajectories of the dynamical system correspond to periodic and solitonic wave solutions of the plasma system, respectively. In addition to the trajectories, one can also check the conservative nature from phase plane of a dynamical system [27]. Figure 1 shows the plots of potential energy function, and phase plane corresponding to Eq. (20) for $\kappa_{se} = \kappa_{ce} = \kappa_H = 2$, $n_{-0} = 0.05$ cm^{-3}, $n_{+0} = 0.5$ cm^{-3}, $n_H = n_{se0} = 4.95$, $n_{d0} = 10^{-5}$ cm^{-3}, $T_- = T_+ = 1.16 \times 10^4$ K, $T_H = 8 \times 10^4$ K, $T_{se} = 2 \times 10^5$ K, $T_{ce} = 2 \times 10^4$ K, $T_d = 2 \times 10^3$ K, $\alpha_- = \alpha_+ = 10^3$, $z_- = 1$, $z_+ = 2$, $Z_{d0} = 10^3$, and $V = 0.1$. Figure 1a illustrates the graph of potential function ψ against $\phi^{(1)}$. Here, local minima in potential plots signify the existence of solitary solutions and the potential dip at $\phi^{(1)} > 0$ corresponds to a compressive solitonic solution whereas the potential dip at $\phi^{(1)} < 0$ signify a rarefactive solitonic solution. Therefore, for the condition $A > 0$, one can obtain only compressive DIAW solution of Eq. (20). Figure 1b displays the phase portrait of the system (21) for the specified set of parameters and shows the existence of two fixed points, namely, saddle point at $E_0(0, 0)$ and center at $E_1(\frac{2V}{A}, 0)$. There also exist two distinct types of phase trajectories, i.e., homoclinic trajectory ($NHT_{1,0}$) and a family of nonlin-

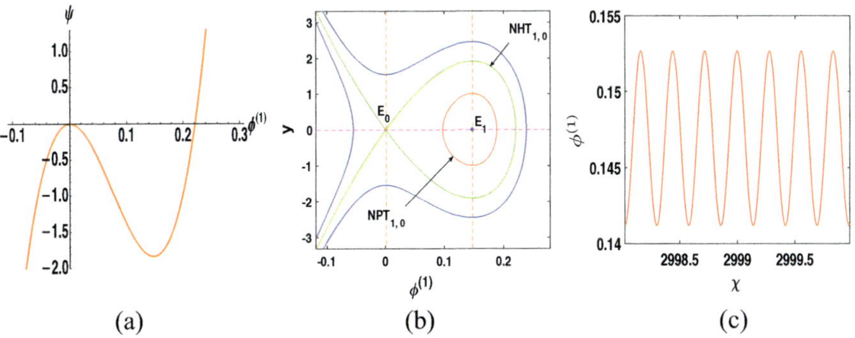

Fig. 1 **a** Potential graph, **b** Phase portrait, and **c** Periodic wave solution of system (21) for $\kappa_{se} = \kappa_{ce} = \kappa_H = 2$, $n_{-0} = 0.05\ \text{cm}^{-3}$, $n_{+0} = 0.5\ \text{cm}^{-3}$, $n_H = n_{se0} = 4.95\ \text{cm}^{-3}$, $n_{d0} = 10^{-5}\ \text{cm}^{-3}$, $T_- = T_+ = 1.16 \times 10^4$ K, $T_H = 8 \times 10^4$ K, $T_{se} = 2 \times 10^5$ K, $T_{ce} = 2 \times 10^4$ K, $T_d = 2 \times 10^3$ K, $\alpha_- = \alpha_+ = 10^3$, $z_- = 1$, $z_+ = 2$, $Z_{d0} = 10^3$, and $V = 0.1$

ear periodic trajectories ($NPT_{1,0}$). The periodic wave solution for initial condition $(0.13, 0)$, corresponding to a periodic trajectory presented in Fig. 1b is displayed in Fig. 1c.

4.1 Solitary Wave Solutions

Considering the transformation $\chi = \xi - V\tau$ and using boundary conditions $\phi^{(1)} \to 0$ and $\frac{d\phi^{(1)}}{d\chi} \to 0$ as $\chi \to \pm\infty$, one can derive compressive solitonic wave solution of equation (20) analogous to the homoclinic trajectory at fixed point E_0 and enclosing point E_1 as

$$\phi^{(1)} = \frac{3V}{A} sech^2\left(\sqrt{\frac{V}{4B}}\chi\right), \tag{22}$$

with the electric field $E = -\nabla\phi^{(1)}$, i.e.,

$$E = \frac{3V}{A}\sqrt{\frac{V}{B}}\, sech^2\left(\sqrt{\frac{V}{4B}}\chi\right) \tanh\left(\sqrt{\frac{V}{4B}}\chi\right). \tag{23}$$

The effects of plasma parameters on the basic properties (amplitude and width) of compressive solitary wave solution of system (20) and its associated electric field are depicted in Figs. 2 and 3, respectively. Both the figures are obtained by varying the mentioned parameters and keeping the other parameter fixed as $\kappa_{se} = \kappa_{ce} = \kappa_H = 2$, $n_{-0} = 0.05\ \text{cm}^{-3}$, $n_{+0} = 0.5\ \text{cm}^{-3}$, $n_H = n_{se0} = 4.95$, $n_{d0} = 10^{-5}$ cm^{-3}, $T_- = T_+ = 1.16 \times 10^4$ K, $T_H = 8 \times 10^4$ K, $T_{se} = 2 \times 10^5$ K, $T_{ce} = 2 \times 10^4$

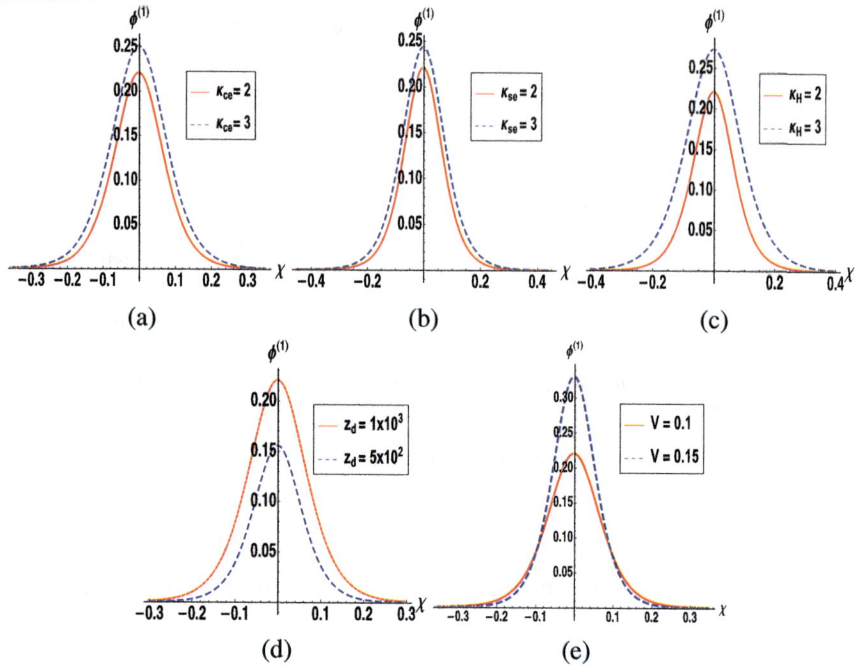

Fig. 2 Dependency of DIA solitary wave solution of equation (3) on mentioned plasma parameters

K, $T_d = 2 \times 10^3$ K, $\alpha_- = \alpha_+ = 10^3$, $z_- = 1$, $z_+ = 2$, $Z_{d0} = 10^3$, and $V = 0.1$. It can be inferred from Fig. 2 that when superthermal plasma particles move far away from the Maxwellian then both amplitude and width of DIA compressive solitary wave also deplete. Now, growing value of dust charge number (z_d) amplifies and broadens the solitary wave. Furthermore, the DIA solitary waves amplify and become narrow as the wave moves towards supersonic region. From Fig. 3, it can be noted that κ_{ce} influences electric field differently as compared to κ_{se} and κ_H. Both amplitude and width of electric field associated with solitary wave solution of the KdV equation (20) enhance with an increase in the value of κ_{ce} i.e., decrease in superthermality of cometary electrons whereas superthermality of solar electrons and hydrogen ions are responsible in the reduction of the amplitude of electric field. The influence of dust charge number and wave velocity (V) is similar to that of the DIA compressive solitary wave.

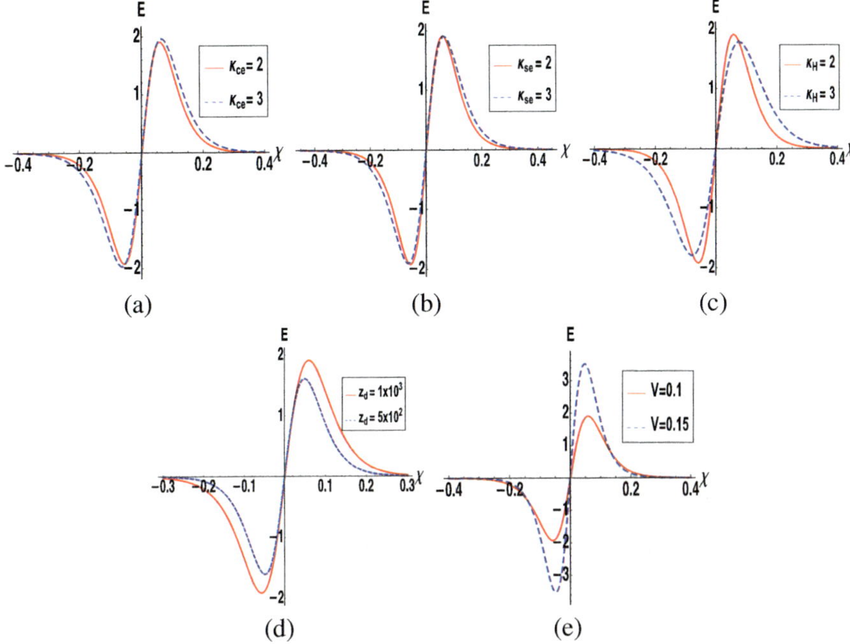

Fig. 3 Effect of plasma parameters on electric field associated with compressive solitary wave solution of Eq. (3)

5 Conclusion

The DIAWs are studied in a six-component cometary tail plasma constituting Oxygen ion-pair, negatively charged dusts with superthermal light hydrogen ions, solar electrons, and cometary electrons. Phase plane analysis of DIAWs in the framework of the KdV equation is carried out. Typical set of plasma parameters observed in the tail region of comet Halley [5, 8, 22, 23] are considered for numerical simulation. The main results of our work are mentioned below:

1. Phase plane corresponding to KdV Eq. (20) only defines a homoclinic orbit on the positive phase of wave potential, which signifies the existence of a compressive solitary wave in the tail region of comet Halley.
2. On analysis of analytical wave solutions of Eq. (20), it is predicted that the DIA solitary wave amplifies when plasma particles tends towards the Maxwellian. On the contrary, electric field enhances with decrease in superthermality of cometary electrons whereas it grows with increase in superthermality of hydrogen ions and solar electrons.
3. The increasing values of dust charge number amplify the DIA solitary wave and its associated electric field. Also, the soliton and its associated electric field amplify and become narrow as the velocity of traveling wave increases.

Acknowledgements The first author is obliged to the Sikkim Manipal Institute of Technology and Sikkim Manipal University for providing research fellowship under TMA Pai University Research Fund (Ref. No. 118/SMU/REG/ UOO/104/2019).

References

1. Rao, N.N., Shukla, P.K., Yu, M.Y.: Dust-acoustic waves in dusty plasmas. Planet. Space Sci. **38**, 543–546 (1990)
2. Shukla, P.K., Silin, V.P.: Dust ion-acoustic wave. Phys. Scripta **45**, 508 (1992)
3. Melandso, F.: Lattice waves in dust plasma crystals. Phys. Plasmas **3**, 3890–3901 (1996)
4. Abulwafa, E.M., AtallaElhanbaly, A.A.M., Bedeir, A.M.: Linear and nonlinear study of six-component dusty cometary plasma. In: Proceedings of the 11th Conference on Nuclear and Particle Physics, vol. 20, p. 23 (2019)
5. Bedeir, A.M., Abulwafa, E.M., Elhanbaly, A.M., Mahmoud, A.A.: A fully nonlinear solitary wave in six-component dusty cometary plasma. Phys. Scripta **96**(9), 095603 (2021)
6. Venugopal, C., Neethu, T.W.: Dust acoustic shock waves in a six component charge varying cometary plasma. J. Phys. Adv. App. **1**(1), 1–12 (2020)
7. Rubin, M., Hansen, K.C., Gombosi, T.I., Combi, M.R., Altwegg, K., Balsiger, H.: Ion composition and chemistry in the coma of Comet 1P/Halley-A comparison between Giotto's ion mass spectrometer and our ion-chemical network. Icarus **199**, 505–519 (2009)
8. Brinca, A.L., Tsurutani, B.T.: Unusual characteristics of electromagnetic waves excited by cometary newborn ions with large perpendicular energies. Astron. Astrophys. **187**, 311–319 (1987)
9. Chaizy, P., Reme, H., Sauvaud, J.A., d'Uston, C., Lin, R.P., Larson, D.E., Mitchell, D.L., Anderson, K.A., Carlson, C.W., Korth, A., Mendis, D.A.: Negative ions in the coma of comet Halley. Nature **349**, 393–396 (1991)
10. Madanian, H., Cravens, T.E., Rahmati, A., Goldstein, R., Burch, J.: Suprathermal electrons near the nucleus of comet 67P/Churyumov-Gerasimenko at 3 AU: model comparison with Rosetta data. J. Geophys. Res. Space Phys. **121**, 5815 (2016)
11. Willington, N.T., Varghese, A., Saritha, A.C., Philip, N.S., Venugopal, C.: Ion acoustic shock waves with drifting ions in a five component cometary plasma. Adv. Space Res. **68**, 4292–4302 (2021)
12. Saini, N.S., Kourakis, I., Hellberg, M.A.: Arbitrary amplitude ion-acoustic solitary excitations in the presence of excess superthermal electrons. Phys. Plasmas **16**, 62903 (2009)
13. Sarkar, J., Chandra, S., Goswami, J., Ghosh, B.: Formation of solitary structures and envelope solitons in electron acoustic wave in inner magnetosphere plasma with suprathermal ions. Contrib. Plasma Phys. **60**(7), e201900202 (2020)
14. Goswami, J., Sarkar, J., Chandra, S., Ghosh, B.: Amplitude-modulated electron-acoustic waves with bipolar ions and kappa-distributed positrons and warm electrons. Pramana J. Phys. **95**, 54 (2021)
15. Samanta, U.K., Saha, A., Chatterjee, P.: Bifurcations of dust ion acoustic travelling waves in a magnetized dusty plasma with a q-nonextensive electron velocity distribution. Phys. Plasmas **20**, 022111 (2013)
16. Selim, M.M., El-Depsy, A., El-Shamy, E.F.: Bifurcations of nonlinear ion-acoustic travelling waves in a multicomponent magnetoplasma with superthermal electrons. Astrophys. Space Sci. **360**, 66 (2015)
17. El-Monier, S.Y., Atteya, A.: Bifurcation analysis for dust-acoustic waves in a four-component plasma including warm ions. IEEE Trans. Plasma Sci. **46**, 815–824 (2018)
18. Rahim, Z., Adnan, M., Qamar, A., Saha, A.: Nonplanar dust-acoustic waves and chaotic motions in Thomas Fermi dusty plasmas. Phys. Plasmas **25**, 083706 (2018)

19. Tamang, J., Saha, A.: Bifurcations of small-amplitude supernonlinear waves of the mKdV and modified Gardner equations in a three-component electron-ion plasma. Phys. Plasmas **27**(1), 012105 (2020)
20. Prasad P.K., Saha A.: Dynamical behavior of ion-acoustic periodic and solitary structures in magnetized solar wind plasma. In: Giri, D., Buyya, R., Ponnusamy, S., De D., Adamatzky, A., Abawajy, J.H. (eds.) Proceedings of the Sixth International Conference on Mathematics and Computing. Advances in Intelligent Systems and Computing, vol. 1262, pp. 419–428. Springer, Singapore (2021). https://doi.org/10.1007/978-981-15-8061-1_33
21. Atteya, A., El-Borie, M., Roston, G., El-Helbawy, A., Prasad, P., Saha, A.: Ion-acoustic stable oscillations, solitary, periodic and shock waves in a quantum magnetized electron-positron-ion plasma. Zeitschrift für Naturforschung A **76**(9), 757–768 (2021)
22. Michael, M., Willington, N.T., Jayakumar, N., Sebastian, S., Sreekala, G., Venugopal, C.: Korteweg-deVries-Burgers (KdVB) equation in a five component cometary plasma with kappa described electrons and ions. J. Theor. Appl. Phys. **10**, 289–296 (2016)
23. Neethu, T.W., Shilpa, S., Saritha, A.C., Philip, N.S., Venugopal, C.: Dust acoustic solitary waves in a five component cometary plasma with dust charge variation. Radiat. Effects Defects Solids **176**, 284–299 (2021)
24. Guckenheimer, J., Holmes, P.J.: Nonlinear Oscillations Dynamical Systems and Bifurcations of Vector Fields. Springer, New York (1983)
25. Chow, S.N., Hale, J.K.: Methods of Bifurcation Theory. Springer, New York (1981)
26. Dubinov, A.E., Kolotkov, D.Y., Sazonkin, M.A.: Nonlinear theory of ion-sound waves in a dusty electron-positron-ion plasma. Tech. Phys. **57**, 585–593 (2012)
27. Strogatz, S.H.: Nonlinear dynamics and chaos: with applications to physics, biology, chemistry, and engineering, 1st Indian edn. CRC Press (2007)

Cylindrical and Spherical Ion-Acoustic Shock and Solitary Waves in a Nonplanar Hybrid q-nonextensive Nonthermal Plasma

Subrata Roy, Santanu Raut, and Rishi Raj Kairi

Abstract In this article we investigate the propagating properties of ion-acoustic wave (IAW) in a plasma comprising positively charged ions and electrons abiding by hybrid q-non-extensive non-thermal velocity distribution equation. The nonplanar KdV-Burger (NKDVB) equation is derived from the basic governing equation. Considering the impact from the ion streaming velocity, inter-particle collisions, and viscosity, a Burgers term is introduced in the present system and using Weighted Residual Method (WRM) and Simplified Hirota bilinear method (SHBM) progressive solitary wave solution and shock wave are derived. Finally, the effect of different physical parameters on solitary and shock wave on the propagation of IAW in the present plasma environment is noticed.

Keywords Ion-acoustic wave · Hybrid q-nonextensive non-thermal distribution · Nonplanar kdV-Burgers equation · Solitary and shock wave

1 Introduction

Investigation on plasma environment becomes an interesting topic in recent times as it is ubiquitously found in space environments as well as laboratory experiments. Sagdeev [1] observed theoretically these types of waves by considering a mechanical analogy whereas, Ikezi [2] in the year 1970 studied the same with an experimental set-up. Subsequently, several authors paid their attention to observe the various types of plasma models. Shukla and Silin [3] investigated theoretically dust ion acoustic wave (DIAW) in a dust plasma and Barkan et al. observe the same experimentally in [4]. Angelo [5] observed high-frequency and low-frequency acoustic waves in a

S. Roy (✉) · R. R. Kairi
Department of Mathematics, Cooch Behar Panchanan Barma University, Cooch Behar 736101, India
e-mail: send2sroy94@gmail.com

S. Raut
Department of Mathematics, Mathabhanga College, Coochbehar 736146, India

S. Banerjee and A. Saha (eds.), *Nonlinear Dynamics and Applications*,
Springer Proceedings in Complexity,
https://doi.org/10.1007/978-3-030-99792-2_11

plasma environment in an experimental setup. Dubuloz et al. [6] observed the solitary IAW as well as DIAW in a plasma medium utilizing the KdV model as well as modified KdV model. It is unfortunate that most of the observation on IAW in plasma systems are considered in linear space only. But, in many cases, the reality in the laboratory, as well as the space environment, is totally different. Sometimes, the physical plasma environment becomes finite and the waves are confined to move in a bounded state. Such type of nonplanar geometrical space arises in Capsule implosions, Supernova explosions, Saturn's magnetosphere, etc. In the year 1974, Maxon and Viecelli [7] investigated the characteristic of IAW in nonplanar geometrical space and found that the nonplanar wave moves faster than the planar counterparts. Gao et al. [8] observed the propagating behaviors of nonplanar DIAW in unmagnetized dusty plasma through the KdV-Burgers model. Further, several authors performed various theoretical and experimental works to observe the behaviors of IAW in different nonplanar plasma systems [9, 10]. Recently, Demiray [11] observed the IAW propagating in a nonplanar plasma system comprising positively charged cold ions and electrons satisfying hybrid q-nonextensive nonthermal velocity distribution. Some experimental on plasma circumstances indicates that the characteristic of wave propagation in a dissipative system significantly depends on interparticle collisions and viscosity etc. [12]. To consider the impact from viscosity a Burgers term is added to Demiray's observation [11] which may produce solitary (in case of the negligible effect of dissipation) as well as shock (in case of strong dissipation) type wave features. The SHBM is employed to find an approximate analytical solution that produces a shock wave. In the present observation, we analyze the significant impact of different physical parameters as well as time parameter (τ) on shock and solitary IAW through the NKDVB framework. The work of this manuscript is formulated as follows: In Sect. 2, the governing equations for the system are stated and the NKDVB equation is derived employing RPM. In Sect. 3 NKDVB model is derived by using RPM. Sect. 4 presents an approximate analytical solitary wave solution for the said equation. In Sect. 5, a shock solution is generated by employing Hirota's bi-linear approach. In Sect. 6, analyzes the numerical structure of the solutions. Section 7 concludes the investigation.

2 Problem Formulation

We consider here a unmagnetized viscous plasma comprising positively charged cold ions along with q-nonextensive nonthermal velocity distributed electrons. The dynamics of the system is described by the normalized basic governing equations:

$$\frac{\partial n}{\partial t} + \nabla.(n\mathbf{u}) = 0 \tag{1}$$

$$\frac{\partial \mathbf{u}}{\partial t} + (\mathbf{u}.\nabla)\mathbf{u} = -\nabla\psi + \delta\nabla^2\mathbf{u} \tag{2}$$

$$\nabla^2\psi = n_e - n \tag{3}$$

where n and n_e denote normalized number densities of cold ion and electron, $\mathbf{u}$ represents velocity of the ion fluid whereas, the electrostatic potential is noted by ψ and δ is the viscocity coefficient. Density of normalized q-nonextensive nonthermal electron is given by

$$n_e = [1 + (q-1)\psi]^{\frac{q+1}{2(q-1)}}(1 + L_1\psi + L_2\psi^2) \tag{4}$$

The coefficients L_1 and L_2 are as

$$L_1 = -\frac{16q\alpha}{(3 - 14q + 15q^2 + 12\alpha)}, \quad L_2 = -(2q-1)L_1. \tag{5}$$

where the parameter α determines nonthermal electrons numbers in the system. The parametric zones of (q, α) and their validity for solitary wave solutions are described by Williams et al. [13]. In the extensive limiting case $(q \to 1)$ and $\alpha = 0$, reduces the distribution to the Maxwell-Boltzmann velocity distribution whereas, it reduces to Cairn distribution [14] for the case $(q \to 1)$ and $\alpha \neq 0$. We consider the field Eqs. (1)–(3) in the following form,

$$\frac{\partial n}{\partial t} + \frac{1}{R^\mu}\frac{\partial(R^\mu n u)}{\partial R} = 0, \tag{6a}$$

$$\frac{\partial u}{\partial t} + u\frac{\partial u}{\partial R} = -\frac{\partial \psi}{\partial R} + \delta\left[\frac{1}{R^\mu}\frac{\partial}{\partial R}\left(R^\mu \frac{\partial u}{\partial R}\right) - \frac{\mu u}{R^2}\right], \tag{6b}$$

$$\frac{1}{R^\mu}\frac{\partial}{\partial R}\left(R^\mu \frac{\partial \psi}{\partial R}\right) = n_e + n \tag{6c}$$

Here the parameter $\mu = 1, 2$ signify the motion of cylindrical and spherical IAWs whereas, the dynamics of planar IAW is represented by $\mu = 0$. For small values of ψ, the electron number density n_e expressed in Eq. (4) can be presented in a power series as,

$$n_e = 1 + p_1\psi + p_2\psi^2 + p_3\psi^3 + \cdots \tag{7}$$

Here the coefficients p_1, p_2 and p_3, are taken as

$$p_1 = L_1 + \frac{q+1}{2}, \quad p_2 = L_2 + L_1\left(\frac{q+1}{2}\right) + \frac{(q+1)(3-q)}{2} \tag{8}$$

$$p_3 = \frac{(q+1)(3-q)(5-3q)}{48} + L_1\frac{(q+1)(3-q)}{8} + L_2\frac{(q+1)}{2} \tag{9}$$

3 Formation of Nonplaner KdV Burgers Equation

To derive the NKDVB equations the standard RPM [15] will be utilized. In order to derive the depending variables n, u, and ψ are stretched in a series of ϵ as [16],

$$\begin{aligned} n &= 1 + \epsilon n_1 + \epsilon^2 n_2 + \epsilon^3 n_3 + \cdots \\ u &= 0 + \epsilon u_1 + \epsilon^2 u_2 + \epsilon^3 u_3 + \cdots \\ \psi &= 0 + \epsilon \psi_1 + \epsilon^2 \psi_2 + \epsilon^3 \psi_3 + \cdots \end{aligned} \tag{10}$$

Now, the new stretched coordinates are taken as,

$$\xi = \epsilon^{1/2}(R - v_p t), \ \tau = \epsilon^{3/2} t. \tag{11}$$

We assume a weak damping in the viscous plasma and write

$$\delta \approx \epsilon^{\frac{1}{2}} \delta_0 \tag{12}$$

Substituting the expression in Eq. (10) along with the coordinates (11) into the (6a)–(6c) and equating the coefficients of different order of ϵ we obtain,

$$n_1 v_p = u_1 \tag{13}$$

$$\frac{\partial n_1}{\partial \tau} - v_p \frac{\partial n_2}{\partial \xi} + \frac{\mu u_1}{v_p \tau} - \frac{\partial}{\partial \tau}(n_1 u_1 + u_2) = 0 \tag{14}$$

$$v_p u_1 = \psi_1 \tag{15}$$

$$\frac{\partial u_1}{\partial \tau} - v_p \frac{\partial u_2}{\partial \xi} - u_1 \frac{\partial u_1}{\partial \xi} + \frac{\partial \psi_1}{\partial \xi} - \delta_0 \frac{\partial^2 u_1}{\partial \xi^2} = 0 \tag{16}$$

$$n_1 = p_1 \psi_1 \tag{17}$$

$$\frac{\partial^2 \psi_1}{\partial \xi^2} - p_1 \psi_2 - p_2 \psi_1^2 - n_2 = 0 \tag{18}$$

Using the results (13), (15) and (17), we find $v_p = \frac{1}{p_1^{1/2}}$ which signifies the phase velocity of the perturbation mode and the parameter ϵ measures the weakness and the dispersion of the perturbation. Using the results (13)–(18) stated above and setting $\psi_1 = \psi$ we obtain the NKDVB equation as

$$\frac{\partial \psi}{\partial \tau} + P\psi \frac{\partial \psi}{\partial \xi} + Q \frac{\partial^3 \psi}{\partial \xi^3} + R \frac{\partial^2 \psi}{\partial \xi^2} + S\psi = 0, \tag{19}$$

where $P = \frac{3}{2}p_1^{1/2} - \frac{p_2}{p_1^{3/2}}$, $Q = \frac{1}{2p_1^{3/2}}$, $R = -\frac{\delta_0}{2}$, $S = \mu/2\tau$. Here, the nonlinear coefficient P is a function of the parameters α and q. The solitary type wave becomes compressive when $P > 0$, whereas, it becomes rarefactive for $P < 0$.

4 Progressive Wave Solution for Nonplanar KdV-Burgers (NKDVB) Equation

Here, we apply the WRM to obtain an approximate analytical solution for the NKDVB equation. In order to employ WRM [10, 11] we utilize the solution of the standard KdV equation. In consideration of negligible small values of δ_0 and large time τ the nonplanar KdV-Bugers Eq. (19) is written as

$$\frac{\partial \psi}{\partial \tau} + P\psi\frac{\partial \psi}{\partial \xi} + Q\frac{\partial^3 \psi}{\partial \xi^3} = 0 \tag{20}$$

which permits the solution

$$\psi = \lambda_0 \mathrm{sech}^2\eta_0, \quad \eta_0 = w(\xi - V_0\tau) \tag{21}$$

with

$$w^2 = \frac{P\lambda_0}{12Q}, \quad V_0 = \frac{P\lambda_0}{3} \tag{22}$$

Here λ_0 presents the constant amplitude of the soliton. Now we assume that amplitude and width of the soliton will be time dependent due to the application of Burgers term and geometric term. Thus, the progressive wave solution of Eq. (21) is proposed as

$$\psi = \lambda(\tau)\mathrm{sech}^2\eta, \quad \eta = w(\xi - V(\tau)) \tag{23}$$

with

$$w^2(\tau) = \frac{P\lambda(\tau)}{12Q}, \quad V'(\tau) = \frac{P\lambda(\tau)}{3} \tag{24}$$

Here the prime stands for presenting differentiation of $V(\tau)$ against τ. Actually, the solution presented in (23) is the same as the solution appeared in (22) except the presence of $\lambda(\tau)$ in (24). However, $\lambda(\tau)$ is to be determined for the later. It is found that though the solution (23) satisfies the Eq. (19) rather a residue term $\mathcal{R}(\eta, \tau)$ presents there as,

$$\mathcal{R}(\eta, \tau) = \left[\lambda' + 4R\lambda Q^2 + S\lambda - \frac{2\lambda w'\eta}{w}\tanh\eta\right]\mathrm{sech}^2\eta - 6Rw^2\lambda\mathrm{sech}^4\eta. \tag{25}$$

Here, the term $\mathcal{R}(\eta, \tau)$ is an even function of η. To obtain a differential equation for finding $\lambda(\tau)$, a strong restriction on $\lambda(\tau)$ is imposed by introducing a weighted function. Here, we choose $sech^2\eta$ as a weighted function. We multiply Eq. (25) by $sech^2\eta$, and integrate it from $\eta = -\infty$ to $\eta = \infty$ and lastly put the result to zero,

$$\lambda' + S\lambda - \frac{w'\lambda}{2w} - \frac{4Rw^2\lambda}{5} = 0. \tag{26}$$

We eliminate w between the Eqs. (24) and (26) and get,

$$\lambda' + \frac{4}{3}S\lambda = \frac{4PR}{45Q}\lambda^2. \tag{27}$$

The solution of Eq. (27) becomes

$$\lambda(\tau) = \left(M\tau^{\frac{2\mu}{3}} + \frac{4PR\tau}{(\frac{2\mu}{3} - 1)45Q}\right)^{-1}, \tag{28}$$

where M is a simple integration constant and λ_0 ($\tau \to \tau_0$, $\lambda(\tau) \to \lambda_0$) is the initial amplitude satisfying Eq. (22). Thus, $\lambda(\tau)$ is written as

$$\lambda(\tau) = \lambda_0 \left(\frac{\tau_0}{\tau}\right)^{\frac{2\mu}{3}} \left(1 + \frac{4PR\lambda_0\tau_0}{(\frac{2\mu}{3} - 1)45Q}\left[\left(\frac{\tau_0}{\tau}\right)^{\frac{2\mu}{3}-1} - 1\right]\right)^{-1}. \tag{29}$$

Using (29) in the expression $V'(\tau)$ and integrating we get

$$V(\tau) = \frac{P\lambda_0}{3} - \frac{45Q}{4R} ln\left[\left(\frac{\tau_0}{\tau}\right)^{\frac{2\mu}{3}}\left(\frac{\lambda_0}{\lambda}\right)\right], \tag{30}$$

Using the result (24) we find

$$w(\tau) = \sqrt{\frac{P\lambda_0}{12Q}}\left(\frac{\tau_0}{\tau}\right)^{\frac{\mu}{3}} \left(1 + \frac{4PR\lambda_0\tau_0}{(\frac{2\mu}{3} - 1)45Q}\left[\left(\frac{\tau_0}{\tau}\right)^{\frac{2\mu}{3}-1} - 1\right]\right)^{-\frac{1}{2}}. \tag{31}$$

Thus, the final solution of equation is

$$\psi(\xi, \tau) = \lambda(\tau)sech^2[w(\tau)(\xi - V(\tau))], \tag{32}$$

where the results (29)–(31) provide the values of $\lambda(\tau)$, $V(\tau)$ and $w(\tau)$ respectively.

This analytical solution (32) can be reduced to the planar solution of Eq. (19) for $\mu = 0$ whereas, the cylindrical and spherical wave solution are given by ($\mu = 1$) and ($\mu = 2$) respectively.

5 Shock Type Wave Solution for Nonplanar KdV-Burgers (NKDVB) Equation

The strong anomalous dissipation may cause to form shock type wave solution. Now, to explore shock type wave solution, we utilize SHBM [17] and rewrite Eq. (19) in the following form:

$$\frac{\partial\psi}{\partial\tau} + P(\tau)\psi\frac{\partial\psi}{\partial\xi} + Q(\tau)\frac{\partial^3\psi}{\partial\xi^3} + R(\tau)\frac{\partial^2\psi}{\partial\xi^2} + S(\tau)\psi = 0. \tag{33}$$

To find a single kink solution, We first use the transformation

$$\psi(\xi,\tau) = e^{\theta}, \quad \text{where } \theta = k\xi - \omega(\tau). \tag{34}$$

We derive the dispersion relation as

$$\omega(\tau) = \int_{\tau_0}^{\tau} (Q(\tau)k^3 + R(\tau)k^2 + S(\tau))d\tau, \tag{35}$$

The shock solution of Eq. (33) is taken as $\psi = A(ln(f))_{\xi}$ where $f(\xi,\tau)$, is determined as

$$f(\xi,\tau) = 1 + e^{\theta} = 1 + e^{k\xi-\omega(\tau)}, \tag{36}$$

Utilizing Eqs. (35) and (36) we find,

$$\psi = A\frac{ke^{\theta}}{1+e^{\theta}}. \tag{37}$$

Substituting Eq. (37) into Eq. (33) we obtain a polynomial equation for $e^{n\theta}$. Putting the coefficient of $e^{n\theta}$, to zero, a system of algebraic equation is obtained. Solving these algebraic equation we get

$$A = \frac{2k^3R(\tau) - S(\tau)}{P(\tau)k^2}. \tag{38}$$

Combining Eqs. (37) and (38) we find the singleton shock solution as,

$$\psi(\theta) = \frac{2k^3R(\tau) - S(\tau)}{2P(\tau)k}\left(1 + \tanh\left(\frac{\theta}{2}\right)\right) \tag{39}$$

$$\text{where } \theta = k\xi - \int_{\tau_0}^{\tau} (Q(\tau)k^3 + R(\tau)k^2 + S(\tau))d\tau.$$

Putting the expression as $P(\tau) = P$, $Q(\tau) = Q$, $R(\tau) = R$, and $S(\tau) = \frac{\mu}{2\tau}$ in Eq. (33), we find the standard non planer KdV-Burgers equation and the shock solution of Eq. (19) can be written as,

$$\psi(\xi, \tau) = \frac{4k^3 R - \frac{\mu}{\tau}}{4Pk}\left(1 + \tanh\left(\frac{\theta}{2}\right)\right), \tag{40}$$

$$\text{where } \theta = k\xi - \left[(Qk^3 + Rk^2)(\tau - \tau_0) + \frac{\mu}{2}\, ln\left(\frac{\tau}{\tau_0}\right)\right].$$

6 Results and Discussion

To detect a suitable parametric domain, the variation of the nonlinear coefficient P and the dispersion coefficient Q due to the change in the parameter α and q are shown in Figs. 1a–d. Figure 2a exhibits the variation of amplitude in planar and nonplanar wave profiles and it is found that the planar wave moves over the cylindrical wave and the spherical wave. Figures 2b, c show the evolution of cylindrical and spherical wave respectively for different time intervals. The significant impact of the Burgers term follows from Fig. 2d. Enhancing δ_0 causes for losing of potential energy of the system and naturally the soliton have negligible backward shifting with declining amplitude. The impacts of the nonextensive parameters q on wave propagation depicted in Fig. 2e. It is found that the wave gets steepened with enhancing α and gets flattened due to the rise in nonthermal electron numbers (see Fig. 2f). For a clear vision of wave propagation, the three-dimensional profiles of the cylindrical and the spherical wave are drawn in Fig. 3a, b. The significant impact of Burgers term in diminishing amplitude is also understood from Fig. 3c. Figures 4a and 5a exhibit the evolution of kink type soliton for different times in cylindrical geometry. Figures 4b and 5b shows that the increasing δ_0 leads the soliton to move dipper. The significant effect of the nonextensive parameter q is shown in Figs. 4c and 5c. It is important to note

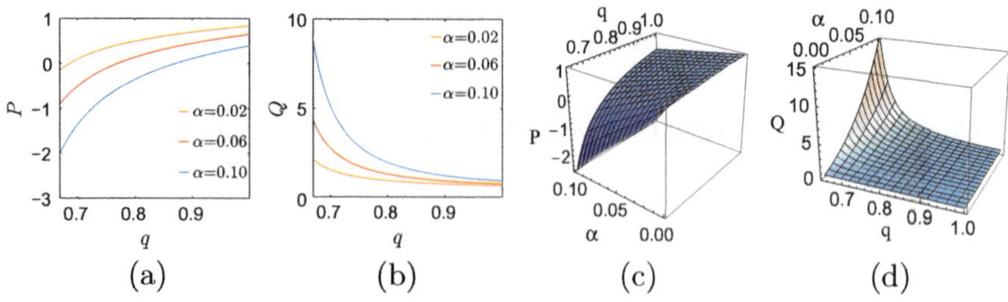

Fig. 1 **a** Variation of nonlinearity coefficient P with q for various values of α, **b** Variation of dispersion coefficient Q with q for various values of α. **c** 3D Variation of nonlinearity coefficient P with q and α, **d** 3D Variation of dispersion coefficient Q with q and α

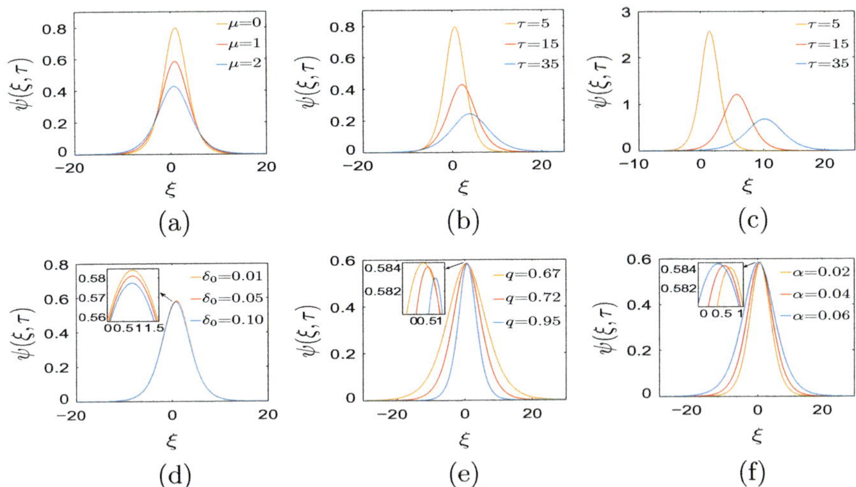

Fig. 2 Profiles of $\psi(\xi, \tau)$ versus ξ of Solution (32), **a** when $q = 0.9$, $\alpha = 0.01$, $\tau_0 = 5$, $\delta_0 = 0.01$, $\tau = 8$, $\lambda_0 = 0.8$, **b** when $q = 0.95$, $\alpha = 0.01$, $\tau_0 = 5$, $\delta_0 = 0.02$, $\mu = 1$, $\lambda_0 = 0.9$, **c** when $q = 0.9$, $\mu = 1$, $\tau_0 = 5$, $\alpha = 0.01$, $\tau = 8$, $\lambda_0 = 0.8$, **d** when $q = 0.9$, $\alpha = 0.02$, $\tau_0 = 4$, $\delta_0 = 0.02$, $\mu = 1$, $\lambda_0 = 3$, **e** $\delta_0 = 0.02$, $\mu = 1$, $\tau_0 = 5$, $\alpha = 0.01$, $\tau = 8$, $\lambda_0 = 0.8$, **f** when $q = 0.9$, $\mu = 1$, $\tau_0 = 5$, $\delta_0 = 0.01$, $\tau = 8$, $\lambda_0 = 0.8$

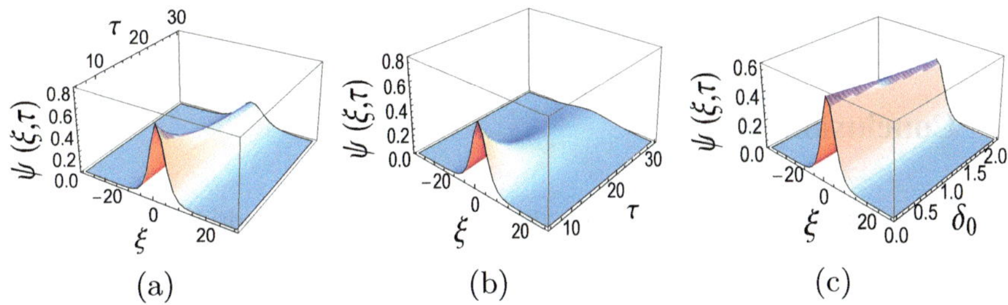

Fig. 3 3D Profiles of Solution (32), **a** when $\mu = 1$, , $q = 0.9$, $\alpha = 0.01$, $\tau_0 = 5$, $\delta_0 = 0.01$, $\tau = 8$, $\lambda_0 = 0.8$, **b** when $\mu = 2$, $q = 0.9$, $\alpha = 0.01$, $\tau_0 = 5$, $\delta_0 = 0.01$, $\tau = 8$, $\lambda_0 = 0.8$, **c** when $\mu = 1$, $q = 0.9$, $\alpha = 0.01$, $\tau_0 = 5$, $\tau = 8$, $\lambda_0 = 0.8$

that there is a critical point of the parameter q for which anti-kink type shocks are found below the values of critical point and oppositely kink type shocks exist for the higher values of q.

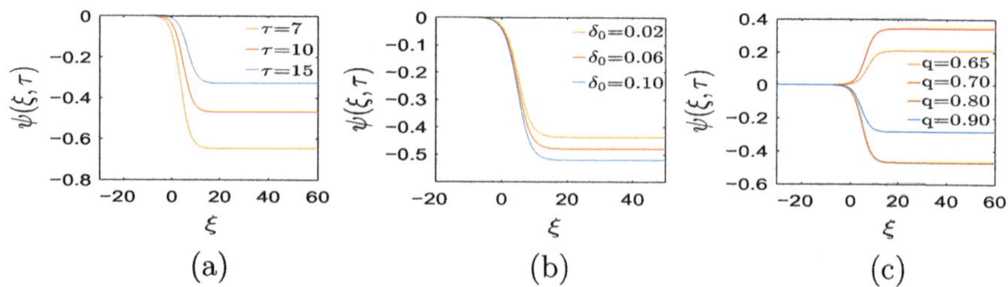

Fig. 4 2D Profiles of Solution (40), **a** when $q = 0.8,\ \alpha = 0.05,\ \tau_0 = 5, k = 0.5,\ \delta_0 = 0.05,\ \mu = 1$, **b** when $\alpha = 0.05,\ \tau_0 = 5,\ k = 0.5,\ \mu = 1,\ q = 0.8,\ \tau = 10$, **c** when $q = 0.8,\ \tau = 10,\ \tau_0 = 5, k = 0.5,\ \delta_0 = 0.05,\ \mu = 1$, **d** when $\alpha = 0.05,\ \tau_0 = 5,\ k = 0.5,\ \mu = 1,\ \delta_0 = 0.05,\ \tau = 10$

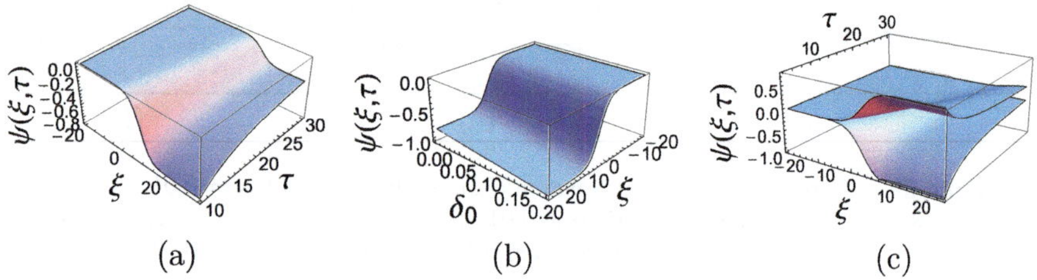

Fig. 5 3D Profiles of Solution (40), **a** when $q = 0.8,\ \alpha = 0.05,\ \tau_0 = 4,\ k = 0.5,\ \delta_0 = 0.05,\ \mu = 0$, **b** when $q = 0.8,\ \alpha = 0.05,\ \tau_0 = 4,\ k = 0.5,\ \mu = 1$, **c** when $q = 0.7, 0.8,\ \alpha = 0.05,\ \tau_0 = 4,\ k = 0.5,\ \delta_0 = 0.05,\ \mu = 1$,

7 Conclusion

Using the basic governing equation the propagation of cylindrical and spherical IAW in unmagnetized plasma comprising positively charge cold ions and electrons describing hybrid q-nonextensive nonthermal velocity distribution are analyzed through the NKDVB model. The WRM and SHBM are employed to obtain a new class of the solitary and shock wave solution for the present system. The significant effects of different physical parameters such as α, q, δ_0 and time parameter τ are observed from a numerical standpoint. Finally, it can be concluded that the results of our investigation may be helpful to observe the propagating behavior of solitary and shock waves in laboratory plasma as well as space plasma environments.

References

1. Sagdeev, R.Z.: Cooperative phenomena and shock waves in collisionless plasmas. Rev. Plasma Phys. **4**, 23 (1966)
2. Taylor, R.J., Baker, D.R., Ikezi, H.: Observation of collisionless electrostatic shocks. Phys. Rev. Lett. **24**(5), 206 (1970)
3. Shukla, P.K., Silin, V.P.: Dust ion-acoustic wave. Phys. Scr. **45**(5), 508 (1992)

4. Barkan, A., D'angelo, N., Merlino, R.L.: Experiments on ion-acoustic waves in dusty plasmas. Planet. Space Sci. **44**(3), 239–242 (1996)
5. D'Angelo, N.: Coulomb solids and low-frequency fluctuations in RF dusty plasmas. J. Phys. D: Appl. Phys. **28**(5), 1009 (1995)
6. Dubouloz, N., Pottelette, R., Malingre, M., Treumann, R.A.: Generation of broadband electrostatic noise by electron acoustic solitons. Geophys. Res. Lett. **18**(2), 155–158 (1991)
7. Maxon, S., Viecelli, J.: Cylindrical solitons. Phys. Fluids **17**(8), 1614–1616 (1974)
8. Gao, D.N., Zhang, Z.R., Wu, J.P., Luo, D., Duan, W.S., Li, Z.Z.: Cylindrical and Spherical Dust-Ion Acoustic Solitary Waves by Damped Korteweg-de Vries-Burgers Equation. Braz. J. Phys. **49**(5), 693–697 (2019)
9. Masood, W., Imtiaz, N., Siddiq, M.: Ion acoustic shock waves in dissipative electron-positron-ion plasmas with weak transverse perturbations. Phys. Scr. **80**(1), 015501 (2009)
10. Roy, S., Saha, S., Raut, S., Das, A.N.: Studies on the effect of kinematic viscosity on electron-acoustic cylindrical and spherical solitary waves in a plasma with trapped electrons. J. Appl. Math. Comput. Mech. **20**(2) (2021)
11. Demiray, H.: Analytical solution for nonplanar waves in a plasma with q-nonextensive nonthermal velocity distribution: weighted residual method. Chaos, Solitons & Fractals **130**, 109448 (2020)
12. Raut, S., Mondal, K.K., Chatterjee, P., Roy, A.: Propagation of dust-ion-acoustic solitary waves for damped modified Kadomtsev-Petviashvili-Burgers equation in dusty plasma with a q-nonextensive nonthermal electron velocity distribution. SeMA J. 1–23 (2021)
13. Williams, G., Kourakis, I., Verheest, F., Hellberg, M.A.: Re-examining the Cairns-Tsallis model for ion acoustic solitons. Phys. Rev. E **88**(2), 023103 (2013)
14. Cairns, R.A., Bingham, R., Dendy, R.O., Nairn, C.M.C., Shukla, P.K., Mamun, A.A.: Ion sound solitary waves with density depressions. Le J. de Phys. IV **5**(C6), C6-43, 072306 (1995)
15. Taniuti, T.: Reductive perturbation method and far fields of wave equations. Prog. Theor. Phys. Suppl. **55**, 1–35, 023103 (1974)
16. Mushtaq, A., Shah, H.A.: Nonlinear Zakharov-Kuznetsov equation for obliquely propagating two-dimensional ion-acoustic solitary waves in a relativistic, rotating magnetized electron-positron-ion plasma. Phys. Plasmas **12**(7), 072306 (2005)
17. Awawdeh, F., Jaradat, H.M., Al-Shara, S.: Applications of a simplified bilinear method to ion-acoustic solitary waves in plasma. Eur. Phys. J. D **66**(2), 1–8, 072306 (2012)

Formation of Shocks in Ionospheric Plasma with Positron Beam

Sunidhi Singla, Manveet Kaur, and Nareshpal Singh Saini

Abstract The nonlinear dynamics of ion acoustic shocks (IAShs) in electron-ion plasma has been studied in the presence of relativistic positron beam as a result of the vital importance in astrophysical environments mainly ionospheric region. The Korteweg-de Vries-Burgers (KdVB) equation and its oscillatory solution is derived by employing reductive perturbation method. From the solution of KdVB equation, existence of oscillatory IAShs and their characteristics are studied under the influence of various plasma parameters such as temperature of ions as well as positron, relativistic factor, viscosity of ions and different parameters of beam. The behaviour of IAShs propagating in a relativistic plasma model is strongly dependent on the ion and positron temperatures, the mass ratio and the relativistic effects. The kinematic viscosity and the equilibrium ion number density play very important roles in the basic features of the produced IAShs.

Keywords Shocks · Ion acoustic · Positron beam · Reductive perturbation

1 Introduction

During the past two decades, many authors have studied the electron-ion plasma in different space and laboratory environments. Electron-ion plasma evokes great interest for many researches due to its existence in pulsar environments, earth's ionosphere, polar regions of neutron stars, white dwarfs [1, 2], pulsar magnetosphere [3–5] etc. It is noteworthy that introduction of positron beam immensely affects the physical properties of plasma. Misra et al. [6] studied the nonlinear propagation and interaction of electron–positron plasma that shows distinctive behavior feature from the usual linear mode. With the increase in the superthermality of electrons and positron beam speed, amplitude and width of the solitons get modified significantly as reported by Shan et al. [7]. Numerous studies has confirmed the influence

S. Singla (✉) · M. Kaur · N. S. Saini
Department of Physics, Guru Nanak Dev University, Amritsar 143005, India
e-mail: singla.sunidhi94@gmail.com

S. Banerjee and A. Saha (eds.), *Nonlinear Dynamics and Applications*,
Springer Proceedings in Complexity,
https://doi.org/10.1007/978-3-030-99792-2_12

and importance of interaction of ion/electron/positron beam in space and laboratory plasmas [8–11]. Sarma et al. studied the propagation properties of ion acoustic non-linear structures (IANSs) in a relativistic plasma containing ion fluid, positron and relativistic electrons. It was reported that different plasma parameters have immense impact on the characteristics of IANSs [12].

For the formation of KdV-Burgers equation reductive perturbation approach was used, and the influence of various plasma physical parameters on IA shock waves was depicted. Many KdV-Burgers equations solutions were examined and in the presence of electron orbital motion excitation from KdV oscillations to the shock solution was described [13]. Very recently, Singh et al. [14] studied the effect of anisotropic term on electron-acoustic shocks by deriving the KdV-Burgers equation in superthermal plasma. To the best of our knowledge, study of oscillatory shocks in a electron-ion plasma in the presence of relativistic positron beam has not been reported yet. In this investigation, main focus of the study is to illustrate the influence of positron beam on ion acoustic shocks by deriving KdV-Burgers equation and its oscillatory solution. The manuscript is organized as follows: in Sect. 2, three fluids beam-plasma model is described. The derivation of KdV-Burgers equation are presented in Sect. 3. Oscillatory solution of KdV-Burgers equation is illustrated in Sect. 4. Parametric analysis is presented in Sect. 5 and conclusions are summarized in Sect. 6.

2 Three-Fluid Beam-Plasma Model

An unmagnetized plasma involving three components specifically cold inertial ions, hot inertial electrons with injection of relativistic positron beam is considered to study the characteristics of IA shock waves. The normalized equations for three fluids model (continuity, momentum, pressure and Poisson equations) are described as follows:

For ions:

$$\frac{\partial n_i}{\partial t} + \frac{\partial}{\partial x}(n_i v_i) = 0, \tag{1}$$

$$\frac{\partial v_i}{\partial t} + v_i \frac{\partial v_i}{\partial x} + \frac{\sigma}{n_i}\frac{\partial p_i}{\partial x} = -\frac{\partial \phi}{\partial x} + \eta_i \frac{\partial^2 v_i}{\partial x^2}, \tag{2}$$

$$\frac{\partial p_i}{\partial t} + v_i \frac{\partial p_i}{\partial x} + 3 p_i \frac{\partial v_i}{\partial x} = 0, \tag{3}$$

For electrons:

$$\frac{\partial n_e}{\partial t} + \frac{\partial}{\partial x}(n_e v_e) = 0, \tag{4}$$

$$\beta_1 n_e \left(\frac{\partial v_e}{\partial t} + v_e \frac{\partial v_e}{\partial x} \right) + \frac{\partial p_e}{\partial x} = n_e \frac{\partial \phi}{\partial x} + \eta_e \frac{\partial^2 v_e}{\partial x^2}, \tag{5}$$

$$\frac{\partial p_e}{\partial t} + v_e \frac{\partial p_e}{\partial x} + 3 p_e \frac{\partial v_e}{\partial x} = 0, \tag{6}$$

For positron beam:

$$\frac{\partial n_b}{\partial t} + \frac{\partial}{\partial x}(n_b v_b) = 0, \tag{7}$$

$$\beta_2 n_b \left(\frac{\partial v_b'}{\partial t} + v_b \frac{\partial v_b'}{\partial x} \right) + \alpha \frac{\partial p_b}{\partial x} = -n_b \frac{\partial \phi}{\partial x}, \tag{8}$$

$$\frac{\partial p_b}{\partial t} + v_b \frac{\partial p_b}{\partial x} + 3 p_b \frac{\partial v_b'}{\partial x} = 0, \tag{9}$$

$$\frac{\partial^2 \phi}{\partial x^2} = n_e(1 + \mu_b) - n_b \mu_b - n_i. \tag{10}$$

where, l = i, e, b, $\sigma = \frac{T_i}{T_e}$, $\beta_1 = \frac{m_e}{m_i}$, $\mu_b = \frac{n_{bo}}{n_{io}}$, $v_b' = v_b \left(1 + \frac{v_b^2}{2c_1^2}\right)$, $\beta_2 = \frac{m_b}{m_i}$ and $\alpha = \frac{T_b}{T_c}$, $c_1 = \frac{c}{c_s}$, Normalisation of the physical quantities are done as follows to make them dimensionless. $t = T'/\omega_{pi}$, $x = X'/\lambda_{Di}$, $v_l = V'_l/c_s$, $n_l = N'_l/N'_{lo}$, $\phi = e\Phi'/k_B T_e$, $p_l = P'_l/n_{lo} k_B T_l$ and the kinematic viscosity η_l is normalized by $\lambda_{Dl}^2 n_{lo} M_l \omega_{pl}$, where $\omega_{pi} = \sqrt{4\pi N_{io} e^2/(k_B T_e)}$ is the ion-plasma oscillation frequency, $c_s = \sqrt{k_B T_e/m_i}$ is the ion acoustic speed, λ_{Dl} represents Debye length, N_{lo} is the unperturbed number density. The temperature of the l-th charged particle is T_l.

3 Derivation of KdV-Burgers Equation and Its Solution

To study the nonlinear propagation of ion acoustic shock waves in a plasma with relativistic positron beam, we introduce the following stretching coordinates [13]:

$$\zeta = \epsilon^{\frac{1}{2}}(x - v_p t), \qquad \tau = \epsilon^{\frac{3}{2}} t. \tag{11}$$

where v_p denotes the phase velocity of the IAShs. The weakness of the perturbation is measured by ϵ, which is small in the range $(0 < \epsilon < 1)$. Expansions of the state variables in power series of ϵ is illustrated as;

$$\begin{aligned}
n_{i,e,b} &= 1 + \epsilon n_{i,e,b1} + \epsilon^2 n_{i,e,b2} + \epsilon^3 n_{i,e,b3} + ... \\
p_{i,e,b} &= 1 + \epsilon p_{i,e,b1} + \epsilon^2 p_{i,e,b2} + \epsilon^3 p_{i,e,b3} + ... \\
v_{i,e} &= \epsilon v_{i,e1} + \epsilon^2 v_{i,e2} + \epsilon^3 v_{i,e3} + ... \\
v_b &= v_{bo} + \epsilon v_{b1} + \epsilon^2 v_{b2} + \epsilon^3 v_{b3} + ... \\
\phi &= \epsilon \phi_1 + \epsilon^2 \phi_2 + \epsilon^3 \phi_3 + ...
\end{aligned}$$

We consider $\eta_{e,i} = \epsilon^{\frac{1}{2}}\eta_{e0}(\eta_{i0})$, where $\eta_{e0}(\eta_{i0})$ is the equilibrium fluid viscosity of electron(ion). Also, scaling of η is taken into account is such a manner that only dissipative term is getting affected and not affecting other terms namely nonlinear and dispersive terms. After using stretching coordinates and expansion of state variable perturbation in Eqs. (1–10) and collecting the coefficients of lowest power of ϵ we get:

$$n_{i1} = R\phi_1, \qquad v_{i1} = Rv_p\phi_1, \qquad p_{i1} = 3R\phi_1 \tag{12}$$

$$n_{e1} = S\phi_1, \qquad v_{e1} = Sv_p\phi_1, \qquad p_{e1} = 3\,S\phi_1 \tag{13}$$

$$n_{b1} = 2c_1^2 PQ\phi_1, \qquad v_{b1} = 2c_1^2 PQ(v_p - v_{bo})\phi_1, \qquad p_{b1} = 3Q\phi_1 \tag{14}$$

where $R = \frac{1}{(V_p^2-3\sigma)}$, $S = \frac{1}{(3-\beta_1 v_p^2)}$, $P = \frac{1}{(1+3v_{bo}^2)}$, $Q = \frac{1}{(\beta_2(v_p-v_{bo})^2-3\alpha)}$;

and Poisson equation leads to the dispersion relation,

$$(1+\mu_b)S - \mu_b 2c_1^2 PQ - R = 0 \tag{15}$$

The next higher order equations from the perturbation theory are as follows:

$$\frac{\partial n_{i1}}{\partial \tau} - v_p\frac{\partial n_{i2}}{\partial \zeta} + \frac{\partial v_{i2}}{\partial \zeta} + \frac{\partial}{\partial \zeta}(n_{i1}v_{i1}) = 0, \tag{16}$$

$$\frac{\partial v_{i1}}{\partial \tau} - v_p\frac{\partial v_{i2}}{\partial \zeta} + v_{i1}\frac{\partial v_{i1}}{\partial \zeta} + \sigma\frac{\partial p_{i2}}{\partial \zeta} - \sigma n_{i1}\frac{\partial p_{i1}}{\partial \zeta} - \eta_i\frac{\partial^2 v_{i1}}{\partial \zeta^2} + \frac{\partial \phi_2}{\partial \zeta} = 0, \tag{17}$$

$$\frac{\partial p_{i1}}{\partial \tau} - v_p\frac{\partial p_{i2}}{\partial \zeta} + v_{i1}\frac{\partial p_{i1}}{\partial \zeta} + 3\frac{\partial v_{i2}}{\partial \zeta} + 3p_{i1}\frac{\partial v_{i1}}{\partial \zeta} = 0, \tag{18}$$

$$\frac{\partial n_{e1}}{\partial \tau} - v_p\frac{\partial n_{e2}}{\partial \zeta} + \frac{\partial v_{e2}}{\partial \zeta} + \frac{\partial}{\partial \zeta}(n_{e1}v_{e1}) = 0, \tag{19}$$

$$\beta_1\frac{\partial v_{e1}}{\partial \tau} - v_p\beta_1\frac{\partial v_{e2}}{\partial \zeta} + \beta_1 v_{e1}\frac{\partial v_{e1}}{\partial \zeta} - \beta_1 v_p n_{e1}\frac{\partial v_{e1}}{\partial \zeta} + \frac{\partial p_{e2}}{\partial \zeta} - \eta_e\frac{\partial^2 v_{e1}}{\partial \zeta^2} - n_{e1}\frac{\partial \phi_1}{\partial \zeta} = \frac{\partial \phi_2}{\partial \zeta}, \tag{20}$$

$$\frac{\partial p_{e1}}{\partial \tau} - v_p \frac{\partial p_{e2}}{\partial \zeta} + v_{e1} \frac{\partial p_{e1}}{\partial \zeta} + 3 \frac{\partial v_{e2}}{\partial \zeta} + 3 p_{e1} \frac{\partial v_{e1}}{\partial \zeta} = 0, \tag{21}$$

$$\frac{\partial n_{b1}}{\partial \tau} - (v_p - v_{bo}) \frac{\partial n_{b2}}{\partial \zeta} + \frac{\partial v_{b2}}{\partial \zeta} + \frac{\partial}{\partial \zeta}(n_{b1} v_{b1}) = 0, \tag{22}$$

$$\frac{\beta_2}{2c_1^2}\left((1+3v_{bo}^2)\frac{\partial v_{b1}}{\partial \tau} - (1+3v_{bo}^2)(v_p - v_{bo})\frac{\partial v_{b2}}{\partial \zeta} + \frac{\partial v_{b1}}{\partial \zeta}\right.$$
$$\left. -2v_{b0}(v_p - v_{b0})v_{b1}\right) - \frac{\beta_2}{2c_1^2}(v_p - v_{b0})(1+3v^2{}_{b0})n_{b1}\frac{\partial v_{b1}}{\partial \zeta}$$
$$+\alpha \frac{\partial p_{b2}}{\partial \zeta} = -\frac{\partial \phi_2}{\partial \zeta} - n_{b1}\frac{\partial \phi_1}{\partial \zeta}, \tag{23}$$

$$\frac{\partial p_{b1}}{\partial \tau} - (v_p - v_{bo})\frac{\partial p_{b2}}{\partial \zeta} + v_{b1}\frac{\partial p_{b1}}{\partial \zeta} + \frac{3(1+3v_{bo}^2)}{2c_1^2}$$
$$\left(\frac{\partial v_{b2}}{\partial \zeta} + p_{b1}\frac{\partial v_{b1}}{\partial \zeta}\right) + \frac{6}{2c_1^2} v_{bo} v_{b1} \frac{\partial v_{b1}}{\partial \zeta} = 0, \tag{24}$$

$$(1+\mu_b)n_{e2} - \mu_b n_{b2} - n_{i1} = 0. \tag{25}$$

By differentiating Eq. (25) w.r.t. ζ, we have;

$$(1+\mu_b)\frac{\partial n_{e2}}{\partial \zeta} - \mu_b \frac{\partial n_{b2}}{\partial \zeta} - \frac{\partial n_{i2}}{\partial \zeta} = 0 \tag{26}$$

Eliminating the second order perturbed quantities from Eqs. (16–25), we obtain the following KdV-Burgers Equation:

$$\frac{\partial \phi_1}{\partial \tau} + A\phi_1 \frac{\partial \phi_1}{\partial \zeta} + B\frac{\partial^3 \phi}{\partial \zeta^3} = C\frac{\partial^2 \phi_1}{\partial \zeta^2} \tag{27}$$

where, nonlinear coefficient, $A = \frac{B'}{A'}$,

dispersion coefficient, $B = \frac{1}{A'}$

and dissipation coefficient, $C = \frac{C'}{A'}$

and

$$A' = -(1+\mu_b)2\beta_1 v_p S^2 - \mu_b 2\beta_2 2c_1^2(v_p - v_{bo})PQ^2 - 2v_p R^2 \tag{28}$$

$$B' = -\mu_b 2c_1^2 P Q^2 (2c_1^2 P(1 - 2v_{bo}(v_p - v_{bo})P) + 9Q) - 3(1 + \mu_b)(4S - 1)S^2 - 3(1 + 4\sigma R)R^2 \quad (29)$$

$$C' = (1 + \mu_b)\eta_e v_p S^2 + \eta_i v_p R^2 \quad (30)$$

Solution of Eq. (27) represents the formation of shock waves with different coefficient A, B and C. In the limiting case, i.e. in the absence of dissipation term, (C = 0) Eq. (27) transforms to KdV equation and it agrees with the results obtained in [12]. Furthermore, we have studied the formation of shocks whereas in authors in ref. [12] have studied the formation of solitons.

4 Solution of KdV-Burgers Equation

By introducing single variable transformation in Eq. (27), we have examined its analytical solution. The solution of Eq. (27) is determined by using the Tanh-method [14],

$$\phi_1(\xi, \tau) = \phi_{max}\left(1 - \frac{1}{4}\left[1 + \tanh\left(\frac{\xi - u\tau}{W}\right)\right]^2\right) \quad (31)$$

where, $\phi_{max} = \frac{12C^2}{25AB}$, $W = \nabla^{-1} = \frac{10B}{C}$ and $u = \frac{6C^2}{25B}$

4.1 Oscillatory Shocks Solution

After dealing with different asymptotic boundary conditions, different type of solution for Eq. (27) is obtained. Using transformation $\xi = \zeta - U\tau$ in the Eq. (27) and assuming $\phi = \phi_0 + \Phi$ with $\phi_0 = 2U/A$ and after linearising with respect to ϕ, we obtain,

$$\frac{d^2\Phi}{d\xi^2} - \frac{C}{B}\frac{d\Phi}{d\xi} - \frac{U}{B}\Phi = 0. \quad (32)$$

The solution of above equation represents damped harmonic oscillator. Thus, the oscillatory shock wave solutions of Eq. (27) is given as [13]

$$\phi_{osc} = \frac{2U}{A} + \Theta exp(-\Omega\xi)\cos(\varpi_1\xi) \quad (33)$$

where,Θ is arbitrary constant, $\Omega = -C/2B$ is the damping factor and $\varpi_1 = \sqrt{\frac{U}{B}(1 - \frac{C^2}{4UB})}$, also $\frac{U}{B}$ represents the natural frequency of the plasma system.

5 Parametric Analysis

The propagation properties of ion acoustic shocks by changing the values of different physical parameters are analyzed numerically. We have considered positron beam density (μ_b) of the range 0.2–0.8, electron to ion temperature ratio(σ) of the range 0.1–0.5 and kinematic viscosity ranges from 0.2–0.8. The considered plasma parameters are from the Earth's upper atmosphere (ionosphere region) [8]. It is numerically that with different plasma parameters, A shows negative value, which implies only negative potential IAShs.

The solution of oscillatory shock waves is plotted against ξ for different plasma parameters. Figure 1 represents the variation of oscillatory shocks with different values of positron beam density ratio (μ_b). It is seen that increase in the value of μ_b leads to the decrease in amplitude of shocks. Thus, with the introduction of positron beam, shocks of smaller amplitude are formed and with increase in the value of μ_b oscillatory behavior of the shocks tends to diminish. Figure 2 depicts the influence of electron to ion temperature ratio (σ), it is remarked that the with increase in the value of σ, amplitude of shocks decreases. Figure 3 represents the variation of profile of oscillatory with viscosity of electrons (η_e), it is depicted that the amplitude is decreased with change in the value of η_e. Furthermore, lesser viscous medium tends to have stronger oscillatory shock waves. Similar behavior is seen for kinematic viscosity of ions η_i. Finally, it is highlighted that the important findings may be useful in describing clearly the propagation properties of IAShs in multi-component plasma with positron beam which is found in Earth's ionospheric region.

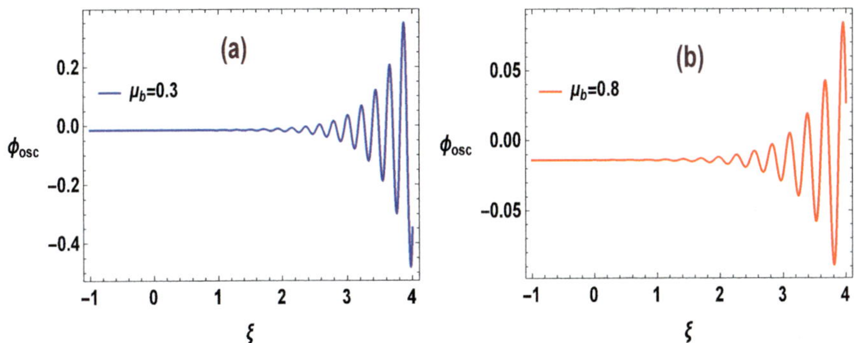

Fig. 1 The variation of ion acoustic oscillatory shocks for different values of positron beam density ratio (μ_b) **a** $\mu_b = 0.3$ **b** $\mu_b = 0.8$, with fixed values of $\sigma = 0.1$ and $\eta_e = 0.2$

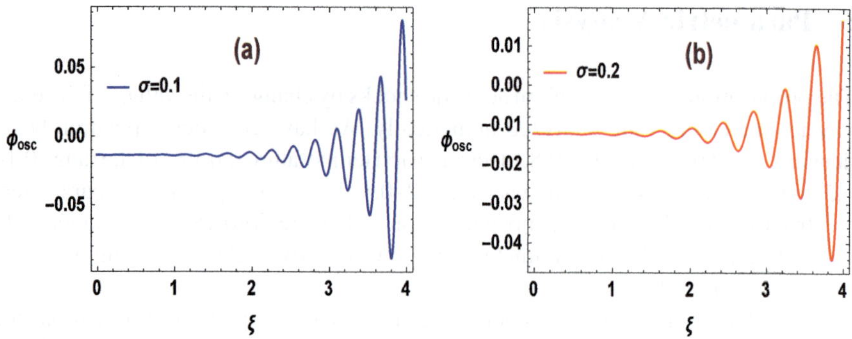

Fig. 2 The variation of ion acoustic oscillatory shocks for different values of $\sigma = 0.1$ and $\sigma = 0.2$ respectively, where $\mu_b = 0.5$ and $\eta_e = 0.2$

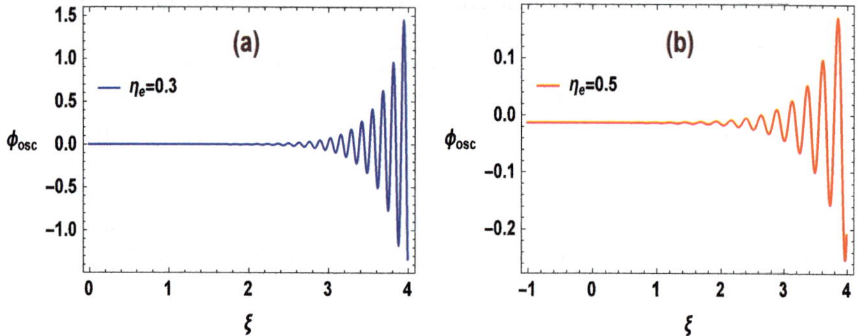

Fig. 3 The variation of ion acoustic oscillatory shocks for different values of $\eta_e = 0.2$ and $\eta_e = 0.5$ respectively, where $\sigma = 0.1$ and $\mu_b = 0.5$

6 Conclusions

In this manuscript, the propagation properties of IAShs in a electron-ion plasma in the presence of beam of relativistic positrons. After the implication of reductive perturbation method, KdV-Burgers equation and its solution are derived. Only negative potential shocks are obtained. Positron beam and other physical parameters play an important role in the formation of different types of oscillatory shocks and their characteristics. The amplitude of oscillatory shocks decreases with the increase in μ_b and η_e. This investigation may be useful to understand the nonlinear phenomena leading to the formation of ion acoustic shocks in astrophysical dense plasma environments, especially in Earth's ionospheric region [8, 12].

Acknowledgements Department of Science and Technology, Govt. of India, New Delhi under DST-SERB project No. CRG/2019/003988 is highly acknowledged to support this investigation.

References

1. Lallement, R., Welsh, B.-Y., Barstow, M.-A., Casewell, S.-L.: High ions towards white dwarfs: circumstellar line shifts and stellar temperature. Astrophys. Astron. **533**(A140), 1–13 (2011)
2. Sabry, R., Moslem, W.-M., Shukla, P.-K.: Freak waves in white dwarfs and magnetars. Phys. Plasmas **19**(12), (122903 (1–7)) (2012)
3. Tribeche, M., Aoutou, K., Younsi, S., Amour, R.: Nonlinear positron acoustic solitary waves. Phys. Plasmas **16**(7), 072103(1–6) (2009)
4. Rahman, M.-M., Alam, M.-S., Mamun, A.-A.: Cylindrical and spherical positron-acoustic Gardner solitons in electron-positron-ion plasmas with nonthermal electrons and positrons. Astrophys. Space Sci. **352**, 193–200 (2014)
5. El-Shamy, E.-F.: Nonlinear propagation of positron-acoustic periodic travelling waves in a magnetoplasma with superthermal electrons and positrons. Chin. Phys. Lett. **34**(6), 065201(1–5) (2017)
6. Misra, A.-P., Roy Chowdhury, A.: Nonlinear interaction of electromagnetic pulses with an electron–positron plasma coupled NLS equation. Chaos, Solitons Fractals **15**(5), 1–10 (2003)
7. Shan, S.-A., El-Tantawy, S.-A., Moslem, W.-M.: On the fully nonlinear acoustic waves in a plasma with positrons beam impact and superthermal electrons. Phys. Plasmas **20**(8), 082104(1–7) (2013)
8. Gsponer, A.: The Physics of high-intensity high-energy Particle Beam Propagation in open Air and outer-space. Plasmas. http://arxiv.org/abs/physics/0409157v3 [physics.plasm-ph] (2004)
9. Popel, S.-I., Vladimirov, S.-V., Shukla, P.-K.: Ion-acoustic solitons in electron-positron-ion plasmas. Phys. Plasmas **2**(716), 1–5 (1995)
10. Shan, S.-A., El-Tantawy, S.-A., Moslem, W.-M.: On the fully nonlinear acoustic waves in a plasma with positrons beam impact and superthermal electrons. Phys. Plasmas, **20**(8), 082104(1–7) (2013)
11. Singla, S., Saini, N.-S.: Head-on collision of ion-acoustic multi-solitons and study of rogue waves in electron-beam superthermal plasma. Results Phys. **22**(10), 103898(1–11) (2021)
12. Sarma, R., Mishra, A.-P., Adhikary, N.-C.: Nonlinear ion–acoustic solitary waves in an electron–positron–ion plasma with relativistic positron beam. Chin. Phys. B **27**(10), 105207(1–6) (2018)
13. Atteya, A., Behery, E.-E., El-Taibany, W.-F.: Ion acoustic shock waves in a degenerate relativistic plasma with nuclei of heavy elements. Eur. Phys. J. Plus **132**(109), 1–8 (2017)
14. Singh, K., Saini, N.-S.: Effect of anisotropic pressure on electron acoustic oscillatory and monotonic shocks in superthermal magnetoplasma. Radio Sci. **54**(12), 1192–1203 (2019)

Nonlinear Propagation of Gaussian Laser Beam in an Axially Magnetized Cold Quantum Plasma

P. P. Nikam, V. S. Pawar, S. D. Patil, and M. V. Takale

Abstract The evolution of self-focusing of Gaussian laser beam in underdense magnetized cold quantum plasma has been studied under parabolic equation approach. We have established beam-width parameter differential equation of Gaussian laser beam using WKB and paraxial approximations. This equation is solved numerically. The results are presented graphically by considering applied magnetic field along (forward) as well as opposite (reverse) to the axis of propagation of laser. It is seen that, the forward magnetization increase the self-focusing effect as compared to the reverse magnetization. In addition, quantum effects enhance the self-focusing behaviour of laser.

Keywords Gaussian beam · Self-focusing · Plasma · Magnetized · Quantum

1 Introduction

Many applications likes fusion by lasers [1], generation of higher order harmonics [2], laser having wavelength in the X-ray region [3] and other applications [4, 5] based on laser-plasma experiments. For such applications, laser should propagate more distance than the Rayleigh length. In plasmas, different nonlinear optical effects are present. Self-focusing (SF) of laser beam is one of them [6, 7]. SF in plasmas is mainly contributed due to three mechanisms such as; relativistic, ponderomotive and thermal. Recently, laser-matter (plasma) interaction in quantum regime has been investigated considerably [8]. Patil et al. [9] highlighted the SF of Gaussian beam in quantum plasma and extended the same to thermal quantum plasma [9]. Aggarwal

P. P. Nikam · S. D. Patil (✉)
Department of Physics, Devchand College, Arjunnagar 591 237, Maharashtra, India
e-mail: sdpatilphy@gmail.com

V. S. Pawar
Department of Physics, Raje Ramrao Mahavidyalaya, Jath 416 404, Maharashtra, India

M. V. Takale
Department of Physics, Shivaji University, Kolhapur 416 004, Maharashtra, India

S. Banerjee and A. Saha (eds.), *Nonlinear Dynamics and Applications*,
Springer Proceedings in Complexity,
https://doi.org/10.1007/978-3-030-99792-2_13

et al. [10] studied Gaussian beam propagation in density ramped magnetized cold plasma. Also, collective consequence of relativistic and ponderomotive mechanisms on SF of Gaussian beam was studied by Aggarwal et al. [11–13] in magnetized RCQP. Recently, Pawar et al. [14] explored effect of laser (aperture coefficient) and plasma (density) parameters on SF of finite AiG beams in RCQP.

In this present paper SF of Gaussian beams in underdense, magnetized RCQP is studied. An external magnetic field is parallel (forward) or antiparallel (reverse) to the propagation vector of beam. In Sect. 2, nonlinear ordinary second order beam-width parameter (BWP) differential equation has been achieved under WKB and paraxial approximations. Sections 3 and 4 contains graphical results and conclusions respectively.

2 Theoretical Formulation

Consider the wave vector of cylindrically distributed Gaussian beam propagating on the z-axis in magnetized RCQP. The total dielectric function ε of plasma is written as,

$$\varepsilon = \varepsilon_0 + \varepsilon_2\big(EE^*\big) \tag{1}$$

Here, ε_0 and ε_2 are intensity independent and dependent parts of ε. The ε_2 has various forms under different physical conditions. The effective plasma electron density n_e for relativistic, magnetized plasma can be given as [12]

$$n_e = \frac{n_0}{\gamma(1 - (\omega_c/\gamma\omega))}$$

here, ω_c and ω are the electron cyclotron frequency and frequency of laser used, γ is the relativistic factor written as;

$$\gamma = \left[1 + A^2 + 2A^2(\sigma\Omega_c)\left(\frac{1}{\sqrt{1+A^2}}\right) + 3A^2(\sigma\Omega_c)^2\left(\frac{1}{1+A^2}\right)\right]^{\frac{1}{2}}$$

where, $A^2 = \frac{e^2|E_0|^2}{m^2c^2\omega^2}$ with c is velocity of light in free space, e and m are electronic charge and electrons rest mass respectively. For collisionless, magnetized cold quantum plasma ε_0 and ε_2 can be written as [12]

$$\varepsilon_0 = 1 - \frac{\Omega_p^2}{1 - \sigma\Omega_c} \tag{2}$$

$$\varepsilon_2(EE^*) = \Omega_p^2\left(1 - \frac{n_e}{n_0}\left(1 - \frac{\beta}{\gamma}\right)^{-1}\right) \tag{3}$$

where, $\Omega_p = \omega_p/\omega$, $\Omega_c = \omega_c/\omega$, $\beta = 4\pi^4 h^2/m^2\omega^2\lambda^4$, $\omega_p = \left(4\pi n_0 e^2/m\right)^{1/2}$ and $\omega_c = eB_0/mc$. Here, ω_p is frequency of plasma electron, h is Planck's constant, λ is the wavelength of laser beam, B_0 is an external magnetic field, σ is magnetization parameter, which decides type of magnetization. For the case of forward (reverse) magnetization $\sigma = +1(\sigma = -1)$. The unmagnetized case of reference corresponds to $\sigma = 0$.

The wave equation for laser beam in plasmas with ε assumed to be as,

$$\nabla^2 \mathrm{E} - \frac{\varepsilon}{c^2}\frac{\partial^2 \mathrm{E}}{\partial t^2} + \nabla\left(\frac{\mathrm{E}\cdot\nabla\varepsilon}{\varepsilon}\right) = 0 \tag{4}$$

By employing WKB approximation, $(c^2/\omega^2)\left|\left(\frac{1}{\varepsilon}\right)\nabla^2 \ln\varepsilon\right| \ll 1$. Therefore, $\nabla\left(\frac{\mathrm{E}\cdot\nabla\varepsilon}{\varepsilon}\right)$ can be ignored. By assuming electric field $E = A(x, y, z)e^{i(\omega t - kz)}$, the evolution of E in magnetized plasma is governing by following nonlinear differential equation,

$$\frac{\partial^2 E}{\partial z^2} - 2ik\frac{\partial E}{\partial z} + \delta\left(\frac{\partial^2 E}{\partial r^2} + \frac{1}{r}\frac{\partial E}{\partial r}\right) + \frac{\omega^2}{c^2}(\varepsilon - \varepsilon_0)E = 0 \tag{5}$$

where, $\delta = \left[1 + (\varepsilon_0/\varepsilon_{0zz})\right]/2$ with $\varepsilon_{0zz} = 1 - \Omega_p^2$. By substituting $E = A_0 \exp(-ikS(r, z))$, where k and S are the wave vector and eikonal of beam. One can write for initially Gaussian beam as,

$$A_0^2 = \frac{E_0^2}{f^2} exp\left(-\frac{r^2}{r_0^2 f^2}\right) \tag{6}$$

$$S = \frac{r^2}{2f}\frac{df}{dz} + \varphi(z) \tag{7}$$

where, E_0 is electric field at the central position $r = z = 0$, r_0 is initial beam radius and f is dimensionless BWP.

In subsequent stages, by succeeding proposal developed by Akhmanov et al. [6] and its modification lead by Sodha et al. [7]. An expression for f under paraxial approximation as

$$\frac{d^2 f}{d\xi^2} = \frac{4\delta^2}{f^3} - \frac{\rho^2 p\chi(-\psi + \delta\sigma\Omega_c)\delta}{f^3(\beta - \sqrt{\psi})^2\sqrt{\psi}(\sqrt{\psi} - \sigma\Omega_c)^2} \tag{8}$$

where,

$\chi = 1 - \frac{p\sigma\Omega_c}{f^2\left(1+\frac{p}{f^2}\right)^{\frac{3}{2}}} + \frac{2\sigma\Omega_c}{\sqrt{1+\frac{p}{f^2}}} - \frac{3p\sigma^2\Omega_c^2}{f^2\left(1+\frac{p}{f^2}\right)^2} + \frac{3\sigma^2\Omega_c^2}{\left(1+\frac{p}{f^2}\right)}$ and $\psi = 1 + \frac{p}{f^2} + \frac{2p\sigma\Omega_c}{f^2\sqrt{1+\frac{p}{f^2}}} + \frac{3p\sigma^2\Omega_c^2}{f^2\left(1+\frac{p}{f^2}\right)}$.

Here, $\xi = z/kr_0^2$ is non-dimensional distance travelled by laser in plasma and $p = \alpha E_0^2$ is initial intensity parameter of beam.

3 Results and Discussion

Equation (8) is an ordinary nonlinear beam width parameter differential equation in magnetized RCQP. The Eq. (8) is solved by considering following laser-plasma parameters as: $\omega = 1.778 \times 10^{20}$rad/s, $r_0 = 20\mu$m, $\Omega_p = 0.1$, $\Omega_c = 0.2$, $p = 1$ and $\sigma = 0, \pm 1$.

Figure 1 shows, the effect of σ on $f(\xi)$. The behaviour of f with ξ is accordance with density gradient which is further responsible for changes in ε of plasma. It is observed that, the SF is more (less) dominant in forward (reverse) magnetization in comparison with unmagnetized case of reference. Figure 2 shows effect of σ on $\varepsilon(\xi)$. From Fig. 2, it is observed that variation of ε is also oscillatory with ξ. The oscillatory peaks of ε in Fig. 2 matches to the oscillatory valleys of f in Fig. 1.

Figure 3, illustrates the effect of plasma density on $f(\xi)$. From this it is clear that, as density parameter Ω_p increases, enhanced SF of beam occurs with decrease in SF length. However, oscillatory behaviour of f with ξ for different σ values is same as in Fig. 1. As usual, SF in plasma is enhanced by addition of quantum effects.

Effect of different values of Ω_c on $f(\xi)$ is plotted in Fig. 4. From this it is clear that, as magnetic field parameter Ω_c increases, SF length decreases. The effect of σ on behavior of f with ξ is same as observed in Fig. 1.

Fig. 1 Effect of σ on $f(\xi)$

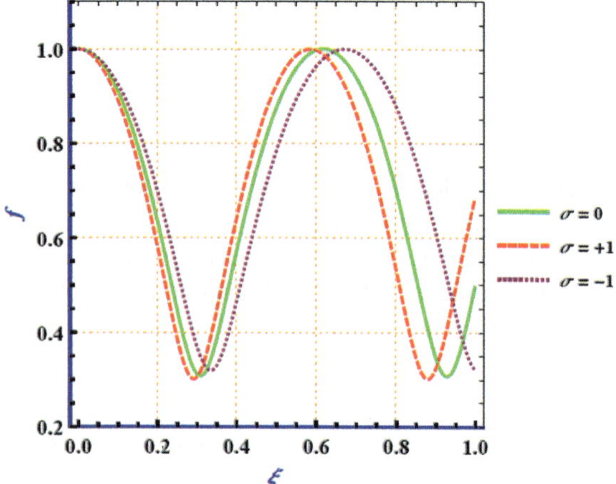

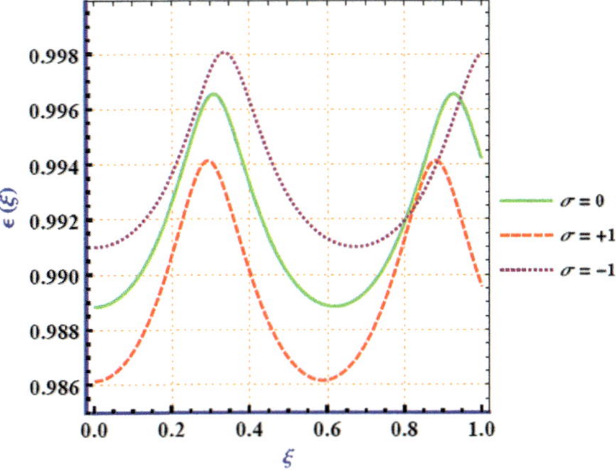

Fig. 2 Effect of σ on $\varepsilon(\xi)$

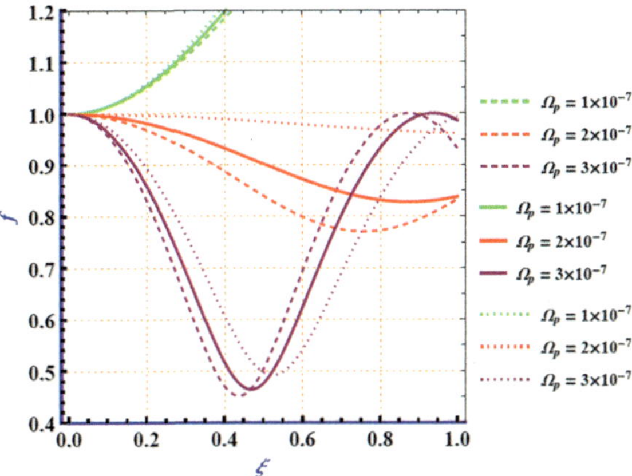

Fig. 3 Effect of Ω_p on $f(\xi)$. Solid curves ($\sigma = 0$), Dashed curves $\sigma = +1$, Dotted curves $\sigma = -1$

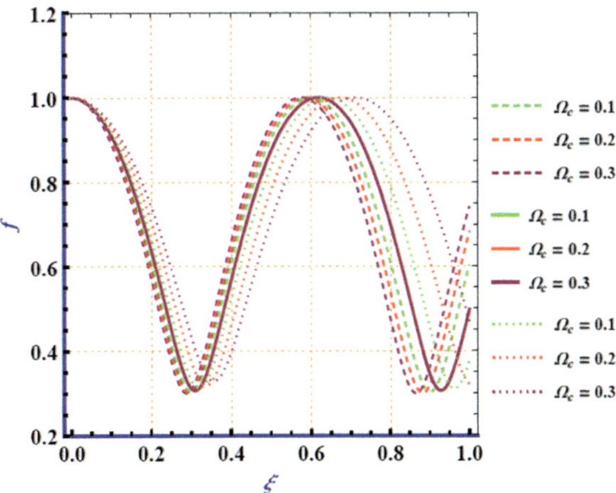

Fig. 4 Effect of Ω_c on $f(\xi)$. Solid curves ($\sigma = 0$), Dashed curves $\sigma = +1$, Dotted curves $\sigma = -1$

4 Conclusions

In this paper, second order nonlinear BWP differential equation of Gaussian beam in magnetized RCQP is obtained. It is found that, strength and direction of external magnetic field affects the SF of the Gaussian beam in plasma. In addition, quantum mechanical effects add the SF behaviour of laser in plasma.

References

1. Lalousis, P., Foldes, I.B., Hora, H.: Ultrahigh acceleration of plasma by picosecond terawatt laser pulses for fast ignition of fusion. Laser Part. Beams **30**, 233–242 (2012)
2. Askari, H.R., Azish, Z.: Effect of a periodic magnetic field on phase matching condition in second harmonic generation at interactions of laser-plasma. Optik **122**, 1159 (2011)
3. Lemoff, B.E., Yin, G.Y., Gordon, C.L., Barthy, C.P.J., Harris, S.E.: Demonstration of a 10-Hz femtosecond-pulse-driven XUV laser at 41.8 nm in Xe IX. Phys. Rev. Lett. **74**, 1574 (1995)
4. Parashar, J., Pandey, H.D., Tripathi. V.K.: Two-dimensional effects in a tunnel ionized plasma, Phys. Plasmas **4**, 3040 (1997)
5. Liu, C.S., Tripathi, V.K.: Self-focusing and frequency broadening of an intense short-pulse laser in plasmas. J. Opt. Soc. Am. A **18**, 1714 (2001)
6. Akhmanov, S.A., Sukhorukov, A.P., Khokhlov, R.V.: Self-focusing and diffraction of light in a Nonlinear medium. Sov. Phys. Usp. **10**, 609–636 (1968)
7. Sodha, M.S., Ghatak, A.K., Tripathi, V.K.: Self-focusing of laser beams in plasmas and semiconductors. Prog. Opt. **13**, 169–265 (1976)
8. Marklund, M., Shukla, P.K.: Nonlinear collective effects in photon-photon and photon-plasma interactions. Rev. Mod. Phys. **78**, 591 (2006)
9. Patil, S.D., Takale, M.V.: Stationary self-focusing of Gaussian laser beam in relativistic thermal quantum plasma. Phys. Plasmas **20**, 072703 (2013)
10. Aggarwal, M., Kumar, H., Kant, N.: Propagation of Gaussian laser beam through magnetized cold plasma with increasing density ramp. Optik **127**, 2212–2216 (2016)
11. Aggarwal, M., Kumar, H., Richa., Gill, T.S.: Self-focusing of Gaussian laser beam in weakly relativistic and ponderomotive cold quantum plasma. Phys. Plasmas **24**, 013108 (2017)
12. Aggarwal, M., Goyal, V., Richa, Kumar, H., Gill, T.S.: Weakly relativistic self-focusing of Gaussian laser beam in magnetized cold quantum plasma. Laser Part. Beams **35**, 699–705 (2017).
13. Aggarwal, M., Goyal, V., Gill, T.S.: Relativistic-Ponderomotive Self-focusing of Gaussian laser beam propagating in magnetized cold quantum plasma. Braz. J. Phys. **51**, 1642–1650 (2021)
14. Pawar, V.S., Nikam, P.P., Kokare, S.R., Patil, S.D., Takale, M.V.: Relativistic self-focusing of finite Airy-Gaussian laser beams in cold quantum plasma. J. Opt. **50**, 403–409 (2021)

Inelastic Soliton Collision in Multispecies Inhomogeneous Plasma

K. Raghavi, L. Kavitha, and C. Lavanya

Abstract In the present study, the multispecies in homogeneous dusty plasma with three constituents electrons, positrons and ions are observed. The Non linear Schr odinger Equation (NLSE) is obtained along the lines of commonly using reductive perturbation technique. We explore the collision of inelastic solitons upon employing the Hirota's bilinearization procedure. The results show that the properties of soliton collision are significantly influenced by the positron embedded in the three component plasma.

Keywords Inhomogeneous plasma · Reductive perturbation technique · Hirota bilinearization

1 Introduction

At present wave phenomena in dusty plasmas have much importance by reason of its part in the analysis of space and astrophysical background which includes tail of comets, ring of planets, interstellar space, asteroid zones etc. [1–3]. In a similar dusty plasma system, the grains of dust are immersed in the surrounding plasma and dispersion background. They interact with plasma particles owing to the charge associated with them. The dust, gain charge because of some conditions such as Collisional charging, Photo electric emisssion etc., and the state of unchanged charged dust grains modify the wave spectra whereas dynamics of dust charge supports the new eigen modes in dusty plasma [4–8]. In early stages, Zabusky and Kruskal observed

K. Raghavi · L. Kavitha (✉)
Department of Physics, School of Basic and Applied Sciences,
Central University of Tamil Nadu, Thiruvarur 610005, India
e-mail: lkavitha@cutn.ac.in

L. Kavitha
The Abdus Salam International Centre for Theoretical Physics, Trieste, Italy

C. Lavanya
Department of Physics, Periyar University, Salem 636 011, Tamil Nadu, India

S. Banerjee and A. Saha (eds.), *Nonlinear Dynamics and Applications*,
Springer Proceedings in Complexity,
https://doi.org/10.1007/978-3-030-99792-2_14

the influence of solitons in the plasma system which is collisionless and found the recurrence of beginning states [9]. S. Gardner et al. derived a technique for solving the Korteweg-de Vries equation which can be applied to forecast exactly the solitary waves that emerges from random initial conditions [10]. The Ion waves in a massively charged dusty plasma was examined by Angelis et al. and they derived an equation explaining low-frequency electrostatic perturbations on a non homogeneous background. The model is used for interpreting noise enhancement in low frequency range in the regions of dust in Halley's comet [5]. T. E. Sheridan investigated the characteristics of large amplitude negative potential solitary wave in multi-component dusty plasma particularly consisting of three components [11]. T. S. Gill and H. Kaur used the Sagdeev Pseudopotential method to study the solitary waves in unmagnetised dusty plasma [12]. Xue investigated the collision between dust acoustic solitary waves in an unmagnetised dusty plasma and established the phase shift modification by varying dust charge [13]. One dimensionally solitary waves in plasma system was reported by many investigators. Sayed and Mamum investigated the solitary waves in dusty plasma system consisting of four components using the reductive perturbation method [14]. Lin and Duan studied DASWs using multi-component dusty plasma consisting of non thermal ions, electrons that obey Boltzmann distribution law and negatively charged dust fluid [15]. Dusty plasma consisting of positive dust grains were investigated by T. K. Baluku et al. and tracked down the results that the DIA solitons are restricted to the positive potential whereas in case of negative dust grains either positive or negative potential may exist [16]. Harvey et al. observed the interaction of two counterpropagating solitons of identical amplitude and found that during collision, the sum of the initial soliton amplitudes was greater than the overlapped soliton amplitude [17]. T. Suji and Oikawa investigated the influence of solitons in a two layer fluid two dimensionally and finally generated a new wave called stem [18]. The collisional properties of dust acoustic solitary waves is investigated by Ghosh et al. using Poincare Lightthill Kuo method [19]. Alfven waves in space and astrophysical regions are studied by V. Jatenco-Pereira et al. and resulted that the existence of superthermal ions and dust modifies the dispersion of Alfven waves [3]. S. K.Sharma et al. investigated the collision and propagation of DASW in a strongly coupled dusty plasma [20]. L. Kavitha et al. investigated the excitations of solitons in low frequency Alfven waves in magnetized dusty plasmas [21]. Najah Kabalan et al. recorded the DA multisolton interactions using numerical simulations in the system of dusty plasma which is strongly coupled [22]. Krishan Kumar et al. reported the reflection aspects of DA solitary waves in dusty plasma in an experimental view [23]. The manuscript is organized as follows. Sect. 1 gives an introduction to the work with related literature. In Sect. 2 we establish the nonlinear equations for ion acoustic solitary waves. We use the reductive perturbation approach and arrive the nonlinear Schrödinger equation in Sect. 3 elucidate the collisional attributes of dust acoustic solitary waves. Sect. 4 contains a breif summary of our investigation.

2 Theoretical Model and Equation of Motion

An inhomogeneous plasma which exhibits less collision is considered. The constituents of such plasma are positrons, electrons and ions [24]. The dynamics of the ions are represented in normalized form by the below equations.

$$\frac{\partial n_i}{\partial t} + \frac{\partial (n_i v_i)}{\partial x} = 0, \tag{1}$$

$$\frac{\partial v_i}{\partial t} + v_i \frac{\partial v_i}{\partial x} = -\frac{\partial \phi}{\partial x}, \tag{2}$$

and

$$\frac{\partial^2 \phi}{\partial x^2} = (n_e - n_i - n_p). \tag{3}$$

The densities of positron, electron and ion are expressed in terms of $n_p, n_e\ n_i$ respectively. The velocity of ion fluid (v_i) and electrostatic potential (ϕ) are the variables of the equation. The debye length λ_{Dd}=$\sqrt{\frac{T_e}{4\pi e^2 n_0}}$, normalizes the space coordinate and ion inverse plasma frequency ω_{pi}^{-1}=$\sqrt{\frac{m_i}{4\pi e^2 n_0}}$, normalizes the time coordintes. Further, the electrostatic potentials is normalized by $\frac{T_e}{e}$ and velocity by $c_{si} = \sqrt{\frac{T_e}{m_i}}.T_p$ and T_e represents the temperature of positron and electron respectively. Moreover, m_i is the ion mass and e designates the magnitude of the electron charge. The evolution equation we arrived is the Nonlinear Schrӧdinger equation, by involving the standard reductive perturbation technique [25, 26] and the nonlinear Schrӧdinger equation we obtained is:

$$i\frac{\partial \phi}{\partial \tau} + P\frac{\partial^2 \phi}{\partial \xi^2} + Q|\phi|^2\phi = 0, \tag{4}$$

In Eq. (4), P is the nonlinearity coefficient and Q is the dispersion coefficient. They are given as

$$\begin{aligned}
P =\ & \frac{1}{k\Big[-k^2 - n_{e0}\big(\frac{e}{T_e}\big) - n_{p0}\big(\frac{e}{T_p}\big)\Big]}\Big[-v_g k A_1 J + 2v_g k^2 + A_2 J k n_{i0} - 2k\omega \\
& +2k^2 v_{i0} - v_g A_1 + n_{i0} A_2 + v_{i0} A_1 + v_{i0} k A_1 J - 2v_{i0} k^2\Big], \\
Q =\ & \frac{-1}{2n_{i0} T_e^2 T_p^2\Big[-k^2 - n_{e0}\big(\frac{e}{T_e}\big) - n_{p0}\big(\frac{e}{T_p}\big)\Big]}\big[-6kA_1^2 A_2 T_e^2 T_p^2 + 8A_1\omega B T_e^2 T_p^2 k^2 \\
& +2A_1\omega B T_e T_p^2 n_{e0} e + 2A_1\omega B T_e^2 T_p n_{e0} e + A_1\omega n_{e0} e^2 T_p^2 - A_1\omega n_{p0} e^2 T_e^2 - A_1\omega \\
& \times n_{p0} T_e^2 T_p^2 - 8A_1 k^3 v_{i0} B T_e^2 T_p^2 - 2A_1 k v_{i0} B T_e T_p^2 n_{e0} e - 2A_1 k v_{i0} B T_e^2 T_p n_{p0} e - 3 \\
& \times A_1 k v_{i0} n_{e0} e^2 T_p^2 + 3A_1 k v_{i0} n_{p0} T_e^2 e^2 + A_1 k v_{i0} n_{p0} T_e^2 T_p^2 + 8k^3 A_2 n_{i0} B T_e^2 T_p^2 + 2kA_2 \\
& \times n_{i0} B T_e T_p^2 n_{e0} e + 2kA_2 n_{i0} B T_e^2 T_p n_{p0} e + 3kA_2 n_{i0} n_{e0} B e^2 T_p^2 - 3kA_2 n_{i0} n_{p0} T_e^2 e^2
\end{aligned}$$

$$-kA_2n_{i0}n_{p0}T_e^2T_p^2 - 2kA_1ev_{i0}CT_eT_p^2n_{e0} - 2kA_1ev_{i0}CT_e^2T_pn_{p0} + 2kA_1ev_gCT_e \times T_p^2n_{p0} + 2kA_1ev_gCT_e^2T_pn_{p0} + 2kA_1v_gn_{e0}e^2T_p^2 - 2kA_1v_gn_{p0}T_e^2e^2 + 2A_2ken_{i0}C \times T_eT_p^2n_{e0} + 2A_2ken_{i0}CT_e^2T_pn_{p0}],$$

where

$$J = \frac{1}{-kn_{i0}(A_2\omega - A_2kv_{i0} - k)}\Big[-2k\omega^2 + 4\omega k^2v_{i0} - \omega v_gA_1 + \omega n_{i0}A_2 + \omega v_{i0}A_1 -2k^3v_{i0}^2 + kv_{i0}v_gA_1 - kv_{i0}^2A_1 - v_gA_2kn_{i0} + kn_{i0}\Big],$$

$$B = \frac{(-\omega + kv_{i0}) - 8ikn_{i0}\Big[(\omega - kv_{i0})\Big(\frac{n_{e0}e^2}{2T_e^2} - \frac{n_{p0}e^2}{2T_p^2} + \frac{1}{2}\Big) - kA_1A_2\Big] - ikA_1A_2}{\frac{-\omega+kv_{i0}}{kn_{i0}}\Big[(4k^2 + \frac{n_{e0}e}{T_e} + \frac{n_{p0}e}{T_p})\Big] + 2ik},$$

$$C = \frac{2A_1A_2 + \frac{(v_{i0}-v_g)^2}{n_{i0}}\Big[\frac{n_{e0}e^2}{T_e^2} - \frac{n_{p0}e^2}{T_p^2}\Big] - A_2A_2}{\frac{-(v_{i0}-v_g)^2}{n_{i0}}\Big[\frac{n_{e0}e}{T_e} + \frac{n_{p0}e}{T_p}\Big] + 1},$$

$$A_1 = k^2 + n_{e0}\frac{n_{e0}e}{T_e} + \frac{n_{p0}e}{T_p},$$

and

$$A_2 = \frac{i\omega k^2 + n_{e0}\frac{n_{e0}e}{T_e} + \frac{n_{p0}e}{T_p} - ikv_{i0}k^2 + n_{e0}\frac{n_{p0}e}{T_p}}{ikn_{i0}}.$$

3 Hirota Bilinearization and Plasmic Solitons

The solution for Nonlinear Wave Equations (NWEs) or Partial Differential Equations (PDEs) became a major challenge in solving NonLinear problems. Though many influential techniques like Inverse scattering transform, Bäcklund transformation, etc., exist, Hirota Direct method [27–31] is one of the capable technique for the construction of Multi-soliton solutions. The above technique not only helps in constructing the multi solitonic solution but also helps in obtaining the solutions of integrable nonlinear evolution equations. We apply rational transformation,to find the solution for Eq. (4).

$$\phi = \frac{g(\xi, \tau)}{f(\xi, \tau)}. \tag{5}$$

Here $g(\xi, \tau)$ and $f(\xi, \tau)$ are complex and real function which is to be determined. Further, using Eq. (5) in place of Eq. (4), we gain the following Hirota's bilinear form

$$\begin{aligned} &[iD_\tau + PD_\xi^2]g \cdot f = 0 \\ &PD_\xi^2(f \cdot f) - Qgg^* = 0, \end{aligned} \tag{6}$$

where $*$ is the symbol of asymmetricity and Bilinear operators introduced by Hirota, D_ξ and D_τ are given by

$$D_\xi^m D_\tau^n(g.f) = \left(\frac{\partial}{\partial \xi} - \frac{\partial}{\partial \xi'}\right)^m \left(\frac{\partial}{\partial \tau} - \frac{\partial}{\partial \tau'}\right)^n g(\xi, \tau) f(\xi', \tau')|_{\xi'=\xi, \tau'=\tau}. \tag{7}$$

using a small expansion parameter ϵ, g and f are expanded, Then we obtain

$$\begin{aligned} g &= \epsilon g_1 + \epsilon^3 g_3 + \epsilon^5 g_5 + ..., \\ f &= 1 + \epsilon^2 f_2 + \epsilon^4 f_4 + \epsilon^6 f_6 + ... \end{aligned} \tag{8}$$

on substituting the above Eq. (8) into Eq. (8), and solving we get the recursion relations.

3.1 *One Soliton Solution*

For the construction of one soliton solution, we assume;

$$\begin{aligned} g &= \epsilon g_1, \\ f &= 1 + \epsilon^2 f_2. \end{aligned} \tag{9}$$

The one-soliton solution is obtained explicitly by substitution of Eq. (9) into Eq. (6) and finally solving the resulting equation. The solution is

$$\phi = \frac{e^{\eta_1}}{1 + f_2}, \tag{10}$$

where

$$f_2 = \frac{Q}{2P(k_1 + k_1^*)^2} e^{\eta_1 + \eta_1^*},$$
$$\eta_1 = k_1 \xi - \omega_1 \tau + \eta_{10},$$
$$\omega_1 = -i k_1^2 P,$$

Here, ω_1 and k_1 are complex parameters whereas η_{10} is a constant.

3.2 Two-Soliton Solution

In order to find solutions of two-soliton, we introduce the following series:

$$g = \epsilon g_1 + \epsilon^3 g_3,$$
$$f = 1 + \epsilon^2 f_2 + \epsilon^4 f_4. \tag{11}$$

and replacing in Eq. (6) and working out the developed group of PDE, the definite two soliton solutions is arrived as follows:

$$\phi = \frac{\epsilon g_1 + \epsilon^3 g_3}{1 + \epsilon^2 f_2 + \epsilon^4 f_4}. \tag{12}$$

$$g_3 = e^{a_1} e^{\eta_1 + \eta_1^* + \eta_2} + e^{b_1} e^{\eta_1 + \eta_2 + \eta_2^*},$$
$$f_2 = \frac{Q}{2P} \left[e^{R_1} e^{\eta_1 + \eta_1^*} + e^{R_2} e^{\eta_1 + \eta_2^*} + e^{R_3} e^{\eta_2 + \eta_1^*} + e^{R_4} e^{\eta_2 + \eta_2^*} \right],$$
$$f_4 = e^{\eta_1 + \eta_1^* + \eta_2 + \eta_2^* + \delta},$$

where

$$e^{a_1} = \frac{Q(k_1 - k_2)^2}{2P(k_2 + k_1^*)^2 (k_1 + k_1^*)^2},$$
$$e^{b_1} = \frac{Q(k_1 - k_2)^2}{2P(k_2 + k_2^*)^2 (k_1 + k_2^*)^2},$$
$$e^{R_1} = \frac{1}{(k_1 + k_1^*)^2},$$

$$e^{R_2} = \frac{1}{(k_1 + k_2^*)^2},$$
$$e^{R_3} = \frac{1}{(k_2 + k_1^*)^2},$$
$$e^{R_4} = \frac{1}{(k_2 + k_2^*)^2},$$
$$e^{\delta} = \frac{Q^2}{2P}\left[\frac{(k_1 - k_2)^2(k_1^* - k_2^*)^2}{(k_1 + k_1^*)^2(k_2 + k_2^*)^2(k_1 + k_2)^2(k_1^* - k_2^*)^2}\right].$$

4 Results and Discussion

We investigated the effects of positron concentration on the plasmic soliton collision in electron positron ion plasma. We discuss graphically the interactions of solitary waves through the solution Eq. (12). Figure 1 indicates the interaction between the two soliton solutions Eq. (12) for arbitrary choices of parameters n_{i0}, T_e, T_p and for different values of density of positron n_{p0}. When $n_{p0} = 0.008$ units, the amplitude of soliton s_1 and s_2 is observed at 45 and 40 units respectively. Similarly by setting $n_{p0} = 0.09, 0.21$ and 3 units, the solitons s_1 and s_2 suffers a fall in amplitude as shown in Fig. 1. The Fig. 2 shows the cumulative intensity plot before collision ($t = -10$) and after collision ($t = 10$). The graphical representation of Fig. 2 clearly indicates that the two solitons s_1 and s_2 undergo a significant inelastic collision in which the amplitude of plasmic soliton decreases with the enhancement of positron density n_{p0}.

5 Conclusions

We have explored the collisional dynamics of solitary waves in an unmagnetized plasma consisting of positrons along with electrons and ions. The dynamics of such wave is described by the celebrated nonlinear Schrödinger equation. By applying the technique of Hirota bilinearization, the interaction between the two soliton solution is presented. It is observed that an increase in the positron density generates the inelastic soliton collision. The results manifested that the presence of positrons display a significant play role on the amplitude of the profile of the plasmic solitons.

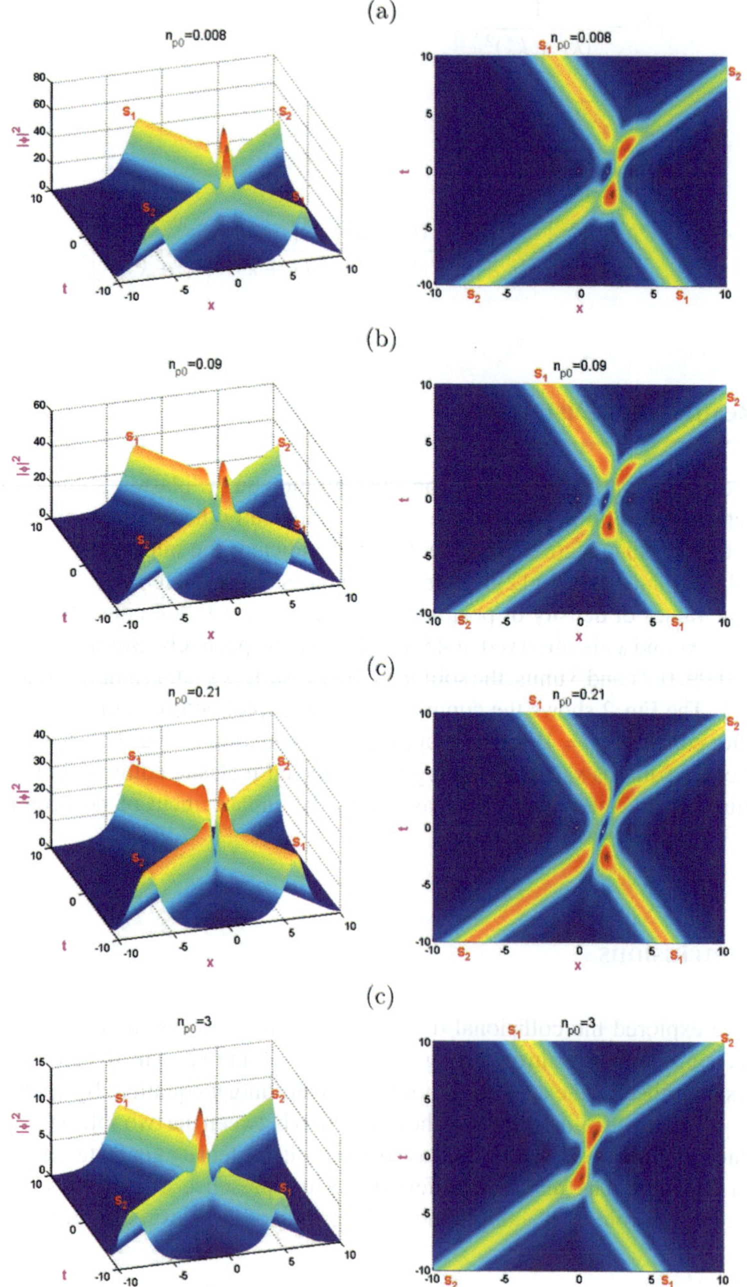

Fig. 1 Snapshots of collision between two soliton to the solution 4.27 with the various choices of parameters $T_e = 1.0, n_{i0} = 6.7, n_{e0} = 0.85, e = -1.6 \times 10^{-19}C, \omega = 9, v_g = 0.79, v_{i0} = 0.7,$ $k = 0.7$, $T_p = 6.0eV$, **a** $n_{p0} = 0.008$, **b** $n_{p0} = 0.09$, **c** $n_{p0} = 0.21$ and **d** $n_{p0} = 3$

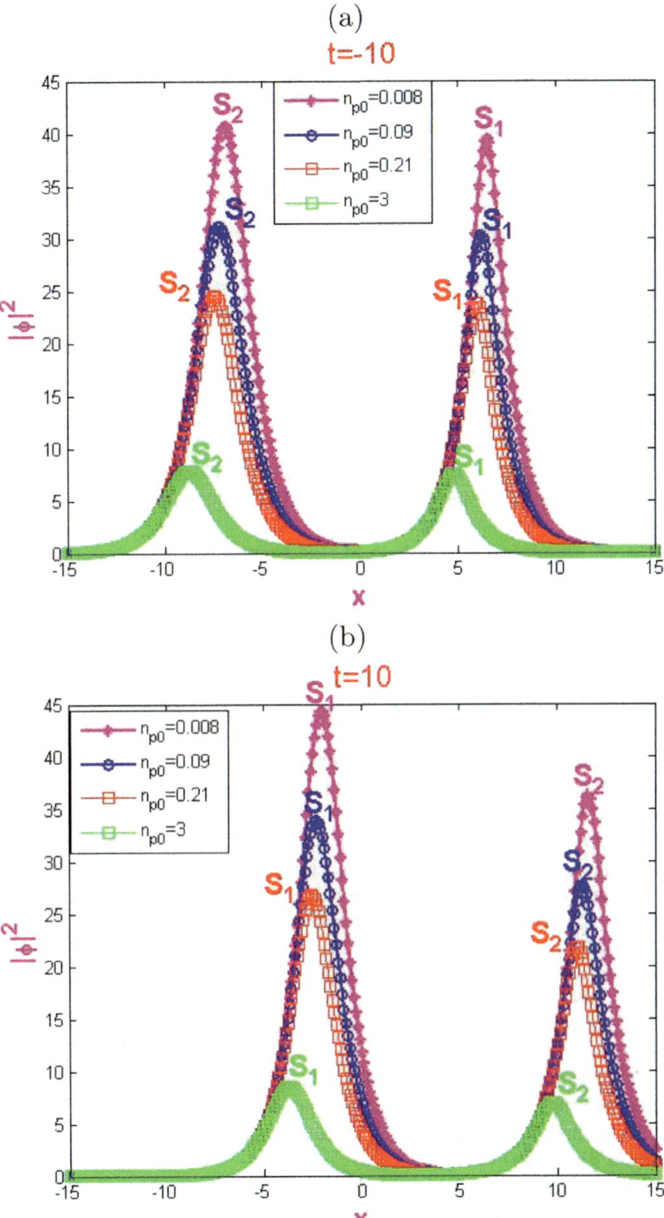

Fig. 2 The cumulative plot showing the intensity profile of two soliton collision for various values of n_{p0} **a** before ($t = -10$) and after **b** ($t = 10$) collision

Acknowledgements K.R. acknowledges the financial support from Central University of Tamil Nadu in the form of URF. L.K. gratefully acknowledges the financial support in the form of Major Research Projects by CSIR (Ref.No.:03(1418)/17/EMR-II), India, UGC-DAE CSR (Ref. No.: CSR-KN/CRS-102/2018-19/1047), DST-SERB (Ref.No.:MTR/2017/000314/MS), India and ICTP, Italy in the form of a Regular Associateship.

References

1. Horanyi, M., Mendis, D.A.: The effects of electrostatic charging on the dust distribution at Halley's Comet. Astrophys. J. **294**, 357–368 (1985)
2. Mendis, D.A., Rosenberg, M.: Some aspects of dust-plasma interactions in the cosmic environment. IEEE Trans. Plasma Sci. **20**, 6 (1992)
3. Jatenco-Pereira, V., Chian, A.C.L., Rubab, N.: Alfvén waves in space and astrophysical dusty plasmas. Nonlin. Proces. Geophys. **21**, 405–416 (2014)
4. Havens, O., Goertz, C.K., Morfill, G.E., Grun, E., Ip, W.: Confinement and structure of electrostatically coupled dust clouds in a direct current plasma-sheath. J. Geophys. Res. **92**, 2281–2287 (1987)
5. de Angelis, U., Formisano, V., Giordano, M.: Electromagnetic wave scattering in dusty plasmas. J. Plasma Phys. **40**, 399–406 (1988)
6. de Angelis, U., Bingham, R., Tsytovich, V.N.: Electromagnetic wave scattering in dusty plasmas. J. Plasma Phys. **40**, 445–456 (1989)
7. de Angelis, U., Forlani, A., Bingham, R., Shukla, P.K., Ponomarev, A., Tsytovich, V.N.: Damping and absorption of high-frequency waves in dusty plasmas. Phys. Plasmas **1**, 236 (1994)
8. Walch, B., Horanyi, M., Robertson, S.: Charging of dust grains in plasma with energetic electrons. Phys. Rev. Lett. **75**, 838–841 (1995)
9. Zabusky, N.J., Kruskal, M.D.: Interaction of "Solitons" in a collisionless plasma and the recurrence of initial states. Phys. Rev. Lett. **15**, 240–243 (1965)
10. Gardner, C.S., Greene, J.M., Kruskal, M.D., Miura, R.M.: Method for Solving the Korteweg-deVries equation. Phys. Rev. Lett. **19**, 1095–1097 (1967)
11. Sheridan, T.E.: On the origin of the ion acoustic soliton. J. Plasma Phys. **60**, 17–28 (1998)
12. Gill, T.S., Kaur, H.: Effect of nonthermal ion distribution and dust temperature on nonlinear dust acoustic solitary waves. Pramana **55**, 855–859 (2000)
13. Xue, J-K.: Head-on collision of dust-acoustic solitary waves. Phys. Rev. E **69**, 016403 (2004)
14. Sayed, F., Mamun, A.A.: Solitary potential in a four-component dusty plasma. Phys. Plasmas, **14**, 014501 (2007)
15. Lin, M., Duan, W.-S.: Envelope solitons in plasma with vortex-like electrons. Chaos, Solitons Fractals **33**, 1189–1196 (2007)
16. Baluku, T.K., Hellberg, M.A., Mace, R.L.: Dust acoustic solitons in plasmas with kappa-distributed electrons and/or ions. Phys. Plasmas **15**, 03701 (2008)
17. Harvey, P., Durniak, C., Samsonov, D., Morfill, G.: Soliton interaction in a complex plasma. Phys. Rev. E **81**, 057401 (2010)
18. Tsuji, H., Oikawa, M.: Two-dimensional interactions of solitons in a two-layer fluid of finite depth. Fluid Dyn. Res. **42**, 065506 (2010)
19. Ghosh, U.N.G., Roy, K., Chatterjee, P.: Head-on collision of dust acoustic solitary waves in a four-component dusty plasma with nonthermal ions. Phys. Plasmas **18**, 103703 (2011)
20. Sharma, S.K., Boruah, A., Bailung, H.: Head-on collision of dust-acoustic solitons in a strongly coupled dusty plasma. Phys. Rev. E **89**, 013110 (2014)
21. Kavitha, L., Lavanya, C., Senthil Kumar, V., Gopi, D., Pasqua, A.: Perturbed soliton excitations of Rao-dust Alfvén waves in magnetized dusty plasmas. Phys. Plasmas, **23**, 043702 (2016)
22. Kabalan, Najah, Ahmad, Mahmoud, Asad, Ali: Study of dust-acoustic multisoliton interactions in strongly coupled dusty plasmas. Adv. Math. Phys. **2020**, 1–9 (2020)

23. Kumar, K., Bandyopadhyay, P., Singh, S., Arora, G., Sen, A.: Reflection of a dust acoustic solitary wave in a dusty plasma. Phys. Plasmas **1**, 12959 (2021)
24. Kaur, H., Gill, T.S., Saini, N.S.: The Kadomstev-Petviashvili equation in dusty plasma with variable dust charge and two temperature ions. Pramana **66**, 1049–1056 (2006)
25. Tiwari, R.S., Mishra, M.K.: Ion-acoustic dressed solitons in a dusty plasma. Phys. Plasmas **13**, 062112 (2006)
26. Kavitha, L., Raghavi, K., Lavanya, C., Kailas, M., Gopi, D.: Propagation of electrostatic solitary waves in the four-component dusty plasma. IEEE Trans. Plasma Sci. **49**, 546 (2021)
27. Kavitha, L., Saravanan, M., Senthilkumar, V., Ravichandran, R., Gopi, D.: Collision of electromagnetic solitons in a weak ferromagnetic medium. J. Mag. Magn. mater. **355**, 37–50 (2014)
28. Kavitha, L., Srividhya, B., Dhamayanthi, S., Senthil Kumar, V., Gopi, D.: Collision and propagation of electromagnetic solitons in an antiferromagnetic spin ladder medium. Appl. Math. Comput. **251**, 643–668 (2015)
29. Kavitha, L., Venkatesh, M., Dhamayanthi, S., Gopi, D.: Nonlinear refractive index induced collision and propagation of nematicons. J. Mol. Liq. **192**, 142–151 (2014)
30. Kavitha, L., Muniyappan, A., Zdravkovic, S., Sataric, M.V., Marlewski, A., Dhamayanthi, S., Gopi, D.: Propagation of kink-antikink pair along microtubules as a control mechanism for polymerization and depolymerization processes. Chin. Phys. B **9**, 098703 (2014)
31. Lakshmanan, M., Rajasekar, S.: Nonlinear Dynamics Integrability. Springer, Chaos and Patterns (2003)

Propagation of Rarefactive Dust Acoustic Solitary and Shock Waves in Unmagnetized Viscous Dusty Plasma Through the Damped Kadomstev-Petviashvili Burgers Equation

Tanay Sarkar, Santanu Raut, and Prakash Chandra Mali

Abstract The non-linear propagation of dust acoustic waves (DAWs) in collisional, unmagnetized, viscous dusty plasma systems containing two temperature ions, electrons, high negatively charged dust grains are investigated. By using the reductive perturbation method (RPM) the damped Kadomtsev-Petviashvili Burgers (dKPB) equation that governs the DAWs is derived. Generally, the impact of viscosity is ignored during the studies of wave dynamics in a plasma medium. In the present investigation, a Burgers term is introduced in order to express the dissipation effect in the viscous plasma circumstance. The strong dissipation due to the presence of Burgers term may cause for rising of a shock solution. However, in a very weak dissipative system, the solitary-like wave solution may arise due to the balance between the dispersion and nonlinearity. Assuming conservation law in the present system, solitary type wave solution is explored, and shock type wave solution is determined by means of Simplified Hirota bilinear method (SHBM). Finally, the effect of the kinematic viscosity, collisional frequency, etc. on wave propagation is demonstrated from numerical understanding.

Keywords Dust acoustic waves · Reductive Perturbation method · Damped KP-Burgers equation · Burgers term · Shock solution · Solitary like wave solution

1 Introduction

Dusty plasma plays a vital role in understanding different nonlinear phenomena appearing in space and astrophysical studies [1, 2]. Due to the presence of different kinds of dust particles of different sizes and masses, dusty plasma gave birth to

T. Sarkar (✉) · P. C. Mali
Department of Mathematics, Jadavpur University, Kolkata 700032, India
e-mail: sarkartanay9@gmail.com

S. Raut
Department of Mathematics, Mathabhanga College, Coochbehar 736146, India

S. Banerjee and A. Saha (eds.), *Nonlinear Dynamics and Applications*,
Springer Proceedings in Complexity,
https://doi.org/10.1007/978-3-030-99792-2_15

various wave modes such as dust acoustic mode [3], dust ion-acoustic mode [4], etc. Over the past few decades, DAW of dusty plasma has come up as a hot research topic and research on this topic is still ongoing. In 1990, Rao et al. [3] first theoretically brought the DAWs to everyone and after five years Barkan et al. [5] confirm the earlier prediction through laboratory experiments.

DAW is the most fundamental content of dusty plasma. DAWs are largely found in the rings of the various planets, mesosphere of the Earth, tails of comets, etc. [6]. A lot of research has been done on dust acoustic waves [7–10].

Among the various nonlinear structures of DAWs, dust acoustic solitary waves (DASWs) and dust acoustic shock waves (DAShWs) are the vastly growing area of research nowadays. Presently DASWs and DAShWs acquired a special place in space plasma research, laboratory plasma research, etc. Solitary waves are produced when there is a balance between nonlinearity and dispersive effects wheres the reason behind the production of shock waves is the dominance of dissipation over the dispersion in nonlinear media [11, 12]. Several researchers have worked for the development of DASWs [13, 14] and DAShWs [15]. Both DASW and DAShW structures may contain in a plasma system where both dispersion and dissipation are present [16].

In the present paper, we analyzed the propagation of DAWs in a collisional, unmagnetized, viscous dusty plasma system comprising of electrons, two temperature ions, and dust particles with a high negative charge. Applying reductive perturbation method, dKPB equation is derived, and using momentum conservation law solitary waves solutions are derived. Shock solution is also obtained for this equation by the SHBM technique. The remaining portion of this paper is arranged as follows: The basic governing equations are considered in Sect. 2. The damped KP-Burgers equation is originated by using RPM in Sect. 3. Solutions of the dKPB equation are given in Sect. 4. The parametric discussion of the solution with diagrams is delivered in Sect. 5 and the conclusions are given in Sect. 6.

2 Basic Model Equations

We consider here an unmagnetized viscous dusty plasma containing of high negatively charged dust grains, electrons and two different temperature ions. For the neutrality of net charge at equilibrium satisfies

$$N_{e0} + N_{d0}Z_{d0} - N_{il0} - N_{ih0} = 0, \tag{1}$$

where N_{e0}, N_{d0}, N_{il0} and N_{ih0} are the values of the number densities of electrons, dust, lower temperature ions and higher temperature ions at equilibrium respectively. Z_{d0} is the unperturbed number of charges on the dust particles. Then the dynamics of DAWs for variable dust charge can be described by the following continuity, motion for the dust and Poisson's equations as

$$\frac{\partial N_d}{\partial t} + \frac{\partial}{\partial x}(N_d U_d) + \frac{\partial}{\partial y}(N_d V_d) = 0, \tag{2}$$

$$\frac{\partial U_d}{\partial t} + U_d \frac{\partial U_d}{\partial x} + V_d \frac{\partial U_d}{\partial y} = Z_d \frac{\partial \phi}{\partial x} - \zeta \left(\frac{\partial^2}{\partial x^2} + \frac{\partial^2}{\partial y^2} \right) U_d - \mu_{id} U, \tag{3}$$

$$\frac{\partial V_d}{\partial t} + U_d \frac{\partial V_d}{\partial x} + V_d \frac{\partial V_d}{\partial y} = Z_d \frac{\partial \phi}{\partial y} - \zeta \left(\frac{\partial^2}{\partial x^2} + \frac{\partial^2}{\partial y^2} \right) V_d - \mu_{id} V, \tag{4}$$

$$\frac{\partial^2 \phi}{\partial x^2} + \frac{\partial^2 \phi}{\partial y^2} = Z_d N_d + N_e - N_{il} - N_{ih}, \tag{5}$$

where N_d denotes the dust number density, Z_d denotes the number of charges on dust particles. The effective temperature T_{eff} satisfies

$$T_{eff} = N_{d0} Z_{d0} \left(\frac{N_{e0}}{T_e} + \frac{N_{il0}}{T_{il}} + \frac{N_{ih0}}{T_{ih}} \right)^{-1}, \tag{6}$$

where T_e, T_{ih} and T_{il} are temperature of electrons, higher temperature ions and lower temperature ions respectively. U_d and V_d are the velocities of the dust flow along the direction of x-axis and y-axis respectively and normalized by the dust acoustic speed $c_d = (\frac{K_B T_{eff} Z_{d0}}{m_d})^{\frac{1}{2}}$ in which K_B is the Boltzmann constant and m_d represents the dust particles mass. The electrostatic potential ϕ is normalized as $\phi = \frac{K_B T_{eff}}{e}$. Space and time variables are scaled over the effective Debye length, $\lambda_d = (\frac{K_B T_{eff}}{4\pi N_{d0} Z_{d0} e^2})^{\frac{1}{2}}$ and the inverse of dust plasma frequency, $\omega_{pd}^{-1} = (\frac{m_d}{4\pi N_{d0} Z_{d0}^2 e^2})^{\frac{1}{2}}$. $\zeta = \frac{\zeta_0}{\omega_{pd} \lambda_d^2}$, here ζ_0 denotes the kinematic viscosity of dust. μ_{id} represents the collisional frequency.

N_e, N_{il} and N_{ih} are respectively the number densities for electrons, lower temperature ions, higher temperature ions and which are given by

$$N_e = \frac{N_{e0}}{N_{d0} Z_{d0}} \exp(\beta_2 s \phi), \tag{7}$$

$$N_{il} = \frac{N_{il0}}{N_{d0} Z_{d0}} \exp(-s \phi), \tag{8}$$

$$N_{ih} = \frac{N_{e0}}{N_{d0} Z_{d0}} \exp(-\beta_1 s \phi), \tag{9}$$

$$\text{where } \beta_1 = \frac{T_{il}}{T_{ih}}, \beta_2 = \frac{T_{il}}{T_e}, \beta_3 = \frac{T_{ih}}{T_e}, s = \frac{T_{eff}}{T_{il}}, \rho_1 = \frac{N_{il0}}{N_{e0}}, \rho_2 = \frac{N_{ih0}}{N_{e0}}. \tag{10}$$

From (1) and (10) it follows

$$\rho_1 + \rho_2 - 1 = 0, \tag{11}$$

$$s = \frac{\rho_1 + \rho_2 - 1}{\rho_1 + \rho_2\beta_1 + \beta_2}. \tag{12}$$

The variable Q_d of dust charge can be calculated from the equation [17]

$$\left(\frac{\partial}{\partial t} + \vec{V}.\vec{\nabla}\right) Q_d = J_e + J_{il} + J_{ih}, \tag{13}$$

where $\vec{V} = (U_d, V_d)$ and J_e, J_{il} and J_{ih} are the electron, lower temperature ions and higher temperature ions currents respectively. We assume that the thermal velocities of electrons and ions are much greater than the streaming velocities, thus $\frac{dQ_d}{dt} << J_e, J_{il}, J_{ih}$ and the equation (13) looks [1]

$$J_e + J_{il} + J_{ih} \approx 0. \tag{14}$$

The electron and ion currents are satisfied the relation [1]

$$J_e = -e\pi r^2 \left(\frac{8T_e}{\pi m_e}\right)^{\frac{1}{2}} N_e \exp\left(\frac{e\Phi}{T_e}\right), \tag{15}$$

$$J_{il} = e\pi r^2 \left(\frac{8T_{il}}{\pi m_i}\right)^{\frac{1}{2}} N_{il} \left(1 - \frac{e\Phi}{T_{il}}\right), \tag{16}$$

$$J_{ih} = e\pi r^2 \left(\frac{8T_{ih}}{\pi m_i}\right)^{\frac{1}{2}} N_{ih} \left(1 - \frac{e\Phi}{T_{ih}}\right), \tag{17}$$

where Φ is the potential of dust particles surface related to the plasma potential ϕ. We gain the dust charge Z_d in normal condition, from $Z_d = \frac{\psi}{\psi_0}$, where $\psi = \frac{\exp^{\Phi}}{T_{eff}}$, and $\psi_0 = \psi(\phi = 0)$. Expressing Z_d in terms of ϕ we get [18]

$$Z_d = 1 + \gamma_1\phi + \gamma_2\phi^2 + \cdots, \tag{18}$$

where $\gamma_1 = \frac{1}{\psi_0}(\frac{d\psi(\phi)}{d\phi})_{\phi=0}$ and $\gamma_2 = \frac{1}{2\psi_0}(\frac{d^2\psi(\phi)}{d\phi^2})_{\phi=0}$.

3 Derivation of Damped KP-Burgers Equation

In this section we derive the damped KP-Burgers equation and for this purpose we have employed the RPM [19]. The stretching co-ordinates are given as

$$\xi = \epsilon(x - \lambda_p t), \ \tau = \epsilon^3 t, \ \eta = \epsilon^2 y, \tag{19}$$

where the small parameter ϵ characterizes the strength of the non-linearity of the system and λ_p stands for presenting the phase velocity of the wave. To get the damped KP-Burgers equation, we expand the perturbation quantities N_d, U_d, V_d, Z_d and ϕ in power series of ϵ as

$$N_d = N_{d0} + \epsilon^2 N_{d1} + \epsilon^4 N_{d2} + \cdots, \tag{20}$$

$$U_d = U_{d0} + \epsilon^2 U_{d1} + \epsilon^4 U_{d2} + \cdots, \tag{21}$$

$$V_d = V_{d0} + \epsilon^3 V_{d1} + \epsilon^5 V_{d2} + \cdots, \tag{22}$$

$$Z_d = Z_{d0} + \epsilon^2 Z_{d1} + \epsilon^4 Z_{d2} + \cdots, \tag{23}$$

$$\phi = \phi_0 + \epsilon^2 \phi_1 + \epsilon^4 \phi_2 + \cdots, \tag{24}$$

$$\zeta = \epsilon \zeta_0, \ \mu_{id} = \epsilon^3 \mu_{id0}. \tag{25}$$

By using the RPM from Eqs. (2)–(5) and using stretching coordinates (19) along with state variables from Eqs. (20)–(25), we gain a evolution equations set. Calculating and considering $\phi_1 = \phi$, we obtain the damped KP-Burgers equation as

$$\frac{\partial}{\partial \xi}\left(\frac{\partial \phi}{\partial \tau} + A\phi\frac{\partial \phi}{\partial \xi} + B\frac{\partial^3 \phi}{\partial \xi^3} + C\frac{\partial^2 \phi}{\partial \xi^2} + D\phi\right) + E\frac{\partial^2 \phi}{\partial \eta^2} = 0, \tag{26}$$

where $A = \frac{\lambda_p^3}{2}\left[(\rho_1 + \rho_2\beta_1^2 - \beta_2^2)\frac{(\rho_1+\rho_2-1)}{(\rho_1+\rho_2\beta_1+\beta_2)^2} - 2\gamma_2\right] + \frac{3}{2}\gamma_1\lambda_p - \frac{3}{2\lambda_p}$, $B = \frac{\lambda_p^3}{2}$, $C = \frac{\zeta_0}{2}$, $D = \frac{\mu_{id0}}{2}$, $E = \frac{\lambda_p}{2}$, and $\lambda_p = (1 + \gamma_1)^{-\frac{1}{2}}$.

4 Solution of Damped KP-Burgers Equation

4.1 Solitary Wave Solutions

In this part we will solve damped KP-Burgers equation (26) using momentum conservation law. Integrating (26) with respect to ξ we have

$$\frac{\partial \phi}{\partial \tau} + A\phi\frac{\partial \phi}{\partial \xi} + B\frac{\partial^3 \phi}{\partial \xi^3} + C\frac{\partial^2 \phi}{\partial \xi^2} + D\phi + E\int \frac{\partial^2 \phi}{\partial \eta^2}\, d\xi = 0. \tag{27}$$

Let $z = \xi + \eta$, then (27) reduces to

$$\frac{\partial \phi}{\partial \tau} + A\phi\frac{\partial \phi}{\partial z} + B\frac{\partial^3 \phi}{\partial z^3} + C\frac{\partial^2 \phi}{\partial z^2} + D\phi + E\frac{\partial \phi}{\partial z} = 0. \tag{28}$$

removing burgers and damping term from the above equation by putting $C = 0$, $D = 0$ then Eq. (28) changes to the equation

$$\frac{\partial \phi}{\partial \tau} + E\frac{\partial \phi}{\partial z} + A\phi\frac{\partial \phi}{\partial z} + B\frac{\partial^3 \phi}{\partial z^3} = 0, \tag{29}$$

which is the KdV type equation and solitary wave solution is of the form

$$\phi(z, \tau) = \phi_m sech^2\left(\frac{z - M\tau}{W}\right), \tag{30}$$

where $\phi_m = \frac{3(M-E)}{A}$ indicates the amplitude, $W = 2\sqrt{\frac{B}{M-E}}$ indicates the width and M indicates the speed of DASW. To find solitary wave solutions of dKPB Eq. (26) we use (30) as a seed solution and so for small values of C (Burgers coefficient) and D (damping coefficient), we assume that the DASW solutions of Eq. (26) with considering amplitude ϕ_m, width W and velocity M are time τ dependent as

$$\phi(z, \tau) = \phi_m(\tau) sech^2\left(\frac{z - M(\tau)\tau}{W(\tau)}\right). \tag{31}$$

Now conserved quantity for KdV type equation (29) is [20]

$$I = \int_{-\infty}^{\infty} \phi^2 \, dz, \tag{32}$$

gives

$$I = 24\frac{\sqrt{B}}{A^2}(M(\tau) - E)^{\frac{3}{2}}. \tag{33}$$

Differentiating (32) w.r.to τ

$$\frac{dI}{d\tau} = -2DI + \frac{32}{15}C\frac{\phi_0^2(\tau)}{W(\tau)}. \tag{34}$$

Again differentiating (33) w.r.to τ

$$\frac{dI}{d\tau} = 36\frac{\sqrt{B}}{A^2}(M(\tau) - E)^{\frac{1}{2}}\frac{dM(\tau)}{d\tau}. \tag{35}$$

From (34) and (35)

$$\frac{d}{d\tau}(M(\tau) - E) + \frac{4D}{3}(M(\tau) - E) = \frac{4C}{15B}(M(\tau) - E)^2. \tag{36}$$

Solving we get

$$M(\tau) = E + \frac{1}{\frac{C}{5BD} + \left(\frac{1}{M_0 - E} - \frac{C}{5BD}\right) e^{\frac{4}{3}D\tau}}, \tag{37}$$

where M_0 represent the value of $M(\tau)$ at $\tau = 0$. Thus the solution of dKPB equation is

$$\phi(\xi, \tau) = \phi_m(\tau) sech^2\left(\frac{\xi + \eta - M(\tau)\tau}{W(\tau)}\right), \tag{38}$$

where $\phi_m(\tau) = \frac{3(M(\tau)-E)}{A}$, $W(\tau) = 2\sqrt{\frac{B}{M(\tau)-E}}$ and (37) provides $M(\tau)$.

4.2 Shock Wave Solutions

Burger's medium produces solitary wave solutions when there is week dissipation. However, the powerful dissipation can result in shock waves. We have already derive solitary solution using conservation law. Now, to investigate shock wave solution, we operate SHBM [21] on dKPB equation

$$\frac{\partial \phi}{\partial \tau} + A\phi\frac{\partial \phi}{\partial \xi} + B\frac{\partial^3 \phi}{\partial \xi^3} + C\frac{\partial^2 \phi}{\partial \xi^2} + D\phi + E\int \frac{\partial^2 \phi}{\partial \eta^2} d\xi = 0. \tag{39}$$

Introducing the potential w, defined by

$$\phi = w_\xi, \tag{40}$$

we may write the Eq.(39) as

$$w_{\tau\xi} + Aw_{\xi\xi}^2 + Bw_{\xi\xi\xi\xi} + Cw_{\xi\xi\xi} + Du_\xi + +Ew_{\eta\eta} = 0. \tag{41}$$

Using the transformation below

$$w = e^\theta, \text{ where } \theta = k\xi + n\eta - \omega\tau. \tag{42}$$

We get the dispersion relation as

$$\omega = \frac{Bk^4 + Ck^3 + En^2 + Dk}{k}. \tag{43}$$

Assumed that the solution of Eq. (41) in the form of shock as

$$w = R(ln(f)), \tag{44}$$

where the function $f(\xi, \eta, \tau)$ is defined as

$$f(\xi, \eta, \tau) = 1 + e^{\theta} = 1 + e^{k\xi + n\eta - \omega\tau}, \tag{45}$$

Substituting Eqs. (44) and (45) into Eq. (41) we obtain a polynomial equation for $e^{n\theta}$. By setting the coefficient of $e^{n\theta}$ to zero, we obtain a system of algebraic equation. Solving we get

$$R = \frac{2(Ck^2 - D)}{Ak^2}. \tag{46}$$

Combining Eqs. (40), (44) and (46) we find the shock solution of dKPB Eq. (39) as,

$$\phi(\xi, \eta, \tau) = \frac{(Ck^2 - D)}{Ak^2} \left(1 + \tanh\left(\frac{k\xi + n\eta - \frac{Bk^4 + Ck^3 + En^2 + Dk}{k}\tau}{2}\right)\right). \tag{47}$$

5 Parametric Discussion

In this part, we will discuss the influence of various physical parameters on solitary and shock wave solutions of the dKPB equation. For the present study, the coefficient B of dispersion is always positive. Thus, the compressive soliton is found for positive values of the nonlinear coefficient A, whereas, the soliton remains rarefactive for negative nonlinear coefficient. From Fig. 1a and b it is obvious that A is negarive and thus, only the rarefactive soliton exists in the present system.

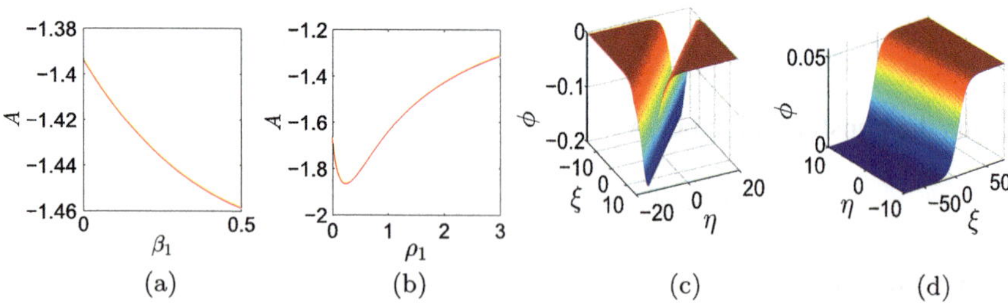

Fig. 1 Variation A vs various parameter, **a** when $\rho_2 = 0.6, \rho_1 = 2, \beta_2 = 0.4, \gamma_1 = 0.3, \gamma_2 = 0.3$. **b** When $\beta_2 = 0.4, \rho_2 = 0.6, \beta_1 = 0.2, \gamma_1 = 0.3, \gamma_2 = 0.3$. **c** 3D profile of $\phi(\tau)$ is plotted against ξ and η with the parameters $\gamma_1 = 0.7$, $\gamma_2 = 0.2$, $\beta_1 = 0.4$, $\rho_2 = 4$, $\rho_1 = 3$, $\beta_2 = 0.4$, $\tau = 2$, $M_0 = 0.5$, $\tau = 2$, $\zeta_0 = 0.04$. **d** 3D profile of $\phi(\tau)$ is plotted against ξ and η with the parameters $\gamma_1 = 0.7$, $\gamma_2 = 0.3$, $\beta_1 = 0.3$, $\rho_2 = 3$, $\mu_{id0} = 0.04$, $\beta_2 = 0.7$, $k = 0.15$, $n = 0.1$, $\zeta_0 = 0.04$, $\tau = 2$

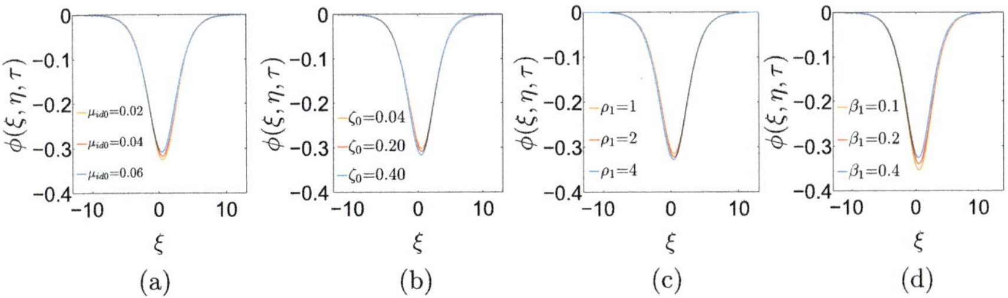

Fig. 2 2D profiles of $\phi(\xi, \eta, \tau)$ for solution (38) is plotted against ξ, **a** when $\gamma_1 = 0.7,\ \gamma_2 = 0.2,\ \zeta_0 = 0.04,\ \rho_2 = 4,\ \rho_1 = 3,\ \beta_1 = 0.4, \beta_2 = 0.4,\ \tau = 2,\ M_0 = 0.5,\ \eta = 0.5$, **b** when $\gamma_1 = 0.7,\ \gamma_2 = 0.2,\ \rho_1 = 3,\ \rho_2 = 4,\ \mu_{id0} = 0.02,\ \beta_1 = 0.4, \beta_2 = 0.4,\ \tau = 2,\ M_0 = 0.5,\ \eta = 0.5$, **c** when $\gamma_1 = 0.7,\ \gamma_2 = 0.2,\ \zeta_0 = 0.04,\ \rho_2 = 4,\ \mu_{id0} = 0.02,\ \beta_1 = 0.4, \beta_2 = 0.4,\ \tau = 2,\ M_0 = 0.5,\ \eta = 0.5$, **d** when $\gamma_1 = 0.7,\ \gamma_2 = 0.2,\ \zeta_0 = 0.04,\ \rho_2 = 4,\ \mu_{id0} = 0.02, \rho_1 = 3,\ \beta_2 = 0.4,\ \tau = 2,\ M_0 = 0.5,\ \eta = 0.5$

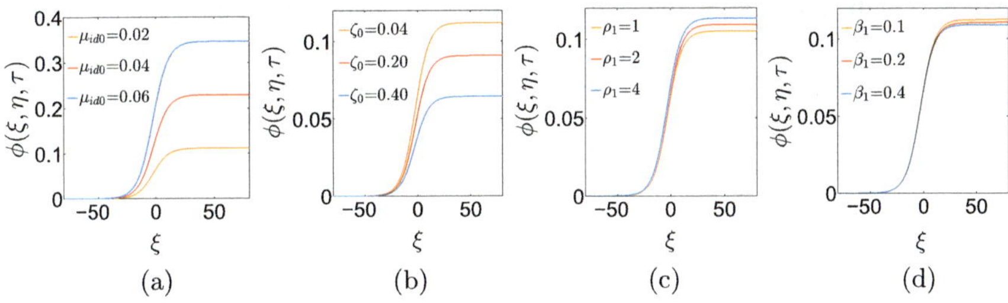

Fig. 3 2D profiles of $\phi(\xi, \eta, \tau)$ for solution is plotted against ξ, **a** when $\gamma_1 = 0.7,\ \gamma_2 = 0.3,\ \beta_1 = 0.3,\ \rho_2 = 3,\ \rho_1 = 3,\ \beta_2 = 0.7,\ k = 0.15,\ n = 0.1,\ \eta = 0.5,\ \zeta_0 = 0.04,\ \tau = 2$, **b** when $\gamma_1 = 0.7,\ \gamma_2 = 0.3,\ \beta_1 = 0.3,\ \rho_2 = 3,\ \rho_1 = 3,\ \beta_2 = 0.7,\ k = 0.15,\ n = 0.1,\ \eta = 0.5,\ \mu_{id0} = 0.04,\ \tau = 2$, **c** when $\gamma_1 = 0.7,\ \gamma_2 = 0.3,\ \beta_1 = 0.3,\ \rho_2 = 3,\ \mu_{id0} = 0.04,\ \beta_2 = 0.7,\ k = 0.15,\ n = 0.1,\ \eta = 0.5,\ \zeta_0 = 0.04,\ \tau = 2$, **d** when $\gamma_1 = 0.7,\ \gamma_2 = 0.3,\ \rho_1 = 3,\ \rho_2 = 3,\ \mu_{id0} = 0.04,\ \beta_2 = 0.7,\ k = 0.15,\ n = 0.1,\ \eta = 0.5,\ \zeta_0 = 0.04,\ \tau = 2$

The effect of the collision frequency (μ_{id0}) and kinematic viscosity (ζ_0) on the solitary wave solutions are displayed in respectively Fig. 2a, b. It is clear from Fig. 2a that the rarefactive soliton goes deeper for higher μ_{id0}. The occurrence of this type of nonlinear phenomenon can be predicted as: As μ_{id0} enhances the value of negative potential energy of the system decrease and naturally, the soliton rises. In respect to Fig. 2a, totally opposite result is seen in Fig. 2b where the rarefactive soliton goes deeper for increment of ζ_0. It is expected because enhancing ζ_0 causes the increase of dissipation in a medium and the soliton becomes wider and deeper. The characteristic of shock waves under the variations of physical parameters μ_{id0} and ζ_0 are depicted respectively in Fig. 3a and b. In Fig. 3a find enhances of μ_{id0} causes rising of the amplitude of shock wave, where as the Fig. 3b demonstrates that the diminishing effect of the amplitude of shock wave due to the increase of ζ_0. Figures 2c and d, 3c and d are depicted to exhibit the effect of the parameters ρ_1 (the ratio of equilibrium values of the number densities of lower temperature ions to the unperturbed number of electrons) and β_1 (ratio of two type of temperature ions) on solitary and shock propagation. For a clear vision of wave propagation in the plasma medium, the three-dimensional profiles of the solitary and shock wave are depicted in Fig. 1a and c respectively.

6 Conclusion

This article demonstrates the propagating behaviors of DASWs and DAShWs in an unmagnetized collisional dusty plasma containing two temperature ion and Maxwellian electrons. The solitary and shock solution are explored employing conservation law and SHBM respectively. Further the solutions are studied from numerical understanding. It is found that the solitary wave rises above due to the enhance in damping coefficient μ_{id0} as damping minimizes the system's negative potential energy. On the other hand, the depth of the shock enhances with the enhance in μ_{id0}. During the evolution of solitary wave, it is also found that soliton goes dipper as viscosity enhances whereas, the shock declines for enhancing ζ_0. Significant impacts from other physical parameters on wave propagation are observed.

References

1. Pakzad, H.R.: Solitary waves of the Kadomstev-Petviashvili equation in warm dusty plasma with variable dust charge, two temperature ion and nonthermal electron. Chaos, Solitons Fractals **42**(2), 874–879 (2009)
2. Raut, S., Mondal, K.K., Chatterjee, P., Roy, A.: Propagation of dust-ion-acoustic solitary waves for damped modified Kadomtsev-Petviashvili-Burgers equation in dusty plasma with a q-nonextensive nonthermal electron velocity distribution. SeMA **78**(4), 571–593 (2021)
3. Rao, N.N., Shukla, P.K., Yu, M.Y.: Dust-acoustic waves in dusty plasmas. Planetary Space Sci. **38**(4), 543–546 (1990)
4. Shukla, P.K., Silin, V.P.: Dust ion-acoustic wave. Physica Scripta **45**(5), 508 (1992)
5. Barkan, A., Merlino, R.L., D'angelo, D.: Laboratory observation of the dust-acoustic wave mode. Phys. Plasma **2**(10), 3563–3565 (1995)
6. Merlino, R.L.: 25 years of dust acoustic waves. J. Plasma Phys. **80**(06), 773–786 (2014)
7. Hartquist, T.W., Havnes, O.: Energy deposition due to the dissipation of dust-acoustic waves. Astrophys. Space Sci. **240**(2), 235–239 (1996)
8. Rosenberg, M., Kalman, G.: Dust acoustic waves in strongly coupled dusty plasmas. Phys. Rev. E **56**(6), 7166–7173 (1997)
9. Yaroshenko, V.V., Verheest, F., Morfill, G.E.: Dust-acoustic waves in collisional dusty plasmas of planetary rings. Astron. Astrophys. **461**(2), 385–391 (2006)
10. Shukla, P.K.: Twisted dust acoustic waves in dusty plasmas. Phys. Plasmas **19**(8) (2012)
11. Shukla, P.K., Mamun, A.A.: Introduction to Dusty Plasma Physics, 1st edn. IOP, London (2002)
12. Mondal, K.K., Roy, A., Chatterjee, P., Raut, S.: Propagation of dust-ion-acoustic solitary waves for damped forced Zakharov Kuznetsov equation in a relativistic rotating magnetized electron-ion-plasma plasma. Int. J. Appl. Comput. Math. **6**(3), 1–17 (2020)
13. Bains, A.S., Saini, N.S., Gill, T.S.: Dust acoustic solitary structures in multidust fluids superthermal plasma. Can. J. Phys. **91**(7), 582–587 (2013)
14. Saha, A., Chatterjee, P.: Bifurcations of dust acoustic solitary waves and periodic waves in an unmagnetized plasma with nonextensive ions. Astrophys. Space Sci. **351**(2), 533–537 (2014)

15. Melandso, F., Shukla, P.K.: Theory of dust-acoustic shocks. Planet. Space Sci. **43**(5), 635–648 (1995)
16. Mamun, A.A., Shukla, P.K.: Electrostatic solitary and shock structures in dusty plasmas. Phys. Scr. **T98**, 107–114 (2002)
17. Melands, F., Aslaksen, T., Havnes, O.: A new damping effect for the dust-acoustic wave. Planet. Space Sci. **41**(4), 321–325 (1993)
18. Gill, T.S., Saini, N.S., Kaur, H.: The Kadomstev-Petviashvili equation in dusty plasma with variable dust charge and two temperature ions. Chaos Solitons Fractals **28**(4), 1106–1111 (2006)
19. Washimi, H., Taniuti, T.: Propagation of ion-acoustic solitary waves of small amplitude. Phys. Rev. Lett. **17**(19), 996–998 (1966)
20. Ali, A., Kalisch, H.: On the formulation of mass, momentum and energy conservation in the KdV equation. Acta Applicandae Mathematicae **133**(1), 113–131 (2014)
21. Awawdeh, F., Jaradat, H.M., Al-Shara, S.: Applications of a simplified bilinear method to ion-acoustic solitary waves in plasma. Eur. Phys. J. D **66**(2), 40 (2012)

Stability of the Dust-Acoustic Solitons in the Thomas-Fermi Dense Magnetoplasma

A. Atteya

Abstract An investigation is presented theoretically for the multi- dimensional instability of dust-acoustic solitary waves in the dense Thomas-Fermi magnetoplasma. The plasma system contains classical negatively charged dust grains with degenerate electrons and ions particles. Based on the reductive perturbation approach, the Zakharov-Kuznetsov (ZK) equation has been formulated. This nonlinear ZK equation is analyzed for its solitary wave solutions. Only rarefactive solitary waves are obtained, those are influenced by the parameters such as the dust temperature and number density, the electrons, and ions Fermi temperatures, and densities. The waves growth rate of the is computed. The previous parameters' effects on the instability are also discussed. The present results are beneficial in understanding the propagation and the instability of nonlinear aspects in dense plasma systems like white dwarfs and high-intensity laser-solid matter interaction experiments where the Thomas-Fermi dense magnetoplasma state may occur.

Keywords Quantum semiconductor plasma · Degenerate holes · Plasma waves · Exchange-correlation forces · Zakharov-Kuznetsov equation · Bright soliton · Dark soliton

1 Introduction

Quantum plasma physics is important for understanding the superdense astrophysical bodies performance [1] (such as, the white dwarfs, neutron stars, and Jupiter interior), ultrasmall electronic devices [2], microplasmas [3], and ultracold plasmas [4], laser-based plasma compression [5], etc. Manfredi investigated the quantum plasma through different approaches [6].

A quintessential plasma contains the electrons and ions, while dusty plasmas are extended by containing massive negative or positive dust grains. The dusty plasmas

---

A. Atteya (✉)
Department of Physics, Faculty of Science, Alexandria University, P.O. 21511, Alexandria, Egypt
e-mail: ahmedatteya@alexu.edu.eg

S. Banerjee and A. Saha (eds.), *Nonlinear Dynamics and Applications*,
Springer Proceedings in Complexity,
https://doi.org/10.1007/978-3-030-99792-2_16

have attracted significant interest in the recent past, and it is associated with industrial applications, and also in astrophysical environments [7–10]. As a result of the charge, size, and mass of the massive dust component, the dynamical profile which characterizes dusty plasmas is complex on comparing to electron-ion plasmas. Furthermore, the electrostatic waves phase speed is affected by the dust-modified charge balance in the case of dust-ion-acoustic (DIA) waves [11], as also confirmed experimentally [12, 13]. On the other side, the dust components take the dust-acoustic (DA) waves aspect, where they are low-frequency oscillations [14–16].

In opposite to classical plasmas, low temperature and the high number density of the particles characterize dense quantum plasmas, and the Wigner-Poisson or the Schrödinger-Poisson treatment are used to model it [6]. Due to quantum corrections, they host various nonlinear structures, and instabilities [6]. Accordingly, fluid transport models have been used in various frameworks of condensed matter physics, such as semiconductors [17], nanoparticles, superfluidity [18], and superconductivity [19]. Fluid transport models with no effects of the quantum diffraction are termed as Thomas-Fermi non-stationary models.

In the Maxwell-Boltzmann statistics, as the temperature and density increase, the ideal quasi-neutral classical gas pressure increases, thus confirm thermodynamical equilibrium at the high temperature case. However, for the very dense plasma, new laws are associated with the distribution of the Fermi-Dirac, at high Fermi temperature. At an absolute zero temperature, the pressure remains nonzero and becomes only a function of the density. The chemical potential μ for such a completely degenerate plasma comes near to the value of the Fermi energy, so we can neglect the quantum diffraction effects. This is applicable only if number density of plasma particle is high degeneracy state, with slightly weak interactions. Hence, the density of plasma is important in the dynamics of collective modes in degenerate dusty plasma, comprising of mobile dust grains and degenerate inertialess ions and electrons [20–24]. Abdelsalam et al. investigated the properties of the dust excitations in the Thomas-Fermi plasma. Later, the extended study [25] derived the magnetized Korteweg-de Vries (KdV), KdV- Burger, Zakharov-Kuznetsov (ZK) and ZK-Burger equations and concluded that DA shock and solitary waves are changed due to the variation of temperatures, concentrations, and viscosity of the dust. The solitary and rogue DA waves formation and propagation are examined in a degenerate thermal Thomas-Fermi dusty plasma through the incorporation of transverse velocity perturbation effects [26]. The Thomas-Fermi density distribution is taken for ions and electrons, whereas the dust is considered as classical and dynamic. Obliquely propagating waves nonlinear properties are studied by Irfan et al. [27] in a dense degenerate cold Thomas–Fermi magnetoplasma, comprising of contains non-degenerate negatively-charged dust species. They derived and analyzed numerically the dust-cyclotron dispersion relation. They also formulated an equation of energy-balance by employing the Sagdeev pseudopotential theory. It was concluded that the soliton existence domain and the wave characteristics depend upon the system parameters. Unidirectional DA waves Overtaking collision in the Thomas-Fermi magnetoplasma has been analyzed [28].

The ZK equation has been derived when nonthermal ions are present in an external magnetic field to analyze the DA solitary waves characteristics [29]. The dispersion of linear DA waves in a dusty plasma have been examined [30, 31], where twisted DA vortex beam creation was reported. The instability of DA waves in a magneto-dusty plasma was checked by the small-k expansion technique [32–36]. The instability of DA waves in a magnetized dusty plasma was checked by Mamun [32]. Unstable DA wave structures were found due to the presence of the external magnetic field. The finite-amplitude DA waves' instabilities in a magnetized three-component dusty plasma with nonthermal particles were also discussed by Mamun et al. [33]. They illustrated that the features of the produced DA waves is modified by temperature of dusts and the nonthermal ions, while the DA wave stability is not affected by the nonthermal parameter variation.

The DIA waves three-dimensional stability have been also studied in a magnetized multicomponent dustyplasma by El-Taibany et al. [34] through using the small-k expansion technique. They showed that the higher growth rate associated with the larger wave amplitude and results in unstable solitary waves that are formed in the presence of negatively-charged ions. Akhter et al. [35] investigated the stability of DA waves in a magnetized dusty plasma. They found that the presence of an external magnetic field and the opposite polarity dust particles modified the DA wave instability-criterion. The obliquely propagating DA waves stability in a magnet0- multicomponent dusty plasma was derived by El-Labany et al. [36]. Saini et al. [37] derived the Zakharov-Kuznetsov (ZK) nonlinear equation for ion-acoustic solitary waves in a magnetized plasma. They also studied the stability analysis and checked the parametric range for the presence of stable and unstable solitons. The ZK equation was derived to study the DA solitary waves propagation in a magnetized dusty plasma containing massive, positive, and negative dust [38]. The wave's growth rate was derived and is affected by the polarization force. It is found also that, the instability is affected by the physical parameters. El-Taibany et al. [39] investigated the multi-dimensional instability in strongly coupled dusty plasma comprising ions and electrons in superthermal distribution. The produced waves growth rate is obtained, which is affected by the weakly and strongly coupling cases and the superthermal distribution of both the ions and electrons. The effects of polarization and trapping on multi-dimensional instability of ion-acoustic solitary waves in a multi-ion plasma system were theoretically investigated by Zedan et al. [40]. The instability and growth rate were found to be dependent on the density ratio between ions and dust, obliqueness, the dust cyclotron frequency, and other system parameters [41]. This manuscript is organized as follows. The governing equations and the derivation of the magnetized ZK equation is provided in Sect. 2. The solitary wave solution is in Sect. 3. The stability analysis for the DA waves is examined in Sect. 4. The numerical investigations and discussion are made in Sect. 5. At last, the conclusions are presented in Sect. 6.

2 Derivation of the ZK Equation

We consider quantum Thomas-Fermi dense magnetized plasma comprising of negatively-charged dust particles with degenerate ions and electrons obeying the Fermi-Dirac distributions. The external magnetic field $\mathbf{B}_0$ confined the plasma system and it is in the z-direction, i.e., $\mathbf{B}_0 = \hat{z}B_0$ where B_0 is the magnetic field strength and $\hat{z}$ is the unit vector along z-axis. The quasineutrality condition is $N_{e0} = N_{i0} - N_{d0}Z_{d0}$ at equilibrium, where Z_{d0} is the dust charge at equilibrium, N_{s0} is the sth species equilibrium density ($s = e, i$, and d for electrons, ions, and negatively charged dust grains respectively). The propagation of the DA for the Thomas-Fermi magnetoplasma is governed by [28]

$$\left.\begin{array}{c}\frac{\partial N_d}{\partial t} + \nabla.(N_d U_d) = 0,\\ \frac{\partial \mathbf{U}_d}{\partial t} + \mathbf{U}_d.\nabla\mathbf{U}_d = \nabla\psi - \Omega\mathbf{U}_d \times \hat{z} - \sigma_d N_d \nabla N_d,\\ \nabla^2\psi = \mu_e N_e - \mu_i N_i + N_d,\\ N_e = (1+\sigma_i\psi)^{3/2},\\ N_i = (1-\psi)^{3/2},\end{array}\right\} \tag{1}$$

where N_s is the normalized number density, U_d dust fluid velocity that normalized by the DA speed $C_d = (2Z_{d0}k_B T_{Fi}/m_d)^{1/2}$, ψ is the wave potential that normalized by $2k_B T_{Fi}/e$. $\Omega = \omega_{cd}/\omega_{pd}$ is the normalized dust gyro-frequency with $\omega_{cd} = eZ_{d0}B/m_d$ and $\omega_{pd} = \left(4\pi Z_{d0}^2 n_{d0}e^2/m_d\right)^{1/2}$. Also, $\sigma_d = T_d/T_{Fi}Z_{d0}$, $\mu_i = n_{i0}/Z_d n_{d0}$, and $\mu_e = n_{e0}/Z_d n_{d0}$, are the dust temperature-to-ion Fermi temperature ratio, the ion concentration, and electron concentration, divided by $n_{d0}Z_{d0}$, respectively, with e is the electronic charge, k_B is the Boltzmann constant. $\sigma_i = T_{Fi}/T_{Fe}$ is the ion-to-electron Fermi temperature ratio. The charge-neutrality condition at equilibrium becomes $\mu_{i=}\mu_{e+1}$. The space variable is normalized by $\lambda_0 = \left(2k_B T_{Fi}/4\pi Z_d n_{d0}e^2\right)^{1/2}$, and the time variable t is normalized by ω_{pd}^{-1}.

The ZK equation is formulated by employing the stretching of the independent variables x, y, and t to be defined as [24]

$$X = \epsilon^{1/2}x, Y = \epsilon^{1/2}y, Z = \epsilon^{1/2}(z - v_0 t), T = \epsilon^{3/2}t, \tag{2}$$

where ϵ is a formal small expansion parameter which indicates strength of the system nonlinearity, v_0 is the phase velocity. The dependent variables can be considered as:

$$\left.\begin{array}{c}N_d = 1 + \epsilon N_d^{(1)} + \epsilon^2 N_d^{(2)} + \epsilon^3 N_d^{(3)} + \dots,\\ U_{dx,y} = \epsilon^{\frac{3}{2}} U_{dx,y}^{(1)} + \epsilon^2 U_{dx,y}^{(2)} + \epsilon^{\frac{5}{2}} U_{dx,y}^{(3)} + \dots,\\ U_{dz} = \epsilon U_{dz}^{(1)} + \epsilon^2 U_{dz}^{(2)} + \epsilon^3 U_{dz}^{(3)} + \dots,\\ \psi = \epsilon\psi^{(1)} + \epsilon^2\psi^{(2)} + \epsilon^3\psi^{(3)} + \dots\end{array}\right\} \tag{3}$$

Putting Eqs. (2) and (3) into Eqs. (1), and the lowest orders perturbed quantities by collecting lowest order of ϵ, we get

$$N_d^{(1)} = \frac{-\psi^{(1)}}{v_0^2 - \sigma_d}, U_{dz}^{(1)} = \frac{v_0\psi^{(1)}}{v_0^2 - \sigma_d}. \tag{4}$$

The propagation phase velocity of the DA waves in the magnetized dusty plasma is

$$v_0 = \sqrt{\frac{2 + 3\mu_i\sigma_d + 3\mu_e\sigma_d\sigma_i}{3\mu_i + 3\mu_e\sigma_i}}. \tag{5}$$

Combining the next higher-orders contributions lead to

$$\left.\begin{aligned} U_{dx}^{(1)} &= \frac{-v_0^2}{\Omega\left(v_0^2-\sigma_d\right)}\frac{\partial\psi^{(1)}}{\partial Y}, \\ U_{dy}^{(1)} &= \frac{v_0^2}{\Omega\left(v_0^2-\sigma_d\right)}\frac{\partial\psi^{(1)}}{\partial X}, \\ U_{dx}^{(2)} &= \frac{-v_0^3}{\Omega\left(v_0^2-\sigma_d\right)}\frac{\partial^2\psi^{(1)}}{\partial X\partial Z}, \\ U_{dy}^{(2)} &= \frac{-v_0^3}{\Omega\left(v_0^2-\sigma_d\right)}\frac{\partial^2\psi^{(1)}}{\partial Y\partial Z}. \end{aligned}\right\} \tag{6}$$

Now, the next higher order of ϵ gives

$$\left.\begin{aligned} \frac{\partial N_d^{(2)}}{\partial Z} = &-\frac{2v_0}{\left(v_0^2-\sigma_d\right)^2}\frac{\partial\psi^{(1)}}{\partial T} + \frac{\left(3v_0^2+\sigma_d\right)}{\left(v_0^2-\sigma_d\right)^3}\psi^{(1)}\frac{\partial\psi^{(1)}}{\partial Z} - \frac{1}{\left(v_0^2-\sigma_d\right)}\frac{\partial\psi^{(2)}}{\partial Z} \\ &-\frac{v_0^4}{\Omega^2\left(v_0^2-\sigma_d\right)^2}\frac{\partial^3\psi^{(1)}}{\partial Y^2\partial Z} - \frac{v_0^4}{\Omega^2\left(v_0^2-\sigma_d\right)^2}\frac{\partial^3\psi^{(1)}}{\partial X^2\partial Z}. \end{aligned}\right\} \tag{7}$$

Substituting in the Poisson's equations, we derive the following equation

$$\frac{\partial\psi^{(1)}}{\partial T} + A\psi^{(1)}\frac{\partial\psi^{(1)}}{\partial Z} + B\frac{\partial^3\psi^{(1)}}{\partial Z^3} + C\left(\frac{\partial^3\psi^{(1)}}{\partial X^2\partial Z} + \frac{\partial^3\psi^{(1)}}{\partial Y^2\partial Z}\right) = 0. \tag{8}$$

This equation is the ZK equation with the following nonlinearity coefficient A, and the dispersive terms B, and C:

$$\left.\begin{aligned} A &= \left(\frac{3v_0^2+\sigma_d-\frac{3}{4}\left(v_0^2-\sigma_d\right)^3\left(\mu_i-\mu_e\sigma_i^2\right)}{2v_0\sigma_d-2v_0^3}\right), \\ B &= \frac{\left(v_0^2-\sigma_d\right)^2}{2v_0}, \\ C &= \frac{1}{2v_0}\left(\left(v_0^2-\sigma_d\right)^2 + \frac{v_0^4}{\Omega^2}\right). \end{aligned}\right\} \tag{9}$$

3 Solitary Wave Analysis

To obtain the solitary wave solution of Eq. (8), we shall follow the transformation of the independent variables [32, 42, 43] rotating by an angle θ about the coordinate axes (X, Z) as

$$\left.\begin{aligned} \xi &= X\sin\theta + Z\cos\theta, \\ \zeta &= X\cos\theta - Z\sin\theta, \\ \eta &= Y, \text{ and } \tau = T. \end{aligned}\right\} \tag{10}$$

Using transformations (10) to the ZK Eq. (8), we get

$$\left.\begin{aligned} &\frac{\partial\psi^{(1)}}{\partial\tau} + S_1\psi^{(1)}\frac{\partial\psi^{(1)}}{\partial\xi} + S_2\frac{\partial^3\psi^{(1)}}{\partial\xi^3} + S_3\psi^{(1)}\frac{\partial\psi^{(1)}}{\partial\zeta} + S_4\frac{\partial^3\psi^{(1)}}{\partial\zeta^3} \\ &+S_5\frac{\partial^3\psi^{(1)}}{\partial\xi^2\partial\zeta} + S_6\frac{\partial^3\psi^{(1)}}{\partial\xi\partial\zeta^2} + S_7\frac{\partial^3\psi^{(1)}}{\partial\xi\partial\eta^2} + S_8\frac{\partial^3\psi^{(1)}}{\partial\zeta\partial\eta^2} = 0, \end{aligned}\right\} \tag{11}$$

where

$$\left.\begin{aligned} S_1 &= A\cos\theta,\ S_2 = B\cos^3\theta + C\sin^2\theta\cos\theta, \\ S_3 &= -A\sin\theta,\ S_4 = -B\sin^3\theta - C\cos^2\theta\sin\theta, \\ S_5 &= 2C(\sin\theta\cos^2\theta - \tfrac{1}{2}\sin^3\theta) - 3B\cos^2\theta\sin\theta, \\ S_6 &= -2C(\sin^2\theta\cos\theta - \tfrac{1}{2}\cos^3\theta) + 3B\sin^2\theta\cos\theta, \\ S_7 &= C\cos\theta,\ S_8 = -C\sin\theta. \end{aligned}\right\} \tag{12}$$

The solution of ZK equation in the steady-state takes the form

$$\psi^{(1)} = \psi_0(\rho),$$

where $\rho = \xi - M\tau$, and Mach number (M) is normalized by DA speed C_d. Thus, Eq. (11) can be written as [44]

$$-M\frac{d\psi_0}{d\rho} + S_1\psi_0\frac{d\psi_0}{d\rho} + S_2\frac{d^3\psi_0}{d\rho^3} = 0. \tag{13}$$

integrating and applying appropriate boundary conditions, we obtain the DA pulse solution as

$$\phi_0(\rho) = \phi_m\,\mathrm{sech}^2\left(\frac{\rho}{W}\right), \tag{14}$$

where W and φ_m are the width and amplitude of the solitary wave, respectively; these are expressed as

$$\phi_m = 3M/S_1 \text{ and } W = 2\sqrt{S_2/M}.$$

The associated electric field is obtained as

$$E_0(\rho) = -\nabla\phi_1 = \frac{2\phi_m}{W}\,\mathrm{sech}^2\left(\frac{\rho}{W}\right)\tanh\left(\frac{\rho}{W}\right). \tag{15}$$

4 Stability Analysis

The small-k expansion perturbation technique [42, 43] is adopted to examine the stability of the DA structures. We consider [32, 36]

$$\phi^{(1)} = \phi_0(\rho) + \Phi(\rho, \zeta, \eta, \tau), \tag{16}$$

where Φ represents an obliquely propagating long-wavelength plane-wave that is given by

$$\Phi(\rho, \zeta, \eta, \tau) = \psi(\rho) \exp i[k(l_\xi \rho + l_\zeta \zeta + l_\eta \eta) - \gamma\tau], \tag{17}$$

in which $l_\xi^2 + l_\zeta^2 + l_\eta^2 = 1$, $\psi(\rho)$ and γ can be expanded by considering small values of k to the form

$$\left.\begin{aligned} \psi(\rho) &= \psi_o + k\psi_1 + k^2\psi_2 + \ldots, \\ \gamma &= k\gamma_1 + k^2\gamma_2 + \ldots \end{aligned}\right\} \tag{18}$$

Putting Eq. (16) into Eq. (11) to obtain the linearized ZK equation as

$$\left.\begin{aligned} \frac{\partial\Phi}{\partial\tau} - M\frac{\partial\Phi}{\partial\rho} + S_1\phi_0\frac{\partial\Phi}{\partial\rho} + S_2\frac{\partial^3\Phi}{\partial\rho^3} \\ +S_3\phi_0\frac{\partial\Phi}{\partial\zeta} + S_4\frac{\partial^3\Phi}{\partial\zeta^3} + S_5\frac{\partial^3\Phi}{\partial\rho^2\partial\zeta} + S_6\frac{\partial^3\Phi}{\partial\rho\partial\zeta^2} + S_7\frac{\partial^3\Phi}{\partial\rho\partial\eta^2} + S_8\frac{\partial^3\Phi}{\partial\zeta\partial\eta^2} = 0. \end{aligned}\right\} \tag{19}$$

Substituting Eqs. (17) and (18) into Eq. (19) and we get for the zeroth-order of k

$$(-M + S_1\phi_0)\psi_o + S_2\frac{d^2\psi_o}{d\rho^2} = C', \tag{20}$$

where C' is the integral constant. Two linearly independent solutions for the homogeneous part of Eq. (20), namely [32],

$$f = \frac{d\psi_0}{d\rho}, g = f\int^\rho \frac{d\rho}{f^2}. \tag{21}$$

Thus, the general solution can be given as

$$\psi_0 = C_1 f + C_2 g - C' f\int^\rho \frac{g}{S_2}d\rho + C' g\int^\rho \frac{f}{S_2}d\rho, \tag{22}$$

where C_1 and C_2 are the integral constants. The Wronskian ($\hat{W}$) is defined by

$$\hat{W} = f(dg/d\rho) - g(df/d\rho).$$

The general solution of zeroth-order equation is simplified as

$$\psi_0 = C_1 f. \tag{23}$$

The first and second-order equation from Eqs. (17)–(19) can be obtained, where their solutions lead to the following dispersion relation:

$$\gamma_1 = \Delta - Ml_\xi + \sqrt{\Delta^2 - \Gamma}, \tag{24}$$

where

$$\left.\begin{array}{c} \Delta = \frac{2}{3}(\mu_1\phi_m - 2\mu_2/W^2), \\ \Gamma = \frac{16}{45}(\mu_1^2\phi_m^2 - 3\mu_1\mu_2\phi_m/W^2 - 3\mu_2^2/W^4 + 12S_2\mu_3/W^4), \\ \mu_1 = (S_1l_\xi + S_3l_\zeta),\ \mu_2 = (3S_2l_\xi + S_5l_\zeta), \\ \text{and } \mu_3 = (3S_2l_\xi^2 + 2S_5l_\xi l_\zeta + S_6l_\zeta^2 + S_7l_\eta^2). \end{array}\right\} \tag{25}$$

Therefore, from Eq. (24), we observe that if the condition $\Gamma - \Delta^2 > 0$ is satisfied then instability occurs. We obtain the instability growth rate, gr, to be represented as

$$gr = \sqrt{\Gamma - \Delta^2}. \tag{26}$$

This instability growth rate depends on the system parameters.

5 Numerical Investigations and Discussion

We have formulated the ZK equation by applying the reductive perturbation approach. The small-k perturbation expansion technique examines the multi-dimensional instability of DAWs that is governed by this ZK equation [8, 40]. The results of this study can be summarized as follows:

The basic properties of the DA waves (polarity, amplitude, width, and speed) are found to be slightly modified by the dust temperature and density beside electron and ions densities and also their Fermi temperatures have a considerable effect.

The impact of slight fluctuation in dust temperature via σ_d on the phase speed v_0 for distinct values of ion-to-electron Fermi temperature ratio, σ_i is presented in Fig. 1. The phase speed enhances with the increase of σ_d, while it shrinks with the increase of σ_i, i.e. the DA wave propagates faster if the dust temperature becomes higher or the ions Fermi temperature becomes lower. The wave steepening is determined by the nonlinear term A. This term determine also the polarity of the DA waves (Fig. 1). Figure 2a illustrates the dependence of the nonlinear coefficient of A on σ_i and σ_d. It is depicted that, A is negative for all values of σ_i and σ_d and its absolute value increases as σ_i and σ_d increase. The effects of σ_i and σ_d on the longitudinal, B, and the transverse, C, dispersion coefficients properties are manifested in Fig. 2b and c, respectively. Both coefficients are positive and decrease as σ_i increases, while B (C) attains lower (higher) values with increasing σ_d. The soliton solution Eq. (14) occurs due to the balance between nonlinearity and dispersion effects, maintains its profile. The electrostatic solitary wave amplitude, ψ_m depends on parameters σ_i, σ_d,

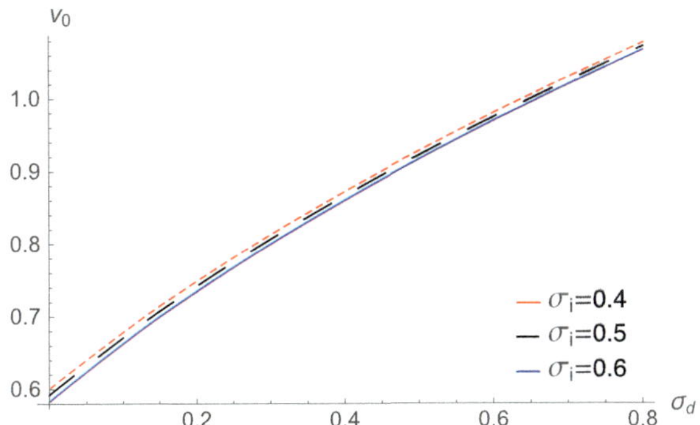

Fig. 1 The phase speed v_0 variation against σ_d at different values of σ_i at $\mu_i = 0.6$

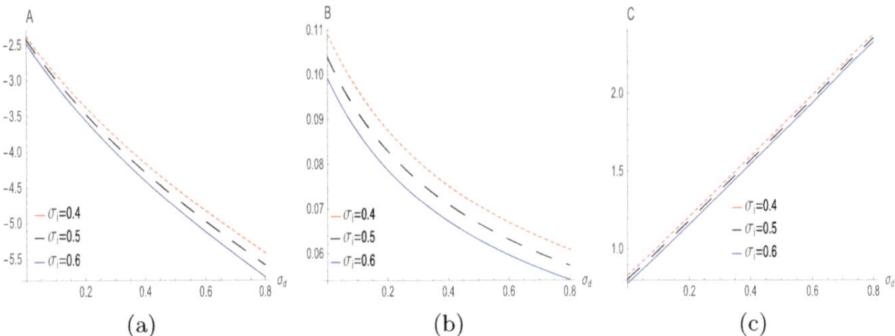

Fig. 2 **a** The variation of the nonlinear term A, **b** the dispersive term B, **c** the dispersive term C represented by Eq. (9) against σ_d for distinct values of σ_i at $\mu_i = 0.6$ and $\Omega = 0.5$

the electron to dust equilibrium densities ratio via μ_e, and obliquity angle θ as shown in Fig. 3. Since the amplitude depends on the nonlinear term, A, the amplitude is negative for all parameters values as depicted in Fig. 2a. The amplitude (in absolute value) becomes larger for smaller values of dust temperature, ion Fermi temperature, and electron equilibrium density or larger values of electron Fermi temperature and obliquity angle. Moreover, width W of the wave is suppressed by increasing values of both σ_i, below a critical value of σ_d and the magnetic field through Ω, as seen in Fig. 4. The angle θ and σ_d larger than the critical value increase lead to increasing W. The results obtained from Figs. 3 and 4 are confirmed through the solitary wave profiles and associated electric field as shown in Figs. 5 and 6, respectively. The variation of the growth rate, gr, against σ_i, σ_d, μ_e, Ω, θ, and frame velocity, M is depicted in Fig. 7, it is obvious that gr goes to zero as σ_d, μ_e, and θ increase. The reduction of gr becomes sharp as σ_i and M increase or as Ω decreases.

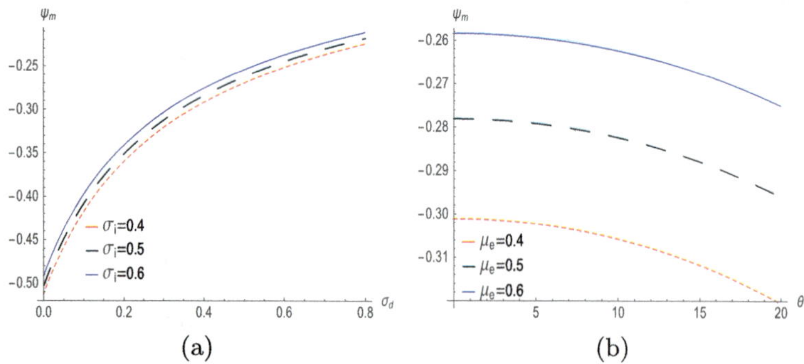

Fig. 3 The effects on the rarefactive DA soliton amplitude ψ_m **a** against σ_d for distinct values of σ_i at $\mu_e = 0.6$ with $\theta = 10$, **b** against θ for different values of μ_e with $\sigma_d = 0.5$ and $\sigma_i = 0.5$

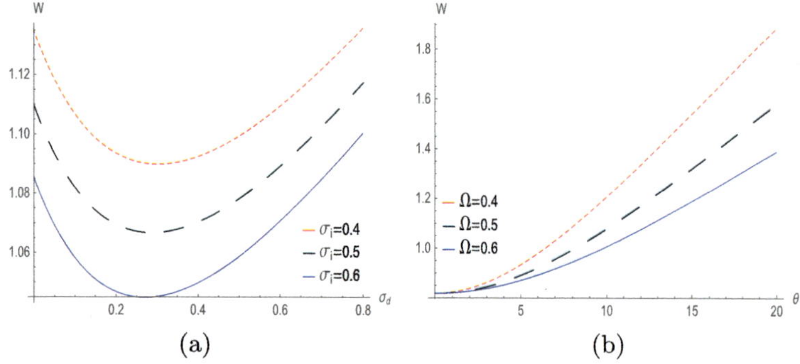

Fig. 4 The effects on the rarefactive DA soliton width W, at $\mu_e = 0.6$ **a** against σ_d for distinct values of σ_i for $\Omega = 0.5$ with $\theta = 10$, **b** against $\theta$for distinct values of Ω with $\sigma_d = 0.5$ and $\sigma_i = 0.5$

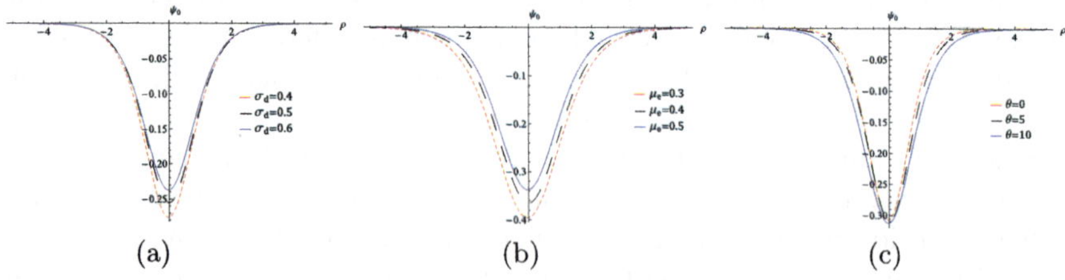

Fig. 5 The evolution of ψ_0 of DA waves that represented by equation (14) with ρ at $\sigma_i = 0.5$ $M = 0.4$ and $\Omega = 0.5$ for distinct values of **a** σ_d with $\mu_e = 0.6$, and $\theta = 10$, **b** μ_e with $\sigma_d = 0.3$, and $\theta = 13$, and **c** $\theta = 13$ with $\sigma_d = 0.3$, and $\mu = 0.6$

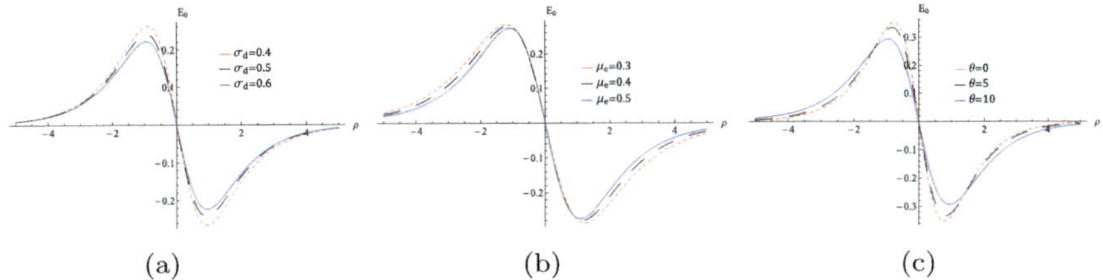

Fig. 6 The evolution of the associated electric field, E_0 of DA waves that represented by Eq. (5) with ρ for the potentials those represented by Fig. 5

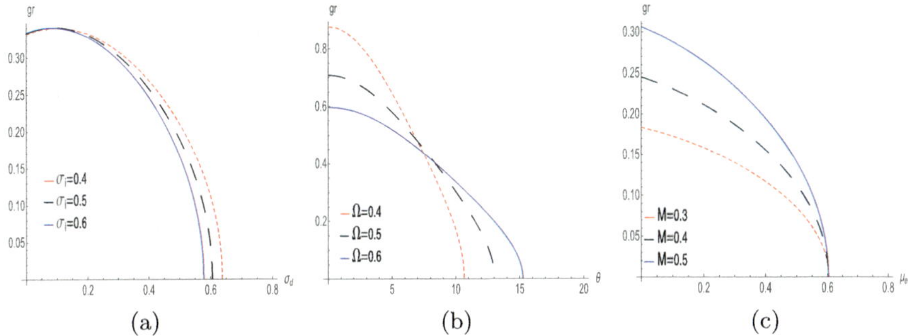

Fig. 7 (Color online) The effects on the growth rate, gr, that represented by Eq. (26) at $l_\xi = 0.7$, $l_\eta = 0.4$ **a** against σ_d for distinct values of σ_i at $\mu_e = 0.6, \theta = 10, M = 0.4$ and $\Omega = 0.5$, **b** against θ for distinct values of Ω at $\mu = 0.6, \sigma_d = 0.3, \sigma_i = 0.5$ and $M = 0.4$, and **c** against μ_e for distinct values of M at $\theta = 13, \sigma_d = 0.3, \sigma_i = 0.5$ and $\Omega = 0.5$

6 Conclusions

The hydrodynamic model has been employed to study a quantum Thomas-Fermi dense plasma consisting of negative dust particles with degenerate ions and electrons obeying the Fermi-Dirac distributions, embedded in a uniform magnetic field. The ZK equation has been formulated by employing the reductive perturbation approach. The solution of ZK equation has been used to explore the characteristics of the solitary wave. Interesting, the soliton amplitude is negative for all values of the ions and electrons densities and Fermi temperatures. This negative polarity structures may occur, depending on the negatively charged dust. The dependence of the soliton width on the electrons densities and Fermi temperatures, on the obliqueness, and the magnetic field (strength) was investigated.

The stability investigation and analysis of the solitary wave solution of ZK equation have been carried out. The influences of various physical parameters on the instability growth rate have been examined. It has been shown that an increase in the dust temperature, magnetic field strength and obliqueness can lessen the instability growth rate. The reduction of the growth rate also obtained by the decrease of the dust equilibrium density.

Our present results can be applied to discern the dynamics of nonlinear localized structures in laboratory and space manners where Thomas-Fermi dense magnetoplasma occurs, e.g. in the high-intensity laser-solid matter interaction experiments, and also in the white dwarfs [26, 27, 40, 45–48].

References

1. Jung, Y.D.: Quantum-mechanical effects on electron-electron scattering in dense high-temperature plasmas. Phys. Plasmas **8**(8), 3842–3844 (2001)
2. Markowich, P.A., Ringhofer, C.A., Schmeiser, C.: Semiconductor Equations, 1st edn. Springer, New York (1990)
3. Becker, K., Koutsospyros, K., Yin, S.M., et al.: Environmental and biological applications of microplasmas. Plasma Phys. Control. Fusion **47**, B513–B523 (2005)
4. Killian, T.C.: Physics-cool vibes. Nature (London) **441**, 297–298 (2006)
5. Malkin, V.M., Fisch, N.J., Wurtele, J.S.: Compression of powerful x-ray pulses to attosecond durations by stimulated Raman backscattering in plasmas. Phys. Rev. E **75**, 026404 (2007)
6. Manfredi, G.: How to model quantum plasmas. Fields Inst. Commun. **46**, 263–287 (2005)
7. Shukla, P.K., Mamun, A.A.: Introduction to Dusty Plasma Physics, 1st edn. Institute of Physics, Bristol (2002)
8. Fortov, V.E., Ivlev, A.V., Khrapak, S.A., Morfill, G.E.: Complex (dusty) plasmas. Phys. Rep. **421**, 1–103 (2005)
9. Atteya, A., El-Borie, M.A., Roston, G.D., El-Helbawy, A.S., Prasad, P.K., Saha, A.: Ion-acoustic stable oscillations, solitary, periodic and shock waves in a quantum magnetize electron-positron-ion plasma. Zeitschrift für Naturforschung A **76**(9), 757–768 (2021)
10. El-Monier, S.Y., Atteya, A.: Dust-acoustic Gardner solitons in cryogenic plasma with the effect of polarization in the presence of a quantizing magnetic field. Zeitschrift für Naturforschung A **76**(2), 121–130 (2021)
11. Shukla, P.K., Silin, V.P.: Dust ion-acoustic wave. Phys. Scr. **45**(5), 508 (1992)
12. Barkan, A., D'Angelo, N., Merlino, R.L.: Experiments on ion-acoustic waves in dusty plasmas. Planet. Space Sci. **44**, 239–242 (1996)
13. Merlino, R.L., Barkan, A., Thompson, C., D'Angelo, N.: Laboratory studies of waves and instabilities in dusty plasmas. Phys. Plasmas **5**, 1607–1614 (1998)
14. Rao, N.N., Shukla, P.K., Yu, M.Y.: Dust-acoustic waves in dusty plasmas. Planet. Space Sci. **38**, 543–546 (1990)
15. Barkan, A., Merlino, R.L., D'Angelo, N.: Laboratory observation of the dust-acoustic wave mode. Phys. Plasmas **2**, 3563–3565 (1995); Pieper, J. B.: J. Goree, Phys. Rev. Lett. **77**, 3137 (1996)
16. Prabhakara, H.R., Tanna, V.L.: Trapping of dust and dust acoustic waves in laboratory plasmas. Phys. Plasmas **3**, 3176–3181 (1996)
17. Banerjee, A., Harbolaa, M.K.: Hydrodynamic approach to time-dependent density functional theory; Response properties of metal clusters. J. Chem. Phys. **113**, 5614–4623 (2000)
18. Domps, A., Reinhard, P.-G., Suraud, E.: Theoretical estimation of the importance of two-electron collisions for relaxation in metal clusters. Phys. Rev. Lett. **81**, 5524–5527 (1998)
19. Loffredo, M., Morato, L.: On the creation of quantized vortex lines in rotating He II. Nuovo Cimento Soc. Ital. Fis., B **108**, 205–215 (1993)
20. Feynman, R.: Statistical Mechanics. A Set of Lectures, 1st edn. Benjamin, Reading (1972)
21. Dubinov, A.E., Dubinova, A.A.: Nonlinear theory of ion-acoustic waves in an ideal plasma with degenerate electrons. Plasma Phys. Rep. **33**, 859–870 (2007)
22. Abdelsalam, U.M., Moslem, W.M., Shukla, P.K.: Localized electrostatic excitations in a Thomas-Fermi plasma containing degenerate electrons. Phys. Plasmas **15**, 052303 (2008)

23. Abdelsalam, U.M., Moslem, W.M., Shukla, P.K.: Ion-acoustic solitary waves in a dense pair-ion plasma containing degenerate electrons and positrons. Phys. Lett. A **372**, 4057–4061 (2008)
24. Atteya, A., El-Borie, M.A., Roston, G.D., El-Helbawy, A.S.: Nonlinear dust acoustic waves in an inhomogeneous magnetized quantum dusty plasma. Waves Random Complex Media 1–16 (2021). https://doi.org/10.1080/17455030.2021.1880030
25. Rahim, Z., Ali, S., Qamar, A.: Dust acoustic solitary and shock excitations in a Thomas-Fermi magnetoplasma. Phys. Plasmas **21**(7), 072305 (2014)
26. Irfan, M., Ali, S., Mirza, A.M.: Dust-acoustic solitary and rogue waves in a Thomas-Fermi degenerate dusty plasma. Astrophys Space Sci. **353**(2), 515–523 (2014)
27. Irfan, M., Ali, S., Ata-ur-Rahman, Mirza, A.M.: Arbitrary amplitude oblique electrostatic solitary waves in a degenerate cold dusty magnetoplasma. IEEE Trans. Plasma Sci. **47**(8), 4151–4158 (2019)
28. Abd-Elzaher, M., Atteya, A.: Obliquely overtaking collisions of electrostatic N-soliton in the Thomas-Fermi dense magnetoplasma. Waves Random Complex Media 1–21 (2021). https://doi.org/10.1080/17455030.2021.1974121
29. El-Taibany, W.F., Sabry, R.: Dust-acoustic solitary waves and double layers in a magnetized dusty plasma with nonthermal ions and dust charge variation. Phys. Plasmas **12**, 082302 (2005)
30. Shukla, P.K.: Twisted dust acoustic waves in dusty plasmas. Phys. Plasmas **19**, 083704 (2012)
31. Shukla, P.K.: Twisted electrostatic ion-cyclotron waves in dusty plasmas. Phys. Rev. E **87**, 015101 (2013)
32. Mamun, A.A.: Instability of obliquely propagating electrostatic solitary waves in a magnetized nonthermal dusty plasma. Phys. Scr. **58**, 505–509 (1998)
33. Mamun, A.A., Russell, S.M., Mendoza-Briceno, C.A., Alam, M.N., Datta, T.K., Das, A.K.: Multi-dimensional instability of electrostatic solitary structures in magnetized nonthermal dusty plasmas. Planet. Space Sci. **48**, 163–173 (2000)
34. El-Taibany, W.F., El-Bedwehy, N.A., El-Shamy, E.F.: Three-dimensional stability of dust-ion acoustic solitary waves in a magnetized multicomponent dusty plasma with negative ions. Phys. Plasmas **18**, 033703 (2011)
35. Akhter, T., Hossain, M.M., Mamun, A.A.: Multi-dimensional instability of dust-acoustic solitary waves in a magnetized plasma with opposite polarity dust. Phys. Plasmas **19**, 093707 (2012)
36. El-Labany, S.K., El-Taibany, W.F., Behery, E.E.: Stability of three-dimensional dust acoustic waves in a dusty plasma with two opposite polarity dust species including dust size distribution. Phys. Rev. E **88**, 023108 (2013)
37. Saini, N.S., Chahal, B.S., Bains, A.S., Bedi, C.: Zakharov-Kuznetsov equation in a magnetized plasma with two temperature superthermal electrons. Phys. Plasmas **21**, 022114 (2014)
38. El-Labany, S.K., El-Taibany, W.F., Behery, E.E., Zedan, N.A.: Stability of three-dimensional obliquely propagating dust acoustic waves in dusty plasma including the polarization force effect. Eur. Phys. J. Plus **130**, 250 (2015)
39. El-Taibany, W.F., Zedan, N.A., Atteya, A.: Stability of three-dimensional dust acoustic waves in a strongly coupled dusty plasma including kappa distributed superthermal ions and electrons. Eur. Phys. J. Plus **134**, 479 (2019)
40. Zedan, N.A., Atteya, El-Taibany, W.F., El-Labany, S.K.: Stability of ion-acoustic solitons in a multi-ion degenerate plasma with the effects of trapping and polarization under the influence of quantizing magnetic field. Waves in Random and Complex Media 1–15 (2020). https://doi.org/10.1080/17455030.2020.1798560
41. Gao, D.-N.: Multi-dimensional instability of dust acoustic waves in magnetized quantum plasmas with positive or negative dust. Brazil. J. Phys. **51**, 66–74 (2021)
42. Allen, M.A., Rowlands, G.: Determination of the growth rate for the linearized Zakharov-Kuznetsov equation. J. Plasma Phys. **50**, 413–424 (1993)
43. Allen, M.A., Rowlands, G.: Stability of obliquely propagating plane solitons of the Zakharov-Kuznetsov equation. J. Plasma Phys. **53**, 63–73 (1995)
44. Haider, M.M., Mamun, A.: Ion-acoustic solitary waves and their multi-dimensional instability in a magnetized degenerate plasma. Phys. Plasmas **19**, 102105 (2012)

45. Jehan, N., Salahuddinin, M., Mahmood, S., Mizra, A.M.: Electrostatic solitary ion waves in dense electron-positron-ion magnetoplasma. Phys. Plasmas **16**, 042313 (2009)
46. Chatterjee, P., Saha, T., Muniandy, S.V., ap, S.L., Wong, C.S.: Solitary waves and double layers in dense magnetoplasma. Phys. Plasmas **16**, 072110 (2009)
47. Rahim, Z., Adnan, M., Qamar, A., Saha, A.: Nonplanar dust-acoustic waves and chaotic motions in Thomas Fermi dusty plasmas. Phys. Plasmas **25**, 083706 (2018)
48. Roy, D., Ghosh, N., Sahu1, B.: Nonlinear modulation of quantum electron acoustic waves in a Thomas–Fermi plasma with effects of exchange-correlation. Indian J Phys. **95**(11), 2479–2490 (2020)

Existence and Stability of Dust-Ion-Acoustic Double Layers Described by the Combined SKP-KP Equation

Sankirtan Sardar and Anup Bandyopadhyay

Abstract Sardar et al. [Phys Plasmas 24:063705 (2017)] have investigated dust-ion-acoustic (DIA) solitons of a combined Schamel's modified Kadomtsev Petviashvili-Kadomtsev Petviashvili (SKP-KP) equation in a dusty plasma consisting of nonthermal electrons which obeys vortex-like velocity distribution. In this paper, we have investigated the existence of double layer solutions and its stability by considering the same combined SKP-KP equation. We have seen that this double layer solution exists when $L = 0$, where L is a function of the parameters. We have analytically discussed the stability of the double layer solutions.

Keywords Stability · Double layers · Combined SKdV-KdV equation · Combined SKP-KP equation

1 Introduction

For the first time, Kadomtsev and Petviashvili [1] attempted to model a soliton in two dimensions known as KP equation. KP equation is generally used to discuss the stability of the Korteweg-de Vries (KdV) solitons. In an unmagnetized plasma, Kako and Rowlands [2] obtained KP equation to discuss the stability of ion acoustic (IA) solitons. Using this KP equation, Infeld et al. [3] investigated the stability of the KdV solitons. They have found that KdV solitons are stable with respect to the transverse perturbation. Employing small-k perturbation expansion method [4–7], Chakraborty and Das [8] derived a modified KP (MKP) equation to study the stability of IA soliton of the MKP equation. Employing multiple-scale perturbation expansion method [9, 10], Chakraborty and Das [11] also studied the higher stability of the same MKP equation in an another paper. Several authors [12–29] have derived the KP equation

S. Sardar (✉)
Department of Mathematics, Guru Ghasidas Vishwavidyalaya, Bilaspur 495009, India
e-mail: sankirtansardar@gmail.com

A. Bandyopadhyay
Department of Mathematics, Jadavpur University, Kolkata 700032, India

S. Banerjee and A. Saha (eds.), *Nonlinear Dynamics and Applications*,
Springer Proceedings in Complexity,
https://doi.org/10.1007/978-3-030-99792-2_17

or different modified KP (MKP) equations and used this KP equation to study the dynamics of IA/dust-acoustic (DA)/DIA waves in different plasmas.

Sardar et al. [30] have investigated dust-ion-acoustic (DIA) solitons of a combined SKP-KP equation in a dusty plasma consisting of warm adiabatic ions, isothermal positrons, immobile dust grains and nonthermal electrons. The electron species obeys vortex-like velocity distribution, where the background distribution is Cairns non-thermal distribution [31]. They have investigated that the alternative soliton of the combined SKP-KP equation exists if and only if $L > 0$, where L is a function of the parameters. They have observed that the alternative soliton cannot define the nonlinear dynamics of DIA waves for $L = 0$ or $L \approx O(\epsilon)$, where ϵ is a small parameter. In this situation, they have reported that further development is required.

In this paper, our aim is to investigate the existence of double layer solutions and its stability by considering the same combined SKP-KP equation. We have seen that this double layer solution exists when $L = 0$. We have analytically discussed the stability of the double layer solutions. Here we have extended previous work of Sardar et al. [30] by cosidering $L = 0$.

2 Evolution Equation

Considering continuity equation, motion equation, pressure equation for ions, Poisson equation, the equations for the velocity distribution functions for positrons and electrons, and the unperturbed charged neutrality condition, Sardar et al. [30] have derived the following equation:

$$\frac{\partial}{\partial \xi}\left[\phi_{\tau}^{(1)} + AB\sqrt{\phi^{(1)}}\phi_{\xi}^{(1)} + AB_1\phi^{(1)}\phi_{\xi}^{(1)} + \frac{1}{2}AC\phi_{\xi\xi\xi}^{(1)}\right] + \frac{1}{2}AD\left(\phi_{\eta\eta}^{(1)} + \phi_{\zeta\zeta}^{(1)}\right) = 0. \tag{1}$$

This equation explains the dynamics of alternative DIA solitons in the present plasma system considered by Sardar et al. [30] when $L > 0$. Here ξ, η, ζ are the stretched spatial coordinates and τ is the stretched time coordinate. The equations of (28), (29), (30), (18) and (34) of Sardar et al. [30] give the expressions of A, B, D, C and B_1 respectively.

In this problem, using the same evolution Eq. (1), we have analyzed DL solution and its stability when $L = 0$.

3 DL Solutions of Combined SKP-KP Equation

For DL solutions of (1), we have considered the following transformations:

$$\eta' = \eta, \zeta' = \zeta, X = \xi - U\tau, \tau' = \tau. \tag{2}$$

Here, U is a constant normalized velocity. Under transformation (2), (1) reduces to the following equation:

$$\frac{\partial}{\partial X}\Big[- U\phi_X^{(1)} + \phi_\tau^{(1)} + AB\sqrt{\phi^{(1)}}\phi_X^{(1)} + AB_1\phi^{(1)}\phi_X^{(1)} + \frac{1}{2}AC\phi_{XXX}^{(1)}\Big] \\ +\frac{1}{2}AD\Big(\phi_{\eta\eta}^{(1)} + \phi_{\zeta\zeta}^{(1)}\Big) = 0. \tag{3}$$

We consider the following equation for travelling waves of (3):

$$\phi^{(1)} = \phi_0(X). \tag{4}$$

Employing (4) in (3), we obtain

$$\frac{d^2}{dX^2}\Big[- U\phi_0 + \frac{2}{3}AB(\phi_0)^{\frac{3}{2}} + \frac{1}{2}AB_1(\phi_0)^2 + \frac{1}{2}AC\frac{d^2\phi_0}{dX^2}\Big] = 0. \tag{5}$$

Now, we use the following boundary condition

$$\phi_0, \frac{d^n\phi_0}{dX^n} \rightarrow 0 \text{ as } X \rightarrow \infty \text{ for all } n = 1, 2, 3, \dots \tag{6}$$

or

$$\phi_0, \frac{d^n\phi_0}{dX^n} \rightarrow 0 \text{ as } X \rightarrow -\infty \text{ for all } n = 1, 2, 3, \dots \tag{7}$$

With the help of (6) or (7), the Eq. (5) can be written as follows:

$$- U\phi_0 + \frac{2}{3}AB(\phi_0)^{\frac{3}{2}} + \frac{1}{2}AB_1(\phi_0)^2 + \frac{1}{2}AC\frac{d^2\phi_0}{dX^2} = 0. \tag{8}$$

Following Das et al. [32], the DL solution of (8) is given as follows:

$$\phi_0 = a^2\Big(1 - \lambda \tanh\frac{X}{W_1}\Big)^2, \text{ where } a = -\frac{2B}{5B_1}, \tag{9}$$

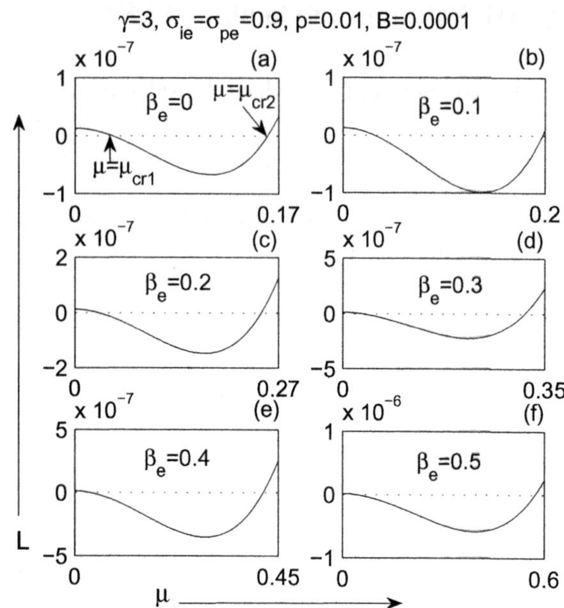

Fig. 1 For different values of β_e, L is drawn with respect to μ for $\gamma = 3$, $\sigma_{ie} = 0.9$, $p = 0.01$, $\sigma_{pe} = 0.9$ and $B = 0.0001$

$$L = B^2 + \frac{75}{2} B_1 C p_1^2 = 0 \text{ with } p_1 = \frac{1}{W_1}. \tag{10}$$

Here $\lambda = \pm 1$ and these two values of λ give two different DL solutions of (1) for the DIA waves corresponding to the boundary conditions (6) and (7) respectively. It is impossible to get alternative soliton of the combined SKP-KP equation by considering only one of the following two conditions:

$$\phi_0, \frac{d^n \phi_0}{dX^n} \to 0 \text{ as } X \to \infty \text{ for } n = 1, 2, 3, \ldots \tag{11}$$

$$\phi_0, \frac{d^n \phi_0}{dX^n} \to 0 \text{ as } X \to -\infty \text{ for } n = 1, 2, 3, \ldots \tag{12}$$

Specifically, using the condition (11), we get a Z-type DL solution whereas considering condition (12), we get a S-type DL solution. But both the conditions are necessary to get an alternative soliton of (1).

Now, we can consider $B = 0.0001$ because $B \neq 0$ but $B \to 0$. Figures 1(a)–1(f) show the variation of L against μ for different values of β_e when values of $p(= 0.01)$, $\sigma_{pe}(= 0.9)$, γ and σ_{ie} are fixed. Each Fig. 1(a)–1(f) has two critical values of μ where $L = 0$.

The solution (9) is the steady state DL solution of (1) along the x-axis. This solution is same as the DL solution of the Schamel's modified Korteweg-de Vries-Korteweg-de Vries (SKdV-KdV) equation, i.e., the steady state DL solution of the equation

$$\phi_{\tau}^{(1)} + AB\sqrt{\phi^{(1)}}\phi_{\xi}^{(1)} + AB_1\phi^{(1)}\phi_{\xi}^{(1)} + \frac{1}{2}AC\phi_{\xi\xi\xi}^{(1)} = 0 \tag{13}$$

is exactly same as the Eq. (9). In the next section, we have considered the stability analysis of DL solution (9) of Eq. (1).

4 Stability Analysis

In the previous section, we have seen that the Eq. (9) gives two different DL solutions for $\lambda = +1$ and $\lambda = -1$. Here, we have analyzed the stability of DL solutions (9) for $\lambda = +1$. Following the same procedure, one can easily analyse the stability of the DL solution (9) for $\lambda = -1$.

To use the multiple-scale perturbation expansion method of Allen and Rowlands [9, 10], we write

$$\phi^{(1)} = \phi_0(X) + q(X, \eta, \zeta, \tau). \tag{14}$$

Here $\phi_0(X)$ is the DL solution (9) of the equation (3) and $q(X, \eta, \zeta, \tau)$ is perturbed component of $\phi^{(1)}$. Substituting (14) into (3) and linearizing the equation, we obtain

$$q_{X\tau} + (Mq)_{XX} + \frac{1}{2}AD(q_{\eta\eta} + q_{\zeta\zeta}) = 0, \tag{15}$$

where

$$M = -U + AB\sqrt{\phi_0} + AB_1\phi_0 + \frac{1}{2}AC\frac{\partial^2}{\partial X^2}. \tag{16}$$

For long-wavelength plane-wave perturbation along a direction having direction cosines (l, m, n), we consider

$$q(X, \eta, \zeta, \tau) = \overline{q}(X)e^{i\{k(lX+m\eta+n\zeta)-\omega\tau\}}. \tag{17}$$

Here k is small and direction cosines follow the relation $l^2 + m^2 + n^2 = 1$. Substituting (17) into (15), we get the following equation:

$$\begin{aligned}&(M\overline{q})_{XX} - i\omega\overline{q}_X + kl\Big\{\omega\overline{q} + 2i(M\overline{q})_X + iAC\overline{q}_{XXX}\Big\}\\&-k^2l^2\Big\{M\overline{q} + \frac{5}{2}AC\overline{q}_{XX} + AD\frac{m^2+n^2}{2l^2}\overline{q}\Big\}\\&-k^3l^3\Big\{2iAC\overline{q}_X\Big\} + k^4l^4\Big\{\frac{1}{2}AC\overline{q}\Big\} = 0,\end{aligned} \tag{18}$$

Following the multiple-scale perturbation expansion method [9, 10], we expand $\overline{q}(X)$ and ω as

$$\overline{q}(X) = \sum_{j=0}^{\infty} k^j q^{(j)}(X, X_1, X_2, X_3, \ldots),\ \omega = \sum_{j=0}^{\infty} k^j \omega^{(j)}, \tag{19}$$

where $\omega^{(0)} = 0,\ X_j = k^j X,\ j = 0, 1, 2, \ldots,$ and each $q^{(j)}(= q^{(j)}(X, X_1, X_2, X_3, \ldots))$ is a function of $X, X_1, X_2, X_3, \ldots$. It is important to note that $X_0 = X$.

Finally, substituting first and second equation of (19) into (18) and comparing the coefficients of different powers of k, we obtain the following equations:

$$\frac{\partial}{\partial X}(Mq^{(j)}) = Q^{(j)},\ \text{where } Q^{(j)} = \int_{\infty}^{X} R^{(j)} dX, \tag{20}$$

and the expressions of $R^{(0)}$ and $R^{(1)}$ are given as follows:

$$R^{(0)} = 0 \tag{21}$$

$$R^{(1)} = i\omega^{(1)} q_0^{(0)} - iACl q_{000}^{(0)} - AC q_{0001}^{(0)} - 2il[Mq^{(0)}]_0 - 2[Mq^{(0)}]_{01}, \tag{22}$$

and we have used the notations: $q_r^{(j)} = \frac{\partial q^{(j)}}{\partial X_r}$, $q_{rs}^{(j)} = \frac{\partial^2 q^{(j)}}{\partial X_r \partial X_s}$, $q_{rst}^{(j)} = \frac{\partial^3 q^{(j)}}{\partial X_r \partial X_s \partial X_t}$, $q_{rsty}^{(j)} = \frac{\partial^4 q^{(j)}}{\partial X_r \partial X_s \partial X_t \partial X_y}$ and $[Mq^{(j)}]_{rs} = \frac{\partial^2 (Mq^{(j)})}{\partial X_r \partial X_s}$.

Considering that $q^{(j)}$ and its first order, second order, and third order derivatives vanish as $X \to \infty$ and from the Eq. (20), we obtain the following general solution:

$$q^{(j)} = A_1^{(j)} f + A_2^{(j)} f \int \frac{1}{f^2} dX + A_3^{(j)} f \int \frac{\phi_0}{f^2} dX + \frac{2}{AC} f \int \frac{\int (f \int Q^{(j)} dX) dX}{f^2} dX, \tag{23}$$

where $f = \frac{d\phi_0}{dX}$, ϕ_0 is given by (9) for $\lambda = 1$ and $A_1^{(j)}, A_2^{(j)}, A_3^{(j)}$ are all arbitrary functions of $X_1, X_2, X_3, \ldots$.

Using MATHEMATICA [33], the solution (23) for $\lambda = +1$ can be written as follows:

$$
\begin{aligned}
q^{(j)} &= A_1^{(j)} f + \frac{3W_1^2}{256a^4}\left(5A_2^{(j)} + 8a^2 A_3^{(j)}\right) f X - \frac{W_1^2}{128a^2}\left(5A_2^{(j)} + 8a^2 A_3^{(j)}\right) \\
&+ \frac{3W_1^2}{128a^2}\left(5A_2^{(j)} + 8a^2 A_3^{(j)}\right)(1-R) + \frac{W_1^2}{128a^2}\left(5A_2^{(j)} + 8a^2 A_3^{(j)}\right) e^{-\frac{2X}{W_1}} \\
&- \frac{W_1^2}{1024a^2}\left(85A_2^{(j)} + 192a^2 A_3^{(j)} + 35A_2^{(j)} R\right) S^2 - \frac{W_1^2}{16a^2} A_2^{(j)} \frac{1}{S^2} \\
&- \frac{W_1^2}{16a^2} A_2^{(j)} (1-R) \frac{1}{S^4} + \frac{2}{AC} f \int \frac{\int (f \int Q^{(j)} dX) dX}{f^2} dX, \qquad (24)
\end{aligned}
$$

where $S = \operatorname{sech}\frac{X}{W_1}$ and $R = \tanh\frac{X}{W_1}$.

4.1 Zeroth Order Equation

For $j = 0$, using (21), one can write (24) as

$$
\begin{aligned}
q^{(0)} &= A_1^{(0)} f + \frac{3W_1^2}{256a^4}\left(5A_2^{(0)} + 8a^2 A_3^{(0)}\right) f X - \frac{W_1^2}{128a^2}\left(5A_2^{(0)} + 8a^2 A_3^{(0)}\right) \\
&+ \frac{3W_1^2}{128a^2}\left(5A_2^{(0)} + 8a^2 A_3^{(0)}\right)(1-R) + \frac{W_1^2}{128a^2}\left(5A_2^{(0)} + 8a^2 A_3^{(0)}\right) e^{-\frac{2X}{W_1}} \\
&- \frac{W_1^2}{1024a^2}\left(85A_2^{(0)} + 192a^2 A_3^{(0)} + 35A_2^{(0)} R\right) S^2 - \frac{W_1^2}{16a^2} A_2^{(0)} \frac{1}{S^2} \\
&- \frac{W_1^2}{16a^2} A_2^{(0)} (1-R) \frac{1}{S^4}, \qquad (25)
\end{aligned}
$$

To make $q^{(0)}$ bounded and consistent at $X = +\infty$, we obtain the following two equations:

$$
-\frac{W_1^2}{16a^2} A_2^{(0)} = 0 \text{ and } -\frac{W_1^2}{128a^2}\left(5A_2^{(0)} + 8a^2 A_3^{(0)}\right) = 0. \qquad (26)
$$

From the Eq. (26), we get $A_2^{(0)} = A_3^{(0)} = 0$. So the expression of $q^{(0)}$ assumed as follows:

$$
q^{(0)} = A_1^{(0)} f = A_1^{(0)} \frac{d\phi_0}{dX}. \qquad (27)
$$

4.2 First Order Equation

Using the Eqs. (22), (27) and MATHEMATICA [33], one can write (24) for $j = 1$ as

$$\begin{aligned} q^{(1)} &= A_1^{(1)} f + \left\{ \frac{3W_1^2}{256a^4}\left(5A_2^{(1)} + 8a^2 A_3^{(1)}\right) + iA_1^{(0)}\frac{\omega^{(1)} - 2lU}{2U} - \frac{\partial A_1^{(0)}}{\partial X_1} \right\} fX \\ &\quad - \frac{W_1^2}{128a^2}\left(5A_2^{(1)} + 8a^2 A_3^{(1)}\right) - \left\{ \frac{3W_1^2}{128a^2}\left(5A_2^{(1)} + 8a^2 A_3^{(1)}\right) + iA_1^{(0)}\frac{a^2\omega^{(1)}}{U} \right\} \\ &\quad \times (1-R) - \left\{ \frac{W_1^2}{128a^2}\left(5A_2^{(1)} + 8a^2 A_3^{(1)}\right) + iA_1^{(0)}\frac{2a^2\omega^{(1)}}{U} \right\} e^{-\frac{2X}{W_1}} \\ &\quad - \left\{ \frac{W_1^2}{1024a^2}\left(85A_2^{(1)} + 192a^2 A_3^{(1)} + 35A_2^{(1)} R\right) + iA_1^{(0)}\frac{a^2\omega^{(1)}}{4U}(7-3R) \right\} S^2 \\ &\quad - \frac{W_1^2}{16a^2} A_2^{(1)} \frac{1}{S^2} - \frac{W_1^2}{16a^2} A_2^{(1)} (1-R)\frac{1}{S^4}, \end{aligned} \tag{28}$$

To make $q^{(1)}$ bounded and consistent at $X = +\infty$, we get $A_2^{(1)} = A_3^{(1)} = 0$. So the expression of $q^{(1)}$ assumed as follows:

$$\begin{aligned} q^{(1)} &= A_1^{(1)} f + \left(iA_1^{(0)}\frac{\omega^{(1)} - 2lU}{2U} - \frac{\partial A_1^{(0)}}{\partial X_1} \right) fX + iA_1^{(0)}\frac{2a^2\omega^{(1)}}{U} e^{-\frac{2X}{W_1}} \\ &\quad + iA_1^{(0)}\frac{a^2\omega^{(1)}}{U}(1-R) - iA_1^{(0)}\frac{a^2\omega^{(1)}}{4U}(7-3R)S^2. \end{aligned} \tag{29}$$

Now, as the first term of $q^{(1)}$ has already been included in $q^{(0)}$, one can remove this term from $q^{(1)}$. One can also remove the ghost secular term of $q^{(1)}$ by choosing

$$\frac{\partial A_1^{(0)}}{\partial X_1} = iA_1^{(0)}\frac{\omega^{(1)} - 2lU}{2U}. \tag{30}$$

Therefore, from the Eq. (29), we obtain

$$q^{(1)} = iA_1^{(0)}\frac{2a^2\omega^{(1)}}{U} e^{-\frac{2X}{W_1}} + iA_1^{(0)}\frac{a^2\omega^{(1)}}{U}(1-R) - iA_1^{(0)}\frac{a^2\omega^{(1)}}{4U}(7-3R)S^2. \tag{31}$$

Now we see that second and third term of $q^{(1)}$ is bounded at $X = \pm\infty$ but the first term of $q^{(1)}$ is not bounded at $X = -\infty$ because of the presence of the term $e^{-\frac{2X}{W_1}}$. To make $q^{(1)}$ bounded at $X = -\infty$, We must have $iA_1^{(0)}\frac{2a^2\omega^{(1)}}{2U} = 0$ and consequently we get

$$\omega^{(1)} = 0. \tag{32}$$

From (32), we see that there is no imaginary part of $\omega^{(1)}$ and consequently DL solution for $\lambda = 1$ is stable at the lowest order of k. The same analysis is true for $\lambda = -1$.

5 Conclusions

In this paper, we have considered the DL solution and its stability described by the combined SKP-KP equation. The form of the DL solution as given by the Eq. (9) suggests that corresponding to a given set of values of the parameters there are two types of DL solutions for $\lambda = +1$ and for $\lambda = -1$. Finally, we have found that DLs are stable at the lowest order of the wave number.

Acknowledgements The authors have used the constructive comments of the reviewer to prepare the manuscript of this paper.

References

1. Kadomtsev, B.B., Petviashvili, V.I.: Sov. Phys. Dokl **15**, 539 (1970)
2. Kako, M., Rowlands, G.: Plasma Phys. **18**, 165 (1976)
3. Infeld, E., Hen, M., Rowlands, G.: Acta Phys. Polon. A **54**, 131 (1978)
4. Rowlands, G.: J. Plasma Phys. **3**, 567 (1969)
5. Infeld, E.: J. Plasma Phys. **8**, 105 (1972)
6. Infeld, E., Rowlands, G.: J. Plasma Phys. **10**, 293 (1973)
7. Zakharov, V.E., Rubenchik, A.M.: Sov. Phys. JETP **38**, 494 (1974)
8. Chakraborty, D., Das, K.P.: J. Plasma Phys. **60**, 151 (1998)
9. Allen, M.A., Rowlands, G.: J. Plasma Phys. **50**, 413 (1993)
10. Allen, M.A., Rowlands, G.: J. Plasma Phys. **53**, 63 (1995)
11. Chakraborty, D., Das, K.P.: J. Plasma Phys. **70**, 89 (2004)
12. Duan, W.S.: Chaos Solitons Fractals **14**, 1315 (2002)
13. Lin, M., Duan, W.S.: Chaos Solitons Fractals **23**, 929 (2005)
14. Pakzad, H.R., Javidan, K.: Chaos Solitons Fractals **42**, 2904 (2009)
15. Pakzad, H.R.: Pramana **74**, 605 (2010)
16. Dorranian, D., Sabetkar, A.: Phys. Plasmas **19**, 013702 (2012)
17. Samanta, U.K., Saha, A., Chatterjee, P.: Phys. Plasmas **20**, 022111 (2013)
18. Samanta, U.K., Saha, A., Chatterjee, P.: Astrophys. Space Sci. **347**, 293 (2013)
19. Sahu, B., Ghosh, N.K.: Astrophys. Space Sci. **343**, 289 (2013)
20. Saha, A., Pal, N., Chatterjee, P.: Phys. Plasmas **21**, 102101 (2014)
21. Saha, A., Chatterjee, P.: Astrophys. Space Sci. **349**, 813 (2014)
22. Saha, A., Pal, N., Chatterjee, P.: Braz. J. Phys. **45**, 325 (2015)
23. Saini, N.S., Kaur, N., Gill, T.S.: Adv. Space Res. **55**, 2873 (2015)
24. Sardar, S., Bandyopadhyay, A., Das, K.P.: Phys. Plasmas **23**, 073703 (2016)
25. Sardar, S., Bandyopadhyay, A., Das, K.P.: Phys. Plasmas **23**, 123706 (2016)
26. Seadawy, A.R., El-Rashidy, K.: Results Phys. **8**, 1216 (2018)
27. Shahein, R.A., Seadawy, A.R.: Indian J. Phys. **93**, 941 (2019)
28. Varghese, A., Saritha, A.C., Willington, N.T., Michael, M., Sebastian, S., Sreekala, G., Venugopal, C.: J. Astrophys. Astr **41**, 1 (2020)
29. Prasad, P.K., Saha, A.: J. Astrophys. Astr **42**, 1 (2021)

30. Sardar, S., Bandyopadhyay, A., Das, K.P.: Phys. Plasmas **24**, 063705 (2017)
31. Cairns, R.A., Mamum, A.A., Bingham, R., Boström, R., Dendy, R.O., Nairn, C.M.C., Shukla, P.K.: Geophys. Res. Lett. **22**, 2709 (1995)
32. Das, J., Bandyopadhyay, A., Das, K.P.: J. Plasma Phys. **74**, 163 (2008)
33. Wolfram, S.: The MATHEMATICA® book, version 4. Cambridge university press (1999)

Dust-ion Collisional and Periodic Forcing Effects on Solitary Wave in a Plasma with Cairns-Gurevich Electron Distribution

Anindya Paul, Niranjan Paul, Kajal Kumar Mondal, and Prasanta Chatterjee

Abstract In this work, our aim is to investigate the effects of dust-ion collision and external periodic force on the ion-acoustic solitary wave (IASW) in the framework of damped forced Korteweg de-Vries (KdV) like Schamel equation. Collisional dusty plasma with Cairns-Gurevich electron distribution has been taken into account to study the propagation of IASW. Reductive perturbation technique (RPT) is employed to derive the damped forced KdV like Schamel equation and its approximate analytical solitary wave solution is determined considering momentum conservation law of KdV like Schamel equation. It also has been shown that, how the other plasma parameters *viz.* non-thermal parameter and ratio of free and trapped electron temperature influence the solitary wave solution in presence of damping and external periodic force. The approximate analytical results of this manuscript may be helpful to understand the basic features of solitary structures in astrophysical plasma where non-thermal and trapped electron distributions are present.

Keywords Damped forced KdV like Schamel equation · Cairns-Gurevich electron distribution · Dust-ion collisional frequency · Non-thermal parameter · RPT · External periodic force

1 Introduction

Analysis of solitary wave structures has become very interesting topic of research to the physicists and mathematicians since last few decades for its versatile application in laboratory and astrophysical plasma [1–3]. Berstein [4] introduced the idea of

A. Paul (✉) · K. K. Mondal
Department of Mathematics, Cooch Behar Panchanan Barma University, Cooch Behar 736101, West Bengal, India
e-mail: paul.anindya20@gmail.com

N. Paul · P. Chatterjee
Department of Mathematics, Siksha Bhavana, Visva-Bharati, Santiniketan, Santiniketan 731235, West Bengal, India

S. Banerjee and A. Saha (eds.), *Nonlinear Dynamics and Applications*, Springer Proceedings in Complexity,
https://doi.org/10.1007/978-3-030-99792-2_18

trapped electrons which was observed in space and laboratory plasma. Plasma species are named as trapped when they move in a finite domain through closed path. Gurevich [5] introduced trapping as a microscopic process where electrons are trapped by potential wells. He showed the effect of trapped electrons in IASWs. To explain characteristic of non-linear electrostatic structures in upper ionosphere observed by Viking and Freja satellites, Cairns et al. [6] considered non-thermal electron distribution and showed that nature of ion-acoustic solitary wave changes significantly in presence of non-thermal electrons. Later Tang [7] and Mamun [18] also worked to show the effect of non-thermal electrons on the IASWs. Abdikian [9] explored dust-ion-acoustic solitary waves in dusty plasma with presence of non-thermal and trapped electrons considering modified Zarakov Kuznetsov equation. Annou et al. [10] investigated the combine effect of trapped and non-thermal electrons on solitons in plasma expansion into vacuum. El-Taibany et al. [11] studied modulated ion-acoustic wave in plasma with electrons obeying Cairns-Gurevich distribution.

Collisions among different plasma constituents occur continuously [12–15]. Due to collisional effects of plasma constituents, damping force is created in plasma medium which influences nonlinear structures in plasma [16–20]. Several works have been studied to investigate the effects of various types of external forces on IASWs. Mainly two kinds of forces have been taken into consideration viz. periodic type source and $sech(\xi, \tau)$ type [21, 22] source that arises from experimental condition or space debris. The results of [23–27] show that the presence of external periodic force brings behavioural changes in IASWs. In [25–27] the effect of frequency & strength of periodic force on solitary wave solution was studied in collisional plasma. The studies [23–27] were done taking external periodic force in the form of $cos(\xi, \tau)$. Chowdhury et al. [24] studied the forced KdV like Schamel equation in a super thermal plasma consisting trapped electron and Paul et al. [20] investigated damped KdV like Schamel equation in a collisional plasma where electrons follow Cairns-Gurevich distribution. To best of our knowledge no work till now has been reported that investigates both dust-ion collisional effect and external periodic forcing effects on solitary wave in plasma where electrons follow Cairns-Gurevich distribution.

The rest of this paper is organized as follows: in Sect. 2, the basic equations and electron which follows the Cairns-Gurevich distribution has been presented. The formation of damped forced KdV like Schamel equation and its time dependent approximate analytical solution has been determined in Sect. 3. In Sect. 4, the effects of different physical parameters on the nonlinear waves have been discussed in detail. Section 5 contains conclusions.

2 Basic Set of Equations and Cairns-Gurevich Electron Distribution

In this work, we consider an unmagnatized, collinsional plasma consisting of cold fluid ions, static dust granules with negative charge, electrons following Cairns-Gurevich distribution. Normalized continuity equation, momentum equation and Poisson's equation are respectively given by,

$$\frac{\partial n_i}{\partial t} + \frac{\partial (n_i u)}{\partial x} = 0, \tag{1}$$

$$\frac{\partial u}{\partial t} + u\frac{\partial u}{\partial x} = -\frac{\partial \phi}{\partial x} - \nu_{id} u, \tag{2}$$

$$\frac{\partial^2 \phi}{\partial x^2} = n_e - n_i + S(x,t). \tag{3}$$

Here, n_i is the ion number density and it is normalized by the unperturbed equilibriam plasma density n_0. u is the ion velocity, normalized to the ion-acoustic speed $C_s = (\frac{T_e}{m_i})^{1/2}$. ϕ is the electrostatic wave potential which is normalized to T_e/e. The time and space variables are in units of the ion plasma frequency $(\omega_{pi})^{-1}$ and the electron Debye radius $\lambda_D = (\frac{T_e}{4\pi n_0 e^2})^{1/2}$. T_e is defined as the temperature of electron. The dust-ion collisional frequency is ν_{id} and $S(x,t)$ is a source term that arises from experimental conditions.

In small amplitude limit of $\phi < 1$, the normalized density of non-thermal electrons obeys the following Cairns-Gurevich distribution [11],

$$n_e = (1 - b\phi + 2b\phi^2)\left(1 + \phi - \frac{4(1-\beta)}{3\sqrt{\pi}}\phi^{3/2} + \frac{1}{2}\phi^2\right), \tag{4}$$

where b is the non-thermal parameter which depends on α and the relation between them is $b = \frac{4\alpha}{1+3\alpha}$. Here, α represents the population of non-thermal electrons and β, the trapping parameter, is the ratio of free electron temperature (T_{fe}) and trapped electron temperature(T_{et}). If $\beta = 0$ then the electron distribution (4) represents plateau like electron distribution and if $\beta = 1$, it becomes Maxwellian.

3 Derivation of Damped Forced KdV Like Schamel Equation and Its Time Dependent Approximate Analytical Solution

To derive the damped forced KdV like Schamel equation, we wish to use Reductive Perturbation Technique (RPT) in a collisional dusty plasma in presence of non-thermal and trapped electrons. The stretched coordinates are taken [20] as follows,

$$\begin{cases} \xi = \epsilon^{1/4}(x - v_0 t), \\ \tau = \epsilon^{3/4} t, \end{cases} \tag{5}$$

where ϵ and v_0 are the strength of nonlinearity and phase velocity of IASW respectively. The expansion of the dependent variables [20, 24] are given by :

$$\begin{cases} n_i = 1 + \epsilon n_i^{(1)} + \epsilon^{3/2} n_i^{(2)} + \dots, \\ u = 0 + \epsilon u_1 + \epsilon^{3/2} u_2 + \dots, \\ \phi = 0 + \epsilon \phi_1 + \epsilon^{3/2} \phi_2 + \dots, \\ \nu_{id} \sim \epsilon^{3/4} \nu_{id0}, \\ S \sim \epsilon^{3/2} S_2. \end{cases} \tag{6}$$

Then, using chain rule, we have,

$$\frac{\partial}{\partial x} \equiv \frac{\partial}{\partial \xi}\frac{\partial \xi}{\partial x} + \frac{\partial}{\partial \tau}\frac{\partial \tau}{\partial x} = \epsilon^{1/4}\frac{\partial}{\partial \xi},$$

$$\frac{\partial^2}{\partial x^2} \equiv \epsilon^{1/2}\frac{\partial^2}{\partial \xi^2},$$

and

$$\frac{\partial}{\partial t} \equiv \frac{\partial}{\partial \xi}\frac{\partial \xi}{\partial t} + \frac{\partial}{\partial \tau}\frac{\partial \tau}{\partial t} = -v_0 \epsilon^{1/4}\frac{\partial}{\partial \xi} + \epsilon^{3/4}\frac{\partial}{\partial \tau}.$$

Using the expansion (6) and (5) into Eqs. (1)–(3) and taking the lowest order coefficient of ϵ, we obtain the following three equations,

$$\epsilon^{5/4} : \ -v_0 \frac{\partial n_i^{(1)}}{\partial \xi} + \frac{\partial u_1}{\partial \xi} = 0, \tag{7}$$

$$\epsilon^{5/4} : \ -v_0 \frac{\partial u_1}{\partial \xi} + \frac{\partial \phi_1}{\partial \xi} = 0, \tag{8}$$

$$\epsilon^{1} : \ -(1 - b)\phi_1 + n_i^{(1)} = 0. \tag{9}$$

Now, from the Eqs. (7), (8) and (9), we obtain the following dispersion relation,

$$v_0 = \frac{1}{\sqrt{1 - b}}. \tag{10}$$

Equating the next higher order coefficient of ϵ, we get the following equations from (1), (2) and (3) respectively as,

$$\epsilon^{7/4}: \frac{\partial n_i^{(1)}}{\partial \tau} - v_0 \frac{\partial n_i^{(2)}}{\partial \xi} + \frac{\partial u_2}{\partial \xi} = 0, \tag{11}$$

$$\epsilon^{7/4}: \frac{\partial u_1}{\partial \tau} - v_0 \frac{\partial u_2}{\partial \xi} + \frac{\partial \phi_2}{\partial \xi} + \nu_{id0} u_1 = 0, \tag{12}$$

$$\epsilon^{3/2}: \frac{\partial^2 \phi_1}{\partial \xi^2} = (1-b)\phi_2 - \frac{4(1-\beta)}{3\sqrt{\pi}} \phi_1^{3/2} - n_i^{(2)} + S_2. \tag{13}$$

Differentiating Eq. (13) partially with respect to ξ, and eliminating the terms $n_i^{(2)}$, u_2 and ϕ_2 from the Eqs. (7)–(13) with the help of dispersion relation, the following nonlinear evolution equation is obtained,

$$\frac{\partial \phi_1}{\partial \tau} + A\sqrt{\phi_1}\frac{\partial \phi_1}{\partial \xi} + B\frac{\partial^3 \phi_1}{\partial \xi^3} + C\phi_1 = B\frac{\partial S_2}{\partial \xi}, \tag{14}$$

where $A = \frac{v_0^3(1-\beta)}{\sqrt{\pi}}$, $B = \frac{v_0^3}{2}$ and $C = \frac{\nu_{id0}}{2}$.

The external periodic force S_2 can be taken as $S_2 = f_0 \xi cos(\omega \tau)$, where f_0 and ω stands for the strength and frequency of the source respectively. Using the above source term in the Eq. (14) we get the following equation,

$$\frac{\partial \phi_1}{\partial \tau} + A\sqrt{\phi_1}\frac{\partial \phi_1}{\partial \xi} + B\frac{\partial^3 \phi_1}{\partial \xi^3} + C\phi_1 = B f_0 cos(\omega \tau), \tag{15}$$

which is the desired damped forced KdV like Schamel equation.

If $C = f_0 = 0$, the Eq. (15) becomes the KdV like Schamel equation,

$$\frac{\partial \phi_1}{\partial \tau} + A\sqrt{\phi_1}\frac{\partial \phi_1}{\partial \xi} + B\frac{\partial^3 \phi_1}{\partial \xi^3} = 0, \tag{16}$$

and its solution is given by,

$$\phi_1 = \phi_m sech^4\left(\frac{\xi - U\tau}{W}\right), \tag{17}$$

where $\phi_m = (\frac{15U}{8A})^2$ and $W = \sqrt{\frac{16B}{U}}$, represent amplitude and width of the solitary wave respectively. Here, U is the speed of the wave.

For small values of f_0 and C, we assume the solution of damped forced KdV like Schamel equation (15) as,

$$\phi_1 = \phi_m(\tau) sech^4\left(\frac{\xi - U(\tau)\tau}{W(\tau)}\right), \tag{18}$$

where $U(\tau)$ is an unknown function of τ, $\phi_m(\tau) = (\frac{15U(\tau)}{8A})^2$ and $W(\tau) = \sqrt{\frac{16B}{U(\tau)}}$.

It is well known that

$$I = \int_{-\infty}^{\infty} \phi_1^2 \, d\xi, \tag{19}$$

is a conserved quantity for a KdV like Schamel equation [20]. Using the momentum conservation law (19) and and with the help of [20] and [24], we get the following differential equation for the unknown function $U(\tau)$,

$$U(\tau)\frac{dU(\tau)}{d\tau} + \frac{4}{7}CU(\tau)^2 = \frac{224}{995}A^2 B f_0 cos(\omega\tau). \tag{20}$$

Solving the differential equation (20) with an initial condition $U(0) = U_0$, we get,

$$U(\tau) = \sqrt{\frac{448\, A^2 B f_0}{135(64C^2 + 49\omega^2)}(8C\cos(\omega\tau) + 7\omega\sin(\omega\tau)) + Ke^{-\frac{8C}{7}\tau}}, \tag{21}$$

where K is given by

$$K = U_0^2 - \frac{3584\, AA^2 BC f_0}{135(64\, C^2 + 49\omega^2)}. \tag{22}$$

Using the value of $U(\tau)$, we subsequently deduce the values of $W(\tau)$ and $\phi_m(\tau)$. Putting these values in Eq. (18), one can obtain the solution of damped forced KdV like Schamel equation (15).

4 Effects of Various Parameters

In this section, the effects of various parameters such as b, β, ν_{id0}, ω and f_0 on solution of Eq. (15) is demonstrated. Figure 1a represents the variation of amplitude in solitons for three distinct values of $b = 0.1, 0.3, 0.5$ when the other parameters are $U_0 = 0.2, \tau = 1, \beta = 0.2, \nu_{id0} = 0.09, f_0 = 0.01, \omega = 0.5$. It is noticed that the amplitude of the solitary waves decreases as b increases and width of the solitary wave increases a little with the increment of b. Figure 1b shows the variation of solitons for different values of trapping parameter (β). When $b = 0.1$ and the other parameters are same as shown in the caption of Fig. 1a. As the value of trapping parameter increases, the amplitude of the solitary wave increase significantly. The width of the solitary waves also increases with the increment of β. In Fig. 1c, the value of ν_{id0} is taken in the interval $(0.01, 0.13)$ and the other parameters are same as in the caption of the figure. From this figure it is clearly observed that as ν_{id0} increases the amplitude of solitary waves increases significantly whereas it is hard to draw conclusion about change in the width of waves from that same figure. Therefore,

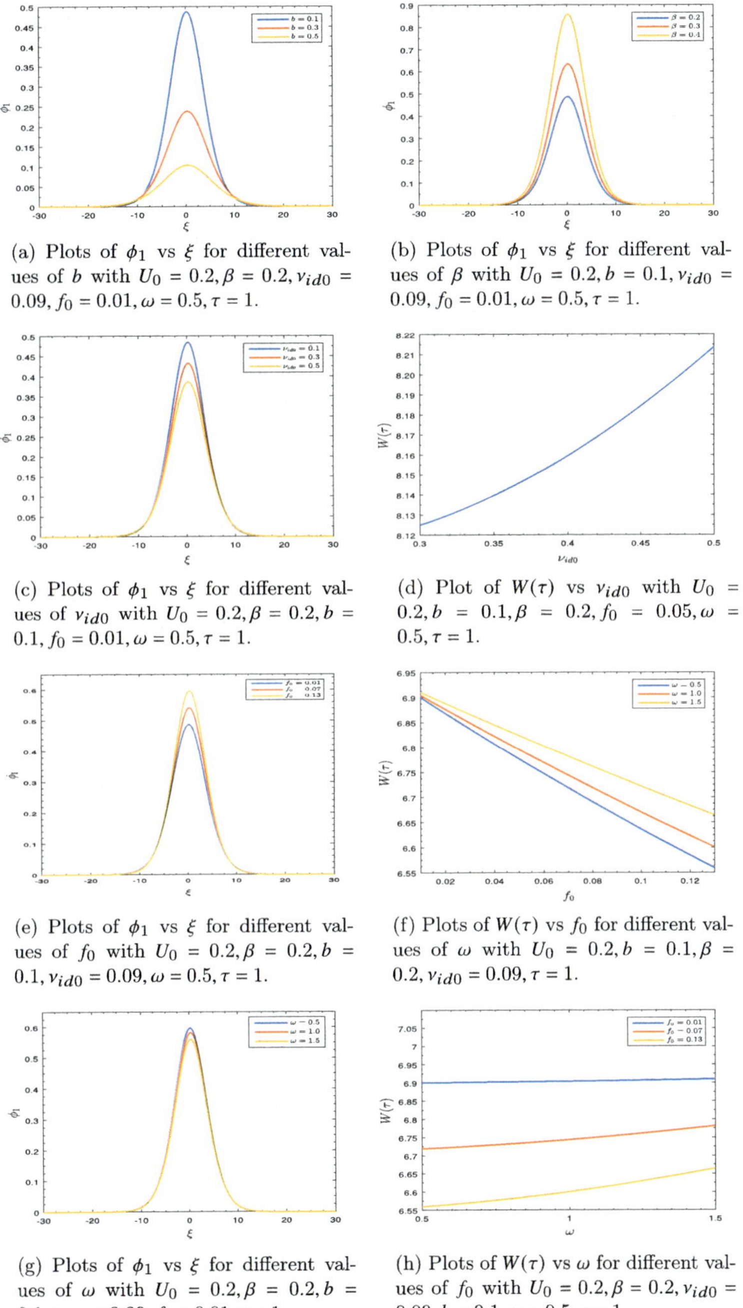

(a) Plots of ϕ_1 vs ξ for different values of b with $U_0 = 0.2, \beta = 0.2, \nu_{id0} = 0.09, f_0 = 0.01, \omega = 0.5, \tau = 1$.

(b) Plots of ϕ_1 vs ξ for different values of β with $U_0 = 0.2, b = 0.1, \nu_{id0} = 0.09, f_0 = 0.01, \omega = 0.5, \tau = 1$.

(c) Plots of ϕ_1 vs ξ for different values of ν_{id0} with $U_0 = 0.2, \beta = 0.2, b = 0.1, f_0 = 0.01, \omega = 0.5, \tau = 1$.

(d) Plot of $W(\tau)$ vs ν_{id0} with $U_0 = 0.2, b = 0.1, \beta = 0.2, f_0 = 0.05, \omega = 0.5, \tau = 1$.

(e) Plots of ϕ_1 vs ξ for different values of f_0 with $U_0 = 0.2, \beta = 0.2, b = 0.1, \nu_{id0} = 0.09, \omega = 0.5, \tau = 1$.

(f) Plots of $W(\tau)$ vs f_0 for different values of ω with $U_0 = 0.2, b = 0.1, \beta = 0.2, \nu_{id0} = 0.09, \tau = 1$.

(g) Plots of ϕ_1 vs ξ for different values of ω with $U_0 = 0.2, \beta = 0.2, b = 0.1, \nu_{id0} = 0.09, f_0 = 0.01, \tau = 1$.

(h) Plots of $W(\tau)$ vs ω for different values of f_0 with $U_0 = 0.2, \beta = 0.2, \nu_{id0} = 0.09, b = 0.1, \omega = 0.5, \tau = 1$.

Fig. 1 Graphical presentations of approximate analytical solitary wave solution with variation of amplitude and width w.r.t to different parameters

Fig. 1d is plotted between $W(\tau)$ versus ν_{id0} for different values of b to understand the variation of the width of solitary waves. It is clear from the graph that the width of the solitary wave increases as the dust-ion collisional frequency grows. Figure 1e represents the variation of solitary wave against ξ where other parameters are same as in the caption. It shows that solitary wave become more and more spiky as the strength of the periodic force f_0 grows. From this figure, it is hard to draw conclusion on the variation of width. Thus, a graph of $W(\tau)$ versus f_0 is plotted in Fig. 1f for three distinct values of $\omega = 0.5, 1.0, 1.5$ when the other parameters are $U_0 = 0.2$, $\tau = 1$, $\beta = 0.2$, $\nu_{id0} = 0.09$, $f_0 = 0.01$, $b = 0.1$. It shows that as the strength of the periodic force f_0 increases the width of the solitary waves decrease. From Fig. 1f represents the profile for ϕ_1 against ξ, keeping all other parameters same as the caption. It is noticed that the height of the solitary wave decreases with the enhancement ω. To understand the effect of the frequency of the force on the width of the solitary wave, a graph is plotted between $W(\tau)$ versus ω in Fig. 1h. a small increment in the width of the solitary wave is observed as ω of the external periodic force grows.

Figure 2a and b show the variance of width and amplitude of solitary wave respectively against τ at different strength of the periodic force when other parameters are same as in the caption. Figure 2a interprets that width of solitary wave decreases as τ increases. Figure 2b explicates that amplitude of solitary wave rises as τ grows. Figure 2c is plotted for $W(\tau)$ against τ for three values of $\nu_{id0} = 0.2, 0.3, 0.4$ when $U_0 = 0.2$, $\beta = 0.2$, $b = 0.1$, $f_0 = 0.05$, $\omega = 0.5$. The plot elucidates that width of the solitary wave surges as τ enhances. Figure 2d depicts a graph between $\phi_m(\tau)$ versus τ at different values of ν_{id0} when $U_0 = 0.2$, $\beta = 0.2$, $b = 0.1$, $f_0 = 0.05$ & $\omega = 0.5$. The graph shows that the amplitude of the IASW reduces with the increment of τ.

5 Conclusions

Using RPT, a damped forced KdV like Schamel equation is derived and the behaviour of its wave propagation has been discussed in a collisional plasma consisting dust granules, trapped and non-thermal electrons which follows Cairns-Gurevich distribution. The study shows that non-thermal parameter (b), trapping parameter (β) strength (f_0) and frequency (ω) of external force, dust-ion collisional frequency (ν_{id0}) have significant effects on the amplitude and the width of solitary waves. It is seen that the amplitude of the solitary waves decreases as the non-thermal parameter (b) increases. This is because, as the value of b increases, the external periodic force helps to decrease the positive potential energy in the solitary wave. It is observed that with the increment of trapping parameter (β), the positive potential energy in the solitary wave increases that helps in the increment of both amplitude and width of the solitary wave. As the dust-ion collisional frequency parameter (ν_{id0}) increases, the collision between the dust granules and ions grows in a more narrow region of the plasma space that results in decay of the internal potential energy of the solitary

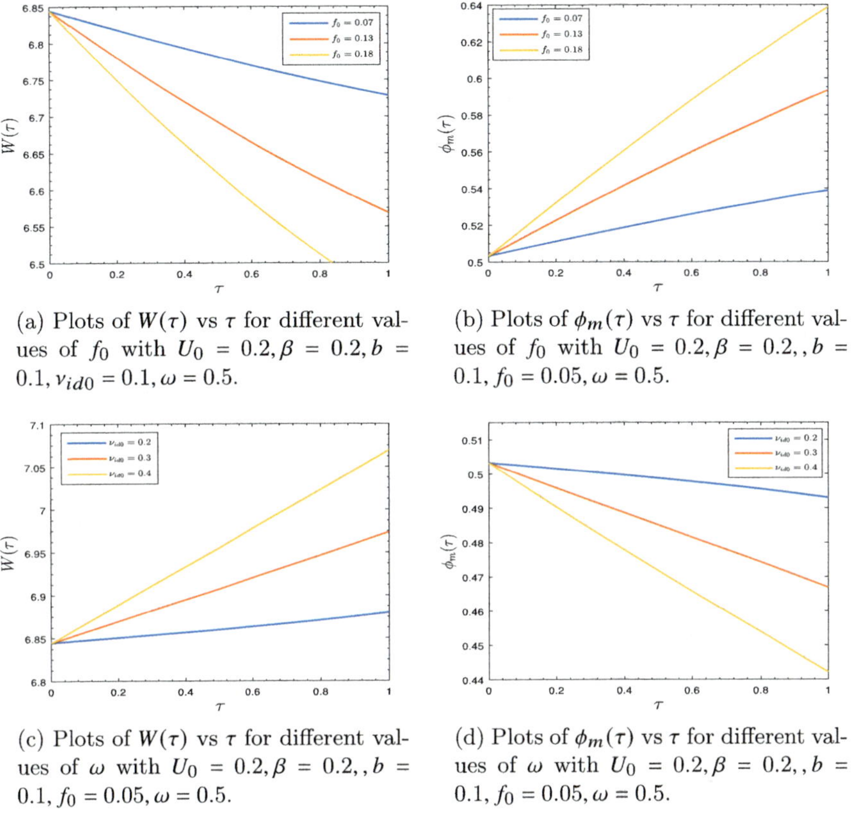

(a) Plots of $W(\tau)$ vs τ for different values of f_0 with $U_0 = 0.2, \beta = 0.2, b = 0.1, \nu_{id0} = 0.1, \omega = 0.5$.

(b) Plots of $\phi_m(\tau)$ vs τ for different values of f_0 with $U_0 = 0.2, \beta = 0.2,, b = 0.1, f_0 = 0.05, \omega = 0.5$.

(c) Plots of $W(\tau)$ vs τ for different values of ω with $U_0 = 0.2, \beta = 0.2,, b = 0.1, f_0 = 0.05, \omega = 0.5$.

(d) Plots of $\phi_m(\tau)$ vs τ for different values of ω with $U_0 = 0.2, \beta = 0.2,, b = 0.1, f_0 = 0.05, \omega = 0.5$.

Fig. 2 Variation of amplitude and width of approximate analytical solitary wave solution w.r.t different parameters

wave causing decrease in the amplitude with almost unaltered width. The increase in strength of the external force (f_0) causes the rise in the positive potential energy of IASW. As a result the peak of the amplitude of the solitary wave increases significantly. The increase in frequency of the external periodic force (ω) causes the insignificant decrement in amplitude of the solitary waves. These results may help in further study of IASWs.

References

1. Ikezi, H., Taylor, R., Baker, D.: Formation and interaction of ion-acoustic solitions. Phys. Rev. Lett. **25**, 11 (1970)
2. Washimi, H., Taniuti, T.: Propagation of ion acoustic waves of small amplitude. Phys. Rev. Lett **77**, 996 (1996)
3. Goertz, C.K.: Dusty plasmas in the solar system. Rev. Geophys. **27**, 271 (1989)

4. Bernstein, I.B., Greene, J.M., Kruskal, M.D.: Exact nonlinear plasma oscillations. Phys. Rev. **108**, 546 (1957)
5. Gurevich, A.V.: Distribution of captured particles in a potential well in the absence of collisions. Sov. Phys. JETP **26**, 575–580 (1968)
6. Cairns, R.A., Mamum, A.A., Bingham, R., Bostrom, R., Dendy, R.O., Nairn, C.M., Shukla, P.K.: Electrostatic solitary structures in nonthermal plasmas. Geophys. Res. Lett. **22**, 2709 (1995)
7. Tang, R.A., Xue, J.K.: Nonthermal electrons and warm ions effects on oblique modulation of ion-acoustic waves. Phys. Plasmas **11**, 3939–3944 (2004)
8. Mamun, A.: Effects of ion temperature on electrostatic solitary structures in nonthermal plasmas. Phys. Rev. E **55**, 1852–1857 (1997)
9. Abdikian, A.: Dust-ion-acoustic solitary waves in a magnetized dusty pair-ion plasma with Cairns-Gurevich electrons and opposite polarity dust particles. Contrib. Plasma Phys. **59**, 20 (2019)
10. Annou, K., Bara, D., Bennaceur-Doumaz, D.: Cairns-Gurevich equation for soliton in plasma expansion into vacuum. J. Plasma Phys. **81** (2015)
11. El-Taibany, W.F., Sabry, R.: Dust-acoustic solitary waves and double layers in a magnetized dusty plasma with nonthermal ions and dust charge variation. Plasma Phys. **12**, 082302 (2005)
12. Yang, X., Wang, C.L., Liu, C.B., Zhang, J.R., Shi, Y.R.: The collision effect between dust grain and ions to the dust ion acoustic waves in a dusty plasma. Phys. Plasmas **19**, 103705 (2012)
13. Popel, S.I., Golub', A.P., Losseva, T.V.: Weakly dissipative dust-ion-solitons. Phys. Rev. E **67**, 056402 (2003)
14. Nakamura., Y., Sarma, A.: Observation of ion-acoustic solitary waves in a dusty plasma. Phys. Plasmas **8**, 3921 (2001)
15. Losseva, T.V., Popel, S.I., Golub', A.P., Izvekova, Yu.N., Shukla, P.K.: Weakly dissipative dust-ion-acoustic solitons in a complex plasmas and the effect of electromagnetic radiation. Phys. Plasmas **19**, 013703 (2012)
16. Misra, A.P., Choudhury, A.R., Choudhury, K.R.: Effect of dust ion collision on dust ion acoustic solitary waves for nonextensive plasmas in the framework of damped Korteweg-de Vries-Burgers equation. Phys. Lett. A **323**, 110 (2004)
17. Maitra, S., Banerjee, G.: Dust ion acoustic solitary waves in a collisional dusty plasma with dust grains having Gaussian distribution. Phys. Plasmas **21**, 113707 (2014)
18. Das, T.K., Ali, R., Chatterjee, P.: Effect of dust ion collision on dust ion acoustic waves in the framework of damped Zakharov-Kuznetsov equation in presence of external periodic force. Phys. Plasmas **24**, 103703 (2017)
19. Paul, N., Mondal, K.K.,Chatterjee, P.: Effect of dust ion collision on dust ion acoustic solitary waves for nonextensive plasmas in the framework of Damped Korteweg–de Vries–Burgers equation. Zeitschrift fur Naturforschung a **74**(10), (2019)
20. Paul, N., Ali, R., Mondal, K.K., Chatterjee, P.: Ion-neutral collisional effect on solitary waves in weakly ionized plasma with Cairns-Gurevich distribution of electrons. Int. J. Appl. Comput. Math. **7**, 172 (2021)
21. Sen, A., Tiwari, S., Misra, S., Kaw, P.: Nonlinear wave excitations by orbiting charged space debris objects. Adv. Space Res. **56**, 429 (2015)
22. Bhowmick, S., Sahu, B.: Propagation of nonlinear excitations of dust acoustic waves by a moving charged object in superthermal plasmas. Indian J. Phys. (2021)
23. Ali, R., Saha, A., Chatterjee, P.: Analytical electron acoustic solitary wave solution for the forced KdV equation in superthermal plasma. Phys. Plasmas **24**, 122106 (2017)
24. Chowdhury, S., Mandi, L., Chatterjee, P.: Effect of externally applied perodic force on ion acoustic waves in superthermal plasmas. Phys. Plasmas **25**, 042112 (2018)
25. Chatterjee, P., Ali, R., Saha, A.: Analytical solitary wave solution of the dust ion acoustic waves for the damped forced Kortewegde Vries equation in superthermal plasmas. Zeitschrift fr Naturforschung A **73**, 151 (2018)

26. Mandi, L., Mondal, K.K., Chatterjee, P.: Analytical solitary wave solution of the dust ion acoustic waves for the damped forced modified Korteweg-de Vries equation in q-nonextensive plasmas. Eur. Phys. J. Spl. Topics **228**, 2753 (2019)
27. Paul, N., Mondal, K.K., Ali, R., Chatterjee, P.: Analytical solitary wave solution of dust ion acoustic waves in nonextensive plasma in the framework of damped forced Korteweg-de Vries-Burgers equation. Indian J. Phys. **95**(12), 2855–2863 (2021)

Electron-Acoustic Solitons in a Multicomponent Superthermal Magnetoplasma

Rajneet Kaur, Geetika Slathia, Kuldeep Singh, and Nareshpal Singh Saini

Abstract In this paper, electron-acoustic solitons (EASs) in magnetized multicomponent plasma having fluid of cold electrons, positrons, superthermal electrons, and positive ions are examined. The nonlinear Zakharov-Kuznetsov (ZK) equation is derived by applying the reductive perturbation method (RPM). The effect of various plasma parameters (concentration of electrons, superthermality of hot electrons/positrons and magnetic field strength) on the characteristic properties of EASs is analysed.

Keywords EA solitons · Superthermal distribution · Zakharov-Kuznetsov equation

1 Introduction

From the past many years, the study of electron-acoustic (EA) has become very fascinating among plasma physicists because of their pivotal role in different plasma environments e.g. astrophysical, laboratory and space plasmas [1]. These waves are evolved due to the existence of two temperature electrons. Due to two distinct temperatures, pressure of hot electrons provide the required restoring force and cold electrons become inertial. Further, due to large mass of ions as compared to that of electrons, ions are considered to form a stationary background. Numerous investigation have already been done to examine the propagation properties of linear and nonlinear EASs. Yu and Shukla [2] reported the characteristics of EASs in a magnetoplasma with multi-temperature electrons. Mace and Hellberg [3] studied the EASs in a fluid model composed of two temperature electrons with magnetized and unmagnetized fluid ions. They discussed the propagation properties of KdV-

R. Kaur (✉) · G. Slathia · K. Singh · N. S. Saini
Department of Physics, Guru Nanak Dev University, Amritsar 143005, India
e-mail: rajneetkaur909@gmail.com

K. Singh
Department of Mathematics, Khalifa University of Science and Technology, Abu Dhabi, UAE

S. Banerjee and A. Saha (eds.), *Nonlinear Dynamics and Applications*,
Springer Proceedings in Complexity,
https://doi.org/10.1007/978-3-030-99792-2_19

ZK equation with plane and multidimensional solitary wave solutions. Danehkar et al. [4] developed a general plasma fluid model to describe the large amplitude EASs in superthermal plasma using pseudopotential method. They showed that only negative potential EASs are formed in a plasma. Devanandhan et al. [5] studied the EASs in a magnetized and plasma composed of hot ions and cold electrons obeying a kappa distribution. They observed that the magnetic field and other plasma parameters have strongly modified the characteristics of EASs. Various researchers have reported the properties of EASs in electron-positron-ion (e-p-i) plasmas. The characteristic properties of positron acoustic solitons in a multicomponent plasma have been examined by Alam et al. [6]. They studied the basic features of Double layer, Gardner solitons with solitary wave solution of mKdV equation. Adnan et al. [7] analysed the ion acoustic waves in a superthermal e-p-i plasma under the influence of magnetic field. They found that the effect of positron concentration and superthermality has modified the ion acoustic solitary waves. Ferdousi et al. [8] studied the ion acoustic solitons in a magnetized plasma composed of nonextensive positrons and electrons. It was observed that nonextensive parameter has altered the propagation properties of ion acoustic solitons. Saha and Tamang [9] analysed the behaviour of positron acoustic waves in a multicomponent plasma containing inertial positrons and Kaniadakis distributed positrons and hot electrons. They observed that the effect of different parameters have modified the nonlinear structures. Bansal et al. [10] examined the characteristic properties of EASs in a magnetoplasma composed of superthermal distributed two temperature electrons, positrons and uniform stationary background ions. The results showed that the nonplanar EASs are significantly modified due to the effect of positron densities as well as positron temperature and other components. The dissipative effects of ion acoustic solitons in a multicomponent collisional e-p-i plasma with non-thermal electrons and isothermal positrons were studied by Gul and Ahmed [11].

Electron-acoustic waves gain more importance when high energy particles in plasmas come into picture. The occurrence of these high energy particles are well explained by kappa distribution function. The superthermal distribution function was first well explained by Vasyliunas [12]. Various researchers have examined the role of high energy superthermal particles in nonlinear dynamics. The characteristic properties of ion acoustic waves with two fluid ions in superthermal plasma were studied by Shahmansouri and Tribeche [13]. The nonlinearity and dispersion properties of ion acoustic solitons are significantly enhanced with change in superthermality parameter. Singh and Sethi [14] studied the characteristic properties of mKdV equation in a collisionless plasma composed of negatively charged dust, two temperature kappa distributed electrons and hot ions. Singh and Saini [15] investigated the EA shock waves in a magnetized multicomponent plasma consists of cold electrons as a fluid, hot positrons and superathermal electrons. They analysed that the strength of EA shocks is increased with increase in superthermality of electrons. The aim of our present work is to study nonlinear dynamics of EASs in an e-p-i superthermal magnetoplasma. The paper is arranged as follows: Sect. 2 presents the basic fluid model equations. The derivation of ZK equation and its solution are illustrated in Sect. 3. Numerical analysis is illustrated in Sect. 4. Conclusions are mentioned in the Sect. 5.

2 Basic Fluid Equations

The dimensionless expressions of densities of superthermal positrons and hot electrons are given as [15]

$$n_p = 1 - \gamma b_1\phi + \gamma^2 b_2\frac{\phi^2}{2} + ... \tag{1}$$

$$n_h = 1 + a_1\phi + a_2\frac{\phi^2}{2} + .. \tag{2}$$

Here, $a_1 = \left(\frac{\kappa_e-\frac{1}{2}}{\kappa_e-\frac{3}{2}}\right)$, $a_2 = \left(\frac{\kappa_e^2-\frac{1}{4}}{(\kappa_e-\frac{3}{2})^2}\right)$, $b_1 = \left(\frac{\kappa_p-\frac{1}{2}}{\kappa_p-\frac{3}{2}}\right)$, $b_2 = \left(\frac{\kappa_p^2-\frac{1}{4}}{(\kappa_p-\frac{3}{2})^2}\right)$. Here, $\kappa_{e,p}$ are the superthermality spectral indices of electrons and positrons. The Maxwellian case can be obtained as $\kappa_{e,p} \to \infty$.

At equilibrium $n_{oh} + n_{oc} = n_{oi} + n_{op}$, where n_{oj}(for $j = c, p, i, h$) are undisturbed number density of cold electrons, hot positrons, stationary ions and hot electrons respectively. The wave is propagating in the x-z plane. We consider the dimensionless equations as [15]:

$$\frac{\partial n_c}{\partial t} + \frac{\partial(n_c u_{cx})}{\partial x} + \frac{\partial(n_c u_{cz})}{\partial z} = 0, \tag{3}$$

$$\frac{\partial u_{cx}}{\partial t} + u_{cx}\frac{\partial u_{cx}}{\partial x} + u_{cz}\frac{\partial u_{cx}}{\partial z} = \frac{\partial \phi}{\partial x} - \Omega u_{cy}, \tag{4}$$

$$\frac{\partial u_{cy}}{\partial t} + u_{cx}\frac{\partial u_{cy}}{\partial x} + u_{cz}\frac{\partial u_{cy}}{\partial z} = \Omega u_{cx}, \tag{5}$$

$$\frac{\partial u_{cz}}{\partial t} + u_{cx}\frac{\partial u_{cz}}{\partial x} + u_{cz}\frac{\partial u_{cz}}{\partial z} = \frac{\partial \phi}{\partial z}, \tag{6}$$

$$\frac{\partial^2 \phi}{\partial x^2} + \frac{\partial^2 \phi}{\partial z^2} = 1 - \alpha + n_c\sigma - \delta + \phi(a_1 + \alpha\gamma b_1) + \frac{\phi^2}{2}(a_2 - \alpha\gamma^2 b_2), \tag{7}$$

The fluid velocity u_c, and electrostatic potential ϕ, are normalized with respect to EA speed, $C_e = (\frac{T_h}{m_e})^{\frac{1}{2}}$, and $\frac{T_h}{e}$, respectively. The space coordinate (x) is normalized by electron Debye length $\lambda_D = (\frac{T_h}{4\pi n_{oh} e^2})^{\frac{1}{2}}$ and time coordinate (t) is scaled by inverse of plasma frequency of electrons, $\omega_{ph} = (\frac{4\pi n_{oh} e^2}{m_e})^{\frac{1}{2}}$. The gyrofrequency of electron, $\omega_c = \frac{eB}{m_e c}$ is scaled with respect to ω_{ph}, $\Omega = \frac{\omega_c}{\omega_{ph}}$. $\alpha = \frac{n_{op}}{n_{oh}}$, $\sigma = \frac{n_{oc}}{n_{oh}}$, $\delta = \frac{n_{oi}}{n_{oh}}$ and $\gamma = \frac{T_h}{T_p}$.

3 Derivation of ZK Equation and Its Solution

To study the dynamics of EASs with weak dispersion and of weak nonlinearity, we assume ω (or k) $<< 1$. All physical quantities vary slowly in space and vary more slowly in time. We have used the RPM to find the ZK equation. The stretched coordinates are given as [16, 17]:

$$\xi = \epsilon^{\frac{1}{2}} (z - Vt), \quad \zeta = \epsilon^{\frac{1}{2}} x, \quad \text{and} \quad \emptyset = {}^{\frac{3}{2}}\mathrm{t} \tag{8}$$

The expansions used are given as:

$$n_c = 1 + \epsilon n_{c1} + \epsilon^2 n_{c2} + \epsilon^3 n_{c3} + ..., \tag{9}$$

$$u_{cx} = \epsilon^{\frac{3}{2}} u_{cx1} + \epsilon^2 u_{cx2} + \epsilon^{\frac{5}{2}} u_{cx3} + ..., \tag{10}$$

$$u_{cy} = \epsilon^{\frac{3}{2}} u_{cy1} + \epsilon^2 u_{cy2} + \epsilon^{\frac{5}{2}} u_{cy3} + ..., \tag{11}$$

$$u_{cz} = \epsilon u_{cz1} + \epsilon^2 u_{cz2} + \epsilon^3 u_{cz3} + ..., \tag{12}$$

$$\phi = \epsilon \phi_1 + \epsilon^2 \phi_2 + \epsilon^3 \phi_3 + ..., \tag{13}$$

using Eqs. (8)–(13) in Eqs. (3)–(7) neutrality condition is obtained as: $(\delta + \alpha) = (1 + \sigma)$. After simplifying, we get the first order equations as:

$$n_{c1} = -\phi_1 Q, \tag{14}$$

$$u_{cy1} = \frac{1}{\sigma} \frac{\partial \phi_1}{\partial \zeta}, \tag{15}$$

$$u_{cz1} = -V\phi_1, \tag{16}$$

$$V = \frac{1}{\sqrt{Q}}, \tag{17}$$

where, V is the phase velocity of EASs and $Q = \frac{(a_1 + \alpha\gamma b_1)}{\sigma}$. By equating the quantities for higher orders of ϵ and doing rigorous calculations, we have obtained the following

ZK equation as:

$$\frac{\partial \phi}{\partial \tau} + A\phi \frac{\partial \phi}{\partial \xi} + B\frac{\partial^3 \phi}{\partial \xi^3} + C\frac{\partial}{\partial \xi}\left(\frac{\partial^2 \phi}{\partial \zeta^2}\right) = 0, \tag{18}$$

where, $\phi_1 = \phi$ and nonlinear coefficient $A = B\left(-3\sigma Q^2 - (a_2 - \alpha\gamma^2 b_2)\right)$, dispersion coefficient $B = \frac{1}{2VQ^2\sigma}$, and transverse dispersion coefficient $C = B\left(1 + \frac{\sigma}{\Omega^2}\right)$.

We consider a transformation $Y = l_x\zeta + l_z\xi - \Lambda\tau$, ($l_x$, l_z are the direction cosines), to evaluate the solution of Eq. (18). Λ denotes the velocity of solitons w.r.t. moving frame scaled with C_e. The solution of ZK equation is obtained as [18]:

$$\phi = \phi_0 sech^2\left(\frac{Y}{\Delta}\right), \tag{19}$$

where $\phi_0 = \frac{3\Lambda}{Al_z}$ is maximum amplitude and $\Delta = \left(\frac{4Fl_z}{\Lambda}\right)^{\frac{1}{2}}$ is the width of EASs. Here, $F = Bl_z^2 + C\left(1 - l_z^2\right)$.

4 Numerical Analysis

To carry out numerical analysis, the range of various physical parameters in laboratory and astrophysical/space plasmas [19] is chosen as: n_{op} ~(1.5–3) cm^{-3}, n_{oc} ~(0.1–0.4) cm^{-3}, T_h ~(200–1,000) eV, n_{oh} ~(1.5–3) cm^{-3}, and T_p ~ (200–1,000) eV. The propagation properties of EASs are strongly influenced by the change in the value of any parameter.

Figure 1, describes the behaviour of phase velocity (V) with superthermality index of positrons (via κ_p) and superthermality index of hot electrons (via κ_e).

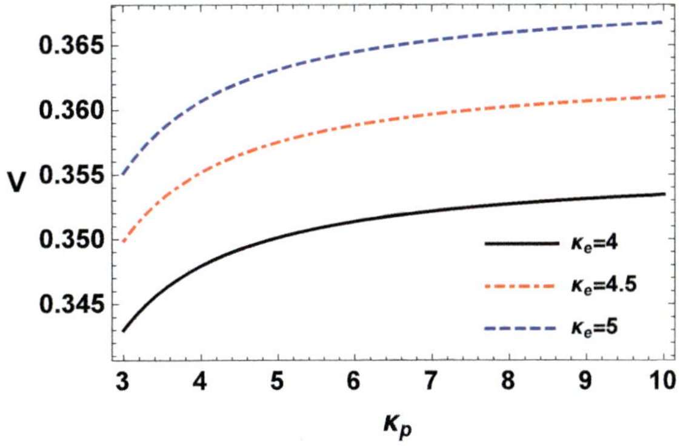

Fig. 1 Plot for the phase velocity (V) with superthermality of positrons (κ_p) and with electrons (κ_e)

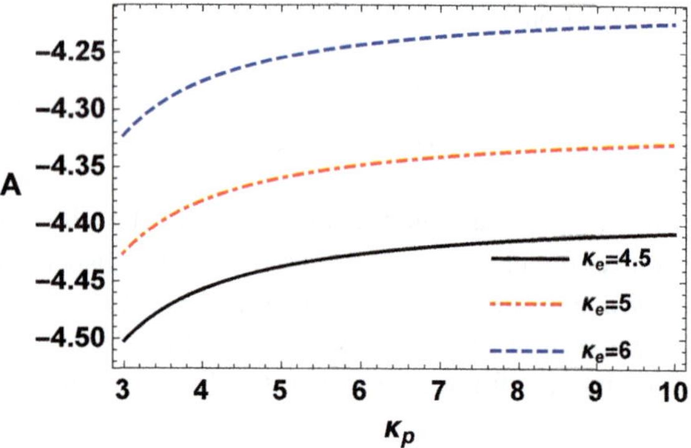

Fig. 2 Plot for the nonlinear coefficient (A) with superthermality index of positrons (via κ_p) and with superthermality index of electrons (κ_e)

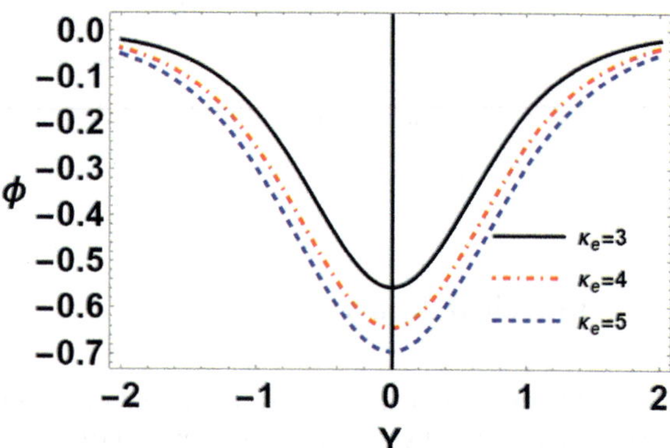

Fig. 3 Plot for EASs profile (ϕ) with superthermality index of electrons (κ_e), here $\gamma = 4$, $\delta = 0.1$, $\kappa_p = 4$, $\sigma = 0.2$, $l_z = 0.6$, $\omega = 0.2$, $\Lambda = 0.3$

It is found that with increment in κ_p and κ_e (i.e., decrease in the superthermality of positrons/electrons), the phase velocity of EASs is enhanced. It is clear that superthermality effects have significantly modified the dispersion properties of EASs and makes the wave to propagate slowly in case of more superthermal charged particles.

Figure 2, describes the nature of coefficient (A) with superthermality index of positrons(κ_p) and superthermal index of hot electrons (via κ_e). It is seen that with increase in κ_p and κ_e, magnitude of A increases. It is found that A is negative, so only negative potential EASs are reported in the considered plasma model.

In Fig. 3, we have analysed the characteristics of EASs profile (ϕ) with superthermality of electrons (via κ_e). It is noticed that with increase in the value of κ_e, the amplitude and width of EASs are increased along negative axis. This variation in the properties of EASs occurs due to the change in nonlinearity and dispersion effects.

In Fig. 4, depicts the nature of EASs profile (ϕ) with number density ratio of positron to hot electron $\alpha(= n_{op}/n_{oh})$ and shows that amplitude(width) of EASs is increased (decreased) with increase in α. It is noteworthy to mention that any change in number density ratios makes variation in the nonlinear coefficient A that further modifies the profile of solitons.

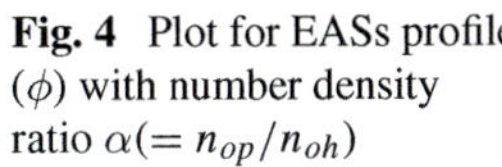

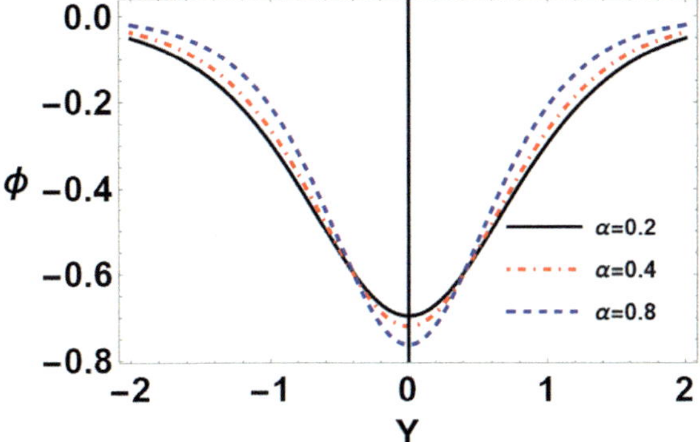

Fig. 4 Plot for EASs profile (ϕ) with number density ratio $\alpha(=n_{op}/n_{oh})$

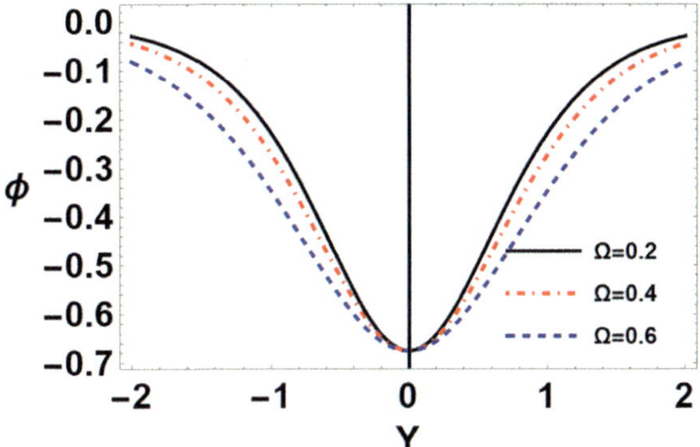

Fig. 5 Plot for EASs profile (ϕ) with magnetic field strength (Ω)

Figure 5, describes the profile of EASs (ϕ) with magnetic field strength (via Ω). The width of EASs increases with the increase in the value of magnetic field strength, whereas the amplitude remains same. It is clear that the dispersion effects are more pronounced with the variation of magnetic field strength.

Figure 6, describes the variation of profile of EASs (ϕ) for multiple values of (κ_p). It is analysed that with rise in superthermality index of positrons (via κ_p), the magnitude of EASs decreases. This emphasizes that superthermality index has strongly influenced the properties of EASs with change in different nonlinear effects.

Figure 7, depicts the variation of EASs profile (ϕ) for multiple values of temperature ratio of hot electrons to positrons $\gamma(=\frac{T_h}{T_p})$ and highlights that width and amplitude of EASs are decreased with increase in the value of $\gamma(=\frac{T_h}{T_p})$. This highly change in width and amplitude of EA solitons is noticed due to the effect of temperature ratio on the nonlinear coefficient A.

Figures 8 and 9, represent the 3D profiles of EASs with magnetic field strength (Ω) and superthermality of positrons (κ_p) respectively. These figures further confirm the modification in the profile of EA solitons with the variation of different parameters simultaneously.

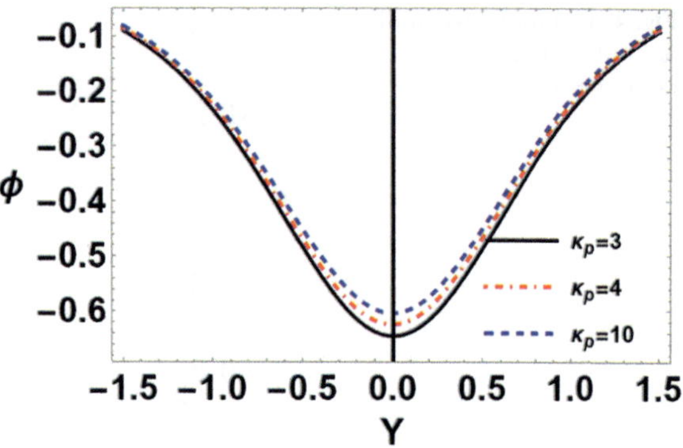

Fig. 6 Plot for EASs profile (ϕ) with superthermality index of positrons (κ_p)

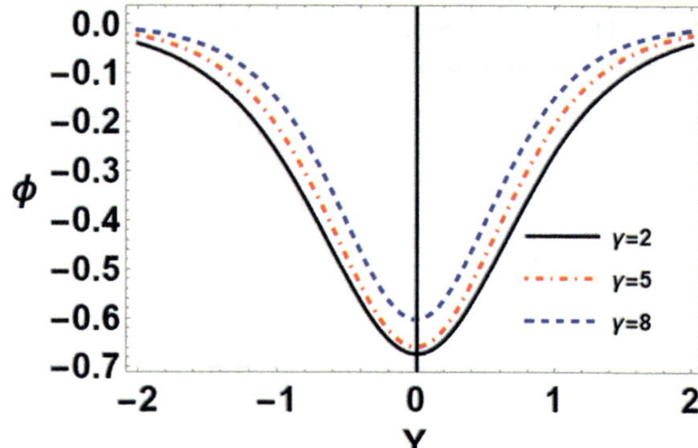

Fig. 7 Plot for EASs profile (ϕ) with temperature ratio of hot electrons to positrons $\gamma(=\frac{T_h}{T_p})$

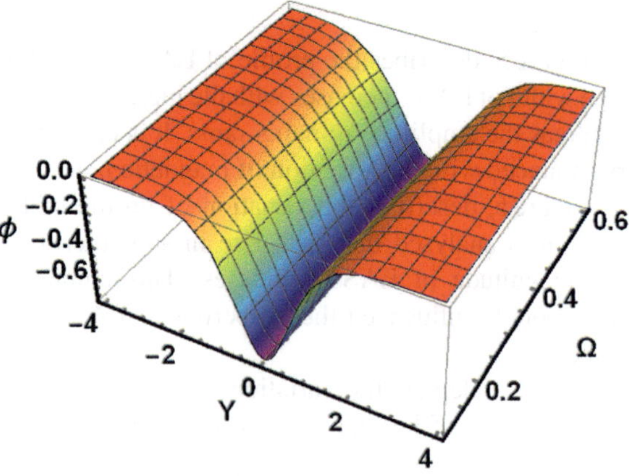

Fig. 8 Plot for 3D EASs profile (ϕ) with magnetic field strength (Ω)

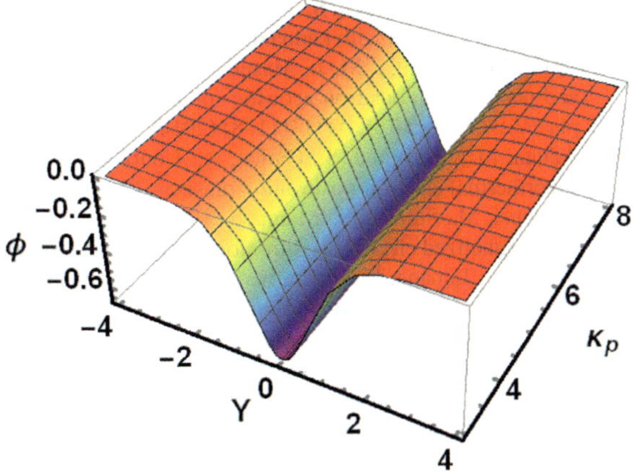

Fig. 9 Plot for 3D EASs profile (ϕ) with superthermality index of positrons (κ_p)

5 Conclusion

We have studied the salient features of EASs in a magnetized multicomponent plasma having inertial cold electrons, inertialess ions and superthermal positrons as well as electrons. The RPM is adopted to develop the nonlinear ZK equation and its solution to describe the dynamics of EASs. Only EASs with negative polarity exist. The effects of various plasma parameters such as κ_p, κ_e, α, γ and Ω have significantly influence the characteristic properties of EASs. The phase velocity of solitons is enhanced with increase in κ_p and κ_e. Nonlinear coefficient (A) flourishes with increase in κ_p and κ_e. The width and amplitude of EASs are increased with increase in κ_e and α. The width of EASs increases with increase in the value of strength of magnetic field (Ω). With the increase in the value of ratio of temperature of hot electrons to positrons $\gamma(=\frac{T_h}{T_p})$, the width and amplitude of EASs are decreased. Negative potential EASs are significantly influenced with the change in dispersion and nonlinearity effects.

The outcome of present study can be beneficial for the indepth understanding of EASs with superthermal positrons and electrons in Van Allen radiation belts, auroral zone, planetary magnetospheres [20–23].

References

1. Mace, R.-L., Baboolal, S., Bharuthram, R., Hellberg, M.-A.: Arbitrary-amplitude electron-acoustic solitons in a two-electron-component plasma. J. Plasma Phys. **45**(1), 323–338 (1991)
2. Yu, M.-Y., Shukla, P.-K.: Linear and nonlinear modified electron-acoustic waves. J. Plasma Phys. **29**(3), 409–413 (1983)
3. Mace, R.-L., Hellberg, M.-A.: The Korteweg-de Vries-Zakharov-Kuznetsov equation for electron-acoustic waves. Phys. Plasmas **8**(6), 2649–2656 (2001)
4. Danehkar, A., Saini, N.-S., Hellberg, M.-A., Kourakis, I.: Electron-acoustic solitary waves in the presence of a suprathermal electron component. Phys. Plasmas **18**(7), 072902(1–10) (2011)

5. Devanandhan, S., Singh, S.-V., Lakhina, G.-S., Bharuthram, R.: Electron acoustic waves in a magnetized plasma with kappa distributed ions. Phys. Plasmas **19**(8), 082314(1–7) (2012)
6. Alam, M.-S., Uddin, M.-J., Masud, M.-M., Mamun, A.-A.: Roles of superthermal electrons and positrons on positron-acoustic solitary waves and double layers in electron-positron-ion plasmas. Chaos **24**(3), 033130(1–8) (2014)
7. Adnan, M., Mahmood, S., Qamar, A.: Coupled ion acoustic and drift waves in magnetized superthermal electron-positronion plasmas. Phys. Plasmas **21**(9), 092119(1–8) (2014)
8. Ferdousi, M., Sultana, S., Mamun, A.-A.: Oblique propagation of ion-acoustic solitary waves in a magnetized electron-positronion plasma. Phys. Plasmas **22**(3), 032117(1–7) (2015)
9. Sahaa, A., Tamang, J.: Qualitative analysis of the positron-acoustic waves in electron-positron-ion plasmas with κ deformed Kaniadakis distributed electrons and hot positrons. Phys. Plasmas **24**(7), 082101(1–10) (2017)
10. Bansal, S., Aggarwal, M., Gill, T.-S.: Planar and nonplanar electron-acoustic solitary waves in the presence of positrons. Plasma Phys. Rep. **46**(7), 715–723 (2020)
11. Gul, N., Ahmad, R.: Dissipative ion-acoustic solitons in electron-positron-ion plasma with non-thermal electrons and iso-thermal positrons. Adv. Space Res. **68**(1), 161–169 (2021)
12. Vasyliunas, V.-M.: A survey of low-energy electrons in the evening sector of the magnetosphere with OGO 1 and OGO 3. J. Geophys. Res. **73**(9), 2839–2884 (1968)
13. Shahmansouri, M., Tribeche, M.: Propagation properties of ion acoustic waves in a magnetized superthermal bi-ion plasma. Astrophys. Space Sci. **350**(2), 623–630 (2014)
14. Saini, N.-S., Sethi, P.: Dust ion-acoustic cnoidal waves in a plasma with two temperature superthermal electrons. Phys. Plasmas **23**(10), 103702(1–10) (2016)
15. Singh, K., Saini, N.-S.: Effect of anisotropic pressure on electron acoustic oscillatory and monotonic shocks in superthermal magnetoplasma. Radio Sci. J. **54**(12), 1192–1203 (2019)
16. Washimi, H., Taniuti, T.: Propagation of ion-acoustic solitary waves of small amplitude. Phys. Rev. Lett. **17**(19), 996–998 (1966)
17. Elwakil, S.-A., El-Shewy, E.-K., Abdelwahed, H.-G.: Solution of the perturbed Zakharov-Kuznetsov (ZK) equation describing electron-acoustic solitary waves in a magnetized plasma. Chin. J. Phys. **49**(3), 732–744 (2011)
18. Singh, K., Sethi, P., Saini, N.-S.: Nonlinear excitations in a degenerate relativistic magneto-rotating quantum plasma. Phys. Plasma **26**(9), 092104 (1–10) (2019)
19. Jilani, K., Mirza, A.-M., Khan, T.-A.: Electrostatic electron acoustic solitons in electron-positron-ion plasma with superthermal electrons and positrons. Astrophys. Space Sci. **349**(1), 255–263 (2015)
20. Voronov, S.-A., Galper, A.-M., Kirilov-Ugryumov, V.-G., Koldashov, S.-V., Popov, A.-V.: Charge composition of the high-energy electrons and positrons in the Van Allen radiation belts. JETP Lett. **43**, 306–307 (1986)
21. Galper, A.-M., Koldashov, S.-V., Mikhailov, V.-V., Voronov, S.-A.: Electrons with energy greater than 20 MeV in the inner radiation belt. Radiat. Meas. **26**(3), 375–378 (1996)
22. Pottelette, R., Ergun, R.-E., Treumann, R.-A., Berthomier, M., Carlson, C., McFadden, J.-P., Roth, I.: Modulated electron acoustic waves in auroral density cavities: FAST observations. Geophys. Res. Lett. **26**(16), 2629–2632 (1999)
23. Plyaskin, V.: Mapping Earth's radiation belts using data from STS91 mission of AMS. Astropart. Phys. **30**(1), 18–27 (2008)

Non-linear Fluctuating Parts of the Particle Distribution Function in the Presence of Drift Wave Turbulence in Vlasov Plasma

Banashree Saikia and P. N. Deka

Abstract We have considered a Vlasov plasma with both the resonant and non-resonant mode waves. The non-resonant mode is considered as a perturbation to plasma where a turbulent field is present which is in resonant mode. The interaction of these waves is characterized by the Vlasov Maxwell set of equations. The evaluation process of the fluctuating parts of the distribution function owing to the presence of resonant mode wave, due to the modulation field, and the nonlinear fluctuating parts of distribution function due to the non-resonant wave is presented in this work.

Keywords Nonlinear wave-particle interaction · Density and temperature gradients · Drift wave turbulence

1 Introduction

In this paper, we have considered electromagnetic Ordinary (O) mode as high-frequency wave and ion cyclotron drift wave as the low-frequency resonant mode wave. In high beta plasmas, Davidson and Wu [1] first discussed the O-mode wave instability, which is exclusively increasing mode. A group of experts studied the formation of unstable ordinary mode waves in the Earth's magnetospheric Auroral region [2, 3]. Increased electromagnetic radiation in the top ionospheric regions has been thoroughly investigated, and this radiation has been labelled as Auroral Kilometric Radiation(AKR) [3]. The Auroral Kilometric Radiation is made up of X and O-mode radiations according to later research [3]. In a series of investigations, Ibscher and Schlickeiser [4–7] investigated the Ordinary mode instability expanding it to the small beta plasma region by using a counter-streaming bi-Maxwellian model [9]. On the other hand, the Ordinary mode instability has mostly been studied in terms of marginal instability criterion rather than numerical dispersion relation solution. In a magnetized non relativistic bi-Maxwellian plasma, [8] temperature anisotropic

B. Saikia (✉) · P. N. Deka
Department of Mathematics, Dibrugarh University, Dibrugarh 786004, Assam, India
e-mail: spinkjht@gmail.com

S. Banerjee and A. Saha (eds.), *Nonlinear Dynamics and Applications*,
Springer Proceedings in Complexity,
https://doi.org/10.1007/978-3-030-99792-2_20

effects on O-mode and its instability were examined. Deka and Borgohain [3] investigated the amplification of O-mode in inhomogeneous plasma using plasma maser theory. In this study, they have investigated that the amplification of electromagnetic wave is achievable in space plasma at the expense of drift wave turbulence. The amplification process and wave energy exchange are also possible at some energy level since the plasma maser effect doesn't require a frequency matching condition. Deka and Deka [10] investigated the amplification of ion-acoustic waves in Burning plasma in presence of drift wave turbulence. Here, they have discussed ion acoustic instabilities while considering the ion distribution function that is consistent in burning plasma. Recently, Senapati and Deka [11] studied the instability of electron Bernstein mode in the presence of drift wave turbulence caused by density and temperature gradients. In their research, they have studied the growth rate of high frequency electron Bernstein mode in Tokamak plasmas as a function of density and temperature gradients. In this paper, we have investigated the formation of O-mode waves in non-uniform plasma media through non-linear wave particle interaction in the context of ion cyclotron drift wave turbulence.

2 Formulation of the Problem

In our problem, we have considered an inhomogeneous plasma which supports drift motion and turbulence. To describe this system, we consider a particle distribution function [12] which involve gradient parameters for temperature and density associated with external force. The density and temperature gradients are taken along the positive y-direction (Fig. 1).

$$f_j(T_j, y, \mathbf{v}) = \left(\frac{m}{2\pi T_{0j}}\right)^{\frac{3}{2}} \left[1 + \mu\left(y + \frac{v_x}{\Omega_j}\right)\right] exp\left[-\left(\frac{m\mathbf{v}^2}{2T_{0j}} - \frac{Fy}{T_{0j}}\right)\right] \quad (1)$$

where $\Omega_j = \frac{eB_0}{mc}$ denotes the cyclotron frequency of the ion, μ denotes the density and gradient smallness parameter which can be derived using equation (1) at y=0 as

$$\mu = \left[\left(\frac{\partial}{\partial T_j}\frac{dT_j}{dy}\right) + \left(\frac{1}{f_0}\frac{df_j}{dy}\right)\right]_{y=0} - \frac{F}{T_{0j}}$$

The particle drift motion owing to the pressure gradient is caused by the force field **F** and the equation of the motion is given by:

$$m\frac{d\mathbf{v}}{dt} = \mathbf{F} - \frac{e}{c}\mathbf{v} \times \mathbf{B}_0$$

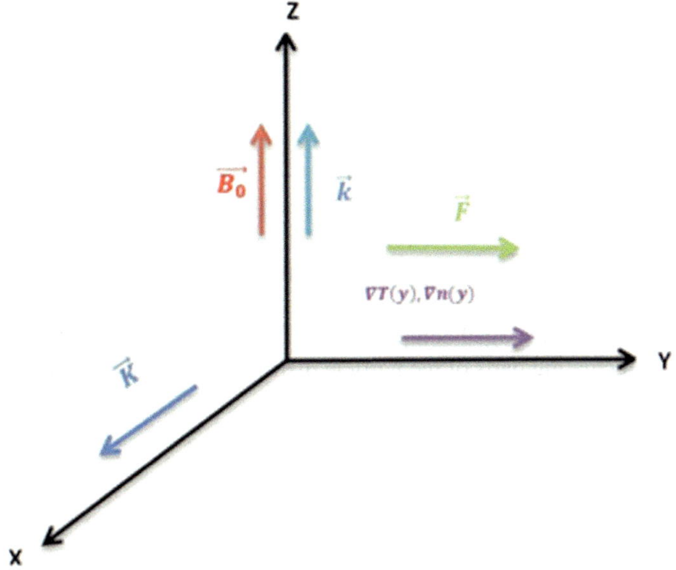

Fig. 1 Geometry of the model: $\mathbf{K} = (K_\perp, 0, 0)$ represents the propagation vector of the Ordinary mode wave $\mathbf{k} = (0, 0, k_\parallel)$ represents the propagation vector of the ion-cyclotron drift wave $\mathbf{B}_0$ represents the magnetic field along the positive z-axes

$$\frac{d\mathbf{v}}{dt} = \frac{\mathbf{F}}{m}\hat{y} - \Omega_j(\mathbf{v} \times \hat{z})$$

The interaction of the high frequency Ordinary mode with ion-cyclotron drift wave is governed by Vlassov–Maxwell set of equations

$$\left[\frac{\partial}{\partial t} + \mathbf{v} \cdot \frac{\partial}{\partial \mathbf{r}} - \left\{\frac{e}{m}\left(\mathbf{E} + \frac{\mathbf{v} \times \mathbf{B}_0}{c}\right) - \frac{\mathbf{F}}{m}\right\} \cdot \frac{\partial}{\partial \mathbf{v}}\right] F_{0j}\,(\mathbf{r}, \mathbf{v}, t) = 0 \tag{2}$$

$$\nabla \times \mathbf{E} = -\frac{1}{c}\frac{\partial \mathbf{B}}{\partial t} \tag{3}$$

$$\nabla \times \mathbf{B} = \frac{1}{c}\frac{\partial \mathbf{E}}{\partial t} + \frac{4\pi}{c}\mathbf{J} \tag{4}$$

$$\mathbf{J} = -en_j \int \mathbf{v} f_{0j}(\mathbf{r}, \mathbf{v}, t) d\mathbf{v} \tag{5}$$

$$\nabla \cdot \mathbf{E} = -4en_j\pi \int f_j(\mathbf{r}, \mathbf{v}, t) d\mathbf{v} \tag{6}$$

The unperturbed particle distribution function and fields are described by the linear response theory [13] of a turbulent plasma.

$$F_{0j} = f_{0j} + \epsilon f_{1j} + \epsilon^2 f_{2j} \tag{7}$$

and

$$\mathbf{E}_{0l} = \epsilon \mathbf{E}_l + \epsilon^2 \mathbf{E}_2 \tag{8}$$

Here, ϵ denotes a smallness parameter related to ion-cyclotron drift wave turbulent field $\mathbf{E}_l = (0, 0, E_{l\parallel})$ having propagation vector $\mathbf{k} = (0, 0, k_\parallel)$, f_{0j} denotes the spatial and time averaged components of the distribution function, f_{1j} and f_{2j} denotes the fluctuating parts of the particle distribution function and $\mathbf{E}_2$ denotes the electric field of second order.

To the order of ϵ, from Eq. (2), we get:

$$\left[\frac{\partial}{\partial t} + \mathbf{v} \cdot \frac{\partial}{\partial \mathbf{r}} - \left\{\frac{e}{m}\left(\frac{\mathbf{v} \times \mathbf{B}_0}{c}\right) - \frac{\mathbf{F}}{m}\right\} \cdot \frac{\partial}{\partial \mathbf{v}}\right] f_{1j}(\mathbf{r}, \mathbf{v}, t) = \frac{e}{m}\left(\mathbf{E}_l \cdot \frac{\partial}{\partial \mathbf{v}} f_{0j}\right) \tag{9}$$

Now to obtain f_{1j}, we apply Fourier transform,

$$A(\mathbf{r}, \mathbf{v}, t) = \Sigma_{k,\omega} A(\mathbf{k}, \omega, \mathbf{v}) \exp[i(\mathbf{k} \cdot \mathbf{r} - \omega t)] \tag{10}$$

The fluctuating parts f_{1j} of the low frequency turbulence field is determined from Eq. (9) by integrating along the unperturbed orbit.

$$f_{1j}(\mathbf{k}, \omega) = \frac{ie}{m} \frac{E_{l\parallel} \cdot \frac{\partial}{\partial v_\parallel} f_{0j}}{\omega - k_\parallel v_\parallel + i \cdot 0^+} \tag{11}$$

We next apply a perturbation $\delta \mathbf{E}_h$ of a high frequency ordinary mode with an electric field $\delta \mathbf{E}_h = (0, 0, \delta E_h)$ having propagation vector $\mathbf{K} = (K_\perp, 0, 0)$, a magnetic field $\delta \mathbf{B}_h = (0, \delta B_h$, and frequency Ω.

We calculate the total perturbed electric field, the magnetic field and the particle distribution function as:

$$\begin{aligned} \delta \mathbf{E} &= \mu' \delta \mathbf{E}_h + \mu' \epsilon \delta \mathbf{E}_{lh} + \mu' \epsilon^2 \Delta \mathbf{E} \\ \delta \mathbf{B} &= \mu' \delta \mathbf{B}_h + \mu' \epsilon \delta \mathbf{B}_{lh} + \mu' \epsilon^2 \Delta \mathbf{B} \\ \delta f &= \mu' \delta f_h + \mu' \epsilon \delta f_{lh} + \mu' \epsilon^2 \Delta f \end{aligned}$$

where $\delta \mathbf{E}_{lh}, \Delta \mathbf{E}, \delta \mathbf{B}_{lh}, \Delta \mathbf{B}$ denotes the modulation fields, δf_h denotes the fluctuating part owing to a high frequency Ordinary mode, δf_{lh} and Δf denotes the particle distribution function corresponding to the modulation field and $\mu' \ll \epsilon$ where μ' denotes the smallness parameter for perturbed field. To the order of μ, $\mu\epsilon$ and $\mu\epsilon^2$, we now have

$$P \delta f_h = \frac{e}{m}\left(\delta \mathbf{E}_h + \frac{\mathbf{v} \times \delta \mathbf{B}_h}{c}\right) \cdot \frac{\partial f_{0j}}{\partial \mathbf{v}} \tag{12}$$

$$P\delta f_{lh} = \frac{e}{m}\delta \mathbf{E}_h \cdot \frac{\partial}{\partial \mathbf{v}} f_{1j} + \frac{e}{m} \cdot \frac{\mathbf{v} \times \delta \mathbf{B}_h}{c} \cdot \frac{\partial}{\partial \mathbf{v}} f_{1j} + \frac{e}{m}\delta \mathbf{E}_{lh} \cdot \frac{\partial}{\partial \mathbf{v}} f_{0j} + \frac{e}{m}\mathbf{E}_l \cdot \frac{\partial}{\partial \mathbf{v}}\delta f_h + \frac{e}{m} \cdot \frac{\mathbf{v} \times \delta \mathbf{B}_{lh}}{c} \cdot \frac{\partial}{\partial \mathbf{v}} f_{0j} \tag{13}$$

$$P\Delta f = \frac{e}{m}\left[\delta \mathbf{E}_{lh} \cdot \frac{\partial f_{1j}}{\partial \mathbf{v}} + \frac{\mathbf{v} \times \delta \mathbf{B}_{lh}}{c} \cdot \frac{\partial f_{1j}}{\partial \mathbf{v}} + \mathbf{E}_l \cdot \frac{\partial}{\partial \mathbf{v}}\delta f_{lh}\right] \tag{14}$$

where $P = \left[\frac{\partial}{\partial t} + \mathbf{v} \cdot \frac{\partial}{\partial \mathbf{r}} - \left\{\frac{e}{m}\left(\frac{\mathbf{v} \times \mathbf{B}_0}{c}\right) - \frac{\mathbf{F}}{m}\right\} \cdot \frac{\partial}{\partial \mathbf{v}}\right]$.

3 Nonlinear Dispersion Relation of Electromagnetic O-Mode Wave

Considering the linear response theory of plasma turbulence, we obtain δf_h, δf_{lh} and Δf by integrating along the unperturbed orbits, using cylindrical co-ordinate system to the velocity space [13, 14].

From equation (12), we have

$$\delta f_h = -\frac{ie}{m}\delta \mathbf{E}_h \left[\left(1 - \frac{K_\perp v_\perp}{\omega} cos\theta\right)\frac{\partial f_{0j}}{\partial v_\parallel} S_{p,q} - f_{0j}\left\{\frac{m v_\parallel}{\Omega T_{0j}} + \left(\frac{m v_\parallel}{T_{0j}} - \frac{\epsilon K_\perp v_\parallel}{\Omega \Omega_j}\right) S_{p,q}\right\}\right] \tag{15}$$

where $S_{p,q} = \Sigma_{a,b} \frac{J_a(\alpha^{'} J_b(\alpha^{'})exp[i(b-a)\theta]}{a\Omega_j - \Omega + K_\perp v_F}$, $\alpha = \frac{K_\perp v_\perp}{\Omega_j}$

From equation (13), we have

$$\delta f_{lh} = I_{lh}^1 + I_{lh}^2 + I_{lh}^3 \tag{16}$$

where

$$I_{lh}^1 = \frac{e}{m}\int_{-\infty}^{0}\left[\mathbf{E}_l \cdot \frac{\partial}{\partial \mathbf{v}}\delta f_h\right] exp\left[i\left\{(\mathbf{K} - \mathbf{k}) \cdot (\mathbf{r}^{'} - \mathbf{r}) - (\Omega - \omega)\tau\right\}\right] d\tau \tag{17}$$

$$I_{lh}^2 = \frac{e}{m}\int_{-\infty}^{0}\left[\delta \mathbf{E}_h + \frac{\mathbf{v} \times \delta \mathbf{B}_h}{c} \cdot \frac{\partial f_{1e}}{\partial \mathbf{v}}\right] exp\left[i\left\{(\mathbf{K} - \mathbf{k}) \cdot (\mathbf{r}^{'} - \mathbf{r}) - (\Omega - \omega)\tau\right\}\right] d\tau \tag{18}$$

$$I_{lh}^3 = \frac{e}{m}\int_{-\infty}^{0}\left[\delta \mathbf{E}_{lh} + \frac{\mathbf{v} \times \delta \mathbf{B}_{lh}}{c} \cdot \frac{\partial f_{0e}}{\partial \mathbf{v}}\right] exp\left[i\left\{(\mathbf{K} - \mathbf{k}) \cdot (\mathbf{r}^{'} - \mathbf{r}) - (\Omega - \omega)\tau\right\}\right] d\tau \tag{19}$$

From equation (14), we have

$$\Delta f = \frac{e}{m}\int_{-\infty}^{0}\left[\delta \mathbf{E}_{lh} + \frac{\mathbf{v} \times \delta \mathbf{B}_{lh}}{c} \cdot \frac{\partial f_{1e}}{\partial \mathbf{v}} + \mathbf{E}_l \cdot \frac{\partial}{\partial \mathbf{v}}\delta f_{lh}\right] exp\left[i\left\{\mathbf{K} \cdot (\mathbf{r}^{'} - \mathbf{r}) - \Omega\tau\right\}\right] d\tau \tag{20}$$

The Maxwell's equation gives the modulated field as

$$\nabla \times \delta \mathbf{B}_{lh} = \frac{1}{c}\frac{\partial}{\partial t}\delta \mathbf{E}_{lh} + \frac{4\pi}{c}\mathbf{J}$$

$$\mathbf{J} = -en_e \int \mathbf{v}\delta f_{lh} d\mathbf{v}$$

$$\nabla \times \delta \mathbf{E}_{lh} = -\frac{1}{c}\frac{\partial}{\partial t}\delta \mathbf{B}_{lh}$$

Therefore, we have

$$\nabla \times \delta \mathbf{B}_{lh} = \frac{1}{c}\frac{\partial}{\partial t}\delta \mathbf{E}_{lh} - \frac{4\pi e n_e}{c}\int \mathbf{v}\delta f_{lh} d\mathbf{v}$$

This can be written in the simplified form as:

$$\delta E_h = \frac{4\pi i e n_e(\Omega - \omega)}{c^2 K_{\perp}^2 - \Omega^2}\int v_{\parallel}(\delta f_h + \Delta f) d\mathbf{v}$$

This equation may be expressed as follows after simplification:

$$\delta E_h \epsilon_h(\mathbf{K}, \Omega) = 0 \tag{21}$$

Here $\epsilon_h(\mathbf{K}, \Omega)$ indicates the non-linear dispersion relation of O-mode wave which is described by

$$\epsilon_h(\mathbf{K}, \Omega) = \epsilon_0(\mathbf{K}, \Omega) + \epsilon_d(\mathbf{K}, \Omega) + \epsilon_p(\mathbf{K}, \Omega). \tag{22}$$

4 Discussions

In our present study, we have been calculating the nonlinear dispersion relation of electromagnetic Ordinary mode waves in context of ion cyclotron drift wave turbulence, that is a frequent characteristics in an inhomogeneous plasma. The fluctuating parts f_{1j} , δf_h, δf_{lh} and Δf which are provided in Eqs. (11) , (15) , (16) and (20) respectively have been obtained. Here, f_{1j} denotes the fluctuating part due to ion-cyclotron turbulent field which is linear in nature. Further, δf_h denotes the fluctuating part of particle distribution function due to perturbed electromagnetic ordinary mode whereas δf_{lh} and Δf denotes the nonlinear fluctuating parts of distribution function. We are focussing on nonlinear fluctuating part Δf for estimating growth rate of O-mode. The expressions of Δf contains polarisation coupling and direct coupling terms. The turbulent field parameters E_l, $k_{|\ |}$, $v_{\parallel}$ are involved besides the parameter associated with O-mode phenomena. Nonlinear dispersion relation will be given by

$$\epsilon_h(\mathbf{K}, \Omega) = \epsilon_0(\mathbf{K}, \Omega) + \epsilon_d(\mathbf{K}, \Omega) + \epsilon_p(\mathbf{K}, \Omega)$$

containing the linear part $\epsilon_0(\mathbf{K}, \Omega)$, the direct coupling part $\epsilon_d(\mathbf{K}, \Omega)$ and the polarisation coupling part $\epsilon_p(\mathbf{K}, \Omega)$.

5 Conclusions

In previous research [3], Ordinary mode was examined in the context of drift wave turbulence in absence of external force causing particle drift. Only density gradient parameters were to emerge in the nonlinear dispersion relation in those circumstances. However, drift motion of particles owing to pressure gradient must exist in practical instances in space and tokamak plasmas, and is connected with $\mathbf{F} \times \mathbf{B}$ drift. As a result, we've incorporated the external force $\mathbf{F}$ in our current research, which ties drift motion to particles.

References

1. Davidson, R.C., Wu, C.S.: Ordinary-mode electromagnetic instability in high-beta plasmas. Phys. Fluids **13**, 1407–1409 (1970)
2. Gurnett, D.A.: The earth as radio source: terrestrial kilometric radiation. J. Geophys. Res. **79**, 4227–4238 (1974)
3. Deka, P.N., Borgohain, A.: On unstable electromagnetic radiation through nonlinear wave-particle interactions in presence of drift wave turbulence. J. Plasma Phys. **78**(5), 515–524 (2012)
4. Ibscher, D., Lazar, M., Schlickeiser, R.: On the existence of Weibel instability in a magnetized plasma. II. Perpendicular wave propagation: the ordinary mode. Phys. Plasmas **19**, 072116-1–072116-10 (2012)
5. Ibscher, D., Lazar, M., Michno, M.J., Schlickeiser, R.: Towards a complete parametrization of the ordinary-mode electromagnetic instability in counterstreaming plasmas. I. Minimizing ion dynamics. Phys. Plasmas **20**, 012103-1–012103-9 (2013)
6. Ibscher, D., Schlickeiser, R.: Towards a complete parametrization of the ordinary-mode electromagnetic instability in counterstreaming plasmas. II. Ion effects. Phys. Plasmas **20**, 042121-1–042121-9 (2013)
7. Ibscher, D., Schlickeiser, R.: Solar wind kinetic instabilities at small plasma betas. Phys. Plasmas **21**, 022110-1–022110-4 (2014)
8. Bashir, M.F., Murtaza, G.: Effect of temperature anisotropy on various modes and instabilities for a magnetized non-relativistic Bi-Maxwellian plasma. Braz. J. Phys. **42**, 487–504 (2012)
9. Hadi, F., Bashir, M.F., Qamar, A., Yoon, P.H., Schlickeiser, R.: On the ordinary mode instability for low beta plasmas. Phys. Plasmas **21**, 052111-1–052111-5 (2014)
10. Deka, P.N., Deka, J.K.: On amplification of ion-acoustic mode in burning plasma in presence of drift wave turbulence. J. Fusion Energy **37**, 301–307 (2018)
11. Senapati, P., Deka, P.N.: Instability of electron bernstein mode in presence of drift wave turbulence associated with density and temperature gradients. J. Fusion Energy **39**, 477–490 (2020)
12. Ichimaru, S.: Basic principles of plasma physics a statistical approach, vol. 1, 1st edn. Addison-Wesley, Boca Raton (1973)
13. Krall, N.A., Trivelpiece, A.W.: Principles of plasma physics, 2nd edn., pp. 403–407. McGraw-Hill, New York (1973)
14. Chen, F.: Introduction to plasma physics and controlled fusion, vol. 1, 2nd edn. Plenum Press, New York (1974)

Effect of Superthermal Charge Fluctuation on Bifurcation of Dust-Ion-Acoustic Waves Under the Burgers Equation in a Magnetized Plasma

Jharna Tamang

Abstract Bifurcation of dust-ion-acoustic waves (DIAWs) in dusty plasmas composed of fluid ions, immobile dust grains, and superthermal charge fluctuations of electrons and ions is examined. The study is done under the framework of the Burgers equation obtained through the reductive perturbation technique. The effect of charge fluctuations holds the responsibility for formation of the shock solution of the Burgers equation. The changes on periodic and shock wave solutions of the Burgers equation are shown varying system parameters. The results of this study, shock and periodic wave solutions of the Burgers equation in plasmas with dust charge fluctuation in magnetized plasmas, are shown for the first time implementing the notion of nonlinear dynamical systems.

Keywords Phase plane profile · Periodic wave · Kink and anti-kink waves · Dynamical system

1 Introduction

Dusty plasmas have wide applications in astrophysics, plasma, biophysics, fusion devices, space science [1, 2]. The charged dust grains produces wave modes, named as dust-ion-acoustic wave (DIAW) and dust-acoustic wave (DAW). Experimentally Barkan et al. [3] discovered DIAWs for the first time. Theoretically, Shukla and Silin [4] were the first to report a observation of low frequency DIAW in a dusty plasma. Many researchers [4–7] studied propagation of DIAW both theoretically and experimentally in various plasma systems while, some researchers [8–11] studied impacts of external magnetic force on the electrostatic waves. Under magnetic effect,

J. Tamang (✉)
Department of Mathematics, Sikkim Manipal Institute of Technology, Sikkim Manipal University, Majitar, East-Sikkim 737136, India
e-mail: jharnatamang.12@gmail.com

Department of Mathematics, Sikkim Alpine University, Kamrang, Namchi, South-Sikkim 737126, India

S. Banerjee and A. Saha (eds.), *Nonlinear Dynamics and Applications*,
Springer Proceedings in Complexity,
https://doi.org/10.1007/978-3-030-99792-2_21

Anowar and Mamun [12] studied the same under the KdV equation in dusty plasmas. In magnetic dusty plasma, El-Labany et al. [13] investigated solitary DIAWs with isothermal electrons. Shalaby et al. [14] reported DIAW in magnetized dusty plasmas under the ZK equation. Very recently, Samanta et al. [15] discussed the formation of rogue waves and examined DIAWs in multicomponent degenerate plasmas.

Many particles of plasmas follow Maxwellian distributions. However, space plasma is usually noticed to follow non-Maxwellian distributions. These distributions can be structured by kappa distribution [16]. In magnetized plasmas, superthermal parameter affects the nature of electrostatic waves [17, 18]. Alinejad et al. [19] examined shock DIAWs in dusty plasmas with effect of superthermal electrons. Recently, Shahmansouri and Alinejad [10] reported the impacts of direction of propagation and superthermal electrons under magnetic field on large amplitude DIAW. The charge on dust grains in dusty plasmas is not definite because of ion and electron currents passing through the grain surface. While, the dissipation may occur due to charge fluctuations of dust in dusty plasmas [20, 21] and is responsible generation of shock structures.

The concept of dynamical systems [22–24] has discovered many characteristics of nonlinear waves in plasmas through phase portrait and time series analysis. Recently, many researchers [25–28] reported the propagation of nonlinear acoustic waves in plasmas. In 2018, Tamang et al. [5] studied DIAWs in collisional dusty plasma under the ionization effect implementing the notion of planar dynamical systems. Chatterjee et al. [29] reported solitary wave solution of DIAWs in superthermal plasmas. Very recently, Sharma et al. [30] studied dynamical features of DIAWs in nonextensive dusty plasma applying the same theory. Using this notion of dynamical systems, the dynamical feature of DIAWs based on the Burgers equation under the magnetic and charge fluctuation is not reported as yet.

The manuscript is organized as: in Sect. 2, the model equations are considered. In Sect. 3, we derive the Burgers equation. In Sect. 4, the dynamical system of the Burgers equation is formed. The potential energy function plot is also displayed. In Subsections of 4, analytical wave solutions are presented. Lastly, conclusions are given in Sect. 5.

2 Model Equations

The traveling of DIAWs in an electron-ion plasma system under magnetic field is represented by the following model equations [11]

$$\frac{\partial n}{\partial t} + \vec{\nabla} \cdot (n\vec{u}) = 0, \tag{1}$$

$$\frac{\partial \vec{u}}{\partial t} + (\vec{u} \cdot \vec{\nabla})\vec{u} = -\vec{\nabla}\phi + \omega_{ci}\vec{u} \times \hat{z}, \tag{2}$$

$$\vec{\nabla}^2\phi = \mu n_e - n + (1-\mu)Z_d. \tag{3}$$

where n and n_e represent ion and electron number densities. Here, velocity component is given by $\overrightarrow{B_0} = B_0\hat{z}$, $\overrightarrow{u} = (u, v, w)$ and $\nabla = (\partial x, \partial y, \partial z)$. Electromagnetic wave potential is given by ϕ. Temperature ratio is given by $\sigma = \frac{T_i}{T_e}$, where $T_i(T_e)$ is temperature of ions (electrons). Dust-acoustic speed is given by $C_s = \sqrt{Te/mi}$ and ion-cyclotron frequency is given by $\omega_{pi} = \sqrt{4\pi n_{i0}e^2/m_i}$. The Debye length is given by $\lambda_D = C_s/\omega_{pi}$, where k_B refers to the Boltzmann constant and m_h is hot electron mass. Here, $\mu = n_{e0}/n_{i0} = 1 - \dfrac{Z_{d0}n_{d0}}{n_{i0}}$.

$$n_e = 1 + C_1\phi + C_2\phi^2, \tag{4}$$

where $C_1 = -\dfrac{-\kappa - 1/2}{\kappa - 3/2}$ and $C_2 = \dfrac{1}{2}\dfrac{(-\kappa - 1/2)(-\kappa + 1/2)}{(\kappa - 3/2)^2}$.

The electron and ion charging currents are considered as:

$$I_e = -e\pi r_d^2 n_{e0}\sqrt{\frac{8T_e}{\pi m_e}}\sqrt{\frac{\kappa - 3/2}{\kappa^4}}\frac{\kappa}{\kappa - 1}\frac{\Gamma(\kappa + 1)}{\Gamma(\kappa - 1/2)}\left(1 + \frac{\alpha Z_d - \phi}{\kappa - 3/2}\right)^{1-\kappa}, \tag{5}$$

$$I_i = e\pi r_d^2 n_{i0}\sqrt{\frac{8T_i}{\pi m_i}}\left(1 + \frac{\alpha Z_d}{\sigma}\right). \tag{6}$$

where r_d is the radius of dust, $\sigma = \frac{T_i}{T_e}$ and $\alpha = Z_{d0}e^2/r_d T_e$.

The normalized dust charging fluctuation is described by the succeeding equation

$$\nu\left(\frac{\partial Z_d}{\partial t} + u_d\frac{\partial Z_d}{\partial x}\right) = \gamma\left(1 + \frac{\alpha Z_d - \phi}{\kappa - 3/2}\right)^{1-\kappa} - \chi n_i\left(1 + \frac{\alpha Z_d}{\sigma}\right), \tag{7}$$

where $\nu = \sqrt{\alpha m_e(1 - \mu)/2m_i}$, $\chi = (\dfrac{r_d}{n_{d0}^{-1/2}})^{3/2}\sqrt{m_e T_i/T_e m_i}$ and

$\gamma = \mu\left(\dfrac{r_d}{n_{d0}^{-1/2}}\right)^{3/2}\sqrt{\kappa - 3/2}\Gamma(\kappa + 1)/\Gamma(\kappa - 1/2)\kappa(\kappa - 1)$.

The normalized dust charging frequency follows as:

$$\nu_{ch} = \frac{1}{e}\frac{\partial}{\partial Z_d}(I_e + I_i)_{|\phi=0, Z_d=1} = \sqrt{\frac{Z_{d0}^2 r_d^2}{2\pi\sigma}\frac{\omega_{pi}}{\lambda_D}}\frac{(\alpha\kappa + \kappa - 3/2) + \sigma(\kappa - 1)}{\sigma + \kappa - 3/2}. \tag{8}$$

3 Derivation of the Burgers Equation

We obtain the Burger equation using the stretching of independent variables as

$$\xi = \epsilon(l_x x + l_y y + l_z z - Vt), \quad \tau = \epsilon^2 t, \tag{9}$$

where ϵ is small parameter ($0 < \epsilon << 1$) and V is phase velocity of the wave. Next, the dependent variables expanded as,

$$\begin{cases} n = 1 + \epsilon n_1 + \epsilon^2 n_2 + \cdots, \\ u = \epsilon u_1 + \epsilon^2 u_2 + \cdots, \\ v = \epsilon v_1 + \epsilon^2 v_2 + \cdots, \\ w = \epsilon w_1 + \epsilon^2 w_2 + \cdots, \\ Z_d = 1 + \epsilon Z_{d1} + \epsilon^2 Z_{d2} + \cdots, \\ \phi = \epsilon \phi_1 + \epsilon^2 \phi_2 + \cdots. \end{cases} \tag{10}$$

By collecting terms of lowest order $\epsilon^{\frac{3}{2}}$, we get the following equations

$$\begin{cases} n_1 = \dfrac{l_z}{V} w_1, \\ u_1 = -\dfrac{l_y}{\omega_{ci}} \dfrac{\partial \phi_1}{\partial \xi}, \\ v_1 = \dfrac{l_x}{\omega_{ci}} \dfrac{\partial \phi_1}{\partial \xi}, \\ w_1 = \dfrac{l_z}{V} \phi_1, \\ \mu C_1 \phi_1 - n_1 + (1-\mu) Z_{d1} = 0, \\ -Z_{d1} + d_1 \phi_1 - d_2 n_1 = 0, \end{cases} \tag{11}$$

where $d_1 = \nu \dfrac{1-\kappa}{\kappa - 3/2} \dfrac{1}{\gamma\alpha \frac{1-\kappa}{\kappa-3/2} - \chi \frac{\alpha}{\sigma}}$ and $d_2 = \dfrac{\chi(1+\alpha/\sigma)}{\chi \frac{\alpha}{\sigma} - \gamma\alpha \frac{1-\kappa}{\kappa-3/2}}$.

Due to $E \times B_0$ drift in magnetized plasma, u_1 and v_1 show up. We obtain the dispersion relation of the DIAW as follows

$$V = l_z \sqrt{\frac{1 + (1-\mu) d_2}{\mu C_1 + (1-\mu) d_1}}. \tag{12}$$

Collecting terms of the next higher order of ϵ, we get

$$\begin{cases} \dfrac{\partial n_1}{\partial \tau} - V\dfrac{\partial n_2}{\partial \xi} + l_z\dfrac{\partial}{\partial \xi}(n_1 w_1) + l_x\dfrac{\partial u_1}{\partial \xi} + l_y\dfrac{\partial v_1}{\partial \xi} + l_z\dfrac{\partial w_2}{\partial \xi} = 0, \\ \dfrac{\partial w_1}{\partial \tau} - V\dfrac{\partial w_2}{\partial \xi} + l_z w_1\dfrac{\partial w_1}{\partial \xi} = -l_z\dfrac{\partial \phi_2}{\partial \xi}, \\ \mu C_1\phi_2 + \mu C_2\phi_1^2 - n_2 + (1-\mu)Z_{d2} = 0 \\ Z_{d2} = d_1\phi_2 + d_4\phi_1^2 - d_2 n_2 + d_3\dfrac{\partial \phi_1}{\partial \xi}, \end{cases} \tag{13}$$

where $d_3 = V\nu\left(\dfrac{-d_1 + d_2\frac{l_z^2}{V^2}}{\gamma\alpha\frac{1-\kappa}{\kappa-3/2} - \chi\frac{\alpha}{\sigma}}\right)$ and $d_4 = \dfrac{\chi\alpha l_z^2}{\sigma V^2}\left(\dfrac{d_1 - d_2\frac{l_z^2}{V^2}}{\gamma\alpha\frac{1-\kappa}{\kappa-3/2} - \chi\frac{\alpha}{\sigma}}\right)$.

The following relation is obtained by comparing the terms of order ϵ^2

$$\mu C_1\phi_2 + \mu C_2\phi_1^2 - n_2 + (1-\mu)Z_{d2} = 0. \tag{14}$$

Differentiating equation (14) and eliminating higher order perturbed terms using equations (11)–(13), we finally acquire the Burgers equation as

$$\frac{\partial \phi_1}{\partial \tau} + A\phi_1\frac{\partial \phi_1}{\partial \xi} = B\frac{\partial^2 \phi_1}{\partial \xi^2}, \tag{15}$$

where A and B are the nonlinear and dispersion coefficients given respectively as $A = \dfrac{l_z^2}{2V} + \dfrac{l_z}{V} - \dfrac{V[\mu C_2 + (1-\mu)d_4]}{2[\mu C_1 + (1-\mu)d_1]}$ and $B = \dfrac{V(1-\mu)d_3}{2[\mu C_1 + (1-\mu)d_1]}$.

4 Dynamical System

The Burgers equation (15) is converted into dynamical system (DS) using the transfiguration

$$\eta = \xi - U\tau, \tag{16}$$

with U as the wave speed. Thus, we obtain the following DS as

$$\begin{cases} \dfrac{d\phi_1}{d\eta} = y, \\ \dfrac{dy}{d\eta} = \dfrac{A^2}{2B^2}\phi_1\left(\phi_1 - \dfrac{U}{A}\right)\left(\phi_1 - \dfrac{2U}{A}\right). \end{cases} \tag{17}$$

In Fig. 1, fixed points (P_0 and P_2) are connected by nonlinear heteroclinic trajectories ($\text{NHeT}_{1,0}$) enclosing center at fixed point (P_1). A trajectory that starts from one fixed point and ends in another fixed point is said to form heteroclinic structure [24, 31, 32]. Such trajectories are associated with shock waves, such as kink and anti-kink waves. A trajectory that encloses one fixed point and has zero separa-

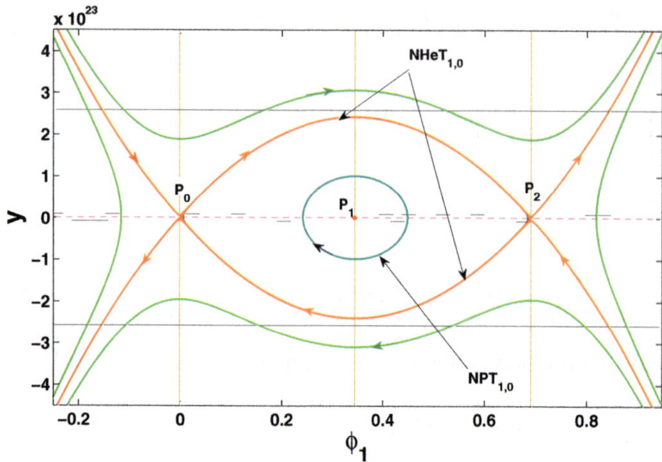

Fig. 1 Phase portrait of equation (17) for $\kappa = 2$, $l_z = 0.1$, $\sigma = 0.1$, $\mu = 0.3$, $\omega_{ci} = 0.5$, $r_d = 2$ μm, $Z_{d0} = 10^4$, $n_{d0} = 1\,\text{m}^{-3}$ and $U = 0.2$

trix is said to form nonlinear periodic trajectory. Here, nonlinear periodic trajectory ($\text{NPT}_{1,0}$) encloses one fixed point (P_1) that corresponds to periodic wave solution [33]. The nonlinear wave trajectories portrayed in Fig. 1 describe nonlinear wave solutions of the Burgers equation through phase plane plots. The nonlinear periodic wave solution shown here are different from supernonlinear periodic wave as supernonlinear periodic waves are characterized by nontrivial topology of their phase portraits. Supernonlinear periodic trajectory encloses more than one fixed points and are separated by at least one separatrix layer [33].

Next, we analyze the occurrence of wave forms by examining potential energy function. Let ψ be the potential energy function such that

$$\frac{d^2\phi_1}{d\eta^2} = -\frac{d\psi}{d\phi_1}, \tag{18}$$

which gives

$$\psi = -\frac{A^2}{2B^2}\left(\frac{1}{4}\phi_1^4 - \frac{U}{A}\phi_1^3 + \frac{U^2}{A^2}\phi_1^2\right). \tag{19}$$

Now, we plot the curve of the potential energy function corresponding to Fig. 1.

From Fig. 2 it is evident that there exist one maximum point and two local minima in the potential energy function graph. The region enclosed by the trajectory from fixed points (P_0 and P_2) are region of the shock wave solution while, the region about maximum point at P_1 describes the periodic wave solution. In Fig. 2, potential dip at P_1 denotes stable fixed point and maxima at points P_O and P_2 represent unstable points [24]. Stable fixed point indicates periodic trajectory and from the other two

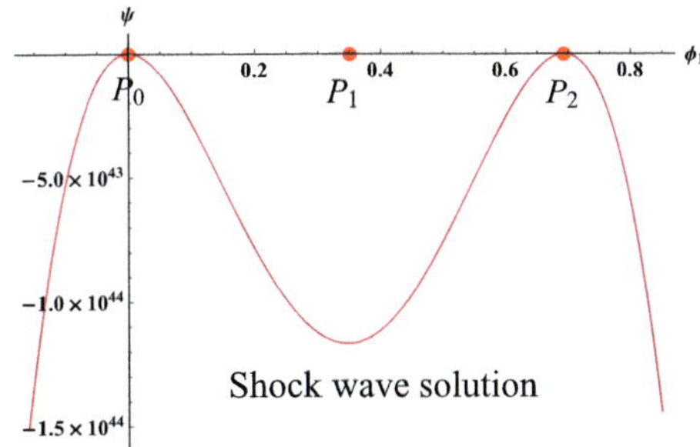

Fig. 2 Potential energy function of equation (19) for same parametric values as Fig. 1

fixed points (P_0 and P_2) the shock wave feature is interpreted from the potential profile that represents dissipation. Here, dissipation leads to the formation of shock structures [20].

4.1 Periodic Wave Solution

The analytical periodic solution of the Burgers equation (15) can be obtained from equation (17) considering the Hamiltonian function $H(\phi_1, y)$ as

$$H(\phi_1, y) = \frac{y^2}{2} - \frac{A^2}{B^2}\left(\frac{\phi_1^4}{4} - \frac{U\phi_1^3}{A} + \frac{U^2\phi_1^2}{A^2}\right) = h, \tag{20}$$

from which one can obtain

$$\frac{dy}{d\eta} = \frac{A}{2B}\sqrt{(r_1 - \psi)(\psi - r_2)(\psi - r_3)(\psi - r_4)}, \tag{21}$$

where r_1, r_2, r_3 and r_4 are roots of $h_i + \dfrac{A^2}{4B^2}\left(\phi_1^4 - \dfrac{4U\phi_1^3}{A} + \dfrac{4U^2\phi_1^2}{A^2}\right) = 0$. Substituting Eq. (21) in Eq. (20), one can obtain the nonlinear periodic dust-ion-acoustic wave (NPDIAW) solution given by

$$\phi_1 = \frac{r_1 + r_4\left\{\frac{r_1 - r_2}{r_2 - r_4} sn^2\left(\frac{A}{2Bg}\eta, z\right)\right\}}{1 + \frac{r_1 - r_2}{r_2 - r_4} sn^2\left(\frac{A}{2Bg}\eta, z\right)}, \tag{22}$$

with sn being the Jacobi elliptic function, $g = \dfrac{2}{\sqrt{(r_1 - r_3)(r_2 - r_4)}}$ and $z = \sqrt{\dfrac{(r_1 - r_2)(r_3 - r_4)}{(r_1 - r_3)(r_2 - r_4)}}$.

In Fig. 3, we show the variation of nonlinear periodic dust-ion-acoustic wave (NPDIAW) solution varying parameters κ, l_z and μ. It is clear from the Fig. 3 that NPDIAW becomes spiky as the values of parameter κ rise, while the NPDIAW becomes smooth for higher values of l_z and μ as the height of NPDIAW diminishes. It is also observed from Fig. 3 that the width of NPDIAW narrows down as parameters κ, l_z and μ are increased.

4.2 *Dust-Ion-Acoustic Kink Wave (DIAKW) and Dust-Ion-Acoustic Anti-kink Wave (DIAAKW)*

In order to acquire DIAKW and DIAAKW solutions analytically, we introduce a new variable ζ as

$$\zeta = c(\xi - \lambda\tau), \tag{23}$$

with wave speed given by λ and $c > 0$.

Solving with $tanh$ method, the DIAKW and DIAAKW solutions of equation (15) are obtained as

$$\phi_1(\xi, \tau) = \pm\frac{\lambda}{A}\left[1 - tanh\left\{\frac{\lambda}{2B}(\xi - \lambda\tau)\right\}\right]. \tag{24}$$

With the help of numerical simulations, the DIAKW and DIAAKW solutions are presented in Fig. 4 varying κ, l_z and μ by keeping $\tau = 1$ with other physical parameters same as Fig. 1. It is noticed that as we increase the values of superthermal

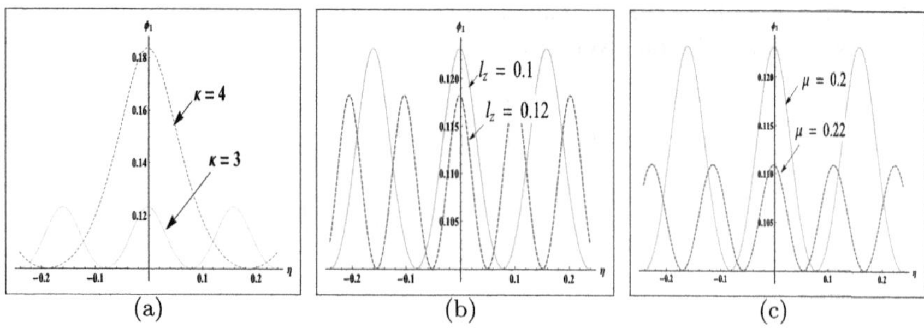

Fig. 3 NPDIAW variations changing parameters κ, l_z and μ with other parameters are same as in Fig. 1

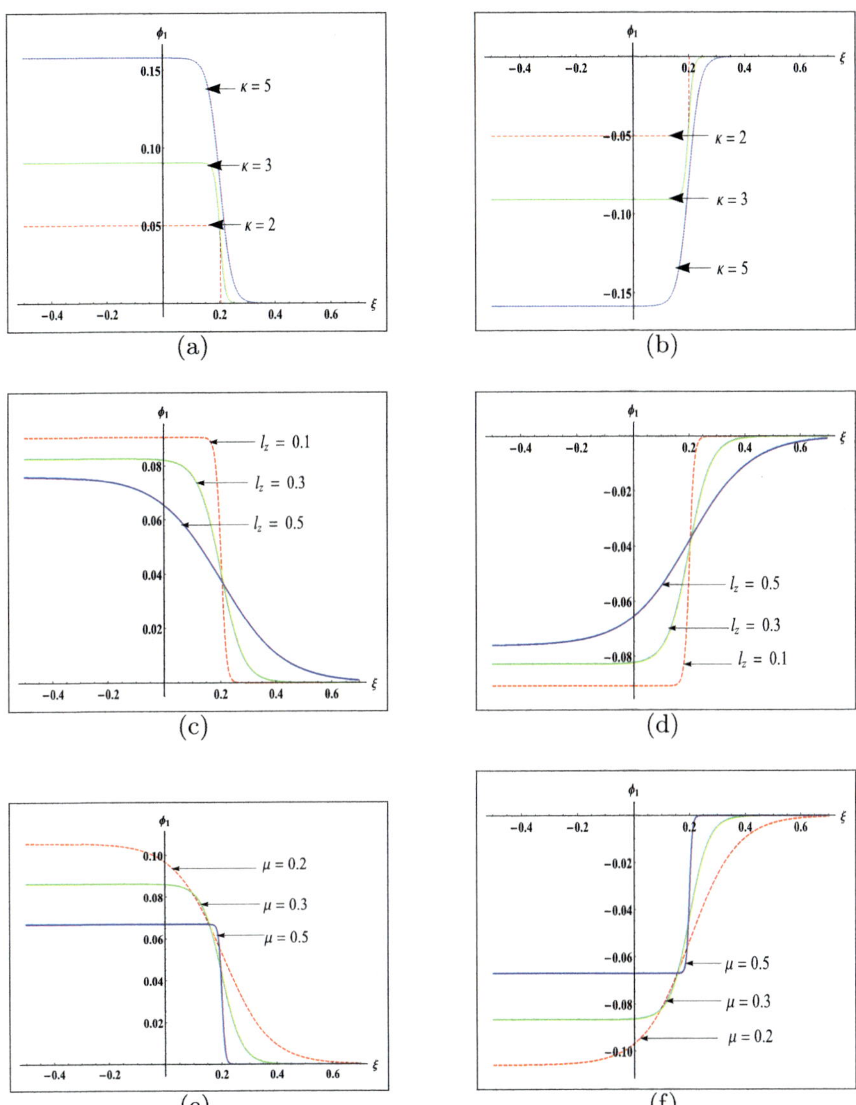

Fig. 4 DIAAKW and DIAKW variations changing parameters κ, l_z and μ with other parameters same as in Fig. 1

parameter (κ), the height of the wave increases. Whereas, the height diminishes while the smoothness of DIAAKW and DIAKW increases when the values of direction cosine (l_z) and number density ratio of electrons and ions (μ).

5 Conclusions

The DIAWs in a superthermal dust charge fluctuation plasma system under the Burgers equation are studied. Using the traveling wave transformations, the Burgers equation is converted into dynamical systems. Through phase plane plot and potential energy function, the shock wave and periodic wave solutions are made evident. Analytical wave solutions of DIAWs are derived using Hamiltonian function. Effects of superthermal charge fluctuation parameters on shock wave and periodic wave solutions of DIAWs are shown. Superthermal parameter (κ), number density ratio (μ) and direction cosine (l_z) have huge impacts on DIAWs. The existence of shock and periodic wave solutions of the Burgers equation for DIAW is shown for the first time in superthermal charge fluctuation in magnetized plasma systems employing the notion of planar dynamical systems. Dusty plasmas are observed in astrophysical environments, namely solar and planetary nebulae and usually with an involvement of magnetic fields [34–36]. Our study may be helpful to understand behaviors of DIAWs existing in magnetized plasma systems where superthermal charge fluctuation occurs, such as in solar wind, mercury, Saturn and Earth's magnetosphere [31].

References

1. Shukla, P.K., Mamun, A.A.: Introduction to Dusty Plasma Physics. Institute of Physics, Bristol (2002)
2. Verheest, F.: Waves in Dusty Space Plasmas, vol. 35. Kluwer Academic, Dordrecht (2000)
3. Barkan, A., D'Angelo, N., Merlino, R.L.: Experiments on ion-acoustic waves in dusty plasmas. Planet. Space Sci. **44**, 239 (1996)
4. Shukla, P.K., Silin, V.P.: Dust-ion-acoustic wave. Phys. Scr. **45**, 508 (1992)
5. Tamang, J., Sarka, K., Saha, A.: Solitary wave solution and dynamic transition of dust ion acoustic waves in a collisional nonextensive dusty plasma with ionization effect. Phys. A Stat. Mech. Appl. **505**, 18–34 (2018)
6. Nakamura, Y., Bailung, H., Shukla, P.K.: Observation of ion-acoustic shocks in a dusty plasma. Phys. Rev. Lett. **83**, 1602 (1999)
7. El-Taibany, W.F., El-Bedwely, N.A., El-Shamy, E.F.: Three-dimensional stability of dust-ion-acoustic solitary waves in a magnetized multicomponent dusty plasma with negative ions. Phys. Plasmas **18**, 033703 (2011)
8. Ghosh, S., Sarkar, S., Khan, M.: Low frequency wave propagation in a cold magnetized dusty plasma. Planet. Space Sci. **47**, 273 (1998)
9. Ghosh, S., Sarkar, S., Khan, M., Gupta, M.R.: Small amplitude nonlinear dust ion acoustic waves in a magnetized dusty plasma with charge fluctuation. Phys. Scr. **63**, 395, 033703 (2001)
10. Shahmansouri, M., Alinejad, H.: Arbitrary amplitude dust ion acoustic solitary waves in a magnetized suprathermal dusty plasma. Phys. Plasmas **19**, 123701 (2012)

11. Shahmansouri, M.: Dynamics of dust-ion acoustic shock waves in a magnetized charge variable superthermal complex plasma. Phys. Scr. **89**, 075604 (2014)
12. Anowar, M.G.M., Mamun, A.A.: Dust ion-acoustic solitary waves in a hot adiabatic magnetized dusty plasma. Phys. Lett. A **372**, 5896, 075604 (2008)
13. El-Labany, S.K., Shalaby, M., El-Shamy, E.F., El-Sherif, L.S.: Effects of two-temperature electrons, external oblique magnetic field, concentration of charged dust grains and higher-order nonlinearity on dust ion-acoustic solitary waves in Saturn's E-ring. Planet. Space Sci. **57**, 1246, 075604 (2009)
14. Shalaby, M., El-Labany, S.K., El-Shamy, E.F., El-Taibany, W.F., Khaled, M.A.: On the stability of obliquely propagating dust ion-acoustic solitary waves in hot adiabatic magnetized dusty plasmas. Phys. Plasmas **16**, 123706 (2009)
15. Samanta, P., De, A., Dey, S., Maity, D., Ghosh, A., Chandra, S.: Nonlinear excitations in dust-ion acoustic waves and the formation of Rogue waves in stable parametric region in a 3-component degenerate plasma. Afr. Rev. Phys. **15**, 10, 123706 (2021)
16. Goswami, J., Sarkar, J., Chandra, S., et al.: Amplitude-modulated electron-acoustic waves with bipolar ions and kappa-distributed positrons and warm electrons. Pramana-J. Phys. **95**, 54, 123706 (2021)
17. Alinejad, H., Mamun, A.A.: Oblique propagation of electrostatic waves in a magnetized electron-positron-ion plasma with superthermal electrons. Phys. Plasmas **18**, 112103 (2011)
18. Sultana, S., Kourakis, I., Hellberg, M.A.: Oblique propagation of arbitrary amplitude electron acoustic solitary waves in magnetized kappa-distributed plasmas. Plasma Phys. Control Fusion **54**, 105016 (2012)
19. Alinejad, H., Tribeche, M., Mohammdi, M.A.: Dust ion-acoustic shock waves due to dust charge fluctuation in a superthermal dusty plasma. Phys. Lett. A **375**, 4183, 105016 (2011)
20. Ghosh, S., Chaudhury, T.K., Sarkar, S., Khan, M., Gupta, M.R.: Collisionless damping of nonlinear dust ion acoustic wave due to dust charge fluctuation. Phys. Rev. E **65**, 037401 (2002)
21. Mamun, A.A., Shukla, P.K.: Electrostatic solitary and shock structures in dusty plasmas. Phys. Scr. **T98**, 107, 037401 (2002)
22. Strogatz, S.H.: Nonlinear Dynamics and Chaos. Levant, Kolkata (2007)
23. Chow, S.N., Hale, J.K.: Methods of Bifurcation Theory, 1st edn. Springer, New York (1982)
24. Saha, A., Banerjee, S.: Dynamical Systems and Nonlinear Waves in Plasmas, 1st edn. CRC Press, Boca Raton (2021)
25. Samanta, U.K., Saha, A., Chatterjee, P.: Bifurcations of dust-ion-acoustic travelling waves in a magnetized dusty plasma with a q-nonextensive electron velocity distribution. Phys. Plasma **20**, 022111 (2013)
26. Selim, M.M., El-Depsy, A., El-Shamy, E.F.: Bifurcations of nonlinear ion-acoustic travelling waves in a multicomponent magnetoplasma with superthermal electrons. Astrophys. Space Sci. **360**, 66, 022111 (2015)
27. El-Monier, S.Y., Atteya, A.: Bifurcation analysis for dust-acoustic waves in a four-component plasma including warm ions. IEEE Trans. Plasma Sci. **46**, 815–824, 022111 (2018)
28. El Ghani, O., Driouch, I., Chatei, H.: Effects of non-extensive electrons on the sheath of dusty plasmas with variable dust charge. Contrib. Plasma Phys **59**, e201900030 (2019)
29. Chatterjee, P., Ali, R., Saha, A.: Analytical solitary wave solution of the dust ion acoustic waves for the damped forced Korteweg-de Vries equation in superthermal plasmas. Zeitschrift für Naturforschung A. **73**, 151–159, e201900030 (2018)
30. Sharma, P., Das, A., Tamang, J., Saha, A.: Dynamical properties of dust-ion-acoustic wave solutions in a nonextensive collisional dusty plasma. J. Taibah Univ. Sci. **15**, 710–720, e201900030 (2021)
31. Saha, A., Chatterjee, P.: Dust ion acoustic travelling waves in the framework of a modified Kadomtsev-Petviashvili equation in a magnetized dusty plasma with superthermal electrons. Astrophys. Space Sci. **349**, 813–820, e201900030 (2014)
32. Tamang, J., Saha, A.: Dynamical properties of nonlinear ion-acoustic waves based on the nonlinear Schrödinger equation in a multi-pair nonextensive plasma. Zeitschrift für Naturforschung A **75**, 687–697, e201900030 (2020)

33. Saha, A., Tamang, J.: Effect of q-nonextensive hot electrons on bifurcations of nonlinear and supernonlinear ion-acoustic periodic waves. Adv. Space Res. **63**, 1596–1606 (2019)
34. Melzer, A., Krüger, H., Maier, D., et al.: Physics of magnetized dusty plasmas. Rev. Mod. Plasma Phys. **5**, 11, e201900030 (2021)
35. Reissl, S., Stutz, A.M., Klessen, R.S., Seifried, D., Walch, S.: Magnetic fields in star-forming systems ii: examining dust polarization, the Zeeman effect, and the faraday rotation measure as magnetic field tracers. Mon. Not. R. Astron. Soc. **500**(1), 153, e201900030 (2020)
36. Li, H.B.: Magnetic fields in molecular clouds-observation and interpretation. Galaxies **9**(2), 41, e201900030 (2021)

Dynamical Aspects of Ion-Acoustic Solitary Waves in a Magnetically Confined Plasma in the Presence of Nonthermal Components

Jintu Ozah and P. N. Deka

Abstract The characteristics of three-dimensional ion-acoustic solitary waves (IASWs) have been investigated in a magnetized plasma including ions, nonthermaly dispersed electrons and positrons. The reductive perturbation technique (RPT) is used to develop the Zakharov–Kuznetsov (ZK) equation for observing ion-acoustic wave structure, and a soliton solution is obtained by using the tangent hyperbolic (*tanh*) method. The influence of various parameters on the soliton profile, such as nonthermal parameters for electrons and positrons, density ratios of positron–electron and ion–electron, and temperature ratio of electron–positron, is presented graphically.

Keywords Solitary waves · Reductive perturbation technique · Magnetized plasma · Nonthermal electrons · Positrons

1 Introduction

Investigation of ion-acoustic solitary waves is an interesting research problem in the field of plasma physics that has been extensively studied by numerous authors [1–4]. For the first time, Washimi and Taniuti [5] observed the distinctive behaviour of ion-acoustic solitary waves in plasma, which can be investigated using the Korteweg-de Vries (K-dV) equation. After that, ion-acoustic solitary waves in both magnetized and unmagnetized plasmas have been studied by a large number of researchers in different theoretical and experimental circumstances during the past few decades. There has been a lot of interest in the investigation of different types of nonlinear solitary waves in plasmas, like magneto-acoustic solitary waves, spherical and cylindrical-acoustic solitary waves, and lower-hybrid solitary waves [6–9]. A study of ion-acoustic solitary waves in magnetized negative ion plasma consisting of nonthermal electrons was carried out by Labany et al. [10]. They observed that the solitary waves are substantially influenced by the positive-to-negative ion mass ratio, the corresponding negative-to-positive ion density ratio, and the parameters of nonthermal electrons.

J. Ozah (✉) · P. N. Deka
Dibrugarh University, Dibrugarh, Assam 786004, India
e-mail: jintuozah@gmail.com

S. Banerjee and A. Saha (eds.), *Nonlinear Dynamics and Applications*,
Springer Proceedings in Complexity,
https://doi.org/10.1007/978-3-030-99792-2_22

For analysing ion-acoustic waves in a magnetized plasma, Zakharov and Kuznetsov developed the nonlinear equation known as the ZK equation. This ZK equation may be found in many branches of physics, such as fluid mechanics, astrophysics, solid state physics, and so on [11, 12]. It is most apparent in the subject of plasma physics. Using the extended *tanh* approach and the direct assumption method, Li et al. derived the ZK equation and got an exact travelling wave solution [13]. Taibany et al. [14] developed the ZK equation to investigate the IASWs in a magnetized multicomponent dusty plasma with negative ions. Recently, many researchers have been showing an interest in studying the impact of nonextensive electron distribution on IASWs in magnetized plasma. Mandi et al. [15] have investigated the effect of the q-nonextensivity of electrons on the characteristics of IASWs. Furthermore, the propagation of solitons in nonthermal plasma has generated much interest among researchers. Because of their practical significance, they continue to pique people's curiosity. Many studies in plasma physics, as well as complex plasma, have focused on ion-acoustic solitary waves and their properties in the field of nonthermal plasma. Pakzad [16] studied the behaviour of soliton structures in a three-component unmagnetized plasma containing cold ions, nonthermal electrons, and positrons. Dev et al. [17] studied the dust IASWs in a magnetized plasma in the presence of nonthermal electrons, positrons and relativistic thermal ions. They discovered that in the absence of nonthermal electron and positron populations, the plasma system behaves in the least nonlinear manner, but the system behaves in the most nonlinear manner when the populations of nonthermal electrons and positrons have the maximal value. In their investigation into three-dimensional ion-acoustic soliton structures, including warm ions, positrons, and nonthermal electrons, Chawla et al. [18] reveal that the presence of nonthermal electrons considerably impacts the amplitude and width of soliton pulses.

In this paper, the effects of nonthermal electrons, nonthermal positrons, and the influence of magnetic fields on the structure of three-dimensional nonlinear IASWs are investigated. We anticipate that the presence of nonthermal electrons and positrons will alter the characteristics of solitons as well as their existence regime.

2 Basic Model Equations

We consider a plasma model with constituent ions, nonthermal electrons and nonthermal positrons, where the magnetic field B_0 is along the z-axis. The following normalized sets of ion continuity equations, momentum equations, and Poisson equations serve as the governing equations for the current plasma model:

$$\frac{\partial n_i}{\partial t} + \frac{\partial (n_i u)}{\partial x} + \frac{\partial (n_i v)}{\partial y} + \frac{\partial (n_i w)}{\partial z} = 0 \tag{1}$$

$$\frac{\partial u}{\partial t} + u\frac{\partial u}{\partial x} + v\frac{\partial u}{\partial y} + w\frac{\partial u}{\partial z} = -\frac{\partial \phi}{\partial x} + \frac{\Omega_i}{\omega_{pi}} v \tag{2}$$

$$\frac{\partial v}{\partial t}+u\frac{\partial v}{\partial x}+v\frac{\partial v}{\partial y}+w\frac{\partial v}{\partial z}=-\frac{\partial \phi}{\partial y}-\frac{\Omega_i}{\omega_{pi}}u \tag{3}$$

$$\frac{\partial w}{\partial t}+u\frac{\partial w}{\partial x}+v\frac{\partial w}{\partial y}+w\frac{\partial w}{\partial z}=-\frac{\partial \phi}{\partial z} \tag{4}$$

$$\frac{\partial^2 \phi}{\partial x^2}+\frac{\partial^2 \phi}{\partial y^2}+\frac{\partial^2 \phi}{\partial z^2}=n_e-\mu_p n_p-\mu_i n_i \tag{5}$$

The Boltzmann distributions [17] for nonthermal electrons and positrons are defined as

$$n_e=\left(1-\beta_e\phi+\beta_e\phi^2\right)\exp(\phi) \tag{6}$$

$$n_p=\left(1+\sigma_{pe}\beta_p\phi+\sigma_{pe}^2\beta_p\phi^2\right)\exp\left(-\sigma_{pe}\phi\right) \tag{7}$$

In the above expressions.
Also $\mu_p=\frac{n_{p0}}{n_{e0}}$, $\mu_i=\frac{n_{i0}}{n_{e0}}$ and $\sigma_{pe}=\frac{T_e}{T_p}$.

$$\beta_e=\frac{4\alpha_e}{1+3\alpha_e},\ \beta_p=\frac{4\alpha_p}{1+3\alpha_p}$$

here, α_e is the nonthermal parameter for electrons and α_p is the nonthermal parameter for positrons, which represent the population of energetic nonthermal electrons and positrons, respectively. T_e and T_p are the temperatures of electrons and positrons. In the above equations, the ion number densities n_i are normalized by n_{i0} and velocities (u, v, w) by the ion-acoustic speed $C_s=(T_e/m_i)^{1/2}$, where m_i is the ion mass. Space coordinates (x, y, z) and time t are normalized in terms of Debye length $\lambda_D=\left(\varepsilon_0 T_e/n_i^0 e^2\right)^{1/2}$ and the inverse of plasma frequency $\omega_{pi}=\left(4\pi e^2 n_i^0/m_i\right)^{1/2}$ respectively. The electric potential ϕ is normalized by T_e/e, where e is the electronic charge. Ω_i and ω_{pi} are the ion-cyclotron frequency and plasma frequency.

3 Reductive Perturbation Method

To derive the ZK equation from the above basic set of equations, we used the reductive perturbation technique. The stretching coordinates [10, 19] are assume as

$$\xi=\varepsilon^{1/2}x,\ \eta=\varepsilon^{1/2}y,\ \zeta=\varepsilon^{1/2}(z-\lambda_0 t)\ \text{and}\ \tau=\varepsilon^{3/2}t \tag{8}$$

where the symbol ε is the expansion parameter that measures the strength of nonlinearity and λ_0 is the phase velocity of IASWs. We express the physical parameters in the power series expansion of ε in the following way:

$$\left.\begin{aligned} n_i &= 1+\varepsilon^1 n^1+\varepsilon^2 n^2+\varepsilon^3 n^3+\ldots.\\ u &= \varepsilon^{3/2}u^1+\varepsilon^2 u^2+\varepsilon^{5/2}u^3+\ldots.\\ v &= \varepsilon^{3/2}v^1+\varepsilon^2 v^2+\varepsilon^{5/2}v^3+\ldots.\\ w &= 0+\varepsilon^1 w^1+\varepsilon^2 w^2+\varepsilon^3 w^3+\ldots.\\ \phi &= 0+\varepsilon^1 \phi^1+\varepsilon^2 \phi^2+\varepsilon^3 \phi^3+\ldots. \end{aligned}\right\} \tag{9}$$

We transform x and t by using the stretch coordinates as

$$\left.\begin{aligned} &\frac{\partial}{\partial x}\equiv\varepsilon^{1/2}\frac{\partial}{\partial\xi},\ \frac{\partial}{\partial y}\equiv\varepsilon^{1/2}\frac{\partial}{\partial\eta},\ \frac{\partial}{\partial z}\equiv\varepsilon^{1/2}\frac{\partial}{\partial\zeta},\ \frac{\partial}{\partial t}\equiv-\lambda_0\varepsilon^{1/2}\frac{\partial}{\partial\zeta}+\varepsilon^{3/2}\frac{\partial}{\partial\tau}\\ &\frac{\partial^2}{\partial x^2}\equiv\varepsilon\frac{\partial^2}{\partial\xi^2},\ \frac{\partial^2}{\partial y^2}\equiv\varepsilon\frac{\partial^2}{\partial\eta^2},\ \frac{\partial^2}{\partial z^2}\equiv\varepsilon\frac{\partial^2}{\partial\zeta^2} \end{aligned}\right\} \tag{10}$$

Using (10), the transformation equations of (1)–(5) may be obtained as

$$-\lambda_0\varepsilon^{1/2}\frac{\partial n_i}{\partial\zeta}+\varepsilon^{3/2}\frac{\partial n_i}{\partial\tau}+\varepsilon^{1/2}\frac{\partial(n_i u)}{\partial\xi}+\varepsilon^{1/2}\frac{\partial(n_i v)}{\partial n}+\varepsilon^{1/2}\frac{\partial(n_i w)}{\partial\zeta}=0 \tag{11}$$

$$-\lambda_o\varepsilon^{1/2}\frac{\partial u}{\partial\zeta}+\varepsilon^{3/2}\frac{\partial u}{\partial\tau}+u\varepsilon^{1/2}\frac{\partial u}{\partial\xi}+v\varepsilon^{1/2}\frac{\partial u}{\partial\eta}+w\varepsilon^{1/2}\frac{\partial u}{\partial\zeta}=-\varepsilon^{1/2}\frac{\partial\phi}{\partial\xi}+\frac{\Omega_i}{\omega_{pi}}v \tag{12}$$

$$-\lambda_o\varepsilon^{1/2}\frac{\partial v}{\partial\zeta}+\varepsilon^{3/2}\frac{\partial v}{\partial\tau}+u\varepsilon^{1/2}\frac{\partial v}{\partial\xi}+v\varepsilon^{1/2}\frac{\partial v}{\partial\eta}+w\varepsilon^{1/2}\frac{\partial v}{\partial\zeta}=-\varepsilon^{1/2}\frac{\partial\phi}{\partial\eta}+\frac{\Omega_i}{\omega_{pi}}u \tag{13}$$

$$-\lambda_o\varepsilon^{1/2}\frac{\partial w}{\partial\zeta}+\varepsilon^{3/2}\frac{\partial w}{\partial\tau}+u\varepsilon^{1/2}\frac{\partial w}{\partial\xi}+v\varepsilon^{1/2}\frac{\partial w}{\partial\eta}+w\varepsilon^{1/2}\frac{\partial w}{\partial\zeta}=-\varepsilon^{1/2}\frac{\partial\phi}{\partial\zeta} \tag{14}$$

$$\varepsilon\frac{\partial^2\phi}{\partial\xi^2}+\varepsilon\frac{\partial^2\phi}{\partial\eta^2}+\varepsilon\frac{\partial^2\phi}{\partial\zeta^2}=n_e-\mu_p n_p-\mu_i n_i \tag{15}$$

Now using (9) in the above equations and then collecting the lowest order terms. in ε, we get

$$\left.\begin{aligned} n^1 &= \frac{w^1}{\lambda_0}, & w^1 &= \frac{\phi^1}{\lambda_0},\\ u^1 &= -\frac{\omega_{pi}}{\Omega_i}\frac{\partial\phi^1}{\partial\eta},\ \phi^1 = & &\frac{\mu_i n^1}{(1-\beta_e+\mu_p\sigma_{pe}-\beta_p\mu_p\sigma_{pe})}\\ v^1 &= \frac{\omega_{pi}}{\Omega_i}\frac{\partial\phi^1}{\partial\xi}, \end{aligned}\right\} \tag{16}$$

After solving for the first order perturbation terms, the dispersion relation of nonlinear IASWs is obtained as

$$\lambda_0 = \sqrt{\frac{\mu_i}{1 - \beta_e + \mu_p \sigma_{pe} - \beta_p \mu_p \sigma_{pe}}} \tag{17}$$

Equation (17) represents the phase velocity of nonlinear IASWs.
The next higher order of ε gives

$$u^2 = \frac{\omega_{pi}\lambda_0}{\Omega_i}\frac{\partial v^1}{\partial \zeta}, \tag{18}$$

$$v^2 = -\frac{\omega_{pi}\lambda_0}{\Omega_i}\frac{\partial u^1}{\partial \zeta}, \tag{19}$$

$$\frac{\partial n_i^1}{\partial \tau} - \lambda_0 \frac{\partial n_i^2}{\partial \zeta} + \frac{\partial u^2}{\partial \xi} + \frac{\partial v^2}{\partial \eta} + \frac{\partial w^2}{\partial \zeta} + \frac{\partial (n_i^1 w^1)}{\partial \zeta} = 0, \tag{20}$$

$$\frac{\partial w^1}{\partial \tau} - \lambda_0 \frac{\partial w^2}{\partial \zeta} + w^1 \frac{\partial w^1}{\partial \zeta} + \frac{\partial \phi^2}{\partial \zeta} = 0, \tag{21}$$

$$\begin{aligned} &\frac{\partial^2 \phi^1}{\partial \xi^2} + \frac{\partial^2 \phi^1}{\partial \eta^2} + \frac{\partial^2 \phi^1}{\partial \zeta^2} - (1 - \beta_e + \mu_p \sigma_{pe} - \beta_p \mu_p \sigma_{pe})\phi^2 \\ &- \frac{1}{2}(1 - \sigma_{pe}^2 \mu_p)(\phi^1)^2 + \mu_i n^2 = 0, \end{aligned} \tag{22}$$

Now, differentiating equation (22) w.r.t ζ, we get

$$\begin{aligned} &\frac{\partial}{\partial \zeta}\left(\frac{\partial^2 \phi^1}{\partial \xi^2} + \frac{\partial^2 \phi^1}{\partial \eta^2}\right) + \frac{\partial^3 \phi^1}{\partial \zeta^3} - (1 - \beta_e + \mu_p \sigma_{pe} - \beta_p \mu_p \sigma_{pe})\frac{\partial \phi^2}{\partial \zeta} \\ &- (1 - \sigma_{pe}^2 \mu_p)\phi^1 \frac{\partial \phi^1}{\partial \zeta} + \mu_i \frac{\partial n^2}{\partial \zeta} = 0. \end{aligned} \tag{23}$$

Now, using the lowest order terms, the Eq. (23) can be written as

$$\frac{\partial}{\partial \zeta}\left(\frac{\partial^2 \phi^1}{\partial \xi^2} + \frac{\partial^2 \phi^1}{\partial \eta^2}\right) + \frac{\partial^3 \phi^1}{\partial \zeta^3} - \frac{\mu_i}{\lambda_0^2}\frac{\partial \phi^2}{\partial \zeta} - (1 - \sigma_{pe}^2 \mu_p)\phi^1 \frac{\partial \phi^1}{\partial \zeta} + \mu_i \frac{\partial n^2}{\partial \zeta} = 0. \tag{24}$$

Now, eliminating the second order quantities from (20), (21) and (24), we obtain the ZK equation in terms of ϕ^1 as

$$\frac{\partial \phi^1}{\partial \tau} + A\phi^1 \frac{\partial \phi^1}{\partial \zeta} + B\frac{\partial^3 \phi^1}{\partial \zeta^3} + C\frac{\partial}{\partial \zeta}\left(\frac{\partial^2 \phi^1}{\partial \xi^2} + \frac{\partial^2 \phi^1}{\partial \eta^2}\right) = 0, \tag{25}$$

where the non-linear coefficient A is given by

$$A = \frac{2}{\lambda_0} - \frac{\lambda_0^3}{2\mu_i}\left(1 - \sigma_{pe}^2 \mu_p\right).$$

B and C are the dispersive and higher order coefficients, expressed as

$$B = \frac{\lambda_0^3}{2\mu_i}, \text{ and } C = \frac{\lambda_0^3}{2\mu_i} + \frac{\lambda_0^3}{2}\frac{\omega_{pi}^2}{\Omega_i^2}.$$

4 Solution of ZK Equation

To analyse the Eq. (25), we use the *tanh* method. We consider the transformation $\chi = \gamma(l\xi + m\eta + n\zeta - U\tau)$, where $\phi(\xi, \eta, \zeta, \tau) = \psi(\chi)$, we can use the following changes:

$$\frac{\partial}{\partial \tau} \equiv -U\gamma \frac{d}{d\chi}, \frac{\partial}{\partial \xi} \equiv l\gamma \frac{d}{d\chi}, \frac{\partial}{\partial \eta} \equiv \gamma m \frac{d}{d\chi}, \frac{\partial}{\partial \zeta} \equiv \gamma n \frac{d}{d\chi},$$
$$\frac{\partial^2}{\partial \xi^2} \equiv \gamma^2 l^2 \frac{d^2}{d\chi^2}, \frac{\partial^2}{\partial \eta^2} \equiv \gamma^2 m^2 \frac{d^2}{d\chi^2}, \frac{\partial^3}{\partial \zeta^3} \equiv \gamma^3 n^3 \frac{d^3}{d\chi^3}$$

Now the Eq. (25) becomes a reduced ordinary differential equation as

$$-U\gamma \frac{d\psi}{d\chi} + \frac{A\gamma n}{2}\frac{d\psi^2}{d\chi} + B\gamma^3 n^3 \frac{d^3\psi}{d\chi^3} + C\gamma n \frac{d}{d\chi}\left[\gamma^2(l^2 + m^2)\frac{d^2\psi}{d\chi^2}\right] = 0. \quad (26)$$

Integrating the above equation, we get

$$-U\psi + \frac{1}{2}An\psi^2 + \gamma^2 n\left[Bn^2 + C\left(l^2 + m^2\right)\right]\frac{d^2\psi}{d\chi^2} = 0 \quad (27)$$

To solve the above equation, we use the *tanh* method. Consider a new independent variable as:

$z = \tan(\chi)$, where $\psi(\chi) = w(z)$.

and we get.

$\frac{d^2}{d\chi^2} = \left(1 - z^2\right)^2 \frac{d^2}{dz^2} - 2z\left(1 - z^2\right)\frac{d}{dz}$.
Now the Eq. (27) becomes

$$-Uw + \frac{1}{2}Anw^2 + \gamma^2 n\left(Bn^2 + C\left(l^2 + m^2\right)\right)\left(1 - z^2\right)^2 \frac{d^2 w}{dz^2}$$

$$-2\gamma^2 n\left(Bn^2 + C\left(l^2 + m^2\right)\right)z\left(1 - z^2\right)\frac{dw}{dz} = 0. \tag{28}$$

In the *tanh* method the series solution of the Eq. (28) can be written as:

$$w(z) = \sum_{i=1}^{m} a_i z^i \tag{29}$$

In Eq. (29), the value of m can be obtained by balancing the highest order linear term with the nonlinear terms. On substitution of Eq. (29) into Eq. (28), we get $m = 2$.

As a result, the solution $w(z) = \sum_{i=1}^{m} a_i z^i$ is of the form

$$w(z) = a_0 + a_1 z + a_2 z^2. \tag{30}$$

Now substituting $w, \frac{dw}{dz}, \frac{d^2w}{dz^2}$ from (30) into (28), then equating different coefficient of z, we get.

$a_0 = -a_2$ and $a_1 = 0$.

Hence Eq. (30) reduce as

$$w(z) = a_0\left(1 - z^2\right). \tag{31}$$

Using (31) into (28) and equating the coefficients of z^2, we get

$$a_0 = \frac{12\gamma^2\left[Bn^2 + C\left(l^2 + m^2\right)\right]}{A}.$$

And hence $\gamma = \sqrt{\frac{U}{4n\left(Bn^2 + C\left(l^2 + m^2\right)\right)}}$.

Using the values of the parameters, Eq. (31) provides a solution as

$$\phi = \phi_m \operatorname{sech}^2\left(\frac{\chi}{W}\right). \tag{32}$$

Here, (32) represents the solution of the Eq. (25), where ϕ_m and W are the amplitude and width of the soliton.

$$\text{Where } \phi_m = \frac{3U}{An} \tag{33}$$

and

$$W = \sqrt{\frac{U}{4n\left(Bn^2 + C\left(l^2 + m^2\right)\right)}}. \tag{34}$$

5 Results and Discussion

For the study of soliton structures due to the existence of nonthermal components of electrons and positrons, we have plotted the variation of nonlinear coefficient A with electron-to-positron temperature ratio (σ_{pe}), positron-to-electron density ratio (μ_p) and ion-to-electron density ratio (μ_i) for different parameters of nonthermal electrons and positrons. The polarity (positive or negative) of the soliton structure completely depends on the sign of the nonlinear coefficient A. The positive polarity (compressive soliton) structure exists for the positive value of the nonlinear coefficient and the negative polarity (rarefactive soliton) structure exists for the negative value of the nonlinear coefficient. Figure 1 shows that the nonlinearity of plasma increases with the electron-to-positron temperature ratio (σ_{pe}). The same result has been observed in Fig. 2, where the nonlinearity changes with the positron-to-electron density ratio (μ_p). The variation of nonlinearity with ion-to-electron density ratio (μ_i) in Fig. 3 shows the existence of rarefactive soliton structures. It can be seen from this graph that the sign of nonlinearity becomes negative after a certain value of μ_i. Therefore, the parameter μ_i is very crucial to obtaining rarefactive soliton structures. The range of the parameter μ_i for the existence of negative polarity can be obtained from Fig. 3b. Further, it is observed that, in all cases, the nonlinearity of plasma decreases as the parameters of nonthermal electrons and positrons are increased.

Figure 4 shows the change of the compressive soliton structure with χ for different values of nonthermal parameters of electrons (α_e) and positrons (α_p). As in equation (34), it shows that the amplitude is the inverse of the nonlinear coefficient A. As a result, as nonthermal parameters are increased, plasma nonlinearity decreases, and hence the amplitude of the soliton structure rises.

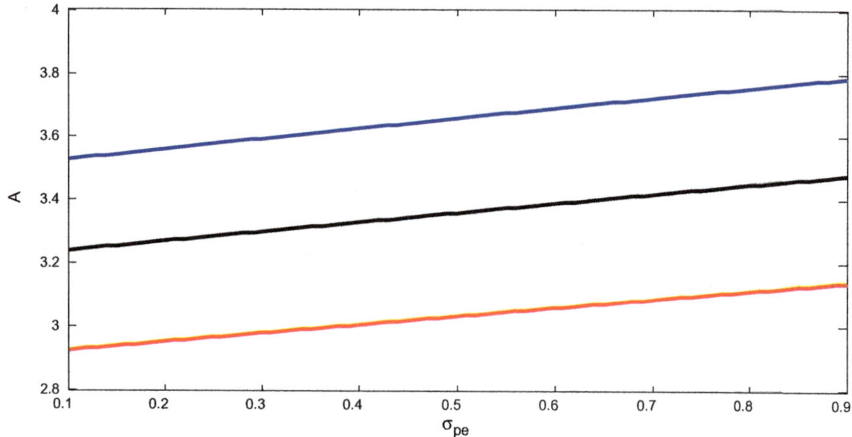

Fig. 1 Variation of nonlinear coefficient against electron-to-positron temperature ratio (σ_{pe}) with $\mu_i = 0.3$ and $\mu_p = 0.2$. Blue line corresponds to $\alpha_e = 0.01$, $\alpha_p = 0.02$; black line corresponds to $\alpha_e = 0.03$, $\alpha_p = 0.04$; red line corresponds to $\alpha_e = 0.05$, $\alpha_p = 0.06$

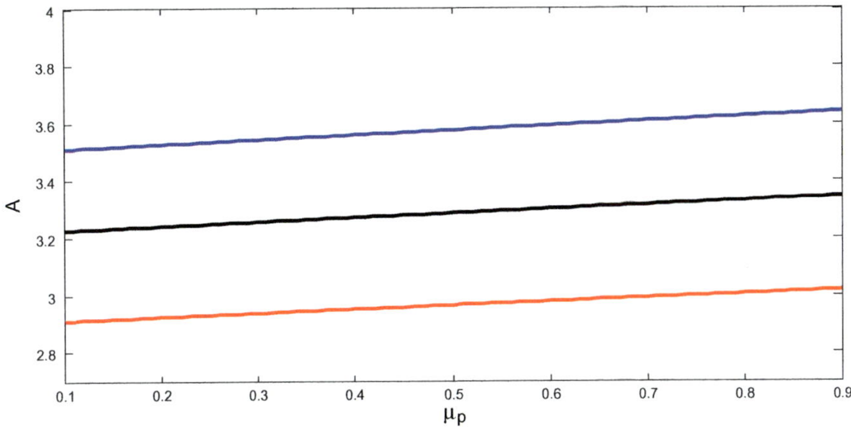

Fig. 2 Variation of nonlinear coefficient against positron-to-electron density ratio (μ_p) with $\sigma_{pe} = 0.1$ and $\mu_i = 0.3$. Blue line corresponds to $\alpha_e = 0.01$, $\alpha_p = 0.02$; black line corresponds to $\alpha_e = 0.03$, $\alpha_p = 0.04$; red line corresponds to $\alpha_e = 0.05$, $\alpha_p = 0.06$

We have also shown the rarefactive soliton structure in Fig. 5 for different values of nonthermal parameters while keeping all other parameters fixed. The amplitude (width) of the solitons decreases (increases) with an increase in the values of α_e and α_p.

A similar kind of variation in the soliton structure is observed in Fig. 6, when we change the positron-to-electron density ratio (μ_p). This observation also provides an information about the change in amplitude and width when the value of (μ_p) changes. Increase in the density ratio, enhances the amplitude of the soliton.

6 Conclusion

In the present work, we have studied the characteristics (amplitude and width) of nonlinear ion-acoustic solitary waves in a **magnetically confined plasma** under the influence of nonthermal electrons and positrons. We have determined the range of the parameters for the existence of both the positive and negative polarity soliton. We have noticed the following main results in our present study:

1. In our investigation, both the positive and negative polarity of soliton exist, whereas in the earlier study, this aspect was not covered by Chawla et al. [18].
2. The parameter ion-to-electron density ratio (μ_i) is very crucial to getting two kinds of soliton structure. A very small change in this parameter, changes the polarity of the soliton structure.
3. We can obtained the parameter range for the existence of a negative polarity soliton from Fig. 3b.

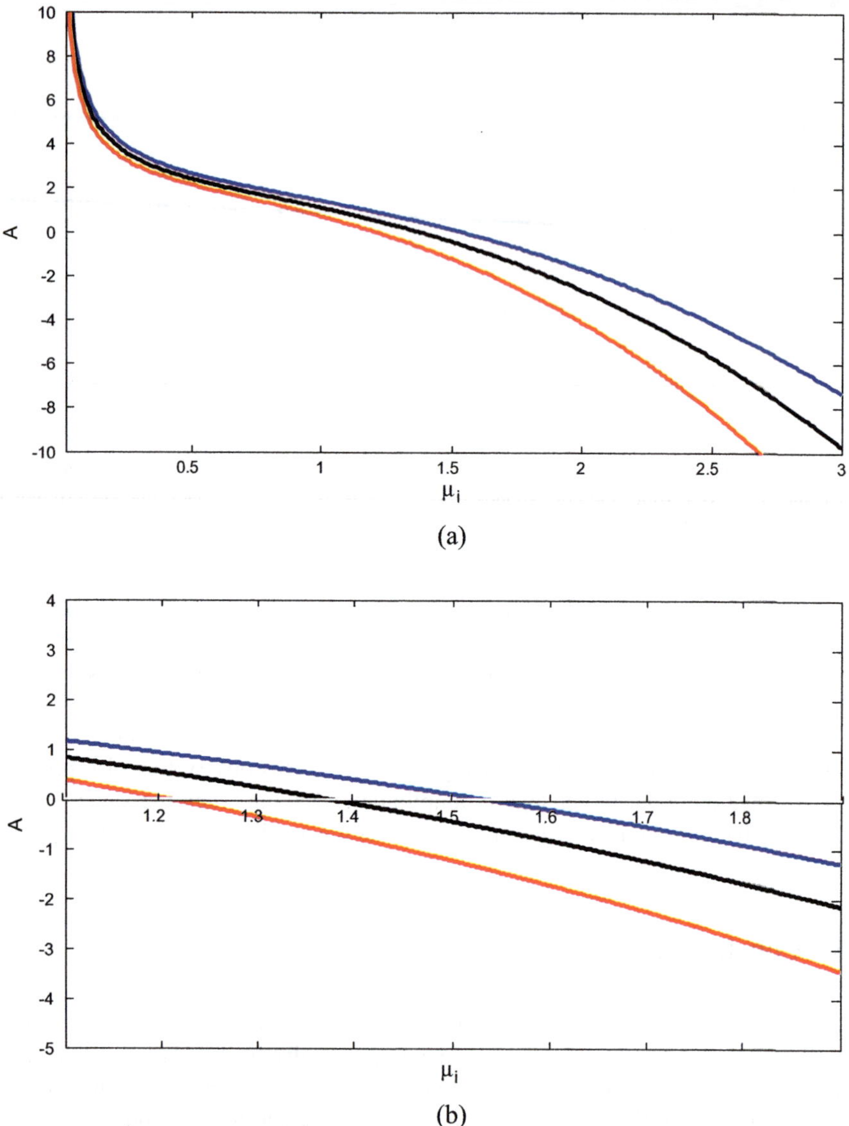

Fig. 3 Variation of nonlinear coefficient against ion-to-electron density ratio (μ_i) with $\sigma_{pe} = 0.1$ and $\mu_p = 0.2$. Blue line corresponds to $\alpha_e = 0.01, \alpha_p = 0.02$; black line corresponds to $\alpha_e = 0.03$, $\alpha_p = 0.04$; red line corresponds to $\alpha_e = 0.05$, $\alpha_p = 0.06$

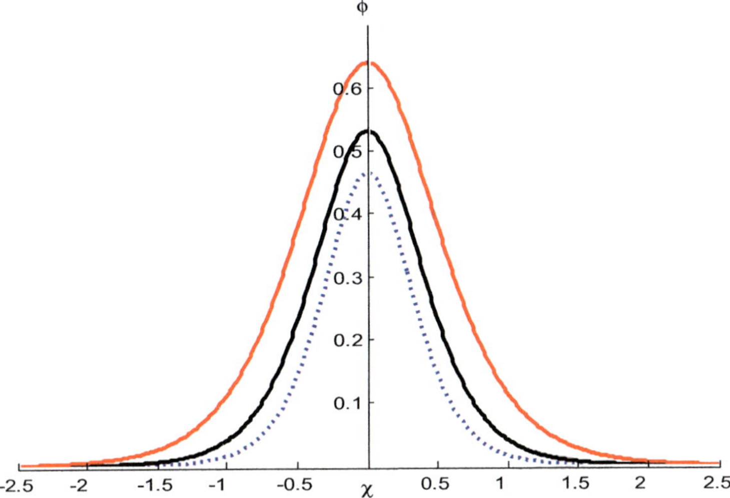

Fig. 4 Variation of compressive soliton wave structure for different values of α_e and α_p, with $\mu_i = 0.3$, $\sigma_{pe} = 0.1$, $\mu_p = 0.2$, $n = 0.6$, $U = 0.9$, $\Omega_i = 0.5$ and $\omega_{pi} = 1.4$. Blue (dotted) curve corresponds to $\alpha_e = 0.01$ and $\alpha_p = 0.03$; black (solid) curve corresponds to $\alpha_e = 0.04$ and $\alpha_p = 0.06$; red (solid) curve corresponds to $\alpha_e = 0.07$ and $\alpha_p = 0.09$

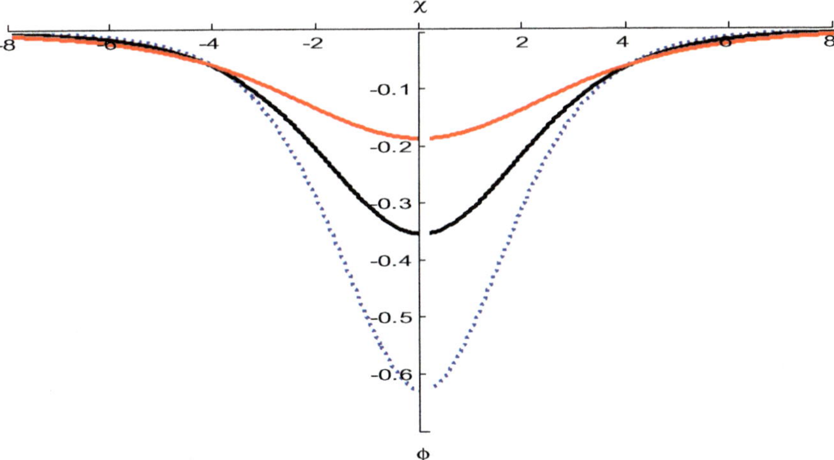

Fig. 5 Variation of rarefactive soliton structure for different values of α_e and α_p, with $\mu_i = 1.8$, $\mu_p = 0.2, \sigma_{pe} = 0.1, n = 0.6, U = 0.9, \Omega_i = 0.5$ and $\omega_{pi} = 1.4$. Blue (dotted) curve corresponds to $\alpha_e = 0.01$ and $\alpha_p = 0.03$; black (solid) curve corresponds to $\alpha_e = 0.04$ and $\alpha_p = 0.06$; red (solid) curve corresponds to $\alpha_e = 0.07$ and $\alpha_p = 0.09$

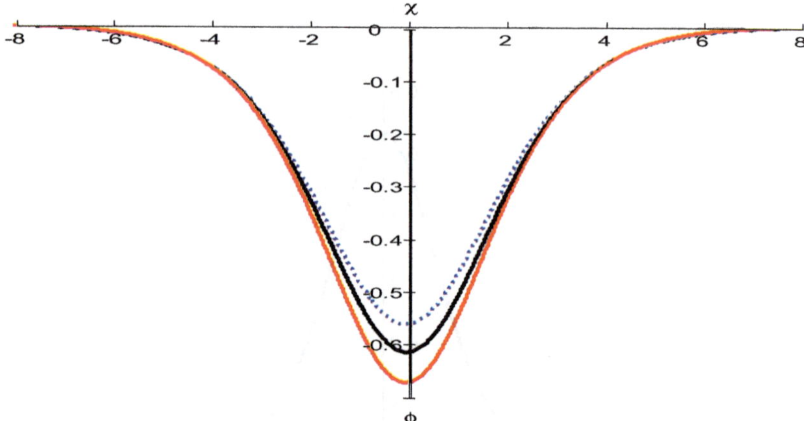

Fig. 6 Variation of rarefactive soliton structure for different values of μ_p, with $\mu_i = 1.8, \sigma_{pe} = 0.1$, $n = 0.6, U = 0.9, \Omega_i = 0.5$ and $\omega_{pi} = 1.4, \alpha_e = 0.05$ and $\alpha_p = 0.06$. Blue (dotted) curve corresponds to $\mu_p = 0.1$; black (solid) curve corresponds to $\mu_p = 0.5$ and; red (solid) curve corresponds to $\mu_p = 0.9$

We hope these results will be incredibly helpful in both space and laboratory plasmas, where solitary wave propagation is quite useful.

References

1. Kamalam, T., Ghosh, S.S.: Ion acoustic super solitary waves in a magnetized plasma. Phys. Plasmas **25**, 122302 (2018)
2. Salmanpoor, H., Sharifian, M., Gholipour, S., Zarandi, M.B., Shokri, B.: Oblique propagation of solitary waves in weakly relativistic magnetized plasma with kappa distributed electrons in the presence of negative ions. Phys. Plasmas **25**, 032102 (2018)
3. Sultana, S., Islam, S., Mamun, A.A., Schlickeiser, R.: Oblique propagation of ion-acoustic solitary waves in a magnetized plasma with electrons following a generalized distribution function. Phys. Plasmas **26**, 012107 (2019)
4. Debnath, D., Bandyopadhyay, A., Das, K.P.: Ion acoustic solitary structures in a magnetized nonthermal dusty plasma. Phys. Plasmas **25**, 033704 (2018)
5. Washimi, H., Taniuti, T.: Propagation of ion-acoustic solitary waves of small amplitude. Phys. Rev. Lett. **17**, 996 (1996)
6. Obregon, M.A., Stepanyants, Y.A.: Oblique magneto-acoustic solitons in a rotating plasma. Phys. Lett. A **249**, 315–323 (1998)
7. Mamun, A.A., Shukla, P.K.: Spherical and cylindrical dust acoustic solitary waves. Phys. Lett. A **290**, 173–175 (2001)
8. Huang, G., Velarde, M.G.: Head-on collision of two concentric cylindrical ion acoustic solitary waves. Phys. Rev. E **53**, 2988–2991 (1996)
9. Schuck, P.W., Bonnell, J.W., Pincon, J.-L.: Properties of lower hybrid solitary structures: a comparison between space observations, a laboratory experiment, and the cold homogeneous plasma dispersion relation. J. Geophys. Res. **109**, A01310 (2004)

10. El-Labany, S.K., Sabry, R., El-Taibany, W.F., Elghmaz, E.A.: Propagation of three-dimensional ion-acoustic solitary waves in magnetized negative ion plasmas with nonthermal electrons. Phys. Plasmas **17**, 042301 (2010)
11. Seadawy, A.R.: Three-dimensional nonlinear modified Zakharov-Kuznetsov equation of ion-acoustic waves in a magnetized plasma. Comput. Math. Appl. **71**, 201–212 (2016)
12. Biswas, A., Zerrad, E.: Solitary wave solution of the Zakharov-Kuznetsov equation in plasmas with power law nonlinearity. Nonlinear Anal. RWA **11**, 3272–3274 (2010)
13. Li, B., Chen, Y., Zhang, H.: Exact travelling wave solutions for a generalized Zakharov-Kuznetsov equation. Appl. Math. Comput. **146**, 653–666 (2003)
14. El-Taibany, W.F., El-Bedwehy, N.A., El-Shamy, E.F.: Three-dimensional stability of dust-ion acoustic solitary waves in a magnetized multicomponent dusty plasma with negative ions. Phys. Plasmas **18**, 033703 (2011)
15. Mandi, L., Mondal, K.K., Chatterjee, P.: Analytical solitary wave solution of the dust ion acoustic waves for the damped forced modified Korteweg-de Vries equation in q-nonextensive plasmas. Eur. Phys. J. Special Topics **228**, 2753–2768 (2019)
16. Pakzad, H.R.: Ion acoustic solitary waves in plasma with nonthermal electron and positron. Phys. Lett. A **373**, 847–850 (2009)
17. Dev, A.N., Deka, M.K., Kalita, R.K., Sarma, J.: Effect of non-thermal electron and positron on the dust ion acoustic solitary wave in the presence of relativistic thermal magnetized ions. Eur. Phys. J. Plus **135**, 843 (2020)
18. Chawla, J.K., Singhadiya, P.C., Tiwari, R.S.: Ion-acoustic waves in magnetised plasma with nonthermal electrons and positrons. Pramana J. Phys. **94**, 13 (2020)
19. Seadawy, A.R., Lu, D.: Ion acoustic solitary wave solutions of three-dimensional nonlinear extended Zakharov-Kuznetsov dynamical equation in a magnetized two-ion-temperature dusty plasma. Results Phys. **6**, 590–593 (2016)

10. El-Labany, S.K., Sabry R., El-Taibany W.F., Elghmaz, E.A.: Propagation of three-dimensional ion-acoustic solitary waves in magnetized negative ion plasmas with nonthermal electrons. Phys. Plasmas 17, 042301 (2010)
11. Seadawy, A.R.: Three-dimensional nonlinear modified Zakharov–Kuznetsov equation of ion-acoustic waves in a magnetized plasma. Comput. Math. Appl. 71, 201–212 (2016)
12. Biswas, A., Zerrad, E.: Solitary wave solution of the Zakharov–Kuznetsov equation in plasmas with power law nonlinearity. Nonlinear Anal. RWA 11, 3272–3274 (2010)
13. Li, B., Chen, Y., Zhang, H.: Exact travelling wave solutions for a generalized Zakharov–Kuznetsov equation. Appl. Math. Comput. 146, 653–666 (2003)
14. El-Labany, S.K., El-Taibany, W.F., [illegible] three-dimensional stability of dust-ion acoustic solitary waves in a magnetized [illegible] plasma with negative ions. Phys. Plasmas 18, 073701 (2011)
15. Mandi, L., Mondal, K.K., Chatterjee, P.: Analytical solitary wave solution of the dust ion acoustic waves for the damped forced modified Korteweg–de Vries equation in q-nonextensive plasmas. Eur. Phys. J. Special Topics 228, 2753–2768 (2019)
16. Rahman, [illegible]: Ion-acoustic solitary waves in plasmas with nonthermal electrons and positrons. Phys. Lett. A [illegible] (2008)
17. [illegible]
18. [illegible]
19. [illegible]

Maxwellian Multicomponent Dusty-Plasma with Fluctuating Dust Charges

Ridip Sarma

Abstract By employing a method known as $\left(\frac{G'}{G}\right)$ expansion method, we rekindle our study on the existences of soliton propagation in a complex plasma consisting of Maxwellian electrons, negative ions, positive ions and positrons, in the presence of dust charge fluctuation. The study involves the augmentation of KdV Equation and then Burger Equation derivable by usual reductive perturbation technique. In this paper we expect to evaluate some important observations which could be of interest in the scientific community. We have investigated our observations in the context of Auroral region, radial spokes of Saturn's ring system and Solar F Corona region.

Keywords Maxwellian multicomponent plasma · Dust charge variation · Nonlinear waves · Soliton and shock dynamics · G′/G method · Space plasmas

1 Introduction

The presence of negative ions in a plasma has been studied in a number of papers. We mention here some of these papers. Tagare and Reddy [1] investigated the ionic temperature effect on ion-acoustic solitons in a warm plasma consisting of positive ions and negative ions and non-isothermal electrons on the basis of a modified KdV equation derived by using the reductive perturbation method. Nakamura et al. [2] experimentally investigated Ion–Acoustic waves in a multicomponent plasma with negative ions. Earlier D'Angelo et al. [3] studied theoretically ion-acoustic waves in a multicomponent plasma with negative ions.

Mamun and Shukla [4] studied the role of negative ions on the charging of dust grains. They have considered two models for negative ions: streaming negative ions and Boltzmannian negative ion distribution. They have found that the effects of negative ion number density, negative ion charge and negative ion streaming speed,

R. Sarma (✉)
Department of Mathematics, Assam Don Bosco University, Tapesia, Sonapur, Assam 782402, India
e-mail: ridip.sarma@dbuniversity.ac.in

S. Banerjee and A. Saha (eds.), *Nonlinear Dynamics and Applications*,
Springer Proceedings in Complexity,
https://doi.org/10.1007/978-3-030-99792-2_23

significantly impact the dust grain charge. Oohura and Hatakeyama [5] have experimentally studied in the laboratory certain aspects of pair-ion (C_{60}^{+} and C_{60}^{-}) plasma. Adhikary et al. [6] experimentally studied in a dusty plasma, the propagation characteristics of rarefactive ion- acoustic solitary waves containing negative ions. They compared their experimental results with the theoretical results of KdV-Burgers equation. In a later paper, Adhikary [7] studied a dusty plasma composed of Boltzmannian distributed singly charged positive ions, singly charged negative ions, electrons, positrons and cold static negatively charged dust particles. He took into account viscous effect and found that viscosity plays an important role in dissipation of the Dust Ion Acoustic (DIA) shock wave propagating through it. In another paper, Adhikary et al. [8] presented an investigation of the properties of Dust Acoustic (DA) solitary wave propagation under the effect of non-thermal ions and trapped electrons in an adiabatic dusty plasma. The plasma is composed of electrons, singly charged positive and negative ions and charged dust particles. They derived a modified KdV equation and obtained a stationary analytical solution. From the study of the solution they observed that both the ions in the dusty plasma play a key role for the formation of both rarefactive and compressive DA solitary waves and that the ion concentration impacts the transformation of negative to positive potentials of the waves. Haider and Nahar [9] studied the nonlinear dynamics of Dust Ion Acoustic (DIA) solitary and shock waves in a plasma composed of positive and negative ions, oppositely charged stationary dust particles and superthermal electrons. There are many researchers [10–12], who studied four component dusty plasma.

We consider in this work a four-component plasma model with negatively charged dust. The constituents are negatively charged dust particles under the influence of Maxwellian distributed positive and negative ions, electrons and positrons. The manuscript is ordered as follows: Sect. 2 illustrates the basic equations governing the plasma dynamics. Sect. 3 describes the derivation of nonlinear equations. To study the existence and propagation of solitons and shocks, a $\left(\frac{G'}{G}\right)$ method [13] has been used. The speciality of the method is it is concise and straightforward and can be used to solve evolutionary equations directly without any requirement for initial/boundary conditions or initial trial function at the outset. Graphical representation includes specific choice of typical parameters of various astrophysical regions playing a vital role in different domains of cosmic dust-laden plasmas. Results and Discussions are in Sect. 4 analysing a comparative study of the different waves in various plasma spaces, followed by Conclusions in Sect. 5.

2 Basic Model Equations Governing Plasma Dynamics

The basic equations governing our plasma dynamics are written as [13, 14]

Equation of continuity

$$\frac{\partial n_d}{\partial t} + \nabla.(n_d u_d) = 0 \tag{1}$$

Equation of motion

$$\frac{\partial n_d}{\partial t} + \nabla.(n_d u_d) = 0 \tag{2}$$

along with the Poisson equation

$$\nabla^2 \varphi = -4\pi \sum q_\beta n_\beta z_\beta \tag{3}$$

where $\beta = i, n, e, p, d$ represents positive ions, negative ions, electrons, positrons and dust particles respectively. q_β represents $+e$ for positive ion, positron and $-e$ for an electron, negative ion and a dust particle. Here, dust particle's mass is represented by m_d, velocity by u_d and the charge density by n_d. Here, u_d is normalised by $c_i = (\frac{k_B T_{e0}}{m_i})^{\frac{1}{2}}$ (k_B represents Boltzmannian constant) and n_β represents the density of charge number, normalized with the equilibrium value $n_{\beta 0}$, where the equilibrium state is represented by suffix 0. The positive ion, negative ion, positron and the dust charge numbers are represented as $z_{i,n,p,d}$, where z_d is the number of electrons residing on a dust particle. φ is the electrostatic potential normalised by $(\frac{k_B T_{e0}}{e})$. The time variable t and space variable x are normalised respectively by $\omega_{pi}^{-1} = (\frac{m_i}{4\pi e^2 n_{i0}})$ and $_D = (\frac{k_B T_{e0}}{4\pi e^2 n_{i0}})^{\frac{1}{2}}$, known as inverse of plasma frequency and Debye length respectively.

We perturb the parameters as $f_\beta = f_{\beta 0} + \tilde{f}_\beta$ $(f = u, n, z)$, where the corresponding equilibrium value is $f_{\beta 0}$ and the perturbed value is $\tilde{f}_\beta$. We assume that electrons and ions are Maxwellian, and therefore we can write the Eq. (3) as

$$\nabla^2 \varphi = \alpha\varphi + \alpha' \varphi^2 + 4\pi e(z_{d0}\tilde{n}_d + \tilde{z}_d\, n_{d0} + \tilde{z}_d \tilde{n}_d) \tag{4}$$

where

$$\alpha = \frac{4\pi e^2}{K_B}\left(\frac{n_{e0}}{T_e} + \frac{n_{i0} z_i^2}{T_i} + \frac{n_{p0}}{T_p} + \frac{n_{n0} z_n^2}{T_n}\right) = \left({}^1/\lambda_D^2\right)$$

and $\alpha' = \frac{2\pi e^3}{{K_B}^2}\left(\frac{n_{e0}}{{T_e}^2} - \frac{n_{i0} z_i^3}{{T_i}^2} + \frac{n_{n0} z_n^3}{{T_n}^2} - \frac{n_{p0}}{{T_p}^2}\right)$

where $T_{i,e,n,p}$ represents the temperatures of positive ions, electrons, negative ions and positrons respectively.

When we immerse the dust particles in a plasma, the basic charging equation is written as

$$\frac{dQ_d}{dt} = \sum I_\gamma, \gamma = i, e, n, p \tag{5}$$

where I_γ is the current [15–17] generated on the dust grains by charged particle γ given by

$$I_i = \pi a^2 e \sqrt{\frac{8T_i}{\pi m_i}} n_i (1 - \frac{eq_d}{aT_i})$$

$$I_n = -\pi a^2 e \sqrt{\frac{8T_n}{\pi m_n}} n_n \left(1 + \frac{eq_d}{aT_n}\right)$$

$$I_e = -\pi a^2 e \sqrt{\frac{8T_e}{\pi m_e}} n_e exp(\frac{eq_d}{aT_e})$$

and

$$I_p = \pi a^2 e \sqrt{\frac{8T_p}{\pi m_p}} n_p exp(-\frac{eq_d}{aT_p})$$

Now following Jana et al. [15], and applying on our current model, we obtain

$$\frac{dQ_d}{dt} + \tilde{Q}_d = |I_{e0}|(\frac{\tilde{n}_i}{n_{i0}} + \frac{\tilde{n}_p}{n_{p0}} - \frac{\tilde{n}_n}{n_{n0}} - \frac{\tilde{n}_e}{n_{e0}}) \tag{6}$$

where

$$Q_d = Q_{d0} + \tilde{Q}_d \tag{7}$$

with Q_{d0} and $\tilde{Q}_d$ considered at equilibrium to be the dust charges and perturbed states respectively. Also, we have.

$$\eta = \left(|\mathrm{eI_{e0}}|/\mathrm{C}\right)\left(\frac{1}{\mathrm{k_B T_e}} + \frac{1}{\mathrm{k_B T_p}} + \frac{1}{\omega_{i0}} + \frac{1}{\omega_{n0}}\right), \tag{8}$$

where C is the capacitance of the dust grains, $\omega_{i0} = k_B T_i - e\varphi_{f0}$ and $\omega_{p0} = k_B T_p - e\varphi_{f0}$ where φ_{f0} is the floating potential.

Substituting in Eq. (6) the values of Eqs. (7) and (8), we get

$$\frac{\partial \tilde{z}_d}{\partial t} + \tilde{z}_d = P_1\varphi + P_2\varphi^2 \tag{9}$$

with

$$P_1 = |I_{e0}|/k_B(\frac{z_i}{T_i} + \frac{z_n}{T_n} + \frac{1}{T_e} + \frac{1}{T_p}) \tag{10}$$

$$P_2 = |I_{e0}|e/k_B^2(\frac{1}{{T_e}^2} + \frac{z_n^2}{T_n^2} - \frac{1}{{T_p}^2} - \frac{z_i^2}{T_i^2}) \tag{11}$$

3 Nonlinear Wave Equations: Derivation and Solution

3.1 *Korteweg De Vries Equation*

Following Washimi and Tanuiti [18], we apply the reductive perturbation method to the basic equations for studying soliton dynamics. The stretching coordinates of the model are:

$$\zeta = \sqrt{\in}(x - v_0 t), \tau = \sqrt[3]{\in} t \tag{12}$$

where, $\in$ measures the small expansion parameter.

We use the perturbed expansion:

$$(S = ud,\ nd,\ \varphi) = \sum_{N=0}^{\infty} \in^N S^{(N)}, \tag{13}$$

having $S^{(0)} = 0$ for φ and u_d.

For the first order in $\in$, we use reductive perturbation scheme and stretching coordinates in the basic equations to get the following relations.

$$u_{d,x}^{(1)} = -\frac{(ez_{d0})}{(m_d v_0)} \varphi^{(1)}, \tag{14}$$

$$n_d^{(1)} = -\frac{\left(\alpha + \frac{p}{\eta}\right)}{4\pi e z_{d0}} \varphi^{(1)}, \tag{15}$$

$$z_d^{(1)} = \frac{P_1}{\ } \varphi^{(1)} \tag{16}$$

Now, the result for phase velocity comes out to be

$$v_0^2 = \omega_{pd}^2 \left(\alpha + \frac{P}{\ }\right)^{-1} \tag{17}$$

where ω_{pd} is the dust charge frequency defined as

$$\omega_{pd} = \left(4\pi e^2 z_{d0}^2 n_{d0}/m_d\right)^{\frac{1}{2}} \tag{18}$$

$$P = 4\pi e n_{d0} P_1 \tag{19}$$

The subsequent higher order term in $\in$ gives the following equations

$$\frac{\partial n_d^{(1)}}{\partial \tau} - v_0 \frac{\partial n_d^{(2)}}{\partial \zeta} + n_{d0} \frac{\partial u_{d,x}{}^{(2)}}{\partial \zeta} + \frac{\partial (n_d^{(1)} u_{d,x}^{(1)})}{\partial \zeta} = 0 \tag{20}$$

$$\frac{\partial \vec{u}_d^{(1)}}{\partial \tau} - v_0 \frac{\partial \vec{u}_d^{(2)}}{\partial \zeta} + u_{d,x}^{(1)} \frac{\partial \vec{u}_{d,x}^{(1)}}{\partial \zeta} = \frac{e}{m_d} (z_{d0} \frac{\partial \varphi^{(2)}}{\partial \zeta} \hat{x} + z_d^{(1)} \frac{\partial \varphi^{(1)}}{\partial \zeta} \hat{x}) \tag{21}$$

where along x-direction, the unit vector is $\hat{x}$.

$$\frac{\partial^2 \varphi^{(1)}}{\partial \zeta^2} = \alpha \varphi^{(2)} + \alpha'(1)2 + 4\pi e \left(z_{d0} n_d^{(2)} + z_d^{(2)} n_{d0} + z_d^{(1)} n_d^{(1)} \right) \tag{22}$$

$$z_d^{(2)} = \frac{P_1}{} \varphi^{(2)} + \frac{P_2}{} \left(\varphi^{(1)}\right)^2 \tag{23}$$

Applying the results of the order ε from Eqs. (14) to (17) in Eqs. (20) to (23) along with some mathematical simplifications, the K-dV equation is obtained as

$$\frac{\partial \varphi^{(1)}}{\partial \tau} + A \varphi^{(1)} \frac{\partial \varphi^{(1)}}{\partial \zeta} + B \frac{\partial^3 \varphi^{(1)}}{\partial \zeta^3} = 0 \tag{24}$$

where the coefficients A and B are

$$A = \frac{3 v_0}{8\pi e z_{d0} n_{d0}} \frac{P}{} - \left(\alpha' + \frac{P'}{} \right) \frac{v_0^3}{\omega_{pd}^2} - \frac{3 e z_{d0}}{2 m_d v_0} \tag{25}$$

$$B = \frac{v_0^3}{2 \omega_{pd}^2} \tag{26}$$

where

$$P' = 4\pi e n_{d0} P_2$$

Applying the $(\frac{G'}{G})$ method, and the transformation $\varphi^{(1)}(\zeta, \tau) = \psi(\xi)$, with $\xi = \zeta - V\tau$ where the frame is moving with the velocity V having boundary conditions

$$(i)\,\varphi^{(1)} \to 0,\ (ii)\,\frac{d\varphi^{(1)}}{d\xi} \to 0,\ (iii)\,\frac{d^2\varphi^{(1)}}{d\xi^2} \to 0\, as\, |\xi| \to \infty \tag{27}$$

enabling us to obtain the solution of the KdV equation as.
$\psi(\xi) = \frac{3V}{A} sech^2(\frac{\sqrt{V}}{2\sqrt{B}}\xi)$, where $\xi = \zeta - V\tau$

$$Or\ \psi(\xi) = \psi_m sech^2\left(\frac{\xi}{\psi_w}\right),\ \text{where } \psi_m = \frac{3V}{A} \text{ and } \psi_w = 2\sqrt{\frac{B}{V}} \tag{28}$$

We leave out two other solutions yielded by the $(\frac{G'}{G})$ method as they are not relevant to the soliton character.

The coefficients A and B of the solution (Eq. 28) are functions of plasma parameters. Now to study the nature of the soliton, we consider some values of plasma parameters [19] as reported in Auroral region, Saturn's radial spokes and Solar F Corona astrophysical region.

It has to be noted that the following expressions are considered while writing the relation Eq. (25):

$$\frac{P}{-} = 4\pi r_d n_{d0}\left[\frac{(T_i + T_e + T_n + T_p)T_{eff}}{(T_i + T_n)(T_e + T_p + T_{eff})}\right]$$

$$\frac{P'}{-} = \frac{4\pi r_d n_{d0}}{k_B}\left[\frac{T_{eff}((T_i + T_n)^2 - (T_e + T_p)^2)}{(T_e + T_p)(T_i + T_n)^2((T_e + T_p) + T_{eff})}\right]$$

$$T_{eff} = (T_i + T_n) - \frac{e^2 Z_d}{K_B r_d}$$

where the values of $v_0, \alpha \& \alpha'$ have been used.

3.2 Burgers Equation

To study the propagation of shock waves, we take the help of pair of stretching coordinates, $\zeta = \in (x - v_0 t)$ and $= \in^2 t$, to the same set of governing Eqs. (1) to (3) and equating first order in$\in$, we obtain:

$$u_{d,x}^{(1)} = -\frac{(ez_{d0})}{(m_d v_0)}\varphi^{(1)}, \tag{29}$$

$$n_d^{(1)} = -\frac{\left(\alpha + \frac{p}{\eta}\right)}{4\pi e z_{d0}}\varphi^{(1)}, \tag{30}$$

$$z_d^{(1)} = \frac{P_1}{}\varphi^{(1)} \tag{31}$$

Now, the result for phase velocity comes out to be

$$v_0^2 = \omega_{pd}^2\left(\alpha + \frac{P}{}\right)^{-1} \tag{32}$$

with the dust plasma frequency defined as

$$\omega_{pd} = \left(4\pi e^2 z_{d0}^2 n_{d0}/m_d\right)^{\frac{1}{2}} \tag{33}$$

and

$$P = 4\pi e n_{d0} P_1 \tag{34}$$

However, for the next higher order of $\in$, we get-

$$\frac{\partial n_d^{(1)}}{\partial \tau} - v_0\frac{\partial n_d^{(2)}}{\partial \zeta} + n_{d0}\frac{\partial u_{d,x}^{(2)}}{\partial \zeta} + \frac{\partial(n_d^{(1)} u_{d,x}^{(1)})}{\partial \zeta} = 0 \tag{35}$$

$$\frac{\partial \vec{u}_d^{(1)}}{\partial \tau} - v_0\frac{\partial \vec{u}_d^{(2)}}{\partial \zeta} + u_{d,x}^{(1)}\frac{\partial \vec{u}_{d,x}^{(1)}}{\partial \zeta} = \frac{e}{m_d}\left(z_{d0}\frac{\partial \varphi^{(2)}}{\partial \zeta}\hat{x} + z_d^{(1)}\frac{\partial \varphi^{(1)}}{\partial \zeta}\hat{x}\right) \tag{36}$$

where along x-direction, $\hat{x}$ is the unit vector.

$$\frac{\partial^2 \varphi^{(1)}}{\partial \zeta^2} = \alpha\varphi^{(2)} + \alpha'\left(\varphi^{(1)}\right)^2 + 4\pi e\left(z_{d0}n_d^{(2)} + z_d^{(2)}n_{d0} + z_d^{(1)}n_d^{(1)}\right) \tag{37}$$

$$z_d^{(2)} = \frac{P_1}{}\varphi^{(2)} + \frac{P_2}{}\left(\varphi^{(1)}\right)^2 + \frac{v_0}{}\frac{\partial z_d^{(1)}}{\partial \zeta} \tag{38}$$

Applying the perturbed relations, with some mathematical simplifications, we derive the Burgers equation

$$\frac{\partial \varphi^{(1)}}{\partial \tau} + A\varphi^{(1)}\frac{\partial \varphi^{(1)}}{\partial \zeta} - C\frac{\partial^2 \varphi^{(1)}}{\partial \zeta^2} = 0 \tag{39}$$

where

$$A = \frac{3v_0}{8\pi e z_{d0} n_{d0}}\frac{P}{} - \left(\alpha' + \frac{P'}{}\right)\frac{v_0^3}{w_{pd}^2} - \frac{3e z_{d0}}{2m_d v_0} \tag{40}$$

$$C = \frac{v_0^4}{2w_{pd}^2}\frac{P}{2} \tag{41}$$

where $P' = 4\pi e n_{d0} P_2$.

Considering the transformation $\varphi(\zeta, \tau) = \varphi(\xi)$, where $= \zeta - V\tau$ and applying the $\left(\frac{G'}{G}\right)$ method, we obtain the shock wave solution

$$\varphi(\xi) = -\varphi_m(1 - \tanh(\frac{\xi}{\varphi_w})) \tag{42}$$

Here, $\varphi_m = \frac{V}{A}$, $\varphi_w = \frac{2C}{V}$ represents the amplitude and width of the shock structure respectively.

Now to study the nature of the soliton, we consider some values of plasma parameters [19] as reported in Auroral region, Saturn's radial spokes and Solar F Corona astrophysical region.

We use the following expressions to write the Eq. (40),

$$\frac{P}{-} = 4\pi r_d n_{d0}\left[\frac{(T_i + T_e + T_n + T_p)T_{eff}}{(T_i + T_n)\big(T_e + T_p + T_{eff}\big)}\right]$$

$$\frac{P'}{-} = \frac{4\pi r_d n_{d0}}{k_B}\left[\frac{T_{eff}\Big((T_i + T_n)^2 - \big(T_e + T_p\big)^2\Big)}{\big(T_e + T_p\big)(T_i + T_n)^2\big(\big(T_e + T_p\big) + T_{eff}\big)}\right]$$

$T_{eff} = (T_i + T_n) - \frac{e^2 Z_d}{K_B r_d}$ have been used.

4 Results and Discussions

Figure 1 exhibits plotting of the Auroral region of Ionosphere where the ion (positive and negative) temperature is taken as (I) $T_{i,n} = 2 \times 10^3 K$, (II) $T_{i,n} = 4 \times 10^3 K$ and (III) $T_{i,n} = 5 \times 10^3 K$. We consider the temperature of electron and positron to be fixed at $T_{e,p} = 10^4 K$. Figure 1a exhibits nonlinear coefficient A with varying dust density n_{d0}. We find that at lower density n_{d0}, there is fast increase in A, then at higher density slows down. Secondly, the rate of slowing down of curve (III) is more than curves (I) and (II) so that curve (III) intersects curve (II) at P and gradually comes closer to curve (I) and may intersect it at a very high density. Thirdly, the rarefactive and compressive regions show up at all the temperatures.

Figure 1b exhibits shock waves, showing that nonlinear term A drops fastly at lower density and then at higher density slows down. However, the curve (I) goes farthest down intersecting the curve (II) at Q and the curve (III) at R. Secondly, the curve (II) slows down more than the curve (III) and meets it. Thirdly, in the density

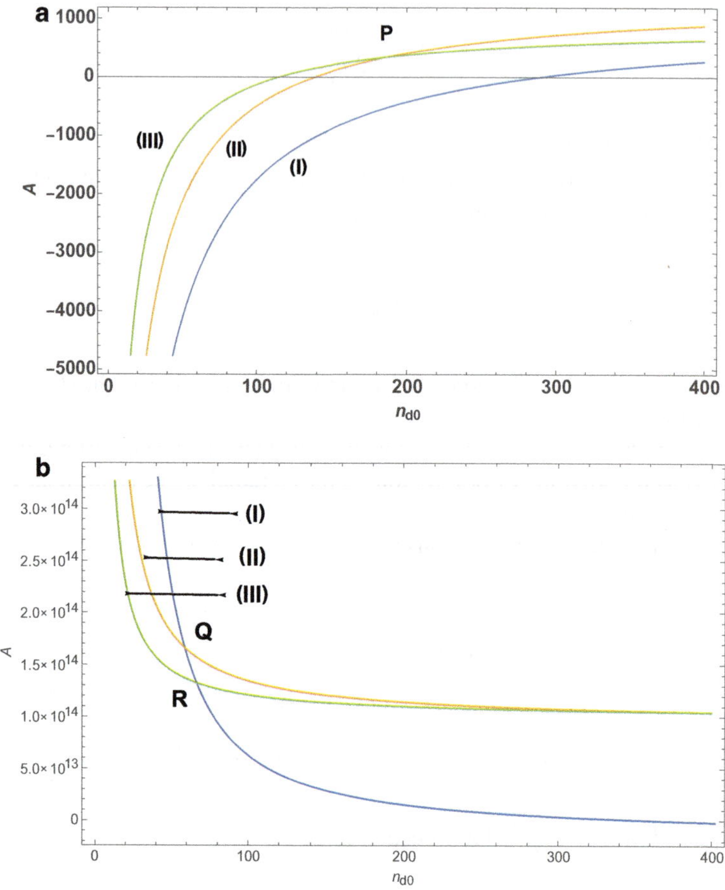

Fig. 1 **a** Plotting of coefficient of nonlinearity A with varying density of dust n_{d0} for solitons found in the Auroral region of Ionosphere where the positive and negative ions temperature is taken as (I) $T_{i,n} = 3 \times 10^3 K$, (II) $T_{i,n} = 5 \times 10^3 K$ and (III) $T_{i,n} = 7 \times 10^3 K$.We have considered $T_{e,p} = 10^4 K$ as the value for electron and positron temperature. **b** Plotting of coefficient of nonlinearity A with varying density of dust n_{d0} for shocks found in the Auroral region of Ionosphere where the positive and negative ions temperature is taken as (I) $T_{i,n} = 3 \times 10^3 K$, (II) $T_{i,n} = 5 \times 10^3 K$ and (III) $T_{i,n} = 7 \times 10^3 K$.We have considered $T_{e,p} = 10^4 K$ as the value for electron and positron temperature.

range under consideration, remembering the comment after Eq. (42), (II) and (III) are in the rarefactive region but (I) is mostly in the rarefactive region but just dips into the compressive region at the higher density end.

If we compare Fig. 1a, b we first observe that with increase in n_{d0}, A rises for solitons and on the other hand A drops with n_{d0} in case of shock waves. We also observe that there is some similar character like at lower density value, A varies fast but at higher values of density, slows down. In Fig. 1a, all the curves are partly in the compressive region and partly in the rarefactive region, in Fig. 1b, according to the comment after Eq. (42), the curve (II) and curve (III) are in the rarefactive region but the curve (I) is mostly in the rarefactive region and just dips in the compressive region at the higher density end. Finally, intersections of curves occur for both types of waves.

Figure 2 exhibits the Saturn's radial spokes where the pair of ion (positive and negative) temperature is taken as (I) $T_{i,n} = 1 \times 10^3 K$, (II) $T_{i,n} = 2 \times 10^3 K$, and

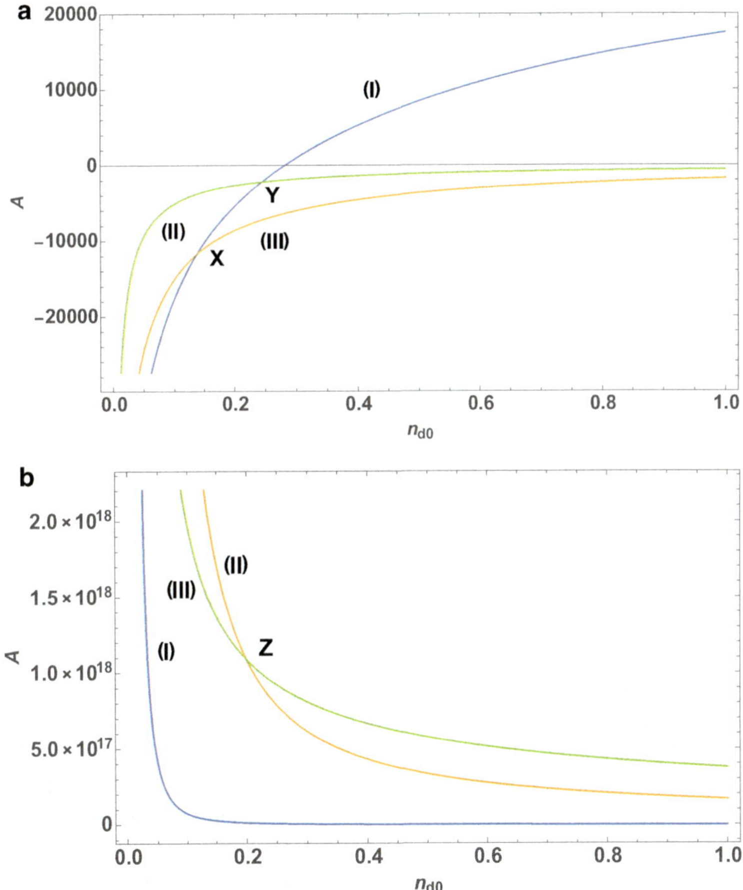

Fig. 2 **a** Plotting of coefficient of nonlinearity A with varying density of dust n_{d0} for solitons found in the Saturn's radial spokes where the positive and negative ions temperature is taken as (I) $T_{i,n} = 1 \times 10^3 K$, (II)$T_{i,n} = 2 \times 10^3 K$ and (III) $T_{i,n} = 4 \times 10^3 K$ We have considered $T_{e,p} = 2{*}10^4 K$ as the value for electron and positron temperature. **b** Plotting of coefficient of nonlinearity A with varying density of dust n_{d0} for shocks found in the Saturn's radial spokes where the positive and negative ions temperature is taken as (I) $T_{i,n} = 1 \times 10^3 K$, (II)$T_{i,n} = 2 \times 10^3 K$ and (III) $T_{i,n} = 4 \times 10^3 K$. We have considered $T_{e,p} = 2{*}10^4 K$ as the value for electron and positron temperature.

(III) $T_{i,n} = 4 \times 10^3 K$. We have considered $2 \times 10^4 K$ as fixed electron and positron temperature. We can observe for Fig. 2a that for lower density A rise fast and and slows down at higher values of density. However, the curve (I) rises higher than the curve (II) and curve (III) so that the curve (I) intersects the curve (II) at X and the curve (III) at Y. Secondly, the curve (I) has both rarefactive and compressive regions. But the curve (II) and curve (III) have only the rarefactive region.

In Fig. 2b for shock waves, at lower density A falls fast and at higher density slows down. However, the curve (I) falls faster than the curve (II) and curve (III) and then curve (I) remains unchanged with change of density. Secondly, the curve (II) falls faster than curve (III) so that curve (II) intersects curve (III) at Z. Thirdly, as per the comment after Eq. (42), the curve (II) and curve (III) have only the rarefactive region but the curve (I) lies first in the rarefactive region and then in the borderline between compressiveness and rarefactivity.

If we compare Fig. 2a, b we first observe that with increase in n_{d0}, A rises for solitons and on the other hand A drops with n_{d0} in case of shock waves. We also observe that there is some similar character like at lower density value, A varies fast but at higher values of density, slows down. Both the curve (II) and (III) are in the rarefactive region (vide comment after Eq. (42)) in both figures. But whereas the curve (I) has both compressive and rarefactive regions in Fig. 2a. the curve (I) is in the rarefactive region at lower density and then lies at the borderline between the rarefactiveness and compressiveness.

In Fig. 3a, we choose the Solar F Corona region having the ion (positive and negative) temperature as (I) $T_{i,n} = 2 \times 10^4 K$, (II) $T_{i,n} = 4 \times 10^4 K$, (III) $T_{i,n} = 8 \times 10^4 K$. We consider the electron temperature and positron temperature to be fixed at $T_{e,p} = 8 \times 10^5 K$. For solitons all the curves (I), (II) and (III) slowly and steadily rise with no tendency to intersect. Secondly, corresponding to any dust density, (I) < (II) < (III). Thirdly, all the curves are in the rarefactive region.

Figure 3b exhibits shock waveform in the Solar F Corona region, the positive and negative ions temperature is taken as (I) $T_{i,n} = 2 \times 10^4 K$, (II) $T_{i,n} = 4 \times 10^4 K$, (III) $T_{i,n} = 8 \times 10^4 K$. We have fixed the electron and positron temperature as $T_{e,p} = 8 \times 10^5 K$. For shock waves, we note first that all the curves for (I), (II), and (III) at lower density falls fast and then at higher density slows down. Secondly, at higher density they show a tendency to come closer to one another and may intersect one another at still higher density. Thirdly, at any density, (II) > (I) > (III). Finally, according to the comment after Eq. (42), all the curves are in the rarefactive region.

We conclude with a comment on the points of intersection P, Q, R, X, Y, and Z in Figs. 1a, b and 2a, b. At such a point, say at P, corresponding to the same dust density which we call a "critical dust density", we have (II) = (III) corresponding to two different temperatures. In other words, at this dust density, the electrostatic potential is the same. This seems to be an interesting role played by the charged dust density!

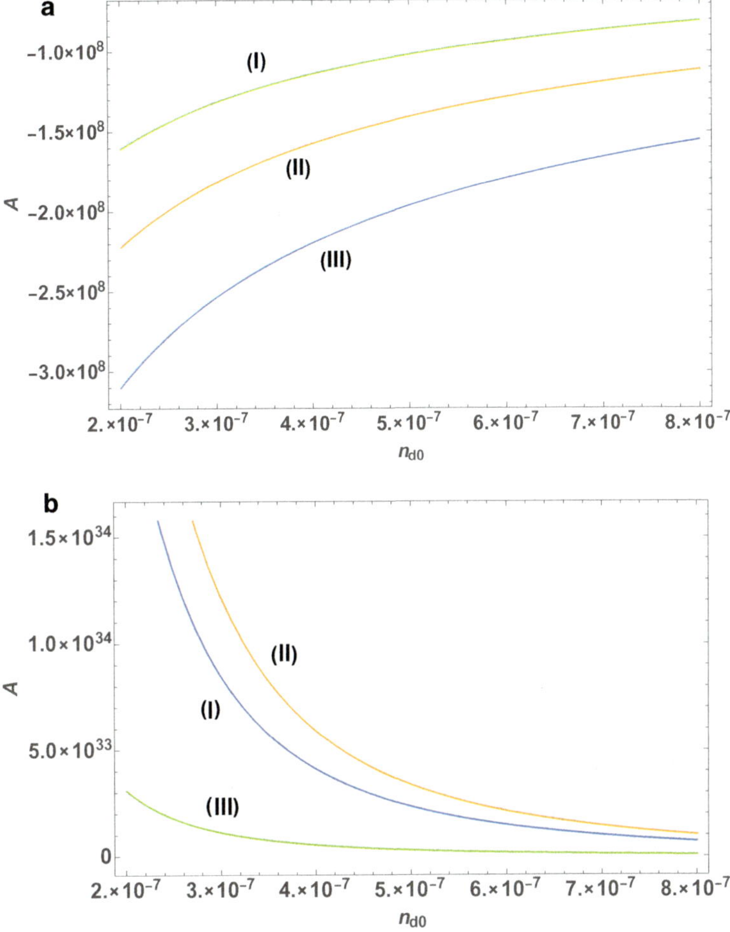

Fig. 3 **a** Plotting of coefficient of nonlinearity A with varying density of dust n_{d0} for solitons found in the Solar Corona where the positive and negative ions temperature is taken as (I) $T_{i,n} = 2\times10^4 K$, (II)$T_{i,n} = 4\times10^4 K$ and (III) $T_{i,n} = 8\times10^4 K$. We have considered $T_{e,p} = 8{*}10^5 K$ as the value for electron and positron temperature. **b** Plotting of coefficient of nonlinearity A with varying density of dust n_{d0} for shocks found in the Solar Corona where the positive and negative ions temperature is taken as (I) $T_{i,n} = 2\times10^4 K$, (II)$T_{i,n} = 4\times10^4 K$ and (III) $T_{i,n} = 8\times10^4 K$. We have considered $T_{e,p} = 8{*}10^5 K$ as the value for electron and positron temperature.

5 Conclusion

In the present paper, we have conducted a systematic study of the effects of nonlinear wave dynamics and dust density in multicomponent plasma on the formation of solitons and shockwaves along with consideration of dusty plasma in different regions and Boltzmann distributed electrons and ions. It is seen that compared to earlier observations, the presence of Maxwellian charges presents different natures of soliton. The present literature exhibited a plasma model to know the effect of Maxwellian force on nonlinear waves with the expectation of new findings. Here, in this paper, by employing reductive perturbation method and the unified method of $\left(\frac{G'}{G}\right)$ expansion method we have derived the salient features of solitons and shockwaves. It is shown that for any dusty plasma model, there exists a critical dust density,n_{dc} at which it is not possible to obtain soliton like structures and at which the nonlinear coefficient becomes zero. It has been observed that at the neighbourhood of the critical dust density both compressive and rarefactive solitons grow in amplitude under the action of Maxwellian force exhibiting generation of high energy which in turn finds the phenomena of soliton radiation.

References

1. Tagare, S.G., Reddy, R.V.: Effect of ionic temperature on ion-acoustic solitons in a two-ion warm plasma consisting of negative ions and non-isothermal electrons. Plasma Phys. Control. Fus. **29**(5), 671–676 (1987)
2. Nakamura, Y., Odagiri, T., Tsukabayashi, I.: Ion acoustic waves in a multicomponent plasma with negative ions. Plasma Phys. Control. Fus. **39**, 105–115 (1997)
3. D'Angelo, N.D., Goeler, S.V., Ohe, T.: Propagation and damping of ion waves in a plasma with negative ions. Phys. Fluids **9**, 1605–1606 (1966)
4. Mamun, A.A., Shukla, P.K.: Solitary potentials in cometary dusty plasmas. Geophys. Res. Letts. **29**(18), 1870–1873 (2002)
5. Oohura, W., Hatakeyama, R.: Basic studies of the generation and collective motion of pair-ion plasmas. Phys. Plasmas **14**, 055704–055710 (2007)
6. Adhikary, N.C., Deka, M.K., Bailung, H.: Observation of rarefactive ion acoustic solitary waves in dusty plasma containing negative ions. Phys. Plasmas **16**, 063701–063707 (2009)
7. Adhikary, N.C.: Effect of viscosity on dust-ion acoustic shock wave in dusty plasma with negative ions. Phys. Lett. A **376**, 1460–1464 (2012)
8. Sarma, R., Misra, A.P., Adhikary, N.C.: Nonlinear ion acoustic solitary waves in an electron positron ion plasma with relativistic positron beam. Chin. Phys. B **27**(10), 105207–105212 (2018)
9. Haider, M.M., Nahar, A.: Dust-ion-acoustic solitary and shock structures in multi-ion plasmas with nonthermal electrons, Z. Naturforsch, **72**(7), 627–635 (2017)
10. Ahmad, M., Ahmad, Z., Farooq, M., Gul, A.: Kinetic treatment of acoustic waves in a four component dusty plasma. Cont. Plasma Phys. **59**(3), 367–380 (2019)
11. Jahan, S., Mannan, A., Chowdhury, N.A., Mamun, A.A.: Kinetic treatment of acoustic waves in a four component dusty plasma. Plasma Phys. Reports **46**, 90 (2020)
12. Raghavi, K., Kavitha L.: Modulational instability analysis of four component dusty plasma system. IEEE Trans. Plasma Sci. https://doi.org/10.1109/TPS.2021.3122447
13. Das, G.C., Sarma, R.: New method for rekindling the nonlinear solitary waves in Maxwellian complex space plasma. Phys. Plasmas **25**, 043703–043711 (2018)
14. Sarma, R., Das, G.C., Das, R., Adhikary, N.C.: On the nonilinear solitary and shock waves in Maxwellian multicomponent space plasma. Phys. Plasmas **25**, 073704–073710 (2018)
15. Jana, M.R., Sen, A., Kaw, P.K.: Collective effects due to charge fluctuation dynamics in a dusty plasma. Phys. Rev. E **48**, 3930–3933 (1993)
16. Bhatt, J.R., Pandey, B.P.: Self consistent charge dynamics and collective modes in a dusty plasma. Phys. Rev. E **50**, 3980–3983 (1994)

17. Shukla, P.K., Mamun, A.A.: Introduction to dusty plasma physics. IOP, Bristol (2002)
18. Washimi, H., Tanuiti, T.: Propagation of ion acoustic solitary waves of small amplitude. Phys. Rev. Lett. **17**, 996–998 (1966)
19. Mendis, D.A.: Progress in the study of dusty plasmas. Plasma Sources Sci. Tech. **11A**, 219–228 (2002)

Effect of Polarization Force on Dust-Acoustic Solitary and Rogue Waves in (r, q) Distributed Plasma

Manveet Kaur, Sunidhi Singla, and Nareshpal Singh Saini

Abstract In this investigation, dust-acoustic solitary and rogue waves are analysed by including the role of polarization force in a dusty plasma in the presence of ions obeying (r, q) distribution velocity. The (r, q) distributed ions lead to change in the expression of polarization force which subsequently modifies the nonlinear structures. Using reductive perturbation technique, the Kadomstev–Petviashvili equation and nonlinear Schrödinger equation (NLSE) are formulated. From the solution of KP equation, we have studied the characteristics of small amplitude dust-acoustic structures under transverse perturbations. We have also presented the stability of the solitary wave solutions of KP equation. Further, from the rational solutions of NLSE, we have analysed the characteristics of dust-acoustic rogue waves by varying different plasma parameters. It is emphasized that all different physical parameters have great influence on the characteristics of dust-acoustic solitary waves and different order rogue waves.

Keywords KP equation · NLSE · Polarization force · (r, q) distribution

1 Introduction

Over the last five decades, researchers have focused on the study of various nonlinear structures (such as solitons, rogue, shocks waves, double layers, and so on) and instabilities in various types of plasmas utilising Maxwellian and non-Maxwellian velocity distributions. Numerous nonlinear equations and energy balance equations have been constructed in the framework of perturbative and non-perturbative methodologies respectively to explore such nonlinear structures (with small or high amplitudes) in space/astrophysical plasmas. These nonlinear equations' solutions have been utilised to explain nonlinear solitary formations in 1-D, 2-D, and 3-D under the influence of different plasma parameters. The Kadomstev–Petviashvili (KP) [1]

M. Kaur (✉) · S. Singla · N. S. Saini
Department of Physics, Guru Nanak Dev University, Amritsar 143005, India
e-mail: manveetsandhu7@gmail.com

S. Banerjee and A. Saha (eds.), *Nonlinear Dynamics and Applications*,
Springer Proceedings in Complexity,
https://doi.org/10.1007/978-3-030-99792-2_24

equation is one of the nonlinear equations used to describe nonlinear systems in two dimensions. It is an extension of the Korteweg-de Vries (KdV) equation. Numerous researchers have studied the characteristics of nonlinear structures by employing KP equation in different type of plasma environments [2–9].

Numerous research articles published over the last three decades have reported the various physical aspects that due to their existence in dusty plasma, the nonlinear structures viz., dust-acoustic waves (DAWs) and dust-ion acoustic waves (DIAWs) are modified. In an unmagnetized plasma, one of such aspect is the forces on charged particles such as electrostatic force, ion drag force, the net plasma pressure force and most importantly polarization force [10]. Although the polarization force is small in comparison to other forces of system but its effects are prominent for low frequency DAWs and incur in a dusty plasma due to deformation of Debye sheath around the particles. It is directed opposite to electric field, direction in which Debye length increases and alters with variation in dust size but independent of charge on the dust particles [11]. Several authors have investigated the role of polarization force on various nonlinear solitary structures utilising different distributions (Maxwellian and non-Maxwellian) in the context of the reductive perturbation methodology and the Sagdeev method [12–22]. In a highly coupled dusty plasma, the influence of polarization force and temperature of dust on DA solitary and shock waves was studied by Mamun and Ashrafi [12]. They observed that the properties of nonlinear structures are highly modified. In an inhomogeneous unmagnetized dusty plasma, Asaduzzaman and Mamun [14] analysed the effect of polarization force owing to dust density inhomogeneity on linear DA waves and reported that the dust grain charge inhomogeneity and grain size distribution along with density inhomogeneity altered the propagation properties of linear and nonlinear waves. Ashrafi et al. [15] investigated the polarization force for non-thermal and trapped ions, adiabaticity of electrons and ions. They explored that magnitude of the polarization force is significantly influenced in different dusty plasma situations. The effect of the polarization force on the dust-acoustic soliton energy was examined by Mayout et al. [16]. They explored that when net force acting on the grains decreases then DA soliton energy shows depletion.

The effect of polarization force and Cairns distributed ions on different DA nonlinear structures was illustrated by Singh and Saini [20]. They reported the influence of polarization parameter for existence domain of such kind of nonlinear structures. Further, the influence of nonextensively modified polarization force on interaction of DA shock waves in a magnetized dusty plasma was illustrated by Saini et al. [22]. They observed that change in values of parameters influence significantly the propagation properties of DA shocks and phase shifts evolved because of interaction of DA shocks.

Satellite data has revealed that particle distributions in different regions of space diverge significantly from Maxwellian distributions. The well-known nonthermal property of space plasmas reveals the presence of superthermal tails and flat tops at low energies. According to observation and theory, some space plasmas can be more successfully described by a generalized (r, q) distribution function. The major aspect of the (r, q) distribution function in space plasmas is that it depicts distributions ranging from flat-topped to kappa distributions in the limiting situation [23]. In general, (r, q) distribution function is written as follows [23, 24]:

$$f_{r,q}(v) = \frac{a_1}{\pi {b_1}^{3/2} {v_{th}}^{3/2}} \left[1 + \frac{1}{q-1} \left(\frac{v^2 - 2e\phi/m_e}{b_1(2T_e/m_e)^{r+1}}\right)\right]^{-q}, \quad (1)$$

$$\text{where} \quad a_1 = \frac{3\Gamma[q](q-1)^{3/(2+2r)}}{4\Gamma[q - \frac{3}{2+2r}]\Gamma[1 + \frac{3}{2+2r}]},$$

$$\text{and} \quad b_1 = \frac{3(q-1)^{\frac{-1}{(1+r)}}\Gamma[q - \frac{3}{2+2r}]\Gamma[\frac{3}{2+2r}]}{\Gamma[q - \frac{5}{2+2r}]\Gamma[\frac{5}{2+2r}]},$$

where the electrostatic potential is represented by ϕ, the electron temperature and mass are denoted as T_e and m_e respectively. The distribution contains two spectral indices, r and q, which regulate the flatness at lower energies and the tail at higher energies [23]. It must establish the $q > 1$ and $(r + 1) > 5/2$ as requirement [23]. The distribution reduces to a Maxwellian distribution when $r = 0$ and $q \to (\kappa + 1)$, and this becomes a generalized Lorentzian distribution, when $r = 0$ and $q \to (\kappa + 1)$ [23]. The generalized (r, q) distribution is suggested by Qureshi et al. [24], which has been studied extensively for various types of plasmas [23–26]. Shah et al. [23] explained the density depletions reported by the Freja and Viking satellites with a generalized (r, q) distribution, formerly illustrated with Cairns distribution function. Ali et al. [25] determined theoretically the potential on the surface of spherical dust in a multi-ion (r, q) distributed dusty plasma. Nasir et al. [26] studied the instability of currentless ion acoustic waves using Boltzmann-Vlasov kinetic model with generalized (r, q) distribution in electrons and ions plasma.

Due to importance of polarization force in dusty plasma and the concept of generalized (r, q) distribution, we have investigated the solitary and rogue waves under the effect of polarization force in a dusty plasma composed of Boltzmann distributed electrons and generalized (r, q) distributed ions. The manuscript is described as follows: The fluid model is presented in Sect. 2. The derivation of Kadomstev–Petviashvili (KP) equation and its solution has been described in Sect. 2.1. The derivation of NLSE and its solution are presented in Sect. 2.2. Parametric analysis is described in Sect. 2.3. Section 2.4 illustrates conclusions.

2 Fluid Model

We have considered a plasma having Maxwell Boltzmann distributed electrons, (r, q) distributed ions, polarization force and negatively charged dust particles as fluid. The expression for polarization force is [20]

$$F_p = -Z_d e F_R C_{r0}\left[1 - C_{r1}\left(\frac{e\phi}{K_B T_i}\right) + C_{r2}\left(\frac{e\phi}{K_B T_i}\right)^2\right]^{\frac{1}{2}} \tag{2}$$

where $F_R = \frac{Z_d e^2}{4K_B T_i \lambda_{Di0}}$, $\lambda_{Di0} = \left(\epsilon K_B T_i / n_i e^2 C_{r1}\right)^{\frac{1}{2}}$, $C_{r0} = (C_{r1} - 2C_{r2}\phi)$,

$$C_{r1} = \frac{(q-1)^{\frac{-1}{(r+1)}}\Gamma\left[\frac{1}{2(r+1)}\right]\Gamma\left[q - \frac{1}{2(r+1)}\right]}{2b_1\Gamma\left[\frac{3}{2(r+1)}\right]\Gamma\left[q - \frac{3}{2(r+1)}\right]},$$

$$C_{r2} = \frac{-(q-1)^{\frac{-2}{1+r}}(1+4r)\Gamma\left[\frac{-1}{2(1+r)}\right]\Gamma\left[q + \frac{1}{2(r+1)}\right]}{8{b_1}^2\Gamma\left[\frac{3}{2(r+1)}\right]\Gamma\left[q - \frac{3}{2(r+1)}\right]},$$

Equation (2) illustrates that polarization force expression is dependent upon (r, q) distributed ions and independent of role of electrons. It approaches to Maxwellian limit when $r = 0$ and $q \to \infty$ and if $r = 0$, $q \to (\kappa + 1)$, then the (r, q) distribution becomes kappa distribution. The normalized fluid equations are written as [20]:

$$\frac{\partial n_d'}{\partial t} + \frac{\partial (n_d' u_d')}{\partial x} + \frac{\partial (n_d' v_d')}{\partial y} = 0 \tag{3}$$

$$\frac{\partial u_d'}{\partial t} + u_d'\frac{\partial u_d'}{\partial x} + v_d'\frac{\partial u_d'}{\partial y} = F_p\frac{\partial \phi'}{\partial x} \tag{4}$$

$$\frac{\partial v_d'}{\partial t} + u_d'\frac{\partial v_d'}{\partial x} + v_d'\frac{\partial v_d'}{\partial y} = F_p\frac{\partial \phi'}{\partial y} \tag{5}$$

$$\frac{\partial^2 \phi'}{\partial x^2} + \frac{\partial^2 \phi'}{\partial y^2} = n_d' + n_e' - n_i' \tag{6}$$

The expressions for number density of (r, q) distributed ions and Maxwellian electrons in dimensionless form are written as [20, 23]:

$$n_i' = \mu_i(1 - C_{r1}\phi' + C_{r2}\phi'^2 - C_{r3}\phi'^3 + \ldots), \tag{7}$$

$$n_e' = \mu_e\left(1 + \theta_{ie}\phi' + \frac{{\theta_{ie}}^2\phi'^2}{2} + \frac{{\theta_{ie}}^3\phi'^3}{6}\right). \tag{8}$$

Also F_p in governing equations is expressed as

$$F_p = \left[1 - F_R(C_{r1} - 2C_{r2}\phi' + \frac{1}{2}{C_{r1}}^2\phi' - ...)\right] \tag{9}$$

where F_R is the polarization force parameter that describes the impact of negatively charged dust particles on (r, q) distributed ions. The number density of ions $n_i{}'$, electrons $n_e{}'$ and electric potential ϕ' are normalized as $n_e{}' = n_e/(Z_{d0}n_{d0})$, $n_i{}' = n_i/(Z_{d0}n_{d0})$ and $\phi' = (e\phi)/(K_B T_i)$. By applying these normalized values and using $\theta_{ie} = T_i/T_e$, $\mu_e = n_{e0}'/(Z_{d0}n_{d0}') = 1/(\mu_{ie}\text{-}1)$, $\mu_i = n_{i0}'/(Z_{d0}n_{d0}') = \mu_{ie}/(\mu_{ie}\text{-}1)$ and $\mu_{ie} = n_{i0}'/n_{e0}'$ is the ratio of number density of ions to electrons in equilibrium such that $\mu_e\text{-}\mu_i = -1$. After substituting the values of ion and electron number density in equation (6), Poisson equation become as a function of potential,

$$\frac{\partial^2\phi'}{\partial x^2} + \frac{\partial^2\phi'}{\partial y^2} = n_d{}' - \mu_i\left[1 - C_{r1}\phi' + C_{r2}\phi'^2 - C_{r3}\phi'^3 + ...\right] + \mu_e\left[1 + \theta_{ie}\phi' + \frac{{\theta_{ie}}^2\phi'^2}{2} + \frac{{\theta_{ie}}^3\phi'^3}{6} + ...\right], \tag{10}$$

For further analysis, we have dropped prime sign from the variables and other quantities for mathematical simplicity.

2.1 Derivation of Kadomstev Petviashvili (KP) Equation

For present fluid model having nonthermal ions under polarization force effect, we have used following stretching co-ordinates [7]:

$$\xi = \epsilon(x - Vt), \quad \tau = \epsilon^3 t \;\; \text{and} \;\; \eta = \epsilon^2 y \tag{11}$$

to derive KP equation, where ϵ is dimensionless quantity which describes role of weak nonlinearity, V represents the phase velocity of solitons. In this investigation, waves having weak dispersion as well as nonlinearity are considered with ω (or k) $<< 1$. Different physical quantities are varying slowly in space and very slowly in time. The stretching coordinates defined in Eq. 11 are essential components of reductive perturbation method for deriving KP equation. The expansion for depended variables is written as:

$$S = S_{d0} + \sum_{j=1}^{\infty}\epsilon^{2j}S_{dj},\; v_d = \sum_{j=1}^{\infty}\epsilon^{2j+1}v_{dj}. \tag{12}$$

The system variables n_d, u_d and ϕ are represented by S at a given position and time. The small deviation of these variables from equilibrium state is represented by S_{d0}, the values of system of variables n_{d0}, u_{d0} and ϕ_0 are 1, 0 and 0 respectively. Thus, applying these stretching co-ordinates and expansions in the set of modified dust model equations (3–5) and (10), yields different kinds of equations. Then, by collecting the lower order quantities of ϵ, we obtain the following:

$$n_{d1} = -\frac{1 - F_R C_{r1}}{V^2}\phi_1, u_{d1} = -\frac{1 - F_R C_{r1}}{V}\phi_1,$$

$$\frac{\partial v_{d1}}{\partial \xi} = \frac{1 - F_R C_{r1}}{V}\frac{\partial \phi_1}{\partial \eta} \quad \text{and} \quad V = \sqrt{\frac{1 - F_R C_{r1}}{C_3}}. \tag{13}$$

where $C_3 = \mu_i C_{r1} + \mu_e \theta_{ie}$. Similarly by collecting coefficients of higher power of ϵ, we obtain

$$\frac{\partial n_{d1}}{\partial \tau} - V\frac{\partial n_{d2}}{\partial \xi} + \frac{\partial (n_{d1}u_{d1})}{\partial \xi} + \frac{\partial u_{d2}}{\partial \xi} + \frac{\partial v_{d1}}{\partial \eta} = 0, \tag{14}$$

$$\begin{aligned}\frac{\partial u_{d1}}{\partial \tau} - V\frac{\partial u_{d2}}{\partial \xi} + u_{d1}\frac{\partial u_{d1}}{\partial \xi} &= 2F_R C_{r2}\phi_1\frac{\partial \phi_1}{\partial \xi} - \frac{F_R {C_{r1}}^2}{2}\phi_1\frac{\partial \phi_1}{\partial \xi} \\ &+ \frac{3F_R}{2}C_{r1}C_{r2}{\phi_1}^2\frac{\partial \phi_1}{\partial \xi} + (1 - F_R C_{r1})\frac{\partial \phi_2}{\partial \xi},\end{aligned} \tag{15}$$

$$\begin{aligned}\frac{\partial v_{d1}}{\partial \tau} - V\frac{\partial v_{d2}}{\partial \xi} + u_{d1}\frac{\partial v_{d1}}{\partial \xi} &= 2F_R C_{r2}\phi_1\frac{\partial \phi_1}{\partial \eta} - \frac{F_R {C_{r1}}^2}{2}\phi_1\frac{\partial \phi_1}{\partial \eta} \\ &+ \frac{3F_R}{2}C_{r1}C_{r2}{\phi_1}^2\frac{\partial \phi_1}{\partial \eta} + (1 - F_R C_{r1})\frac{\partial \phi_2}{\partial \eta},\end{aligned} \tag{16}$$

$$\frac{\partial^2 \phi_1}{\partial \xi^2} = n_{d2} + \mu_i C_{r1}\phi_2 - \mu_i C_{r2}{\phi_1}^2 + \mu_e \theta_{ie}\phi_2 + \frac{\mu_e {\theta_{ie}}^2}{2}{\phi_1}^2 \tag{17}$$

After long mathematical calculations in above different equations, we have obtained the KP equation as [7]

$$\frac{\partial}{\partial \xi}\left[\frac{\partial \phi}{\partial \tau} + A\phi\frac{\partial \phi}{\partial \xi} + B\frac{\partial^3 \phi}{\partial \xi^3}\right] + C\frac{\partial^2 \phi}{\partial \eta^2} = 0 \tag{18}$$

where

$$A = \frac{-V^3}{2(1 - F_R C_{r1})}\left[\frac{3(1 - F_R C_{r1})^2}{V^4} - \frac{2F_R C_2}{V^2} + \frac{F_R {C_{r1}}^2}{2V^2} + 3C_4\right], \tag{19}$$

$$B = \frac{V^3}{2(1 - F_R C_{r1})}, C = \frac{V}{2} \quad \text{and} \quad C_4 = -\mu_i C_{r2} + \mu_e\frac{{\theta_{ie}}^2}{2} \tag{20}$$

The nonlinear coefficient is represented by A, B represents the dispersion coefficient and C is higher order coefficient.

To find the steady solution of KP equation (18), we have transformed this equation with $\Lambda = \xi - \eta - u\tau$ as an appropriate transformation, where u represents the velocity of frame. By substituting this single variable transformation, we get KP equation in ordinary differential form. Thus, by integration with proper boundary conditions ($\phi(\Lambda), \phi'(\Lambda) \to 0$ as $\mid \Lambda \mid \to \infty$), we obtain

$$\phi = \phi_m sech^2\left(\frac{\Lambda}{w}\right), \tag{21}$$

where $\phi_m = 3\left(\frac{u-C}{A}\right)$ represents peak amplitude and $w = 2\sqrt{\frac{B}{u-C}}$ is width of the KP solitons.

2.2 The Derivation of NLSE and Its Solution

To explore the characteristics of dust-acoustic waves in the (r, q) distributed plasma under the impact of polarization force, we find the solutions of Eq. (18) by incorporating the following new variables as $\phi(\xi, \eta, \tau) = \phi(\varrho, \tau)$ and $\varrho = F\xi + G\eta - m\tau$. Here, F and G are direction cosines, so that $F^2 + G^2 = 1$. m is a which may be considered like the Mach number. Now, by substituting new variable in (18), we get:

$$\frac{\partial}{\partial\tau}\phi(\varrho,\tau) + FA\phi(\varrho,\tau)\frac{\partial}{\partial\varrho}\phi(\varrho,\tau) + F^3B\frac{\partial^3}{\partial\varrho^3}\phi(\varrho,\tau) = 0, \tag{22}$$

$m = \frac{G^2C}{F} = \frac{(1-F^2)C}{F}$. The solution of this equation is assumed as [27, 28]:

$$\phi(\varrho,\tau) = \sum_{n=1}^{\infty}\epsilon^n \sum_{l=-\infty}^{\infty} \phi_l{}^n(\varrho,\tau)exp\left[il(k\varrho - \omega\tau)\right], \tag{23}$$

we consider $\xi = \epsilon(\varrho - u_g\tau)$ and $t = \epsilon^2\tau$ as stretching coordinates, where u_g denotes the group velocity.

Now, consider that all perturbed states rely solely upon rapid scales (through the phase $(k\varrho - \omega\tau)$). Whereas, the prolonged scale (ξ, t) enter the argument of the lth harmonic amplitude $\phi_l{}^n$. If the condition $\phi_{-l}{}^n = \phi_l{}^{n*}$ (* defines complex conjugate) is satisfied only then $\phi(\varrho, \tau)$ will be real. The derivative operator are written as [27, 28]:

$$\frac{\partial}{\partial\varrho} \equiv \frac{\partial}{\partial\varrho} + \epsilon\frac{\partial}{\partial\xi}, \tag{24}$$

$$\frac{\partial}{\partial\tau} \equiv \frac{\partial}{\partial\tau} - u_g\epsilon\frac{\partial}{\partial\xi} + \epsilon^2\frac{\partial}{\partial t}. \tag{25}$$

For $l = 1, n = 1$, the different expressions yields as $\omega = -F^3k^3B$, and from $l = 1$, $n = 2$, we find the group velocity is $u_g = -3F^3k^2B$. The zeroth harmonic with $l = 0, n = 2$ gives

$$\phi_2^{(0)} = \left(\frac{A}{u_g}\right) \mid \phi_1^{(1)} \mid^2. \tag{26}$$

From second harmonic, we get

$$\phi_2^{(2)} = \left(\frac{A}{6F^2k^2B}\right) \mid \phi_1^{(1)} \mid^2. \tag{27}$$

From $n = 3, l = 1$, and doing rigorous analytical calculations, we get the NLSE as [29]

$$i\frac{\partial\phi}{\partial t} + \frac{\mathcal{P}}{2}\frac{\partial^2\phi}{\partial\xi^2} + \mathcal{Q} \mid \phi \mid^2 \phi = 0, \tag{28}$$

for simplicity $\phi_1^{(1)} = \phi$. $\mathcal{Q} = \frac{A^2F^2}{\mathcal{P}}$ represents nonlinear coefficient and $\mathcal{P} = 6F^3Bk$ denoted as dispersion coefficient. The different solutions of Eq. (28) are determined respectively as [30]

$$\mid \phi_1(\xi, t) \mid = \left(\frac{4(1+2it)}{1+4t^2+4\frac{\xi^2}{\mathcal{P}}} - 1\right)\frac{e^{it}}{\sqrt{\mathcal{Q}}}, \tag{29}$$

$$\mid \phi_2(\xi, t) \mid = \sqrt{\frac{\mathcal{P}}{\mathcal{Q}}}\left(1 + \frac{m_1 + im_2}{m_3}\right)e^{(i\mathcal{P}t)}, \tag{30}$$

where, $m_1 = \left[\frac{3}{8} - \frac{\xi^4}{2} - \frac{3\xi^2}{2} - 6(\mathcal{P}\xi t)^2 - 10(\mathcal{P}t)^4 - 9(\mathcal{P}t)^2\right]$,
$m_2 = -\mathcal{P}t\left(-\frac{15}{4} + \xi^4 - 3\xi^2 + 4(\mathcal{P}\xi t)^2 + 4(\mathcal{P}t)^4 + 2(\mathcal{P}t)^2\right)$ and
$m_3 = \frac{3}{32} + \frac{\xi^6}{12} + \frac{\xi^4}{8} + \frac{1}{2}\xi^4(\mathcal{P}t)^2 + \frac{9\xi^2}{16} + \xi^2(\mathcal{P}t)^4 - \frac{3(\mathcal{P}\xi t)^2}{2} + \frac{2(\mathcal{P}t)^6}{3} + \frac{9(\mathcal{P}t)^4}{2} + \frac{33(\mathcal{P}t)^2}{8}$.
$\mid \phi_1(\xi, t) \mid$ ($\mid \phi_2(\xi, t) \mid$) describes the characteristics of first (second) order rogue waves.

2.3 *Parametric Analysis*

The role of variety of physical parameters such as polarization parameter (via F_R), ratio of number density of ions to electrons (via μ_{ie}), ratio of temperature of ions to electrons (via θ_{ie}), spectral indices r, and q on properties of DA solitary and rogue waves. For numerical analysis, we have considered the data from the planetary rings [31]. The parametric ranges are $n_{e0} = 4 \times 10^7, n_d = 10^7, n_{i0} = 5 \times 10^7, Z_d = 3 \times 10^3$, $F_R = 0 - 0.14$.

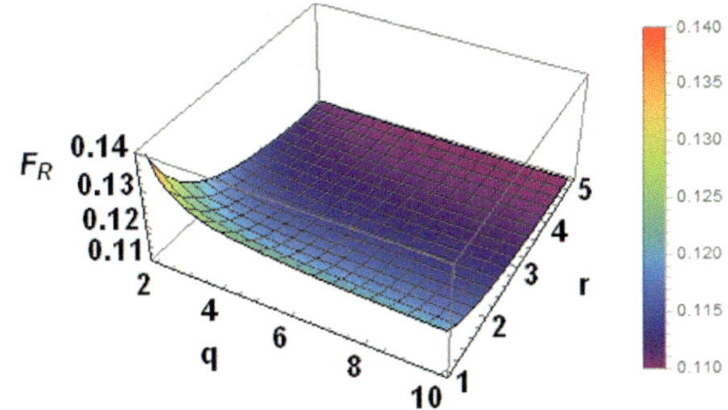

Fig. 1 The polarization force parameter (F_R) versus spectral indices (**a**) r; (**b**) q

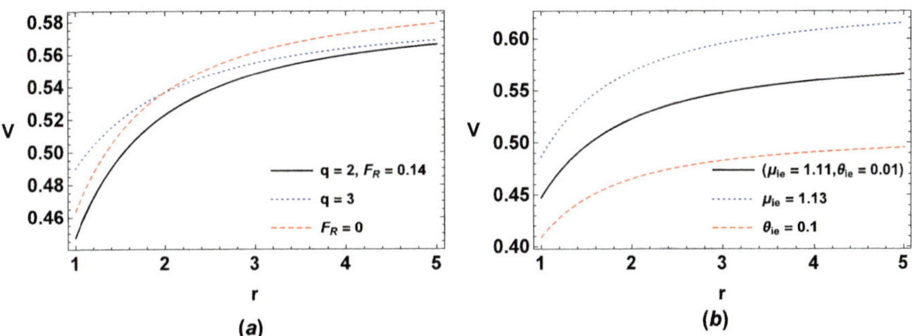

Fig. 2 The phase velocity (V) versus index r for different values of (a) index q, polarization force parameter (F_R); (b) ratio of number density of ions to electrons (μ_{ie}) and ratio of temperature of ions to electrons (θ_{ie})

The Fig. 1 shows the polarization force parameter (F_R) versus indices r and q. It is shown that when the values of r and q are increased, F_R decreases. In other words, when the indices r and q decrease, the polarization parameter increases. This means that indices r and q have a big influence on the polarization parameter.

The variation of phase velocity (V) with index r for different values of index q, polarization force parameter (F_R), ratio of number density of ions to electrons (μ_{ie}) and ratio of temperature of ions to electrons (θ_{ie}) is portrayed in Fig. 2. It is depicted that with increase in r, q, ratio of number density of ions to electrons and without polarization force parameter (F_R), the phase velocity is increased and it is reduced with enhancement in ratio of temperature of ions to electrons. This implies that polarization force suppresses the phase velocity of solitary waves.

Figure 3a, b elucidate the solitary profile (ϕ) with Λ for the change in values of r and q, polarization force parameter (F_R), ratio of number density of ions to electrons (μ_{ie}) and ratio of temperature of ions to electrons (θ_{ie}). It is described that with increase in r, q indices, ratio of number density of ions to electrons and without polarization force parameter (F_R), the nonlinearity effect decreases and dispersion effect increases, so the amplitude of solitary waves is decreased on negative axis and

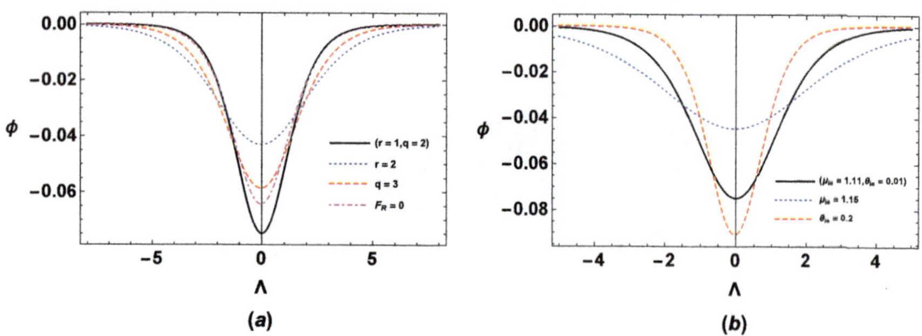

Fig. 3 The soliton profile (ϕ) for different values of (a) r, q indices, polarization force parameter (F_R); (b) ratio of number density of ions to electrons (μ_{ie}) and ratio of temperature of ions to electrons (θ_{ie})

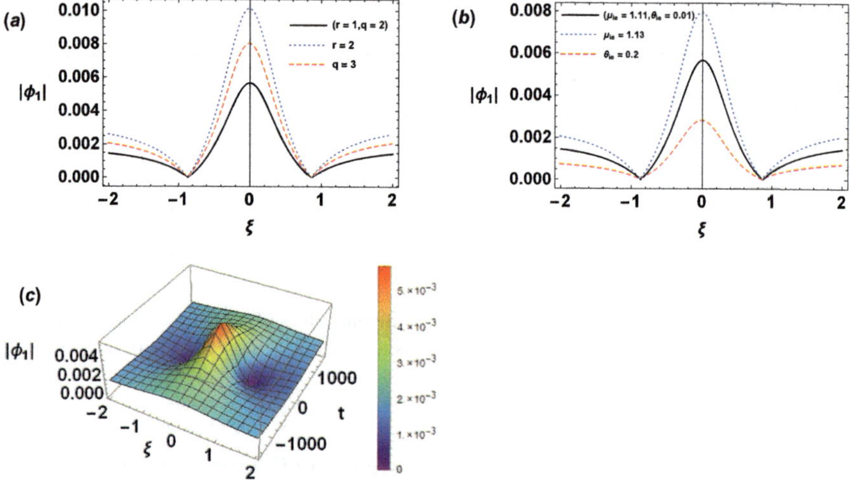

Fig. 4 The variation of first order rogue wave profile with (**a**) r, q indices; (**b**) ratio of number density of ions to electrons (μ_{ie}) and ratio of temperature of ions to electrons (θ_{ie}); (**c**) 3-D representation

width is increased. The amplitude of solitary wave is enhanced and width is reduced with change in the value of ratio of temperature of ions to electrons, this occurs due to increase in nonlinearity and decrease in dispersion effects. This also implies that contribution of polarization force enhances, the amplitude of solitary waves along negative axis.

The variation of first order profile ($|\phi_1|$) and second order profile ($|\phi_2|$) of rogue waves with ξ for change in r, q indices, μ_{ie} and θ_{ie} is presented in Figs. 4 and 5. It is noticed that with increase in r, q indices and μ_{ie} (i.e., ratio of number density of ions to electrons) the nonlinearity increases so the amplitude of both orders rogue waves is increased. The amplitude is reduced with increase in θ_{ie} (i.e., ratio of temperature

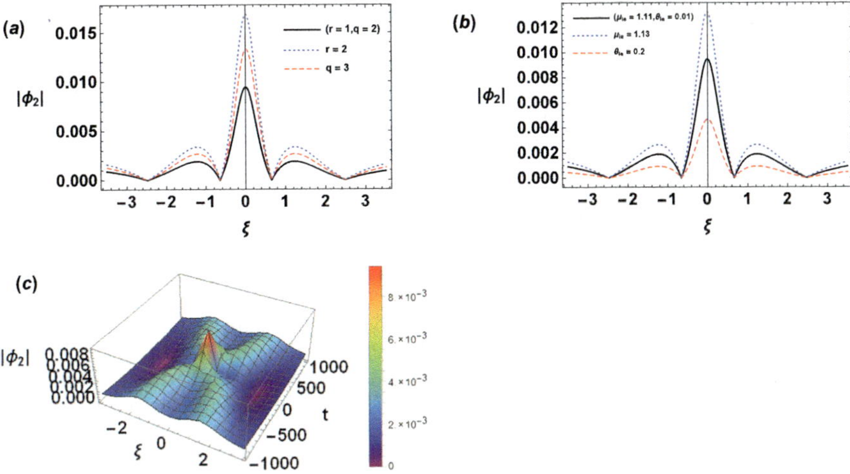

Fig. 5 The second order rogue wave profile with (**a**) r, q indices; (**b**) ratio of number density of ions to electrons (μ_{ie}) and ratio of temperature of ions to electrons (θ_{ie}); (**c**) 3-D representation

of ions to electrons) due to decrease in nonlinearity. Its 3-dimensional presentation is described in Figs. 4c and 5c.

From the whole numerical analysis we can conclude that all physical parameters such as r, q indices, polarization force parameter (F_R), ratio of number density of ions to electrons (μ_{ie}) and ratio of temperature of ions to electrons (θ_{ie}) have great influence on propagation properties of dust-acoustic solitary and rogue waves.

2.4 Conclusion

The influence of polarization force on characteristics of dust-acoustic (DA) solitary and rogue waves in negatively charged dusty plasma containing Maxwellian electrons and ions with generalized (r, q) distribution is studied. The KP and NLS equations have been derived. From their respective solutions characteristics of solitary and rogue waves have also been examined. We observe negative polarity DA KP solitons. The polarization parameter (F_R) is suppressed with increase r, q indices. The propagation properties of DA solitary and rogue waves are highly affected by variation in r, q indices, ratio of number density of ions to electrons, ratio of temperature of ions to electrons, and with polarization force parameter. It is emphasized that the implication of this work may be of great importance to analyse the nonlinear phenomena from physics point of view in planetary rings [31].

References

1. Kadomtsev, B.-B., Petviashivili, V.-I.: On stability of solitary waves in weakly dispersive media. Sov. Phys. Doklady **15**(4), 753–756 (1970)
2. Mushtaq, A., Khan, S.-A.: Ion acoustic solitary wave with weakly transverse perturbations in quantum electron-positron-ion plasma. Phys. Plasmas **14**(5), 052307(1–6) (2007)
3. Pakzad, H.-R.: The Kadomstev–Petviashvili (KP) equation ion acoustic waves in weakly relativistic plasma with nonthermal electron, positron and warm ion. J. Phys.: Conf. Series **96**, 012146(1–6) (2008)
4. Jian-Hong, C.: Effects of dust size distribution on nonlinear waves in a dusty plasma. Chinese Phys. B **18**(6), 2121–2128 (2009)
5. Dorranian, D., Sabetkar, A.: Dust acoustic solitary waves in a dusty plasma with two kinds of nonthermal ions at different temperatures. Phys. Plasmas **19**(1), 013702–6 (2012)
6. Elwakil, S.-A., El-hanbaly, A.-M., El-Shewy, E.-K., El-Kamash, I.-S.: Electron acoustic soliton energy of the Kadomtsev–Petviashvili equation in the Earth's magnetotail region at critical ion density. Astrophys. Space Sci. **349**(1), 197–203 (2014)
7. Saini, N.-S., Kaur, N., Gill, T.-S.: Dust acoustic solitary waves of Kadomstev–Petviashvili (KP) equation in superthermal dusty plasma. Adv. Space Res. **55**(12), 2873–2882 (2015)
8. Kohli, R., Kaur, N., Singh, M., Saini, N.-S.: Effect of ion beam on dust-acoustic waves under transverse perturbations in dusty plasma. IEEE Trans. Plasma Sci. **46**(4), 782–789 (2018)
9. Ruderman, M.-S.: Kadomtsev–Petviashvili equation for magnetosonic waves in Hall plasmas and soliton stability. Phys. Scripta **95**(9), 095601(1–9) (2020)
10. Hamaguchi, S., Farouki, R.-T.: Plasma-particulate interactions in nonuniform plasmas with finite flows. Phys. Plasmas **1**(7), 2110–2118 (1994)
11. Khrapak, S.-A., Ivlev, A.-V., Yaroshenko, V.-V., Morfill, G.-E.: Influence of a polarization force on dust acoustic waves. Phys. Rev. Lett. **102**(24), 245004-3 (2009)
12. Mamun, A.-A., Ashrafi, K.-S., Shukla, P.-K.: Effects of polarization force and effective dust temperature on dust-acoustic solitary and shock waves in a strongly coupled dusty plasma. Phys. Rev. E **82**(2), 026405–6 (2010)
13. Shukla, N., Shukla, P.-K.: Polarization-force-induced dust grain acceleration and intrinsic magnetization of dusty plasmas. J. Plasma Phys. **76**(5), 677–680 (2010)
14. Asaduzzaman, M., Mamun, A.-A.: Roles of polarization force and nonthermal electron on dust-acoustic waves in an inhomogeneous dusty plasma with positively charged dust. Phys. Plasmas **19**(9), 093704–6 (2012)
15. Ashrafi, K.-S., Mamun, A.-A., Shukla, P.-K.: Polarization force for different dusty plasma situations. J. Plasma Phys. **80**(1), 1–7 (2014)
16. Mayout, S., Bentabet, K., Tribeche, M.: Effect of the polarization force on the dust-acoustic soliton energy. Contribut. Plasma Phys. **56**(2), 99–103 (2016)
17. Bentabet, K., Tribeche, M.: Dust-acoustic solitons in a polarized dusty plasma with nonthermal ions. IEEE Trans. Plasma Sci. **45**(4), 736–741 (2017)
18. Abbasi, A., Vaziri, M.-R.: Effect of polarization force on the Jeans instability in collisional dusty plasmas. Plasma Sci. Technol. **20**(3), 035301(1–4) (2018)
19. Singh, K., Ghai, Y., Kaur, N., Saini, N.-S.: Effect of polarization force on dust-acoustic cnoidal waves in dusty plasma. European Phys. J. D **72**(9), 160(1–8) (2018)
20. Singh, K., Saini, N.-S.: Breather structures and peregrine solitons in a polarized space dusty plasma. Front. Phys. **8**, 602229(1–12) (2020)
21. Dolai, B., Prajapati, R.-P.: Effects of dust-charge gradient and polarization forces on the waves and Jeans instability in strongly coupled dusty plasma. Phys. Lett. A **384**(25), 126462(1–7) (2020)
22. Saini, N.-S., Singh, K., Sethi, P.: Interaction of dust-acoustic shock waves in a magnetized dusty plasma under the influence of polarization force. Laser Particle Beams **2021**, 6679085(1–8) (2021)
23. Shah, K.-H., Qureshi, M.-N.-S., Masood, W., Shah, H.-A.: An alternative explanation for the density depletions observed by Freja and Viking satellites. AIP Adv. **8**(8), 085010–12 (2018)

24. Qureshi, M.-N.-S., Shah, H.-A., Murtaza, G., Schwartz, S.-J., Mahmood, F.: Parallel propagating electromagnetic modes with the generalized (r, q) distribution function. Phys. Plasmas **11**(8), 3819–3829 (2004)
25. Ali, S., Abid, A.-A., Du, J., Mamun, A.-A.: Dust surface potential in a generalized (r, q)-distributed multi-ion dusty plasma. Contribut. Plasma Phys. **58**(10), 976–984 (2018)
26. Nasir, W., Ehsan, Z., Qureshi, M.-N.-S., Shah, H.-A.: Solar wind driven electrostatic instabilities with generalized (r, q) distribution function. Contribut. Plasma Phys. **59**(9), e201800159(1–8) (2019)
27. El-Labany, S.-K., Moslem, W.-M., El-Bedwehy, N.-A., Abd El-Razek, H.-N.: Rogue wave in Titan's atmosphere. Astrophys. Space Sci. **338**(1), 3–8 (2012)
28. El-Shewy, E.-K., Abdelwahed, H.-G., Abdo, N.-F., Shahein, R.-A.: Rogue waves for Kadomstev–Petviashvili solutions in a warm dusty plasma with opposite polarity. Chem. Phys. Phys. Kinet. Plasma Phys. **71**, 284–291 (2016)
29. Saini, N.-S., Singh, M., Bains, A.-S.: Dust kinetic Alfvén solitary and rogue waves in a superthermal dusty plasma. Phys. Plasmas **22**(11), 113702–10 (2015)
30. Peregrine, D.-H.: Water waves, nonlinear Schrödinger equations and their solutions. J. Australian Math. Soc. Series B-Appl. Math. **25**, 16–43 (1983)
31. Goertz, C.-K., Linhua-Shan, S., Havnes, O.: Electrostatic forces in planetary rings. Geophys. Res. Lett. **15**(1), 84–87 (1988)

Dust-Ion-Acoustic Multisoliton Interactions in the Presence of Superthermal Particles

Dharitree Dutta and K. S. Goswami

Abstract The propagation and interaction of dust-ion-acoustic solitons in the plasma composed of superthermal electrons and positrons, positively charged inertial ion, and static dust particles is examined in this work. The reductive perturbation method for small amplitude is adopted to derive the KdV equation. Hirota's bilinear method has been employed to calculate the multisoliton solutions of the KdV equation. The role of various plasma parameters on the soliton has been studied. It has been observed that the superthermality of the electrons and positrons and their concentration can alter the nature of the solitons. The presence of dust particles also influence the solitary structures. These results can be used to understand the nonlinear structures in different space and atmospheric environment, e.g. in the Van Allen radiation belt.

Keywords DIA soliton · Reductive perturbation method · Hirota's bilinear method

1 Introduction

In recent years, there is plenty of research [1–8] in the fascinating field of nonlinear structures because of their occurrence in different space and atmospheric environment. Soliton is one of those nonlinear structures which receive a huge amount of interests from the researcher worked in different fields of plasma physics. The distinguishing feature of a soliton is that it can maintain the size and shape even after its interaction with another soliton or solitons. These nonlinear structures are generated because of the balance between nonlinearity and dispersion. Initially it was stud-

D. Dutta (✉)
Department of Physics, Anandaram Dhekial Phookan College, Nagaon, Assam 782002, India
e-mail: dutta3dharitree@gmail.com

K. S. Goswami
Centre of Plasma Physics—Institute for Plasma Research, Nazirakhat, Sonapur, Kamrup(M), Assam 782402, India

S. Banerjee and A. Saha (eds.), *Nonlinear Dynamics and Applications*,
Springer Proceedings in Complexity,
https://doi.org/10.1007/978-3-030-99792-2_25

ied in electron -ion plasma. However, dust being omnipresent at different space and ionospheric plasma, it has been later observed that the presence of dust particles in a plasma introduces various low frequency waves and nonlinear structures [9]. Dust-ion-acoustic wave is one of the distinct normal modes observed in unmagnetized dusty plasma. Shukla and Silin had introduced this dust-ion-acoustic wave through their theoretical work, later which was proved experimentally by Barkan et al. [9]. Since then, this nonlinear structure has attained considerable amount of interests from the researchers.

Recently, the plasma physicists have proclaimed the presence of electron-positron-ion plasma in various space and atmospheric environment. In addition, in some other environment e.g., the interior region of accretion disks near neutron stars and magnetars, in the Milky Way, in the Saturn's and Jupiter's magnetosphere, the electron-positron-ion-dust (e-p-i-d) plasma have been observed [10]. The presence of positron in the e-i-d plasma brings significant changes in the nature of the nonlinear waves. This is because the intrinsic symmetry between the electron (e^-) and positron (e^+) within the plasma brings a dynamical change from that of an electron-ion-dust plasma. The symmetry in the mass of the electron and positron originates different fluid and kinetic instabilities in the plasma, and controls the possibility of the generation of nonlinear acoustic waves [11]. In view of the vital role of electron-positron-ion-dust plasma in space and atmospheric plasma environment, in this work, the formation of a DIA soliton in an unmagnetized dusty plasma comprise of superthermal electrons and positrons, cold fluid ions and static dust grains have been studied. The KdV equation is derived using the reductive perturbation method. Another salient feature of this work is to incorporate the soliton interactions. Different space-based observations proved the existence of a series of solitons and double layers in different space and atmospheric environment. NASA's Van Allen Probe spacecraft had shown recently the existence of a series of low frequency electrostatic solitons in the Earth's magnetosphere [12, 13]. The interaction of multiple solitons can be studied with the help of Hirota's bilinear method [14]. This method was first published in 1971 by Hirota. The multi soliton solutions of the KdV equation have been derived using this method. The effect of different plasma parameters on the soliton interactions have also been discussed. The manuscript is organized as: the Sect. 2 contains theoretical formulation. In Sect. 3, the Hirota's bilinear method has been employed. In Sect. 4, the results have been discussed. Finally it has been concluded with a brief summery in Sect. 5.

2 Theoretical Formulation

A homogeneous, collisionless, unmagnetized plasma comprise of superthermal electrons and positrons, cold fluid ions along with stationary dust particles in the background have been considered here. The set of normalized equations [15] describing the plasma model have been given bellow.

$$\frac{\partial N_i}{\partial t} + \frac{\partial (N_i U_i)}{\partial x}, \tag{1}$$

$$\frac{\partial U_i}{\partial t} + U_i \frac{\partial U_i}{\partial x} + 3\sigma N_i \frac{\partial N_i}{\partial x} = -\frac{\partial \phi}{\partial x}, \tag{2}$$

$$\frac{\partial^2 \phi}{\partial x^2} = \mu_e N_e + \mu_d - \mu_p N_p - N_i. \tag{3}$$

Here, N_i and U_i are the number densities and velocities of ions normalized by equilibrium ion density (n_{i0}) and ion-acoustic velocity ($c_i = \sqrt{k_B T_e/m_i}$), respectively, where T_e is the electron temperature and m_i is the ion mass. ϕ is the normalized potential (normalized by $k_B T_e/e$, e being the electric charge). σ is the ratio of ion and electron mass. μ_e, μ_d, and μ_p are the normalized equilibrium densities of electrons, dust, and positrons, respectively, i.e., $\mu_e = n_{e0}/n_{i0}$, $\mu_d = Z_d n_{d0}/n_{i0}$, and $\mu_p = n_{p0}/n_{i0}$. The equilibrium condition states: $\mu_e = 1 - \mu_d + \mu_p$. The time variable is normalized by the inverse of ion plasma frequency ($\omega_i = \sqrt{4\pi n_{i0} e^2/m_i}$), whereas the space variable is normalized by the ion Debye length ($\lambda_{D_i} = \sqrt{k_B T_e/4\pi n_{i0} e^2}$). The superthermal electrons and positrons can be described with the $\kappa-$ distribution and the expression for their respective normalized number densities are,

$$N_e = \left(1 - \frac{\phi}{\kappa_e - 3/2}\right)^{-\kappa_e + 1/2}, N_p = \left(1 + \frac{\phi}{\beta(\kappa_p - 3/2)}\right)^{-\kappa_p + 1/2}. \tag{4}$$

Here, κ_e and κ_p are superthermality index of electrons and positrons respectively. β is the positron to electron temperature ratio ($\beta = T_p/T_e$).

The KdV equation is derived using equations 1–4 and the stretched coordinates [16] used here are $\xi = \epsilon^{1/2}(x - Mt)$, and $\tau = \epsilon^{3/2} t$, where ϵ is the smallness parameter that measures the weakness of the amplitude and M is the Mach number. The variable N_i, U_i and ϕ can be expanded about the unperturbed states in the power series of ϵ as,

$$\begin{aligned} N_i &= N_i^{(0)} + \epsilon N_i^{(1)} + \epsilon^2 N_i^{(2)} + \epsilon^3 N_i^{(3)} + \dots \\ U_i &= \epsilon U_i^{(1)} + \epsilon^2 U_i^{(2)} + \epsilon^3 U_i^{(3)} + \dots \\ \phi &= \epsilon \phi^{(1)} + \epsilon^2 \phi^{(2)} + \epsilon^3 \phi^{(3)} + \dots \end{aligned} \tag{5}$$

After following the reductive perturbation method, and solving the set of equations of of first order perturbed quantities the dispersion relation for linear dust-ion-acoustic waves has been calculated.

$$M = \left[\frac{1 + 3\sigma(\mu_e P_1 + \mu_p Q_1)}{\mu_e P_1 + \mu_p Q_1}\right]^{1/2}. \tag{6}$$

Here, $P_1 = (2\kappa_e - 1)/(2\kappa_e - 3)$, and $Q_1 = (2\kappa_p - 1)/\beta(2\kappa_p - 3)$. Further manipulation of the the second order perturbed quantities finally results the KdV equation,

$$\frac{\partial \psi}{\partial \tau} + A\psi \frac{\partial \psi}{\partial \xi} + B\frac{\partial^3 \psi}{\partial \xi^3} = 0, \tag{7}$$

where, $\psi = \phi^{(1)}$. A, and B represents the nonlinearity and dispersion coefficients, and their expression are,

$$A = \frac{(\mu_p Q_2 - \mu_e P_2)(M^2-3\sigma)^2}{M} + \frac{3(M^2+\sigma)}{2M(M^2-3\sigma)}, \; B = \frac{(M^2-3\sigma)^2}{2M}. \tag{8}$$

Here, $P_2 = \left(4\kappa_e^2 - 1\right)/2(2\kappa_e - 3)^2$, and $Q_2 = \left(4\kappa_p^2 - 1\right)/2\beta^2(2\kappa_p - 3)^2$.

3 Multisoliton Solution of the KdV Equation

To employ Hirota's bilinear method, the dependent variable ψ of the KdV equation has been transformed to a new form which is a logarithmic transformation of an auxiliary function $f(\xi, \tau)$ [14],

$$\psi(\xi, \tau) = \frac{12B}{A}\frac{\partial^2[\ln f(\xi, \tau)]}{\partial \xi^2}. \tag{9}$$

Substituting this transformation for Eq. 7 and integrating once with respect to ξ,

$$\left(\frac{f_{\xi,\tau}}{f} - f_\xi f_\tau\right) + 3B\left(\frac{f_{\xi,\xi}}{f}\right)^2 - 4B\frac{f_\xi f_{\xi,\xi,\xi}}{f^2} + B\frac{f_{\xi,\xi,\xi,\xi}}{f} = 0. \tag{10}$$

Multiplying Eq. 10 with f^2 results in the bilinear form as,

$$ff_{\xi,\tau} - ff_\xi f_\tau + 3B(f_{\xi,\xi})^2 - 4Bf_\xi f_{\xi,\xi,\xi} + Bff_{\xi,\xi,\xi,\xi} = 0. \tag{11}$$

From Eq. 11, it has been observed that the transformation leads to the disappearance of the nonlinearity coefficient (A) from the bilinear form of the KdV equation. The bilinear equation can be simplified and converted to an ODE in terms of Hirota derivative as

$$D_\xi(D_\tau + BD_\xi^3)(f(\xi, \tau) \cdot f(\xi, \tau)) = 0. \tag{12}$$

Here, D_ξ, D_τ are binary operators and on a paired function it can be expressed as,

$$D_\xi^m D_\tau^n(f.f) = \left(\frac{\partial}{\partial \xi} - \frac{\partial}{\partial \xi'}\right)^m \left(\frac{\partial}{\partial \tau} - \frac{\partial}{\partial \tau'}\right)^n (f(\xi, \tau) \cdot f(\xi', \tau'))|_{\xi=\xi', \tau=\tau'} \tag{13}$$

The function $f(\xi, \tau)$ can be expanded in the power series of ι as

$$f(\xi, \tau) = 1 + \iota f_1 + \iota^2 f_2 + \iota^3 f_3 + \ldots \tag{14}$$

If the original equation (KdV equation in this case) admits a N-soliton solution, then Eq. 14 will truncate at the n = N term provided f is the sum of precisely N simple exponential terms.

3.1 One and Two Soliton Solutions

For the one soliton solution of Eq. 7, it has been considered that $f_1 = e^{\eta}$, where $\eta = k\xi + \omega\tau$. For $\iota = 1$,

$$f(\xi, \tau) = 1 + e^{\eta} \tag{15}$$

Substituting Eqs. 13–15 the dispersion relation for the DIA wave can be derived and is, $\omega = -Bk^3$.
Therefore, the known form of a single soliton has been recovered as,

$$\psi(\xi, \tau) = \frac{12B}{A} \frac{\partial^2[\ln(1 - exp(k\xi - Bk^3\tau))]}{\partial\xi^2} = \frac{3B}{A} k^2 sech^2\left(\frac{k\xi - Bk^3\tau}{2}\right). \tag{16}$$

Here, k is the propagation vector, and $3Bk^2/A$ represents the amplitude of the soliton.
For two soliton solutions of Eq. 7, the function $f(\xi, \tau)$ can be expressed as,

$$f(\xi, \tau) = 1 + e^{\eta_1} + e^{\eta_2} + a_{12}e^{\eta_1+\eta_2}, \tag{17}$$

where $\eta_i = k_i\xi + \omega_i\tau + \gamma_i$, k_i are the propagation vectors, γ_i are the phase shifts, $\omega_i = -Bk_i^3$ and i = 1,2 for the first and second soliton respectively. a_{12} is an interaction parameter of the two solitons and it depends upon the propagation vectors k_1, and k_2 and can be expressed as, $a_{12} = (k_1 - k_2/k_1 + k_2)^2$.
The two soliton solution can be derived using the transformation (9) to the function $f(\xi, \tau)$.

$$\psi(\xi, \tau) = \frac{12B}{A} \frac{k_1^2 e^{\eta_1}(1 + a_{12}e^{2\eta_2}) + k_2^2 e^{\eta_2}(1 + a_{12}e^{2\eta_1}) + 2(k_1 - k_2)^2 e^{\eta_1+\eta_2}}{(1 + e^{\eta_1} + e^{\eta_2} + a_{12}e^{\eta_1+\eta_2})^2}. \tag{18}$$

Equation 18 represents the two solitons solution which indicates the interaction of two solitons. If initially $\tau \to -\infty$, the larger soliton is behind the smaller one, then the phase shifts [17] after interaction of the two solitons are

$$\gamma_1 = \frac{2B^{1/3}}{k_1} \ln \frac{k_1 + k_2}{k_1 - k_2}, \gamma_2 = -\frac{2B^{1/3}}{k_2} \ln \frac{k_1 + k_2}{k_1 - k_2}, \tag{19}$$

for larger and smaller solitons respectively. The phenomenon can easily be understood from Fig. 3, and is explained in the next section.

4 Results and Discussions

Figure 1 shows how the plasma parameters alter the nonlinearity and dispersion of the solitary wave. In Fig. 1a, b, the variation of 'A', and 'B' with the superthermality of the electrons and positrons have been shown. From the figure, it has been observed that both the nonlinearity and dispersion of the soliton are strongly influenced by the superthermality of the electrons and positrons. The nonlinearity coefficients are higher for higher superthermality (i.e. lower values of κ_e, and κ_p), whereas the dispersion coefficients are lower for higher superthermality. The electron and positron densities also have impact on the nonlinearity and dispersion of the soliton. The nonlinearity decrease with increasing electron density and after reaching the minimum value, it increases with increasing electron density, as shown in Fig. 1c. On the other hand, the nonlinearity increases with increasing positron density (Fig. 1d). However, the dispersion is higher for lower electron and positron densities. From Fig. 1, it has been observed that for all values of κ_e, κ_p, μ_e, and μ_p only compressive soliton will generate.

The change in the potential of the single compressive soliton at $\tau = 0$ for different values of the plasma parameters have been shown in Fig. 2. The potential of the soliton is maximum for Maxwellian electrons and positrons ($\kappa_e = \kappa_p \to \infty$), and is minimum for highly superthermal electrons and positrons ($\kappa_e = \kappa_p = 2$), as shown in Fig. 2a. This result can be elucidated as the increasing spectral indices cause increase in the electron and positron pressure. The increased pressure enhance the restoring force, due to which the potential of the soliton is higher for larger values of κ_e and κ_p. The potential of the soliton also depends upon the densities of the component. In Fig. 2b, the amplitude of the soliton for different dust densities has been shown. It is observed that the amplitude of the soliton is minimum for ion-acoustic mode ($\mu_d = 0$) and it increases on increase of the dust density (μ_d). The increase in dust density indicates a decrease in electron density, which leads to the increase of the

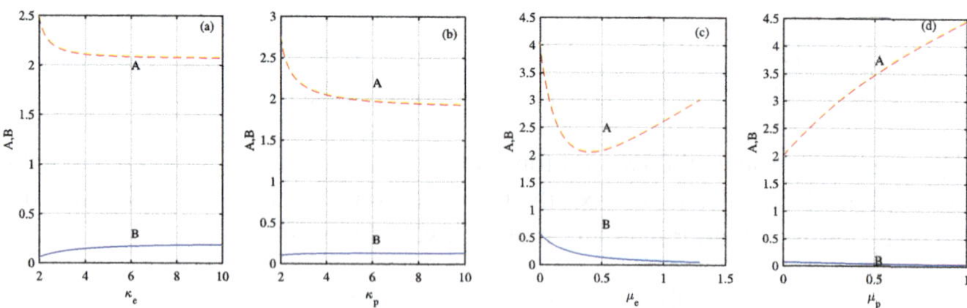

Fig. 1 The nonlinear (A) and dispersion coefficient (B) of the KdV equation for the variation of (**a**) the superthermal index of electron (κ_e) with $\kappa_p = 2$, $\sigma = 0.01$, $\mu_e = 1.27$, $\mu_p = 0.3$, $\beta = 1$, (**b**) the superthermal index of positron (κ_p) with $\kappa_e = 2$, $\sigma = 0.01$, $\mu_e = 1.27$, $\mu_p = 0.3$, $\beta = 1$, (**c**) the electron density (μ_e) with $\kappa_p = \kappa_e = 2$, $\sigma = 0.01$, $\mu_p = 0.3$, $\beta = 1$, and (**d**) the positron density (μ_p) with $\kappa_p = \kappa_e = 3$, $\sigma = 0.01$, $\mu_e = 1.27$, $\beta = 1$

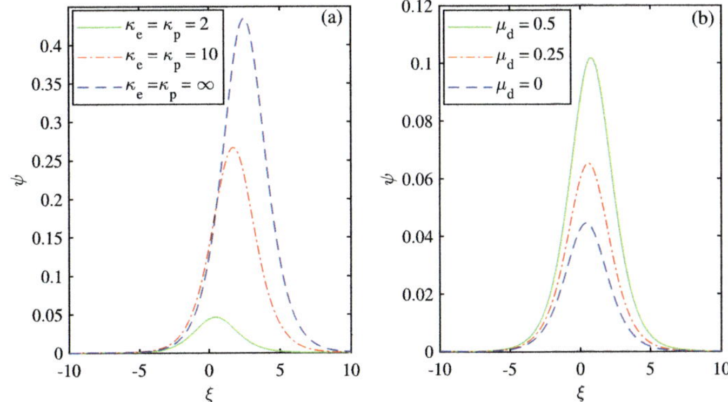

Fig. 2 Variation of ψ versus ξ with (**a**) $\sigma = 0.01$, $k = 1$, $\mu_e = 1.27$, $\mu_p = 0.3$, $\beta = 1$, and (**b**) $\sigma = 0.01$, $k = 1$, $\mu_p = 0.3$, $\kappa_e = \kappa_p = 2$, and $\beta = 1$

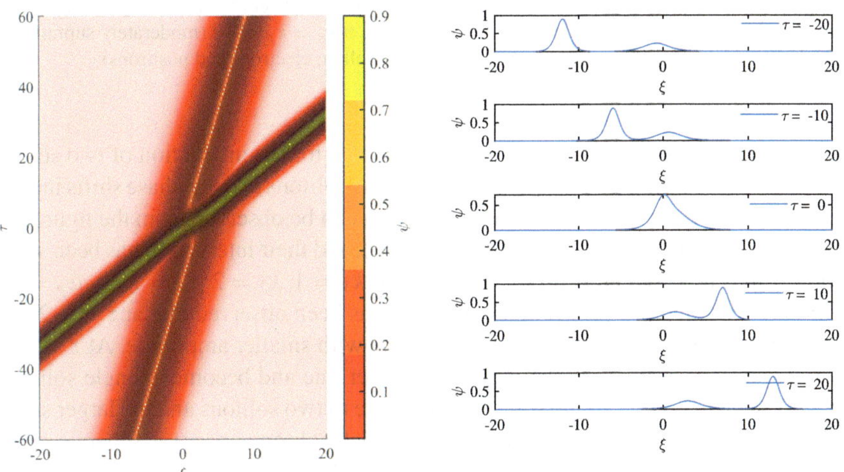

Fig. 3 (LHS) Surface plot of the electrostatic potential (ψ) of two interacting solitary waves with $\sigma = 0.01$, $k_1 = 1$, $k_2 = 2$, $\mu_e = 1.27$, $\mu_p = 0.3$, $\kappa_e = \kappa_p = 2$, and $\beta = 1$, (RHS) Variation of the electrostatic potential (ψ) of two interacting solitary waves versus ξ

dispersion as well as nonlinearity of the solitary waves. Therefore, the width and amplitude of the soliton increase with increased dust density.

The time evolution of the compressive DIA soliton in the plasma has been shown in Fig. 3. Two compressive DIA solitons, one with larger amplitude and narrower width than the second one, travelling in the same direction, undergone an interaction. At the interaction region, a single soliton is formed whose amplitude and width lie within the intermediate range of their respective amplitudes and widths. Then the single soliton breaks into two solitary structures and return to their initial amplitudes and widths to

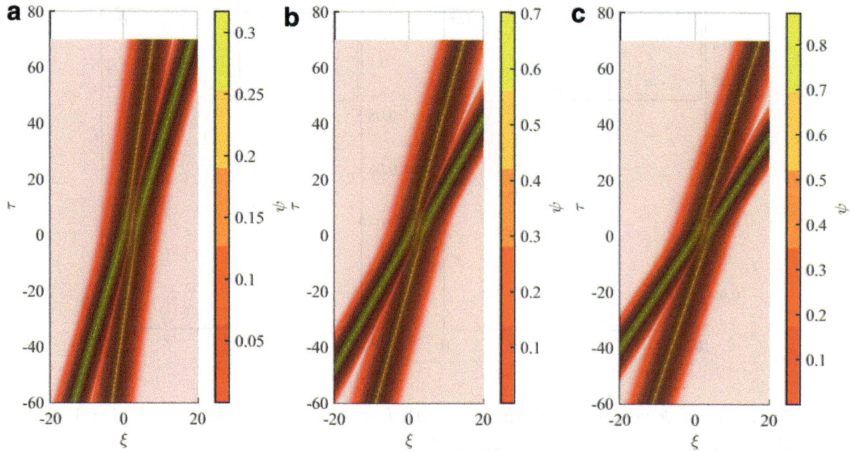

Fig. 4 Surface plot of the electrostatic potential (ψ) of two interacting solitary waves with $\sigma = 0.01$, $k_1 = 1$, $k_2 = 2$ $\mu_e = 1.27$, $\mu_p = 0.3$, $\beta = 1$ for three different set of values of κ_e and κ_p: **a** $\kappa_e = \kappa_p = 2$ (highly suprathermal electrons and positrons), **b** $\kappa_e = \kappa_p = 10$ (moderately suprathermal electrons and positrons), and **c** $\kappa_e = \kappa_p \to \infty$ (Maxwellian electrons and positrons)

regain their shape at a later time. The surface plot for the interaction of two solitons has been shown in Fig. 3a. The trajectories of the solitons and the phase shifts induced in the trajectories of the solitons after collision can be observed from the figure. The propagation of the solitons at different time (τ), and their interaction has been shown in Fig. 3b. The parameters used are $\sigma = 0.01$, $k_1 = 1$, $k_2 = 2$ $\mu_e = 1.27$, $\mu_p = 0.3$, $\kappa_e = \kappa_p = 2$, and $\beta = 1$. From the figure, it has been observed that at $\tau = -20$, the soliton with large amplitude is behind the one with smaller amplitude. At $\tau = -10$, the larger one starts to interact with the smaller one and become a single soliton at $\tau = 0$. At $\tau = 10$, the single soliton decompose to two solitons and the larger soliton overtake the smaller one. Finally at $\tau = 20$, each soliton appears as seperate soliton acquiring their initial size and shape. The phase shifts of the two solitons can also be clearly observable from the figure (RHS). Both of the solitons have positive phase shifts. However, the larger soliton has a higher phase shift than the smaller soliton.

Figure 4 depicts the surface plots of the interacting DIA solions for three cases of superthermality (κ): (a) highly superthermal electrons and positrons ($\kappa_e = \kappa_p = 2$), (b) moderately superthermal electrons and positrons ($\kappa_e = \kappa_p = 10$), and (c) Maxwellian electrons and positrons ($\kappa_e = \kappa_p \to \infty$). The figure shows that the amplitude of the two solitons increase with the increased values of spectral indices. The influence of the superthermal particles on the phase shift of the solitons after collision can be observed from the figure. The phase shift, i.e., the temporal change of the position of the two solitons is minimum for highly superthermal electrons and positrons, and is maximum for Maxwellian electrons and positrons. It can be explained with the help of the results shown in Fig. 3. From that figure we have observed that the larger soliton moves with a faster speed. Moreover, the increase

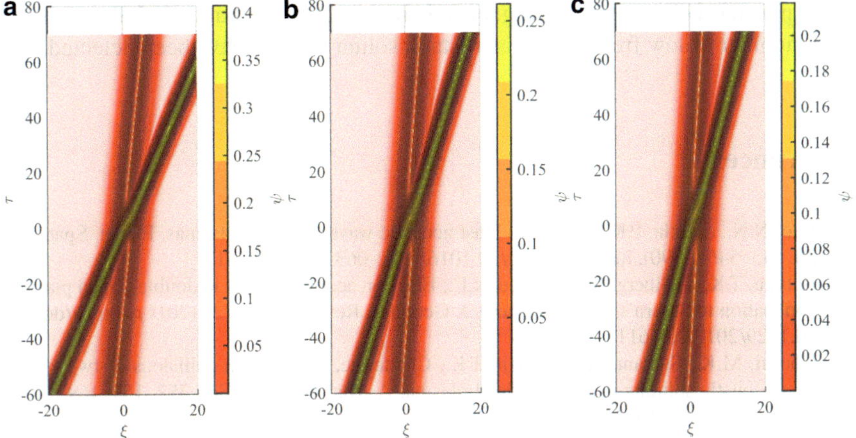

Fig. 5 Surface plot of the electrostatic potential (ψ) of two interacting solitary waves with $\sigma = 0.01$, $k_1 = 1$, $k_2 = 2$, $\mu_p = 0.3$, $\beta = 1$, $\kappa_e = \kappa_p = 2$ for three different set of values of μ_d: **a** $\mu_d = 0.5$, **b** $\mu_d = 0.25$, and **c** $\mu_d = 0$

in spectral index leads to the soliton with higher amplitude. So, on increase of the spectral index, the speeds of the solitons increase. This cause a faster change in the position of the two solitons and hence an increased phase shift. On the other hand, with a decrease in the superthermality, the temporal scale of interaction is reduced. This is because the solitary waves with Maxwellian particles have higher amplitudes as compared to superthermal particles. The larger amplitude solitons move with higher speed than the smaller amplitude solitons and, therefore, interact in a less time interval. The interaction of the solitons in presence of different dust concentrations have been shown in Fig. 5. In Fig. 5a, b the normalized dust densities are 0.5, and 0.25 respectively. In the absence of dust (i.e. $\mu_d = 0$), the interaction of the ion-acoustic solitons are shown in Fig. 5c. From the figure, it has been observed that the increased dust density enhances the phase shift. Similarly, the time of interaction of the solitons is reduced with reduced dust density.

5 Conclusion

Here the generation and interaction of two DIA solitons in the plasma containing superthermal electrons and positrons, inertial positive ions and negatively charged static dust have been studied. It has been observed that the superthermal particles play a vital role on the nature of the solitons. The superthermalility of electrons and positrons, and dust concentrations also play a crucial role on the nature of DIA solitons, and on the interaction of two dust-ion-acoustic solitons. The outcome of this work would helped to understand the collision of solitons in different laboratory and space plasma such as, Earth's magnetosphere, Van Allen radiation belt etc., where a large number of low frequency electrostatic solitary waves have been detected.

References

1. Rao, N.N., Shukla, P.K., Yu, M.Y.: Dust acoustic waves in dusty plasmas. Planet. Space Sci. **38**, 543–546 (1990). https://doi.org/10.1016/0032-0633(90)90147-I
2. Baluku, T.K., Hellberg, M.A., Mace, R.L.: Electron acoustic waves in double-kappa plasmas: application to Saturn's magnetosphere. J. Geophys. Res. **116**, A04227 (2011). https://doi.org/10.1029/2010JA016112
3. Ghorui, M.K., Samanta, U.K., Maji, T.K., Chatterjee, P.: Head-on collisions of two type of dust-acoustic solitons in a magnetized plasma. Astrophys. Space Sci. **352**, 159–169 (2014). https://doi.org/10.1007/s10509-014-1812-3
4. Dutta, D., Goswami, K.S.: Dust ion acoustic double layer in the presence of superthermal electrons. Indian J. Phys. **93**, 257–265 (2019). https://doi.org/10.1007/s12648-018-1279-0
5. Dutta, D., Goswami, K.S.: Electron acoustic double layers in a magnetized plasma in the presence of superthermal particles. J. Plasma Phys. **85**, 905850308 (2019). https://doi.org/10.1017/S0022377819000424
6. Dutta, D., Adhikari, S., Moulick, R., Goswami, K.S.: Evolution of dust ion acoustic soliton in the presence of superthermal electrons. Phys. Scr. **94**, 125210 (2019). https://doi.org/10.1088/1402-4896/ab3a5b
7. Sahu, B., Roychoudhury, R.: Two-soliton solution of ion acoustic solitary waves in nonplanar geometry. Astrophys. Space Sci. **345**, 91–98 (2013). https://doi.org/10.1007/s10509-013-1378-5
8. Jahangir, R., Masood, W.: Interaction of electron acoustic waves in the presence of superthermal electrons in terrestrial magnetosphere. Phys. Plasmas **27**, 042105 (2020). https://doi.org/10.1063/1.5143400
9. Shukla, P.K., Mamun, A.A.: Introduction to dusty plasma physics, 1st edn. IOP, Bristol (2002)
10. Banerjee, G., Maitra, S.: Arbitrary amplitude dust ion acoustic solitons and double layers in the presence of nonthermal positrons and electrons. Phys. Plasmas **23**, 123701 (2016). https://doi.org/10.1063/1.4971223
11. Sarri, G., et al.: Generation of neutral and high-density electron-positron pair plasmas in the laboratory. Nat. Commun. **6**, 6747 (2015). https://doi.org/10.1038/ncomms7747
12. Mozer, F.S., Bale, S.D., Bonnell, J.W., Chaston, C.C., Roth, I., Wygant, J.: Megavolt parallel potentials arising from double-layer streams in the earth's outer radiationbelt. Phys. Rev. Lett. **111**, 1–5 (2013). https://doi.org/10.1103/PhysRevLett.111.235002
13. Dillard, C.S., Vasko, I.Y., Mozer, F.S., Agapitov, O.V., Bonnell, J.W.: Electron-acoustic solitary waves in the Earth's inner magnetosphere. Phys. Plasmas **25**, 022905 (2018). https://doi.org/10.1063/1.5007907
14. Hirota, R.: Exact solution of the Korteweg-de Vries equation for multiple collisions of solitons. Phys. Rev. Lett. **27**, 1192 (1971). https://doi.org/10.1103/PhysRevLett.27.1192
15. Saini, N.S., Singh, K.: Head-on collision of two dust ion acoustic solitary waves in a weakly relativistic multicomponent superthermal plasma. Phys. Plasmas **23**, 103701 (2016). https://doi.org/10.1063/1.4963774
16. Taniuti, T., Wei, C.C.: Reductive perturbation method in nonlinear wave propagation. J. Phys. Soc. Jpn. **24**, 941–946 (1968). https://doi.org/10.1143/JPSJ.24.941
17. Marchant, T.R., Smyth, N.F.: Soliton interaction for the extended Kortewege-de Vries equation. IMA J. Appl. Math. **56**, 157–176 (1996). https://doi.org/10.1093/imamat/56.2.157

Fluid Dynamics and Nonlinear Flows

Numerical Study of Shear Flow Past Two Flat Inclined Plates at Reynolds Numbers 100, 200 Using Higher Order Compact Scheme

Rajendra K. Ray and Ashwani

Abstract In this study, an incompressible two-dimensional flow across two flat inclined plates is investigated numerically using a Higher-order compact (HOC) finite difference scheme. The shear parameter values $P = 0.0, 0.1$ are used to simulate simulations for two Reynolds numbers (Re), 100 and 200. Each plate is of length "d" and the shortest distance between the plates is exactly half of the plate's length (i.e., $0.5d$). Plate-1 and Plate-2 are inclined with angles of attack (i.e., with x-axis), $\alpha = (\pi - 45°)$ and $-\alpha$, respectively. The impact of the Reynolds numbers and shear rate on the process of vortex shedding is investigated from the perspective of stream function, vorticity contours, center-line velocity fluctuation, and phase diagrams. The numerical findings show not only the influence of vortex shedding from two flat inclined plates in shear flow but also several important flow generating properties with P and Re. This is the first time, to our knowledge, a numerical investigation has been performed to study the vortex shedding phenomena for two flat inclined plates with angles of attacks, $\alpha = (\pi - 45°)$ and $-\alpha$ respectively.

Keywords Shear flow · Flat inclined plates · HOC Scheme · Streamlines · Vorticity contours · Phase diagram

1 Introduction

The flow around bluff bodies has attracted much interest for for well over a century because of its engineering utility as well as its scientific significance in fluid dynamics. Flow through offshore platforms, heat exchangers and around bridge piers are just a few examples of mechanical, civil, and marine engineering applications. As a result, over the preceding century, a number of successful numerical, experimental,

R. K. Ray · Ashwani (✉)
School of Basic Sciences, Indian Institute of Technology Mandi, Mandi, Himachal Pradesh 175005, India
e-mail: mr.punia11@gmail.com
URL: https://www.iitmandi.ac.in/

S. Banerjee and A. Saha (eds.), *Nonlinear Dynamics and Applications*, Springer Proceedings in Complexity,
https://doi.org/10.1007/978-3-030-99792-2_26

and analytical research were carried out. In a uniform flow, the majority of study on flow through flat plates, square and circular cylinders has already been published in the literature [1–11]. At a Reynolds number of 250, Najjar and Balachandar [19] conducted a 3D numerical study of the flow behind a standard flat plate. It was noticed that the wake had small unsteadiness, which affected the flow field while also global integral characteristics like the drag coefficient and Strouhal number. Fage and Johansen [20] demonstrated that the Strouhal number had an almost constant value of 0.148 at angles of attack α ranging from 30° to 90° in their early experiments using a plate at 18 distinct angles of incidence. Vortex shedding takes place at the two corners of a plate that is slanted, resulting in uneven mean velocity profile in the recirculation region's very nearby wake. According to Lam [21], The wake is regulated by a sequence of counterclockwise vortices shed from the top edge of the plate at an impact angle of 30°. Breuer and Jovicic [22] and Breuer et al. [23] studied the flow across an inclined plate at an angle of 18° with a Reynolds number of 20000, and their large-eddy simulation (LES) findings clearly revealed that the trailing edge vortices were significantly dominated by the wake. There's also no frequent vortex shedding right at the front edge, according to them. Turki [24] presented a numerical analysis of the control of vortex shedding behind a square cylinder in a laminar channel flow with a plate in the $110 <= Re <= 200$ range. At the plate with a critical length (L_C), the author presented that the vortex shedding completely stopped. When the inlet flow is sheared instead than uniform, wake exhibits a different flow behavior, as demonstrated from both numerical and experimental analysis by Ayukawa et al. [15], Kiya et al. [13], Hayashi and Yoshino [27], Adachi and Kato [12], Cheng et al. [17], Kwon et al. [14], Cao et al. [16], Kumar and Ray [18].

Kalita et al. [25] worked on higher-order compact schemes (HOC) for the unsteady 2D convection-diffusion equation with variable convection coefficients in 2002 and they demonstrated that in an area with a greater level of complexity, HOC scheme can produce extremely precise results with a small selection of grid points. The HOC approach for incompressible viscous flows in the polar coordinate system (r, θ) was further developed by Ray and Kalita [26] in 2009. As a result, the Higher Order Compact (HOC) finite difference methods for the calculation of incompressible viscous flows have been slowly gaining traction in recent years because of their excellent precision and advantages over compact difference stencils.

The flow behaviour of laminar shear flow over two flat inclined plates is investigated in this work. The angle positions of the flat plate with respect to the x-axis are $\alpha = (\pi - 45°)$ and $-\alpha$. The problem is solved numerically by using the higher-order compact (HOC) finite difference technique. In time variable, it's second-order accurate, while in space variables, it's fourth-order accurate. Figure 1 shows a graphic representation of the problem.

The rest of the work is structured in the following manner. The mathematical description of the physical problem as well as the discretization technique of governing equations is described in Sect. 2. Section 3 i.e Results and Discussion section deals with numerical investigation of the flow through two flat inclined plates. Finally, we explain our findings in Sect. 4 (conclusion section), which is followed by references.

2 Mathematical Modelling

An incompressible, unsteady shear flow past two flat inclined plates is considered here. The length of both plates is exactly d and the shortest distance between two plates is exactly half of the length of the plate as depicted in the problem's schematic diagram Fig. 1. The incompressible, two-dimensional (2D) Navier–Stokes equations govern the flow. In Cartesian coordinate (x, y), the stream-function vorticity (F - ϑ) formulation of the 2D, incompressible Navier–Stokes equation in non-dimensional form is as follows:

$$\frac{\partial^2 \vartheta}{\partial x^2} + \frac{\partial^2 \vartheta}{\partial y^2} = Re\left(\frac{\partial \vartheta}{\partial t} + u\frac{\partial \vartheta}{\partial x} + v\frac{\partial \vartheta}{\partial y}\right) \tag{1}$$

$$\frac{\partial^2 F}{\partial y^2} + \frac{\partial^2 F}{\partial x^2} = -\vartheta \tag{2}$$

where vorticity and stream-function are represented by ϑ and F, respectively.

v = component of velocity in the y-direction, u = component of velocity in the x-direction

In terms of stream function (F), the velocity component u, v may be represented as follows:

$$v = -\frac{\partial F}{\partial x}, u = \frac{\partial F}{\partial y} \tag{3}$$

so vorticity,

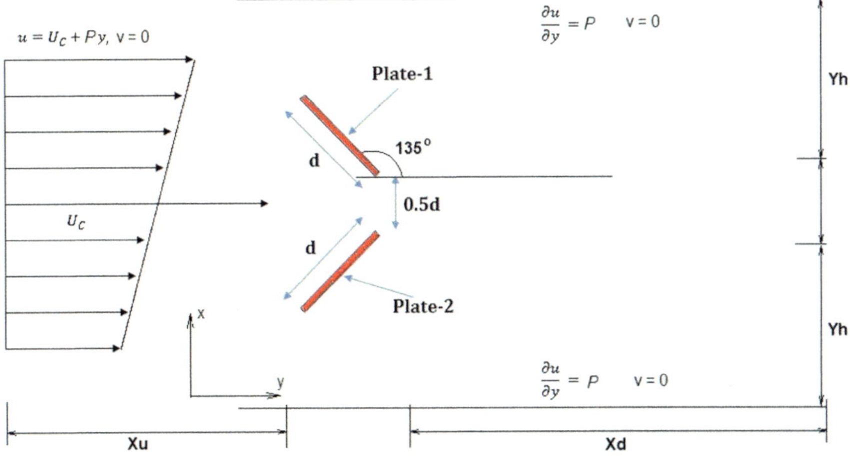

Fig. 1 Schematic Diagram of flow across two flat inclined plates

$$\vartheta = \frac{\partial v}{\partial x} - \frac{\partial u}{\partial y} \tag{4}$$

Although the physical domain is infinite, we consider a finite domain for computational purposes, where downstream boundaries, as well as the boundaries at top and bottom of the domain, are maintained far enough away from the plates so that the near wake instability should not be influenced by these boundaries of the computational domain. At the top and bottom boundaries, we use slip boundary conditions, and at the downstream border, we use convective boundary conditions [9, 18, 18]. On the surface of the flat inclined plate, the no-slip boundary condition is utilized. The linear shear flow is considered at the inlet boundary, i.e.,

$$u = U_c + Py, \quad v = 0 \tag{5}$$

where, U_c = Inflow velocity at the center-line, P = Shear rate

2.1 *Numerical Discretization*

Higher order compact (HOC) finite difference technique on uniform Cartesian grid [9, 18, 25] is used to discretize the governing equations. HOC scheme has already shown the ability to reproduce complex flow phenomena very efficiently and very accurately [9, 18, 26, 28, 29]. The governing equations (1) are discretized using HOC scheme at the $(i, j)^{th}$ node as follows:

$$\begin{aligned} &[Re + P_{ij}{\delta_x}^2 + Q_{ij}{\delta_y}^2 + R_{ij}\delta_x + S_{ij}\delta_y + T_{ij}\delta_x\delta_y + \\ &\quad U_{ij}\delta_x{\delta_y}^2 + V_{ij}{\delta_x}^2\delta_y + W_{ij}{\delta_x}^2{\delta_y}^2]\vartheta_{ij}^{n+1} \\ &= [Re + P_{ij}{\delta_x}^2 + Q_{ij}{\delta_y}^2 + R_{ij}\delta_x + S_{ij}\delta_y + T_{ij}\delta_x\delta_y + \\ &\quad U_{ij}\delta_x{\delta_y}^2 + V_{ij}{\delta_x}^2\delta_y + W_{ij}{\delta_x}^2{\delta_y}^2]\vartheta_{ij}^{n} \end{aligned} \tag{6}$$

Likewise, (2) has the following HOC discretization:

$$[\delta_x^2 + \delta_y^2 - (K1 + L1)\delta_x^2\delta_y^2]F_{i,j} = [-1 + K1\delta_x^2 + L1\delta_y^2]\vartheta_{i,j} \tag{7}$$

where,

$$\begin{aligned} P_{i,j} &= -M2Re - 0.5\Delta t X1_{i,j} \\ Q_{i,j} &= -N2Re - 0.5\Delta t X2_{i,j} \\ R_{i,j} &= -M1Re - H12u_{i,j}(Re)^2 - 0.5\Delta t X3_{i,j} \\ S_{i,j} &= -N1Re - K12v_{i,j}(Re)^2 - 0.5\Delta t X4_{i,j} \\ T_{i,j} &= -0.5\Delta t X5_{i,j} \\ U_{i,j} &= -0.5\Delta t X6_{i,j} \\ V_{i,j} &= -0.5\Delta t X7_{i,j} \\ W_{i,j} &= -0.5\Delta t X8_{i,j} \end{aligned}$$

and,

$X1_{i,j} = 1 + M1Reu_{i,j} + M2(Re)^2u_{i,j}^2 + 2M2Re(u_x)_{i,j}$

$X2_{i,j} = 1 + N1Rev_{i,j} + N2(Re)^2v_{i,j}^2 + 2N2Re(v_y)_{i,j}$

$X3_{i,j} = -Reu_{i,j} + M1Re(u_x)_{i,j} + N1Re(u_y)_{i,j} + M2Re^2u_{i,j}(u_x)_{i,j} + M2Re(u_{xx})_{i,j} + N2Re(u_{yy})_{i,j} + N2Re^2v_{i,j}(u_y)_{i,j}$

$X4_{i,j} = -Rev_{i,j} + M1Re(v_x)_{i,j} + N1Re(v_y)_{i,j} + M2Re^2u_{i,j}(v_x)_{i,j} + M2Re(v_{xx})_{i,j} + N2Re(v_{yy})_{i,j} + N2Re^2v(v_y)_{i,j}$

$X5_{i,j} = M1Rev_{i,j} + N1Reu_{i,j} + M2Re^2u_{i,j}v_{i,j} + 2M2Re(v_x)_{i,j} + 2N2Re(u_y)_{i,j} + N2Re^2u_{i,j}v_{i,j}$

$X6_{i,j} = -M1 - M2Reu_{i,j} + N2Reu_{i,j}$

$X7_{i,j} = -N1 + M2Rev_{i,j} - N2Rev_{i,j}$

$X8_{i,j} = -M2 - N2$

$K1 = -h^2/12, \quad L1 = -H^2/12,$

$M1 = N1 = Reu_{i,j}h^2/6,$

$M2 = N2 = -h^2/12,$

Where, The grid spacings in space variables and time variables are h and k, respectively. δ_x and δ_y are the first order central difference operators and δ_x^2 and δ_y^2 are the seconder central difference operators in x and y direction, respectively. More information on the HOC discretization may be found in [9, 25].

3 Results and Discussion

3.1 Grid and Time Independence Test

Uniform grids are used to discretize the computing domain. For three distinct grid sizes (400 × 137), (800 × 275) and (1600 × 550), a grid independence test was performed with a constant time increment of 0.01, and the results are presented in Table 1 for a representative point (0.8, –0.5). Table 1 indicates that a grid size of (800 × 275) is sufficient to generate acceptable results. For the grid sizes (800 × 275) and (1600 × 550), the relative error is only 0.394%. Table 2 presents the time independence test at a representative location (0.80, −0.50) behind the two inclined plates for 3 time increments 0.002, 0.006, and 0.01 values, on a grid of size (800 × 275). One can easily see that the variable's values are not changing much. So, the grid (800 × 275) with a time increment of 0.01 is adequate to depict the flow phenomena properly, based on the previous findings. We used a grid of size (800 × 275) and Δt = 0.01 in our computations.

Table 1 Velocities, Stream function, vorticity readings at (0.80, −0.5) behind the two inclined plates with $\Delta t = 0.01$ at $P = 0.1$, $Re = 100$ at various grid sizes

t	$(L \times B)$	u	v	F	ϑ	Max. relative error (%)
12.0	(400 × 137)	−0.20132	0.03171	−051582	−1.83485	8.433%
	(800 × 275)	−0.27180	0.08238	−0.61532	−2.00384	0.394%
	(1600 × 550)	−0.27150	0.08218	−0.61588	−2.01175	–

Table 2 The influence of time increment on the outputs for $Re = 100$, $P = 0.1$, $t = 5.0$ at point (0.80, −0.5)

Δt	u	v	F	ϑ
0.002	−0.41745	0.05158	−0.56567	0.13175
0.006	−0.41723	0.05234	−0.56602	0.12851
0.01	−0.41710	0.05268	−0.56622	0.13413

3.2 Numerical Study

The influence of shear rate (P) on the vortex shedding phenomena of flow across two flat inclined plates is investigated here for Reynolds numbers $Re = 100, 200$ and shear parameters $P = 0.0, 0.1$. Figure 2a and b plot the streamline pattern for $P = 0.0$ and $P = 0.1$ respectively at Reynold number (Re) = 100 for fully developed flow. The uniform flow is represented by $P = 0.0$. Fluid particles have distinct relative velocities on both the upper and lower surface of the Plates for positive P values.

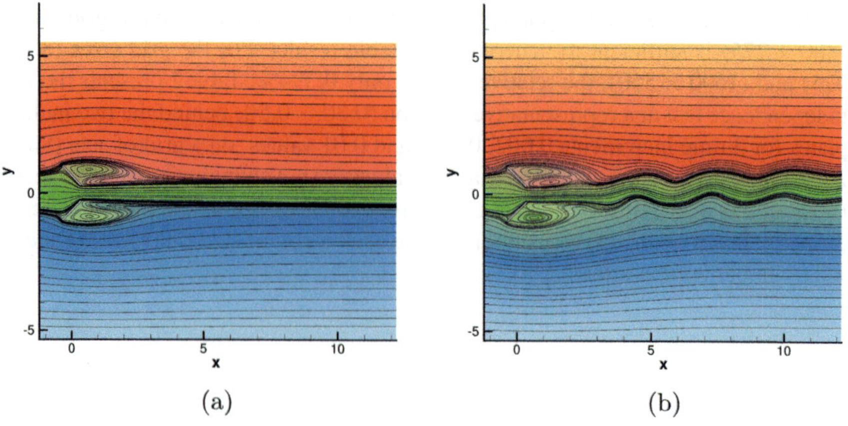

Fig. 2 Streamline flow at $Re = 100$, **a** $P = 0.0$, **b** $P = 0.1$

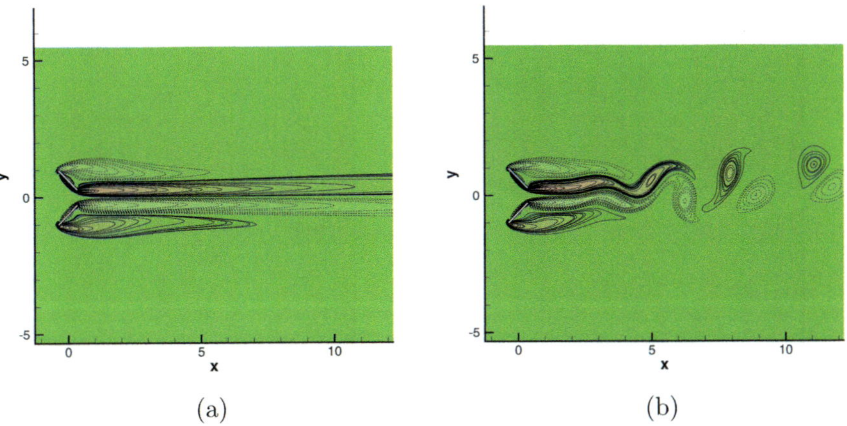

Fig. 3 Contours of vorticity for $Re = 100$: **a** $P = 0.0$, **b** $P = 0.1$

This causes an asymmetric shear gradient at the plate's surface in the boundary layer. Asymmetrical vortices are created from the plate surface because the initial vortex is always generated from the surface with the greater relative velocity. The strengths of these asymmetrical vortices vary, resulting in uneven transverse and longitudinal vortex spacing. As a result, shear flows passing between two inclined plates display more intricate vortex shedding than uniform flows. For $P = 0.0$, the flow behind the two plates ultimately becomes fully steady, while for $P = 0.1$, the flow becomes unsteady from the start. The wake is symmetric along the centerline when $P = 0.0$ and does not change with time. Figure 3a and b show the shear influence on vortex shedding more clearly for $Re = 100$. It is clear from Fig. 3a that flow has stabilised for $P = 0.0$. Because of the asymmetry in the entering freestream, the vortex shedding phenomena behind the two plates is different when $P = 0.1$ than when $P = 0.0$. The Kármán vortex street is maintained by the periodic vortex shedding occurrence, although the positive and negative vortices are varied in size and intensity for $P = 0.1$. The vortices that emerge from the lower side of the top plate are larger than those that emerge from the top side of the lower plate.

The flow phenomena are next investigated for $Re = 200$ with $P = 0.0$ and 0.1. The streamlines contours for $P = 0.0$ and 0.1 are plotted in Figs. 4a and b, respectively. The streamline pattern for the uniform flow ($P = 0.0$) is different from the preceding ($Re = 100$). The flow behind the two plates does not become stable at $Re = 200$ for uniform flow. As demonstrated in Fig. 4a, the flow pattern behind the plates is symmetric. Because of the shear impact, the streamlines pattern differs from $P = 0.0$ to $P = 0.1$ (Fig. 4b). Figure 5a and b shows the vorticity contours for $P = 0.0$ and $P = 0.1$ for $Re = 200$. Our findings in Fig. 4a and b are confirmed in these graphs. For $Re = 200$, the vortex shedding phenomenon is observed to be different for $Re = 100$ and $P = 0.0$. Here, The flow-pattern is symmetric and periodic behind the two plates. The periodic vortex shedding phenomena still exists for $P = 0.1$, but the positive and negative vortices are distinct in size, shape, and inten-

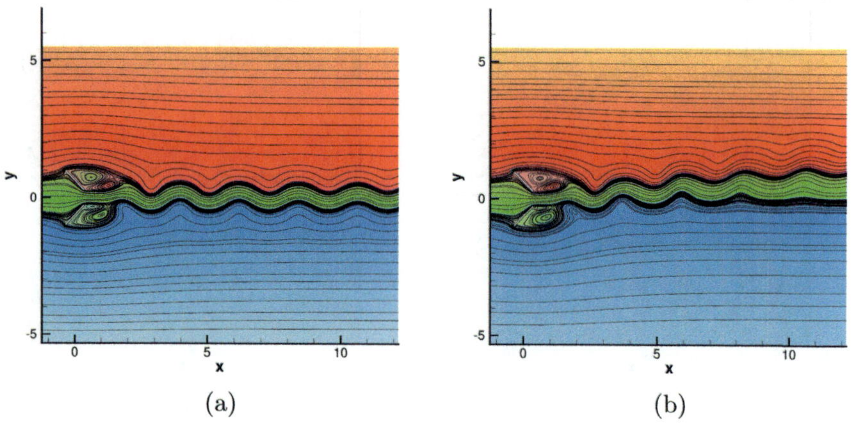

Fig. 4 Streamline flow at $Re = 200$, **a** $P = 0.0$, **b** $P = 0.1$

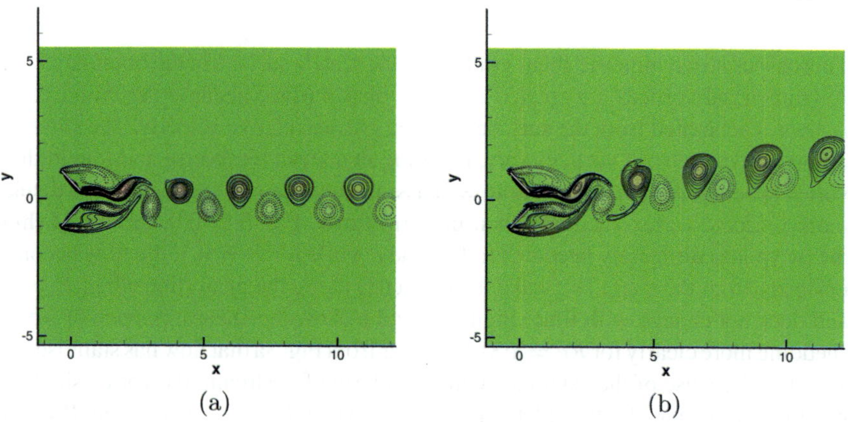

Fig. 5 Contours of vorticity for $Re = 200$: **a** $P = 0.0$, **b** $P = 0.1$

sity. The vortices on the bottom side of the Plate-1 are larger than the vortices on the top side of the Plate-2 ($P = 0.1$). The frequency of vortex shedding is significantly greater for $Re = 200$ and $P = 0.1$, but the vortex size and intensity are lower than for $Re = 100$ and $P = 0.1$.

We plot the axial velocity profiles of v versus x at $y = 0$ in Fig. 6a and u over y at $x = 0$ in Fig. 6b for $Re = 100$ after the flow has fully developed into its periodic condition. These figures show that there is no fluctuation in v velocity for $P = 0.0$ as the flow accelerates equally from both surface of the two inclined plates and v velocity decrease along the x-axis for $P = 0.1$. whereas the amount of v variation is greater in the nearby wake region for $P = 0.1$. Figure 6a shows that as P rises, u rises in the upper part of the plates while falling in the lower part. Centeral-line velocity fluctuation also plotted for different values of P at $Re = 200$ as shown in Fig. 7a

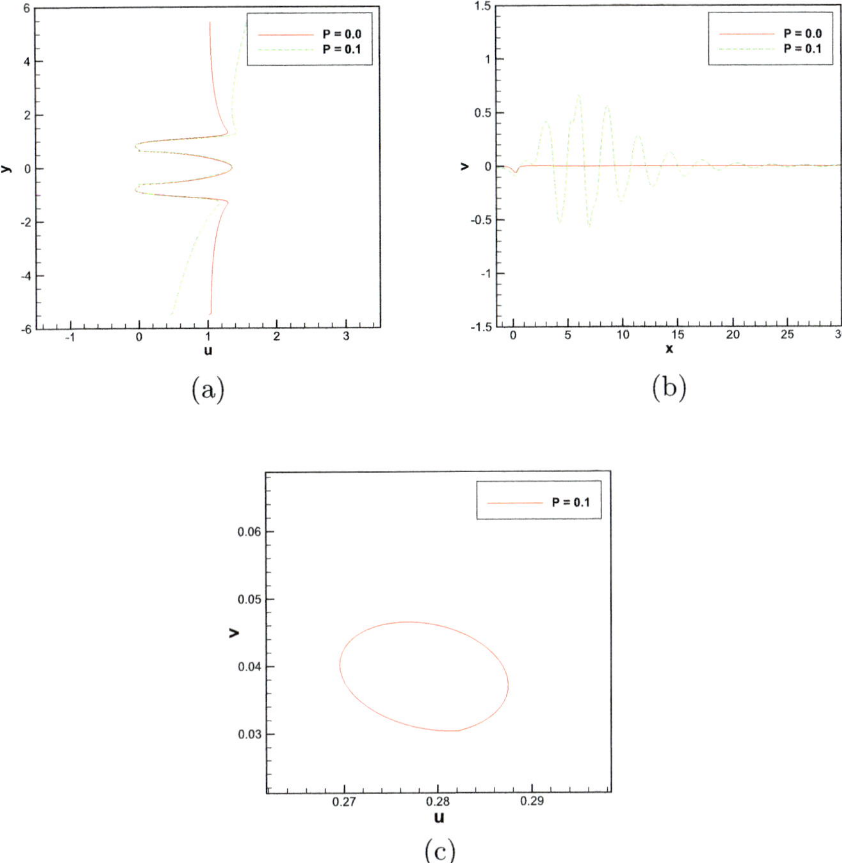

Fig. 6 Centerline velocity variation at $Re = 100$ and various P values: **a** u along the y-axis, **b** v along the x-axis, **c** (u-v) phase diagram

and b. We can observe that the u velocity profile behaves similarly to the previous $Re = 100$ value, but v-velocity varies with the increased height of oscillations along the x-axis only in the near wake. When the flow has reached a stable periodic condition phase diagram (Figs. 6c, 7c) are also drawn between u versus v at a measuring point (0.80, −0.5) behind the both flat inclined plates for $Re = 100, 200$ at various values of P. A periodic solution is depicted from these graphs.

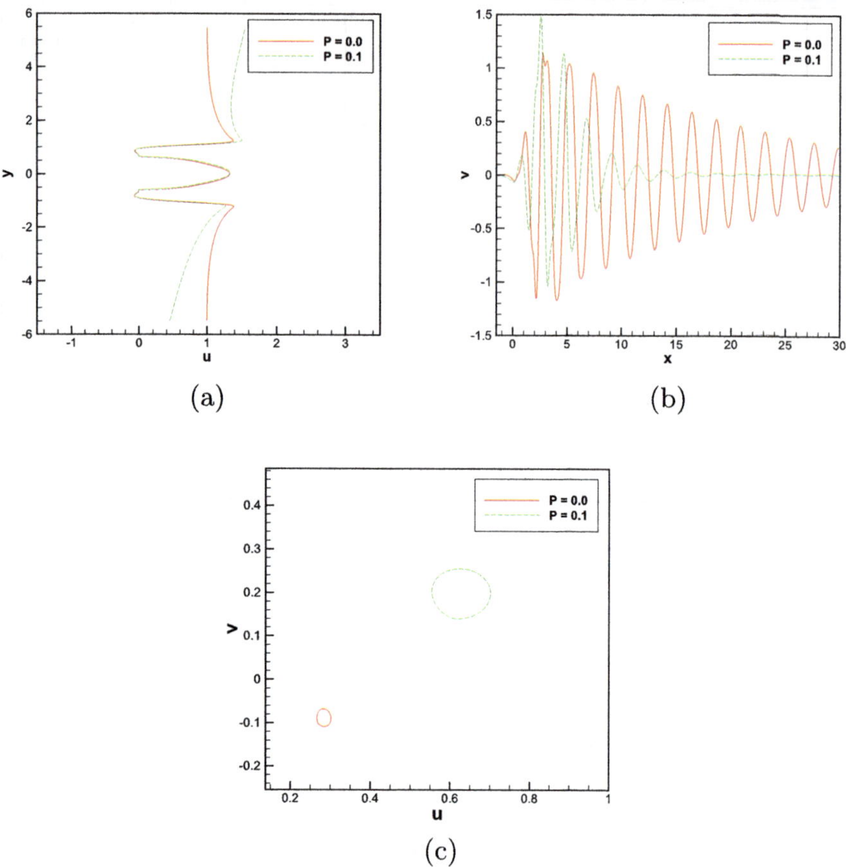

Fig. 7 Centerline velocity variation at $Re = 200$ and various P values: **a** u along the y-axis, **b** v along the x-axis, **c** (u-v) phase diagram

4 Conclusion

The HOC finite difference simulation of incompressible, linear shear flow across two flat inclined plates at $Re = 100, 200$ and various P values is presented in this paper. The current simulation yields some new and useful details on flow near the two flat inclined plates at various shear parameter values ($P = 0.0$, $P = 0.1$). The development of the wake behind inclined plates and the vortex shedding phenomena are substantially influenced by shear rate and Reynolds number, according to our findings. For all parameter values examined here, the flow that is completely developed is investigated in terms of streamline flow, vorticity contours, centerline velocity fluctuation, and phase diagrams. For different Reynolds numbers, certain intriguing flow phenomena have been observed. At $Re = 200$, the vorticity contours are symmetric about $y = 0$ for uniform flow ($P = 0.0$), whereas at $Re = 100$, the flow becomes almost steady-state for uniform flow ($P = 0.0$). The variations in the size and intensity of the alternatively shedding vortices become much more noticeable with shear flow. The frequency of vortex shedding is significantly greater for $Re = 200$ and $P = 0.1$, but the vortex size and intensity are lower than for $Re = 100$ and $P = 0.1$. Although the positive vortices have a narrow form and Over the positive vortices, the negative vortices are rounded for both $Re = 100$ and $Re = 200$ at $P = 0.1$. We can observe that for $Re = 200$, u-velocity behaves similarly to the preceding Re. But, On the other hand, the v-velocity profile has a periodic characteristic. For both P values, the amplitudes are bigger than those of $Re = 100$.

References

1. Yang, D., Pettersen, B., Andersson, H., Narasimhamurthy, V.: Vortex shedding in flow past an inclined flat plate at high incidence. Phys. Fluids **24**, 084103 (2012)
2. Xu, L., Nitsche, M.: Start-up vortex flow past an accelerated flat plate. Phys. Fluids **27**, 033602 (2015)
3. Xu, L., Nitsche, M., Krasny, R.: Computation of the Starting Vortex Flow Past a Flat Plate. Procedia IUTAM **20**, 136–143 (2017)
4. Sohankar, A., Norberg, C., Davidson, L.: Simulation of three-dimensional flow around a square cylinder at moderate Reynolds numbers. Phys. Fluids **11**, 288–306 (1999)
5. Ranjan, R., Dalal, A., Biswas, G.: A numerical study of fluid flow and heat transfer around a square cylinder at incidence using unstructured grids. Numer. Heat Transf. Part A: Appl. **54**, 890–913 (2008)
6. A transformation-free HOC scheme for incompressible viscous flow past a rotating and translating circular cylinder. J. Sci. Comput. **46**, 265–293 (2010)
7. Kalita, J., Sen, S.: Triggering asymmetry for flow past circular cylinder at low Reynolds numbers. Comput. Fluids **59**, 44–60 (2012)
8. Mushyam, A., Bergada, J.: A numerical investigation of wake and mixing layer interactions of flow past a square cylinder. Meccanica **52**, 107–123 (2016)
9. Ray, R., Kumar, A.: Numerical study of shear rate effect on unsteady flow separation from the surface of the square cylinder using structural bifurcation analysis. Phys. Fluids **29**, 083604 (2017)
10. Kalita, J., Sen, S.: Unsteady separation leading to secondary and tertiary vortex dynamics: the sub- and sub-phenomena. J. Fluid Mech. **730**, 19–51 (2013)
11. Shademan, M., Naghib-Lahouti, A.: Effects of aspect ratio and inclination angle on aerodynamic loads of a flat plate. Adv. Aerodyn. 2 (2020)
12. Yoshioka, N., Adachi, K., Nakamura, A., Ishimura, H.: An experimental investigation of viscoplastic flow past a circular cylinder at high Reynolds numbers. Rheol. Acta **14**, 993–1000 (1975)
13. Kiya, M., Tamura, H., Arie, M.: Vortex shedding from a circular cylinder in moderate-Reynolds-number shear flow. J. Fluid Mech. **101**, 721–735 (1980)
14. Kwon, T., Sung, H., Hyun, J.: Experimental investigation of uniform-shear flow past a circular cylinder. J. Fluids Eng. **114**, 457–460 (1992)
15. Ayukawa, K., Ochi, J., Kawahara, G., Hirao, T.: Effects of shear rate on the flow around a square cylinder in a uniform shear flow. J. Wind Eng. Ind. Aerodyn. **50**, 97–106 (1993)
16. Cao, S., Hirano, K., Ozono, S., Wakasugi, Y.: On the vortex shedding from a circular cylinder in a linear shear flow. J. Wind. Eng. **25**, 53–62 (2000)
17. Cheng, M., Tan, S., Hung, K.: Linear shear flow over a square cylinder at low Reynolds number. Phys. Fluids **17**, 078103 (2005)

18. Kumar, A., Ray, R.: Numerical study of shear flow past a square cylinder at Reynolds numbers 100, 200. Proc. Eng. **127**, 102–109 (2015)
19. Najjar, F., Balachandar, S.: Low-frequency unsteadiness in the wake of a normal flat plate. J. Fluid Mech. **370**, 101–147 (1998)
20. Fage, A., Johansen, F.: On the Flow of Air Behind an Inclined Flat Plate of Infinite Span. HMSO, London (1927)
21. Lam, K.: Phaselocked eduction of vortex shedding in flow past an inclined flat plate. Phys. Fluids **8**, 1159–1168 (1996)
22. Breuer, M., Jovicic, N.: Separated flow around a flat plate at high incidence: An LES investigation. J. Turbul. **2**, N18 (2001)
23. Breuer, M., Jovicic, N., Mazaev, K.: Comparison of DES, RANS and LES for the separated flow around a flat plate at high incidence. Int. J. Numer. Meth. Fluids **41**, 357–388 (2003)
24. Turki, S.: Numerical simulation of passive control on vortex shedding behind square cylinder using splitter plate. Eng. Appl. Comput. Fluid Mech. **2**, 514–524 (2008)
25. Kalita, J., Dalal, D., Dass, A.: A class of higher order compact schemes for the unsteady two-dimensional convection-diffusion equation with variable convection coefficients. Int. J. Numer. Meth. Fluids **38**, 1111–1131 (2002)
26. Kalita, J., Ray, R.: A transformation-free HOC scheme for incompressible viscous flows past an impulsively started circular cylinder. J. Comput. Phys. **228**, 5207–5236 (2009)
27. Hayashi, T., Yoshino, F.: On the evaluation of the aerodynamic forces acting on a circular cylinder in a uniform shear flow. Trans. Jpn. Soc. Mech. Eng. Ser. B **56**, 289–294 (1990)
28. Mittal, H., Al-Mdallal, Q., Ray, R.: Locked-on vortex shedding modes from a rotationally oscillating circular cylinder. Ocean Eng. **146**, 324–338 (2017)
29. Ray, R.: A transformation-free HOC scheme for incompressible viscous flow past a rotating and translating circular cylinder. J. Sci. Comput. **46**, 265–293 (2010)

On Transport Phenomena of Solute Through a Channel with an Inclined Magnetic Field

Susmita Das and Kajal Kumar Mondal

Abstract Under the effect of an angled magnetic field and the constant gradient of pressure, the present study investigates the solute dispersion in a Magneto-Hydrodynamics (MHD) flow between two infinite parallel plates, with the upper plate moving at a constant speed while the lower plate remains stationary. The unsteady advection-diffusion equation is solved by Aris's moments method with aid of a finite-difference scheme. It is shown that with the enhancement of absorption parameter, inclination angle of magnetic field and Hartmann number, the dispersion of the solute decreases. It is observed that after a certain critical time, the coefficient of dispersion asymptotically comes to a stationary circumstance for all cases. The present result may be applied for separation of matter from the fluids. The reaction parameter (β), inclination of an angle of the magnetic field (α), the Hartmann number (M), and the dispersion time (t) all have a significant impact on the solute's mean concentration profiles.

Keywords Inclined magnetic field · Dispersion · Method of moments · Channel · Absorption · Dispersion · Distribution of mean Concentration

1 Introduction

The study of the behavior of electrically conducting fluids and their magnetic properties is known as magneto-hydrodynamics. Due to its several applications in a variety of fields, including MHD power generation, oil reservoir engineering, separation of matter from fluids, aerodynamics, astrophysics and environmental mechanics, the study of dispersion phenomena in MHD fluids flows is highly important. Taylor [1] first studied the basic mechanism of dispersion of tracers in a laminar Poiseuille

Supported by Science and Technology and Biotechnology Department, Government of West Bengal, India.

S. Das (✉) · K. K. Mondal
Cooch Behar Panchanan Barma University, Cooch Behar 736101, India
e-mail: rimidas2105@gmail.com

S. Banerjee and A. Saha (eds.), *Nonlinear Dynamics and Applications*,
Springer Proceedings in Complexity,
https://doi.org/10.1007/978-3-030-99792-2_27

flow through a pipe. Aris [2] developed Taylor's work by removing some restrictions using his method of moments.

Gupta and Chatterjee [3] explained the solute transport in a MHD flow which is flowing through a channel with the aid of a transverse magnetic field analytically. They revealed that the coefficient of dispersion reduces as the magnetic field enhances. Annapurna and Gupta [4] extended the work and showed that the fluctuations in the dispersion coefficient reduces with enhancement of the Hartmann number. Many researchers [5–9] explores this area successfully.

Mazumder and Das [10] investigated that the first order boundary absorption on the dispersion process when the fluid is moving through a tube. They showed that the coefficient of dispersion reaches to its stationary circumstance after a certain critical time. Using a semi analytical approach, Sebastian and Nagarani [11] analyzed the dispersion of the contaminant through an annulus with an reaction parameter at the outer wall.

The dispersion of solute has been studied by many researchers over the last two decades in different flow geometry. But till now, no work has been discussed to study the mass transport phenomena of solute through a channel with absorption boundary consisting of an inclined magnetic field. The primary objective of the present research work is to present the effects of Hartmann number, the angle of inclination of the magnetic field and absorption parameter on the dispersion process of tracers through a channel of electrically conducting fluid. The inclined magnetic field is a magnetic field with nonzero inclination and it is the angle between the direction of the vector $\overrightarrow{B}$ with the perpendicular to the flow direction. Since, it significantly effects the velocity profile and consequently, on the dispersion process of the solute, the recent study is highly important for investigation of the basic mechanism of the tracers in a MHD flow. The time-dependent advection-diffusion equation with recommended initial and boundary conditions is solved using Aris' method of moments followed by a finite difference implicit scheme. The coefficient of dispersion and the distribution of the mean concentration of the contaminant are presented for all time period.

2 Formulation of the Problem

Consider a steady, laminar, fully developed, incompressible, viscous, two dimensional electrically conducting fluid, flowing through a infinite parallel plates caused by a constant gradient of pressure along the x^*-direction and y^*-axis is taken perpendicular to the direction of the channel flow. The stationary lower plate is situated at $y^* = -h$. The upper plate is at $y^* = h$ and it moves at a constant speed U (Fig. 1). A magnetic field $\overrightarrow{B}$ is applied at an angle α with the vertical y^*-direction. The interaction between magnetic and velocity fields give rise to an electric field $\overrightarrow{E}$ and it satisfies the relation $\overrightarrow{E} = \overrightarrow{V} \times \overrightarrow{B}$. In this research, the flow is considered along the x^*-direction only and consequently, the velocity and magnetic flux profiles are given by $\overrightarrow{V} = (u, 0, 0)$ and $\overrightarrow{B} = (0, B \sin \alpha, 0)$. It is also assumed no electric

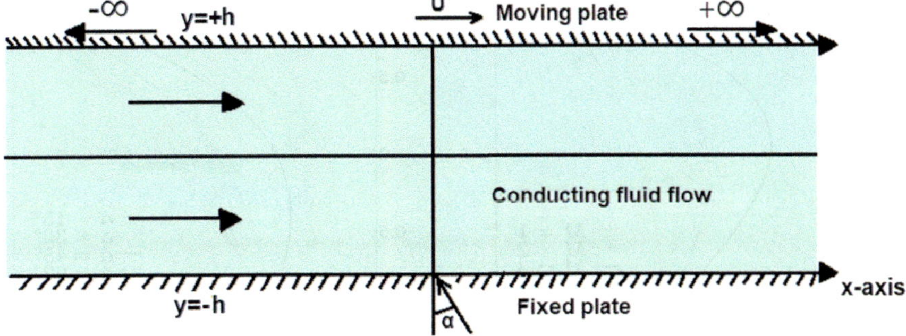

Fig. 1 MHD flow through a infinite parallel plates in presence of an inclined magnetic field

field is applied in the above mentioned flow. In this flow situation, the dimensional $\overrightarrow{x}$-momentum equation is

$$0 = -\frac{1}{\rho}\frac{\partial p^*}{\partial x^*} + \frac{\mu}{\rho}\left(\frac{\partial^2 u^*}{\partial x*^2}\right) + \frac{\sigma B^2 \sin^2\alpha}{\rho}u^* \tag{1}$$

and consequently, the dimensionless form of the Eq. (1) can be expressed as

$$\frac{d^2u}{dy^2} - M^2\sin^2\alpha u = -P \tag{2}$$

where $P(= -\frac{dp}{dx})$ is the pressure gradient which is constant, in the longitudinal x-direction, α is an inclination of angle of the magnetic field and $M = Bh\sqrt{(\frac{\sigma}{\rho\nu})}$ is the Hartmann number which is directly proportional to the the Magnetic field B. Also, M represents the relative significance between the magnetic force and the viscous force. Solving (1) with respect to the non-dimensional boundary conditions $u = 0$ at $y = -1$ and $u = 1$ at $y = 1$, the dimensionless velocity distribution is given by

$$\begin{aligned} u(y) = &\frac{1}{\sinh(2M\sin\alpha)}\left[\sinh(M\sin(y+1)) - \frac{P}{M^2\sin^2\alpha}\sin h(M\sin\alpha)\right. \\ &\left.\times\cosh(M\sin\alpha)y\right] + \frac{P}{M^2\sin^2\alpha} \end{aligned} \tag{3}$$

Figure 2a and b represent the velocity distribution for different values of Hartmann number (M) and inclination of an angle (α) of the magnetic field respectively. From the figures, it is seen that velocity significantly decreases with the increment of M and α. This is because, the applied magnetic field produces a Lorentz force which acts against the flow. In a similar manner, the inclination of an angle of the magnetic field resists the flow.

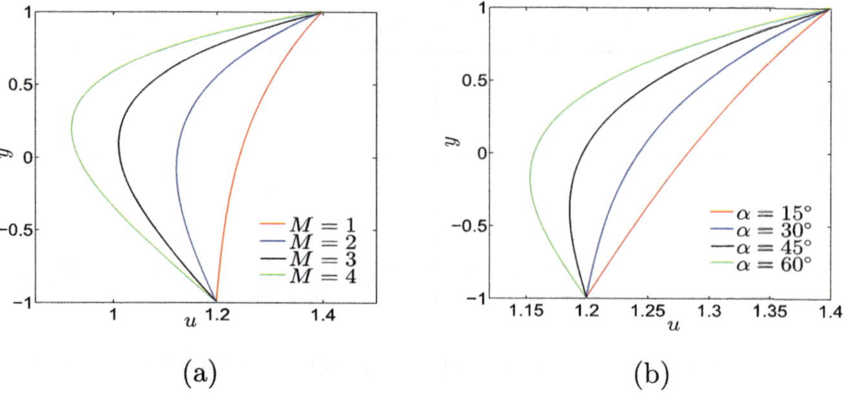

Fig. 2 Velocity distribution for various values of **a** M when $\alpha = 30°$, **b** α when $M = 1$

When a passive solute with an invariable molecular diffusivity D is introduced in the above mentioned flow in existence of a first-order reaction at the inner and outer walls of the channel, the mean concentration $C(x, y, t)$ of the tracers satisfies the following dimensionless advection-diffusion equation,

$$\frac{\partial C}{\partial t} + S_c u(y)\frac{\partial C}{\partial x} = \frac{\partial^2 C}{\partial x^2} + \frac{\partial^2 C}{\partial y^2}, \tag{4}$$

Here, the dimensionless parameters are $t = \frac{Dt^*}{h^2}, x = \frac{x^*}{h}, y = \frac{y^*}{h}, u = \frac{u^* h}{\nu}$ where $\nu = \frac{\mu}{\rho}$ and $S_c = \frac{\nu}{D}$ is the Schmidt number expressed as the ratio of momentum diffusivity (kinematic viscosity) and mass diffusivity. The Schmidt number is used to characterize the fluid flow in which momentum and mass diffusion processes occur simultaneously. The corresponding initial and boundary conditions are considered as,

$$C(x, y, 0) = \delta(y) \tag{5a}$$

$$\left[\frac{\partial C}{\partial y} + \beta C\right]_{y=1} = 0 \qquad \text{at } y = 1 \tag{5b}$$

$$\left[\frac{\partial C}{\partial y} - \beta C\right]_{y=-1} = 0 \qquad \text{at } y = -1 \tag{5c}$$

Also, it is assumed that at all points, the value of C is finite,

$$\frac{1}{2}\int_{-1}^{1}\int_{-\infty}^{\infty} C(x, y, 0)dxdy = 1 \tag{6}$$

where $\beta = (\beta^* h)$ defines the first-order reaction rate or reaction parameter corresponding to the catalytic reaction at both walls. If $\beta = 0$ then one can suggest the dispersion process of solute is chemically inert or neutrally buoyant. From Aris's moment method, the kth integral moment of the mean concentration distribution is as follows,

$$C_k(y,t) = \int_{-\infty}^{\infty} x^k C(x,y,t)dx \tag{7}$$

The concentration of the solute is distributed along the channel's cross section as follows,

$$M_k(t) = \overline{C}_k = \frac{\int_{-1}^{1} C_k(y,t)dy}{2} \tag{8}$$

Using (7), the diffusion Eq. (4) with initial and boundary conditions become

$$\frac{\partial C_k}{\partial t} - \frac{\partial^2 C_k}{\partial y^2} = ku(y)S_c C_{k-1} + k(k-1)C_{k-2} \tag{9}$$

with

$$C_k(y,0) = \delta(y) \tag{10}$$

where

$$C_k(y,0) = 1 \quad for \quad k = 0 \tag{11}$$

$$C_k(y,0) = 0 \quad for \quad k > 0 \tag{12}$$

$$\left[\frac{\partial C_k}{\partial y} + \beta C_k\right]_{y=1} = 0 \tag{13a}$$

$$\left[\frac{\partial C_k}{\partial y} - \beta C_k\right]_{y=-1} = 0 \tag{13b}$$

and

$$\frac{dM_k}{dt} = kS_c\overline{u(y)C_{k-1}} + k(k-1)\overline{c}_{k-2} - \frac{1}{2}\beta[C_k(-1,t) + C_k(+1,t)] \tag{14}$$

Where

$$M_k(0) = 1 \; for \; k = 0 \tag{15}$$

$$M_k(0) = 0 \quad for \; k > 0 \tag{16}$$

The cross-sectional mean concentration of the tracers is indicated by the over-bar. For a concentration distribution, the kth integral moment about the mean is given by

$$\nu_k(t) = \frac{1}{2M_0} \int_{-1}^{1} \int_{-\infty}^{\infty} (x - x_g) C dx dy \tag{17}$$

where

$$x_g = \frac{1}{2M_0} \int_{-1}^{1} \int_{-\infty}^{\infty} x C dx dy = \frac{M_1}{M_0} \tag{18}$$

x_g represents the first moment or the centroid of the solute which defines the initial location of the slug's centre of gravity with mean fluid velocity and M_0 is the total mass of the contaminant. The variance, skewness, and kurtosis of a contaminant's distribution are designated as ν_2, ν_3 and ν_4 respectively. The non-zero values of skewness suggest the deviation from the Gaussianity of the concentration distribution and when the kurtosis is greater than 3, the peak of the mean concentration of the contaminant becomes sharper.

3 Numerical Procedure

when $k \geqslant 1$ and $\beta \neq 0$, a finite difference implicit scheme is used for solving the Eq. (9), with above initial (10) and boundary [(13a)–(13b)] conditions, due to the analytical complexity. Here, the mesh point (p, q) indicates a point where $t_p = (p - 1) \times \Delta t$ and $y_q = -1 + (q - 1) \times \Delta y$. The increments along the time t direction and the space y directions are represented by $\Delta t = t_{p+1} - t_p$ and $\Delta y = y_{p+1} - y_p$ respectively. Using forward difference for $\frac{\partial C_k}{\partial t}$ and three-point averaged central difference for $\frac{\partial^2 C_k}{\partial y^2}$, the resulting system of linear algebraic equation is given by,

$$E_q \Omega_k(p+1, q+1) + F_q \Omega_k(p+1, q) + G_q \Omega_k(p+1, q-1) = H_q \tag{19}$$

where E_q, F_q, G_q and H_q are the matrix elements. The finite difference schemes for the initial and boundary conditions are

$$C_k(1, q) = \begin{cases} 1 \text{ for } k = 0, \\ 0 \text{ for } k \geqslant 1 \end{cases} \tag{20}$$

and for $k \geqslant 0$

$$C_k(p+1, 0) = C_k(p+1, 2) - 2\beta \Delta y C_k(p+1, 1) \quad \text{at } y = -1 \tag{21a}$$

$$C_k(p+1, M+1) = C_k(p+1, M-1) - 2\beta \Delta y C_k(p+1, M) \quad \text{at } y = 1 \tag{21b}$$

In this work, $M(=31)$ is the value of the index q at the outer wall of the channel. The index p which represents the dispersion time is lies between 1 and N where N_{max}

is taken as 12000. The Thomas algorithm [12] is employed to solve the resultant tri-diagonal coefficient matrix. Simpson's one-third rule is used to compute the values of M_k using the known values of $u(y)$ and C_k.

Since the diffusion of the contaminant along the longitudinal direction is negligibly small in compare to that of in the lateral direction, the effective longitudinal dispersion coefficient is taken as [13]

$$D_a = \frac{1}{2P_e^2}\frac{d\nu_2}{dt} \tag{22}$$

Also, the coefficients of skewness and kurtosis of the tracer distribution are represented as,

$$\beta_2 = \frac{\nu_3}{\nu_2^{\frac{3}{2}}} \tag{23a}$$

$$\beta_3 = \frac{\nu_4}{\nu_2^2} - 3 \tag{23b}$$

To find the distribution of the mean concentration $C_m(x, t)$ along the longitudinal direction, the Hermite polynomial expression [14] and the central moments ν_2, β_2, β_3 are used. The Hermite polynomial expression for the distribution of the mean concentration is of the form,

$$C_m(t, x) = M_0(t)e^{-X^2}\sum_{r=0}^{\infty} a_r(t)H_r(x) \tag{24}$$

where $X = \frac{x-x_g}{\sqrt{2\nu_2}}$, $x_g = \frac{M_1}{M_0}$ and the Hermite polynomials $H_r(x)$ that satisfy the recurrence relation [14]

$$H_{r+1}(x) = 2xH_r(x) - 2rH_{r-1}(x), \qquad r = 1, 2, 3, \ldots \tag{25}$$

with $H_0(x) = 1$. The coefficients a_r are found [14] from Eq. (24),

$$a_0 = \frac{1}{\sqrt{2\pi\nu_2}}, \; a_1 = a_2 = 0, \; a_3 = \frac{\sqrt{2}a_0\beta_2}{24}, \; a_4 = \frac{a_0\beta_3}{96}.$$

4 Results and Discussions

In this section, the different flow characteristic such as dispersion coefficient and mean concentration distribution of the solute are studied with respect to the various flow parameters namely, Hartmann number (M), absorption parameter (β) and inclination of an angle (α) of the magnetic field. Figure 3a, b and c represent the

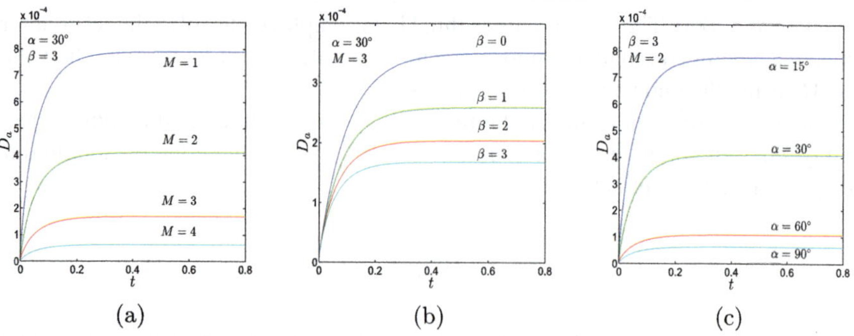

Fig. 3 Coefficient of Dispersion D_a for $S_c = 1000$ and different values of **a** M when: $\beta = 3$ and $\alpha = 30°$. **b** β when: $\alpha = 30°$ and $M = 3$. **c** α when: $\beta = 3$ and $M = 2$

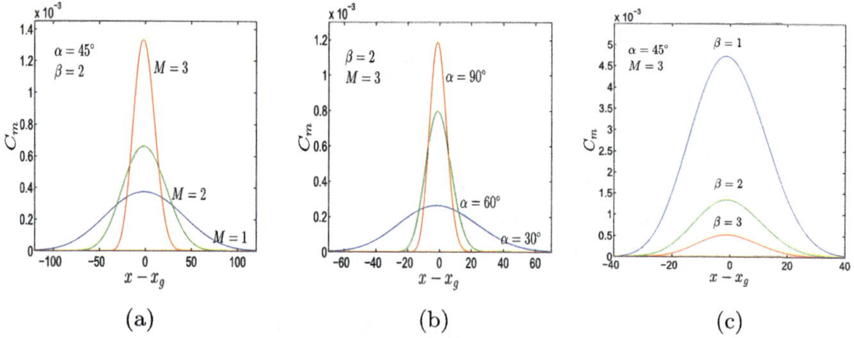

Fig. 4 Distribution of Mean concentration $C_m(t, x)$ for $S_c = 1000$ at time t = 1.1 for various values of **a** M when $\alpha = 45°$ and $\beta = 2$ **b** α when $M = 3$ and $\beta = 2$ **c** β when $\alpha = 45°$ and $M = 3$

variations of the dispersion coefficient for different values of M, β and α against the dispersion time t, where the other flow parameter are described in the caption of the figure. Figure 3a shows that the coefficient of dispersion D_a of the solute reduces if the Hartmann number enhances. The cause behind this, as M increases, the resisting Lorentz force enhances and consequently, there is a drop in D_a. The reduction in the coefficient of dispersion is also observed with the enhancement of both β and α [see Fig. 3b and c]. It is seen that for all cases the D_a reaches to a stationary circumstance after certain dispersion time. It is also observed from the Fig. 3b that in absence of reaction parameter at the walls, the D_a reaches to its steady state after a longer instant of time in compare to that of the dispersion coefficient when β is present.

Figure 4a exhibits the variation of the distribution of mean concentration $C_m(t, x)$ at a particular time ($t = 1.1$) against the axial distance $x - x_g$ for different M when $\alpha = 45°$, $\beta = 2$ and $S_c = 1000$. It is shown that the enhancement of M leads to increment of the peak of the distribution of the mean concentration of the tracer molecules but the axial expansion of the distributions diminishes. The reason behind this, as M increases the flow velocity as well the dispersion coefficient reduces along

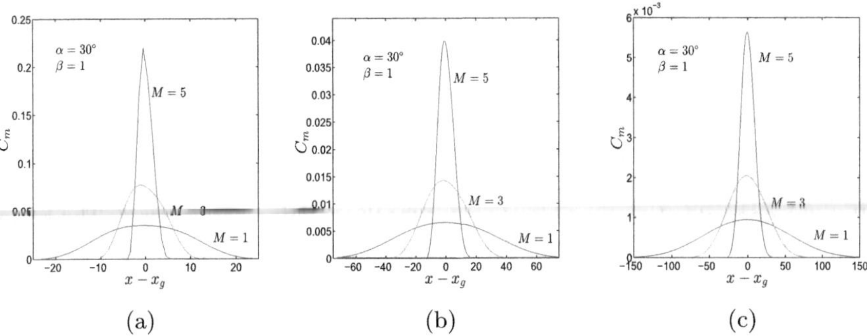

Fig. 5 Distribution of Mean concentration $C_m(t, x)$ for $S_c = 1000$, $\beta = 1$ and $\alpha = 30°$ for various M at time **a** t = 0.08 **b** t = 0.4 **c** t = 1.2

the axial direction and hence there is an increment in the amplitude of the mean concentration. The similar behavior is observed for the variation of α in the mean concentration distribution [see Fig. 4b]. Figure 4c illustrates the mean concentration distribution for various values of β where the other parameters are $\alpha = 45°$, $M = 3$ and $S_c = 1000$. An opposite phenomena is found in the distribution of the mean concentration of the traces in compared to that of the Fig. 4a and b, as β increases. As β increases, the tracer molecules are attracted by the reaction parameter and it is depleted along the walls. Thus, the distribution of the concentration mean becomes flatter.

Figure 5a–c represent the mean concentration distribution $C_m(t, x)$ for different values of Hartmann number with the dispersion times $t = 0.08, 0.4, 1.2$ respectively, where the fixed parameter are $\alpha = 30°$, $\beta = 1$ and $S_c = 1000$. It is observed for all cases that, as the Hartmann number enhances, the amplitude of the mean concentration profiles enhances. But, the strength of the concentration of the solute decreases prominently as time proceeds. It is significantly note that the mean concentration distribution of the tracers shows asymmetry when the dispersion time is small (see Fig. 5a) and it tends to become symmetric when t is large (see Fig. 5c).

5 Conclusions

The present investigation addresses an analysis on the solute dispersion through an incompressible magneto-hydrodynamics flow governed by an unsteady advection-diffusion equations, by employing Aris's method of moments with aid of a finite difference implicit method. The effects of Hartmann number M, angle of inclination of magnetic field α and reaction parameter β on the coefficient of dispersion and profiles of the solute's mean concentration are investigated. It is shows that the coefficient of Dispersion D_a reduces with the enhancement of β, M and α. When the absorption parameter β enhances at the both boundary, an enhances the number of molecules of the solute undergoes in the absorption and hence there is a reduction in D_a. Again, when M and α increases, the resisting Lorentz force enhances and thus D_a decreases. It is seen that the peak of the mean concentration profiles increase with the increment of both M and α. This is because, with the increment of M and α, the flow velocity reduces and consequently, the mixing of the molecules with the flow decreases. Thus, the amplitude of distribution of the mean concentration enhances. But, the opposite phenomena is observed for the reaction parameter β. The reason behind this, the total amount of the tracers reduces in the channel flow due to the absorption effect of β at the walls. It is remarkable to note that as time proceeds the strength of distribution of the mean concentration decreases significantly, as the rate of the mixing of the solute material is large in compare to that of in smaller times. It is seen that the profiles of distribution of the mean concentration contain an asymmetric behavior for small dispersion time and it becomes symmetric when the dispersion time reaches in a Taylor regime.

Acknowledgements The work is financially supported by **Science & Technology and Biotechnology Department, Government of West Bengal, India** under the project grant number 342 (Sanc.)—ST/P/S & T/16G-27/2018.

References

1. Taylor, G.I.: Dispersion of soluble matter in solvent flowing slowly through a tube. Proc. R. Soc. A **219**, 186–203 (1953)
2. Aris, R.: On the dispersion of a solute in a fluid flowing through a tube. Proc. R. Soc. A **235**, 67–77 (1956)
3. Gupta, A.S., Chatterjee, A.S.: Dispersion of soluble matter in the hydro-magnetic laminar flow between two parallel plates. Math. Proc. Camb. Phil. Soc. **64**(4), 1209 (1968)
4. Annapurna, N., Gupta, A.S.: Exact analysis of unsteady m.h.d convective diffusion. Proc. R. Soc. A **367**, 281–289 (1979)
5. Mondal, K.K., Mazumder, B.S.: On solute dispersion in pulsatile flow through a channel with absorbing walls. Int. J. Non-Linear Mech. **40**(1), 69–81 (2004)
6. Roy, A.K., Saha, A.K.: Mathematical model on magneto-hydrodynamic dispersion in a porous medium under the influence of bulk chemical reaction. Korea-Aust. Rheol. J. **32**(4), 287–299 (2020)
7. Gupta, P.S.: Effect of conducting walls on the dispersion of soluble matter in MHD channel flow. Chem. Eng. Commun. **7**(4–5), 301–307 (1980)
8. Mondal, K.K., Dhar, S., Mazumder, B.S.: On dispersion of solute in steady flow through a channel with absorption boundary: an application to sewage dispersion. Theor. Comput. Fluid Dyn **34**, 643–658 (2020)
9. Dhar, S., Poddar, N., Mondal, K.K., Mazumder, B.S.: On Dispersion of solute in a hydro-magnetic flow between two parallel plates with boundary absorption. Phys. Fluids **33**, 083609 (2021)
10. Mazumder, B.S., Das, S.K.: Effect of boundary reaction on solute dispersion in pulsatile flow through a tube. J. Fluid Mech **239**, 523 (1992)
11. Sebastian, B.T., Nagarani, P.: Convection-diffusion in unsteady non-Newtonian fluid flow in an annulus with wall absorption. Korea-Aust. Rheol. J. **30**(4), 261–271 (2018)
12. Anderson, D.A., Tanehill, J.C., Pletcher, R.H.: Computational Fluid Mechanics and Heat Transfer, 599 (1984)

13. Mondal, K.K., Mazumder, B.S.: On the solute dispersion in a pipe of annular cross-section with absorption boundary. ZAMM. Z. Angew. Math. Mech. **85**, 422–430 (2005)
14. Mehta, R.V., Merson, R.L., Mccoy, B.J.: Hermite Polynomial representation of chromatography elution curves. J. Chromatogr. **88**, 1 (1974)

Unsteady MHD Hybrid Nanoparticle (Au-Al_2O_3/Blood) Mediated Blood Flow Through a Vertical Irregular Stenosed Artery: Drug Delivery Applications

Rishu Gandhi and Bhupendra K. Sharma

Abstract The current study investigates the influence of hybrid nanoparticles (Au & Al_2O_3) on blood flow through a vertical artery with irregular stenosis with two-dimensional pulsatile blood flow, an inclined external magnetic field, viscous dissipation, and Joule heating. The blood flow is assumed to be unsteady, laminar, viscous, and incompressible, and the artery walls are considered permeable. The Reynolds temperature-dependent viscosity model is used to determine the variable viscosity effects. The governing momentum and energy equations are solved using Crank–Nicolson finite difference method by employing an appropriate coordinate transformation to build an accurate mesh using rectangular mesh units. Outcomes of the work are represented graphically for non-dimensional velocity, wall shear stress, flow rate, and non-dimensional temperature, respectively. The recent findings could be useful to biological researchers looking into the therapy of different cardiovascular disorders.

Keywords Irregular-shaped stenosis · Heat transfer · Joule heating · Temperature-dependent viscosity · Hybrid nanoparticles

1 Introduction

Cardiovascular diseases (CVDs) have become a significant global public health issue, with the highest morbidity and mortality rates among all conditions. A growing body of evidence in the scientific literature indicates vascular fluid dynamics plays a vital role in the onset and progression of arterial disorders. Stenosis is a term used to describe narrowing an artery segment's lumen. This is caused by the deposition of numerous chemicals on the endothelium of the artery wall, such as cholesterol. Many researchers have studied blood circulation dynamics via stenosed arteries, both

R. Gandhi (✉) · B. K. Sharma
Department of Mathematics, Birla Institute of Technology and Science, Pilani, Rajasthan, India
e-mail: rishugandhi155@gmail.com
URL: https://www.bits-pilani.ac.in

S. Banerjee and A. Saha (eds.), *Nonlinear Dynamics and Applications*, Springer Proceedings in Complexity,
https://doi.org/10.1007/978-3-030-99792-2_28

theoretically and experimentally. Ponalagusamy et al. [12] investigated blood flow through an artery with tapered stenosis considering an unsteady two-fluid model of blood. Basri et al. [2] employed Fluid-Structure Interaction to perform a transient study of three pulse cycles to examine hemodynamic parameters in normal and single stenosed renal arteries. Chandgar et al. [4] considered an inclined multiple stenosed artery and analyzed drug transport utilizing spherical gold nanoparticles by employing a single-phase model as well as a discrete-phase model. Mensah et al. [10] highlighted the 15 most prevalent causes of death from cardiovascular disorders and the 13 risk factors associated with them. Ponalagusamy [11] depicted a four-layered mathematical model of blood flow through a mildly stenosed artery. To determine how varying core viscosity influences blood flow control, they compared the outcomes for variable core viscosity with constant core viscosity. By considering blood as a biomagnetic fluid, Sharma et al. [15] explored Soret and Dufour's effects in an artery with tapering effects. Chen et al. [5] examined the influence of coexisting cardiovascular disease on the severity of COVID-19 and discovered that COVID-19 could have a significant effect on heart function and lead to myocardial damage. Majee et al. [9] conducted a comprehensive study on targeted drug delivery with magnetic nanoparticles with the goal of better understanding the blood flow in an atherosclerotic artery. Shahzadi et al. [13] addressed the importance of permeability in hybrid nanofluid blood flow through a bifurcated stenosed artery. She discovered that permeability effects are more efficient at reducing hemodynamic effects in atherosclerotic arteries with bifurcation effects. Tripathi et al. [17] investigated the impact of heat and mass transmission through a stenosed artery considering the variable magnetic field and varying viscosity. Kumawat et al. [8] investigated two-phase blood flow in a curved artery with time-variant stenosis, taking into account variable viscosity in the core and plasma region, respectively. Sharma et al. [14] investigated heat transmission and entropy generation in a multi-stenosed artery with tapering effects, viscous dissipation, radiation, and Joule heating using hybrid nanoparticles (gold and alumina).

Nanotechnology's application in biomedicine is a rapidly growing field with promising prospects for improving human illness diagnosis and therapy. The capacity to incorporate medications into a functionalized nanoparticle marks a new beginning in which drugs can be delivered to tissues or cells selectively. Consistent drug delivery to a target produces a more substantial therapeutic effect with lower drug levels than conventional dosing approaches. This research work is also motivated by the benefits of the nanoparticle drug delivery system. Having surveyed the literature, it is visualized that the inclined external magnetic field effects on hybrid nanoparticles (Au & Al_2O_3) via a vertical artery with irregular stenosis having permeable walls is not investigated yet. Most studies have overlooked the effects of Joule heating, viscous dissipation, and variable viscosity in blood flow problems. In this work, the Reynolds model for temperature-dependent viscosity is employed. The form of the stenosis or the fatty buildup of particles on the artery wall cannot be precisely described so, we've considered an irregular geometry of stenosis. The study's objective is to explore inclined magnetic field effects through an irregular stenosed artery consider-

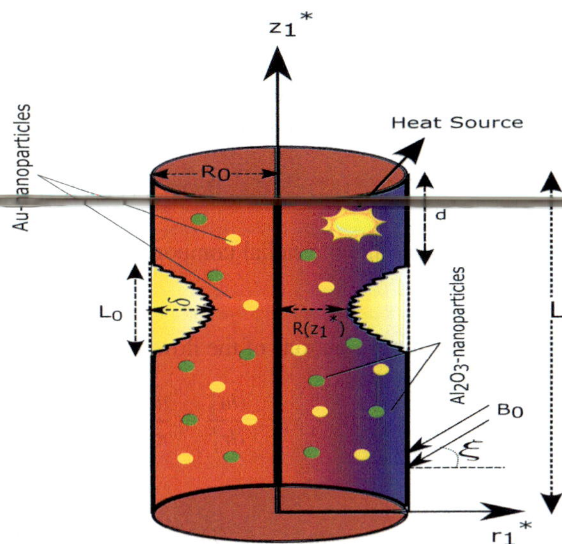

Fig. 1 The geometrical representation of an artery with irregular stenosis

ing permeable wall conditions in the presence of Joule heating, viscous dissipation, and variable viscosity.

2 Mathematical Modeling

The cylindrical coordinate system $(r_1^*, \tilde{\theta}, z_1^*)$ is considered to analyze the blood flow behavior through a vertical artery with irregular stenosis by assuming blood as a Newtonian fluid. The flow is in the radial (i.e., r_1^*-axis) and axial (z_1^*-axis) directions respectively. The mild stenotic assumption reduces the bi-directional flow to uni-directional flow. The impact of an inclined external magnetic field, Joule heating, pulsatile blood flow, and viscous dissipation is the subject of this study. The induced magnetic field is considered meager compared to the applied magnetic field. Figure 1 represents the geometry of the artery with suspended nanoparticles.

The geometry of the stenosis (irregular-shaped) [18] is assumed as

$$R(z_1^*) = \begin{cases} R_0 - 2\delta\left[cos\left(\frac{2\pi}{L_0}\left(\frac{z_1^*-d}{2} - \frac{L_0}{4}\right)\right) - \frac{7}{100}cos\left(\frac{32\pi}{L_0}\left(z_1^* - d - \frac{L_0}{2}\right)\right)\right] & d \le z_1^* \le d + L_0, \\ R_0 & \text{otherwise.} \end{cases} \quad (1)$$

where R(z_1^*) represents the radius of the artery's stenosed portion, d indicates the stenosis location, R_0 represents the radius of the artery in the non-stenotic area, δ represents the critical stenosis height, and L_0 represents the length of the stenosis.

2.1 Governing Equations

The velocity and temperature fields for the present study are defined as follows in the case of unsteady and axisymmetric hybrid nanoparticle-doped blood flow:

$$\tilde{V}^* = \tilde{V}^*[u_1^*(r_1^*, z_1^*, t_1^*),\ 0,\ w_1^*(r_1^*, z_1^*, t_1^*)]\,, \qquad \tilde{T}^* = \tilde{T}^*(r_1^*, z_1^*, t_1^*)$$

where u_1^* represents the radial component and w_1^* represents the axial component respectively.

The governing equations of the flow are represented as:

$$\frac{\partial u_1^*}{\partial r_1^*} + \frac{u_1^*}{r_1^*} + \frac{\partial w_1^*}{\partial z_1^*} = 0 \tag{2}$$

$$\begin{aligned} \rho_{hnf}\left[\frac{\partial u_1^*}{\partial t_1^*} + u_1^*\frac{\partial u_1^*}{\partial r_1^*} + w_1^*\frac{\partial u_1^*}{\partial z_1^*}\right] &= -\frac{\partial p_1^*}{\partial r_1^*} + \frac{1}{r_1^*}\frac{\partial}{\partial r_1^*}\left[\mu_{hnf}(\tilde{T}^*)\frac{\partial u_1^*}{\partial r_1^*}\right] \\ &+ \frac{\partial}{\partial z_1^*}\left[\mu_{hnf}(\tilde{T}^*)\left(\frac{\partial u_1^*}{\partial z_1^*} + \frac{\partial w_1^*}{\partial r_1^*}\right)\right] \\ &- 2\mu_{hnf}(\tilde{T}^*)\frac{u_1^*}{r_1^{*2}} - \sigma_{hnf}B_0^2 cos^2\xi w_1^* \end{aligned} \tag{3}$$

$$\begin{aligned} \rho_{hnf}\left[\frac{\partial w_1^*}{\partial t_1^*} + u_1^*\frac{\partial w_1^*}{\partial r_1^*} + w_1^*\frac{\partial w_1^*}{\partial z_1^*}\right] &= -\frac{\partial p_1^*}{\partial z_1^*} + \frac{1}{r_1^*}\frac{\partial}{\partial r_1^*}\left[r_1^*\mu_{hnf}(\tilde{T}^*)\left(\frac{\partial u_1^*}{\partial z_1^*} + \frac{\partial w_1^*}{\partial r_1^*}\right)\right] \\ &+ \frac{\partial}{\partial z_1^*}\left[2\mu_{hnf}(\tilde{T}^*)\frac{\partial w_1^*}{\partial z_1^*}\right] \\ &+ (\rho\gamma)_{hnf}g(\tilde{T}^* - \tilde{T}_1^*) - \sigma_{hnf}B_0^2 sin^2\xi w_1^* \end{aligned} \tag{4}$$

$$\begin{aligned} (\rho C_p)_{hnf}\left[\frac{\partial \tilde{T}^*}{\partial t_1^*} + u_1^*\frac{\partial \tilde{T}^*}{\partial r_1^*} + w_1^*\frac{\partial \tilde{T}^*}{\partial z_1^*}\right] &= k_{hnf}\left[\frac{\partial^2 \tilde{T}^*}{\partial r_1^{*2}} + \frac{1}{r_1^*}\frac{\partial \tilde{T}^*}{\partial r_1^*} + \frac{\partial^2 \tilde{T}^*}{\partial z_1^{*2}}\right] \\ &+ \sigma_{hnf}B_0^2 sin^2\xi w_1^{*2} + \phi^* \end{aligned} \tag{5}$$

where

$$\phi^* = 2\mu_{hnf}\left[\left(\frac{\partial u_1^*}{\partial r_1^*}\right)^2 + \left(\frac{u_1^*}{r_1^*}\right)^2 + \left(\frac{\partial w_1^*}{\partial z_1^*}\right)^2 + \frac{1}{2}\left(\frac{\partial u_1^*}{\partial z_1^*} + \frac{\partial w_1^*}{\partial r_1^*}\right)^2\right] \tag{6}$$

The boundary conditions for the flow are:

Thermophysical Properties of Hybrid Nanofluid	
Viscosity	$\mu_{hnf} = \frac{\mu_f}{(1-\phi_1)^{2.5}(1-\phi_2)^{2.5}}$
Density	$\rho_{hnf} = [(1-\phi_2)\{(1-\phi_1)\rho_f + \phi_1\rho_{s_1}\}] + \phi_2\rho_{s_2}$
Heat Capacity	$(\rho C_p)_{hnf} = [(1-\phi_2)\{(1-\phi_1)(\rho C_p)_f + \phi_1(\rho C_p)_{s_1}\}] + \phi_2(\rho C_p)_{s_2}$
Thermal Conductivity	$\frac{k_{hnf}}{k_{bf}} = \frac{k_{s_2} + (n-1)k_{bf} - (n-1)\phi_2(k_{bf}-k_{s_2})}{k_{s_2} + (n-1)k_{bf} + \phi_2(k_{bf}-k_{s_2})}$ where $\frac{k_{bf}}{k_f} = \frac{k_{s_1}+2k_f-(n-1)\phi_1(k_f-k_{s_1})}{k_{s_1}+(n-1)k_f+\phi_1(k_f-k_{s_1})}$
Electrical Conductivity	$\frac{\sigma_{hnf}}{\sigma_{bf}} = \frac{\sigma_{s_2} + (n-1)\sigma_{bf} - (n-1)\phi_2(\sigma_{bf}-\sigma_{s_2})}{\sigma_{s_2} + (n-1)\sigma_{bf} + \phi_2(\sigma_{bf}-\sigma_{s_2})}$ where $\frac{\sigma_{bf}}{\sigma_f} = \frac{\sigma_{s_1}+(n-1)\sigma_f-(n-1)\phi_1(\sigma_f-\sigma_{s_1})}{\sigma_{s_1}+(n-1)\sigma_f+\phi_1(\sigma_f-\sigma_{s_1})}$
Thermal Expansion Coefficient	$\Upsilon_{hnf} = [(1-\phi_2)\{(1-\phi_1)\Upsilon_f + \phi_1\Upsilon_{s_1}\}] + \phi_2\Upsilon_{s_2}$

(a)

Thermophysical Properties	Blood	Gold (Au)	Alumina (Al_2O_3)
Density$[\rho(kg/m^3)]$	1063	19320	3970
Thermal Conductivity $[(W/mK)]$	0.492	314	40
Thermal Expansion Coefficient $[\Upsilon \times 10^{-5}(K^{-1})]$	0.18	1.4	0.85
Heat Capacitance $[C_p(J/kgK)]$	3594	129	765
Electrical Conductivity $[\sigma(S/m)]$	6.67×10^{-1}	4.52×10^{7}	3.5×10^{7}

(b)

Fig. 2 **a** Thermophysical parameters of hybrid nanofluid [6], **b** Thermophysical properties of blood and nanoparticles

$$\frac{\partial w_1^*}{\partial r_1^*} = 0, \quad \frac{\partial \tilde{T}^*}{\partial r_1^*} = 0 \text{ at } r_1^* = 0; \tag{7}$$

$$w = w_s, \quad \frac{\partial w_1^*}{\partial r_1^*} = \frac{\alpha}{\sqrt{k_1^*}}(w_s - w_{porous}), \quad \tilde{T}^* = \tilde{T}_w^* \text{ at } r_1^* = R, \tag{8}$$

where w_{porous} is the velocity in the permeable boundary, w_s is the slip velocity, Da is the Darcy number, α (called the slip parameter) is a dimensionless quantity depending on the material parameters which characterize the structure of the permeable material within the boundary region.

The initial conditions are assumed as:

$$w_1^* = 0, \tilde{T}^* = 0 \text{ at } t_1^* = 0 \tag{9}$$

Figure 2a depicts the thermophysical parameters of hybrid nanofluid whereas Fig. 2b shows the thermophyiscal properties of blood and nanoparticles.

Blood flows through the cardiovascular system due to the heart's pumping motion, causing a pressure gradient across the vascular network. The pressure gradient is separated into two parts: non-fluctuating (continuous) and fluctuating (pulsatile) [3]:

$$-\frac{\partial p_1^*}{\partial z_1^*} = A_0 + A_1 cos(w_p t_1^*), t_1^* > 0 \tag{10}$$

where $w_p = 2\pi f_p$, f_p denotes the heart pulse frequency, A_0 signifies the amplitudes of the steady-state component, and A_1 represents the pulsatile components of the pressure gradient, respectively.

The preceding governing equations (2)–(5) are non-dimensionalized to achieve a numerical solution by introducing the following transformation variables:

$$\bar{r}_1^* = \frac{r_1^*}{R_0}, \bar{w}_1^* = \frac{w_1^*}{U_0}, \bar{u}_1^* = \frac{L_0 u_1^*}{\delta^* U_0}, \bar{t}_1^* = \frac{U_0 t_1^*}{R_0}, \bar{z}_1^* = \frac{z_1^*}{L_0}, \bar{p}_1^* = \frac{R_0^2 p_1^*}{U_0 L_0 \mu_0}, \tilde{\theta} = \frac{\tilde{T}^* - \tilde{T}_1^*}{\tilde{T}_w^* - \tilde{T}_1^*},$$
$$\bar{R} = \frac{R}{R_0}, \bar{d} = \frac{d}{L_0}, \bar{w}_s = \frac{w_s}{U_0}, \bar{w}_{porous} = \frac{w_{porous}}{U_0}, Q = \frac{Q_0 R_0^2}{(\tilde{T}_w^* - \tilde{T}_1^*) k_f}, Re = \frac{U_0 \rho_f R_0}{\mu_f},$$
$$Gr = \frac{\rho_f R_0^2 g \gamma_f (\tilde{T}_w^* - \tilde{T}_1^*)}{\mu_f U_0}, M^2 = \frac{\sigma_f B_0^2 R_0^2}{\mu_f}, Ec = \frac{U_0^2}{C_p(\tilde{T}_w^* - \tilde{T}_1^*)}, Pr = \frac{\mu_f C_p}{k_f},$$
$$Da = \frac{k_1^*}{R_0^2}, Br = EcPr = \frac{\mu_f U_0^2}{k_f(\tilde{T}_w^* - \tilde{T}_1^*)}. \tag{11}$$

After inserting the non-dimensionalized variables, the normalized form of the pressure gradient is:

$$-\frac{\partial p_1^*}{\partial z_1^*} = B_1[1 + ecos(c_1 t_1^*)] \tag{12}$$

where

$$e = \frac{A_1}{A_0}, B_1 = \frac{A_0 R_0^2}{\mu_0 U_0}, c_1 = \frac{2\pi R_0 f_p}{U_0} \tag{13}$$

With the substitution of variables given in (11), ignoring the bars, assuming that in comparison to the radius of the artery, the maximal height of stenosis is lesser, i.e., $\delta(= \delta^*/R_0) << 1$ and the radius of the artery and the length of the stenotic area have comparable magnitudes, i.e., $\epsilon(=R_0/L_0) = O(1)$, and further utilizing the radial coordiante transformation given by $\left(x_1^* = \frac{r_1^*}{R(z_1^*)}\right)$, the governing (2)–(5) become:

$$Re\frac{\rho_{hnf}}{\rho_f}\frac{\partial w_1^*}{\partial t_1^*} = B_1[1 + ecos(c_1 t_1^*)] + \left(\frac{\mu_{hnf}(\tilde{\theta})}{\mu_0}\right)\left(\frac{1}{R^2}\right)\left[\frac{\partial^2 w_1^*}{\partial x_1^{*2}} + \frac{1}{x_1^*}\frac{\partial w_1^*}{\partial x_1^*}\right]$$
$$+ \frac{(\rho\gamma)_{hnf}}{(\rho\gamma)_f} Gr\tilde{\theta} - \frac{\sigma_{hnf}}{\sigma_f} M^2 w_1^* \tag{14}$$

Table 1 The values of emerging parameters

Parameters	ϕ_1	ϕ_2	d	B_1	c_1	e	δ	β_0	w_s	α	ξ	M	Gr	Da	Re	Pr	Ec
Value	0.03	0.03	0.56	1.41	1	0.2	0.1	0.5	0.1	0.1	$\pi/4$	$\sqrt{3}$	0.5	0.1	2	21	0.1

$$\frac{(\rho C_p)_{hnf}}{(\rho C_p)_f}\frac{\partial\tilde{\theta}}{\partial t_1^*} = \frac{1}{RePr}\frac{k_{hnf}}{k_f}\left(\frac{1}{R^2}\right)\left[\frac{\partial^2\tilde{\theta}}{\partial x_1^{*2}} + \frac{1}{x_1^*}\frac{\partial\tilde{\theta}}{\partial x_1^*}\right] + \frac{\sigma_{hnf}}{\sigma_f}\frac{B_1 M^2}{Re}w_1^{*2} + \frac{\mu_{hnf}(\tilde{\theta})}{\mu_0}\frac{Ec}{Re}\left(\frac{1}{R^2}\right)\left(\frac{\partial w_1^*}{\partial x_1^*}\right)^2 \tag{15}$$

Here, Reynolds viscosity model [7] is considered for the temperature-dependent viscosity, which is as follows:

$$\mu_f(\tilde{\theta}) = \mu_0 e^{-\beta_0\tilde{\theta}} = \mu_0[1-\beta_0\tilde{\theta}] \text{ where } \beta_0 << 1 \tag{16}$$

The associated boundary conditions (7), (8) and the initial conditions (9) becomes:

$$\left.\frac{\partial w_1^*}{\partial x_1^*}\right|_{x_1^*=0} = 0,\ \left.\frac{1}{R}\frac{\partial w_1^*}{\partial x_1^*}\right|_{x_1^*=1} = \frac{\alpha}{\sqrt{Da}}(w_s - w_{porous}),\ w_1^*|_{x_1^*=1} = w_s,\ \left.\frac{\partial\tilde{\theta}}{\partial x_1^*}\right|_{x_1^*=0} = 0,\ \tilde{\theta}|_{x_1^*=1} = 1 \tag{17}$$

$$w_1^*|_{t_1^*=0} = 0,\ \tilde{\theta}|_{t_1^*=0} = 0 \tag{18}$$

The wall shear stress (WSS) and the flow rate are expressed as:

$$\tau_w = -\frac{1}{R}\left(\frac{\partial w_1^*}{\partial x_1^*}\right)_{x_1^*=1} \tag{19}$$

$$Q_1 = 2\pi R^2\int_0^1 w_1^* x_1^* dx_1^* \tag{20}$$

The equations given by (14) and (15) are coupled partial differential equations, and numerical schemes are often used to solve these equations as finding an exact solution is a tedious work. The Crank–Nicolson scheme is based on the implicit finite difference method. The fact that this method is unconditionally stable is one of the key reasons for its use [16]. Furthermore, the order of convergence is two in time and space. The step size in the spatial direction is x = 1/N+1, discretizing the spatial variable in N+1 grid points. The value of t^k given as $t^k = (k-1)dt$, determines the time instant, and dt signifies a small increment in time. The method being implicit is stable for any value of dt and dx, we have chosen the values very precisely as $dt = 0.0001$ and $dx = 0.0001$. Also, no further change is noticed in the results with a decrement in these values. The reduced system of Eqs. (14) and (15) are solved using the Tri-diagonal Matrix Algorithm (TDMA) [1] as they form a tri-diagonal system of equations.

3 Results and Graphical Analysis

The goal of this study is to investigate the hemodynamic features under the effect of pulsatile blood flow via an artery with irregular stenosis, an inclined external magnetic field, viscous dissipation, and Joule heating incorporating hybrid nanoparticles. The validation of the work done is depicted by Fig. 3. Streamline contours for different flow parameters are depicted in Figs. 4 and 5. The results for velocity, wall shear stress, volumetric flow rate, and temperature are illustrated graphically in Figs. 6, 7, 8 and 9. The computational work has been carried out by using the data illustrated in the Table 1.

3.1 Validation of the Numerical Results

The results obtained are compared with previously published work to authenticate the results obtained in the present study. Figure 3a, b are used for validating velocity and temperature profiles of the present work with previous study done by Tripathi et al. [18]. The results are compared for Au-nanoparticles, which is common in both the research work and the effect of the Darcy number (Da), velocity slip (α), and Eckert number (Ec) has been ignored. Tripathi et al. [18] used the FTCS scheme to solve the dimensionless governing equations. Therefore, in this comparison, the FTCS scheme is employed for [18] work, and the Crank–Nicolson scheme is used in the current study. These figures show a good agreement between our study for velocity and temperature with the previous research [18].

The velocity contours depict the actual blood flow patterns as these correctly illustrate the research efforts. Figures 4 and 5 represent these contours for different influential parameters. The impact of w_s is depicted in Fig. 4. The number of trapped boluses and their size increases with increasing w_s values implying that the velocity profile elevates with an increase in w_s. The slip effect amplifies the acceleration in the axial flow, which is the reason for elevation in velocity profiles with increasing w_s values. The velocity contours for different Re values are illustrated in Fig. 5. With increasing Re values, the trapped boluses become fewer and eventually vanish, suggesting a drop in velocity values. The regime becomes viscous-dominated at low Re values. As a result, a drop in velocity is correlated with a rise in Re values.

The velocity profiles for Darcy number (Da), wall slip velocity (w_s), inclination parameter (ξ), and varying concentrations of Au & Al_2O_3 nanoparticles (ϕ_1, ϕ_2) are illustrated in Fig. 6. Figure 6a shows the influence of Da on non-dimensional velocity profiles in the presence and absence of a magnetic field. The medium's permeability increases as the value of Da rises, resulting in a declination of the velocity profile.

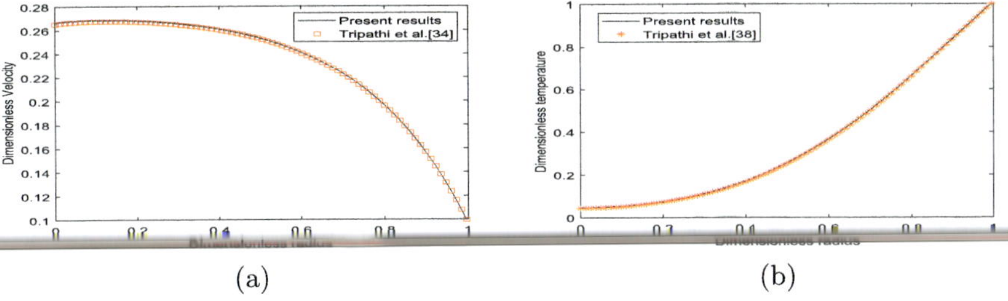

(a) (b)

Fig. 3 **a** Comparison of velocity profile for $M^2 = \sqrt{3}$, **b** Comparison of temperature profile for Pr = 21

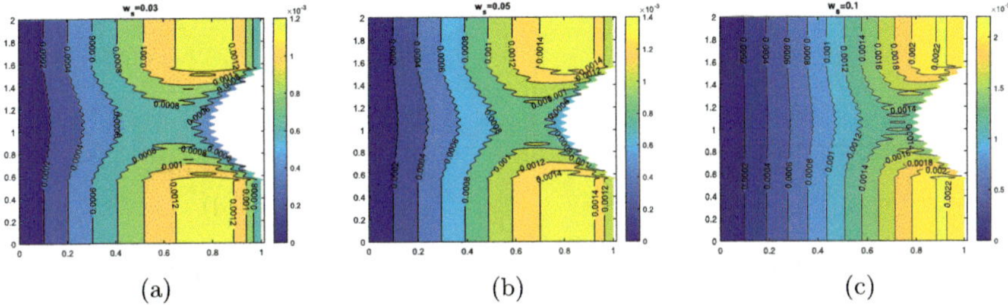

(a) (b) (c)

Fig. 4 Velocity contours for wall slip velocity **a** $w_s = 0.03$, **b** $w_s = 0.05$, **c** $w_s = 0.1$

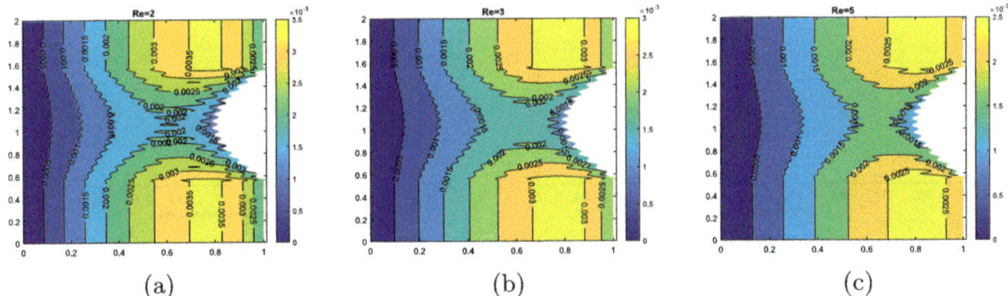

(a) (b) (c)

Fig. 5 Velocity contours for Reynolds number **a** Re = 2, **b** Re = 3, **c** Re = 5

It's also worth noting that when M^2 rises, the velocity profile declines slightly. This demonstrates that using a magnetic field to reduce blood velocity is advantageous. The velocity profiles for different values of w_s is represented by Fig. 6b. The hydrodynamic wall slip effect causes the axial flow to accelerate as the wall slip velocity increases. The inclusion of slip results in a momentum increases at the application zone, which improves the velocity profile. The Au-Al_2O_3/blood hybrid nanofluid consistently reaches lower magnitudes than the Au/blood nanofluid, showing the presence of Al_2O_3 nanoparticles decelerates the blood flow. Figure 6c highlights the influence of ξ on the dimensionless velocity profile. With increasing values of ξ, a decreasing trend in velocity profiles is observed. This is because as the angle of inclination increases, the influence of the magnetic field on fluid particles increases. Thus,

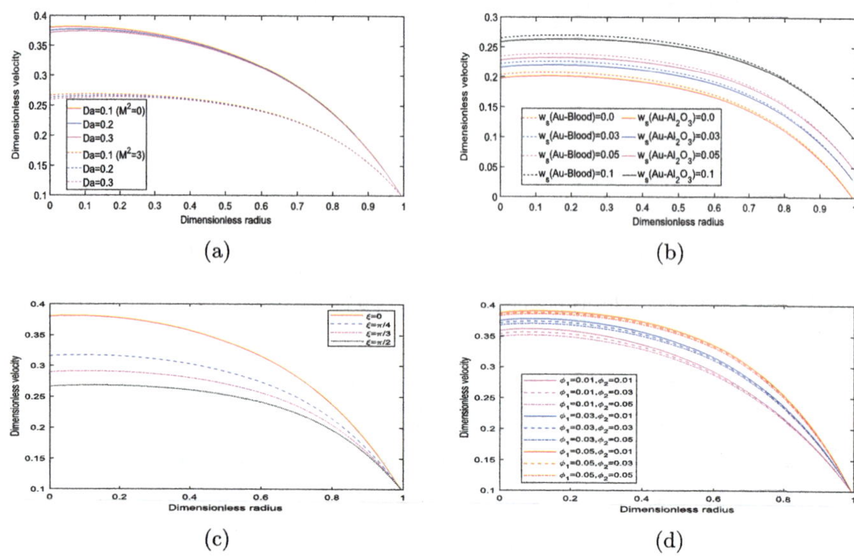

Fig. 6 Effect of **a** Darcy number (Da), **b** wall slip velocity (w_s), **c** inclination parameter (ξ), and **d** varying concentrations of Au & Al_2O_3 nanoparticles (ϕ_1, ϕ_2) on velocity at $z_1^* = 1.06$ and $t_1^* = 1.2$

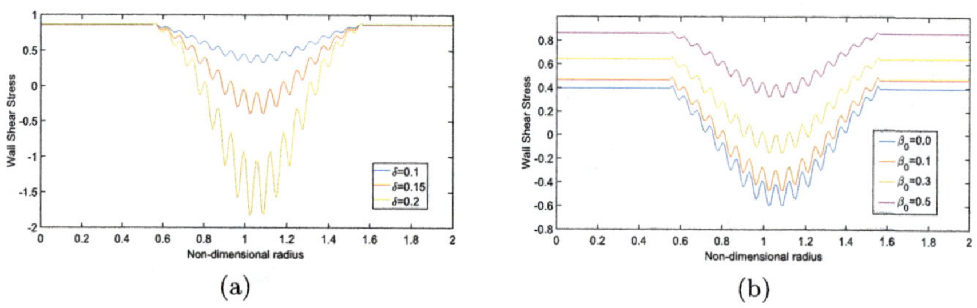

Fig. 7 Influence of **a** stenotic depth (δ), **b** viscosity parameter (β_0), on wall shear stress at $t_1^* = 1.2$

the Lorentz force enhances, which causes a reduction in velocity values. The influence of varying concentrations of Au & Al_2O_3 nanoparticles (ϕ_1, ϕ_2) on velocity profile is shown in Fig. 6d. The value $\phi_1 = 0.01$, $\phi_2 = 0.05$ shows minimal velocity, whereas the maximum velocity is shown for $\phi_1 = 0.05$, $\phi_2 = 0.01$. This demonstrates that Au & Al_2O_3 nanoparticles show opposite effects on velocity. The velocity profiles show an enhancement with an increase in the concentration of Au-nanoparticles, whereas a decrement in velocity profile is analyzed with an increase in Al_2O_3-nanoparticles. This benefits bringing blood velocity under control and allowing surgeons to make adjustments as needed.

The profiles corresponding to wall shear stress are demonstrated in Fig. 7 for stenotic depth (δ) and viscosity parameter (β_0). The wall shear stress profiles highlighting the impact of δ are represented by Fig. 7a. As shown by the profiles, the

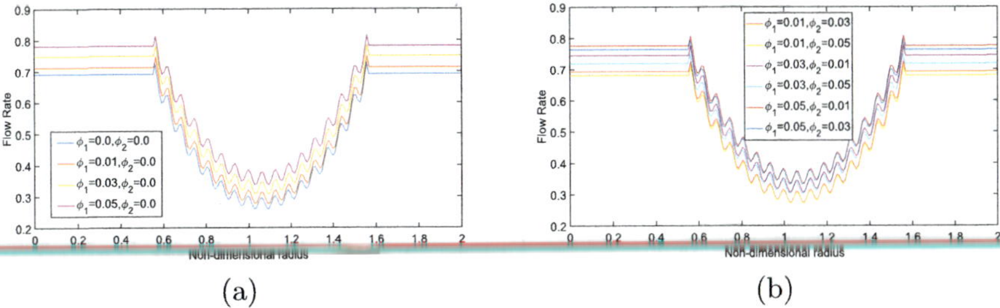

Fig. 8 Effect of **a** different concentrations of Au-nanoparticles (ϕ_1), **b** varying concentrations of Au & Al_2O_3 nanoparticles (ϕ_1, ϕ_2) on Flow rate at $t_1^* = 1.2$

values of wall shear stress show a decrement with increasing δ values. This implies that there are lower shear stress values with an increment in the stenotic depth. These lower values are more dangerous, as suggested by Zhang et al. [19] in their study. Figure 7b signifies the wall shear stress profiles for different values of β_0. With an increase in the β_0 values corresponding increase in the wall shear stress values is observed. The increment in wall shear stress denotes that the particles' internal resistance has decreased as the magnitude of the viscosity parameter has increased.

The volumetric flow rate is the amount of fluid(blood) that flows in a given amount of time. The flow rate profiles for varying concentrations of Au-nanoparticles (ϕ_1) and varying concentrations of Au & Al_2O_3 nanoparticles (ϕ_1, ϕ_2) are depicted in Fig. 8. Figure 8a represents the influence of Au-nanoparticles concentration on volumetric flow rate. According to the findings, the flow rate profiles demonstrate an increasing trend with increasing concentration of Au-nanoparticles. The flow rate profiles for the influence of varying concentrations of Au & Al_2O_3 nanoparticles is shown in Fig. 8b. It can be interpreted that there is a declination in flow rate on increasing the concentration of Al_2O_3 nanoparticles, whereas it enhances with increment in Au-nanoparticles. So, it can be concluded from Fig. 8a and b that the Al_2O_3 nanoparticles reduce the flow rate.

The non-dimensional temperature profiles for Eckert number (Ec) and Prandtl number (Pr) are illustrated by Fig. 9. Figure 9a depicts the impact of Ec on the non-dimensional temperature profile. The Ec describes how work against viscous fluid stress converts kinetic energy into internal energy. The heat energy develops in the fluid due to this conversion which causes the fluid's (blood) temperature to rise. As a result, a rise in Ec denotes a higher temperature distribution. Figure 9b portrays the non-dimensional temperature profiles for different values of Pr. With increasing Pr, the temperature profiles show a decreasing trend. The Pr is a dimensionless quantity that correlates viscosity to thermal conductivity in a fluid. As a result, it determines the link between the motion of a fluid and its heat transfer capacity. The thickness of the thermal boundary layer reduces, which means the transfer of heat from the artery decreases with increasing Pr values leading to a declination of arterial temperature.

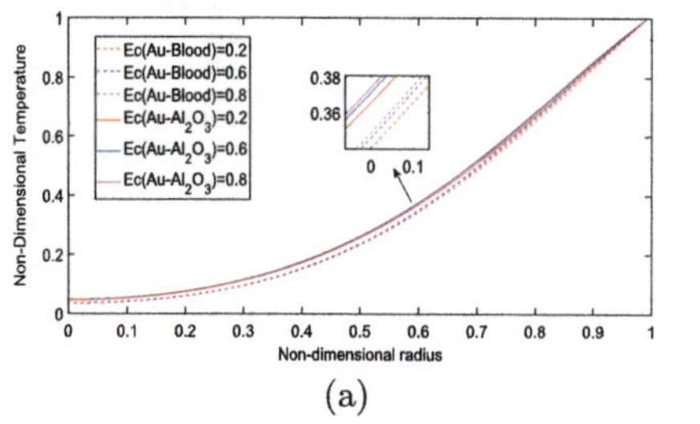

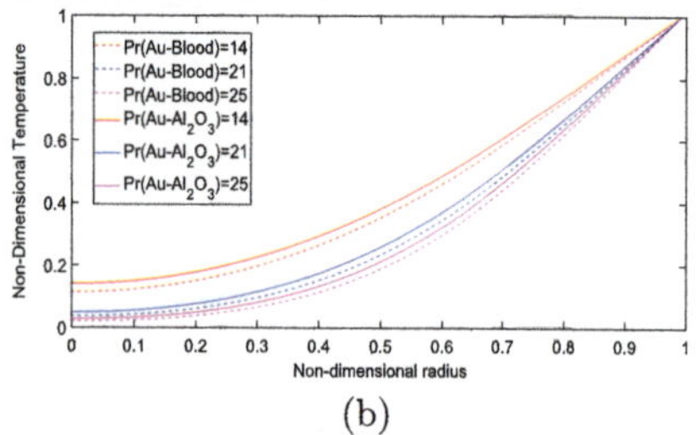

Fig. 9 The non-dimensional temperature profiles for **a** Eckert number (Ec), **b** Prandtl number (Pr) at $z_1^* = 1.06$ and $t_1^* = 1.2$

4 Conclusions

This study focuses on the influence of hybrid nanoparticles on blood flow through an artery with mild irregular stenosis. This model aims to improve nanoparticle drug delivery to a desired location in the human body. The partial differential equations are solved using an implicit finite difference scheme, i.e., the Crank–Nicolson scheme. The effects of a wide range of physical parameters have been evaluated on axial blood velocity, temperature profile, volumetric flow rate, and artery wall shear stress (WSS). The following are some of the study's key findings: The streamline contours illustrate that as w_s increases, so does the velocity. The velocity profile diminishes as Da and ξ are increased. The profiles for wall shear stress show an enhancement with increasing β_0 values, whereas an increase in δ leads to lower profiles. The flow rate increases as ϕ_1 increases but decreases as ϕ_2 increases. The temperature profiles increase as the Ec values rise but decrease as the Pr values rise.

References

1. Anderson, J.D., Wendt, J.: Computational fluid dynamics, vol. 206. Springer, Berlin (1995)
2. Basri, A.A., Khader, S.M.A., Johny, C., Pai, R., Zuber, M., Ahmad, K.A., Ahmad, Z.: Numerical study of haemodynamics behaviour in normal and single stenosed renal artery using fluid-structure interaction. J. Adv. Res. Fluid Mech. Therm. Sci. **51**(1), 91–98 (2018)
3. Burton, A.C.: Physiology and biophysics of the circulation. Acad. Med. **40**(8), xxx–xxxvi (1965)
4. Changdar, S., De, S.: Investigation of nanoparticle as a drug carrier suspended in a blood flowing through an inclined multiple stenosed artery. Bionanoscience **8**(1), 166–178 (2018)
5. Chen, C., Yan, J., Zhou, N., Zhao, J., Wang, D.: Analysis of myocardial injury in patients with covid-19 and association between concomitant cardiovascular diseases and severity of covid-19. Zhonghua xin xue guan bing za zhi, pp. 567–571 (2020)
6. Devi, S.A., Devi, S.S.U.: Numerical investigation of hydromagnetic hybrid cu-al2o3/water nanofluid flow over a permeable stretching sheet with suction. Int. J. Nonlinear Sci. Numer. Simul. **17**(5), 249–257 (2016)

7. Ellahi, R., Raza, M., Vafai, K.: Series solutions of non-newtonian nanofluids with reynolds' model and vogel's model by means of the homotopy analysis method. Math. Comput. Modell. **55**(7–8), 1876–1891 (2012)
8. Kumawat, C., Sharma, B., Mekheimer, K.: Mathematical analysis of two-phase blood flow through a stenosed curved artery with hematocrit and temperature dependent viscosity. Phys. Scripta **96**(12), 125277 (2021)
9. Majee, S., Shit, G.: Modeling and simulation of blood flow with magnetic nanoparticles as carrier for targeted drug delivery in the stenosed artery. Europ. J. Mech. B/Fluids **83**, 42–57 (2020)
10. Mensah, G.A., Roth, G.A., Fuster, V.: The global burden of cardiovascular diseases and risk factors: 2020 and beyond (2019)
11. Ponalagusamy, R., Manchi, R.: A four-layered model for flow of non-newtonian fluid in an artery with mild stenosis. Sādhanā **44**(7), 1–14 (2019)
12. Ponalagusamy, R., Priyadharshini, S.: Numerical investigation on two-fluid model (micropolar-newtonian) for pulsatile flow of blood in a tapered arterial stenosis with radially variable magnetic field and core fluid viscosity. Comput. Appl. Math. **37**(1), 719–743 (2018)
13. Shahzadi, I., Bilal, S.: A significant role of permeability on blood flow for hybrid nanofluid through bifurcated stenosed artery: drug delivery application. Comput. Methods Programs Biomed. **187**, 105248 (2020)
14. Sharma, B., Gandhi, R., Bhatti, M.: Entropy analysis of thermally radiating mhd slip flow of hybrid nanoparticles (au-al2o3/blood) through a tapered multi-stenosed artery. Chem. Phys. Lett. 139348 (2022)
15. Sharma, M., Sharma, B.K., Gaur, R., Tripathi, B.: Soret and dufour effects in biomagnetic fluid of blood flow through a tapered porous stenosed artery. J. Nanofluids **8**(2), 327–336 (2019)
16. Smith, G.D., Smith, G.D., Smith, G.D.S.: Numerical solution of partial differential equations: finite difference methods. Oxford University Press, Oxford (1985)
17. Tripathi, B., Sharma, B.K.: Influence of heat and mass transfer on two-phase blood flow with joule heating and variable viscosity in the presence of variable magnetic field. Int. J. Comput. Methods **17**(03), 1850139 (2020)
18. Tripathi, J., Vasu, B., Bég, O.A., Gorla, R.S.R.: Unsteady hybrid nanoparticle-mediated magneto-hemodynamics and heat transfer through an overlapped stenotic artery: Biomedical drug delivery simulation. Proc. Inst. Mech. Eng. Part H: J. Eng. Med. 09544119211026095 (2021)
19. Zhang, B., Gu, J., Qian, M., Niu, L., Zhou, H., Ghista, D.: Correlation between quantitative analysis of wall shear stress and intima-media thickness in atherosclerosis development in carotid arteries. Biomed. Eng. Online **16**(1), 1–17 (2017)

An Analytical Approach to Study the Environmental Transport of Fine Settling Particles in a Wetland Flow

Subham Dhar, Nanda Poddar, and Kajal Kumar Mondal

Abstract The present research deals with an analytical solution of convection-diffusion equation which represents the water phase based superficial concentration in a vegetated wetland. Gill's series expansion method is used to obtain the dispersion coefficient and mean concentration distribution of the settling particles. For the limiting case of vegetation factor, tortuosity and settling velocity, the dispersivity is compared with the earlier research work and an excellent agreement is achieved with them. Effects of settling velocity, tortuosity and vegetation factor, on the dispersion coefficient and mean concentration are observed. A physical application is performed to predict the critical length of the concentration cloud beyond the ecological safe level. It is seen that with the increment of settling velocity, the duration for safe level increases. Moreover, the study provides two important criteria for wastewater treatment namely critical length and duration of solute cloud in the wetland. The work may also be applicable to investigate the sedimentation process and flood damage control.

Keywords Vegetated wetland · Gill's series expansion method · Dispersion · Settling velocity · Critical length

1 Introduction

The study of solute dispersion phenomena was initially introduced by Taylor [1]. Following his work, Aris [2] employed the method of moments and removed some restrictions of Taylor's theory. Later, Gill and Sankarasubramanian [3, 4] used the series expansion method to find the solution of the convection-diffusion equation. Several researchers [5–7] investigated the free surface effect on the dispersion of

Supported by CSIR India and UGC India.

S. Dhar (✉) · N. Poddar · K. K. Mondal
Cooch Behar Panchanan Barma University, Cooch Behar 736101, India
e-mail: dharsuvam94@gmail.com

S. Banerjee and A. Saha (eds.), *Nonlinear Dynamics and Applications*,
Springer Proceedings in Complexity,
https://doi.org/10.1007/978-3-030-99792-2_29

tracer in a wetland flow. The influences of bed absorption and bulk degradation on mass transfer phenomena are studied by numerous researchers [8, 9]. Various analytical methods such as multi-scale analysis [5, 10–12], method of moments [13, 14], Taylor's classical technique [15] were used to investigate the environmental dispersivity and mean concentration distribution. Hammer [16] and Zhou et al. [17] showed how the constructed wetland is useful for wastewater treatment. Wu et al. [18] observed the tidal effect on environmental dispersion coefficient for a depth dominated wetland flow. Recently, Dhar et al. [19] and Poddar et al. [20] explored the dispersion of settling particles in a wetland flow using finite difference technique and the method of moments respectively.

The current research presents an analytical approach to investigate the environmental transport of fine particles with settling velocity in a depth dominated wetland flow. The influences of linear and nonlinear boundary reactions on the solute dispersion was studied by many researchers but the transport of fine sediment particles in a wetland flow paid low attention though it has a wide range of applications in the direction of environmental engineering. Gill's series expansion method is used to explore the environmental dispersivity and mean concentration in this study. Moreover, the critical length and the duration of the concentration cloud are obtained with the help of the mean concentration profile of the tracers. It is seen that, the solute concentration is conglomerated near the source when the fall velocity of the particles in the wetland flow increase significantly. The maximum critical length and corresponding duration of the cloud of settling particles are obtained for 5 day Biochemical Oxygen Demand (BOD_5), Total Phosphorus (TP) and Total Nitrogen (TN) which relates the wastewater treatment in the wetland.

2 Mathematical Formulation

2.1 Velocity Distribution

Consider the dispersion of settling particles in a fully developed and unidirectional flow with constant porosity (ϕ), vegetation force (F), tortuosity (κ), momentum dispersivity (L) and mass dispersivity K in a depth dominated wetland channel of depth H. A Cartesian coordinate system is used where x^*-axis is along the longitudinal direction, y^*-axis along the vertical direction and the origin is situated at the bed surface. For a depth dominated wetland channel, effects of width are considered negligible as the first approximation, and so related terms are taken as width-averaged (see Fig. 1). For a constant pressure gradient, the momentum equation reduces as [15],

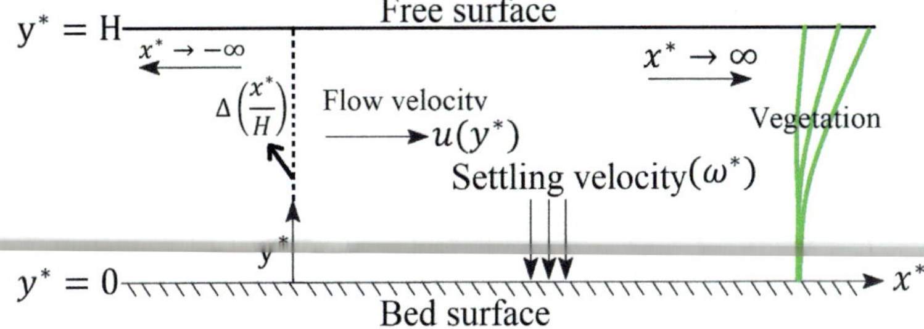

Fig. 1 Sketch of the present model

$$\kappa(\mu + L_{yy})\frac{d^2u}{dy^{*2}} = \mu F u + \frac{dp}{dx^*} \tag{1}$$

with stationary conditions,

$$u(y^*)|_{y^*=0} = u'(y^*)|_{y^*=H} = 0 \tag{2}$$

Considering the characteristic velocity $u_c = -\frac{dp}{dx^*} \cdot \frac{H^2}{\mu+L_{yy}}$ and the dimensionless parameter $y = \frac{y^*}{H}$, Eq. (1) can be rewritten as,

$$\kappa\frac{d^2u}{dy^2} - \alpha^2 u = -u_c, \tag{3}$$

where, $\alpha = \sqrt{\frac{\mu F H^2}{\mu+L_{yy}}}$ is the vegetation factor which depends on vegetation force (F), fluid viscosity (μ) and vertical momentum dispersivity (L_{yy}). The corresponding stationary condition is given by,

$$u(y)|_{y=0} = u'(y)|_{y=1} = 0. \tag{4}$$

Solving (3) and (4), the normalized velocity profile is found as,

$$\psi = \frac{u}{u_c} = \frac{1}{\alpha^2}\left(1 - \frac{\cosh(\frac{\alpha}{\sqrt{\kappa}}(y-1))}{\cosh(\frac{\alpha}{\sqrt{\kappa}})}\right). \tag{5}$$

It is amusing to observe that, without vegetation factor and tortuosity (i.e., $\alpha \to 0, \kappa = 1$), the velocity profile becomes an open channel flow as

$$\lim_{\alpha\to 0,\kappa\to 1} \psi = \frac{y(2-y)}{2} \tag{6}$$

The effects of vegetation factor and tortuosity on the velocity profile are shown in Fig. 2. It is observe that, when the amount of vegetation in wetland increases, the flow velocity significantly decreases as the vegetation factor resists the flow. On the

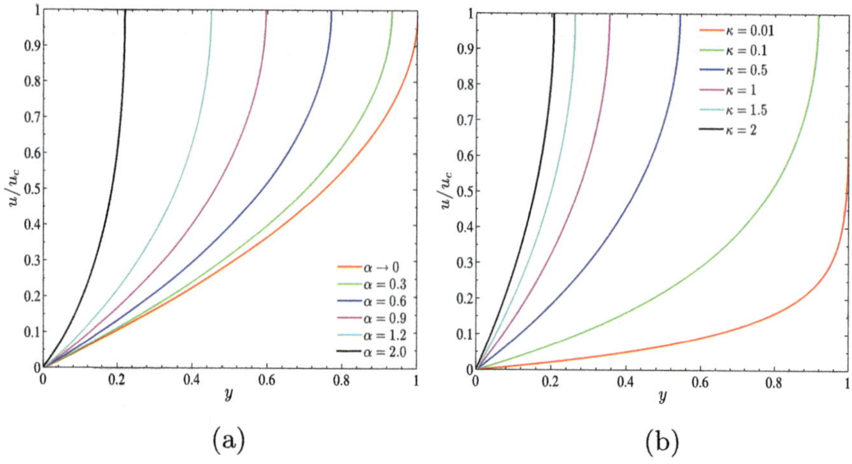

Fig. 2 Variation of velocity profile $u(y)/u_c$: **a** $\alpha \to 0$, $\alpha = 0.3, 0.6, 0.9, 1.2, 2$ with $\kappa = 0.5$ and **b** $\kappa = 0.01, 0.1, 0.5, 1, 1.5, 2$ with $\alpha = 1$

other hand, as tortuosity increases the flow becomes tortous and consequently speed of the flow decreases.

2.2 *Mass Transport Equation*

When the fine particles of settling velocity ω^* with mass diffusivity λ is injected into the above mentioned vegetated wetland flow, the water phase based concentration C^* satisfies the following mass transfer equation

$$\frac{\partial C^*}{\partial t^*} + \frac{u}{\varphi}\frac{\partial C^*}{\partial x^*} - \omega^*\frac{\partial C^*}{\partial y^*} = \kappa\left(\lambda + \frac{K}{\varphi}\right)\left(\frac{\partial^2 C^*}{\partial x^{*2}} + \frac{\partial^2 C^*}{\partial y^{*2}}\right). \tag{7}$$

with the corresponding initial and boundary conditions

$$C^*(x^*, y^*, t^*)|_{t^*=0} = \Delta\left(\frac{x^*}{H}\right) \tag{8a}$$

$$\left[\kappa\left(\lambda + \frac{K}{\varphi}\right)\frac{\partial C^*}{\partial y^*} + \omega^* C^*\right]_{y^*=0,H} = 0 \tag{8b}$$

$$C^*(x^*, y^*, t^*)|_{x^*\to\pm\infty} = 0. \tag{8c}$$

Introducing the non-dimensional parameters $t = \frac{\kappa\left(\lambda+\frac{K}{\varphi}\right)t^*}{H^2}$, $y = \frac{y^*}{H}$, $C = \frac{C^*}{Q/H^2}$, $x = \frac{\kappa\left(\lambda+\frac{K}{\varphi}\right)}{H^2\bar{u}}(x^* - \bar{u}t^*)$, $\Psi = \frac{u}{\bar{u}}$, $\omega = \frac{\omega^* H}{\kappa\left(\lambda+\frac{K}{\varphi}\right)}$, the Eq. (7) becomes

$$\frac{\partial C}{\partial t} + (\Psi - 1)\frac{\partial C}{\partial x} - \omega\frac{\partial C}{\partial y} = \frac{1}{P_e^2}\frac{\partial^2 C}{\partial x^2} + \frac{\partial^2 C}{\partial y^2} \tag{9}$$

where $P_e = \frac{\bar{u}H}{\kappa\left(\lambda+\frac{K}{\varphi}\right)\varphi}$ is the Péclet number which measures the ratio of the convection rate to diffusion rate. Also, the respective non-dimensional initial and boundary conditions are given by

$$C(x, y, t)|_{t=0} = \Delta(x) \tag{10a}$$

$$\left[\frac{\partial C}{\partial y} + \omega C\right]_{y=0,1} = 0 \tag{10b}$$

$$C(x, y, t)|_{x\to\pm\infty} = 0 \tag{10c}$$

3 Mean Concentration Expansion

The solution of Eq. (9) can be formulated using Gill's series expansion method as [4],

$$C = \bar{C} + \sum_{k=1}^{\infty} f_k(t, y)\frac{\partial^k \bar{C}}{\partial x^k} \tag{11}$$

where $\bar{C} = \int_0^1 C dy$. Putting (11) in Eq. (9), one can get

$$\begin{aligned}\frac{\partial \bar{C}}{\partial t} + (\Psi - 1)\frac{\partial \bar{C}}{\partial x} - \frac{1}{P_e^2}\frac{\partial^2 \bar{C}}{\partial x^2} + \sum_{k=1}^{\infty}\Bigg[\left(\frac{\partial f_k}{\partial t} - \frac{\partial^2 f_k}{\partial y^k} - \omega\frac{\partial f_k}{\partial y}\right)\frac{\partial^k \bar{C}}{\partial x^k} \\ +(\Psi - 1)f_k\frac{\partial^{k+1}\bar{C}}{\partial x^{k+1}} - \frac{1}{P_e^2}\frac{\partial^{k+2}\bar{C}}{\partial x^{k+2}} + f_k\frac{\partial^{k+1}\bar{C}}{\partial x^k \partial t}\Bigg].\end{aligned} \tag{12}$$

To obtain the time dependent dispersion coefficient, the dispersion model can be expressed as

$$\frac{\partial \bar{C}}{\partial t} = \sum_{i=1}^{\infty} K_i(t)\frac{\partial^i \bar{C}}{\partial x^i}, \tag{13}$$

and consequently one can have

$$\frac{\partial^{k+1}\bar{C}}{\partial x^k \partial t} = \sum_{i=1}^{\infty} K_i(t) \frac{\partial^{i+k}\bar{C}}{\partial x^{i+k}}. \tag{14}$$

Using (13) and (14) in (12), we get

$$\left(K_1(t) + \Psi - 1 + \frac{\partial f_1}{\partial t} - \frac{\partial^2 f_1}{\partial y^2} - \omega \frac{\partial f_1}{\partial y}\right) \frac{\partial \bar{C}}{\partial x} + \left(K_2(t) - \frac{1}{P_e^2} + (\Psi - 1) f_1 \right.$$
$$\left. + f_1 K_1 + \frac{\partial f_2}{\partial t} - \frac{\partial^2 f_2}{\partial y^2} - \omega \frac{\partial f_2}{\partial y}\right) \frac{\partial^2 \bar{C}}{\partial x^2} + \cdots = 0. \tag{15}$$

On comparing the terms associated with $\frac{\partial \bar{C}}{\partial x}$ and $\frac{\partial^2 \bar{C}}{\partial x^2}$, we have,

$$K_1(t) + \Psi - 1 + \frac{\partial f_1}{\partial t} - \frac{\partial^2 f_1}{\partial y^2} - \omega \frac{\partial f_1}{\partial y} = 0, \tag{16}$$

$$K_2(t) - \frac{1}{P_e^2} + (\Psi - 1) f_1 + f_1 K_1 + \frac{\partial f_2}{\partial t} - \frac{\partial^2 f_2}{\partial y^2} - \omega \frac{\partial f_2}{\partial y} = 0. \tag{17}$$

Similarly, putting Eq. (11) in (10b), we get

$$\left[\frac{\partial f_k}{\partial y} + \omega f_k = 0\right]_{y=0,1} \qquad \text{for } k = 1, 2, \ldots. \tag{18}$$

Taking average operation on Eqs. (16) and (17), it is found that

$$K_1(t) = 0, \tag{19}$$

$$K_2(t) = \frac{1}{P_e^2} - \int_0^1 \Psi f_1 dy. \tag{20}$$

Since, mean concentration satisfies the one dimensional diffusion like equation, one can neglect other terms of Eq. (13) and it becomes

$$\frac{\partial \bar{C}}{\partial t} = K_2(t) \frac{\partial^2 \bar{C}}{\partial x^2}. \tag{21}$$

Now, solving Eq. (16) with the help of (18), f_1 can be found as

$$f_1 = \frac{\tanh K_a}{K_a - \tanh K_a} \left(\frac{1 + \omega - \omega y}{\omega^2}\right) + \left(\frac{K_a \cosh(K_a(y-1)) - \omega \sinh(K_a(y-1))}{(K_a - \tanh K_a)({K_a}^2 - \omega^2) \cosh K_a}\right)$$
$$+ A e^{-\omega y} + e^{-\frac{\omega y}{2}} \sum_{n=1}^{\infty} \frac{1 - \cos(n\pi)}{n\pi} e^{-(n^2\pi^2 + \frac{\omega^2}{4})t} \left(\cos(n\pi y) - \frac{\omega}{2n\pi} \sin(n\pi y)\right). \tag{22}$$

where $A = \frac{\omega}{e^{-\omega}-1}\left[\frac{(\omega+2)\tanh K_a}{2\omega^2(K_a-\tanh K_a)} + \frac{K_a \sinh K_a+\omega(\cosh K_a-1)}{K_a(K_a^2-\omega^2)(K_a-\tanh K_a)\cosh K_a}\right]$ and $K_a = \frac{\alpha}{\sqrt{\kappa}}$. Now, using (22) in (20), one can get $K_2(t)$ as

$$\begin{aligned} K_2(t) =& \frac{1}{P_e^2} + \frac{K_a}{K_a \cosh K_a - \sinh K_a}\left[\frac{(1+\omega)\tanh K_a \sinh K_a}{\omega^2 K_a(K_a - \tanh K_a)} - \frac{\tanh K_a(\cosh K_a - 1)}{\omega K_a^2(K_a - \tanh K_a)}\right. \\ &+ \frac{1}{(K_a - \tanh K_a)(K_a^2 - \omega^2)\cosh K_a}\left[\frac{\sinh 2K_a}{4} + \frac{K_a}{2} + \frac{\omega \sinh^2 K_a}{2K_a(K_a^2 - \omega^2)}\right] \\ &\left. - A\left(\frac{\omega(\cosh K_a - e^{-\omega}) - K_a \sinh K_a}{K_a^2 - \omega^2}\right) + \sum_{n=1}^{\infty} B_n e^{-(n^2\pi^2+\frac{\omega^2}{4})t}\right], \end{aligned} \tag{23}$$

where $B_n = \frac{1-cos(n\pi)}{n\pi}\left[\left(1 - \frac{\omega}{2n\pi}\right)B_{n_1} + \left(1 + \frac{\omega}{2n\pi}\right)B_{n_2}\right]$, $B_{n_1} = \frac{2K_a-\omega}{(2K_a-\omega)^2+4n^2\pi^2}(e^{-\omega/2}\cos n\pi - e^{-K_a})$ and $B_{n_2} = \frac{2K_a+\omega}{(2K_a+\omega)^2+4n^2\pi^2}(-e^{-\omega/2}\cos n\pi + e^{-K_a})$. Thus, the dispersion coefficient becomes $D_T = P_e^2 K_2(t)$. Now, for large time evolution, without vegetation, tortuosity and settling velocity (i.e., $t \to \infty, \alpha \to 0, \kappa = 1, \omega \to 0$), the dispersion coefficient becomes

$$\lim_{t\to\infty}\lim_{\alpha\to 0}\lim_{\omega\to 0} D_T = 1 + \frac{2}{945}P_e^2 \tag{24}$$

which is exactly equal with the steady dispersion coefficient of Bandyopadhyay and Mazumder [24], and it validates the present solution. Taking the average operation on the initial condition (10a), the solution of the Eq. (13) becomes longitudinal Gaussian distribution as

$$\bar{C} = \frac{1}{\sqrt{4\pi K_2(t)}}\exp\left(-\frac{x^2}{4K_2(t)}\right). \tag{25}$$

Now, let us consider a free surface wetland with porosity $\phi = 0.9$, mean stem diameter $d = 10^{-2}$ m and $H = 1$ m as in [21]. From the Bruggemann equation [22], one can obtain the tortuosity as $\kappa = \sqrt{\phi}$. The vegetated shear force F can be obtained from the Ergun equation [23] as $F = \frac{150(1-\phi^2)}{d^2\phi^3}$. One can take the ambient properties of water such as density $\rho = 10^3$ kgm^{-3}, viscosity $\mu = 10^{-3}$ kgm^{-1}s^{-1}, and diffusivity $\lambda = 10^{-5}$m^2 s^{-1} as in [22]. Also, the mass diffusivity and momentum dispersivity can be taken as $K = 2.86 \times 10^{-3}$m^{2}s^{-1} and $L_{yy} = 1.28$ kgm^{-1}s^{-1} respectively [21]. In case of wastewater emission into the wetland, a sediment cloud would appear in the flow and the mean concentration decays under the combined action of fall velocity and hydraulic transport. For an environmental or ecological risk assessment, related with the solute cloud moving downwards with the mean flow, an influenced region where the involved concentration is beyond the standard level can be obtained. From Eq. (25), the critical length of the influenced region can be found as

$$S(t) = 4\sqrt{-K_2 \log\left(\frac{C_0\phi H^2\sqrt{4\pi K_2}}{Q}\right)}. \tag{26}$$

For an instantaneous emanation of unit quantity per unit width is taken as $Q = 1\,\text{kgm}^{-1}$ in this work. The basic limited values of C_0 for BOD_5, TN and TP are taken as $4 \times 10^{-3}\,\text{kgm}^{-3}$, $1 \times 10^{-3}\,\text{kgm}^{-3}$ and $2 \times 10^{-4}\,\text{kgm}^{-3}$ respectively [22].

4 Discussion of Results

In this section, the effects of Péclet number, vegetation factor, tortuosity and settling velocity on the dispersion coefficient and on the mean concentration distribution are described. Also, a physical application of the current work is shown. From Fig. 3a, it is seen that the dispersion coefficient of the tracers increases with the increment of Péclet number because it increases the convection rate. Enhancement in vegetation results the flow resistance in the wetland and thus the dispersion coefficient decreases gradually (see Fig. 3b). From Fig. 3c, it is observed that, if tortuosity $\kappa \leq 0.1$, a boundary layer is created near the bed surface and consequently, the dispersion coefficient enhances with the enhancement of it. On the other hand, if $\kappa > 0.1$, the flow becomes tortuous and thus dispersivity reduces significantly. The concentration of the fine particles conglomerated near the bed surface as the settling velocity increases and consequently, dispersion of the contaminants decreases prominently and it is presented in Fig. 3d.

Figure 4a shows the temporal evolution of the dispersion coefficient of the solute. It is seen from the figure that, as time progresses dispersion increases for small dispersion times and it reaches its stationary state after a certain critical time. Since, the vegetation factor and tortuosity resists the flow, it is seen from Fig. 4b, c that with the increment of vegetation factor and tortuosity, the concentration agglomerated near the source and hence the peak of the mean concentration profile increases. Figure 4d depicts the mean concentration of the tracers for different values of settling velocity. It is seen that as the settling velocity increases, the peak of the mean concentration distribution enhances. This is because, the deposition of the settling particles

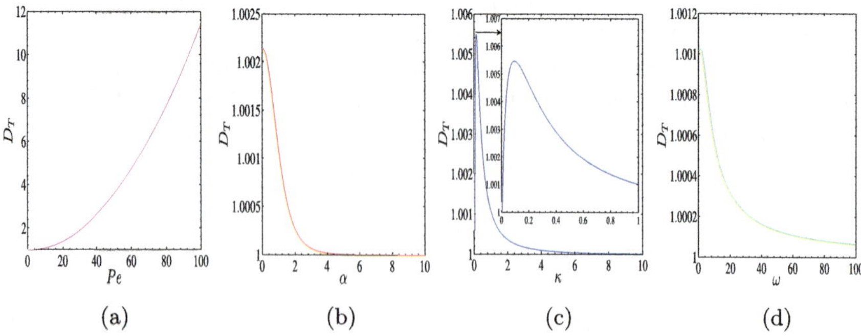

Fig. 3 Variation of dispersion coefficient with **a** Peclet number P_e, **b** vegetation factor α, **c** tortuosity κ, **d** settling velocity ω

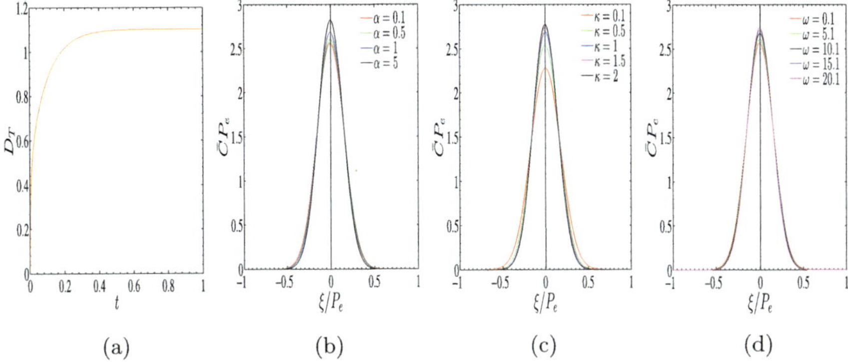

Fig. 4 **a** Temporal evolution of dispersion coefficient; Variation of mean concentration with **b** vegetation factor (α), **c** tortuosity (κ), **d** settling velocity (ω)

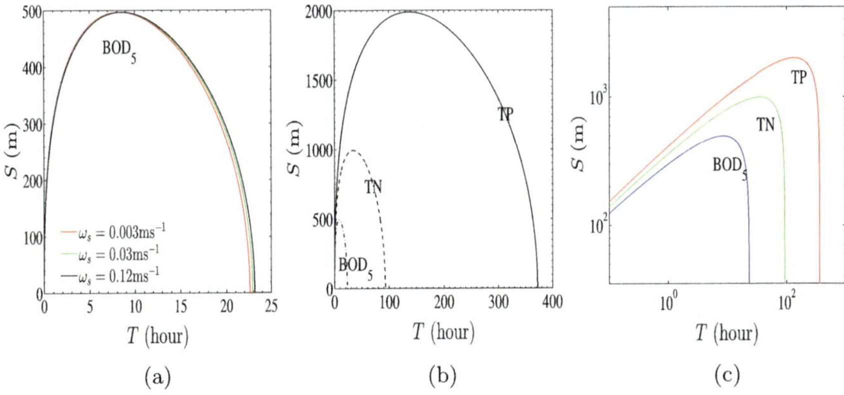

Fig. 5 **a** Variation of influenced region (S) with settling velocity; **b** influenced region for BOD_5, TN, TP; **c** influenced region for BOD_5, TN, TP (log-log)

increases with the increasing values of settling velocity and thus the distribution of mean concentration enhances.

From Fig. 5a, it is observed that the length of the influenced region increases as time proceeds and after a certain time it reaches its maximum and finally it decreases to zero. The maximum critical length for 5 day Biochemical Oxygen Demand (BOD_5), Total Nitrogen (TN) and Total Phosphorus (TP) are 500 m, 1000 m and 2000 m corresponding to the duration 8.6 h, 30 h and 135 h respectively. It is interesting to observe that, the duration of the influenced region of BOD_5 increases with the increment of the settling velocity of the fine particles although the maximum length remains unchanged. This is because, the fine particles settle at the bed of the wetland and consequently the solute particles stay in the wetland for longer time (see Fig. 5a). The removal of BOD_5 also depends on the settling velocity of the particles. It is seen that the duration of the particle cloud enhances though the maximum length remains unaltered with the enhance of settling velocity. It is seen from Fig. 5b and c that, when the releasing quantity is unaltered, BOD_5 has the weakest and TP has the strongest detrimental effect, because the metabolism due to suspended or vegetation supported bacteria may accelerate the removal of soluble organism. Again, since the removal of TP is mainly depends on the decomposition of chemical molecule, it has much stronger damage compare to other constituents.

5 Conclusions

In the present study, Gill's series expansion method is employed to study the transport of fine particles having settling velocity in a wetland flow. Moreover, analytical expressions for Taylor dispersivity and the mean concentration profile of tracer particles are obtained. The present research shows that the dispersion of settling particles decreases with the enhancement of vegetation factor, tortuosity and settling velocity in the wetland flow. It is seen that the concentration of settling particles agglomerated near the source when the settling velocity of particles, tortuosity of the flow and vegetation of the wetland increases. It is also observed that the duration of the settling particles for 5 day Biochemical Oxygen demand (BOD_5) increases with the enhancement of settling velocity. The results may play a significant role to study the wastewater treatment and sedimentation process.

References

1. Taylor, G.I.: Dispersion of soluble matter in solvent flowing slowly through a tube. Proc. R. Soc. Lond. A **219**, 186–203 (1953). https://doi.org/10.1098/rspa.1953.0139
2. Aris R.: On the dispersion of a solute in a fluid flowing through a tube. Proc. R. Soc. Lond. A **235**, 67–77 (1956). https://doi.org/10.1098/rspa.1956.0065
3. Gill, W.N.: A Note on the Solution of Transient Dispersion Problems. Proc. R. Soc. A: Math. Phys. Eng. Sci. **298**(1454), 335–339 (1967). https://doi.org/10.1098/rspa.1967.0107
4. Gill, W.N., Sankarasubramanian, R.: Exact analysis of unsteady convective diffusion. Proc. R. Soc. Lond. A **316**, 341–350 (1970). https://doi.org/10.1098/rspa.1970.0083
5. Wu, Z., Li, Z., Zeng, L., Shao, L., Tang, H., Yang, Q., Chen, G.: Environmental dispersivity in free-water-surface-effect dominated wetland: multi-scale analysis. Front. Environ. Sci. & Eng. China **5**(4), 597–603 (2011). https://doi.org/10.1007/s11783-011-0311-9
6. Zeng, L., Wu, Y.H., Ji, P., Chen, B., Zhao, Y.J., Chen, G.Q., Wu, Z.: Effect of wind on contaminant dispersion in a wetland flow dominated by free-surface effect. Ecol. Model. **237–238**, 101–108 (2012). https://doi.org/10.1016/j.ecolmodel.2012.04.020
7. Wang, P., Wu, Z., Chen, G.Q., Cui, B.S.: Environmental dispersion in a three-layer wetland flow with free-surface. Commun. Nonlinear Sci. Numer. Simul. **18**(12), 3382–3406 (2013). https://doi.org/10.1016/j.cnsns.2013.04.027
8. Zeng, L., Chen, G.Q.: Ecological degradation and hydraulic dispersion of contaminant in wetland. Ecol. Modell. **222**, 293–300 (2011). https://doi.org/10.1016/j.ecolmodel.2009.10.024
9. Wang, H., Zhu, Z., Li, S., Huai, W.: Solute dispersion in wetland flows with bed absorption. J. Hydrol. **579**, 124149 (2019). https://doi.org/10.1016/j.jhydrol.2019.124149

10. Wu, Z., Li, Z., Chen, G.Q.: Multi-scale analysis for environmental dispersion in wetland flow. Commun. Nonlinear Sci. Numer. Simul. **16**, 3168–3178 (2011). https://doi.org/10.1016/j.cnsns.2010.12.002
11. Zhi, L., Ping, W., Tao, S., Yiran, A., Xodong, W.: Critical length of contaminant cloud in a three-layer wetland: multi-scale analysis for environmental dispersivity. Wetlands **36**, 193–203 (2016). https://doi.org/10.1007/s13157-015-0663-1
12. Wu, Z., Zeng, L., Chen, G.Q.: Analytical modeling for environmental dispersion in Wetland. Ecolog. Modell. Eng. Lakes Wetl. **26**, 251–274 (2014). https://doi.org/10.1016/b978-0-444-63249-4.00011-7
13. Wu, Z., Chen, G.Q., Zeng, L.: Environmental dispersion in a two-zone wetland. Ecol. Model. **222**(3), 456–474 (2011). https://doi.org/10.1016/j.ecolmodel.2010.10.026
14. Wang, H., Huai, W.: Analysis of environmental dispersion in a wetland flow under the effect of wind: extended solution. J. Hydrol. **557**, 83–96 (2018). https://doi.org/10.1016/j.jhydrol.2017.12.029
15. Zeng, L., Chen, G.Q., Tang, H.S., Wu, Z.: Environmental dispersion in wetland flow. Commun. Nonlinear Sci. Numer. Simul. **16**(1), 206–215 (2011). https://doi.org/10.1016/j.cnsns.2010.02.019
16. Hammer, D.A.: Constructed wetlands for Wastewater Treatment: Municipal, Industrial, and Agricultural. Lewis Publishers, Chelsea, MI, USA (1989). https://doi.org/10.1201/9781003069850
17. Zhou, J.B., Jiang, M.M., Chen, B., Chen, G.Q.: Emergy evaluations for constructed wetland and conventional wastewater treatments. Commun. Nonlinear Sci. Numer. Simul. **14**(4), 1781–1789 (2009). https://doi.org/10.1016/j.cnsns.2007.08.010
18. Wu, Z., Zeng, L., Chen, G.Q., Li, Z., Shao, L., Wang, P., Jiang, Z.: Environmental dispersion in a tidal flow through a depth-dominated wetland. Commun. Nonlinear Sci. Numer. Simul. **17**(12), 5007–5025 (2012). https://doi.org/10.1016/j.cnsns.2012.04.006
19. Dhar, S., Poddar, N., Kairi, R.R., Mazumder, B.S., Mondal, K.K.: Numerical study on dispersion of fine settling particles in a depth dominated wetland flow. Commun. Nonliear Sci. Numer. Simulat. **96**, 105707 (2021). https://doi.org/10.1016/j.cnsns.2021.105707
20. Poddar, N., Das, S., Dhar, S., Mondal, K.K.: Semi-analytical study on environmental dispersion of settling particles in a width-independent wetland flow. Environ. Fluid Mech. **21**(4), 1–22 (2021). https://doi.org/10.1007/s10652-021-09809-2
21. CPCB, Annual Water Quality Statistics of India, from 2019. Central Pollution Control Board, Government of India (2019). http://www.cpcbenvis.nic.in/waterpollution/2019/Water_Quality_Canals_Sea_Water_Drains_STPs_2019.pdf
22. Liu, S., Masliyah, J.H.: Dispersion in porous media. In: Vafai, K. (ed.) Handbook of Porous Media. CRC Press, Boca Ratton, FL, USA, pp. 81–140 (2005), https://www.routledge.com/Handbook-of-Porous-Media/Vafai/p/book/9781439885543
23. Ergun, S.: Fluid flow through packed columns. Chem. Eng. Prog. **48**, 89–94 (1952). http://dns2.asia.edu.tw
24. Bandyopadhyay, S., Mazumder, B.S.: Unsteady convective diffusion in a pulsatile flow through a channel. Acta Mech. **134**, 1–16 (1999). https://doi.org/10.1007/BF01170300

Effects of Radiation and Chemical Reaction on MHD Mixed Convection Flow over a Permeable Vertical Plate

C. Sowmiya and B. Rushi Kumar

Abstract In this paper, we investigate the mixed convective flow of a viscous fluid in a vertical plate fixed in a porous medium under radiation, Dufour effect. Consider an incompressible, two-dimensional steady-state fluid subjected to flow over a semi-infinite plate influence of viscous dissipation, magnetic field. The non-linear PDEs convert into non-linear ODEs by using non-dimensionalization. The problem solved analytically by employing perturbation method. A study of the effects of velocity, temperature, concentration is significantly determined by existence of viscous dissipation, chemical reaction, radiation, Dufour effect are obtained. In addition, the effects of skin friction, the rate of energy, mass transfer are determined. It is notice, radiation parameter increases with temperature declines, Diffusion mass decreases with chemical reaction constraints.

Keywords Vertical plate · Slip condition · Buoyancy effect · Heat and mass transfer

1 Introduction

Several fields of science and engineering contain significant applications of hydromagnetic incompressible viscous flow as it involves heat, mass transfer under chemical reaction regions. Observing a field, we realize its influence, mixed convection arising flow from heated vertical plate is a significant problem from a theoretical and practical perspective. It is one of the most common topics of interest recently, it has become a separate focus of intensive research since heat and mass are constantly being transferred through porous media, which occur as part of a variety of technological processes. Heat and mass transfer have made considerable progress in magnetohydrodynamic flows over the past few years the MHD power generators and

C. Sowmiya · B. R. Kumar (✉)
Department of Mathematics, SAS, Vellore Institute of Technology, Vellore 632014, Tamil Nadu, India
e-mail: rushikumar@vit.ac.in

S. Banerjee and A. Saha (eds.), *Nonlinear Dynamics and Applications*, Springer Proceedings in Complexity,
https://doi.org/10.1007/978-3-030-99792-2_30

Hall accelerators are a few examples of their use. Mixed convection proceeding from fluid flow through a upright plate under the impact of magnetic field was studied by Yih [1], Barletle [2], Chin [3], Chamkha [4]. Micropolar fluids move convectively past vertical porous plates when they heated or absorbed by heat to which magnetic field applied was investigated by Rahman et al. [5]. Motsa [6] studied based on magnetohydrodynamic of thermal and mass transfer by convection under radiation and chemical effect, through a vertical permeable plate. The hydromagnetic flow of a heat-generating or absorbing vertical surfaces, effects of chemical and hall current had examined by Patil et al. [7]. Makinde et al. [8] examined a diagonal magnetic field, radiation, and heat transfer from variable viscosity fluid observed without accounting for chemical reactions. Impact of buoyancy, radiation on mass transfer across semi-infinite surfaces explored by Shateyi [9]. Mansour et al. [10] described the MHD convection flow and chemical reaction. Palani et al. [11] discussed about heat transfer which lies between vertical plates convective MHD flow effects. Talukdar et al. [12] investigated boundary layer slip condition with thermal radiation, chemical reaction is characterized by mass and energy transfer across a vertical permeable plate. Pal et al. [13] analyzed thermodynamics of mixed convection heat transfer in porous media. Makinde et al. [14] studied dufour and soret effects of a convective boundary layer through a porous medium. Chemical reactions, heat sources have attracted attention in the flow of electrically conductive fluid in various design was evaluated by Bisht et al. [15]. Olanrewaju et al. [17] analysed flat plate subjected to an thermal boundary layer produced by exponentially declining internal heat generation, convective boundary conditions. Makinde et al. [16] analyzed the interaction between MHD mixed convection, radiation, higher order chemical reactions. The boundary layer of an incompressible fluid travels past a semi-infinite vertical plate investigated by Rushi kumar et al. [18]. Sheea Juilet et al. [19] Studied flow over a porous medium over a surface subject to exponential stretching under free convection. Nalivela Nagi Reddy et al. [20] determined effect of mass movement, chemical transmission on the natural movement of mass through endless perpendicular plates, dissipative and radiative liquids glide in convection.

Motivated by the aforementioned literature review, we investigate the effects on steady convection boundary layer through a permeable vertical plate influence of viscous dissipation, buoyancy, heat source, radiation, Dufour effect, chemical reaction, nonlinear PDEs transform into nonlinear ODEs by using non-dimensionalization. The problem solved analytically by employing perturbation method, impact of pertinent parameters on the flow rate of temperature, concentration is graphically depicted.

2 Formulation of the Problem

We consider an incompressible two-dimensional steady flow, viscous, heat-source fluid past a vertical permeable plate with slip flow. There exists a first-order chemical reaction with a fixed rate. B_0 is a magnetic field applied in the existence of radiation and buoyancy effects. The x-axis is taken to the vertical direction along the pla-

Fig. 1 Sketch of flow geometry

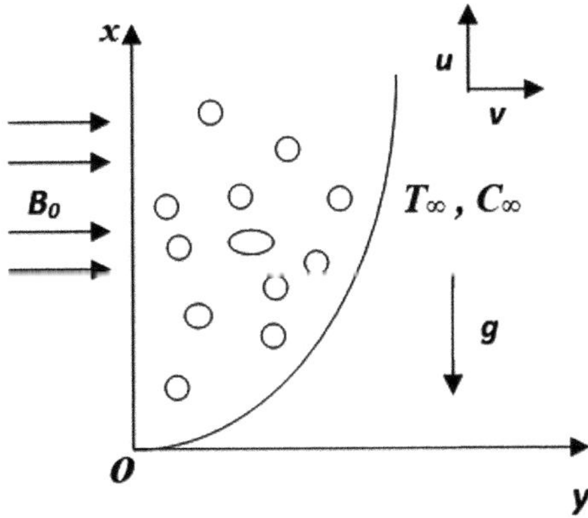

nar surface, y-axis taken surface normal to plate. According to the principle of the immeasurable plane surface, the fluid flow variables are the functions of y, t only (Fig. 1).

Governing equations of fluid flow as follows:

$$\frac{\partial u}{\partial x} + \frac{\partial v}{\partial y} = 0, \tag{1}$$

$$\left(u\frac{\partial u}{\partial x} + v\frac{\partial u}{\partial y}\right) = \nu\left(\frac{\partial^2 u}{\partial x^2} + \frac{\partial^2 u}{\partial y^2}\right) - \frac{\sigma B_0^2}{\rho}u + g\beta_T(T - T_\infty) + g\beta_c(C - C_\infty) - \frac{\nu}{k}u, \tag{2}$$

$$\left(u\frac{\partial T}{\partial x} + v\frac{\partial T}{\partial y}\right) = \frac{k}{\rho C_p}\left(\frac{\partial^2 T}{\partial x^2} + \frac{\partial^2 T}{\partial y^2}\right) + \frac{\nu}{\rho C_p}\left(\frac{\partial u}{\partial x} + \frac{\partial u}{\partial y}\right)^2 - \frac{Q(T - T_\infty)}{\rho C_p}$$
$$- \frac{1}{\rho C_p}\frac{\partial q_r}{\partial y} + \frac{D_m k_T}{C_s C_p}\frac{\partial^2 C}{\partial y^2}, \tag{3}$$

$$\left(u\frac{\partial C}{\partial x} + v\frac{\partial C}{\partial y}\right) = D\left(\frac{\partial^2 C}{\partial x^2} + \frac{\partial^2 C}{\partial y^2}\right) + k_r(C - C_\infty). \tag{4}$$

Where (u, v) be the velocity components of (x, y), C-mass concentration, T-temperature, ν-kinematic viscosity, g-gravity acceleration, σ-conductivity of fluid, β_T-thermal expansion coefficient, k_T-thermal conductivity, C_∞-ambient concentration, T_∞-ambient temperature, C_p-specific heat.

The boundary conditions are:

$$\begin{aligned} &u = U_{slip}, \quad T = T_w \quad C = C_w \quad at \quad y = 0; \\ &u \to u_\infty = 0, \quad T_w \to T_\infty, \quad C_w \to C_\infty \quad as \quad y \to \infty. \end{aligned} \tag{5}$$

A two-dimensional motion and an adequate length of the plate produce a situation, the x-axis is independent of all physical variables,

$$\frac{\partial u}{\partial x} = 0, \tag{6}$$

From equation of continuity, the suction velocity is either a function of t or constant. Hence integrate (1), we get the following form,

$$v = -V_0, \tag{7}$$

where V_0 is velocity for scale of suction which has positive constant other than zero.

Energy flux due to radiation is follows,

$$\frac{\partial q_r}{\partial y} = -4a\sigma(T_\infty^4 - T^4), \tag{8}$$

In the fluid flow region, the temperature difference is small enough, so if Taylor series are expanded around T_∞, T_4 can be obtained by omitting the coefficients of higher order,

$$T^4 = 4T_\infty^3 T - 3T_\infty^4,$$

Non-dimensional transformation are described below:

$$U = \frac{u}{U_0}, \quad V = \frac{v}{V_0}, \quad Y = \frac{V_0 y}{\nu}, \quad \theta = \frac{T - T_\infty}{T_w - T_\infty}, \quad \phi = \frac{C - C_\infty}{C_w - C_\infty}. \tag{9}$$

In non-dimensional form, the above transformation (2) to (4) obtain as follows,

$$-\frac{\partial U}{\partial Y} = \frac{\partial^2 U}{\partial Y^2} + Gr\theta + Gc\phi - \left(M^2 + \frac{1}{\lambda}\right)U, \tag{10}$$

$$-\frac{\partial \theta}{\partial Y} = \frac{1}{Pr}\frac{\partial^2 \theta}{\partial Y^2} + E\left(\frac{\partial U}{\partial Y}\right)^2 + N\frac{\partial^2 \theta}{\partial Y^2} + Du\frac{\partial^2 \phi}{\partial Y^2} - \alpha\theta, \tag{11}$$

$$-\frac{\partial \phi}{\partial Y} = \frac{1}{Sc}\frac{\partial^2 \phi}{\partial Y^2} + K\phi. \tag{12}$$

Combined boundary conditions are;

$$\begin{aligned} U = U_{slip}, \quad \theta = 1, \quad \phi = 1 \quad at \quad Y = 0, \\ U \to 0, \quad \theta \to 0, \quad \phi \to 0 \quad at \quad Y \to \infty. \end{aligned} \tag{13}$$

Since the flow will be super imposed on main flow due to joules dissipation, the Eckart number is always less than 1 ($E << 1$). where $Gr = \frac{\gamma g \beta_T (T_w - T_\infty)}{V_0^2 U_0}$ is the thermal Grashof number, $Gc = \frac{g \beta_C \gamma (C_w - C_\infty)}{V_0^2 U_0}$ is the solutal Grashof number,

$M = \frac{\gamma\sigma B_0^2}{\rho V_0^2}$,-magnetic field, $Pr = \frac{C_p\mu}{k}$-Prandtl number, $N = \frac{16\sigma T_\infty\gamma}{V_0^2\rho C_p 3k}$-radiation heat-flux, $Du = \frac{(C_w-C_\infty)D_m k_T}{(T_w-T_\infty)\gamma C_s C_p}$ is the Dufour number, $Sc = \frac{\nu}{D}$ is the schmidt number, $K = \frac{k_r\gamma}{V_0^2}$ is the chemical reaction, $E = \frac{U_0^2}{\rho C_p(T_w-T_\infty)}$, $\lambda = \frac{V_0^2 K}{\gamma^2}$-porous permeability, $\alpha = \frac{Q}{\rho C_p V_0^2}$-heat absorption.

3 Method of Solution

A closed form solution cannot be found for this set of partial differential equation (10)–(13). In an analytical form, it is possible to solve the problems by using ordinary differential equations, we can solve these equations in which velocity, energy, and mass can be represented as dimensionless form,

$$U(y) = U_0(y) + EU_1(y) + O(E^2), \tag{14}$$

$$\theta(y) = \theta_0(y) + E\theta_1(y) + O(E^2), \tag{15}$$

$$\phi(y) = \phi_0(y) + E\phi_1(y) + O(E^2). \tag{16}$$

Applying (14) to (16) in (10) to (13), and comparing the balanced and unbalanced terms, ignoring higher power of $O(E^2)$, we obtained,

$$U_0^{''} + U_0^{'} - \left(M^2 + \frac{1}{\lambda}\right) U_0 = -Gr\theta_0 - Gc\phi_0, \tag{17}$$

$$U_1^{''} + U_1^{'} - \left(M^2 + \frac{1}{\lambda}\right) U_1 = -Gr\theta_1 - Gc\phi_1, \tag{18}$$

$$\theta_0^{''}(1 + PrN) + \theta_0^{'}Pr - \alpha\theta_0 Pr = -DuPr\phi_0^{''}, \tag{19}$$

$$\theta_1^{''}(1 + PrN) + \theta_1^{'}Pr - \alpha\theta_1 Pr = -DuPr\phi_1^{''} - U_0^{'2}Pr, \tag{20}$$

$$\phi_0^{''} + Sc\phi_0^{'} + KSc\phi_0 = 0, \tag{21}$$

$$\phi_1^{''} + Sc\phi_1^{'} + KSc\phi_1 = 0. \tag{22}$$

Boundary condition are transformed by,

$$\begin{aligned} &U_0 = \gamma U_0^{'}, \quad U_1 = \gamma U_1^{'}, \quad \theta_0 = 1, \quad \theta_1 = 1, \quad \phi_0 = 1 \quad \phi_1 = 1 \quad as \quad y = 0, \\ &U_0 = 0, \quad U_1 = 0, \quad \theta_0 \to 0, \quad \theta_1 \to 0, \quad \phi_0 \to 0 \quad \phi_1 \to 0 \quad as \quad y \to \infty. \end{aligned} \tag{23}$$

Solving the (17)–(22) along with (23) we get,

$$U_0 = A_2 e^{-m_1 y} + A_3 e^{-m_2 y} + A_4 e^{-m_3 y} \tag{24}$$

$$U_1 = A_{13} e^{-m_1 y} + A_{14} e^{-m_2 y} + A_{21} e^{-m_3 y} + A_{15} e^{-2m_1 y} + A_{16} e^{-2m_2 y} + A_{17} e^{-2m_3 y} + A_{18} e^{-(m_1+m_2)y} + A_{19} e^{-(m_1+m_3)y} + A_{20} e^{-(m_3+m_2)y}, \tag{25}$$

$$\theta_0 = (1 - A_1) e^{-m_2 y} + A_1 e^{-m_1 y}, \tag{26}$$

$$\theta_1 = A_5 e^{-m_1 y} + A_{12} e^{-m_2 y} + A_6 e^{-2m_1 y} + A_7 e^{-2m_2 y} + A_8 e^{-2m_3 y} + A_9 e^{-(m_1+m_2)y} + A_{10} e^{-(m_3+m_2)y} + A_{11} e^{-(m_1+m_3)y}, \tag{27}$$

$$\phi_0 = e^{-m_1 y}, \tag{28}$$

$$\phi_1 = e^{-m_1 y}. \tag{29}$$

The density, temperature, and concentration distributions were determined by substituting (24)–(29) in (14) to (16)

$$U(Y,t) = \left[A_2 e^{-m_1 y} + A_3 e^{-m_2 y} + A_4 e^{-m_3 y}\right] + E\left[A_{13} e^{-m_1 y} + A_{14} e^{-m_2 y} + A_{21} e^{-m_3 y} + A_{15} e^{-2m_1 y} + A_{16} e^{-2m_2 y} + A_{17} e^{-2m_3 y} + A_{18} e^{-(m_1+m_2)y} + A_{19} e^{-(m_1+m_3)y} + A_{20} e^{-(m_3+m_2)y}\right], \tag{30}$$

$$\theta(Y,t) = \left[(1 - A_1) e^{-m_2 y} + A_1 e^{-m_1 y}\right] + E\left[A_5 e^{-m_1 y} + A_{12} e^{-m_2 y} + A_6 e^{-2m_1 y} + A_7 e^{-2m_2 y} + A_8 e^{-2m_3 y} + A_9 e^{-(m_1+m_2)y} + A_{10} e^{-(m_3+m_2)y} + A_{11} e^{-(m_1+m_3)y}\right], \tag{31}$$

$$\phi(Y,t) = e^{-m_1 y}\left[1 + E\right]. \tag{32}$$

The physical parameter of wall shear stress τ_w is given by,

$$\tau_w = \mu \left(\frac{\partial u}{\partial y}\right)_{y=0} = \frac{U_0 V_0}{\rho} \left(\frac{\partial U}{\partial Y}\right)_{Y=0},$$

Local skin friction coefficient τ is given by,

$$\tau = \frac{\tau_w \rho}{U_0 V_0}, \tag{33}$$

$$\tau = U'(0) = -\,(A_2 m_1 + A_3 m_2 + A_4 m_3) - E\big(A_{13} m_1 + A_{14} m_2 + A_{21} m_3 + 2A_{15} m_1 + 2A_{16} m_2 + 2A_{17} m_3 + A_{18}(m_1 + m_2) + A_{19}(m_3 + m_2) + A_{20}(m_1 + m_3)\big), \tag{34}$$

Local surface heat flux is given by,

$$q_w = -K \frac{\partial T}{\partial y} = \frac{-K\,(T_w - T_\infty)\,V_0}{\gamma} \left(\frac{\partial \theta}{\partial Y}\right)_{Y=0},$$

$$
\begin{aligned}
Nu = \frac{Nu_x}{Re_x} &= \theta'(0) = ((1 - A_1)m_3 + A_1 m_1) \\
&- \epsilon e^{nt}\big(A_5 m_2 + A_{12} m_5 + 2A_6 m_1 + 2A_7 m_3 \\
&+ 2A_8 m_4 + A_9(m_1 + m_3) + A_{10}(m_3 + m_4) + A_{11}(m_1 + m_4)\big), \quad (35)
\end{aligned}
$$

K—effective heat conductivity.

Local surface mass flux is given by,

$$
Sh = \frac{Sh_x}{Re_x} = \phi'(0) = -m_1(1 + E). \quad (36)
$$

Here, $Re_x = \frac{V_0 K}{\nu}$ is Reynolds Number.

4 Results and Discussion

The problem consider a two-dimensional steady convective fluid over a vertical plate submerged in a porous medium. It include radiation, chemical reaction, Dufour effect in the occurrence of suction. The numerical values of various parameters $Gr, Gc, M, \lambda, \alpha, Pr, Sc, N, K, Du$. have been computed with boundary condition. The results are depicted as graphs in Figs. 2, 3, 4, 5, 6, 7, 8, 9, 10, 11, 12, 13, 14 and 15.

The most important fluids are atmospheric in air, H_2O, results are limited for Prandtl's number ($Pr = 0.71, 7.0$) representing air, H_2O at 20 °C, were chosen Schmidt numbers of ($Sc = 0.21, 0.62, 0.78, 2.62$) representing chemical species of H_2, O_2, CO_2, C_9H_{12} diffusing in the air. In the velocity boundary layer, thermal Grashof number is equal to the relation of buoyant to viscous forces acting on a fluid. For the following graphs we have used $Gr = 1, Gc = 1, E = 0.2, \alpha = 0.1, \lambda = 2, M = 2, N = 1, \gamma = 1, K = 1, Du = 0.5, Sc = 0.6$.

The velocity distribution in Fig. 2 shows that velocity decreases as the values of M increase because of the existence of magnetic field sets in Lorentz force which induces the retarding force on velocity field. Figure 3 depicts velocity rising with an rise the values λ since porosity reduce the drag force help fluid to go fast. The effects of thermal, solutal Grashof number depicted in Figs. 4, 5, from these figures shown the velocity increases with raise the values of Gr and Gc it is due to the presence of thermal buoyancy boost the velocity distribution.

In Fig. 6 temperature distribution diminishes as the value of heat absorption increases, that has a tendency to lower thermal buoyancy effects. Pr increased then fluid temperature distribution is decreases displayed in Fig. 7. By increasing Pr, thermal conductivities will decreases, which leads to faster diffusion of heat apart from heated plate. Figure 8 shown Dufour number rises temperature distribution falls. Radiation heat-flux N causes the fluid temperature distribution to increase as displayed in Fig. 9. When N increases thickness of boundary layer, heat transfer rate also increase. Figure 10 presented the mass distribution is decreasing while increas-

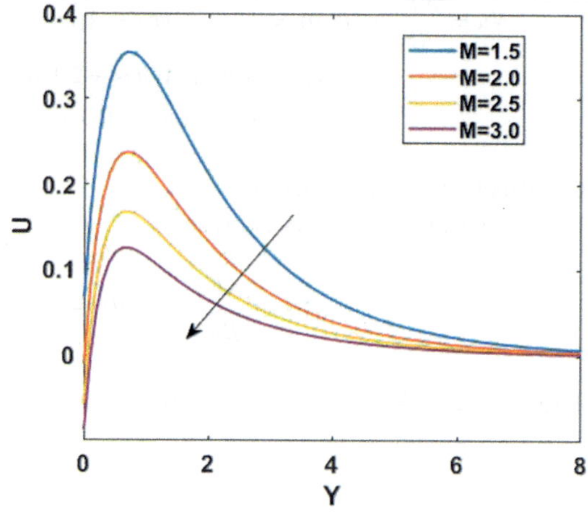

Fig. 2 Effect of M on velocity distribution

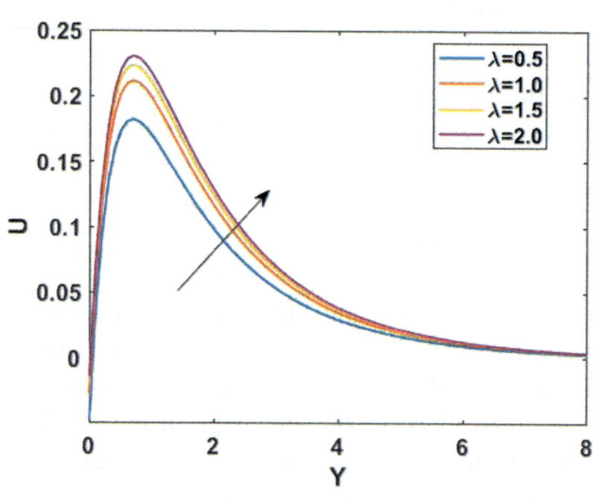

Fig. 3 Effect of λ on velocity distribution

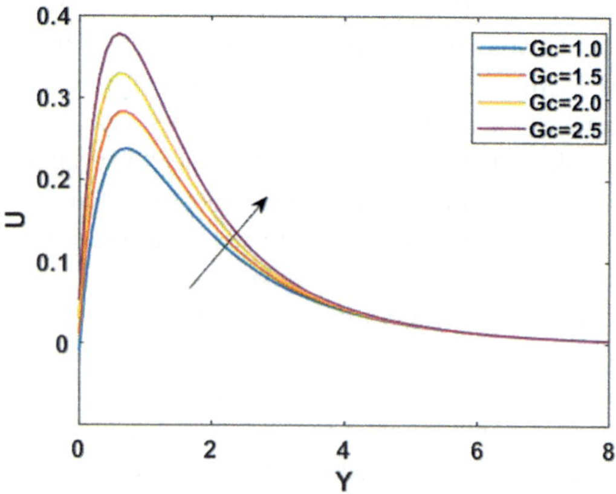

Fig. 4 Effect of Gc on velocity distribution

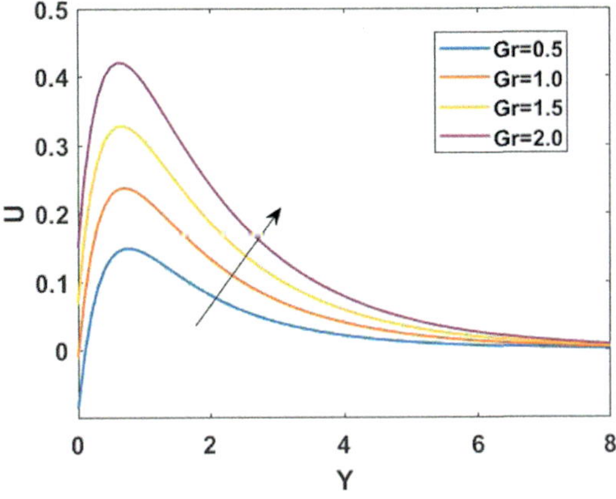

Fig. 5 Effect of Gr on velocity distribution

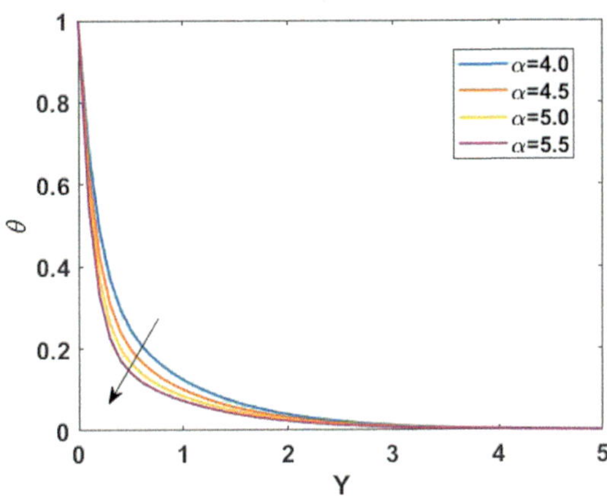

Fig. 6 Effect of α on temperature distribution

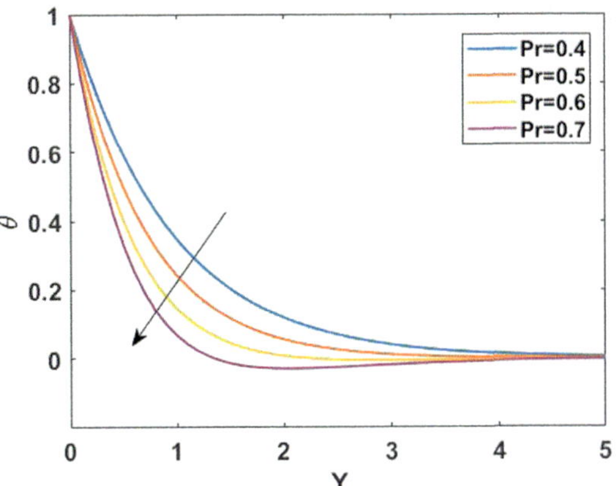

Fig. 7 Effect of Pr on temperature distribution

Fig. 8 Effect of *Du* on temperature distribution

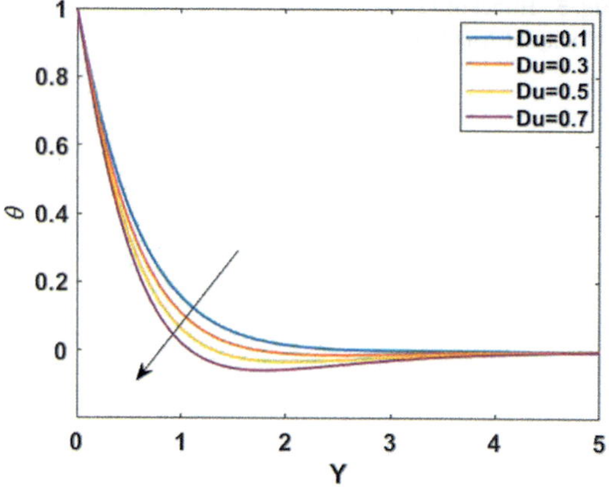

Fig. 9 Effect of radiation on temperature distribution

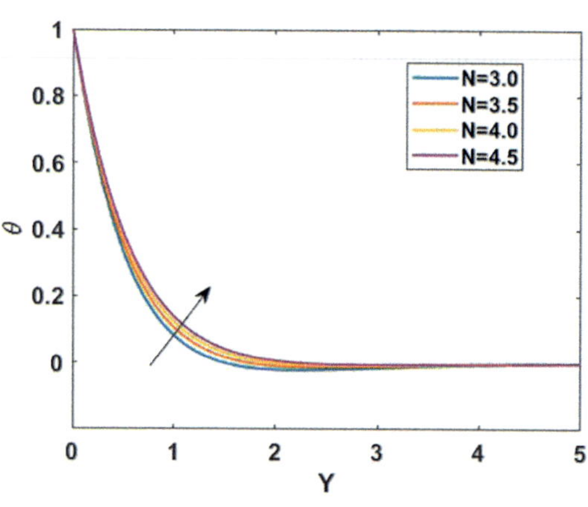

Fig. 10 Effect of *Sc* on concentration distribution

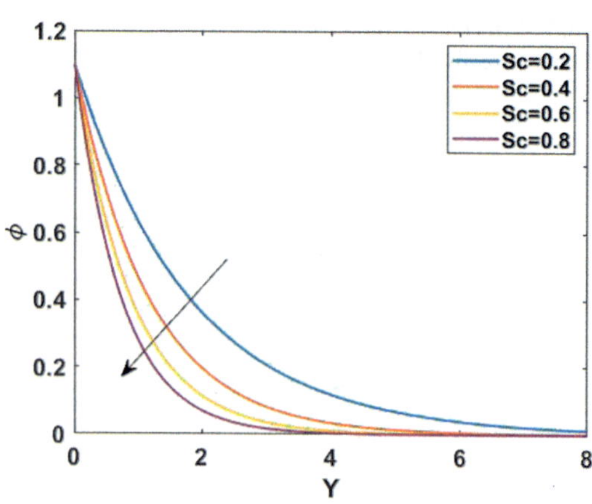

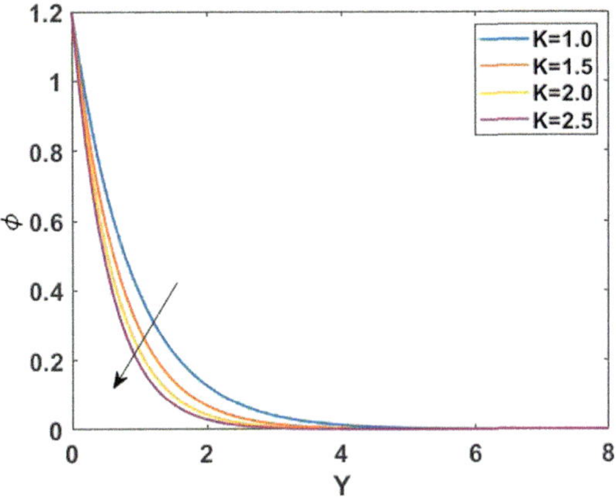

Fig. 11 Effect of K on concentration distribution

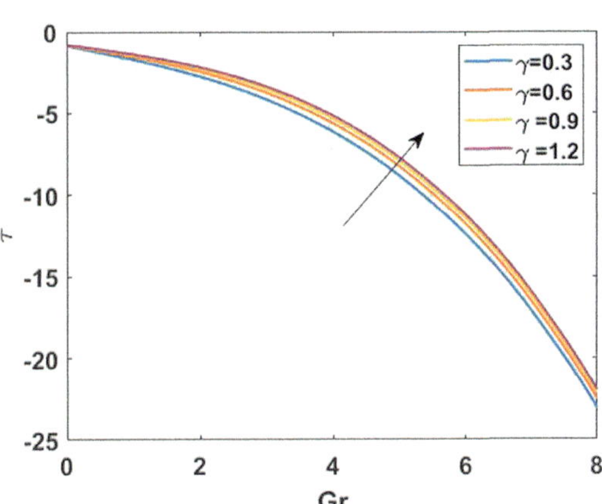

Fig. 12 Effect of γ on skin friction distribution

ing the values of Sc, because Sc is small, it leads to smaller diffusivity which results fall in mass. By increasing chemical reaction values K, the concentration decreases as shown in Fig. 11. The reverse trend is noticed due to appearance of chemical reaction at constant rate. When γ increases, the skin friction also increase and increase the value Gc, skin friction decreases as shown in Fig. 13. Nu against M, various values of Pr is increases in heat transfer monotonically, which is shown in Fig. 14. Sh against the M for various values K, resulting in mass transfer being reduced as represented in Fig. 15.

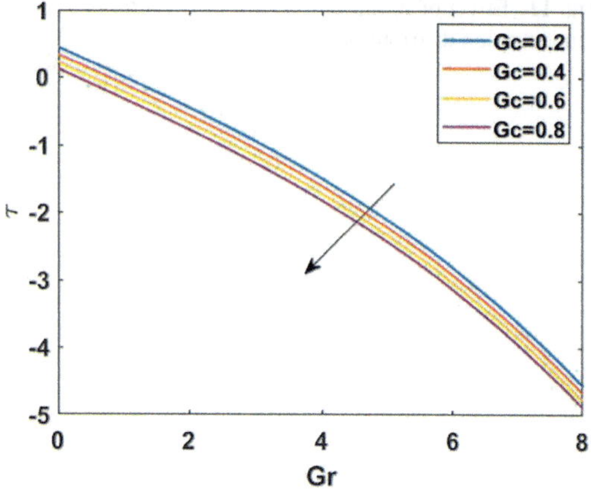

Fig. 13 Effect of Gc on skin distribution

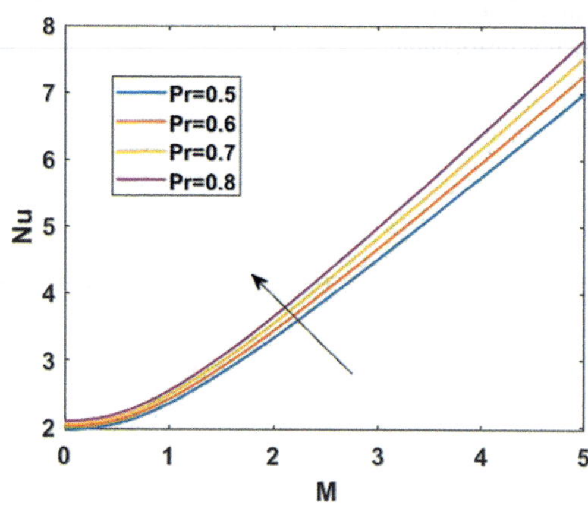

Fig. 14 Effect of Du on Nusselt number distribution

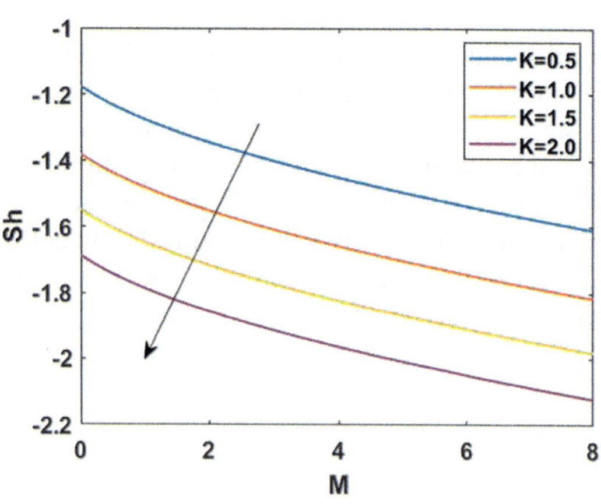

Fig. 15 Effect of K on Sherwood number distribution

5 Conclusion

We have investigate steady state mixed convectional heat, mass transfer in a permeable vertical plates with Dufour and chemical reaction. By using perturbation method, we have obtain an analytical solution to this problem and numerical results are depicted as graphs. Several important parameters are analyzed in this study. Following are some highlights.

- If magnetic field is increases, velocity distribution is decreases.
- If λ, Gc, Gr increases, there is an enhancement in velocity profile.
- An temperature distribution is diminished by increasing heat absorption, Prandtl number, Dufour effects.
- Raising radiation parameter 'N', increases temperature distribution.
- Concentration distribution decreases while Sc, K are increases.

Appendix

$$m_1 = \frac{Sc + \sqrt{Sc^2 - 4KSc}}{2}, \quad m_2 = \frac{-pr}{1 + PrN}, m_3 = \frac{1 + \sqrt{1 + 4N}}{2},$$

$$A_1 = \frac{-Du.Pr.m_1^2}{(1 + PrN)m_1^2 - m_1 pr - \alpha pr}, A_2 = \frac{-(GrA_1 + Gc)}{m_1^2 - m_1 - N}, A_3 = \frac{Gr(1 - A_1)}{m_3^2 - m_3 - N},$$

$$A_4 = \frac{-A_2[1 + \gamma m_1] - A_3[1 + \gamma m_2]}{1 + \gamma m_3}, A_5 = \frac{-pr.Du.m_1^2}{m_1^2(1 + Npr) - m_1 Pr - \alpha pr},$$

$$A_6 = \frac{-pr.A_2^2 m_1^2}{4m_1^2(1 + Npr) - 2m_1 Pr - \alpha pr}, A_7 = \frac{-pr.A_3^2 m_2^2}{4m_2^2(1 + Npr) - 2m_2 Pr - \alpha pr},$$

$$A_8 = \frac{-pr.A_4^2 m_3^2}{4m_3^2(1 + Npr) - 2m_3 Pr - \alpha pr}, A_9 = \frac{-2pr.A_2 A_3 m_1 m_2}{(m_1 + m_2)(1 + Npr) - (m_1 + m_2)Pr - \alpha pr},$$

$$A_{10} = \frac{-2pr.A_4 A_3 m_2 m_3}{(m_2 + m_3)(1 + Npr) - (m_2 + m_3)Pr - \alpha pr},$$

$$A_{11} = \frac{-2pr.A_2 A_4 m_1 m_3}{(m_1 + m_3)(1 + Npr) - (m_1 + m_3)Pr - \alpha pr}, A_{12} = -A_5 - \dots - A_{11},$$

$$A_{13} = \frac{GrA_5 + Gc}{m_1^2 - m_1 - N_1}, A_{14} = \frac{GrA_{12}}{m_2^2 - m_2 - N_1}, A_{15} = \frac{GrA_6}{4m_1^2 - 2m_1 - N_1},$$

$$A_{16} = \frac{GrA_7}{4m_2^2 - 2m_2 - N_1}, A_{17} = \frac{GrA_8}{4m_3^2 - 2m_3 - N_1}, A_{18} = \frac{GrA_9}{(m_1 + m_2)^2 - (m_1 + m_2) - N_1},$$

$$A_{19} = \frac{GrA_{10}}{(m_2 + m_3)^2 - (m_2 + m_3) - N_1}, \quad A_{20} = \frac{GrA_{11}}{(m_1 + m_3)^2 - (m_1 + m_3) - N_1},$$

$$A_{21} = -(A_{13}[1 + \gamma m_1] + A_{14}[1 + \gamma m_2] + A_{15}[1 + \gamma 2m_1] + A_{16}[1 + \gamma 2m_2]$$
$$+A_{17}[1 + \gamma 2m_3] + A_{18}[1 + \gamma(m_1 + m_2)] + A_{19}[1 + \gamma(m_2 + m_3)]$$
$$+A_{20}[1 + \gamma(m_1 + m_3)])/(1 - \gamma m_3).$$

References

1. Yih, K.A.: The effect of transpiration on coupled heat and mass transfer in mixed convection over a vertical plate embedded in a saturated porous medium. Int. Commun. Heat Mass Trans. **24**, 265–275 (1997)
2. Barletle, A.: Combined forced and free convection with viscous dissipation in a vertical duct. Int. J. Heat Mass Transfer **42**, 2243–2253 (1999)
3. Chin, K.E., Nazar, R., Arifin, N.M., Pop, I.: Effect of variable viscosity on mixed convection boundary layer flow over a vertical surface embedded in a porous medium. Int. Commun. Heat Mass Trans. **34**, 464–473 (2007)
4. Chamkha, A.J.: Hydromagnetic three-dimensional free convection on a vertical stretching surface with heat generation or absorption. Int. J. Heat Fluid Flow **20**(1), 84–92 (1999)
5. Rahman, M.M., Sattar, M.A.: Magnetohydrodynamic convective flow of a micropolar fluid past a continuously moving vertical porous plate in the presence of heat generation/absorption. J. Heat Trans. **128**(2), 142–152 (2006)
6. Motsa, S.S.: The effects of thermal radiation, hall currents, soret, and dufour on MHD flow by mixed convection over a vertical surface in porous media. SAMSA J. Pure Appl. Math. **3**, 58–65 (2008)
7. Patil, P.M., Kulkarni, P.S.: Effects of chemical reaction on free convective flow of a polar fluid through a porous medium in the presence of internal heat generation. Int. J. Therm. Sci. **47**(8), 1043–1054 (2008)
8. Makinde, O.D., Ogulu, A.: The effect of thermal radiationon the heat and mass transfer flow of a variable viscosity fluidpast a vertical porous plate permeated by a transverse magneticfield. Chem. Eng. Commun. **195**(12), 1575–1584 (2008)
9. Shateyi, S.: Thermal radiation and buoyancy effects on heat and mass transfer over a semi-infinite stretching surface with suction and blowing. J. Appl. Math. Article ID 414830 (2008)
10. Mansour, M.A., El-Anssary, N.F., Aly, A.M.: Effect of chemical reaction and viscous dissipation on MHD natural convection flows saturated in porous media with suction or injection. Int. J. Appl. Math. Mech. **4**(2), 60–70 (2008)
11. Palani, G., Srikanth, U.: MHD flow past a semi-infinite vertical plate with mass transfer. Nonlinear Anal.: Modell. Control **14**(3), 345–356 (2009)
12. Pal, D., Talukdar, B.: Perturbation analysis of unsteady magnetohydrodynamic convective heat and mass transfer in a boundary layer slip flow past a vertical permeable plate with thermal radiation and chemical reaction. Commun. Nonlinear Sci. Numer. Simul. **15**, 1813–1830 (2010)
13. Pal, D., Talukdar, B.: Buoyancy and chemical reaction effects on MHD mixed convection heat and mass transfer in a porous medium with thermal radiation and Ohmic heating. Commun. Nonlinear Sci. Numer. Simul. **15**, 2878–2893 (2010)
14. Makinde, O.D.: On MHD heat and mass transfer over a moving vertical plate with a convective surface boundary condition. Can. J. Chem. Eng. **88**(6), 983–990 (2010)
15. Bisht, V., Kumar, M., Uddin, Z.: Effect of variable thermal conductivity and chemical reaction on steady mixed convection boundary layer flow with heat and mass transfer inside a cone due to a point. J. Appl. Fluid Mech. **4**(4), 59–63 (2011)
16. Makinde, O.D.: MHD mixed-convection interaction with thermal radiation and nth order chemical reaction past a vertical porous plate embedded in a porous medium. Chem. Eng. Commun. **198**(4), 590–608 (2011)
17. Olanrewaju, P.O., Arulogun, O.T., Adebimpe, K.: Internal heat generation effect on thermal boundary layer with a convective surface boundary condition. Amer. J. Fluid Dyn. **2**(1), 1–4 (2012)
18. Kumar, B.R., Sivaraj, R.: MHD mixed convection flow over a permeable vertical plate with Buoyancy and Soret effects. Int. J. Appl. Mech. Eng. **17**, 51–76 (2012). https://doi.org/10.12725/mjs.22.3 https://doi.org/10.12725/mjs.22.3
19. Juilet, S.S., Vidhya, M., Govindarajan, A.: Effect of mass transfer with chemical reactions on MHD free convective flow of dissipative and radiative fluid past an infinite vertical plate. AIP Conf. Proc. **2277**, 030017 (2020)

20. Reddy, N.N., Rao, V.S., Ravindra Reddy, B.: Chemical reaction impact on MHD natural convection flow through porous medium past an exponentially stretching sheet in presence of heat source/sink and viscous dissipation. Case studies Thermal Engineering, vol. 25 (2021)

Note on the Circular Rayleigh Problem

G. Chandrashekhar and A. Venkatalaxmi

Abstract We consider stability of axial flows of an incompressible, inviscid homogeneous fluid to axisymmetric disturbances known as circular Rayleigh problem of hydrodynamic stability. For circular Rayleigh problem, we derived instability region for a class of flows and which intersect with Batchelor and Gill semicircle under some conditions. Also, we derived a necessary condition for stability.

Keywords Inviscid fluid · Axial flows · Incompressible · Axisymmetric disturbances

1 Introduction

The study of linear stability of axial flows of an inviscid, incompressible, homogeneous fluid to axisymmetric disturbances is an important part of fluid dynamics and geophysical fluid dynamics (cf. [1, 3]). Circular Rayleigh problem is a inviscid case of Orr-Sommerfeld problem (cf. [2]) which deals with incompressible inviscid homogeneous axial flows to axisymmetric disturbances. Many analytical results have been proved for this problem. Reference [1] derived a necessary condition for instability which states that $r\left(\frac{W'}{r}\right)'$ should changes its sign at least once. Also, he proved that

Supported by Organization X

G. Chandrashekhar (✉) · A. Venkatalaxmi
Department of Mathematics, Osmania University, Hyderabad, TG, India
e-mail: chandu.724@gmail.com

A. Venkatalaxmi
e-mail: akavaramvlr@gmail.com

S. Banerjee and A. Saha (eds.), *Nonlinear Dynamics and Applications*,
Springer Proceedings in Complexity,
https://doi.org/10.1007/978-3-030-99792-2_31

$K(r) = \frac{-\psi(r)}{r(W-W_s)} > 0$, where $\psi(r) = r\left(\frac{W'}{r}\right)'$. For instability [4] derived a necessary condition which is $r\left(\frac{W'}{r}\right)'(W - W_s) < 0$ at least once. They also proved a semi-circle know as Batchelor & Gill semi-circle which gives the location of eigen values. Reference [5] proved Howard's conjecture namely growth rate $k\,c_i$ approaches to zero as wave number k approaches to infinity. Reference [7] derived a sufficient condition for stability and short-wave stability. Reference [6] derived a parabolic instability regions which intersects with Batchelor and Gills semi-circle [4] by following the approach of [9]. The parabolic instability region derived by [6] depends on $\phi(r) > 0$ or $\psi(r) < 0$.

In this present work, we derived parabolic instability regions for a class of flows and which intersects with Bachelor and Gill semicircle under some condition. Also, we derived a necessary condition for stability, namely $k^2 > k_c^2$ where k_c is the critical value of the wave number then the flow is stable. This has been illustrated with an example.

2 Circular Rayleigh Problem

The circular Rayleigh Problem (cf. p. 361 of [1]) is given by

$$(W - c)\left[DD_* - k^2\right]u - rD\left[\frac{DW}{r}\right]u = 0, \tag{1}$$

with boundary conditions

$$u(R_1) = 0 = u(R_2). \tag{2}$$

Where $D_* = D + \frac{1}{r},\ D = \frac{d}{dr}$, W is the velocity function, $c = c_r + i\,c_i$ (Complex eigen value), u is the eigen function and k is the wave number .

Apply the transformation $u = (W - c)^{\frac{1}{2}}\,G$, we get

$$D\left[(W - c)\,D_*G\right] - \frac{1}{2}rD\left[\frac{DW}{r}\right]G - \frac{\frac{(W')^2}{4}}{W - c}G - k^2\,(W - c)\,G = 0, \tag{3}$$

$$\textit{With boundary conditions}\ \ G(R_1) = 0 = G(R_2). \tag{4}$$

3 Instability Region

Theorem 1 *If* $W_{\min} > 0,\ c_i > 0$, *then* $c_i^2 \leq \lambda\left[c_r - (W_{\max} - W_{\min})\right]$, *where,* $\lambda = \frac{(W')^2{}_{\max}}{4\,W_{\min}\left[\frac{R_1\,\pi^2}{R_2\,(R_2-R_1)^2} + k^2 R_1\right]}$.

Proof Multiplying (3) by $r\,G^*$, integrating, using (4), we get

$$\int_{R_1}^{R_2} (W - c)\left[|D_* G|^2 + k^2\,|G|^2\right] r\,dr + \int_{R_1}^{R_2} \frac{1}{2} r \left(\frac{W'}{r}\right)' r\,|G|^2\,dr + \int_{R_1}^{R_2} \left(\frac{\frac{(W')^2}{4}}{(W - c)}\right) r\,|G|^2\,dr = 0\,. \tag{5}$$

Equating real and imaginary parts, we have

$$\int_{R_1}^{R_2} (W - c_r)\left[|D_* G|^2 + k^2\,|G|^2\right] r\,dr + \frac{1}{2}\int_{R_1}^{R_2} r \left(\frac{W'}{r}\right)' r\,|G|^2\,dr + \int_{R_1}^{R_2} \left(\frac{\frac{(W')^2}{4}(W - c_r)}{|W - c|^2}\right) r\,|G|^2\,dr = 0\,. \tag{6}$$

Since $c_i > 0$, we have

$$\int_{R_1}^{R_2} \left[|D_* G|^2 + k^2\,|G|^2\right] r\,dr - \int_{R_1}^{R_2} \left(\frac{\frac{(W')^2}{4}}{|W - c|^2}\right) r\,|G|^2\,dr = 0\,. \tag{7}$$

Multiplying (7) by $[W_{\min} - W_{\max}]$ and adding with (6), we get

$$\int_{R_1}^{R_2} (W - c_r + W_{\min} - W_{\max})\left[|D_* G|^2 + k^2\,|G|^2\right] r\,dr + \frac{1}{2}\int_{R_1}^{R_2} r \left(\frac{W'}{r}\right)' r\,|G|^2\,dr + \int_{R_1}^{R_2} \left(\frac{\frac{(W')^2}{4}(W - c_r - W_{\min} + W_{\max})}{|W - c|^2}\right) r\,|G|^2\,dr = 0\,. \tag{8}$$

Since $W_{\min} < c_r < W_{\max}$ this implies that $(W - c_r + W_{\min} - W_{\max})$ is negative hence dropping the first integral term in the above equation, we get

$$\frac{1}{2}\int_{R_1}^{R_2} r\left(\frac{W'}{r}\right)' r\,|G|^2\,dr \geq \int_{R_1}^{R_2}\left(\frac{\frac{(W')^2}{4}\,(W_{\min}-W_{\max}+c_r-W)}{|W-c|^2}\right) r\,|G|^2\,dr\,. \tag{9}$$

Multiplying (7) by c_r and adding with (6), we get

$$\int_{R_1}^{R_2} W\,\left[|D_*G|^2+k^2\,|G|^2\right] r\,dr + \frac{1}{2}\int_{R_1}^{R_2} r\left(\frac{W'}{r}\right)' r\,|G|^2\,dr$$

$$+\int_{R_1}^{R_2}\left(\frac{\frac{(W')^2}{4}\,(W-2c_r)}{|W-c|^2}\right) r\,|G|^2\,dr = 0\,. \tag{10}$$

Substituting (9) in (10), we have

$$\int_{R_1}^{R_2} W\,\left[|D_*G|^2+k^2\,|G|^2\right] r\,dr \;\leq\; \int_{R_1}^{R_2}\left(\frac{\frac{(W')^2}{4}\,(c_r-W_{\min}+W_{\max})}{|W-c|^2}\right) r\,|G|^2\,dr\,.$$

Using Rayleigh Ritz inequality and $\frac{1}{|W-c|^2} \;\leq\; \frac{1}{c_i^2}$, we have

$$c_i^2 \;\leq\; \lambda\,[c_r - W_{\min} + W_{\max}]\,, \tag{11}$$

where $\lambda \;=\; \frac{\frac{(W')^2_{\max}}{4}}{W_{\min}\left[\frac{R_1\,\pi^2}{R_2\,(R_2-R_1)^2}+R_1\,k^2\right]}$.

Theorem 2 *If* $\lambda < \lambda_c$, *where critical value* $\lambda_c = 3\,W_{\max} - W_{\min} - 2\sqrt{W_{\max}\,(2\,W_{\max} - W_{\min})}$, *then the parabola* $c_i^2 \leq \lambda\,[c_r - W_{\min} + W_{\max}]$ *intersect with Batchelor and Gill semicircle.*

Proof Batchelor and Gill semicircle (cf. [4]) is given by

$$\left[c_r - \frac{W_{\min}+W_{\max}}{2}\right]^2 + c_i^2 \;\leq\; \left[\frac{W_{\max}-W_{\min}}{2}\right]^2. \tag{12}$$

Substituting (11) in (12), we get

$$c_r^2 + [\lambda - (W_{\min}+W_{\max})]\;c_r\; + [W_{\min}W_{\max} + \lambda\,(W_{\max}-W_{\min})] \;\leq 0\,.$$

Above equation is a quadratic equation in c_r, for real roots, its discriminant part should be greater than or equal to zero, we have

$$\lambda^2 - 2\,[3\,W_{\max} - W_{\min}] + [W_{\max} - W_{\min}]^2 \;\geq\; 0\,.$$

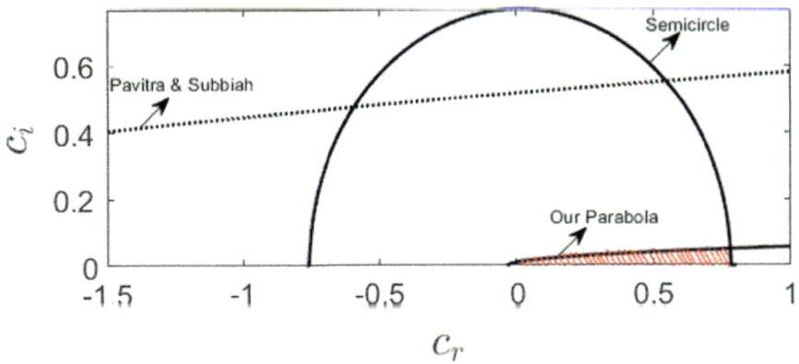

Fig. 1 c_r *vs* c_i (The instability region is the shaded region)

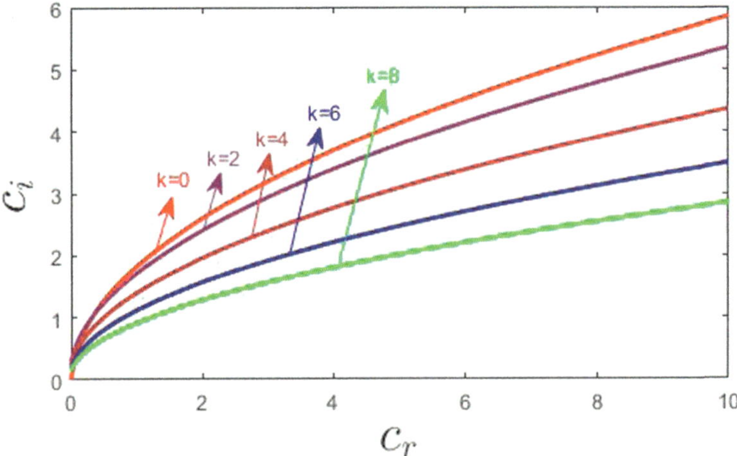

Fig. 2 c_r *vs* c_i (Parabolic instability region)

Solving the above equation, we get $\lambda = 3\,W_{\max} - W_{\min} \pm 2\sqrt{W_{\max}\,(2\,W_{\max} - W_{\min})}$. $\lambda = 3\,W_{\max} - W_{\min} + 2\sqrt{W_{\max}\,(2\,W_{\max} - W_{\min})}$ will lead to $c_r < W_{min}$ and hence

$$\lambda_c = (3\,W_{\max} - W_{\min}) - 2\sqrt{W_{\max}\,(2\,W_{\max} - W_{\min})}\,,$$

then the parabola (11) intersect with Batchelor and Gill semicircle (12).

Example 1 Let us consider example $W(r) = \sin(r^2)$, $R_1 = 1$, $R_2 = 2$. $W_{min} = 0.03925 > 0$, $W_{max} = 0.069756$ and $\lambda = 0.003$, $\lambda_c = 0.00308$. so, $\lambda < \lambda_c$ which implies that the parabola intersects the semi circle (Fig. 1).

Figure 2 illustrate the reduction of parabolic instability region for different values of k, as wave number k increases, the instability region reduces.

Example 2 Let us consider example $W(r) = 1 + (r - \frac{1}{2})$, $R_1 = 1, R_2 = 2$. $W_{min} = 1.5 > 0$, $W_{max} = 2.5$ and $\lambda = 0.03376$, $\lambda_c = 0.0839$. so, $\lambda < \lambda_c$ which implies that the parabola intersects the semi circle.

Theorem 3 *If* $c_i > 0$ *and* $\left(\frac{W'}{r}\right)'_{\min} > 0$ *then* $c_i^2 \leq \lambda' \,[c_r - 2\,W_{\min} + W_{\max}]$, *where* $\lambda' = \frac{(W')^2_{\max}}{2\left(\frac{W'}{r}\right)'_{\min} . R_1}$.

Proof Multiplying (7) by $(W_{\max} - W_{\min})$ and adding with (6) and since $W_{\min} < c_r < W_{\max}$, this implies that $(W - c_r + W_{\max} - W_{\min})$ is positive and hence dropping the first integral term, we get

$$\frac{1}{2}\int_{R_1}^{R_2} r\left(\frac{W'}{r}\right)' r\,|G|^2\,dr + \int_{R_1}^{R_2}\left(\frac{\frac{(W')^2}{4}(W - c_r - W_{\max} + W_{\min})}{|W - c|^2}\right) r\,|G|^2\,dr \leq 0\,.$$

Since $\frac{1}{|W-c|^2} \leq \frac{1}{c_i^2}$, we have

$$R_1\left(\frac{W'}{r}\right)'_{\min}\int_{R_1}^{R_2} r\,|G|^2\,dr \leq \frac{1}{2}\left(\frac{(W')^2_{\max}(c_r - W_{\min} + W_{\max} - W_{\min})}{c_i^2}\right)\int_{R_1}^{R_2} r\,|G|^2\,dr\,.$$

$$i.e., c_i^2 \leq \lambda'\,[c_r - 2\,W_{\min} + W_{\max}]\,, \qquad (13)$$

where $\lambda' = \frac{(W')^2_{\max}}{2\left(\frac{W'}{r}\right)'_{\min} . R_1}$.

Theorem 4 *If* $\lambda' < \lambda'_c$, *where* $\lambda'_c = \left(3 - 2\sqrt{2}\right)(W_{\max} - W_{\min})$, *then the parabola* $c_i^2 \leq \lambda'\,[c_r - 2W_{\min} + W_{\max}]$ *intersect with Batchelor and Gill semicircle.*

Proof Proceeding in the same way as in Theorem 3.2, we get the proof.

4 Short Wave Stability

Theorem 5 *For an unstable mode* $(c_i > 0)$, *it is necessary that* $k^2 < k_c^2$, *where* $k_c^2 = \frac{-\left[r\left(\frac{W'}{r}\right)'\right]^2}{2\,r\left(\frac{W'}{r}\right)'(W - W_s)}$.

Proof Multiplying (1) by $r\left[\frac{(r\,u^*)'}{r}\right]'$, integrating and applying (2), we get

$$\int_{R_1}^{R_2}\left|\left[\frac{(r\,u)'}{r}\right]'\right|^2 r\,dr-\int_{R_1}^{R_2}\left[k^2+\frac{r\left(\frac{W'}{r}\right)'}{W-c}\right]ur\left(\frac{(r\,u^*)'}{r}\right)'dr=0\,. \quad (14)$$

From (1), taking complex conjugate, we have

$$\left(\frac{(r\,u^*)'}{r}\right)'=\left[k^2+\frac{r\left(\frac{W'}{r}\right)'}{W-c^*}\right]u^*. \quad (15)$$

Substituting (15) in (14), we have

$$\int_{R_1}^{R_2}\left|\left[\frac{(r\,u)'}{r}\right]'\right|^2 r\,dr-k^4\int_{R_1}^{R_2}|u|^2\,rdr-k^2\int_{R_1}^{R_2}\left[\frac{r\left(\frac{W'}{r}\right)'}{W-c}\right]$$
$$|u|^2\,rdr-k^2\int_{R_1}^{R_2}\left[\frac{r\left(\frac{W'}{r}\right)'}{W-c^*}\right]|u|^2\,rdr-\int_{R_1}^{R_2}\frac{\left[r\left(\frac{W'}{r}\right)'\right]^2}{|W-c|^2}|u|^2\,rdr=0\,.$$

Equating real parts, we get

$$\int_{R_1}^{R_2}\left|\left[\frac{(r\,u)'}{r}\right]'\right|^2 r\,dr-k^4\int_{R_1}^{R_2}|u|^2\,rdr-2k^2\int_{R_1}^{R_2}\left[\frac{r\left(\frac{W'}{r}\right)'(W-c_r)}{|W-c|^2}\right]|u|^2\,rdr$$
$$-\int_{R_1}^{R_2}\frac{\left[r\left(\frac{W'}{r}\right)'\right]^2}{|W-c|^2}|u|^2\,rdr=0\,. \quad (16)$$

Multiplying (1) by $r\,u^*$, integrating, applying (2) and equating imaginary part, we get

$$-c_i\int_{R_1}^{R_2}\left[\frac{r\left(\frac{W'}{r}\right)'}{|W-c|^2}\right]r\,|u|^2\,dr=0\,. \quad (17)$$

Multiplying (17) by $2k^2\left(\frac{c_r-W_s}{c_i}\right)$ and adding with (16), we get

$$\int_{R_1}^{R_2}\left|\left[\frac{(r\,u)'}{r}\right]'\right|^2 r\,dr - k^4\int_{R_1}^{R_2}|u|^2\,r dr - \int_{R_1}^{R_2}\left[\frac{2k^2 r\left(\frac{W'}{r}\right)'(W-W_s)+\left[r\left(\frac{W'}{r}\right)'\right]^2}{|W-c|^2}\right]|u|^2\,r dr = 0\ .$$

From the above equation, we have

$$2k^2 r\left(\frac{W'}{r}\right)'(W-W_s)+\left[r\left(\frac{W'}{r}\right)'\right]^2 > 0\,.$$

Which implies that, $k^2 < \left[\frac{\left[r\left(\frac{W'}{r}\right)'\right]^2}{-2r\left(\frac{W'}{r}\right)'(W-W_s)}\right]_{r=r_p}$.

Hence $k^2 < {k_c}^2$, where

$$ {k_c}^2 = \frac{\left[r\left(\frac{W'}{r}\right)'\right]^2}{-2r\left(\frac{W'}{r}\right)'(W-W_s)}. \tag{18}$$

Theorem 6 *If* $k^2 > k_c^2$, *where* ${k_c}^2 = \frac{\left[r\left(\frac{W'}{r}\right)'\right]^2}{-2r\left(\frac{W'}{r}\right)'(W-W_s)}$ *then the flow is stable.*

Proof Follows from (18).

Example 3 Let us consider example $W(r) = \sin(r^2),\ R_1 = 1, R_2 = 2.$ $r\left(\frac{W'}{r}\right)' = -4r^2\sin(r^2)$ and $r\left(\frac{W'}{r}\right)'$ changes its sign at $r_s = \sqrt{\pi}$, hence the value of W_s is zero. Which implies ${k_c}^2 = 8$. Therefore If $k > 2.828$ then the flow is stable (Fig. 3).

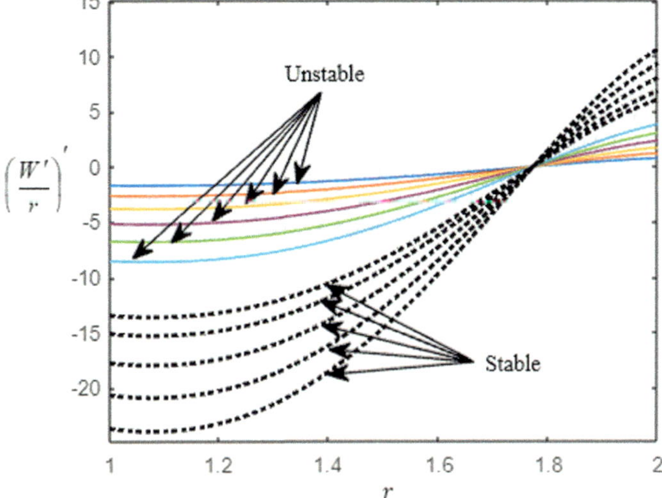

Fig. 3 r *vs* $\left(\frac{W'}{r}\right)'$ (The Stable and Unstable modes for different values of k_c)

5 Concluding Remarks

In this paper, we consider linear stability of axial flows of an inviscid, incompressible, homogeneous fluid to axisymmetric disturbances. For this problem, we derived parabolic instability region for a class of flows, and which intersect with Batchelor and Gill semicircle under some conditions. Also, we derived a necessary condition for stability, namely if the wave number k is greater than or equal to critical wave number k_c then the flow is stable.

Acknowledgements We are thankful to the reviewer for his suggestions in improving the presentation of our paper.

References

1. Chandrasekhar, S.: Hydrodynamic and Hydromagnetic Instability. Clarendon Press, Oxford (1961)
2. Walton, A.G.: Stability of circular Poiseuille-coutte flow to axisymmetric disturbances. J. Fluid. Mech. **500**, 169–210 (2004)
3. Drazin, P.G., Reid, W.H.: hydrodynamic Stability. Cambridge University Press, Cambridge (2004)
4. Batchelor, G.K., Gill, A.E.: On the hydrodynamic and hydromagnetic stability of swirling flows. J. Fluid. Mech. **14**, 529–551 (1962)
5. Iype, M.S.A., Subbiah, M.: On the hydrodynamic and hydromagnetic stability of inviscid flows between coaxial cylinders. Inter. J. Fluid. Mech. Res. **37**, 1–15 (2010)

6. Pavithra, P., Subbiah, M.: Note on instability regions in the circular Raleigh problem of hydrodynamic stability. Proc. Natl. Aca. Sci., Sect. A: Phys. Sci. **91**, 49–54 (2021)
7. Pavithra, P., Subbiah, M.: On sufficient conditions for stability in the circular Rayleigh problem of hydrodynamic stability. J. Anal. **27**, 781–795 (2019)
8. Banerjee, M.B., Gupta, J.R., Subbiah, M.: on reducing Howard's semicircle for homogeneous shear flows. J. Math. Anal. Appl. **130**, 398–402 (1988)
9. Gupta, J.R., Shandil, R.G., Rana, S.D.: On the limitations of the complex wave velocity in the instability problem of heterogeneous shear flows. J. Math. Anal. Appl. **144**, 367–376 (1989)
10. Howard, L.N.: Note on a paper of J W Miles. J Fluid Meh. **10**, 509–512 (1961)

Soret and Chemical Reaction Effects on Heat and Mass Transfer in MHD Flow of a Kuvshinski Fluid Through Porous Medium with Aligned Magnetic Field and Radiation

Raghunath Kodi and Mohana Ramana Ravuri

Abstract This article explores the free convective outflow of radiativekuvshinski fluid through an inclined upward permeable platter implanted in the porous object in the proximity of Radiation absorption, thermal-diffusion (Soret) and aligned magnetic field. When a crosswise magnetic domain exists, it is known that the base is immersed in an identical permeable medium and travels at a fixed velocity in the movement regulation. The nonlinear systems of partial differential equations are changed to the ordinarily differential equalities by utilizing similarities transformations. The changed systems of equations were then solved making use of the perturbation methodology. The graphical representations of velocity, temperature, and concentration distributions obtained from mathematical solutions. Additionally, examine the effect of several specifications of the skin interference, the rate of warmth transference in the frame of the Nusselt number, and the mass transference rate at the surface as measured by the Sherwood number. It was observed that progressing the buoyancy and Soret specifications increased the velocity distribution while diminishes magnetic field specification. Temperature declined when the Prandtl number and Radiation absorption has raised, whereas concentration diminished when the Schmidt number and chemical reaction specification were progressed.

Keywords Radiation absorption · Chemical reaction · Soret effect

R. Kodi (✉)
Department of Humanities and Sciences (Mathematics), Bheema Institute of Technology and Science, Adoni, AP 518301, India
e-mail: kraghunath25@gmail.com

M. R. Ravuri
Department of Basic Science and Humanities (Mathematics), Narasaraopeta Engineering College, Narasaraopeta, AP 522601, India

S. Banerjee and A. Saha (eds.), *Nonlinear Dynamics and Applications*,
Springer Proceedings in Complexity,
https://doi.org/10.1007/978-3-030-99792-2_32

1 Introduction

The flows of the non-Newtonian fluids are increasing considerably owing to plentiful practical submissions in manufactures as well as industrially procedures. Examples of non Newtonian fluids are molten plastic, blood, ketchups, grease, artificially fiber, paints, and certain oily liquids, in addition to numerous others. These fluids violate Newton's law of viscosity. Those liquids were extremely glutinous those did expose their important property of elasticity. Those models of liquids are important in compositely process, polymers depolarization, bubbles absorptions, as well as boiling points, etc. The second-grade fluid exhibited the effects of the natural stress as well as cannot predict the shear thin in addition to shear thicken phenomenon. But a model of the third ordered liquids can predicts together the natural stresses as well as the shear thin as well as the shear thicken phenomenon evenly the constitutive governing equations had more complexity. The resolutions of the specialized third graded liquid toward the stagnation points of the unstable absorbent stretched and/or shrinking surfaces are obtained by Naganthran et al. [1]. The phenomenon's of increment of the thermally conductance of the liquid through scattering nano-particles were studied by Masuda et al. [2]. Buongiorno [3] found that the Brownians movement as well as thermophoresis impacts of nanofluids gave an outstanding increment in fluids thermally conductivities. The computational resolution of nanofluids over the stretched sheets utilizing Buongiornos modeling as well as analyzing the Brownians movement as well as the thermophoresis impacts on the temperature transportation rates at the surface are obtained through Khan as well as Pop [4]. Khan et al. [5] considered the effects of non-linearly radiating on a MHD flows of the Carreau's liquids over the non-linearly stretched surfaces by the convection frontier conditions.

The term "non-Newtonian" is most commonly used to describe fluids that can be used in such diverse areas as chemical, mechanical, and biotechnological systems. Two teams of physicists, mathematicians, and engineers (and several other individuals in various locations) had differing viewpoints on the primary concept of the Kuvshinski fluid flow was examined due to a multitude of factors. Given the plethora of factors to be considered in magnetic flow, the authors Seth and Bhattacharya [6] recently addressed the new Statistical Simulation of hydromagnetic Convection in Porous Media with higher Order Chemical reactions and Newtonian warming. As Newtonian flow is combined with additional diffusion in non-expanding porous media, as hypothesized by Seth et al. [7], they found a new system double-diffusive MHD flow in the pore fluids. Seth et al. [8] decentralized the Flow warmth and magnitude variations are crucial in designing vertical-screw turbines that rotate against fluid or natural convection. Seth et al. [9] study a consequence of Hall movement on MHD Natural Convection in a revolving Fluid/Solid Coupling System on Study of Past-slip non-mode of Viscoelastic Flow past. Seth et al. [10] have tackled the concept of diffusion-controlled convection, which has imperfectly approximated the motion of viscous fluid on rough interfaces and discovered time-dependent hydromagnetic unsteady flow with the Hall Effect and ocean stream expands and contracts with a fast turning mechanism. Hari and Patel [11] have done warmth and expanse variation tests in MHD capillary-coupled capillary tube movement beyond a revolving upright platter within a penetrable object wall at ramped temperature differences. Akhil and Harshad [12] have recently discovered the significance of MHD on mixed convection, a breakthrough in understanding it. Krishna et al. [13] expressed MHD Convection outpour of Kuvshinski liquid movement an indefinite Upright permeable Platter with thermic dispersal and radiation impacts. Krishna et al. [14] conferred the radiating and absorbing on the MHD convection flows of nanoliquids over a vertically travels porous plate.

The attending work is affected by the aligned magnetic domain, Soret consequence, Radiation and Kuvshinski liquid specification on MHD convective warmth and mass transfer outpour of an unstable, incompressible electrically manipulating fluid beyond a semi-unbounded inclined upright permeable platter in the existence of chemical response and Radiation absorption. The governing equations of movement are translated analytically by utilizing the perturbation approach. In this deconstruction, we have generalized the outcome done by Raghunath et al. [15] by evaluating Radiation absorption and Kuvshinski liquid specification. This contemplation may be generous in several industrial applications, such as polymer exhibition, ceramics or glassware edibles processing, and so forth.

2 Mathematical Formulation

Regard an unstable bi-dimensional outpour of a laminar, dense, isentropic, eclectically functioning, emitting, and chemically responding Kuvshinski liquid via a permeable object beyond a semi-unbounded abrupt moving platter. In accord with the coordinate approach, * x-axis is carried along the upright permeable platter in the upward directive and * y-axis expected to it. The liquid is supposed to be a gray, immersing emitting but non dispersion object. The radiative warmth instability in the * x-supervision is regarded inconsequential analogized to that in the * y-supervision. An invariant magnetic specialization is spread perpendicular to the liquid gush approach and presumed that an influenced magnetic area is ignored. Viscous and Darcy antagonism spans are carried into account. The liquid belongings are presumed to be invariant, excluding thickness deviation with temperature and concentration in the body significance stint. The level of foreign mass is assumed to be low So that the Dufour effect is neglected. Also, believe that the overindulgence consequences are overlooked. The physical structure of the problem followed by Obulesu et al. [16]. Beneath the above speculations and gathering the Boussinesq expression, the border coating equations controlling the warmth and mass transmission of a viscoelastic liquid can be composed as (Fig. 1).

$$\frac{\partial v*}{\partial y*} = 0 \rightarrow v^* = -v_0(v_0 > 0) \tag{1}$$

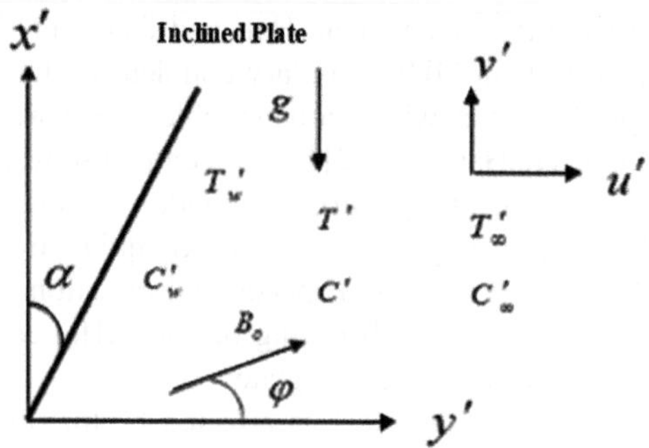

Fig. 1 Physical structure of the problem

$$\left(1+\lambda\frac{\partial}{\partial t}\right)\frac{\partial u*}{\partial t*}+v*\frac{\partial u*}{\partial y*}=-\frac{1}{\rho}\frac{\partial p*}{\partial x*}+\vartheta\frac{\partial^2 u*}{\partial y*^2}+g\beta_T(T*-T^*_\infty)\,\text{Cos}\alpha+ \\ g\beta_C(C*-C^*_\infty)\,\text{Cos}\alpha-\frac{\sigma B_0^2}{\rho}\text{Sin}^2\gamma\, u*-\frac{\vartheta u*}{k*} \tag{2}$$

$$\frac{\partial T*}{\partial t*}+v*\frac{\partial T*}{\partial y*}=\frac{K}{\rho C_p}\frac{\partial^2 T*}{\partial y*^2}-\frac{1}{\rho C_p}\frac{\partial q_r*}{\partial y*}+\frac{Q*}{\rho C_p}(C*-C^*_\infty) \tag{3}$$

$$\frac{\partial C*}{\partial t*}+v*\frac{\partial C*}{\partial y*}=D\frac{\partial^2 C*}{\partial y*^2}-K^*(C*-C^*_\infty)+D_1\frac{\partial^2 T*}{\partial y*^2} \tag{4}$$

With the connected primary along with border stipulations were,

$$u^*=u^*_p,\, T^*=T^*_w+\varepsilon\left(T^*_w-T^*_\infty\right)e^{n^*t^*},\, C^*=C^*_w+\varepsilon\left(C^*_w-C^*_\infty\right)e^{nt} \text{ at } y^*=0 \\ u^*\to u^*_\infty=U_0\left(1+se^{n^*t^*}\right),\quad T^*\to T^*_\infty\quad C^*\to C^*_\infty \text{ as } y^*\to\infty \tag{5}$$

The continuity equation renders that V* is either an invariant or some position of time, hence supposing that.

$$v^*=-\text{V}_0\left(1+\text{A}\,\varepsilon\,\text{e}^{\text{n}^*\text{t}^*}\right) \tag{6}$$

where A is a real optimistic invariant, ε and Aε are smallish than unity, V_0 is the scale of the suction velocity, which has a non-zero optimistic invariant.

Beyond the border layer, Eq. (2) gives

$$-\frac{1}{\rho}\frac{\partial p*}{\partial x*}=\frac{\partial U_\infty}{\partial t*}+\frac{\upsilon}{k*}U^*_\infty+\frac{\sigma B_0^2}{\rho}U^*_\infty\,\text{Sin}^2\gamma \tag{7}$$

We assess an arithmetical prototype for an optically delicate limit gray gas near equaliser in the layout presented by Cramer and Pai [17]. Later Grief et al. [18]

$$\frac{\partial q_r*}{\partial y*} = 4\left(T^* - T_w^*\right) I \tag{8}$$

where $I = \int_0^\infty K_{\lambda\omega}\left(\frac{\partial_{eb\lambda}}{\partial T}\right)_\omega d\lambda,\ K_{\lambda\omega}$ the absorption specification at the wall and ebλ is Planck's specification.

To formalize the arithmetical embodiment of the physical concern, familiarise the subsequent non-dimensional amounts and specifications

$$u = \frac{u*}{U_0}, y = \frac{U_0 y*}{\vartheta}, T = \frac{T*-T_\infty^*}{T_w^* - T_\infty^*}, C = \frac{C*-C_\infty^*}{C_w^* - C_\infty^*}, \Pr = \frac{\mu\rho C_p}{k_T}, Sc = \frac{\vartheta}{D}, M = \frac{\sigma B_0^2\vartheta}{\rho U_0^2},$$

$$Gr = \frac{\vartheta g\beta(T_w^* - T_\infty^*)}{U_0^3}, Gm = \frac{\vartheta g\beta^*(C_w^* - C_\infty^*)}{U_0^3},\quad K = \frac{U_0^2 K_0^*}{\vartheta^2}, t = \frac{t*U_0^2}{\vartheta},$$

$$K_r = \frac{\vartheta K_C^*}{U_0^2}, R = \frac{16a^* v^2 \sigma T_\infty^{*3}}{kU_0^2}, Q = \frac{Q_1 v}{U_0^2}, S_0 = \frac{D_1(T_w^* - T_\infty^*)}{\vartheta(C_w^* - C_\infty^*)}, R = \frac{4\vartheta n^*}{U_0^2} \tag{9}$$

The non-dimensional form of the equalizations (2), (3) and (4) are

$$\left(1+\lambda\frac{\partial}{\partial t}\right)\frac{\partial u}{\partial t} - \frac{\partial u}{\partial y} = \frac{\partial^2 u}{\partial y^2} + Gr\,\theta\ \mathrm{Cos}\alpha + Gm\ \phi\ \mathrm{Cos}\alpha - \left(1+\lambda\frac{\partial}{\partial t}\right)\xi\, u \tag{10}$$

$$\frac{\partial\theta}{\partial t} - \frac{\partial\theta}{\partial y} = \frac{1}{\Pr}\frac{\partial^2\theta}{\partial y^2} + Q\phi - R\theta \tag{11}$$

where $\xi = \left(\frac{1}{k} + M\, u\mathrm{Sin}^2\gamma\right)$

$$\frac{\partial\phi}{\partial t} - \frac{\partial\phi}{\partial y} = \frac{1}{Sc}\frac{\partial^2\phi}{\partial y^2} + S_0\frac{\partial^2\phi}{\partial y^2} - K_r\phi \tag{12}$$

The connected border circumstances are given by

$$\begin{aligned} &u = U_p\quad \theta = 1+\varepsilon e^{nt}\quad \phi = 1+\varepsilon e^{nt},\quad at\quad y = 0\\ &U \to U_\infty = 1+\varepsilon e^{nt},\quad \theta \to 0,\quad \phi \to 0\quad as\quad y \to \infty \end{aligned} \tag{13}$$

3 Method of Solution

The equalizations (11)–(13) are the partial differential arrangement that can't be decrypted in sealed conditions. Regardless, these can be decoded by consolidating them into regular differential equations utilizing the subsequent perturbation approach. Now describe the velocity, temperature, and concentration disbandments in stints of harmonic and non-harmonic processes as

$$\begin{aligned} U(y,t) &= u_0(y) + \varepsilon u_1(y)e^{nt} + O(\varepsilon^2) \\ T(y,t) &= \theta_0(y) + \varepsilon \theta_1(y)e^{nt} + O(\varepsilon^2) \\ C(y,t) &= \phi_0(y) + \varepsilon \phi_1(y)e^{nt} + O(\varepsilon^2) \end{aligned} \tag{14}$$

Covering Equalization (15) into equalizations (11)–(13), and correlating the harmonic and non-harmonic stints, and overlooking the more elevated mandate stints of ε, acquire the subsequent teams of equations of ordering zero and one.

3.1 Zero Order Terms

$$u_0'' + u_0' - \xi\, \mathrm{u}_0 = -\mathrm{Gr}\, \cos\alpha\, \theta_0 - \mathrm{Gm}\, \cos\alpha\, \phi_0 \tag{15}$$

$$\theta_0'' + \mathrm{Pr}\, \theta_0' + \mathrm{Pr}\ \mathrm{Q}\, \phi_0 - R\, \mathrm{Pr}\, \theta_0 = 0 \tag{16}$$

$$\phi_0^{ll} + Sc\, \phi_0^{l} - \mathrm{Sc}\, \mathrm{K}\phi_0 = -\mathrm{Sc}\, \mathrm{Sr}\, \theta_0^{\mathrm{ll}} \tag{17}$$

3.2 First Order Terms

$$u_1'' + u_1' - (\xi + n)\, \mathrm{u}_1 = -\mathrm{Gr}\, \cos\alpha\, \theta_1 - \mathrm{Gm}\, \cos\alpha\, \phi_1 \tag{18}$$

$$\theta_1'' + \mathrm{Pr}\, \theta_1' - n\ \mathrm{Pr}\ \theta_1 - \mathrm{Pr}\ Q\, \phi_1 + \mathrm{Pr}\ R\, \theta_1 = -\mathrm{Pr}\, \mathrm{A}\, \theta_0' \tag{19}$$

$$\phi_1^{ll} + Sc\, \phi_1^{l} - \mathrm{Sc}\, (\mathrm{K} + n)\, \phi_1 = -\mathrm{A}\, \mathrm{Sc}\, \phi_0^{1} - \mathrm{Sc}\, \mathrm{Sr}\, \theta_1^{\mathrm{ll}} \tag{20}$$

The corresponding boundary conditions are

$$\begin{aligned} &u_0 = U_p, u_1 = 0, \theta_0 = 1, \theta_1 = 1, \quad C_0 = 1, \quad C_1 = 1 \quad at \quad y = 0 \\ &u_0 = 1, \quad u_1 = 1, \theta_0 \to 0, \theta_1 \to 0, C_0 \to 0, C_1 \to 0 \quad as \quad y \to \infty \end{aligned} \tag{21}$$

Solving Eqs. (15)–(20) under the boundary conditions (21), the following solutions are obtained

$$\phi_0 = A_3 \exp(-m_1 y) + A_4 \exp(-m_3 y) \tag{22}$$

$$\theta_0 = \exp(-m_1 y) \tag{23}$$

$$u_0 = 1 + A_9 \exp(-m_5 y) + A_{10} \exp(-m_3 y) + A_{11} \exp(-m_5 y) \tag{24}$$

$$\phi_1 = A_5 \exp(-m_1 y) + A_6 \exp(-m_2 y) + A_7 \exp(-m_3 y) + A_8 \exp(-m_4 y) \tag{25}$$

$$\theta_1 = A_1 \exp(-m_1 y) + A_2 \exp(-m_2 y) \tag{26}$$

$$\begin{aligned} u_1 = 1 &+ A_{12} \exp(-m_1 y) + A_{13} \exp(-m_2 y) + A_{14} \exp(-m_3 y) + A_{15} \exp(-m_4 y) \\ &+ A_{16} \exp(-m_3 y) + A_{17} \exp(-m_6 y) \end{aligned} \tag{27}$$

Substituting Eqs. (22)–(27) in Eq. (14), obtain the velocity, temperature and concentration distribution in the boundary layer as follows

$$\begin{aligned} u =& \left(1 + A_9 \exp(-m_1 y) + A_{10} \exp(-m_3 y) + A_{11} \exp(-m_5 y)\right) + \varepsilon e^{nt} (1 + A_{12} \exp(-m_1 y) + \\ & A_{13} \exp(-m_2 y) + A_{14} \exp(-m_3 y) + A_{15} \exp(-m_4 y) + A_{16} \exp(-m_5 y) + A_{17} \exp(-m_6 y)) \end{aligned} \tag{28}$$

$$\theta = \exp(-m_1 y) + \varepsilon e^{nt} (A_1 \exp(-m_1 y) + A_2 \exp(-m_2 y)) \tag{29}$$

$$\begin{aligned} \phi = & A_3 \exp(-m_1 y) + A_4 \exp(-m_3 y) + \varepsilon e^{nt} (A_5 \exp(-m_2 y) + \\ & A_6 \exp(-m_2 y) + A_7 \exp(-m_3 y) + A_8 \exp(-m_4 y)) \end{aligned} \tag{30}$$

3.3 Skin Friction

$$\begin{aligned} \tau = & -(m_1 A_9 + m_3 A_{10} + m_5 A_{11}) \\ & -\varepsilon\, e^{nt} (m_1 A_{12} + m_2 A_{13} + m_3 A_{14} + m_4 A_{15} + m_5 A_{16} + m_6 A_{17}) \end{aligned} \tag{31}$$

3.4 Nusselt Number

$$Nu = -\left(\frac{\partial \theta}{\partial y}\right)_{y=0} = -m_1 - \varepsilon\, e^{nt} (m_1 A_1 + m_2 A_2) \tag{32}$$

3.5 Sherwood Number

$$Sh = m_1 A_3 + m_3 A_4 + \varepsilon\, e^{nt} (m_1 A_5 + m_2 A_6 + m_4 A_7 + m_5 A_8) \tag{33}$$

4 Results and Discussion

The consequence of the Grashof numeral on the velocity silhouettes is caught in Fig. 2. A proliferation in Gr contributes to an upsurge in velocity when all different specifications in the velocity domain are held unchanging. Also, it is detected that as we push away from the platter, the consequence of Gr is not that influential. The adjusted Grash of numeral Gm on the velocity silhouettes is memorialized in Fig. 3. A proliferation in Gm is seen to exploit the velocity to rise. Also, it is noticed that as we push far away from the platter, it is caught that the outcome of Gm is discovered to be not that influential.

The consequences of the magnetic domain specification on the velocity dispersal silhouettes across the border coating are exemplified in Fig. 4. The outcome of

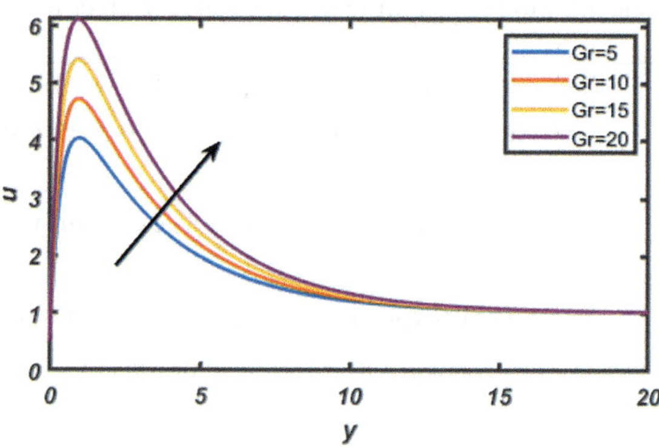

Fig. 2 The consequence of thermal Grash of number (Gr) specification on velocity

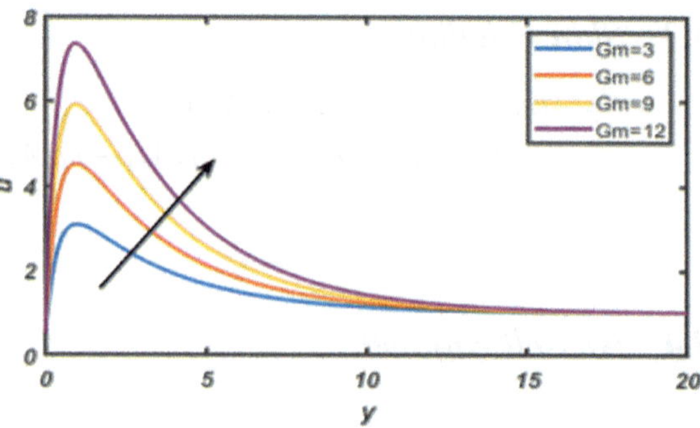

Fig. 3 The consequence of Mass Grash of number (Gr) specification on velocity

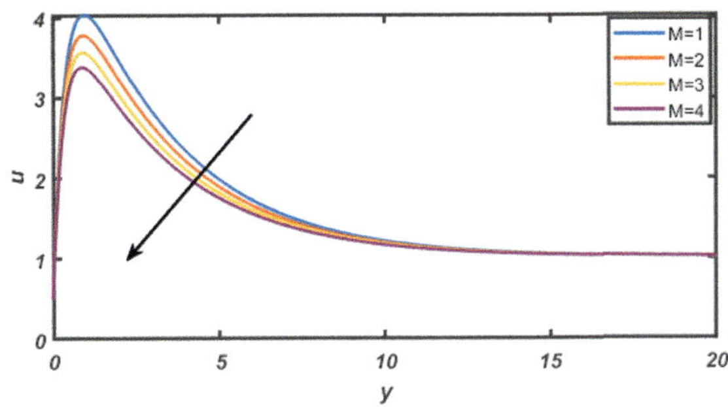

Fig. 4 The consequence of Magnetic field (M) specification on velocity

raising magnetic domain specification M diminishes the velocity issuance across the bordering coat. This is because the intro of a transverse magnetic domain standard to the outpour movement compels the trend to create a drag squad due to Lorentz power, retarding the velocity silhouettes. It can be glimpsed in Fig. 5 that gradient of inclination (α) diminishes the consequence of the buoyancy power due to thermal disbandment. Hence, the enterprising significance to the liquid diminishes as a outcome velocity of the liquid declines. The impact of the aligned magnetic domain specification in the velocity silhouette is portrayed in Fig. 6. It is regarded that the velocity lessens with an enlargement aligned magnetic domain (γ) specification. Figure 7 depicts the consequence of the permeability specification (k) on the velocity disbandment silhouettes from which it evolves transparent that as the permeability specification (k) raises, the velocity gains along with the border layer consistency, which is anticipated since when the hollows of the permeable medium evolve more expansive, the resistivity of the medium may be overlooked. Figure 8 exhibits the consequences of Soret numeral (Sr) on the velocity domain; it is encountered that the velocity expands with an expansion in Sr.

Figure 9 exemplifies the radiation absorption specification consequence on the border layer's temperature silhouettes. As the radiation absorption specification

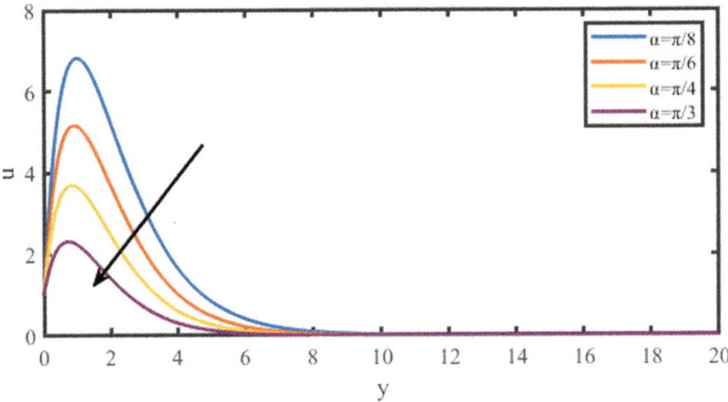

Fig. 5 The consequence of inclined angle (α) specification on velocity

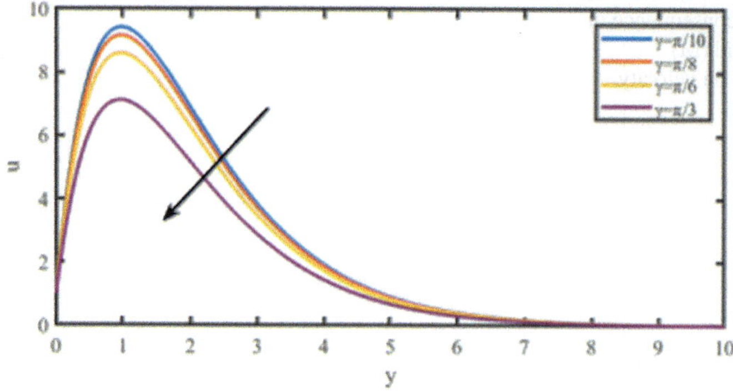

Fig. 6 The consequence of Aligned magnetic field (γ) specification on velocity outlines

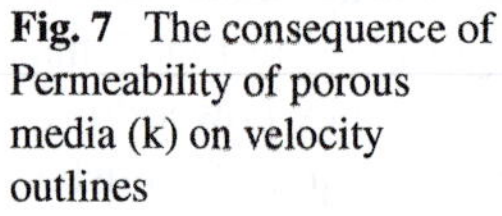

Fig. 7 The consequence of Permeability of porous media (k) on velocity outlines

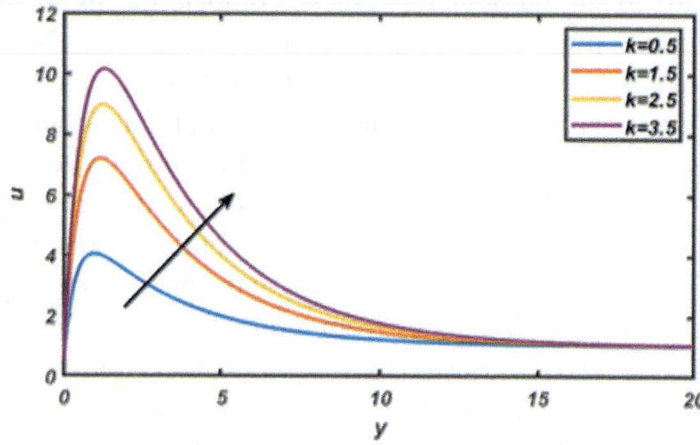

Fig. 8 The consequence of Permeability Soret (Sr) specification on velocity outlines

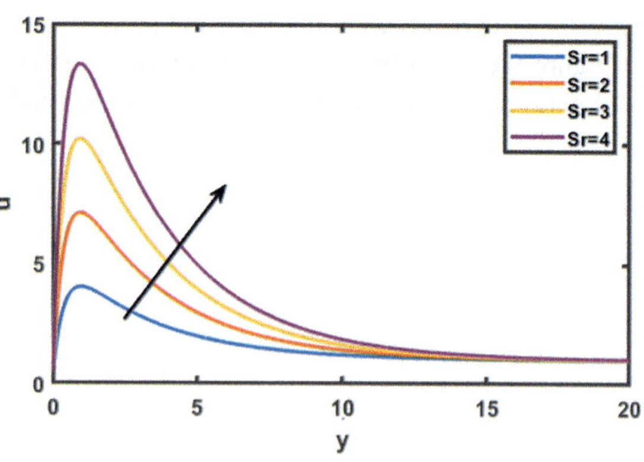

expansions, temperature dispersals rise when the additional physical specifications are designated. Figure 10 portrays the consequence of Prandtl numeral (Pr) on temperature silhouettes in existence of some established fluids such as Hydrogen ($Pr = 0.68$), Air ($Pr = 0.71$), Carbon dioxide ($Pr = 0.76$) and Electrolytic solution

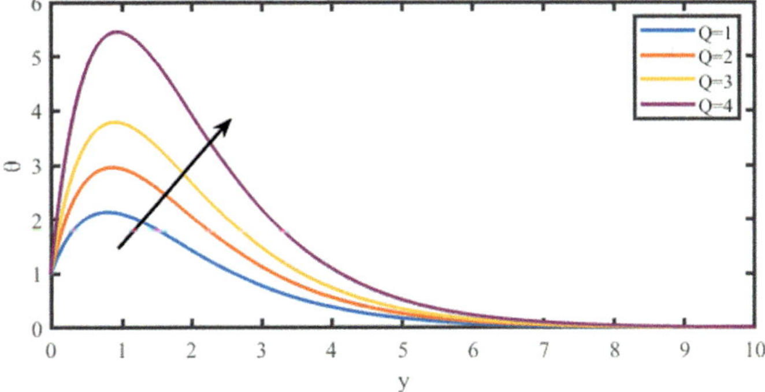

Fig. 9 The consequence of radiation absorption (Q) specification on temperature outlines

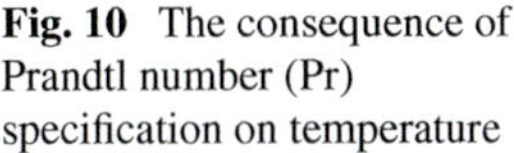

Fig. 10 The consequence of Prandtl number (Pr) specification on temperature

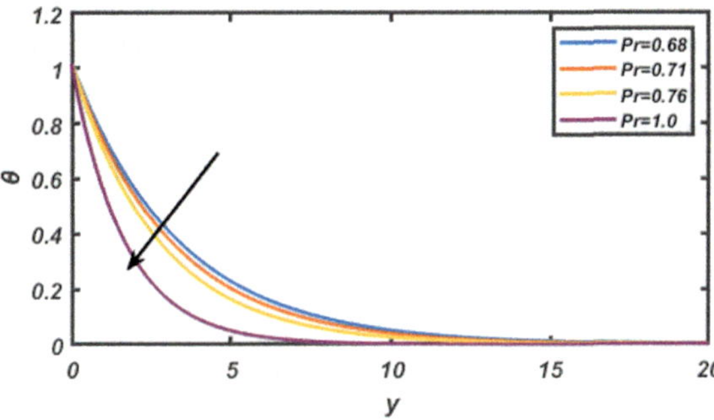

(Pr = 1). This sculpture regarded that enlargement in the Prandtl numeral lessens the temperature of the outpour domain at all matters. Due to the proportion of swiftness diffusivity to thermic diffusivity. Figure 11 depicts the temperature disbandment on the radiation specification (F). This sculpture exhibits that the temperature lessens with an accumulation in the Radiation Specification.

Figure 12 depicts Concentration silhouette for distinct significances of Schmidt

Fig. 11 The consequence of radiation parameter (F) specification on temperature

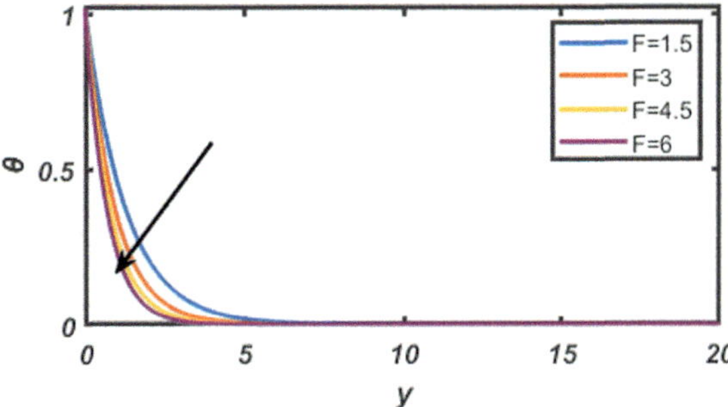

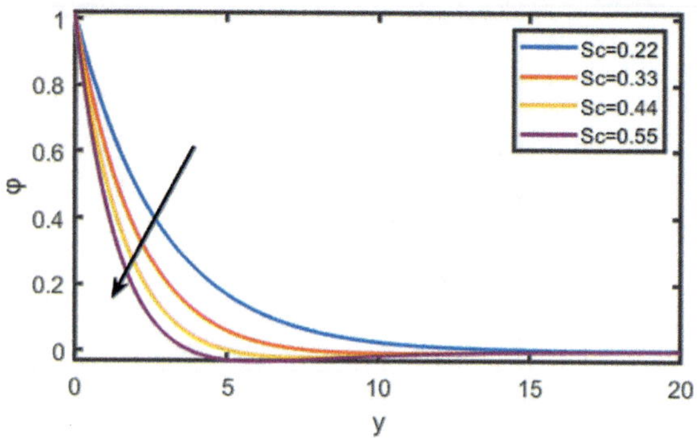

Fig. 12 The consequence of Schmidt number (Sc) specification on concentration

numeral Sc, delivers that growth in Sc declines the concentration silhouette. Physically, this is accurate because the wetness hazes can possess a definitive concentration area, whereas hydrogen can be utilized to preserve an adequate concentration area. Figure 13 exemplifies the concentration silhouettes for distinct matters of Soret numeral Sr. As witnessed from this chart, the concentration of species declines with the importance of the accumulation of the Soret number. Figure 14 depicts the

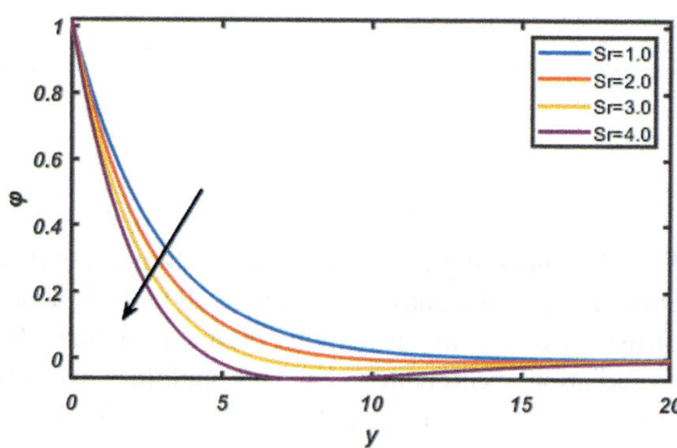

Fig. 13 The consequence of Soret number (Sr) specification on concentration

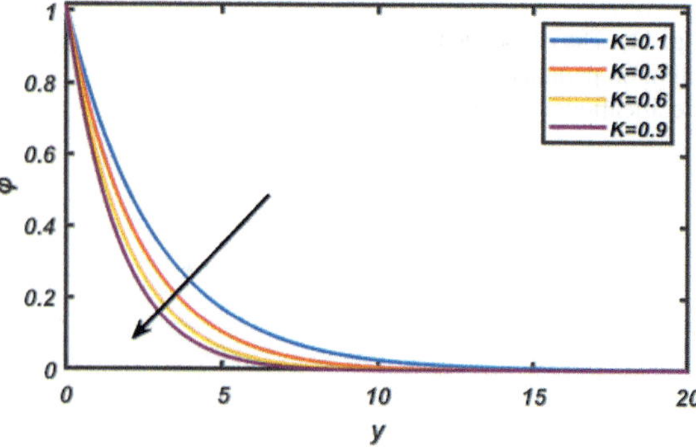

Fig. 14 The consequence of Chemical reaction (K) specification on concentration outlines

convincing of the chemical response influence on concentration. This mannequin viewer's concentration declines with an accumulation in the significances of the chemical retort specification.

The interpretation in skin-friction specification, the rate of warmth transmission in the physique of Nusselt numeral, and the rate of mass transmission in the physique of Sherwood numeral for miscellaneous specification are dissected via Tables 1, 2, 3. For

Table 1 Skin friction

M	γ	Gr	Gm	Q	Kr	τ (previous) Obulesu et al. [16]	τ (previous) Raghunath et al. [15]	τ
1	$\pi/6$	5	5	0.5	0.8	0.0985	0.0927	0.0827
1.5						0.1745	0.1680	0.1370
2						0.1985	0.1851	0.1881
	$\pi/4$					0.1247	0.1254	0.1881
	$\pi/3$					0.2876	0.2897	0.2819
		7.5				1.2479	1.2654	1.2862
		10				2.5478	2.4785	2.4898
			7.5			−1.7854	−1.6754	−1.6192
			10			−3.4578	−3.3324	−3.3212
						−0.6478	−0.6782	−0.6470
						0.0457	0.0467	0.0827
				0.2		−0.3547	−0.3457	−0.3673
				0.4		−2.4785	−2.4578	−2.4396
					1.0	0.2478	0.2547	0.1968
					1.5	0.3245	0.3245	0.2660
						16.1475	16.1547	16.5628
						20.4789	20.4578	20.5053
						24.4521	24.4577	24.1451

Table 2 Nusselt number

Q	Pr	R	Nu (previous) Obulesu et al. [16]	Nu (previous) Raghunath et al. [15]	Nu (present)
0.2	0.71	2	−1.7575	−1.6765	−1.6897
0.4	0.71	2	−1.6785	−1.6093	−1.6797
0.6	0.71	2	−1.5782	−1.5381	−1.5370
0.1	0.25	2	−0.2478	−0.2512	−0.2896
0.1	0.71	2	−1.7210	−1.7088	−1.7820
0.1	7.0	2	−9.3547	−9.3225	−9.4271
0.1	0.71	1	−1.3478	−1.3372	−1.3823
0.1	0.71	3	−1.9874	−1.9969	−1.9856
0.1	0.71	5	−2.4007	−2.4559	−2.5880

Table 3 Sherwood Number

Q	Kr	Sr	Sh (previous) Obulesu et al. [16]	Sh (previous) Raghunath et al. [15]	Sh (present)
0.2	0.5	2	0.4785	0.5868	0.5457
0.4	0.5	2	0.5478	0.5586	0.4577
0.6	0.5	2	0.5367	0.5286	0.5247
0.1	2	2	−0.0874	−0.0722	−0.0687
0.1	4	2	0.6712	0.6003	0.6154
0.1	6	2	3.6874	3.6683	3.7857
0.1	05	1	0.2101	0.2501	0.2155
0.1	0.5	3	0.9875	0.9504	0.9851
0.1	0.5	5	1.6780	1.6507	1.7514

the reality of our work, to analogize our consequences with the existent consequences of Raghunath et al. [15] in the scarcity of permeable object and heat conception. Our result arises to be in exceptional arrangement with the present consequences.

5 Conclusion

From the current examination, the subsequent findings can be illustrated:

1. The liquid velocity supplements when the Grashof numeral (Gr), altered Grashof numeral (Gm), the Soret numeral (Sr), the permeable media (k) proliferation.
2. The liquid velocity diminishes with an accumulation aligned magnetic domain specification (γ), chemical reaction specification (K), and the angle of inclination specification (α).
3. The liquid temperature diminishes with the consequence of the Prandtl numeral (Pr), radiation specification (R), and enhanced when increases of radiation absorption specification (Q).
4. The concentration declines with enhancing of Soret numeral (Sr) chemical reaction specification (Kr) and the Schmidt number (Sc).

References

1. Naganthran, K., Nazar, R., Pop, I.: Unsteady stagnation-point flow and heat transfer of a special third-grade fluid past a permeable stretching/shrinking sheet. Sci. Rep. **6**, 1–13 (2016)
2. Masuda, H., Teramae, A.E.K., Hishinuma, N.: Alteration of thermal conductivity and viscosity of liquid by dispersing ultrane particles. NetsuBussei **7**, 227–233 (1993)
3. Buongiorno, J.: Convective transport in nanofluids. ASME J. Heat Transf. **128**, 240–250 (2006)
4. Khan, W.A., Pop, I.: Boundary-layer flow of a nanofluid past astretching sheet. Int. J. Heat Mass Transf. **53**, 2477–2483 (2010)

5. Khan, M., Hashim, M., Hussain, M., Azam, M.: Magnetohydrodynamic flow of Carreau fluid over a convectively heated surface in the presence of non-linear radiation. J. Magn. Mater. **412**, 63–68 (2016)
6. Seth, G.S., Bhattacharyya, A., Kumar, R., Mishra, M.K.: Modelling and numerical simulation of hydromagnetic natural convection Casson fluid flow with nth-order chemical reaction and Newtonian heating in porous medium. J. Porous Media **22**(9), 1141–1157 (2019)
7. Seth, G.S., Kumar, R., Tripathi, R., Bhattacharyya, A.: Double diffusive MHD Casson fluid flow in a non-Darcy porous medium with Newtonian heating and thermo-diffusion effects. Int. J. Heat Technol. **36**(4), 1517–1527 (2019). https://doi.org/10.18280/ijht.360446
8. Seth, G.S., Bhattacharyya, A., Tripathi, R.: Effect of hall current on MHD natural convection heat and mass transfer flow of rotating fluid past a vertical plate with ramped wall temperature. Front. Heat Mass Transf. (FHMT) **9**, 21 (2017). https://doi.org/10.5098/hmt.9.21
9. Seth, G.S., Bhattacharyya, A., Mishra, M.K.: Study of partial slip mechanism on free convection flow of viscoelastic fluid past a nonlinearly stretching surface. Comput. Therm. Sci.: Int. J. 105–117 (2019). https://doi.org/10.1615/ComputThermalScien.2018024728
10. Seth, G.S., Mahto, N., Tripathi, R., Bhattacharyya, A.: Unsteady hydromagnetic flow formation with hall effect due to time-dependent free stream in a rotating medium. J. Nat. Sci. Sustain. Technol. **11**(3), 197–211 (2018)
11. Hari, R.K., Harshad, P.R.: Heat and mass transfer in magnetohydrodynamic (MHD) Casson fluid flow past over an oscillating vertical plate embedded in porous medium with ramped wall temperature. Propuls. Power Res. **7**(3), 257–267 (2018)
12. Akhil, S., Harshad, P.R.: Influence of thermophoresis and Brownian motion on mixed convection two dimensional MHD Casson fluid flow with non-linear radiation and heat generation. Phys. A **537**, 122710 (2020). https://doi.org/10.1016/j.physa.2019.122710
13. Krishna, R.V., Reddy, G.V., Kiran Kumar, R.V.M.S.S., Varma, S.V.K.: MHD convection flow of kuvshinski fluid past an infinite vertical porous plate with thermal diffusion and radiation effects. Chem. Mater. Res. **8**(2), 18–31 (2016)
14. Krishna, M.V., Ahamad, N.A., Chamkha, A.J.: Radiation absorption on MHD convective flow of nanofluids through vertically travelling absorbent plate. Ain Shams Eng. J. **11**, 1–14 (2021). https://doi.org/10.1016/j.asej.2020.10.028
15. Raghunath, K., Gulle, N., Vaddemani, R.R., Mopuri, O.: Unsteady MHD fluid flow past an inclined vertical porous plate in the presence of chemical reaction with aligned magnetic field, radiation, and Soret effects. Heat Trans. **51**, 1–19 (2022). https://doi.org/10.1002/htj.22423
16. Mopuri, O., Kodi, R., Ganteda, C., Srikakulapu, R., Lorenzini, G.: MHD heat and mass transfer steady flow of a convective fluid through a porous plate in the presence of diffusion thermo and aligned magnetic field. J. Adv. Res. Fluid Mech. Therm. Sci. **89**(1), 62–76 (2022). https://doi.org/10.37934/arfmts.89.1.6276
17. Cramer, K.P., Pai, S.I.: Magneto Fluid Dynamics for Engineers and Applied Physics. McGraw-Hill Book Co, New York (1973)
18. Grief, G., Habib, I.S., Lin, L.C.: Laminar convection of a radiating gas in a vertical channel. J. Fluid Mech. **45**, 513–520 (1971)

5. Khan, M., Hashim, M., Hussain, M., Azam, M.: Magnetohydrodynamic flow of Carreau fluid over a convectively heated surface in the presence of non-linear radiation. J. Magn. Magn. Mater. 412, 63–68 (2016)
6. Seth, G.S., Bhattacharyya, A., Kumar, R., Mishra, M.K.: Modelling and numerical simulation of hydromagnetic natural convection Casson fluid flow with nth-order chemical reaction and Newtonian heating in porous medium. J. Porous Media 22(9), 1141–1157 (2019)
7. Seth, G.S., Kumar, R., Tripathi, R., Bhattacharyya, A.: Double diffusive MHD Casson fluid flow in a non-Darcy porous medium with Newtonian heating and thermo-diffusion effects. Int. J. Heat Technol. 36(4), 1517–1527 (2018). https://doi.org/10.18280/ijht.360447
8. Seth, G.S., Bhattacharyya, A., Tripathi, R.: Effect of Hall current on MHD natural convection heat and mass transfer flow of rotating fluid past a vertical plate with ramped wall temperature. Front. Heat Mass Transf. 9(21) (2017). https://doi.org/10.5098/hmt.9.21
9. Seth, G.S., Bhattacharyya, A., Mishra, M.K.: Study of partial slip mechanism on free convection flow of viscoelastic fluid past a nonlinearly stretching surface. Comput. Therm. Sci. Int. J. 11(1–2), 105–117 (2019). https://doi.org/10.1615/ComputThermalScien.2018024728
10. [illegible]
11. [illegible]
12. [illegible]
13. [illegible]
14. [illegible]
15. [illegible]
16. [illegible]
17. [illegible]
18. [illegible]

Effect of Reversible Reaction on Concentration Distribution of Solute in a Couette Flow

Nanda Poddar, Subham Dhar, and Kajal Kumar Mondal

Abstract A multiple-scale homogenization technique is employed in the current research to show the dispersion phenomena in a Couette flow where the solute may undergoes a reversible phase exchange between the immobile phase (stationary boundary bed phase) and mobile phase (fluid phase). Analytical solutions are obtained to view the influences of retardation factor and phase exchange kinetics on transport coefficient as well as in the two-dimensional longitudinal and transverse concentration distributions. Effects of several transversal position and dispersion time on longitudinal real concentration and the impact of different downstream stations on transverse concentration distributions are also determined. It is seen that with the increment of Damkohlar number the effective dispersivity reduces.

Keywords Homogenization technique · Dispersion · Couette flow · Reversible reaction · Phase exchange kinetics

1 Introduction

The dispersion of solute affected by reversible reaction is motivated for its huge applications in the fields of biological, environmental and chemical engineering. Due to its practical importance, the researchers has paid attention to study the influence of reversible reaction on the transport of tracer in flows in recent days. Sir Taylor [1] presented solute dispersion process through a tube in his pioneer work. He introduced an approximate solution which was valid for large dispersion time under some certain limiting conditions. Then after, Aris [2] dispelled those restrictions employing the method of moments for the solution of advection-diffusion equation. Later, Chatwin [3] accuired an asymptotic series solution of advection-diffusion equation

Supported by UGC India and CSIR India

N. Poddar (✉) · S. Dhar · K. K. Mondal
Cooch Behar Panchanan Barma University, Cooch Behar, WB 736101, India
e-mail: nandapoddarcr7@gmail.com

S. Banerjee and A. Saha (eds.), *Nonlinear Dynamics and Applications*,
Springer Proceedings in Complexity,
https://doi.org/10.1007/978-3-030-99792-2_33

for a pipe. Mei et al. [4] introduced a multi-scale asymptotic homogenization method to determine the dispersion coefficient which is also valid for longer dispersion time in compare to diffusion time. By applying this technique, to investigate the solute transport phenomena, it is observed that longitudinal real concentration, transverse concentration and mean concentration distributions can be found simultaneously. Bandyopadhyay and Mazumder [5] analyzed the scalar transport phenomena in generalized Couette flow by the method of moments. Afterwards, Ng and Yip [6] and Ng and Bai [7], Mazumder and Paul [13], Barik and Dalal [14] studied the effect of reversible sorptive exchange in open channel flow and oscillatory Couette flow respectively. In recent time, several researchers such as Wu and Chen [8], Barik and Dalal [9], Poddar et al. [10], Dhar et al. [11] and Das et al. [12] applied multiple-scale homogenization theory for study the dispersion phenomena of solute through various flow geometry to show the impacts of different flow parameters.

The study of dispersion process of tracers with reversible reaction in steady Couette flow is investigated using homogenization technique for the first time. The main purpose of the present research is to view the effect of reversible phase exchange kinetics between the fluid phase and the stationary boundary bed in a simple Couette flow by employing the multi-scale homogenization technique. In view of the previous literature survey, it is found that, this is the first time in which the study on dispersion of solute is performed for a simple Couette flow for finding the dispersion coefficient, longitudinal real concentration and transverse concentration together. The analytical expressions for dispersion coefficient and real concentration distribution are obtained to inspect the impacts of reversible phase exchange kinetics, retardation factor, time of dispersion and other flow parameters on them. The current analytical result is compared with the results of Bandyopadhyay and Mazumder [5] and Ng and Bai [7] for confirmation of validation. It is seen that, in absence of the retardation factor, the present result of dispersion coefficient of the solute is exactly equal with those of [5, 7].

2 Formulation of the Problem

2.1 Velocity Profile

For the current research problem an one-dimensional laminar, viscous, incompressible Couette flow is considered between two infinite parallel plates. The plates are separated by a distance h. The $\bar{x}$ axis and $\bar{y}$ axis are taken along the longitudinal and transversal directions respectively in a Cartesian coordinate system. It is assumed that the lower plate is situated at $\bar{y} = 0$ and it is at rest, however the upper plate oscillates with a constant characteristic velocity U in its own plane. Neglecting the pressure gradients, momentum equation reduces to

$$\frac{d^2\bar{u}}{d\bar{y}^2} = 0, \tag{1}$$

with boundary conditions

$$\bar{u}(\bar{y})|_{\bar{y}=0} = 0 \quad \text{and} \quad \bar{u}(\bar{y})|_{\bar{y}=h} = U. \tag{2}$$

The analytical solution for velocity distribution in Couette flow is

$$\bar{u} = U\frac{\bar{y}}{h}, \tag{3}$$

2.2 Governing Equation and Boundary Conditions

Consider the dispersion of reactive species through the above mentioned flow (see Fig. 1). It is assumed that the tracer material is fully mixable in the flow of fluid. A portion of tracer concentration is flows with the fluid and the rest stay at the the boundary during the flow. The mobile phase or fluid phase is the phase where concentration moves with the flowing fluid and the immobile phase is that in which the tracer retains at the boundary.

The dispersion problem of concentration of the solute, when it released into the above mentioned Couette flow, is given by:

$$\frac{\partial C}{\partial \bar{t}} + \bar{u}\frac{\partial C}{\partial \bar{x}} = D\left(\frac{\partial^2 C}{\partial \bar{x}^2} + \frac{\partial^2 C}{\partial \bar{y}^2}\right), \quad 0 < \bar{y} < h, \tag{4}$$

where, $C(\bar{x}, \bar{y}, \bar{t})$ is the solute concentration (mass of reactive contaminant dissolved per bulk volume of the fluid) of the fluid phase $\bar{t}$ is the time and D is the molecular diffusivity.

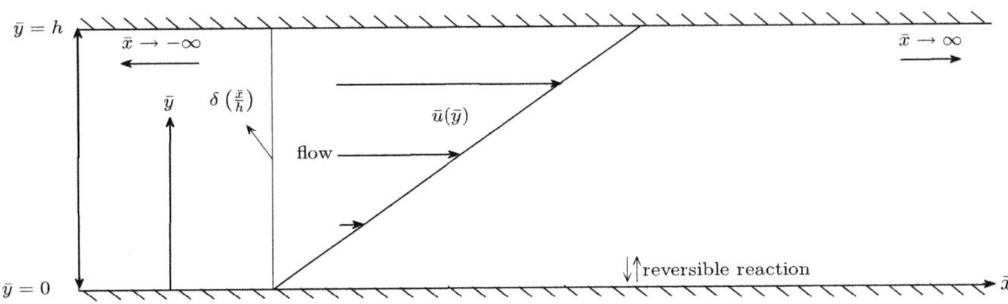

Fig. 1 Systematic diagram of Couette flow

Consider a uniform and instantaneous release of solute with mass Q at the cross-section of $\bar{x} = 0$ at time $\bar{t} = 0$, the initial condition can be taken as

$$C(\bar{x}, \bar{y}, 0) = \frac{Q}{h}\delta\left(\frac{\bar{x}}{h}\right) \tag{5}$$

where, $\delta(\bar{x})$ is the Dirac delta function.

The boundary conditions respectively given as

$$D\frac{\partial C}{\partial \bar{y}} = 0, \quad \text{at} \quad \bar{y} = h \tag{6}$$

$$D\frac{\partial C}{\partial \bar{y}} = \frac{\partial C_s}{\partial \bar{t}} = \gamma(\bar{\theta}C - C_s), \quad \text{at} \quad \bar{y} = 0 \tag{7}$$

where, C_s is the immobile phase concentration (mass of reactive contaminant retained per unit area of channel bed surface), γ is the reversible reaction rate and $\bar{\theta}$ is the retention factor or partition coefficient that relates the concentrations of mobile and immobile phases.

Since, the tracer material can not move at infinity, the upstream and downstream conditions are given by

$$C(\bar{x}, \bar{y}, \bar{t})|_{\bar{x}+\rightarrow\pm\infty} = 0 \tag{8}$$

3 Multi-scale Method of Homogenization

3.1 Scales Selection

In order to find the solution of convection diffusion equation three different time scales are taken as Mei's multi-scale homogenization technique, related with two scales of length h (the channel width) and L (relative length of the tracer cloud). These are T_0, T_1 and T_2, where, $T_0 = h^2/D$ as the diffusion time across the channel width, $T_1 = L/U$ as the convection time across the characteristic length and $T_2 = L^2/D$ is the diffusion time along longitudinal direction. The general expression of the ratio of three distinct time scales is given by

$$T_0 : T_1 : T_2 = 1 : \frac{1}{\epsilon} : \frac{1}{\epsilon^2} \tag{9}$$

where $\epsilon = \frac{h}{L}(\ll 1)$ can be taken as perturbation parameter.

3.2 Dimensionless Governing Equation and Velocity Profile

On introduction of the dimensionless parameters

$$x = \frac{\bar{x}}{L}, y = \frac{\bar{y}}{h}, u = \frac{\bar{u}}{U}, t = \frac{D\bar{t}}{h^2}, Pe = \frac{Uh}{D}, Da = \frac{\gamma h^2}{D}, \theta = \frac{\bar{\theta}}{h}, \tag{10}$$

where Pe, Da, θ are the Peclet number, Damkohler number and dimensionless retention parameter respectively.

Then the concentration transport equation and the associate stationary conditions is reduces to

$$\frac{\partial C}{\partial t} + \epsilon Peu\frac{\partial C}{\partial x} = \epsilon^2 \frac{\partial^2 C}{\partial x^2} + \frac{\partial^2 C}{\partial y^2}, \quad 0 < y < 1, \tag{11}$$

$$\frac{\partial C}{\partial y} = 0, \quad \text{at} \quad y = 1, \tag{12}$$

$$\frac{\partial C}{\partial y} = \frac{\partial C_s}{\partial t} = Da(\theta C - C_s), \quad \text{at} \quad y = 0. \tag{13}$$

The dimensionless velocity is defined as

$$u = y. \tag{14}$$

3.3 Homogenization

Mei's homogenization technique [4] is employed for the asymptotic analysis. The expansion of mobile and immobile concentration into multi-scale are

$$C = C^{(0)} + \epsilon C^{(1)} + \epsilon^2 C^{(2)} + O(\epsilon^3), \tag{15}$$

and

$$C_s = C_s^{(0)} + \epsilon C_s^{(1)} + \epsilon^2 C_s^{(2)} + O(\epsilon^3), \tag{16}$$

On the basis of discussion of Sect. 3.1 the fast, medium and slow time variables are taken as

$$t_0 = t, t_1 = \epsilon t, t_2 = \epsilon^2 t. \tag{17}$$

The original time derivative becomes, according to the chain rule

$$\frac{\partial}{\partial t} \equiv \frac{\partial}{\partial t_0} + \epsilon\frac{\partial}{\partial t_1} + \epsilon^2\frac{\partial}{\partial t_2} \tag{18}$$

Substitution of Eqs. (15) and (18) into Eqs. (11)–(13) results in

$$\left(\frac{\partial C^{(0)}}{\partial t_0}-\frac{\partial^2 C^{(0)}}{\partial y^2}\right)+\epsilon\left(\frac{\partial C^{(0)}}{\partial t_1}+\frac{\partial C^{(1)}}{\partial t_0}+uP_e\frac{\partial C^{(0)}}{\partial x}-\frac{\partial^2 C^{(1)}}{\partial y^2}\right)+\epsilon^2\left(\frac{\partial C_{(0)}}{\partial t_2}\right.$$
$$\left.+\frac{\partial C^{(1)}}{\partial t_1}+\frac{\partial C^{(2)}}{\partial t_0}+uP_e\frac{\partial C^{(1)}}{\partial x}-\frac{\partial^2 C^{(0)}}{\partial x^2}-\frac{\partial^2 C^{(2)}}{\partial y^2}\right)+O(\epsilon^3)=0,\quad 0<y<1, \tag{19}$$

$$\frac{\partial C^{(0)}}{\partial y}+\epsilon\frac{\partial C^{(1)}}{\partial y}+\epsilon^2\frac{\partial C^{(2)}}{\partial y}+O(\epsilon^3)=0 \text{ at } y=1, \tag{20}$$

and

$$\frac{\partial C^{(0)}}{\partial y}+\epsilon\frac{\partial C^{(1)}}{\partial y}+\epsilon^2\frac{\partial C^{(2)}}{\partial y}+O(\epsilon^3)=\frac{\partial C_s^{(0)}}{\partial t_0}+\epsilon\left(\frac{\partial C_s^{(0)}}{\partial t_1}+\frac{\partial C_s^{(1)}}{\partial t_0}\right)$$
$$+\epsilon^2\left(\frac{\partial C_s^{(0)}}{\partial t_2}+\frac{\partial C_s^{(1)}}{\partial t_1}+\frac{\partial C_s^{(2)}}{\partial t_0}\right)+O(\epsilon^3)=Da\left(\theta C^{(0)}-C_s^{(0)}\right)$$
$$+\epsilon\left(Da\left(\theta C^{(1)}-C_s^{(1)}\right)\right)+\epsilon^2\left(Da\left(\theta C^{(2)}-C_s^{(2)}\right)\right)+O(\epsilon^3) \text{ at } y=0. \tag{21}$$

Using this perturbation analysis, for leading order $(O(1))$ the general solution of $C^{(0)}$ is given by

$$C^{(0)}=C_0^{(0)}(x,t_1,t_2)+\sum_{n=1}^{\infty}\mathrm{Re}\left[C_n^{(0)}(x,t_1,t_2)e^{in\pi y}\right]e^{-n^2\pi^2 t_0}. \tag{22}$$

The solution can be considered

$$C^{(0)}=C_0^{(0)}(x,t_1,t_2) \tag{23}$$

On omitting the reliance of $C^{(0)}$ on y, the stationary condition from (21) (leading order $(O(1))$) gives

$$C_s^{(0)}=\theta C_0^{(0)} \tag{24}$$

Since, the time scale t_0 is larger than the another time scale t_1, the derivative with respect to t_0 is negligible, for first order $(O(\epsilon))$ the perturbation problem becomes

$$\frac{\partial C^{(0)}}{\partial t_1}+uP_e\frac{\partial C^{(0)}}{\partial x}=\frac{\partial^2 C^{(1)}}{\partial y^2},\quad 0<y<1, \tag{25}$$

we define section average function $\langle f\rangle$ of a function f with respect to y as

$$\langle f\rangle=\int_0^1 f dy \tag{26}$$

Now taking this section average of Eqs. (25) subject to the condition from perturbation analysis, we get

$$\frac{\partial C^{(0)}}{\partial t_1} + P_e \frac{\langle u \rangle}{R} \frac{\partial C^{(0)}}{\partial x} = 0, \tag{27}$$

here $R = 1 + \theta$ is the retardation parameter.

Subtracting (27) from (26), which suggest the following substitutions

$$C^{(1)} = P_e A(y) \frac{\partial C_0}{\partial x} \tag{28}$$

$$C_s^{(1)} = P_e A_s \frac{\partial C_0}{\partial x} \tag{29}$$

On comparing terms related with $\frac{\partial C^{(0)}}{\partial x}$, the function $A(y)$ is found to be governed by

$$\frac{d^2 A}{dy^2} = u - \frac{\langle u \rangle}{R}, \quad 0 < y < 1, \tag{30}$$

with the stationary conditions

$$\frac{dA}{dy} = 0, \quad \text{at} \quad y = 1, \tag{31}$$

$$\frac{dA}{dy} = \frac{\theta \langle u \rangle}{R} = Da(\theta A - A_s), \quad \text{at} \quad y = 0, \tag{32}$$

and

$$\langle A \rangle = 0. \tag{33}$$

Solving the above equations, we get

$$A = \frac{y^3}{6} - \frac{y^2}{4R} - \frac{1}{2}\left(1 - \frac{1}{R}\right) y + \frac{5}{24} - \frac{1}{6R} \tag{34}$$

$$A_s = \frac{5R}{24} - \frac{3}{8} - \frac{1}{6R} + \frac{1}{2Da}\left(1 - \frac{1}{R}\right) \tag{35}$$

From second order $(O(\epsilon^2))$ perturbation analysis the effective transport equation is given by

$$\frac{\partial C^{(0)}}{\partial t} + \epsilon P_e \frac{\langle u \rangle}{R} \frac{\partial C^{(0)}}{\partial x} = \epsilon^2 \left(\frac{1}{R} - P_e^2 \frac{\langle uA \rangle}{R} + P_e^2 \frac{\langle u \rangle A_s}{R^2}\right) \frac{\partial^2 C^{(0)}}{\partial x^2}, \tag{36}$$

using (14) and (34) one can easily get

$$\langle uA \rangle = \frac{1}{30} - \frac{1}{16R} - \frac{1}{6}\left(1 - \frac{1}{R}\right) + \frac{1}{2}\left(\frac{5}{24} - \frac{1}{6R}\right). \tag{37}$$

In terms of new variables $\tau = T, \xi = \frac{\bar{x}}{h} - P_e \frac{\langle u \rangle R}{T}$ with the help of initial and boundary conditions, the solution of (36) is given by

$$C^{(0)} = \frac{1}{\sqrt{4\pi D_T T}} \exp\left(-\frac{\xi^2}{4D_T T}\right), \tag{38}$$

where

$$D_T = \frac{1}{R} - P_e^2 \frac{\langle uA \rangle}{R} + P_e^2 \frac{\langle u \rangle A_s}{R^2} \tag{39}$$

is the dispersion coefficient.

The explicit form of the dispersion coefficient is

$$D_T = \frac{1}{R} - \frac{P_e^2}{R}\left(\frac{-7}{240} + \frac{1}{48R}\right) + \frac{P_e^2}{2R^2}\left(\frac{5R}{24} - \frac{3}{8} + \frac{1}{6R}\right) + \frac{1}{2Da}\left(1 - \frac{1}{R}\right) \tag{40}$$

In the similar manner of (28) and (29) one can easily find

$$C^{(2)} = P_e^2 B(y) \frac{\partial^2 C^{(0)}}{\partial x^2}, \tag{41}$$

$$C_s^{(2)} = P_e^2 B_s \frac{\partial^2 C^{(0)}}{\partial x^2}. \tag{42}$$

On comparing the terms related with $\frac{\partial^2 C^{(0)}}{\partial x^2}$, we have

$$\frac{d^2 B}{dy^2} = \left(u - \frac{\langle u \rangle}{R}\right) A - \frac{\langle uA \rangle}{R} + \frac{\langle u \rangle A_s}{R^2}, \quad 0 < y < 1, \tag{43}$$

with the boundary conditions

$$\frac{dB}{dy} = 0, \quad \text{at} \quad y = 1, \tag{44}$$

$$\frac{dB}{dy} = \frac{\theta \langle uA \rangle}{R} - \frac{\langle u \rangle A_s}{R^2} = Da(\theta A - A_s), \quad \text{at} \quad y = 0, \tag{45}$$

and

$$\langle B \rangle = 0. \tag{46}$$

Solving the above equations, we get

$$B = \frac{y^6}{30} - \frac{y^5}{60R} - \frac{1}{24}\left(1 - \frac{1}{R} - \frac{1}{4R^2}\right)y^4 + \frac{1}{24}\left(\frac{5}{6} + \frac{1}{3R} - \frac{1}{R^2}\right)y^3 + \left(\frac{7}{240} - \frac{1}{8R} + \frac{1}{12R^2}\right)y - \frac{691}{20160} - \frac{177}{1440R} \tag{47}$$

$$B_s = -\frac{691R}{20160} + \frac{1769}{20160} - \frac{25}{288R} + \frac{1}{30R^2} - \frac{1}{Da}\left(\frac{7}{240} - \frac{1}{8R} - \frac{1}{12R^2}\right) \tag{48}$$

when $P_e > 100$ the longitudinal diffusion is usually disregarded. Therefore the the dispersion coefficient becomes

$$D_T \approx D_{Ta} \tag{49}$$

where

$$D_{Ta} = \frac{\langle uA\rangle}{R} + \frac{\langle u\rangle A_s}{R^2} \tag{50}$$

is called the apparent dispersion coefficient, which is depending on θ and Da. Also the real concentration distribution is obtained in new P_e independent system $\{\eta/P_e, CP_e\}$.

For validation the result, the limiting case ($\theta = 0$ or $R = 1$) of apparent dispersion coefficient is compared with the previous result of Ng & Bai and Bandyopadhyay & Mazumder.

In the case of $\theta = 0$ or $R = 1$ the dispersion coefficient becomes

$$D_{Ta} = \frac{1}{120}, \tag{51}$$

which is Taylor dispersion coefficient for simple Couette flow. It is exactly same result as of limiting case of dispersion coefficient as Bandyopadhyay and Mazumder [5] and Ng and Bai [7].

4 Results and Discussion

How the dispersion coefficient changes with the sorptive partition coefficient (θ) and kinetics (depending on Da) are sketched in Fig. 2. It is observed that lower kinetics of sorptive exchange i.e. when Da is larger, dispersion coefficient is smaller. It is remarkable to note that if the phase exchange kinetics is mild i.e. $Da \ll 1$, the dispersion coefficients are large in comparison to that of the phase exchange kinetics is fast i.e. $Da > 1$. The reason behind this as the phase exchange kinetics is slow that is rate of reversible reaction is less so the solute restrained with the flow as a result dispersion coefficient is large, on the other way when the phase exchange kinetics is large, the rate of reversible reaction is high which absorbs the solute more so the dispersion coefficient is small. It is also clearly seen that with

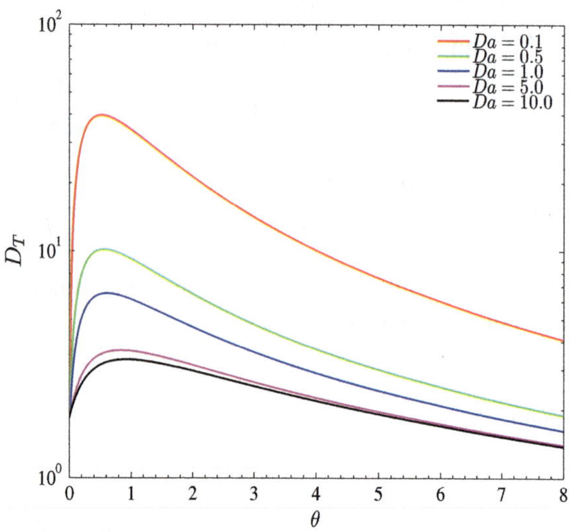

Fig. 2 Evolution of effective dispersivity with retention factor θ for different values of Damkohler number Da

the increases of retardation factor θ the dispersion coefficient initially increases and after a certain value of θ it decreases. The dispersion coefficient attains their highest values when the values of θ in between 0.2–0.4. After a critical value of θ, with the enhancement of θ transport coefficient decreases due to the larger retardation parameter, which reduces the tracer spreading with the flow. In the Fig. 3a longitudinal real concentration distribution is sketched for different transversal position with $\theta = 1$, $Da = 1$, $T = 1$. It is observed that as transversal width increases from the lower plate the concentration distribution move away through the longitudinal direction. The cause behind this with the increases of transversal width from the lower plate the velocity is increases as a result convection also increases. Effect of several dispersion time on real concentration distribution are depicted in Fig. 3b with $\theta = 1$, $Da = 1$, $y = 0.5$. It is clearly obtained that as dispersion time progresses the peak of the real concentration distribution decreases and becomes more flatter. This is because as time proceeds the dispersion in the longitudinal direction increases prominently due to the combined effect of convection, diffusion of the solute. Variations of real concentration distribution for different Damkohlar number with $\theta = 1$, $T = 1$, $y = 0.5$ are shown in Fig. 3c. With the enhancement of Da the peak of the real concentration distribution increases. It occurs when the diffusion rate is much slower than the reversible reaction rate i.e. $Da \gg 1$. As a result from the immobile phase (i.e. the lower plate) the solute moves quickly to the flow and the tracer concentration enhances in the mobile phase. The opposite phenomena happens when molecular diffusion rules the rate of reversible reaction i.e. $Da \ll 1$ and it makes the real concentration distribution more flatter also blunt. Longitudinal real concentration distribution for several values of θ are observed in Fig. 3d with $Da = 1$, $T = 1$, $y = 0.5$. It is interesting to note that the peak of the real concentration distribution of the solute decreases with the increase of θ and after a certain value it also increases.

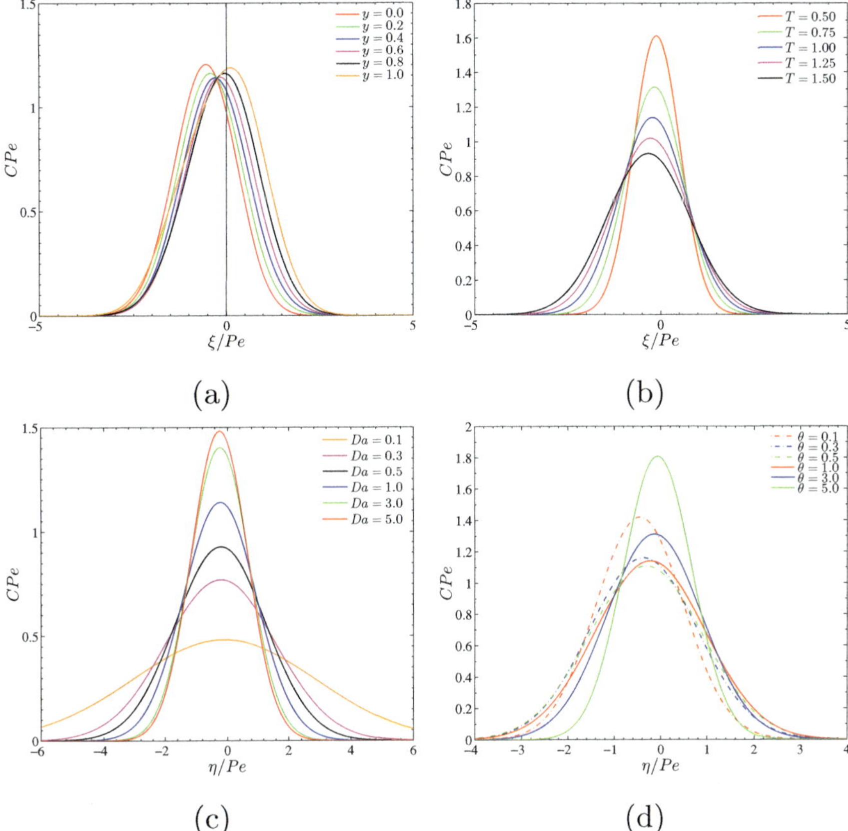

Fig. 3 Longitudinal real concentration distribution for different **a** transversal position y, **b** dispersion time T, **c** different Damkohlar number Da, **d** different retardation factor θ

Transverse concentration distribution for various values of retardation factor is plotted in Fig. 4a. It is seen that initially with the enhancement of retardation factor the transverse concentration decreases and after a certain value as of real concentration distribution it also enhances. Figure 4b shows the transverse concentration of the solute at different downstream locations. It is seen that the concentration of the solute decreases significantly as it is moving towards the downstream directions. For all the downstream stations, the concentration of the tracer is maximum at $y = 1$ and it becomes minimum at the lower plate. The reason behind it, the flow velocity is maximum at $y = 1$ and minimum at $y = 0$.

Effect of reversible reaction on solute transport in Couette flow is sketched in Fig. 5. It is clearly observed that in mobile phase as the Damkohlar number Da increases the concentration of the solute agglomerated near the source. The reason may be the faster rate of exchange between the lower plate of the channel and fluid phase enhances the contaminant concentration in the phase of fluid. Also, the

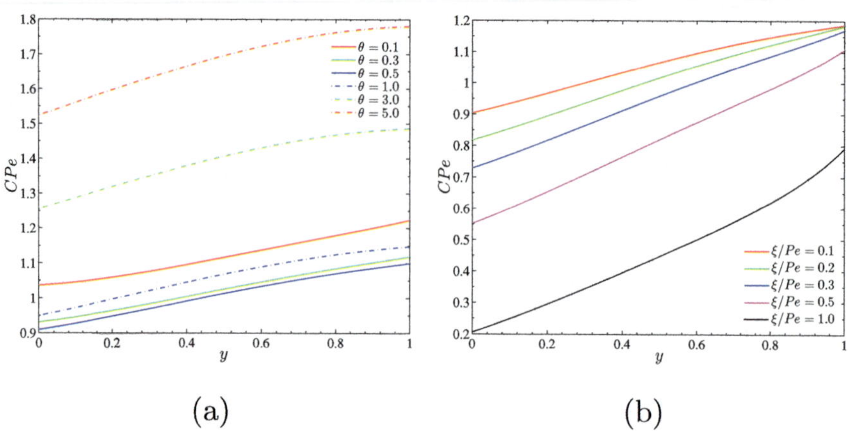

Fig. 4 Transverse variation of concentration profile for different values of **a** retardation factor, **b** downstream stations

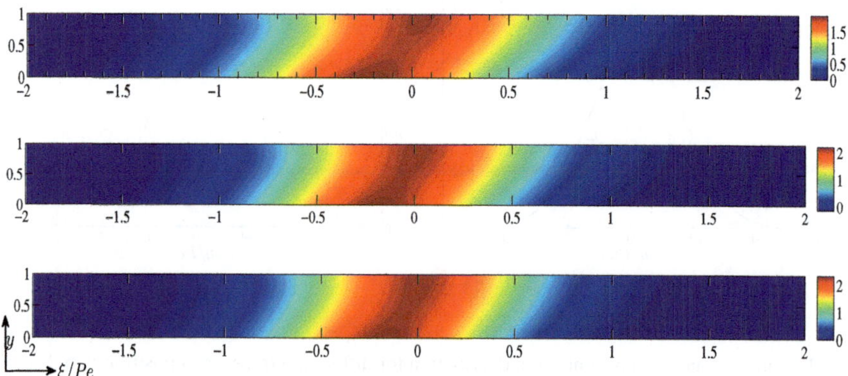

Fig. 5 Mobile phase concentration contours for different $Da = 0.5, 2, 10$

Damkohlar number disproportionate with the diffusivity i.e. with the increment of reversible phase exchange kinetics (Damkohlar number), diffusion decreases, as a result concentration of the solute conglomerated near the source.

5 Conclusion

The solute transport phenomena in simple Couette flow with reversible reaction which relate the phase exchange between the mobile phase and immobile phase is investigated analytically by multi-scale homogenization method. The limiting case of the result is compared with the established analytical result and achieved an excellent agreement. The results shows the effect of Damkohlar number (phase exchange

kinetics), retention factor, dispersion time on dispersion coefficient, transverse and longitudinal real concentration distributions. Some major outcomes are as follows:
(I) with the enhancement of Da the dispersion coefficient decreases.
(II) With the increment of θ the dispersion coefficient increases and after a certain value it is decreases.
(III) When $Da \gg 1$ i.e. the diffusion rate is much slower than the reversible reaction rate, the real concentration distribution of the solute increases. Opposite phenomena occurs for $Da \ll 1$, also peak of the real concentration become blunt.
(IV) As dispersion time progresses the peak of the concentration decreases and become flatter.

References

1. Taylor, G.I.: Dispersion of soluble matter in solvent flowing slowly through a tube. Proc. R. Soc. Lond. A **219**(1137), 186–203 (1953). https://doi.org/10.1098/rspa.1953.0139
2. Aris, R.: On the dispersion of a solute in a fluid flowing through a tube. Proc. Royal Soc. Lond. A. **235**(1200), 67–77 (1956). https://doi.org/10.1098/rspa.1956.0065
3. Chatwin, P.C.: The approach to normality of the concentration distribution of a solute in a solvent flowing along a straight pipe. J. Fluid Mech. **43**(02), 321 (1970). https://doi.org/10.1017/s0022112070002409
4. Mei, C.C., Auriault, J.L., Ng, C.O.: Some applications of the homogenization theory. Adv. Appl. Mech. **32**, 277–348 (1996). https://doi.org/10.1016/s0065-2156(08)70078-4
5. Bandyopadhyay, S., Mazumder, B.S.: On contaminant dispersion in unsteady generalised Couette flow. Int. J. Eng. Sci. **37**(11), 1407–1423 (1999). https://doi.org/10.1016/s0020-7225(98)00132-3
6. Ng, C.O., Yip, T.L.: Effects of kinetic sorptive exchange on solute transport in open-channel flow. J. Fluid Mech. **446**, 321–345 (2001). https://doi.org/10.1017/S0022112001005791
7. Ng, C.O., Bai, Y.C.: Dispersion in oscillatory Couette flow with sorptive boundaries. Acta Mech. **178**(1–2), 65–84 (2005). https://doi.org/10.1007/s00707-005-0230-6
8. Wu, Z., Chen, G.Q.: Approach to transverse uniformity of concentration distribution of a solute in a solvent flowing along a straight pipe. J. Fluid Mech. **740**, 196–213 (2014). https://doi.org/10.1017/jfm.2013.648
9. Barik, S., Dalal, D.C.: Multi-scale analysis for concentration distribution in an oscillatory Couette flow. Proc. R. Soc. Lond. A **475**(2221), 20180483 (2018). https://doi.org/10.1098/rspa.2018.0483
10. Poddar, N., Dhar, S., Mazumder, B.S., Mondal, K.K.: An exact analysis of scalar transport in hydromagnetic flow between two parallel plates: a multi-scale approach. Proc. R. Soc. Lond. A **477**(2248), 20200830 (2021). https://doi.org/10.1098/rspa.2020.0830
11. Dhar, S., Poddar, N., Mondal, K.K., Mazumder, B.S.: On dispersion of solute in a hydromagnetic flow between two parallel plates with boundary absorption. Phys. Fluids **33**, 083609 (2021). https://doi.org/10.1063/5.0060404
12. Das, D., Poddar, N.,Dhar S., Kairi, R.R., Mondal, K. K.: Multi-scale approach to analyze the dispersion of solute under the influence of homogeneous and inhomogeneous reactions through a channel. Int. Commun. Heat Mass Transf. **129**, 105709 (2021). https://doi.org/10.1016/j.icheatmasstransfer.2021.105709
13. Mazumder, B.S., Paul, S.: Dispersion of reactive species with reversible and irreversible wall reactions. Heat Mass Transf. **48**(6), 933–944 (2011). https://doi.org/10.1007/s00231-011-0920-7

14. Barik, S., Dalal, D. C.: Analytical solution for concentration distribution in an open channel flow with phase exchange kinetics. Acta Mech. Sin. **1** (2021). http://ams.cstam.org.cn/EN/abstract/abstract157644.shtml

Mathematical Analysis of Hybrid Nanoparticles ($Au - Al_2O_3$) on MHD Blood Flow Through a Curved Artery with Stenosis and Aneurysm Using Hematocrit-Dependent Viscosity

Poonam and Bhupendra K. Sharma

Abstract The current study deals with hybrid nanoparticles ($Au - Al_2O_3/blood$) to explore the impact of hemodynamic parameters (such as wall shear stress and resistive impedance) on unsteady MHD blood flow via a curved artery in the presence of stenosis and aneurysm. The governing momentum equation is solved using the Crank-Nicolson method. Velocity contours for numerous parameters have been provided to study the overall behavior of flow patterns. Comprehensive solutions for gold and gold-aluminum oxide hybrid blood flow are presented using medically relevant hemodynamic data. The investigation shows that hybrid nanoparticles ($Au - Al_2O_3$) have lower hemodynamic characteristics such as WSS (wall shear stress) and resistive impedance. The findings could aid in identifying and treating cancer, plaque rupture, the clearance of blood clots, infections, and brain aneurysms.

Keywords Curved artery · Hematocrit-dependent viscosity · Hybrid nanoparticles · Aneurysm · Stenosis

1 Introduction

The addition of nanoparticles to blood flow in a curved stenotic artery significantly impacts hemodynamical variables (such as WSS and impedance). Because of the low cytotoxicity of gold nanoparticles and their inert nature, localized SPR, and distinctive optical properties, these nanoparticles are frequently utilized as precise cancer killers. Hybrid nanoparticles are gaining popularity due to innovations in the treatment of numerous cardiovascular problems. Hybrid nanoparticles possess a range of uses in biosciences, including angioplasty, cancer therapy, angiography, and bio-nanopolymer coatings of surgical equipment. Zaman et al. [1] investigated the effect of silver-alumina hybrid nanoparticles on blood flow through an artery with stenosis and an aneurysm. Das et al. [2] looked into the hemodynamic and rhe-

Poonam (✉) · B. K. Sharma
Department of Mathematics, BITS Pilani, Pilani Campus, Pilani 333031, Rajasthan, India
e-mail: aggarwalpurnima123@gmail.com

S. Banerjee and A. Saha (eds.), *Nonlinear Dynamics and Applications*,
Springer Proceedings in Complexity,
https://doi.org/10.1007/978-3-030-99792-2_34

ological changes generated by blood-mediated hybrid nanoparticles $Cu - Al_2O_3$. Jayanti Tripathi et al. performed simulation studies with hybrid nano blood (Au-Ag/blood) via irregular stenosis [3]. Cardiovascular disorders are the leading cause of death worldwide, accounting for approximately 30% of all deaths. According to the literature, the effects of stenosis on the hemodynamics of blood going beyond and through the tapered arterial section have been studied in many theoretical and experimental studies. These researches can contribute to the detection and treatment of a wide range of cardiovascular disorders. The post-stenotic blood flow is induced as the stenosis advances due to the increased wall shear stress caused by the stenosis. This post-stenotic flow contributes to arterial wall weakening and post-stenotic dilation (aneurysm). When compared to a single segment (with stenosis only), the combined impact of various anomalies necessarily raises the chances of rupturing these stenotic segments. Arterial curvature also affects the growth of the abnormal segment of the atherosclerotic artery. As a result, it is a crucial geometrical parameter to research. With this objective in mind, Zaman et al. [4, 5] developed several mathematical models to study the unsteady flow of blood via a curved artery with associated stenosis and aneurysm. The majority of prior studies assumed constant viscosity; however, viscosity is influenced by several factors, including hematocrit, temperature, and shear rate. The hematocrit has a considerable impact on whole blood viscosity. Several mathematical models have been used to investigate the effects of altering viscosity on blood flow hemodynamics. The study mentioned above did not include hematocrit-dependent viscosity and curved arterial flow.

Nomenclature			
x	Axial direction	B_1	Pressure gradient parameter
r	Radial direction	Re	Reynold's Number
t	Time	B	Uniform Magnetic Field
R^*	Radius of the curved channel	R_c	dimensionless radius of curvature of the artery
u	Radial Velocity component	v	Axial velocity component
M^2	Magnetic Number	Q	volumetric flow Rate
u_0	Reference velocity	P	Pressure
e	Systolic to diastolic pressure ratio	R_0	Radius of normal artery
A_1	Amplitude of pulsatile component	***Abbreviation***	
A_0	Amplitude of pressure gradient	WSS	Wall Shear Stress
Greek Letters			
τ_w	Shear stress at the wall	δ	Stenosis depth
σ	Electrical conductivity	ρ_{hnf}	Density of hybrid nano-fluid
μ_0	Reference viscosity	ϕ_1, ϕ_2	Nanoparticles Concentration
μ_f	Blood's viscosity	λ	Impedance, ω_p Circular frequency

Based on our literature study, no attempt has been made to investigate the effects of magnetic field and body acceleration on the flow of a hybrid nanofluid $(Au - Al_2O_3/blood)$ via a curved channel with stenosis and aneurysm using hematocrit-dependent viscosity. We developed a mathematical model to examine curvature effects, $Au - Al_2O_3$ hybrid nanoparticles, body acceleration, hematocrit-dependent viscosity, external magnetic field on pulsatile blood flow through a curved artery with stenosis and aneurysm. The objective of this research is to learn more about how to cure atherosclerosis without surgery to save money on health expenses.

2 Model Formulation

2.1 Geometry of the Model

Consider blood as an unsteady, Newtonian, laminar, viscous, incompressible fluid moving through a stenosed curved artery with aneurysm.The uniform magnetic field B is applied perpendicular to the axial direction. The diseased segment's geometry (Fig. 1) can be mathematically characterized as follows [5]:

$$R(x) = \begin{cases} (\alpha_i^* x + R_0)\left(1 - \left(\frac{\delta_i^*}{2R_0}\left(1 + \cos\frac{2\pi}{\lambda_i}\left(x - \sigma_i^* - \frac{\lambda_i}{2}\right)\right)\right)\right), \\ \qquad\qquad \sigma_i^* \le x \le \sigma_i^* + \lambda_i, i = 1, 2 \\ (\alpha_i^* x + R_0) \qquad \text{Otherwise}, \end{cases} \tag{1}$$

$$-R(x) = \begin{cases} (\alpha_i^* x - R_0)\left(1 - \left(\frac{\delta_i^*}{2R_0}\left(1 + \cos\frac{2\pi}{\lambda_i}\left(x - \sigma_i^* - \frac{\lambda_i}{2}\right)\right)\right)\right), \\ \qquad\qquad \sigma_i^* \le x \le \sigma_i^* + \lambda_i, i = 1, 2 \\ (\alpha_i^* x - R_0) \qquad \text{Otherwise}, \end{cases} \tag{2}$$

where $R(x)$ denotes the upper wall geometry of a curved artery and $-R(x)$ denotes the lower wall geometry. L is the length of the arterial channel, $\alpha = \tan\psi$ represents the constriction of the diseased artery, ψ represents the tapering angle, λ_i denotes the length of the diseased segment, and σ_i^* is the length of the ith abnormal section from the origin and δ_i^* stands for the critical height of the ith diseased section occurring at two explicit locations given by:
$x = \sigma_1^* + \lambda_1/2$, and $x = \sigma_2^* + \lambda_2/2$, where the value of δ_i^* takes a positive value for stenosis and a negative value for an aneurysm.

2.2 Governing Equations

The current flow model is illustrated by two-dimensional orthogonal curvilinear coordinates (r, x). The flow velocity vector is defined as $V = (u(r, x, t), v(r, x, t))$, with u and v representing the radial and axial velocity components, respectively. Figure 1 shows the geometry of a curved artery with stenosis and aneurysm.

The equations for continuity, momentum, and energy are written as:

Continuity equation

$$\frac{\partial u}{\partial r} + \frac{u}{R^* + r} + \frac{R^*}{R^* + r}\frac{\partial v}{\partial x} = 0. \tag{3}$$

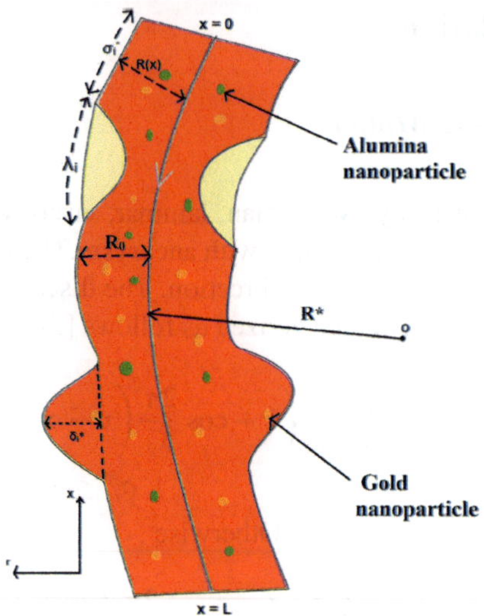

Fig. 1 Geometrical representation of diseased artery

Momentum equations
r- direction

$$\rho_{hnf}\left[\frac{\partial u}{\partial t}+u\frac{\partial u}{\partial r}+\frac{R^{*}v}{R^{*}+r}\frac{\partial v}{\partial x}-\frac{v^{2}}{R^{*}+r}\right]=-\frac{\partial P}{\partial r}+\mu_{hnf}\left[\frac{\partial^{2}u}{\partial r^{2}}+\frac{1}{R^{*}+r}\frac{\partial u}{\partial r}+\left(\frac{R^{*}}{R^{*}+r}\right)^{2}\frac{\partial^{2}u}{\partial x^{2}}-\frac{u}{(R^{*}+r)^{2}}-\frac{2R^{*}}{(R^{*}+r)^{2}}\frac{\partial v}{\partial x}\right]+\left(\frac{4}{3}\frac{\partial u}{\partial r}-\frac{2}{3}\left(\frac{R^{*}}{R^{*}+r}\frac{\partial v}{\partial x}+\frac{u}{R^{*}+r}\right)\right)\frac{\partial \mu_{hnf}}{\partial r},$$

x-direction

$$\rho_{hnf}\left[\frac{\partial v}{\partial t}+u\frac{\partial v}{\partial r}+\frac{R^{*}v}{R^{*}+r}\frac{\partial v}{\partial x}+\frac{uv}{R^{*}+r}\right]=-\frac{R^{*}}{R^{*}+r}\frac{\partial P}{\partial x}+G(t)+\mu_{hnf}\left[\frac{\partial^{2}v}{\partial r^{2}}+\frac{1}{R^{*}+r}\frac{\partial v}{\partial r}+\left(\frac{R^{*}}{R^{*}+r}\right)^{2}\frac{\partial^{2}v}{\partial x^{2}}-\frac{v}{(R^{*}+r)^{2}}+\frac{2R^{*}}{(R^{*}+r)^{2}}\frac{\partial u}{\partial x}\right]+\left(\frac{R^{*}}{R^{*}+r}\frac{\partial u}{\partial x}+\frac{\partial v}{\partial r}-\frac{v}{R^{*}+r}\right)\frac{\partial \mu_{hnf}}{\partial r}-\sigma_{hnf}B^{2}v. \tag{4}$$

The following is the equation for the axial pressure gradient:

$$-\frac{\partial \bar{P}}{\partial \bar{x}}=A_0+A_1\cos(2\pi\omega_p t),\quad t>0,$$

where, A_1 = Amplitude of the pulsatile component, A_0 = Mean pressure gradient, $\omega_p = 2\pi f_p$.
The initial and boundary conditions that the flow is subjected to are as follows:

$$\begin{cases} v = 0 & \text{at} \quad t = 0, \\ v = 0, & \text{at} \quad r = R \quad and \quad r = -R. \end{cases} \tag{5}$$

Here, hematocrit-dependent viscosity is considered which is illustrated as:

$$\mu_f = \mu_0[1 + \beta_1 h(r)], \tag{6}$$

where, $h(r) = h_m\left[1 - \left(\frac{r}{R_0}\right)^m\right]$, h_m = maximum hematocrit at the center of the artery.

$\beta_1 = 2.5$, m = exact shape of velocity profile, $m \geq 2$.

2.3 Non-dimensionalization of Governing Equations

The previous flow Eqs. (3)–(4) is non-dimensionalized using the non-dimensional parameters listed below to evaluate numerical solutions:

$$\bar{r} = \frac{r}{R_0}, \quad \bar{x} = \frac{x}{\lambda_i}, \quad \bar{u} = \frac{\lambda_i u}{\delta^* u_0}, \quad \bar{v} = \frac{v}{u_0}, \quad \bar{t} = \frac{u_0 t}{R_0},$$

$$R_c = \frac{R^*}{R_0}, \epsilon = \frac{R_0}{\lambda_i}, \quad Re = \frac{\rho_f u_0 R_0}{\mu_0}, \quad \delta = \frac{\delta^*}{R_0},$$

$$\bar{P} = \frac{R_0^2 P}{\mu_0 u_0 \lambda_i}, \quad M^2 = \frac{\sigma_f B^2 R_0^2}{\mu_0}.$$

After using non-dimensional parameter and mild stenotic condition, i.e., $\delta(= \frac{\delta^*}{R_0}) << 1, \epsilon(= \frac{R_0}{\lambda_i}) = O(1)$, Eqs. (3)–(4) will be reduced as:

Non–dimensional equations

$$\frac{\partial \bar{P}}{\partial \bar{r}} = 0, \tag{7}$$

$$\begin{aligned} \frac{\rho_{hnf}}{\rho_f} Re \frac{\partial \bar{v}}{\partial \bar{t}} = &-\frac{R_c}{R_c + \bar{r}} \frac{\partial \bar{P}}{\partial \bar{x}} + \frac{\mu_{hnf}}{\mu_0}\left[\frac{\partial^2 \bar{v}}{\partial \bar{r}^2} + \frac{1}{R_c + \bar{r}} \frac{\partial \bar{v}}{\partial \bar{r}} - \frac{\bar{v}}{(R_c + \bar{r})^2}\right] \\ &- \left(\frac{\partial \bar{v}}{\partial \bar{r}} + \frac{\bar{v}}{R_c + \bar{r}}\right) \frac{m\beta_1 h_m \bar{r}^{m-1}}{(1 - \phi_1)^{2.5}(1 - \phi_2)^{2.5}} - \frac{\sigma_{hnf}}{\sigma_f} M^2 \bar{v}, \end{aligned} \tag{8}$$

Pressure gradient and body acceleration are obtained as follows after non-dimensionalization:

$\frac{\partial \bar{P}}{\partial \bar{x}} = B_1(1 + e\cos(c_1 \bar{t})$, where $B_1 = \frac{A_0 R_0^2}{\mu_0 u_0}$, $e = \frac{A_1}{A_0}$, $c_1 = \frac{2\pi R_0 w_p}{u_0}$.

Non-dimensionalized form of associated initial and boundary conditions subjected to the flow are rewritten as:

$$\begin{cases} \bar{v} = 0 & \text{at } t = 0, \\ \bar{v} = 0 & \text{at } r = R \; and \; r = -R. \end{cases} \tag{9}$$

Now, the non-dimensionalized form of geometry of curved arterial channel with stenosis and aneurysm is obtained as:

$$R(x) = \begin{cases} (1+\alpha x)\left(1 - \left(\frac{\delta}{2}\left(1 + \cos 2\pi\left(x - \sigma_i - \frac{1}{2}\right)\right)\right)\right), \\ \qquad\qquad \sigma_i \le x \le \sigma_i + 1, i = 1, 2 \\ (1+\alpha x) \qquad \text{Otherwise}, \end{cases} \tag{10}$$

with $\sigma_i = \frac{\sigma_i^*}{\lambda_i}$; $\alpha = \frac{\alpha^* \lambda_i}{R_0}$.

In Eq. (10), the tapering parameter is $\alpha = \tan(\psi)$, and the associated taper angle is ψ. The following are the mathematical formulae for hemodynamical factors such as wall shear stress, volumetric flow rate, and resistive impedance:

$$\tau_w = \left(\frac{\partial \bar{v}}{\partial \bar{r}}\right)_{\bar{r}=R}, \quad Q = \int_{-R}^{R} \bar{v}\bar{r}d\bar{r}, \quad \lambda = \frac{L\left(\frac{\partial \bar{P}}{\partial \bar{x}}\right)}{Q}. \tag{11}$$

Hybrid nano fluid equation

$$\mu_{hnf} = \frac{\mu_f}{(1-\phi_1)^{2.5}(1-\phi_2)^{2.5}},$$

$$\rho_{hnf} = (1-\phi_2)[(1-\phi_1)\rho_f + \phi_1\rho_{s1}] + \phi_2\rho_{s2},$$

$$\sigma_{hnf} = \sigma_{bf}\left[\frac{\sigma_{s2}(1+2\phi_2) + 2\sigma_f(1-\phi_2)}{\sigma_{s2}(1-\phi_2) + \sigma_f(2+\phi_2)}\right],$$

where

$$\sigma_{bf} = \sigma_f\left[\frac{\sigma_{s1}(1+2\phi_1) + 2\sigma_f(1-\phi_1)}{\sigma_{s1}(1-\phi_1) + \sigma_f(2+\phi_1)}\right].$$

3 Solution Process

Since the governing equations (8) is a non-linear partial difference equation, so a robust numerical method is used to solve the resulting dimensionless boundary value problem. An implicit Crank-Nicolson (an unconditionally stable) scheme based on a finite difference approach is used for the current blood flow problem. Furthermore, in both space and time, this technique is second-order convergent. The partial spatial and temporal derivatives used in this method are stated as:

$$\frac{\partial \bar{v}}{\partial \bar{r}} = \frac{\bar{v}_{i+1}^{k} - \bar{v}_{i-1}^{k}}{2\Delta \bar{r}}, \quad \frac{\partial^2 \bar{v}}{\partial \bar{x}^2} = \frac{\bar{v}_{i+1}^{k} - 2\bar{v}_{i}^{k} + \bar{v}_{i-1}^{k}}{(\Delta \bar{x})^2}, \quad \frac{\partial \bar{v}}{\partial \bar{t}} = \frac{\bar{v}_{i}^{k+1} - \bar{v}_{i}^{k}}{\Delta \bar{t}}. \tag{12}$$

3.1 Discretization of Governing Equations

By using the partial derivatives as given in Eq. (12), we discretize the governing equations (8).

The following are the discretized boundary and initial conditions related with the governing equations:

$$\bar{v}_{1}^{k+1} = 0, \quad \bar{v}_{N+1}^{k+1} = 0, \quad \bar{v}_{i}^{1} = 0. \tag{13}$$

The spatial variable is now uniformly discretized into $N + 1$ discrete grid points x_i, $(i = 1, 2, ..., N + 1)$, with $\Delta x = 1/(N + 1)$ as the step size. $t^k = (k - 1)\Delta t$ indicates the time levels, with Δt representing a small increment in time. Despite the fact that this approach is unconditionally stable for all values of Δt and Δx, we have chosen $\Delta t = 10^{-4}$ and $\Delta x = 10^{-4}$ as the choices for step sizes. As previously stated, the Crank-Nicolson approach is an implicit one, hence the governing equation (8) is reduced to Eq. (14). It form a tri-diagonal system of equations that can be simplified using the Tri-diagonal Matrix Algorithm.

Equation (8) corresponds to a tri-diagonal system, which can be calculated as follows:

$$S_i^k \bar{v}_{i-1}^{k+1} + T_i^k \bar{v}_i^{k+1} + U_i^k \bar{v}_{i+1}^{k+1} = S_i^{'k} \bar{v}_{i-1}^{k} + T_i^{'k} \bar{v}_i^{k} + U_i^{'k} \bar{v}_{i+1}^{k} + F_i^k, \tag{14}$$

where S_i^k, T_i^k, U_i^k, $S_i^{'k}$, $T_i^{'k}$, $U_i^{'k}$, F_i^k are the corresponding coefficient matrices.

4 Results and Graphical Analysis

This mathematical investigation aims to see how hematocrit-dependent viscosity and nanofluid hemodynamics affect blood flow via a curved artery with two aberrant segments (stenosis and aneurysm). As hybrid nanoparticles, Au and Al_2O_3 nanoparticles are combined. The effect of emergent characteristics such as hematocrit-dependent viscosity (h_m), magnetic field (M^2), on WSS and impedance profiles of hybrid blood ($\phi_1 = 0.01, \phi_2 = 0.01$) is investigated. The Crank-Nicolson method has been used to simplify dimensionless governing equations. The wall shear stress and impedance profiles are computed using MATLAB algorithms for the C-N approach. The flow patterns for different values of R_c and volume fraction of nanoparticles are studied using velocity contours.

The computational work has been done by utilizing the default values of parameters as illustrated in (Table 1b).

Table 1 Thermophysical properties and Physical parameters's values table

(a) Thermophysical properties of blood, Au and Al_2O_3 nanoparticles

Thermophysical properties	Gold	Alumina	Blood
Thermal Conductivity $[K(W/mK)]$	314	40	0.492
Electrical Conductivity $[\sigma(S/m)]$	4.10×10^7	3.5×10^7	0.667
Density $[\rho(kg/m^3)]$	19320	3970	1063
Thermal Expansion Coefficient $[\beta \times 10^{-5}(K^{-1})]$	1.4	0.85	0.18
Heat Capacitance $[C_p(J/kgK)]$	129	765	3594

(b) Physical parameters's values with their sources

Parameters	Ranges	Sources
Magnetic Number	0–4	[6]
Maximum Hematocrit	0–2	[7]
Thermal Grashof number	0–6	[8]

4.1 *Validation of the Numerical Results*

The validation of our study is consummated with published work Zaman [5] for the curved artery with stenosis and aneurysm, which is common in both the study.The graphs (Fig. 2a, b) for dimensionless velocity profiles have been plotted for authentication using the following set of emergent parameters: $Pr = 14$, $Gr = 0.8$, $Re = 0.5, Nr = 0, h_m = 0, M^2 = 0, s = 0, R_c = 3$, slip parameter ($\alpha = 0$), thermal slip parameter ($\gamma = 0$), heat source or sink parameter ($\beta = 0$), $B_2 = 0$, $B_1 = 1.41$,

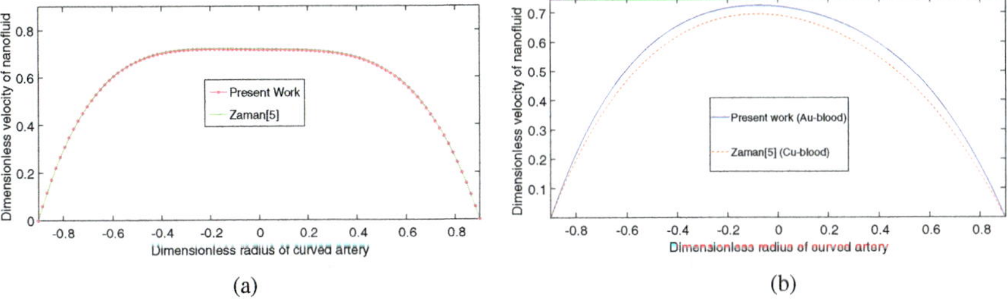

Fig. 2 Variation of velocity profile patterns for **a** Cu-blood, **b** Cu and Au-blood at $t = 0.86$ and $x = 0.7$

$\psi = 0, \phi_1 = 0.05, \phi_2 = 0$. The published work Zaman [5] for the curved artery with stenosis and aneurysm (which is prevalent in both studies), completes the confirmation of our research. The Cu-blood velocity profiles in both studies are shown in Fig. 2a. The current paper's velocity profile matches the published paper [5] quite well.

The velocity profiles of Au-blood (present work) and Cu-blood [5] are compared in Fig. 2b. There is a great match between the velocity trend in our study and the velocity profile trend in Ref. [5]. There is a little difference in velocity profiles due to the use of gold nanoparticles instead of copper nanoparticles since Au-blood velocity is greater than Cu-blood velocity [9].

4.2 Velocity Contours

The varied blood flow patterns in terms of velocity contours are incorporated the research efforts more correctly. The influence of the curved artery's non - dimensional radius (R_c) on the flow patterns are shown in Fig. 3a–c, which shows that the pattern of streamlines and circulating bolus shrinks to symmetric form as the value of R_c increases.

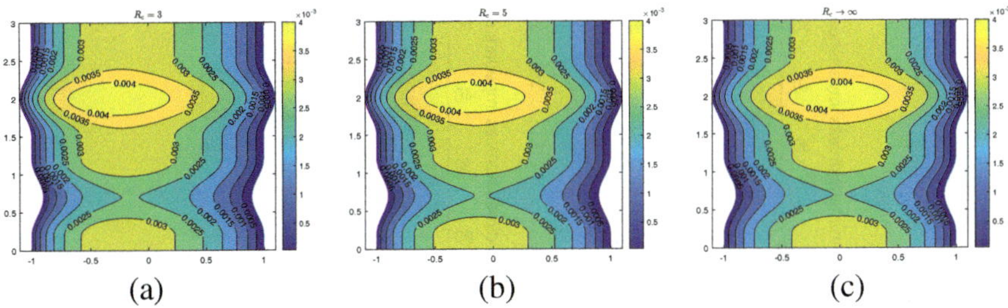

Fig. 3 Variation in blood flow patterns for different values of dimensionless radius of curved channel, **a** $R_c = 3$, **b** $R_c = 5$, **c** $R_c \to \infty$

4.3 Resistance Impedance

The ratio of pressure drop to the flow rate is known as resistive impedance. The study of variation in resistive impedance is quite helpful in controlling blood flow during surgery. Figure 4a shows the variation of resistive impedance in the axial direction for different values of the hematocrit parameter (h_m). It can be observed from the figure that flow impedance increases as the hematocrit parameter increases. Blood viscosity increases as h_m increases, blocking blood flow. The impedance variation at the stenotic neck is greater than at the aneurysm neck for varied h_m values. Figure 4b shows the temporal variation of resistive impedance for different values of dimensionless radius of curvature of the artery. It can be noted from the figure that impedance varies periodically as time passes. The resistive impedance decreases as R_c grows, illustrating that the resistance to the flow in a curved artery is greater than in a straight arterial channel. In addition, when R_c increases, the impedance profiles become close.

The relationship between frictional resistance and maximal stenosis height for varying lengths of the curved artery is shown in Fig. 4c. As the artery lengthens, the resistive impedance rises. This indicates that resistive impedance in the large arteries is higher than in the smaller arteries figure also depicts that the frictional resistance increases (almost linearly) as the stenotic depth grows (owing to increased flow obstruction, which raises flow resistance). It should also be noted that frictional resistance is insignificant in the absence of stenosis, i.e., for ($\delta = 0$). The impedance variation (λ) with axial direction for different values of magnetic number is shown

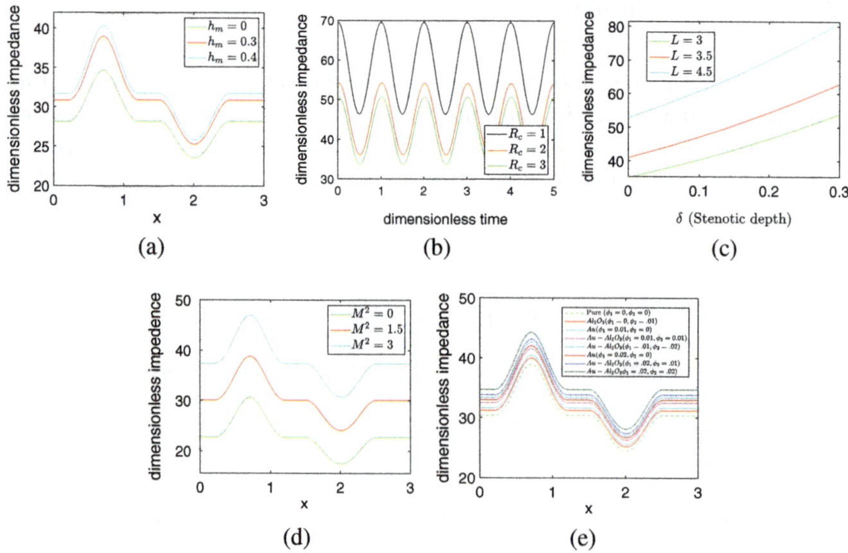

Fig. 4 Variation in impedance profile for, **a** h_m, **b** time and R_c, **c** L with δ, **d** M^2, **e** ϕ_1, ϕ_2

in Fig. 4d. The figures show that when the magnetic number increases, flow resistance increases. When the magnetic field intensity increases, the Lorentz force is generated, slowing the flow and increasing the flow impedance. The influence of the nanoparticle's volume fraction on the flow resistance for $Au - Al_2O_3/blood$ in the axial direction is depicted in Fig. 4e. The resistive impedance increases as the volume fraction of gold and alumina nanoparticles increases. This is very helpful the controlling the flow rate during surgical processes.

4.4 Wall Shear Stress

Wall shear stress in arterial blood flow is described as the force per unit area applied to the blood by the artery's walls parallel to the local tangent plane. As we know, arterial sections with low wall shear stress or significantly oscillating WSS are the most susceptible to atherogenesis disorders. The atherogenic process is influenced by pulsatile blood flow in the arterial system. WSS values can be determined using velocity patterns along the artery's walls. In Fig. 5a, the variation of WSS in the axial direction for different values of magnetic number is illustrated. In the absence of an applied magnetic field, i.e., $M^2 = 0$, the wall shear stress is greatest, and when M^2 increases,a fall in wall shear stress can be observed. For a given value of M^2, WSS in the aneurysm region is higher than WSS in the stenotic region because WSS changes with stenotic depth. The variation in wall shear stress with the depth of the stenosis for various values of magnetic number is depicted in Fig. 5b. The flow is subjected to an opposing Lorentz force as M^2 increases, reducing velocity and wall shear stress. It is also noted that when maximum stenotic depth rises, WSS also reduces. WSS decreases when blood lipids increase, which is consistent with Zhang's experimental findings [10]. Figure 5c shows WSS time series plots in stenotic and aneurysm sites for various values of R_c. The radius of curvature has a significant impact on the WSS profile towards the artery's outer wall. It is revealed that if the channel has a small radius of curvature or a high curvature, the WSS drops, which is consistent with the findings of the [11]. The curved artery's shape is reduced to a straight channel ($R_c = \infty$ or curvature of channel $= 0$) when the value R_c grows, meaning that the straight artery's WSS remains higher than the curved artery ($R_c < \infty$). Figure 5d depicts the effect of volume fraction of nanoparticles on the axial variation of WSS profile for $Au - Al_2O_3/blood$. The WSS decreases as the volume fraction of gold and alumina nanoparticles increases indicating that the effects of both nanoparticles on WSS are the same.

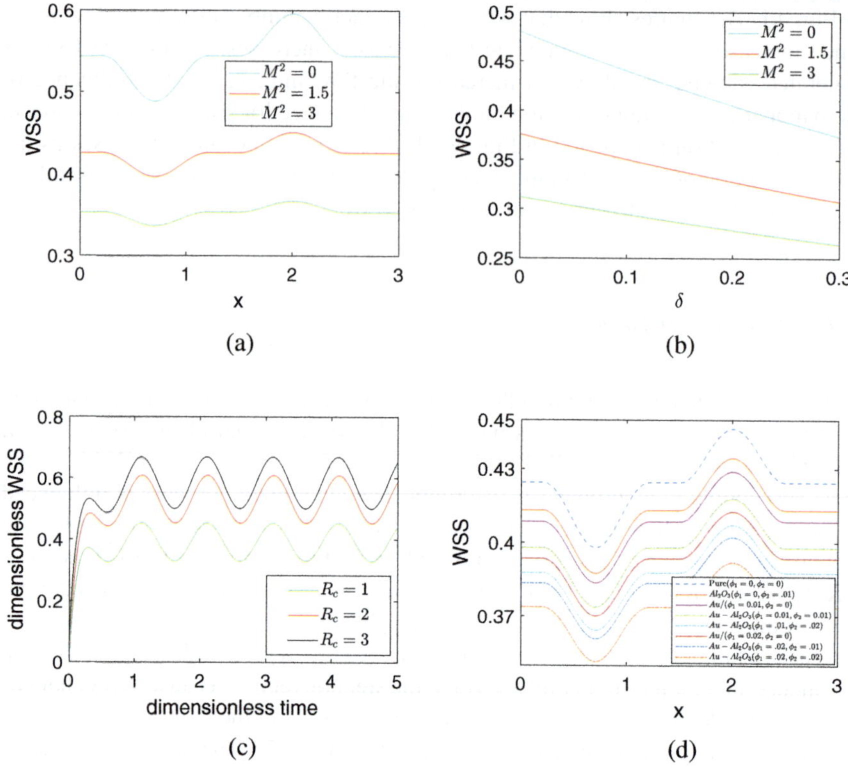

Fig. 5 Variation in WSS for, **a** M^2 with x, **b** M^2 with δ, **c** R_c with time, **d** ϕ_1, ϕ_2

5 Conclusion

The current study examined the effects of addition of nanoparticle on blood flow through a curved artery with minor stenosis and aneurysm circumstances having variable viscosity. This model aims to increase the delivery of medicine in a confined curved artery using nanoparticles. The results shows that Impedance to the flow decreases with increment in R_c, which signifies that resistive impedance is greater for curved artery than for straight channel. WSS decreases when the concentration of gold and alumina nanoparticles increases. Impedance increases when the concentration of gold and alumina nanoparticles increases. It is found that the high curvature or small radius of curvature of the channel leads to a decrease in the WSS.

References

1. Zaman, A., Ali, N., Khan, A.A.: Computational biomedical simulations of hybrid nanoparticles on unsteady blood hemodynamics in a stenotic artery. Math. Comput. Simul. **169**, 117–132 (2020)
2. Das, S., Pal, T., Jana, R.: Outlining impact of hybrid composition of nanoparticles suspended in blood flowing in an inclined stenosed artery under magnetic field orientation. BioNanoScience **11**(1), 99–115 (2021)
3. Tripathi, J., Vasu, B., Bég, O.A.: Computational simulations of hybrid mediated nano-hemodynamics (Ag-Au/Blood) through an irregular symmetric stenosis. Comput. Biol. Med. **130**, 104213 (2021)
4. Zaman, A., Ali, N., Sajjad, M.: Effects of nanoparticles (Cu, TiO2, Al2O3) on unsteady blood flow through a curved overlapping stenosed channel. Math. Comput. Simul. **156**, 279–293 (2019)
5. Zaman, A., Ali, N., Ali, I.: Effects of nanoparticles (Cu (Copper), Silver (Ag)) and slip on unsteady blood flow through a curved stenosed channel with aneurysm. Therm. Sci. Eng. Prog. **5**, 482–491 (2018)
6. Abdel-wahed, M.S.: Magnetohydrodynamic Ferro-Nano fluid flow in a semi-porous curved tube under the effect of hall current and nonlinear thermal radiative. J. Magn. Magn. Mater. **474**, 347–354 (2019)
7. Tripathi, B., Sharma, B.K.: Influence of heat and mass transfer on two-phase blood flow with joule heating and variable viscosity in the presence of variable magnetic field. Int. J. Comput. Methods **17**(03), 1850139 (2017)
8. Tripathi, B., Sharma, B.K., Sharma, M.: MHD pulsatile two-phase blood flow through a stenosed artery with heat and mass transfer **474** (2017). arXiv:1705.09794
9. Elnaqeeb, T., Shah, N.A., Mekheimer, K.S.: Hemodynamic characteristics of gold nanoparticle blood flow through a tapered stenosed vessel with variable nanofluid viscosity. BioNanoScience **9**(2), 245–255 (2019)
10. Zhang, B., Gu, J., Qian, M., Niu, L., Zhou, H., Ghista, D.: Correlation between quantitative analysis of wall shear stress and intima-media thickness in atherosclerosis development in carotid arteries. Biomed. Eng. Online **16**(1), 1–17 (2017)
11. Shahzadi, I., Nadee, S.: Analysis of Ag/blood-mediated transport in curved annulus with exclusive nature of convective boundary. Physica Scripta **94**(11), 115011 (2019)

Response Behavior of a Coaxial Thermal Probe Towards Dynamic Thermal Loading

Anil Kumar Rout, Niranjan Sahoo, Pankaj Kalita, and Vinayak Kulkarni

Abstract An in-house fabricated fast responsive coaxial thermal probe has been utilized to capture the transient temperature response in the exhaust of an internal combustion (IC) engine. The high temperature coaxial thermal probe (ht-CTP) is fabricated from chromel and constantan in the laboratory scale. A constantan wire (of 0.91 mm diameter and 15 mm length) is placed inside the chromel wire (of 3.25 mm diameter and 10 mm length) and clubbed together with an epoxy which also acts as an insulator separating the two. The junction between them is prepared at the surface using abrasion method. The sensor is flush mounted at the exhaust pipe of an IC engine at a distance of 20 mm from the exhaust manifold. In a 4 stroke engine, there is only one exhaust stroke in a complete cycle; therefore, the valve at the exhaust also allows gases one time in a cycle. Therefore, the heat load (contained in the exhaust gas) is also imparted on the sensor in a periodic manner. Hence, by measuring the consecutive temperature response, the cycle time duration has been measured and the engine RPM has been calculated. The RPM is compared with the recorded RPM by the RPM sensor. Along with this, the initial temperature signal is processed for heat flux estimation through analytical method which provides an estimate about the magnitude of transient heat flux imparted on the sensor by the exhaust gas of the engine.

Keywords Thermocouple · Transient temperature · Coaxial thermal probe · IC engine · Cycle time · Transient heat flux

---

A. K. Rout (✉) · N. Sahoo · P. Kalita · V. Kulkarni
School of Energy Science and Engineering, Indian Institute of Technology Guwahati, Guwahati, India
e-mail: anil.rout@iitg.ac.in

S. Banerjee and A. Saha (eds.), *Nonlinear Dynamics and Applications*,
Springer Proceedings in Complexity,
https://doi.org/10.1007/978-3-030-99792-2_35

1 Introduction

The information about transient thermal parameters such as "Transient temperature" and "Transient heat flux" are required for the efficient design and modeling of many engineering systems and sub-systems. Very limited sensors are used to capture instantaneous temperatures. One prominent sensor amongst them is coaxial thermal probe (CTP). Due to fast response, CTP is capable to record instantaneous events in tough conditions like shock tubes, shock tunnels, Gas turbines etc. [1]. Here, the capability of the sensor has been evaluated in other tough conditions like the exhaust flow of internal combustion engine. Nevertheless, the temperature information in combustion chamber of IC engines, the temperature at the exhaust is the vital parameter necessary for upgrading the IC engines.

The information about the temperature in combustion chamber of IC engines is very important which was commonly measured by using sound waves. A transient sound pulse was the main medium for the measurement. The transient sound pulse was applied inside the test chamber and the difference in time between the generated pulsed signal and transmitted pulse was calculated [2]. Another method was preferred for the measurement of average gas temperature where, a steady temperature state was allowed to establish between a thin metallic wire and the gas temperature [3, 4]. Fast response thermocouples were employed for the measurement of transient temperature as well as heat flux through different types of thermocouples [5–10]. The information about both the input and output helps in proper thermodynamic analysis of the engine. Hence, the temperature information at the outlet of engine is also an important aspect in the learning of IC engine features [7, 8 and 11]. Some information about the engine phenomena can be obtained from the exhaust gas analysis and also from the exhaust gas temperature [9]. The unsteadiness of the temperature signal can imply some information about the combustion phenomena taking place inside the combustion chamber [10, 12]. It is also observed that the surface temperature and heat flux obey similar pattern as observed inside the cylinder but, the trend varied when it is measured away from the exhaust [13].

The exhaust valve of an internal combustion engine opens periodically after a certain interval of time corresponding to the engine speed. These CTPs (if mounted at any place in the exhaust line of the engine) can respond to the exhaust hot gas that goes outside through the exhaust manifold after the completion of combustion process inside the engine cylinder. The time difference between the successive responses can be measured and the cycle time corresponding to the engine RPM can be calculated. Therefore, the present work focus on in-house fabrication of a high temperature thermal probe (ht-CTP) for engine and its application in the engine exhaust to capture transient temperature. Subsequently, the information of cycle time is obtained from the temperature response and compared with the cycle time calculated using the value from RPM sensor of the engine. The heat flux imparted on the sensor by the exhaust gas is also estimated using the analytical modeling of the sensor.

2 Fabrication and Calibration of the Thermal Probe

A coaxial probe (ht-CTP) has been fabricated in-house using chromel and constantan as primary elements. For the present case, a (chromel-constantan) thermocouple (CTP) is used due to its high sensitivity, called as "thermo-power" sensor. Constantan wire (0.91 mm diameter, 15 mm length) and chromel wire (3.25 mm diameter, 10 mm length) are considered for fabrication of E-type CTP probe. The inner element (constantan wire) is put into the outer element (chromel) concentrically and symmetrically along with a minimal insulation thickness (~20 μ) in between them. An adequate thickness of epoxy is maintained for its whole length leaving the sensing surface unaffected. The epoxy acts as both electrical insulator as well as binder for both the thermo-elements providing strength to the sensor. Proper care has been taken to make sure that the inner element is straight, co-axial and there is no linkage between the thermo-elements throughout the length to avoid any possibility of multiple connections. A suitable length (10 mm) of the probe is chosen for proper operation and experimentation. An alumina based adhesive is used as insulating material which can sustain a temperature of 1500 °C. The expected temperature at the exhaust of the IC engine for the present experimental condition is nearly 400 °C and an E-type probe can record a temperature up to 700 °C. Therefore, these probes can be used to capture the response confidently. The linkage amongst the two elements is prepared at the measuring surface using scratching technique. Same wires (Chromel and constantan) are spot welded for further instrumentation purpose. The schematic of the fabricated probe is mentioned in Fig. 1. The output from the probe is in the form of electromotive force (EMF) which is generated in response to the change in temperature obeying Seebeck effect. Therefore, to convert voltage into temperature, a correlation is required which is commonly known as the sensitivity of the probe. A furnace based set up is used to calibrate the probe where the probe is placed along with a reference temperature probe in a heating environment and both the temperature and EMF are noted for the change in temperature. The slope of the EMF and Temperature plot provides the sensitivity value which for the present case found to be 59 μV/°C [1].

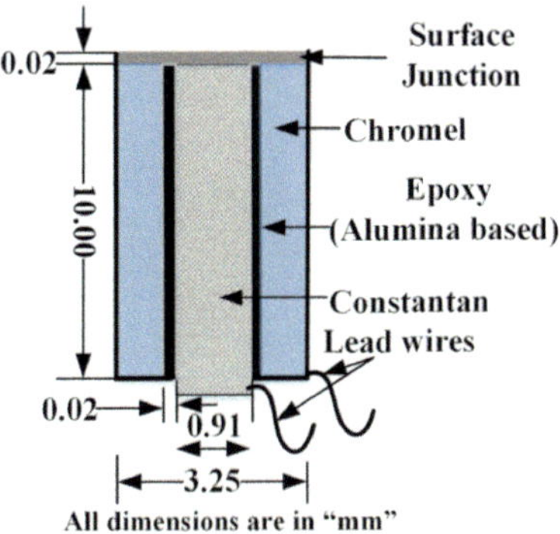

Fig. 1 Schematic of the high temperature probe

3 Experimental Engine Set Up

The sensor is mounted at the exhaust of a four stroke research engine running with petrol. The engine is equipped with eddy current type water cooled dynamometer attached to a loading unit. Piezo type pressure transducer is attached for pressure measurement and Pt-100 RTD for temperature measurement. All the outputs from the sensors equipped with the engine are directly linked to DAS (Make: NI USB-6210, 16-bit). It is a water cooled Research Engine setup-single cylinder, four stroke, Multi-fuel VCR engine with bore and stroke dimensions of 87.5 mm * 110 mm. the capacity is 661 cc and connecting rod length as 234 mm. It has a compression ratio which can be adjusted in between 6–10. The engine has a power rating of 4.5 KW at 1800 rpm, Range of speed: 1200–1800 rpm. Along with the mentioned characteristics, the engine test bench was equipped with many general purpose attachments.

- Fuel tank with a capacity of 15 lit. attached to a measuring tube
- Orifice meter and mano-meter attached air box.
- Rotameters for water flow measurement
- Load sensor indicator
- Calorimeter and its attachments

3.1 Experimental Procedure

The in-house fabricated CTP is exposed to the hot engine by product (hot air) at the exhaust of the engine. The CTP is mounted at 20 mm distance from the exhaust manifold (Fig. 2). The sensor is flush mounted so that the flow is not disturbed by the presence of the sensor and the sensing surface is nicely exposed to the heat load. The sensor is linked to a DAS (Make: NI, Model: NI-9223/ cDAQ-9178) to record high frequency voltage signal at a rate of 100,000 samples per second. The compression ratio is set to 10 which is almost one end for petrol engine. The engine is run for

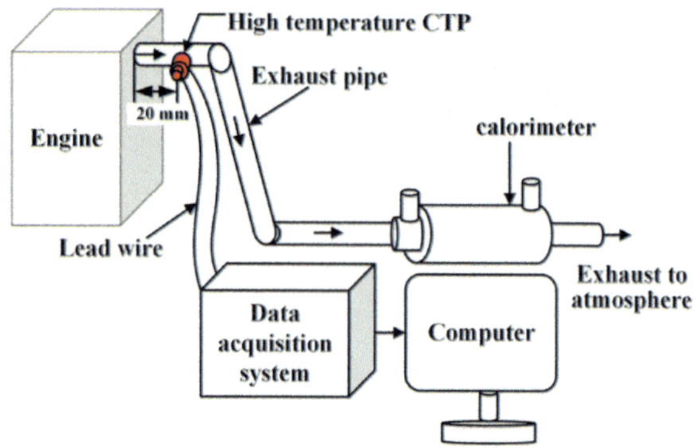

Fig. 2 Schematic highlighting the sensor location

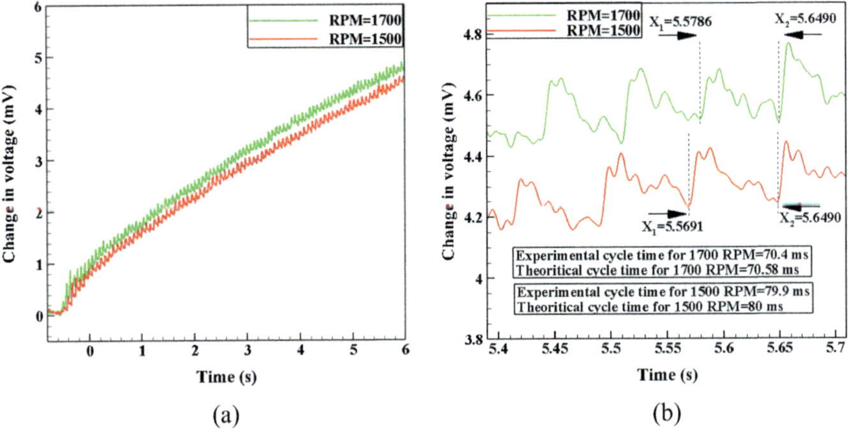

Fig. 3 Temperature response from the probe; **a** voltage signal; **b** Cycle time calculation

two different RPMs (1500 & 1700) with full throttle condition and responses are captured (Fig. 3).

4 Results and Discussion

The voltage signals are recorded using a NI based data acquisition system with the help of lab view interface. The signal is filtered with low pass filter and the results are plotted for 1500 and 1700 RPM in Fig. 3a. The sensor indicates a sharp rise at the arrival of the hot gases and then the subsequent cooling phase. This cycle repeats itself as the exhaust valve of the engine opens after a certain interval of time corresponding to the RPM of the engine. According to the speed of the crank shaft, the time required for 4 stroke cycles can be evaluated using Eq. 1. The experimental cycle time is calculated using the time span between two sharp rise in signal and is compared with the theoretical cycle time (Fig. 3b). The experimental and theoretical cycle time is tabulated in Table 1. A deviation of 0.12 and 0.25% are observed in 1500 RPM and 1700 RPM, respectively.

$$Cycletime, t\,(ms) = \frac{2 \times 60}{RPM} \times 1000 \tag{1}$$

Table 1 Comparison of experimental and theoretical cycle time

RPM	Experimental cycle time (ms)	Theoretical cycle time (ms)	Deviation (%)
1500	79.9	80	0.12
1700	70.4	70.58	0.25

4.1 Transient Heat Flux Calculation Using Analytical Approach

Owing to the unavailability of any straight relation for the transient heat-flux calculation for short duration experiments, it is calculated through appropriate modeling of the sensor. The time-varying temperature obtained from the experiment is employed for the estimation of heat-flux. The process is carried out by considering one-dimensional heat conduction in a semi-infinite substrate. With appropriate assumptions, and boundary conditions, the heat conduction equations take a shape as mention in Eq. 2.

$$\dot{q}_s(t) = \frac{\beta}{\sqrt{\pi}} \int_0^t \frac{1}{\sqrt{t-\tau}} \frac{d\{T_s(\tau)\}}{dt} d\tau \ ; \ \beta = \sqrt{\rho c k} \tag{2}$$

The form presented in Eq. (2) need to be discretized for the temperature. Here, a cubic spline of third order is used to discretize the temperature signal as explained below [6, 7]:

$$[T_s(\tau)]_{CS} = C_{1,i} + C_{2,i}(\tau - \tau_i) + \frac{1}{2}C_{3,i}(\tau - \tau_i)^2 + \frac{1}{6}C_{4,i}(\tau - \tau_i)^3; \quad \tau_i \le \tau \le \tau_{i-1}, i = 1, 2, \ldots M \tag{3}$$

T where, $\tau = S_t\, t$ is the scaled time and S_t is the scaling factor that is considered as unity in this case. The coefficients are given by Eq. 4.

$$C_{n,i} = \frac{dT_s^{n-1}(\tau_i)}{d\tau} \tag{4}$$

Finally, the surface heat flux is estimated using Eq. 5.

$$\dot{q}_s(\tau_{M+1}) = \left\{ 2\frac{\beta}{\sqrt{\pi}} \sum_{i=1}^{M-1} \begin{matrix} V_i\left(P_i^{\frac{1}{2}} - R_i^{\frac{1}{2}}\right) - \frac{W_i}{3}\left(P_i^{\frac{3}{2}} - R_i^{\frac{3}{2}}\right) + \frac{C_{4,i}}{10}\left(P_i^{\frac{5}{2}} - R_i^{\frac{5}{2}}\right) \\ +2\sqrt{\frac{\rho c k}{\pi}}\left(V_M P_M^{\frac{1}{2}} - \frac{W_M}{3} P_M^{\frac{3}{2}} + \frac{C_{4,M}}{10} P_M^{\frac{5}{2}}\right) \end{matrix} \right\} \tag{5}$$

$\sqrt{S_t}$; $M = 1, 2, .J - 1$ *where*, $P_i = \tau_{M-1} - \tau_i$; $R_i = \tau_{M+1} - \tau_{i+1}$; $F_i = C_{1,i} + C_{2,i} P_i + \frac{C_{3,i}}{2} P_i^2 + \frac{C_{4,i}}{6} P_i^3$; $V_i = \frac{dF_i}{d\tau_{M+1}}$; $W_i = \frac{d^2 F_i}{d\tau_{M+1}^2}$

The heat flux signals for all the inputs are acquired from the discretized temperature signals using Eq. (5). More information regarding discretization can be obtained from the reference [14]. Referring to Eq. 2; the magnitude of thermal product (β) for a sensor is required for the calculation of the heat-flux. Researchers adopted various magnitudes for the thermal product. It is tough to access the actual percentage from

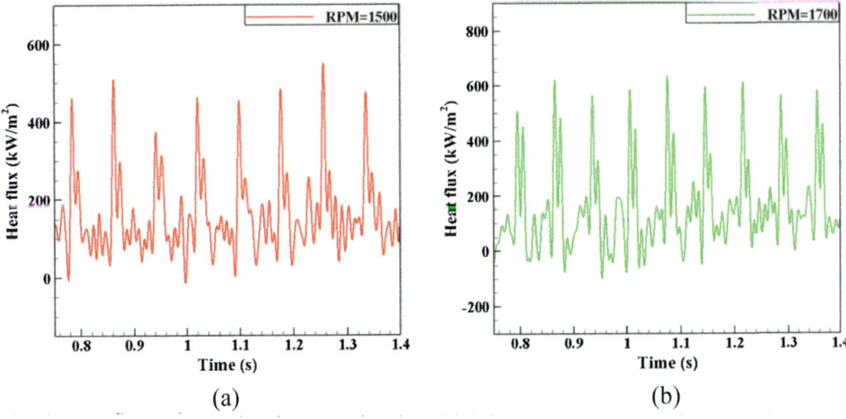

Fig. 4 Heat flux estimation for 0.5 s after the initial rise; **a** RPM = 1500; **b** RPM = 1700

individual thermo-element in the preparation of the junction. Hence, here the thermal product magnitude is considered with 50% stake from individual component and the magnitude calculated is 8650.2 $Jm^{-2}\ K^{-1}\ s^{-0.5}$ [15]. The heat-flux signals are drawn in Fig. 4. Considering the restrictions in the validity of one dimensional heat conduction and semi-infinite substrate assumption, the signal is considered for 0.5 s after from the initial rise of the temperature for the heat flux estimation. As inferred from Fig. 4, the peak magnitude of heat flux is in the range of 500 ± 50 kW/m^2 for 1500 RPMand 550 ± 50 kW/m^2 for 1700 RPM. The exhaust valve opens once in a cycle (4 strokes) and the sensor receives an instantaneous heat load. Therefore, the peak heat flux of nearly similar magnitude indicates the nearly equivalent combustion process.

5 Conclusion

The coaxial probe is exposed to the exhaust gas coming out from the engine exhaust manifold. The cyclic nature of heat load is well captured by the probe. From the experiment it is clear that the sensor is able to capture cycle time correctly. The transient temperature data is used to estimate the heat content getting lost as by product in terms of heat flux. As expected, the peak magnitude of cyclic heat flux are nearly equal and are in the range of 500 ± 50 kW/m^2 for 1500 RPM and 550 $\pm$ 50 kW/m^2 for 1700 RPM. The aim of the study can be extended to engines in moving vehicles where a sensor can be mounted somewhere in the line of exhaust and some amount of exhaust air may be blown over it. From the response, the driver can get the on time engine health monitoring.

References

1. Rout, A.K., Sahoo, N., Kalita, P.: Effectiveness of coaxial surface junction thermal probe for transient measurements through laser based heat flux assessment. Heat Mass Transf. **56**(4), 1141–1152 (2020)
2. Livengood, J.C., Rona, T.P., Baruch, J.J.: Ultrasonic temperature measurement in internal combustion engine chamber. J. Acoust. Soc. Am. **26**(5), 824–830 (1954)
3. Nagao, et al.: Measurement of cylinder gas temperature of internal combustion engines. Bull. JSME **13**(64), 1240–1246 (1970)
4. Touloukian, Y.S.: Specific heat metallic elements and alloys. In: Touloukian, Y.S. (ed.) Thermo-Physical Properties of Matter; TPRC Data Series, vol. 4. IFI/Plenum Press, New York (1970).
5. Assanis, D.N., Badillo, E.: On heat transfer measurements in diesel engines using fast-response coaxial thermocouples. J. Eng. Gas Turbines Power, Trans. ASME **111**, 458–465 (1989)
6. Rakopoulos, C.D., Mavropoulos, G.C.: Experimental instantaneous heat fluxes in the cylinder head and exhaust manifold of an air-cooled diesel engine. Energy Convers. Manage. **41**, 1265–1281 (2000)
7. Kar, et al.: Instantaneous exhaust temperature measurements using thermocouple compensation techniques. SAE Technical Paper Series 2004-01-1418 (2004)
8. Kee, et al.: Fast response exhaust gas temperature measurement in IC engines. SAE Technical Paper Series 2006-01-1319 (2006)
9. Wang, X., Stone, C.R.: A study of combustion, instantaneous heat transfer, and emissions in a spark ignition during warm-up. J. Automob. Eng., IMechE, Part-D **222**, 607–618 (2008)
10. Hotta, S.K., Sahoo, N., Mohanty, K., Kulkarni, V.: Ignition timing and compression ratio as effective means for the improvement in the operating characteristics of a biogas fueled spark ignition engine. Renew. Energy **150**, 854–867 (2020)
11. Sujith, R.I., Unni, V.R.: Dynamical systems and complex systems theory to study unsteady combustion. Proc. Combust. Inst. **38**(3), 3445–3462 (2021)
12. Marr, et al.: A fast response thermocouple for internal combustion engine surface temperature measurements. Exp. Therm. Fluid Sci. **34**, 183–189 (2010)
13. Mavropoulos, G.C.: Unsteady heat conduction phenomena in internal combustion engine chamber and exhaust manifold surfaces. Heat Transf.-Eng. Appl. 283–308 (2011)
14. Taler, J.: Theory of transient experimental techniques for surface heat transfer. Int. J. Heat Mass Transf. **39**(17), 3733–3748 (1996)
15. Agarwal, S., Sahoo, N., Singh, R.K.: Experimental techniques for thermal product determination of coaxial surface junction thermocouples during short duration transient measurements. Int. J. Heat Mass Transf. **103**, 327335 (2016)

Soret and Dufour Effects on Thin Film Micropolar Fluid Flow Through Permeable Media

G. Gomathy and B. Rushi Kumar

Abstract In this study, thin films of micropolar fluid flowing through porous media are analysed over a stretching sheet with heat effect subjected to thermophoresis. We assume that micropolar fluid is the base fluid and that the plate moves linearly and is subjected to reference temperature and concentration variations. An analysis of fluid flow over a steady stretching sheet utilizes the Soret and Dufour effects has been conducted in this study. Through the use of similarity variables, basic fluid flow equations are transformed to a nonlinear set of coupled equations with boundary conditions. MATLAB bvp4c solver is used to solve the problem. Graphs are used to illustrate the impact of physical parameters on the flow profiles that demonstrate the velocity, concentration, and temperature of thin fluid films. A series of tables illustrate the effects of dimensionless parameters on skin friction coefficients, Nusselt and Sherwood numbers. It is observed that increasing permeability parameter results in the decline of the velocity profile. With increasing Dufour and Soret number, temperature and concentration profiles are observed to increase, respectively.

Keywords Micropolar fluid · Dufour-Soret effect · Thin film · Thermophoresis

1 Introduction

In recent decades, micropolar fluids have gained significant attention within the engineering community due to the limitations of Newtonian fluids. Particles in suspension prevent Newtonian fluids from describing fluid flow. In contrast, micropolar fluids can simulate fluid in the presence of dust particles. Micropolar fluids are present in animal blood, dumbell-shaped molecules, liquid suspensions, and polymer fluids. The subject of thin film flow has recently become prominent in research. In chemical techniques, thin film fluids are used to make heat exchangers, requiring

G. Gomathy · B. Rushi Kumar (✉)
Department of Mathematics, School of Advanced Sciences, Vellore Institute of Technology, Vellore, Tamilnadu, India
e-mail: rushikumar@vit.ac.in

S. Banerjee and A. Saha (eds.), *Nonlinear Dynamics and Applications*, Springer Proceedings in Complexity,
https://doi.org/10.1007/978-3-030-99792-2_36

in-depth expertise in motion. Coating of wires and fibres, polymer preparation are possible applications. Tiny particles moved from a hot surface onto a cold one during the extrusion process, is called thermophoresis. As dust particles move along the temperature gradient in gases, they exert thermophoretic force, and the acceleration achieved is referred to as thermophoretic velocity.

Eringen's [4] theory of micropolar fluids took into consideration the microscopic influence caused by the local structure and micro-rotation of fluid particles and believed to have a mathematical model explaining non-Newtonian fluid behaviour. Thermophoresis involves transfer of small particles to cold surfaces, while hot surfaces resist particle transfer; thus, a layer of free particles occurs around hot surface, as determined by Goldsmith and May [7]. Kim and Lee [10] analytically studied a semi-infinite vertical rotating porous plate with electrically conducting viscous incompressible micropolar fluid. They looked at the effect of different thermophysical and flow parameters and how they affected the temperature and flow fields in the boundary layer. A thermophoresis effect is observed in the free convective flow of boundary layer over a permeable barriers was reported by Chamka et al. [5, 9]. The aim of Rashidi et al. [15] was to develop a complete and accurate method to predict thermal transfer of micropolar fluid across a porous media with radiations based on Homotopy analysis. Several investigations regarding micropolar fluids with different physical structures and thermal radiations were discussed in [1, 3, 6, 8, 16].

Omowaye et al. [13] developed an analytical method to flow of an incompressible viscous fluid that flows past a semi-infinite plate that is embedded in porous media. They presumed fluid properties would remain unchanged except for the viscosity of a fluid, which is inversely related to temperature. Shah et al. [17] conducted a study over an unsteady stretched surface that included flow of thin film with effects of Dufour and Soret, which is reported to be the first to incorporate such effects in flow of thin film fluid. Tripathy et al. [18] numerically studied the interaction of a chemical reaction on the convective flow, mass transfer and heat capacity of a micropolar fluid over a stretched sheet embedded in porous media in the presence of a volumetric non-uniform heat source. Palwasha et al. [14] investigated thin film flow of fluid in three dimensions with different thermophysical properties of boundary layers. Vakkar Ali et al. [2] examined the flow of a thin film of micropolar fluid through porous layers subjected to thermophoresis and heat effect past a stretching plate under the assumption that the micropolar fluid itself is the base fluid. Usman et al. [19] investigated variations in Brownian motion, Hall current, thermophoresis and couple stress within the steady convection MHD flow of micropolar nanofluid to understand non-isothermal heat transfer from non-linear walls. Megahed et al. [11] addressed the laminar boundary layer flow, heat transfer problem for MHD fluid caused by a stretched sheets which is unsteady with prolonged heat flux. Naseem et al. [12] computed Soret and Dufour effects applied on radiated material on a porous stretched surface with thermal conductivity that depends on temperature.

To the author's knowledge, there are no studies in the literature that analyze the combined effects of Soret and Dufour of micropolar fluid in porous media over a stretching sheet in the presence of thermal radiation and thermophoresis along with magnetic field. Thus in the current work, we have considered thin films of micropolar fluid flowing through a porous media over a stretching sheet with combined effects of Soret and Dufour in the presence of thermal radiation and thermophoresis along with magnetic field. Once the boundary layer PDE are transformed into ODE, they are solved numerically using shooting technique along with RK4 method. Graphs are made to illustrate the impact of physical parameters like permeability, inertia coefficient, microrotation, thermophoretic, radiation, and dimensionless numbers such as Prandtl, Dufour and Soret on the flow profiles that demonstrate velocity, concentration and temperature of thin fluid flow.

2 Mathematical Formulation

Let us consider the flow of thin micropolar fluids on a stretched plate, stretched at a linear velocity $U_w = ax$. The constant $a > 0$ indicates the stretching rate, while x represents the flow direction. The film is chosen to have a uniform thickness δ, while the nature of the medium is porous, as indicated by Fig. 1. During stretching, the temperature of the stretching plate is set at T_w and the concentration at C_w. On the surface, $T_w = T_0 - T_{ref}\left(\frac{U_w x}{2\nu}\right)$ and $C_w = C_0 - C_{ref}\left(\frac{U_w x}{2\nu}\right)$ are presume to differ with distance x relative to the plate. In addition to the temperature T_0 and concentration C_0 at the plate, there are constant reference temperature T_{ref} $(0 \leq T_{ref} \leq T_0)$ and concentration C_{ref} $(0 \leq C_{ref} \leq C_0)$ as well. Moreover, it is speculated that the liquid film grips and releases radiation. On the x-axis the radiative flux is taken into account, while the y-axis is ignored.

Our suggested model's basic flow equations are:

$$\frac{\partial u}{\partial x} + \frac{\partial v}{\partial y} = 0, \tag{1}$$

$$u\frac{\partial u}{\partial x} + v\frac{\partial u}{\partial y} = \nu\frac{\partial^2 u}{\partial y^2} + k_c\frac{\partial \sigma}{\partial y} + \frac{\nu\varphi}{K}(U_w - u) + Cr\varphi(U_w^2 - u^2), \tag{2}$$

$$G_1\frac{\partial^2 \sigma}{\partial y^2} - 2\sigma\frac{\partial u}{\partial y} = 0, \tag{3}$$

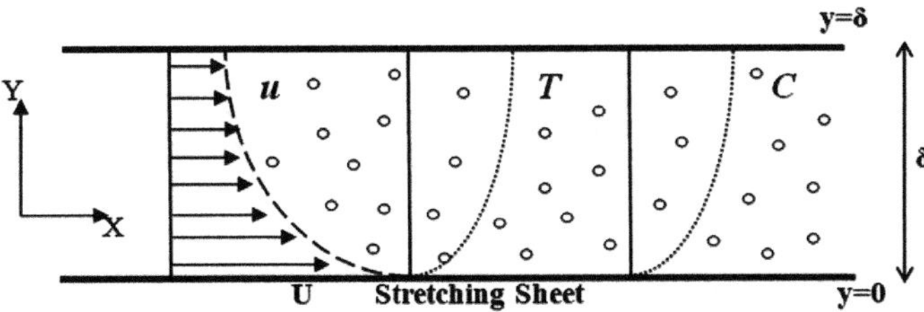

Fig. 1 A schematic diagram of the physical system

$$u\frac{\partial T}{\partial x} + v\frac{\partial T}{\partial y} = \frac{k}{\rho c_p}\frac{\partial^2 T}{\partial y^2} - \frac{1}{\rho c_p}\frac{\partial q_r}{\partial y} + \frac{D_m k_T}{c_s c_p}\frac{\partial^2 C}{\partial y^2}, \tag{4}$$

$$u\frac{\partial C}{\partial y} + v\frac{\partial C}{\partial y} = D_m\frac{\partial^2 C}{\partial y^2} + \frac{D_m k_T}{T_m}\frac{\partial^2 T}{\partial y^2} - \frac{\partial (V_T C)}{\partial y}. \tag{5}$$

For the two-dimensional flow of a liquid film, these are the boundary conditions:

$$\begin{aligned} &u = U_w = ax, v = 0, \sigma = 0, T = T_w, C = C_w \text{ at } y = 0 \\ &\frac{\partial u}{\partial y} = 0, \frac{\partial T}{\partial y} = 0, \frac{\partial \sigma}{\partial y} = 0, \frac{\partial C}{\partial y} = 0, v = \delta x \text{ at } y = \delta \end{aligned} \tag{6}$$

where thermophoretic velocity V_T can be taken as $V_T = -\frac{k_1 \nu}{T_{ref}}\frac{\partial T}{\partial y}$, k_1 represents thermophoretic coefficient.

As defined by Rosseland, the Rosseland approximation is:

$$q_r = -\frac{4\sigma^*}{3k^*}\frac{\partial T^4}{\partial y} \tag{7}$$

As a result of Taylor's series, T^4 can be rewritten as follows:

$$T^4 = 4T_1^3 T - 3T_1^4 \tag{8}$$

Equation (4) can be reduced using Eqs. (7) and (8) as follows:

$$u\frac{\partial T}{\partial x} + v\frac{\partial T}{\partial y} = \frac{k}{\rho c_p}\frac{\partial^2 T}{\partial y^2} + \frac{16\sigma^* T_1^3}{3\rho c_p k^*}\frac{\partial^2 T}{\partial y^2} \tag{9}$$

Incorporating transformations for the non-dimensional variables f, g, θ, ϕ and the similarity variable 'η' as,

$$u = U_w f'(\eta), \quad v = -\left(\frac{\nu U_w}{2x}\right)^{\frac{1}{2}} f(\eta), \quad \sigma = \left(\frac{U_w}{2\nu x}\right)^{\frac{1}{2}} U_w g(\eta), \quad \eta = \left(\frac{a}{2\nu}\right)^{\frac{1}{2}} y \tag{10}$$

$$T = T_0 - T_{ref}\left(\frac{U_w x}{2\nu}\right)\theta(\eta), \quad C = C_0 - C_{ref}\left(\frac{U_w x}{2\nu}\right)\phi(\eta) \tag{11}$$

When Eq. (10) and (11) are substituted into Eqs. (1)–(6), the basic governing equations with boundary conditions can be written as follows:

$$f''' + ff'' - 2f'^2 + \Delta g' + \frac{1}{Mr}(1 - f') + Nr(1 - f'^2) = 0, \tag{12}$$

$$Grg'' - 2(2g + f'') = 0, \tag{13}$$

$$\theta''(1 + \frac{4}{3}R) - Pr(2f'\theta - f\theta') + 2PrDu\phi'' = 0, \tag{14}$$

$$\phi'' + Sc(Sr - \tau\phi)\theta'' + Sr(f - \tau\theta')\phi' - 2Sc\phi f' = 0. \tag{15}$$

$$\begin{gathered} f(0) = 0, g(0) = 0, f'(0) = 1, \theta(0) = 1, \phi(0) = 1, \\ f''(1) = 0, f(1) = 0, g'(1) = 0, \theta'(1) = 0, \phi'(1) = 0. \end{gathered} \tag{16}$$

here f, g, θ, and ϕ represents dimensionless velocity, micro-rotation angular velocity, temperature and concentration function respectively. β represents fluid film non-dimensional thickness. Also Δ, Mr, Nr, Gr, R, τ represents parameter of vortex-viscosity, permeability, inertia coefficient micro-rotation, radiation, thermophoretic respectively, and Pr, Du, Sc, Sr, represents Prandtl, Dufour, Schmidt, and Soret number.

From a mathematical perspective, these parameters can be expressed as follows:

$$\Delta = \frac{k_c}{\nu}; \quad Mr = \frac{Ka}{2\varphi\nu}; \quad Nr = \frac{2\varphi C_r U_w}{a}; \quad Gr = \frac{G_1 a}{\nu}; \quad R = \frac{4\sigma^* T_1^3}{kk^*}$$

$$Du = \frac{D_m k_T}{c_s c_p \nu}\frac{C_w - C_0}{T_w - T_0}; \quad Sc = \frac{\nu}{D_m}; \quad Sr = \frac{D_m k_T}{\nu T_m}\frac{T_w - T_0}{C_w - C_0}; \quad \tau = \frac{k_1 U_w^2}{2\nu a}$$

Skin friction coefficient, Nusselt number and Sherwood number are defined as follows $C_f = \dfrac{\mu\left(\dfrac{\partial u}{\partial y}\right)_{y=0}}{\dfrac{1}{2}\rho U_w^2}$, $Nu = \dfrac{-kx\left(\dfrac{\partial T}{\partial y}\right)_{y=0}}{k(T_w - T_0)}$, $Sh = \dfrac{-D_m x\left(\dfrac{\partial C}{\partial y}\right)_{y=0}}{D_m(C_w - C_0)}$ with variables in Eq. (10), we can determine expressions for non-dimensional skin friction, Nusselt number and Sherwood number as follows:

$$C_f\left(\frac{Re}{2}\right)^{\frac{1}{2}} = f''(0), \quad Nu\left(\frac{Re}{2}\right)^{-\frac{1}{2}} = -\theta'(0), \quad Sh\left(\frac{Re}{2}\right)^{-\frac{1}{2}} = -\phi'(0) \tag{17}$$

$Re = \dfrac{U_w x}{\nu}$ stands for Reynold's number based on stretching velocity.

3 Solution Approach

A finite-difference algorithm in MATLAB, which is called the bvp4c solver, is used to solve boundary value problems (12)–(15), using boundary conditions (16). As an adaptive mesh solver, this solver utilizes residual control to select the mesh and control the error. Using this solver, we can create differential equations by using the `odefun` function, boundary conditions by using the `bcfun` function, and initial guesses by using the `solinit` function. By converting the boundary value problem (BVP) to an initial value problem (IVP) and reducing the higher order ODE's to first order ODE's, this code could be applied to the boundary value problem (BVP). First order ODE's are transformed as follows:

$$f = f(1);\quad f' = f(2);\quad f'' = f(3);$$

$$f''' = -f(1)f(3) + 2(f(2))^2 - \Delta f(5) - \frac{1}{Mr}(1 - f(2)) - Nr(1 - (f(2))^2);$$

$$g = f(4);\quad g' = f(5);$$

$$g'' = \frac{1}{Gr}[2(2f(4) + f(3))];$$

$$\theta = f(6);\quad \theta' = f(7);$$

$$\theta'' = \frac{1}{(1 + \frac{4}{3}R) - 2PrDuSc(Sr - \tau f(8))}[Pr(2f(2)f(6)) + 2PrDuSr(f(1) - \tau f(7))f(9) - 4PrDuScf(2)];$$

$$\phi = f(8);\quad \phi' = f(9);$$

$$\phi'' = \frac{1}{2PrDuSc(Sr - \tau f(8)) - (1 + \frac{4}{3}R)}[PrSc(Sr - \tau f(8))(2f(2)f(6) - f(1)f(7) + (1 + \frac{4}{3}R)Sr(f(1) - \tau f(7))f(9) - 2Sc(1 + \frac{4}{3}R)f(2)f(8)]$$

4 Results and Discussion

An investigation has been conducted on the thin film motion of a micropolar fluid through porous media. It includes the effects of energy radiation and thermophoresis, as well as Dufour-Soret effect through a stretching plate. Numerical computations are performed for $Mr = Nr = Gr = R = Sr = Du = Sc = \Delta = \tau = 1$, $Pr = 0.71$.

Table 1 Values of $f''(0)$ with Mr, Nr and Δ variation

Mr	Nr	Δ	$f''(0)$
0.8	0.3	0.3	2.302991
0.9	0.3	0.3	2.330854
1.0	0.3	0.3	2.353966
0.8	0.3	0.3	2.353966
0.8	0.4	0.3	2.317195
0.8	0.5	0.3	2.282024
0.8	0.3	0.3	2.282024
0.8	0.3	0.4	2.265702
0.8	0.3	0.5	2.249314

Table 2 Values of $\theta'(0)$ with R, Pr and Du variation

R	Pr	Du	$-\theta'(0)$
0.3	0.3	0.3	1.562135
0.4	0.3	0.3	1.221797
0.5	0.3	0.3	1.003682
0.3	0.3	0.3	1.003682
0.3	0.4	0.3	1.292026
0.3	0.5	0.3	1.561992
0.3	0.3	0.3	1.561992
0.3	0.3	0.4	1.471936
0.3	0.3	0.5	1.381899

Tables 1, 2 and 3, shows the computed value of skin friction coefficient, local Nusselt number, and Sherwood number. Figures 2, 3, 4, 5, 6, 7, 8, 9 and 10 illustrates velocity, temperature, and concentration profiles of flow of thin film with variation over physical parameters.

Based on different values of permeability, Fig. 2 displays the variation in velocity distribution. The presence of a porous medium causes higher fluid flow restriction, which subsequently causes deceleration of fluid. Thus, increasing permeability decreases fluid velocity due to fluid motion resistance. According to Fig. 3, higher values of the inertia coefficient lead to increased fluid velocity, as it is directly related to fluid motion. Figure 4 depicts the inertia parameter varies on the profile of microrotation, which is dimensionless. A decrease in the microrotation profile is observed when the inertia parameter increases. From Fig. 5, the microrotation profile of fluid film decreases with increasing its parameter due to the inverse relationship between the microrotation and viscosity parameter is depicted. Thus, as Gr increases, the viscosity of the fluid decreases.

Table 3 Values of $\phi'(0)$ with Sc, Sr and τ variation

Sc	Sr	τ	$-\phi'(0)$
0.3	0.3	0.3	1.267877
0.4	0.3	0.3	1.624919
0.5	0.3	0.3	1.957370
0.3	0.3	0.3	1.957370
0.3	0.4	0.3	1.896043
0.3	0.5	0.3	1.835538
0.3	0.3	0.3	1.197621
0.3	0.3	0.4	1.178786
0.3	0.3	0.5	1.160595

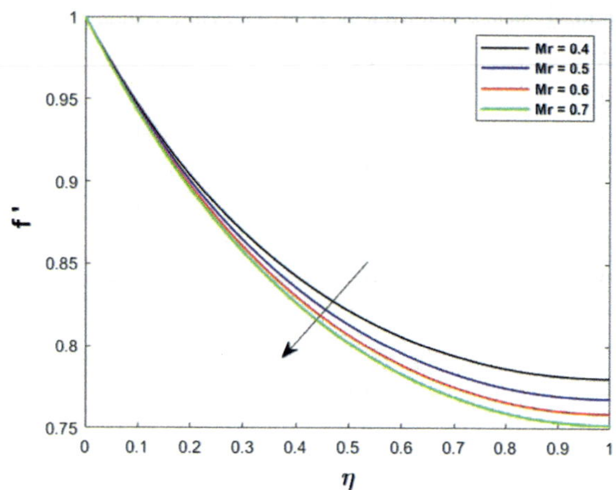

Fig. 2 Velocity profile with Mr variation

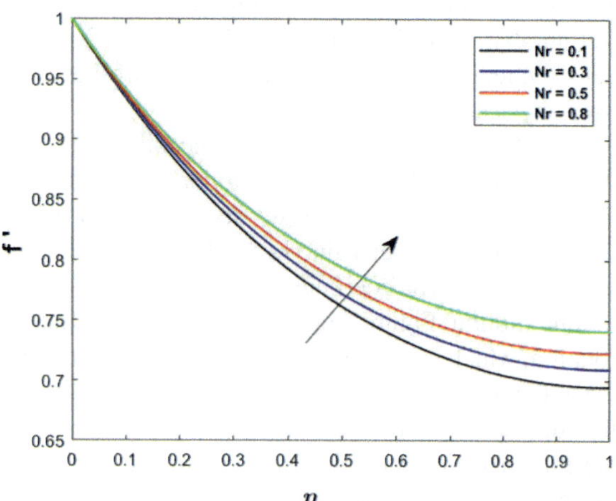

Fig. 3 Velocity profile with Nr variation

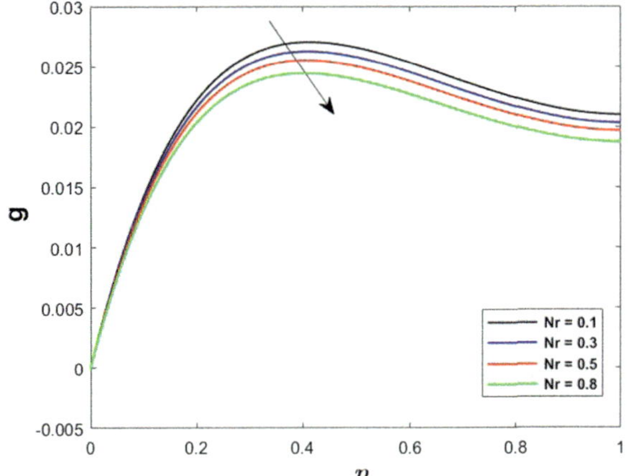

Fig. 4 Microrotation profile with Nr variation

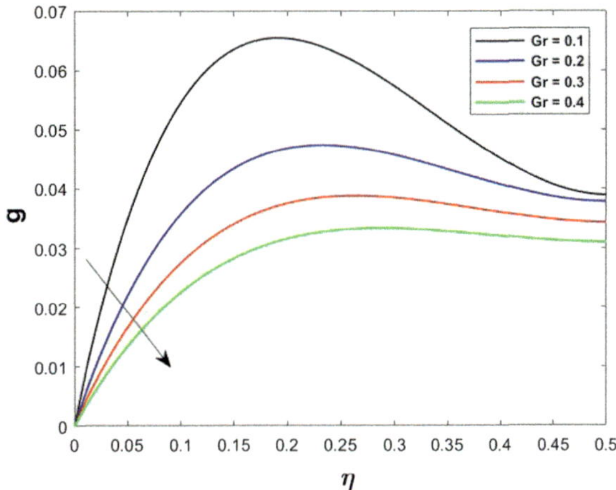

Fig. 5 Microrotation profile with Gr variation

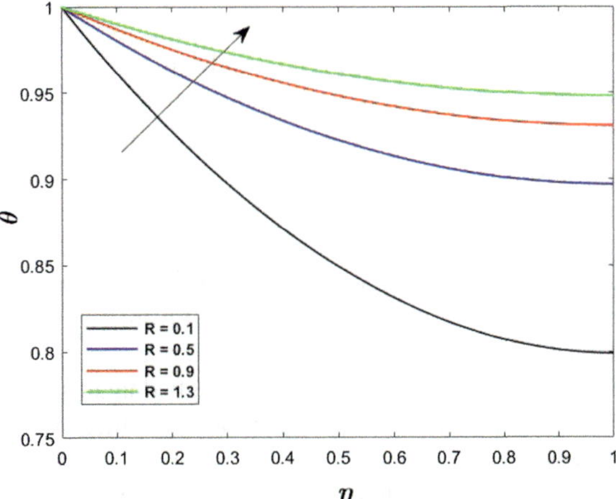

Fig. 6 Temperature profile with R variation

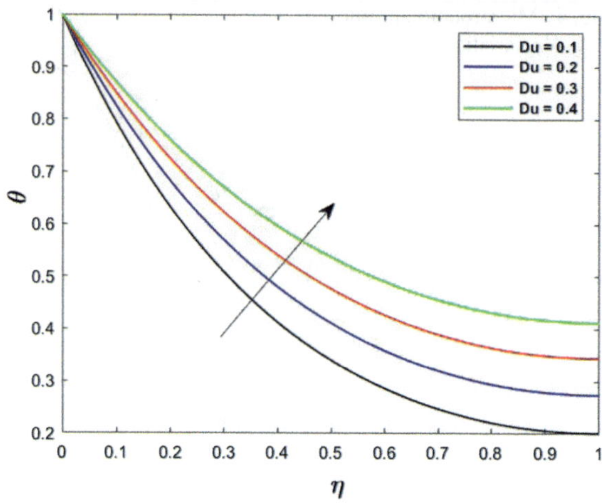

Fig. 7 Temperature profile with *Du* variation

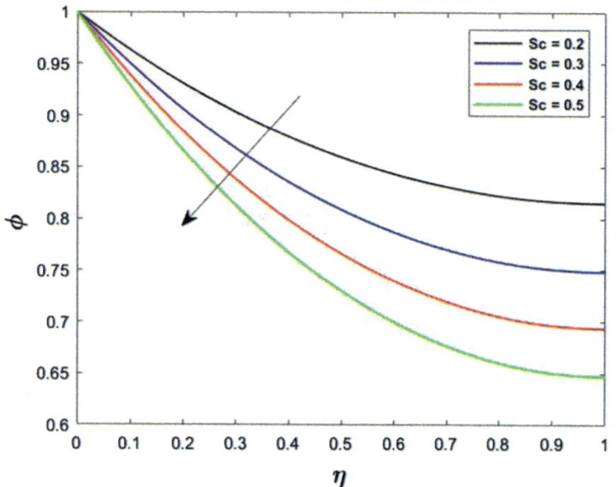

Fig. 8 Concentration profile with *Sc* variation

Radiation from thermal sources is always included as a special case in the energy equation. Temperature rise with higher thermal radiation parameter is illustrated in Fig. 6. Physically, a rise in temperature can be attributed to the rate at which energy is transported in the fluid. As shown in Fig. 7, the Dufour number is influenced by the temperature profile. Dufour number refers to how much thermal energy is transferred in a flow due to concentration gradients. Consequently, concentration gradients become larger at higher Dufour number. Therefore, mass diffusion takes place faster, and energy is transferred from one particle to another at a higher rate. This causes the temperature profile to rise.

In Fig. 8, Schmidt number is weighed against concentration. Due to the inverse relationship between Sc and molecular diffusivity, increasing Schmidt numbers decrease the concentration field. Schmidt numbers characterize the flow of fluids where momentum dissipation is concurrent with mass dissipation. Figure 9 illus-

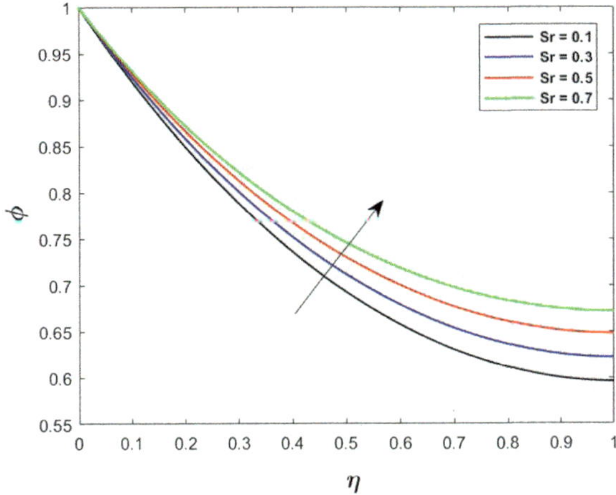

Fig. 9 Concentration profile with Sr variation

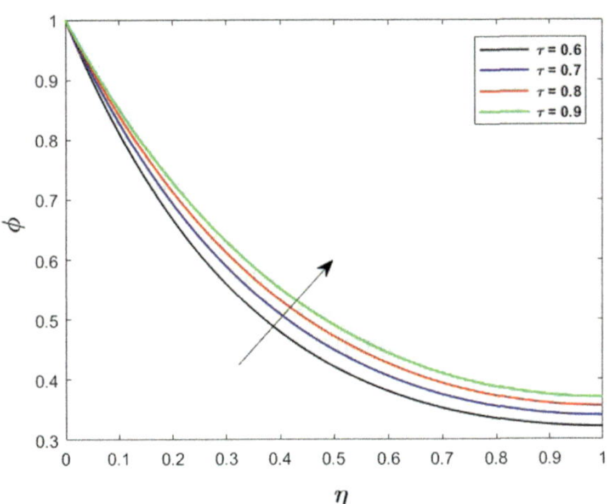

Fig. 10 Concentration profile with τ variation

trates the effect of the Soret number on ϕ. A higher Soret number causes the viscosity to increase, and ϕ to accelerate. Due to the increase in Soret number, the difference in temperature between hot and the surrounding fluids increases, so the temperature rises, which results in a higher concentration. Figure 10 illustrates the relation between concentration and thermophoretic parameter. As the thermophoretic parameter increases, the nondimensional concentration profile also increases.

5 Conclusion

In this study, the flow of thin films past a stretched sheet through a permeable medium is investigated. A micropolar fluid served as base fluid, influenced by thermal radiation and thermophoresis. Bvp4c solver is used to solve coupled nonlinear differential equations modelled in this study. We have displayed and discussed how the physical parameters affect the profile of velocity, concentration, and temperature. Accordingly the problem outcomes are:

- Increase in permeability causes a decrease in fluid velocity due to fluid motion resistance.
- Higher values of the inertia coefficient lead to increased fluid velocity.
- A decrease in the microrotation profile is observed when the inertia parameter increases.
- Due to the growth of energy and transport, the temperature field increases as the thermal radiation parameter increases. This elevates the temperature profile.
- It was found that temperature field enhancement occurs when Dufour number increases.
- It has been found that, as Schmidt number increases, the mass diffusivity decreases, which decreases the concentration.
- Observations shows that an increase in Soret number boosts concentration field.

The subject of thin film flow has recently become prominent in research. In chemical techniques, thin film fluids are used to make heat exchangers, which necessitate an in-depth expertise in motion. This study has possible applications in coating of wires, fibres, and polymer preparation industries.

References

1. Abo-Eldahab, E.M., Ghonaim, A.F.: Radiation effect on heat transfer of a micropolar fluid through a porous medium. Appl. Math. Comput. **169**(1), 500–510 (2005)
2. Ali, V., Gul, T., Afridi, S., Ali, F., Alharbi, S.O., Khan, I.: Thin film flow of micropolar fluid in a permeable medium. Coatings **9**(2) (2019)
3. Bhattacharyya, K., Mukhopadhyay, S., Layek, G., Pop, I.: Effects of thermal radiation on micropolar fluid flow and heat transfer over a porous shrinking sheet. Int. J. Heat Mass Transf. **55**, 2945–2952 (2012)
4. C., E.A.: Theory of micropolar fluids. J. Math. Mech. **16**, 1–18 (1966)
5. Chamkha, A.J., Al-Mudhaf, A.F., Pop, I.: Effect of heat generation or absorption on thermophoretic free convection boundary layer from a vertical flat plate embedded in a porous medium. Int. Commun. Heat Mass Transf. **33**(9), 1096–1102 (2006)
6. Das, K.: Effect of chemical reaction and thermal radiation on heat and mass transfer flow of mhd micropolar fluid in a rotating frame of reference. Int. J. Heat Mass Transf. **54**(15), 3505–3513 (2011)
7. Goldsmith, P., May, F.G.: Diffusiophoresis and Thermophoresis in Water Vapour Systems, pp. 163–194. Aerosal Science, Academic Press, London (1966)

8. Heydari, M., Loghmani, G., Hosseini, S.M.: Exponential bernstein functions: an effective tool for the solution of heat transfer of a micropolar fluid through a porous medium with radiation. Comput. Appl. Math. **36**, 647–675 (2015)
9. Chamkha, J.A., Pop, I.: Effect of thermophoresis particle deposition in free convection boundary layer from a vertical flat plate embedded in a porous medium. Int. Commun. Heat Mass Transf. **31**(3), 421–430 (2004)
10. Kim, Y.J., Lee, J.C.: Analytical studies on mhd oscillatory flow of a micropolar fluid over a vertical porous plate. Surf. Coat. Technol. **171**(1), 187–193 (2003)
11. Megahed, A.M., Reddy, M.G., Abbas, W.: Modeling of MHD fluid flow over an unsteady stretching sheet with thermal radiation, variable fluid properties and heat flux. Math. Comput. Simul. (MATCOM) **185**(C), 583–593 (2021)
12. Naseem, T., Nazir, U., El-Zahar, E.R., Algelany, A.M., Sohail, M.: Numerical computation of dufour and soret effects on radiated material on a porous stretching surface with temperature-dependent thermal conductivity. Fluids **6**(6), 2311–5521 (2021)
13. Omowaye, A., Fagbade, A., Ajayi, A.: Dufour and soret effects on steady mhd convective flow of a fluid in a porous medium with temperature dependent viscosity: Homotopy analysis approach. J. Niger. Math. Soc. **34**(3), 343–360 (2015)
14. Palwasha, Z., Khan, N.S., Shah, Z., Islam, S., Bonyah, E.: Study of two-dimensional boundary layer thin film fluid flow with variable thermo-physical properties in three dimensions space. AIP Adv. **8**(10), 105318 (2018)
15. Rashidi, M., Mohimanian Pour, S.: A novel analytical solution of heat transfer of a micropolar fluid through a porous medium with radiation by dtm-padé. Heat Transf.-Asian Res. **39**(8), 575–589 (2010)
16. Rashidi, M., Mohimanian pour, S., Abbasbandy, S.: Analytic approximate solutions for heat transfer of a micropolar fluid through a porous medium with radiation. Commun. Nonlinear Sci. Numer. Simul. **16**(4), 1874–1889 (2011)
17. Shah, Q., Gul, T., Mamat, M.B., Khan, W., Tofany, N.: Soret and dufour effect on the thin film flow over an unsteady stretching surface. AIP Conf. Proc. **1775**(1), 030088 (2016)
18. Tripathy, R., Dash, G., Mishra, S., Hoque, M.M.: Numerical analysis of hydromagnetic micropolar fluid along a stretching sheet embedded in porous medium with non-uniform heat source and chemical reaction. Eng. Sci. Technol. Int. J. **19**(3), 1573–1581 (2016)
19. Usman, A.H., Shah, Z., Humphries, U.W., Kumam, P., Thounthong, P.: Soret, dufour, and activation energy effects on double diffusive convective couple stress micropolar nanofluid flow in a hall mhd generator system. AIP Adv. **10**(7), 075010 (2020)

Effects of Slip Velocity and Bed Absorption on Transport Coefficient in a Wetland Flow

Debabrata Das, Subham Dhar, Nanda Poddar, Rishi Raj Kairi, and Kajal Kumar Mondal

Abstract In this research work, an investigation on solute transport in a width independent wetland flow with the appearance of the vegetation, bed absorption and slip velocity is explored. The equations of moment are formed from the governing convection-diffusion equation with the help of method of moments. A finite difference implicit technique is imposed to find the solution of the resultant moment equations. The behavior of the dispersion coefficient and skewness is analyzed graphically for the various values of slip parameter, inhomogeneous reaction at the bed surface and for vegetation parameter in the wetland. It is found that the slip velocity reduces the dispersion of the tracers in the wetland flow. The uplifted values of slip parameter, inhomogeneous reaction rate and vegetation factor shorten the critical time to reach the stationary state of the transport coefficient. The slip effect reduces the asymmetry of solute distribution, while the enlarged absorption introduces asymmetry in the tracer distribution.

Keywords Wetland flow · Slip velocity · Bed absorption · Moment method · Dispersivity

1 Introduction

Wetland caused by the flood of the water permanently or seasonally got remarkable attention in last two decades for its huge application in different ecosystems. Wetlands are benefiting human society in different ways, such as through irrigation, flood control, water purification, ecological restoration and biodiversity conservation etc. Also, its play a vital role for the supply of water, climate regulation as well as for contaminant degradation. Aris [1] was the first who introduce the method of moments to

Supported by CSIR INDIA and UGC INDIA.

D. Das (✉) · S. Dhar · N. Poddar · R. R. Kairi · K. K. Mondal
Cooch Behar Panchanan Barma University, Cooch Behar 736101, India
e-mail: debabratadas359548@gmail.com

S. Banerjee and A. Saha (eds.), *Nonlinear Dynamics and Applications*,
Springer Proceedings in Complexity,
https://doi.org/10.1007/978-3-030-99792-2_37

observed the dispersion of the substance in fluid flowing through a tube. The erosion of average mean concentration due to the environmental dispersion was obtained by Zeng et al. [2] with the help of Aris's moment method. Later, to observe the variation of solute concentration in a width dominated flow, Wu et al. [3] applied the multi-scale technique. Zeng et al. [4] discussed a theoretical analysis for the degeneration of depth-averaged concentration with the combined effect of hydraulic dispersion and ecological degradation. To examine the behavior of the vertical distribution of solute concentration in a wetland flow, Wu et al. [5] used mean concentration expansion technique. In this direction, the transport of contaminant in a width independent wetland flow with the presence of absorption and bulk degradation was analyzed by Wang and Chen [6]. They showed that solute cloud becomes more shrunk and reformed with the enhancement of damping factor in presence of the bed absorption. The homogenization method was employed by Wang et al. [7] to find the solution for spatial concentration distributions of solute and reported that non-uniformity occurs in vertical concentration profiles for vegetation shear. In recent time, Dhar et al. [8] investigated environmental dispersion of settling particles numerically and revealed the effects of some important parameters on sediment transport. Further, Poddar et al. [9] introduced integral moment method to notice the dispersion phenomena in a time independent wetland flow and employed Hermite polynomial representation to determine the mean concentration distribution. Ng [10] showed the influence of wall slippage on hydrodynamic dispersion in a parallel-plate channel and observed that keeping slip length constant in the channel, the mixing of a contaminant cloud reduces with the boundary slip. Munoz et al. [11] employed homogenization technique to study the slippage effect on hydrodynamic dispersion of a passive tracers in a oscillatory electro-osmotic flow.

In the present investigation, the main focus is to explore the effect of slip velocity in a width independent wetland flow. A finite difference technique is employed to find the solution of the moment equations. Further, the behavior on dispersion coefficient and skewness for the values of slip parameter, inhomogeneous reaction and vegetation parameter are analyzed. The novelty of the present research is to observe the slip effect on environmental dispersion which has a wide range of application.

2 Mathematical Formulation

In general, near the phase average scale, the basic momentum equation for a wetland flow can be considered as

$$\rho\left(\frac{\partial \mathbf{U}}{\partial \tau}+\nabla\cdot\frac{\mathbf{U}\mathbf{U}}{\phi}\right)=-\nabla p-\mu F\mathbf{U}+\kappa\mu\nabla^2\mathbf{U}+\kappa\nabla\cdot(\mathbf{L}\cdot\nabla\mathbf{U}), \tag{1}$$

where U, ρ, ϕ, μ, τ, p, κ, F are the velocity, density, porosity, dynamic viscosity, time, pressure, tortuosity, shear factor respectively and L is taken as momentum dispersivity. For the shallow wetland flow, the momentum equation is expressed

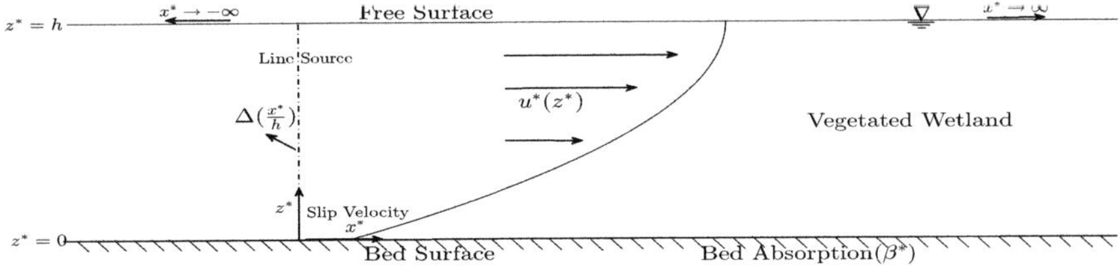

Fig. 1 Schematic diagram for the present model

with the combination of Navier-Stokes equation (single-phase flow), Darcy's law (porous media) and a second order derivative term for the momentum dispersion.

Here, the contaminant transport phenomena is considered in a fully exhibited unidirectional flow, where ϕ, κ, F, and L are constants in the wetland flow of height h. In the above discussed flow, with the help of Cartesian coordinate system, x^*-axis and z^*-axis are taken along the longitudinal and transversal direction of the flow, respectively. Where the origin is considered at one of the channel banks (See Fig. 1). For a width independent wetland flow, width effects are not significant and the concerned quantities are assumed as width-averaged. Equation 1 is transformed due to the pressure gradient as

$$\kappa(\mu + L_z)\frac{d^2u^*}{dz^{*2}} - \mu F u^* = \frac{dp}{dx^*}, \tag{2}$$

where ρ is the fluid density. For the wetland flow, the slip condition at bed surface ($z^* = 0$) with slip length L_s and the no flux boundary condition at the free surface ($z^* = h$) are respectively as,

$$\left[L_s\frac{du^*(z^*)}{dz^*} - u^*(z^*)\right]_{z^*=0} = \left[\frac{du^*(z^*)}{dz^*}\right]_{z^*=h} = 0. \tag{3}$$

The characteristic velocity is defined as $u_c = -\frac{dp}{dx}\frac{W^2}{\mu+L_z}$. Introducing the dimensionless parameters $\zeta = \frac{z^*}{h}$, $l_s = \frac{L_s}{h}$, $u = \frac{u^*}{u_c}$, the general solution of Eq. (2) with the help of boundary conditions (3) becomes,

$$u(z) = \frac{1}{\alpha^2}\left(1 - \frac{\cosh\{\alpha(z-1)\}}{l\alpha\sinh\alpha + \cosh\alpha}\right), \tag{4}$$

where $\alpha \left(= \sqrt{\frac{\mu F h^2}{\kappa(\mu+L_Z)}}\right)$ is the vegetation parameter which illustrates the amalgamated action of the depth of the channel wetland, fluid viscosity, effective vegetation force, tortuosity of the flow and the momentum dispersivity towards the vertical direction. The space variables x^* and dispersion time t^* is normalized as follows,

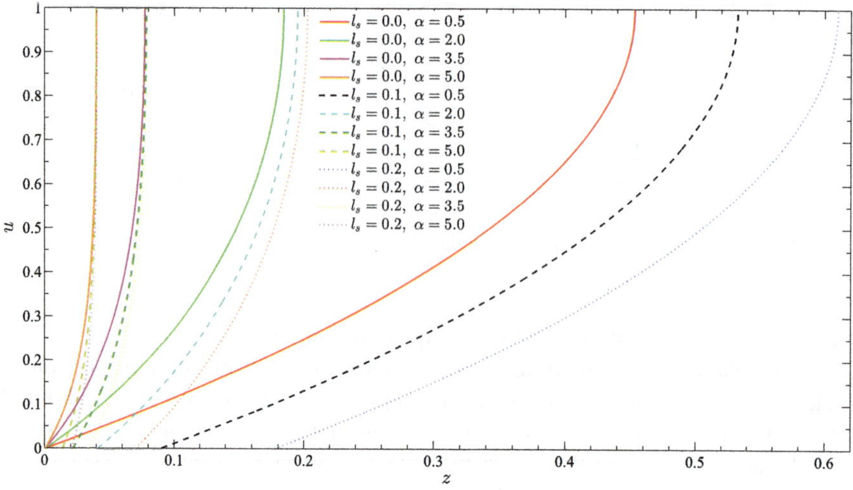

Fig. 2 Velocity profile for different values of vegetation and slip parameters

$$x = \frac{x^*}{h},\ t = \frac{t^* h^2}{\kappa(\lambda + K_z/\phi)}.$$

The velocity distribution in a wetland flow due to different values of vegetation parameter and slip length is shown in Fig. 2. It is noticed that with the increment of vegetation α, velocity deceases and for high vegetation, velocity profiles behave like a log-law profile. Moreover, when slip velocity is present in the flow, it is seen that, near the bed surface, velocity profile slips from the initial point towards the longitudinal direction. It is fascinated to note that as vegetation increases in the wetland, the slip effect diminishes. Whereas for high vegetation, the effects of slip velocity is negligible near the free surface.

When a solute with mass concentration diffusivity λ is injected in the above mentioned flow, the concentration $C^*(x^*, z^*, t^*)$ satisfies the following advection-diffusion equation,

$$\frac{\partial C^*}{\partial t^*} + \frac{u^*}{\phi}\frac{\partial C^*}{\partial x^*} = \kappa\left(\lambda + \frac{K_x}{\phi}\right)\frac{\partial^2 C^*}{\partial x^{*2}} + \kappa\left(\lambda + \frac{K_z}{\phi}\right)\frac{\partial^2 C^*}{\partial z^{*2}}, \tag{5}$$

where K_x and K_z are the constant value known as longitudinal and transversal mass dispersivity. The non-dimensional concentration is $C = \frac{C^*}{Q_m/\phi h}$, where Q_m is the amount of instantaneous released mass at the cross section of $x^* = 0$ and time $t = 0$. The governing Eq. (5) becomes

$$\frac{\partial C}{\partial t} + Peu\frac{\partial C}{\partial x} = R\frac{\partial^2 C}{\partial x^2} + \frac{\partial^2 C}{\partial z^2}, \tag{6}$$

where $Pe = \frac{u_c h}{\kappa(\lambda\phi+K_z)}$ and $R = \frac{\phi\lambda+K_x}{\phi\lambda+K_z}$. Here, Peclet number, Pe is the ratio of the characteristic time for diffusion process $\left(\frac{\phi h^2}{\kappa(\phi\lambda+K_z)}\right)$ to the time of the convection process $\left(\frac{h}{u_c}\right)$, and R describes the ratio of total mass dispersivity and diffusivity along the longitudinal and transversal direction. The boundary conditions are given by

$$\left[\kappa\left(\lambda+\frac{K_z}{\phi}\right)\frac{\partial C}{\partial z^*}-\beta^* C^*\right]_{z^*=0}=\left[\kappa\left(\lambda+\frac{K_z}{\phi}\right)\frac{\partial C}{\partial z^*}\right]_{z^*=h}=0, \tag{7}$$

and their dimensionless forms are,

$$\left[\frac{\partial C}{\partial z}-\beta C\right]_{z=0}=\left[\frac{\partial C}{\partial z}\right]_{z=1}=0, \tag{8}$$

The corresponding dimensionless initial input condition is

$$C(x,z,t)|_{t=0}=\delta(x), \tag{9}$$

where δ is the Dirac delta function. Since, the concentration of the solute particles cannot move at infinity, the normalized additional condition can be taken as

$$C(x,z,t)|_{x\to\pm\infty}=0. \tag{10}$$

Using p-th order integral moment of $C(x, z, t)$ with respect to x according to the method of moments (Aris [1]), we have

$$C_p(z,t)=\int_{-\infty}^{\infty} x^p C(x,z,t)dx. \tag{11}$$

Taking depth average of $C_p(z, t)$, one can obtain

$$M_p(t)=\overline{C}_p=\int_0^1 C_p(z,t)dz. \tag{12}$$

Using (11)–(12) in Eq. (6), with the initial condition (10) and boundary conditions (8), one can get

$$\frac{\partial C_p}{\partial t}=puPeC_{p-1}+Rp(p-1)C_{p-2}+\frac{\partial^2 C_p}{\partial z^2}, \tag{13}$$

with

$$C_p(z,0)=\begin{cases}1 & \text{for } p=0\\ 0 & \text{for } p>0\end{cases}, \tag{14}$$

and

$$\left[\frac{\partial C_p}{\partial z} - \beta C_p\right]_{z=0} = \left[\frac{\partial C_p}{\partial z}\right]_{z=1} = 0, \tag{15a}$$

and therefore,

$$\frac{dM_p}{dt} = pPe\overline{uC}_{p-1} + p(p-1)R\overline{C}_{p-2} - \beta C_p(0,t), \tag{16}$$

with

$$M_p(0) = \begin{cases} 1 & \text{if p} = 0 \\ 0 & \text{if } p > 0 \end{cases}, \tag{17}$$

for $p = 0$, it is found that

$$M_0(t) = 1 - \beta \int C_p(0,t)dt. \tag{18}$$

3 Description of Statistical Components

The p-th order integral moment about the distribution of mean concentration is as follows

$$\nu_p(t) = \frac{1}{M_0}\int_0^1 \int_{-\infty}^{\infty} (x - x_g)^p C dx dz, \tag{19}$$

where

$$x_g = \frac{1}{M_0}\int_0^1 \int_{-\infty}^{\infty} x C dx dz. \tag{20}$$

Here, M_0 defines the total amount of mass of the solute in the whole volume of the wetland. x_g is the centroid of the tracers, it determines the location of the center of gravity of the movement of concentration cloud with the mean velocity of the fluid. Also, variance (ν_2), skewness (ν_3) and kurtosis (ν_4) are represented by the second, third and fourth moments of contaminant distribution, respectively.

Taylor's approximation for the diffusion coefficient $D_a = Pe/192 + 1/Pe$ is developed by Aris [1], where the dispersion time is characterized asymptotically large. In compare to the vertical diffusion, longitudinal diffusion is insignificant and so, the effective transport coefficient may be express as

$$D_a = \frac{1}{2Pe^2}\frac{d\nu_2}{dt}, \tag{21}$$

D_a describes the relation between diffusion and convection process. For a wetland flow, D_a depends on the absorption parameter β, vegetation parameter α and dispersion time t.

To determine the sharpness and symmetricity of the concentration distribution of the solute, the coefficients of skewness and kurtosis are the important criteria. The skewness and kurtosis are given respectively as

$$\beta_2 = \frac{\nu_3}{\nu_2^{\frac{3}{2}}}, \tag{22a}$$

$$\beta_3 = \frac{\nu_4}{\nu_2^2} - 3. \tag{22b}$$

When $\beta_2 = \beta_3 = 0$, the concentration distribution of solute is normal and their nonzero values indicate the deviations from the Gaussianity.

4 Numerical Computation

When $p > 1$, the analytical solution of moment equations with concerned initial and boundary conditions is quite complicated. Thus, using a finite difference implicit scheme based on Crank-Nicholson approach is employed to study the transport of solute particles in the wetland. The grid point (m, n) indicates a point where $z_m = (m-1) \times \Delta z$ and $t_n = (n-1) \times \Delta t$. The enhancements of z and t are $\Delta z = z_{m+1} - z_m$ and $\Delta t = t_{n+1} - t_n$ respectively. All the elements are determined at the grid point $(m, n+1)$, where $m = 1$ relates the bottom of the wetland $z = 0$ and $n = 1$ relates the dispersion time $t = 0$. The two point forward difference technique is adopted for $\frac{\partial C_p}{\partial t}$ at the $(n+1)$-th mesh point

$$\frac{\partial C_p}{\partial t}\Big|_{(m,n+1)} = \frac{C_p(m, n+1) - C_p(m, n)}{\Delta t}. \tag{23}$$

To discretize $\frac{\partial C_p}{\partial z}$, two point averaged central difference technique is used and it is given by

$$\frac{\partial C_p}{\partial z}\Big|_{(m,n+1)} = \frac{C_p(m+1, n+1) - C_p(m-1, n+1) + C_p(m+1, n) - C_p(m-1, n)}{4\Delta z}. \tag{24}$$

The discretization for $\frac{\partial^2 C_p}{\partial z^2}$ is taken as three-point averaged central difference scheme and it is written as

$$\frac{\partial^2 C_p}{\partial z^2}\Big|_{(m,n+1)} = \frac{1}{2\Delta z^2}\big[C_p(m+1,n+1) - 2C_p(m,n+1) + C_p(m-1,n+1) + C_p(m+1,n) - 2C_p(m,n) + C_p(m-1,n)\big]. \tag{25}$$

For the term uC_{p-1}, the discretization is as follows,

$$uC_{p-1}|_{(m,n+1)} = u(m,n+1)C_{p-1}(m,n+1). \tag{26}$$

The finite difference schemes for the initial condition is

$$C_p(m,1) = \begin{cases} 1 \text{ for } p = 0, \\ 0 \text{ for } p \geqslant 1 \end{cases}, \tag{27}$$

and the boundary condition are

$$C_p(0,n+1) = C_p(2,n+1) - 2\beta\Delta z C_p(1,n+1), \quad \text{at the bed surface} \tag{28a}$$

$$C_p(M+1,n+1) = C_p(M-1,n+1). \quad \text{at the free surface} \tag{28b}$$

A tri-diagonal matrix is obtained by using the the above mentioned discretization and it is given by

$$P_m C_p(m+1,n+1) + Q_m C_p(m,n+1) + R_m C_p(m-1,n+1) = S_m, \tag{29}$$

where P_m, Q_m, R_m and S_m are the matrix elements and they are describes as follow

$$\left.\begin{aligned} P_m &= -\frac{\Delta t}{2\Delta z^2}, \\ Q_m &= \left[1 + \frac{\Delta t}{\Delta z^2}\right], \\ R_m &= -\frac{\Delta t}{2\Delta z^2}, \\ S_m &= \left[\frac{\Delta t}{2\Delta z^2}\right] C_p(m+1,n) + \left[1 - \frac{\Delta t}{\Delta z^2}\right] C_p(m,n) + \left[2\frac{\Delta t}{\Delta z^2}\right] C_p(m-1,n) \\ &\quad + \left[puPeC_{p-1}(m,n+1) + p(p-1)RC_{p-2}(m,n+1)\right]\Delta t. \end{aligned}\right\} \tag{30}$$

The index m relates the wetland space whose range is from the value 1 to M. At the wetland free surface, the value of m is considered as M. To obtain a optimal result, in a steady flow through a wetland, the time grid and the space grid size are taken as $\Delta t = 0.0001$ and $\Delta z = 0.02$. The mesh size for steady flow is taken as $\Delta t = 0.0001$. The index related to the time is n and it is expanded from 1 to N. The tri-diagonal coefficient matrix is constructed and solved with the help of initial and boundary condition applying the Thomas algorithm. Simpson's one-third formula is used to compute the values of M_p with the known values of $u(z)$ and C_p at the corresponding mesh points.

5 Discussion of Results

In this section, the temporal variations of dispersion and skewness are illustrated for various values of slip parameter (l_s), absorption parameter (β), vegetation factor (α) and R. From Fig. 3a, it is seen that the increment in the slip length minimizes the dispersion in the flow. Slip on the bed surface reduces the velocity gradient along the transverse direction of the flow and therefore, it abates the dispersion process of the solute. Figure 3b is described that the presence of bed absorption declines the solute dispersion. Further, dispersion coefficient prominently reduces with enlarge values of the bed absorption strength, because enhanced value of β, increases the depletion rate of the solute material at the bed surface. Also, due to the increase of vegetation force in the wetland, reduction in the dispersion coefficient is found (Fig. 3c). This is because of the resisting force in the flow, imposed by the vegetation in the wetland. Moreover, it is interesting to observe from the Fig. 3a–c that the critical time for reaching to the stationary state decreases with the rise in the values of l_s, β, α. However, the opposite phenomena is noticed in Fig. 3d i.e., the coefficient of dispersion of the solute enhances with the increment of R. The reason behind this, the longitudinal diffusion increases as R enlarges.

The influence of different crucial flow parameters on skewness is depicted in Fig. 4. The variation of skewness of the distribution of solute material is shown in Fig. 4a for different slip length. The increment in slip parameters result skewness coefficient to move towards zero i.e. it reduces the asymmetry of the solute distribution and moves towards the Gaussianity. Whereas, the reverse phenomena is found in Fig. 4b for various values of the reaction parameter β. It illustrates that enhanced value of absorption parameter introduces asymmetry in the tracer distribution. From Fig. 4c, it is observed that the enhancement of α, the distribution of tracers gradually becomes symmetric. Moreover, when $\alpha \to 2$, the skewness of the concentration vanishes and it follows Gaussian distribution. It is remarkable to note that the effect of R is similar to the slip parameter l_s (See Fig. 4d).

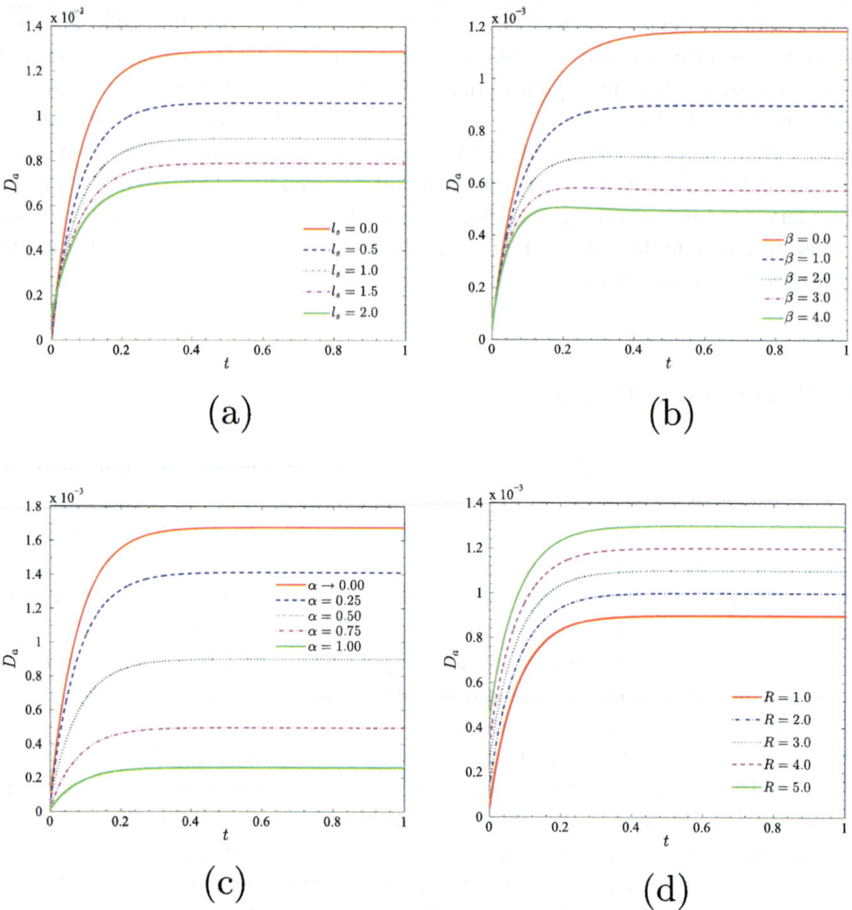

Fig. 3 Dispersivity D_a with time for different values of **a** slip parameter l_s, **b** bed absorption β, **c** vegetation factor α **d** R; when $\alpha = 0.5$, $l_s = 1$, $\beta = 1$, $P_e = 100$ and $R = 1$

The variation of the kurtosis is delineated in Fig. 5. The kurtosis of the tracer distribution is presented in Fig. 5a for various values of slip length. Due to the enhancement in slip parameter, the kurtosis of the tracer distribution moves toward zero for large dispersion time. The similar behaviour is observed in the kurtosis for various values of β and R (See Fig. 5b, d). From the figures, it is concluded that for small dispersion time ($t < 0.2$), the distribution of the tracer material deviates from the Gaussianity and for large dispersion time with large l_s, β, and R, the distribution tends to reach the normal distribution. From Fig. 5c, it is seen that, when $\alpha \to 0$, the distribution approaches towards Gaussian distribution. Moreover, when α increases in the wetland, the flow of the tracer decreases, which increases the peak of the concentration.

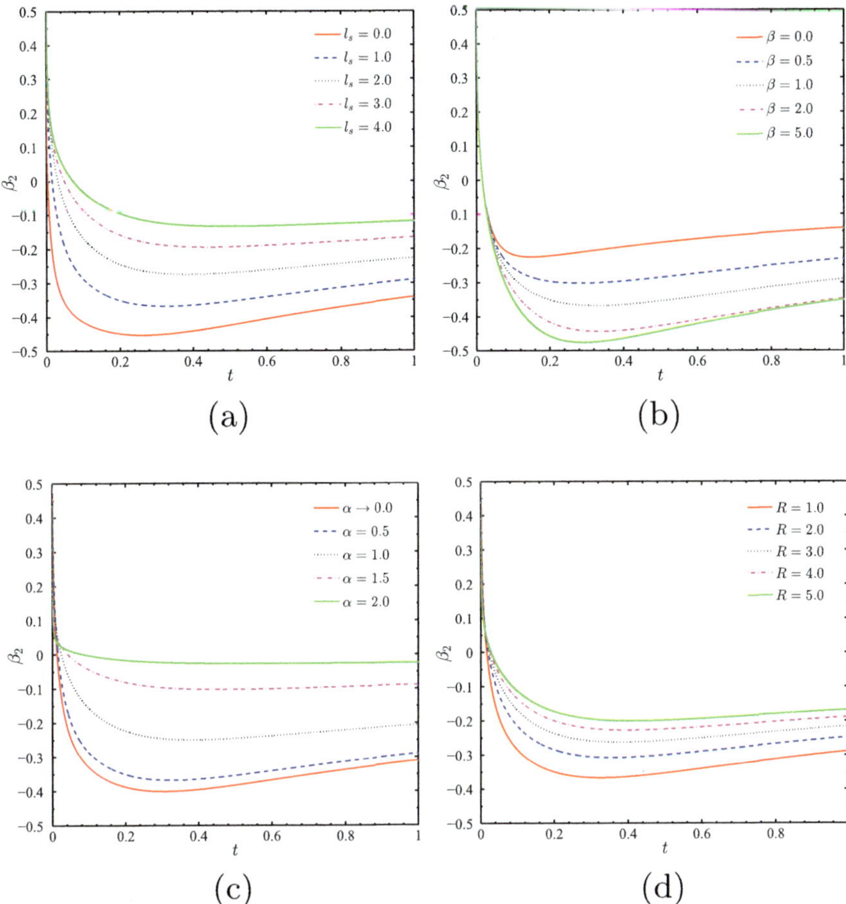

Fig. 4 Skewness β_2 of solute distribution with time for different values of **a** slip parameter l_s, **b** bed absorption β, **c** vegetation factor α **d** R; when $\alpha = 0.5$, $l_s = 1$, $\beta = 1$, $P_e = 100$ and $R = 1$

6 Conclusion

An investigation on solute transport in a width independent vegetated wetland channel flow is presented with the appearance of slip condition and absorption at the bed surface. A finite difference implicit method is used to find the solution of the moment equations. Some crucial observations are noticed in this work:

(a) The Enhancement of slip parameter reduces the dispersion coefficient in the flow.
(b) The uplifted values of slip parameter, inhomogeneous reaction rate and vegetation factor shorten the critical time to reach the stationary state of the transport coefficient.

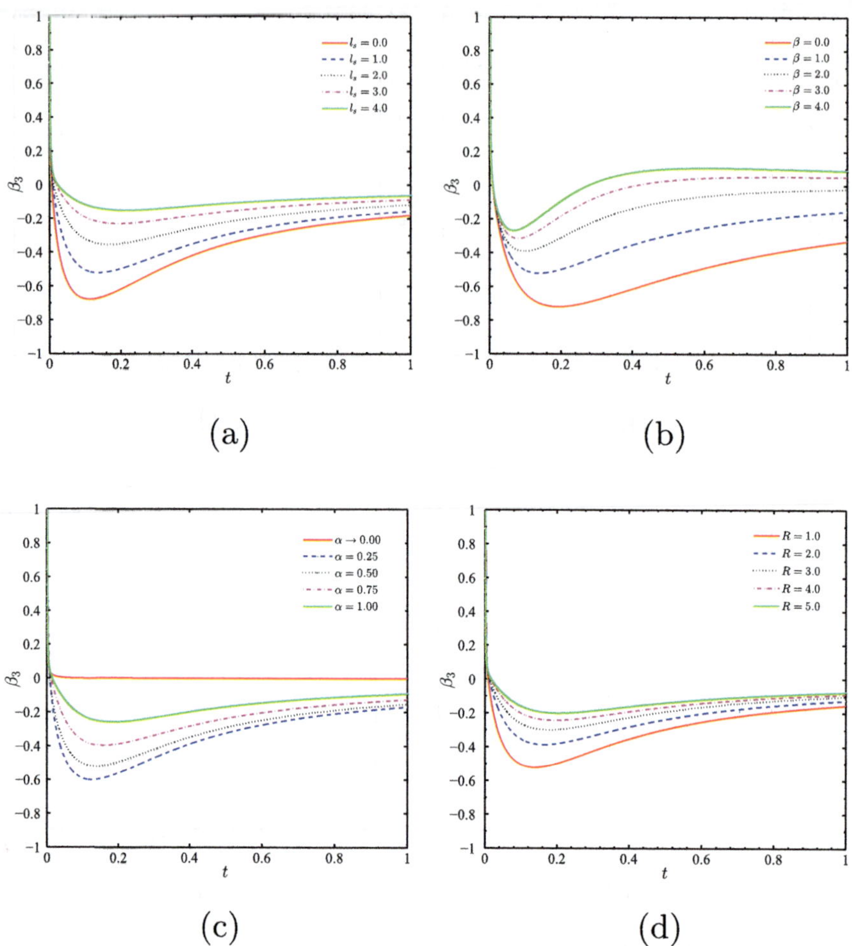

Fig. 5 Kurtosis β_3 of solute distribution with time for different values of **a** slip parameter l_s, **b** bed absorption β, **c** vegetation factor α **d** R; when $\alpha = 0.5$, $l_s = 1$, $\beta = 1$, $P_e = 100$ and $R = 1$

(c) The slip effect reduces the asymmetry of solute distribution, while the enlarged absorption introduces asymmetry in the tracer distribution.

(d) At small dispersion time, the distribution of tracer deviates from Gaussianity. However, for large dispersion time with large l_s, β, and R, the distribution tends to reach the normal distribution.

References

1. Aris, R.: On the dispersion of a solute in a fluid flowing through a tube. Proc. R. Soc. Lond. **235**, 67–77 (1956)
2. Zeng, L., Chen, G.Q., Tang, H.S., Wu, Z.: Environmental dispersion in wetland flow. Commun. Nonlinear Sci. Numer. Simulat. **16**, 206–215 (2010)
3. Wu, Z., Li, Z., Chen, G.Q.: Multi-scale analysis for environmental dispersion in wetland flow. Commun. Nonlinear Sci. Numer. Simulat. **16**(8), 3168–3178 (2011)
4. Zeng, L., Chen, G.Q.: Ecological degradation and hydraulic dispersion of contaminant in wetland. Ecol. Modell. **222**, 293–300 (2011)
5. Wu, Z., Fu, X., Wang, G.: Concentration distribution of contaminant transport in wetland flows. J. Hydrol. **525**, 335–344 (2015)
6. Wang, P., Chen, G.Q.: Contaminant transport in wetland flows with bulk degradation and bed absorption. J. Hydrol. **552**, 674–683 (2017)
7. Wang, H., Zhu, Z., Li, S., Huai, W.: Solute dispersion in wetland flows with bed absorption. J. Hydrol., pp. 124–149 (2019)
8. Dhar, S., Poddar, N., Kairi, R.R., Mazumder, B.S., Mondal, K.K.: Numerical study on dispersion of fine settling particles in a depth dominated wetland flow. Commun. Nonliear Sci. Numer. Simulat. **96**, 105707 (2021)
9. Poddar, N., Das, S., Dhar, S., Mondal, K.K.: Semi-analytical study on environmental dispersion of settling particles in a width-independent wetland flow. Environ. Fluid Mech. **21**(4), 1–22 (2021)
10. Ng, C.O.: How does wall slippage affect hydrodynamic dispersion? Microfluidics Nanofluidics **10**(1), 47–57 (2010)
11. Munoz, J., Arcos, J., Bautista, O., Mendez, F.: Slippage effect on the dispersion coefficient of a passive solute in a pulsatile electro-osmotic flow in a microcapillary. Phys. Rev. Fluids **3**(8) (2018)

Entropy Analysis for MHD Flow Subject to Temperature-Dependent Viscosity and Thermal Conductivity

Umesh Khanduri and Bhupendra K. Sharma

Abstract This research aimed to figure out how to optimise the entropy of MHD flow past a continuously stretching surface. The effect of temperature-dependent variables viscosity and electric conductivity has been taken into account. The fluid region is subjected to a uniform magnetic field. By using similarity analysis, the governing coupled partial differential equations (PDEs) that describe the model are turned into non-linear ordinary differential equations and then computed by employing "BVP4C" in MATLAB software. The effect of various pertinent parameters like Magnetic field parameter M, radiation parameter R, Grashof number Gr, Brinkman number Br, Reynold number Re, and a variation of variables viscosity ϵ_1 and electric conductivity ϵ_2 is analysed and presented graphically on velocity, temperature, entropy, and concentration profile. The comparison is based on previously published studies, and there is a considerable deal of agreement.

Keywords Entropy · Variable viscosity · Stretching sheet · MHD · Viscous dissipation

1 Introduction

Numerous researchers and analysts studied the effect of MHD flow across the stretching sheet due to its wide applications in industrial fields such as machine design, magnetic drug targeting, glass blowing, electronic chips and astrophysics sensors. Hayat et al. [1] demonstrated the influence of magnetic field and porous media on the flow velocity and heat transfer profile under the exponential stretching sheet condition.

The first author's work is supported by CSIR, India, which provides the Ph.D. fellowship and CSIR file No. is 09/719(0105)/2019-EMR-I.

U. Khanduri (✉) · B. K. Sharma
Department of Mathematics, Birla Institute of Technology and Science, Pilani, Rajasthan, India
e-mail: umeshkhanduri09@gmail.com
URL: https://www.bits-pilani.ac.in

S. Banerjee and A. Saha (eds.), *Nonlinear Dynamics and Applications*,
Springer Proceedings in Complexity,
https://doi.org/10.1007/978-3-030-99792-2_38

This study also considered the velocity and thermal slip conditions and performed it by adopting the "Homotopy analysis method". Further, motivated by the applications of boundary layer flow, Sandeep et al. [2] extended the above study on the thin stretching surface. Reddy et al. [3] considered the non-flat sheet and studied the impact of Williamson nanofluid with variable thermal conductivity. In this investigation, they considered the flow viscosity as temperature-dependent instead of constant. They observed that non-dimensional heat and mass transfer profiles rise significantly due to temperature-dependent viscosity inflow. The impact of heat source, radiation and chemical reaction under the convective boundary conditions has been studied by Ram et al. [4] and state that the magnitude of heat and mass transfer rates at the wall improves with the positive add in thermal and solute Biot numbers. Recently, Sharma et al. [5] studied a mathematical model of blood flow to identify the effects of Ohmic heating on the stretched arterial surface, claiming that heat transfer at the surface diminishes as the Ohmic heating parameter increases. Reddy et al. [6] studied the effect of different parameters on the blood flow inside the permeable wall with heat source and viscous dissipation. Tripathi et al. [7] employed the "Homotopy perturbation method" to study the arterial blood flow using the variable viscosity. They considered mild stenosis and hematocrit-dependent viscosity to analyze the blood flow behaviour. Further, they [8] examined the temperature-dependent viscosity and two-layer fluid model in blood flow with elliptical shaped stenosis.

Entropy generation is one of the fundamental phenomena associated with a level of irreversibility that occurs in any thermal process. Several researchers have examined the effect of entropy generation in MHD flow due to its application in many engineering systems. In order to reduce the irreversibility process, one can employ the second law of thermodynamic for designing better thermal equipment. Bejan [9] first introduced this concept by introducing the Bejan number, which signifies the thermal irreversibility to the total heat loss due to fluid frictions. Bhatti et al. [10] discussed the entropy generation on MHD Casson fluid with radiation and the effect of nanoparticles over the porous surface. Mandal et al. [11] studied the properties of nanoparticles by considering Buongiorno's model and delineated the radiation and viscous dissipation effect on unsteady MHD flow. By considering viscous dissipation and the magnetic field effect, Afridi et al. [12] explored entropy generation in boundary layer flows. They perceived irreversible conduction as the principal source of entropy generation far beyond the boundary. Several other researchers [13, 14] have investigated the effect of Entropy generation on MHD flow.

In the present study, we have studied the effect of temperature-dependent variable viscosity and thermal conductivity on the stretching sheet, including entropy generation. The governing equations are converted into ordinary differential equations and then solved using MATLAB software by implementing "BVP4C". The effect of various pertinent parameters had been analyzed and presented graphically on different profile like velocity, temperature, entropy, and concentration.

2 Mathematical Formulation

An unsteady, incompressible, two-dimensional MHD flow over a time-dependent stretching sheet in the X-Y plane is considered as depicted in Fig. 1. The sheet is aligned at an angle α with the vertical and uniform inclined magnetic field $B(t)$ is applied at an angle ξ to the sheet. At time $t = 0$, the sheet is stretched along the x-axis with velocity $\tilde{U}_w(x,t) = \frac{dx}{(1-\chi t)}$. Here, d and χ are constant with conditions $d > 0$, $\chi \geq 0$ and $\chi t < 1$. The velocity components $\tilde{u_1}$ and $\tilde{v_1}$ are taken along x and y axis. Here, $\tilde{T}$ and $\tilde{T}_\infty$ denotes the temperature and ambient temperature of the fluid, σ is the electric conductivity, β is coefficient of temperature expansion, C_p^* is the specific heat, $\tilde{\beta}^*$ is coefficient of concentration expansion, $\tilde{C}$ and $\tilde{C}_\infty$ denotes the concentration and ambient concentration of the fluid, D_B^* is mass diffusion coefficient, and g represent acceleration due to gravity. Temperature-dependent viscosity $\mu(\tilde{T})$ is assumed to change linearly with temperature as follows:

$$\mu(\tilde{T}) = \mu_\infty \left[1 + \frac{\epsilon_1}{\Delta \tilde{T}} (\tilde{T}_w - \tilde{T}) \right] \tag{1}$$

Here $\Delta \tilde{T} = \tilde{T}_w - \tilde{T}$, ϵ_1 is the temperature difference and viscosity variation parameter respectively. And, temperature-dependent thermal conductivity is given as:

$$K(\tilde{T}) = \kappa_\infty \left[1 - \frac{\epsilon_2}{\Delta \tilde{T}} (\tilde{T} - \tilde{T}_\infty) \right] \tag{2}$$

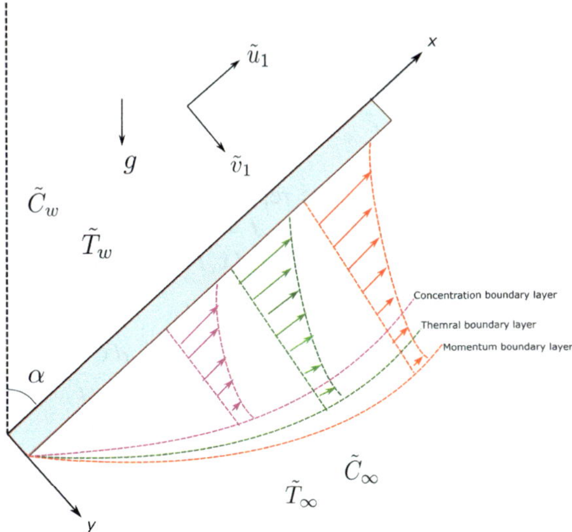

Fig. 1 Pictorial representation of the problem

Here ϵ_2 denotes the thermal conductivity variation parameter. The magnetic Reynold number is assumed to be less than unity, so the induced magnetic field effect is negligible. The fluid properties are considered to be constant. Based on the order of magnitude along with Boussinesq approximation and using the above assumption, the governing equations are:

Continuity:

$$\frac{\partial \tilde{u_1}}{\partial x} + \frac{\partial \tilde{v_1}}{\partial y} = 0 \,, \tag{3}$$

Momentum:

$$\begin{aligned}\frac{\partial \tilde{u_1}}{\partial t} + \tilde{u_1}\frac{\partial \tilde{u_1}}{\partial x} + \tilde{v_1}\frac{\partial \tilde{u_1}}{\partial y} &= g\beta(\tilde{T} - \tilde{T_\infty})\cos\alpha + g\tilde{\beta}^*(\tilde{C} - \tilde{C_\infty})\cos\alpha + \frac{1}{\rho}\frac{\partial}{\partial y}\left(\mu(\tilde{T})\frac{\partial \tilde{u_1}}{\partial y}\right) \\ &\quad - \frac{\sigma B^2(t)}{\rho}\sin^2\xi\tilde{u_1} - \frac{\mu(\tilde{T})}{\rho l_1(t)}\tilde{u_1} \,, \end{aligned} \tag{4}$$

Energy:

$$\frac{\partial \tilde{T}}{\partial t} + \tilde{u_1}\frac{\partial \tilde{T}}{\partial x} + \tilde{v_1}\frac{\partial \tilde{T}}{\partial y} = \frac{1}{\rho C_p^*}\frac{\partial}{\partial y}\left(K(\tilde{T})\frac{\partial \tilde{u_1}}{\partial y}\right) - \frac{1}{\rho C_p^*}\frac{\partial q_r}{\partial y} + \frac{\mu(\tilde{T})}{\rho C_p^*}\left(\frac{\partial \tilde{u_1}}{\partial y}\right)^2 + \frac{\sigma B^2(t)}{\rho C_p^*}\sin^2\xi\tilde{u_1}^2 \,, \tag{5}$$

Concentration:

$$\frac{\partial \tilde{C}}{\partial t} + \tilde{u_1}\frac{\partial \tilde{C}}{\partial x} + \tilde{v_1}\frac{\partial \tilde{C}}{\partial y} = D_B^*\frac{\partial^2 \tilde{C}}{\partial y^2} - \Gamma(t)(\tilde{C} - \tilde{C_\infty}), \tag{6}$$

The associate boundary conditions are:

$$\begin{cases} \tilde{u_1} = \tilde{U_w} + H\mu\frac{\partial \tilde{u_1}}{\partial y}, \;\; \tilde{C} = \tilde{C_w} + J\frac{\partial \tilde{C}}{\partial y}, \;\; \tilde{v_1} = \tilde{V_w}, \;\; \tilde{T} = \tilde{T_w} + P\frac{\partial \tilde{T}}{\partial y}, & \text{at } y = 0 \\ \tilde{u_1} \to 0, \;\; \tilde{C} \to \tilde{C_\infty}, \;\; \tilde{T} \to \tilde{T_\infty} & \text{as } y \to \infty. \end{cases} \tag{7}$$

Where, $H = H_0(1 - \chi t)^{\frac{1}{2}}$ represent the velocity slip factor, $P = P_0(1 - \chi t)^{\frac{1}{2}}$ represent the thermal slip factor, $J = J_0(1 - \chi t)^{\frac{1}{2}}$ represent the concentration slip factor. The surface temperature $\tilde{T_w}(x, t)$, stretching velocity $\tilde{U_w}(x, t)$, and the concentration $\tilde{C_w}(x, t)$ are given as:

$$\tilde{T_w} = \tilde{T_\infty} + \frac{cx}{1 - \chi t}, \qquad \tilde{U_w} = \frac{dx}{1 - \chi t}, \qquad \tilde{C_w} = \tilde{C_\infty} + \frac{ex}{1 - \chi t}.$$

Here, c and e are constant and satisfies the condition $c, e \geq 0$ and $\chi t < 1$. In Eq. (7), $\tilde{V_w}$ represent the injection/suction velocity:

$$\tilde{V_w} = -\sqrt{\frac{\nu \tilde{U_w}}{x}} f(0) \,. \tag{8}$$

The mass transfer at the wall is represented by Eq. (7) with velocity $\tilde{V_w} > 0$ in which $\tilde{V_w} > 0$ represents injection and $\tilde{V_w} < 0$ represents the suction case. The expression $l_1(t)$ used in the Eq. (4) represents the time-dependent permeability and it satisfies the equation $l_1(t) = l_2(1 - \chi t)$. Let us now consider the magnetic field $B(t) = B_0(1 - \chi t)^{-\frac{1}{2}}$ and chemical reaction parameter as $\Gamma(t) = \Gamma_0(1 - \chi t)^{-1}$. Here B_0 represent the magnetic field at initial time, and Γ_0 as constant. In Eq. (5), the radiative heat flux [6] is approximate by Rosseland mean approximation and it is simplified as:

$$q_r = -\frac{4\sigma*}{3k*}\frac{\partial \tilde{T}^4}{\partial y}. \tag{9}$$

Here $k*$ and $\sigma*$ are the mean absorption coefficient and the Stefan-Boltzman constant, respectively. We linearize the term $\tilde{T}^4$ using a Taylor series about the free steam function $\tilde{T_\infty}$, ignoring higher order terms by assuming small temperature differences.

$$\tilde{T}^4 \cong 4\tilde{T}\tilde{T}_\infty^3 - 3\tilde{T}_\infty^4. \tag{10}$$

Now, similarity transformations are defined as:

$$\begin{cases} \eta = d^{\frac{1}{2}}\nu^{-\frac{1}{2}}(1-\chi t)^{-\frac{1}{2}}y, \\ \psi = d^{\frac{1}{2}}x\nu^{\frac{1}{2}}(1-\chi t)^{-\frac{1}{2}}f(\eta), \\ \theta(\eta) = \left(\frac{\tilde{T}-\tilde{T}_\infty}{\tilde{T}_w-\tilde{T}_\infty}\right), \\ \phi(\eta) = \left(\frac{\tilde{C}-\tilde{C}_\infty}{\tilde{C}_w-\tilde{C}_\infty}\right). \end{cases} \tag{11}$$

Here, η represent the independent similarity variable, ψ denotes the stream function and it is defined as $\tilde{u}_1 = \frac{\partial\psi}{\partial y}$, $\tilde{v}_1 = -\frac{\partial\psi}{\partial x}$, which automatically satisfies Eq. (3). Now, substitute Eqs. (10) and (11) into Eqs. (4)–(6),to get differential equations as:

$$(1+e_1(1-\theta))f''' + ff'' - A\left(f' + \frac{1}{2}\eta f''\right) - (f')^2 + Gr\theta\cos\alpha + Gc\phi\cos\alpha - e_1\theta' f'' \\ -M^2 f'\sin^2\xi - (1+e_1(1-\theta))\frac{f'}{K} = 0, \tag{12}$$

$$\frac{(1+R+e_2\theta)}{Pr}\theta'' - A\left(\theta + \frac{1}{2}\eta\theta'\right) + f\theta' - f'\theta + Ec(1+e_1(1-\theta))(f'')^2 + Ec*M^2(f')^2\sin^2\xi \\ + e_2\frac{(\theta')^2}{Pr} = 0, \tag{13}$$

$$\frac{1}{Sc}\phi'' + f\phi' - f'\phi - A\left(\phi + \frac{1}{2}\eta\phi'\right) - \gamma\phi = 0, \tag{14}$$

The associated boundary conditions are:

Table 1 Various dimensionless parameters involved in Eqs. (12)–(14)

$M = B_0\sqrt{\frac{\sigma}{\rho d}}$	$Gr = \frac{g\beta x(\tilde{T}_w - \tilde{T}_\infty)}{\tilde{U_w}^2}$	$A = \frac{\chi}{d}$	$Ec = \frac{\tilde{U_w}^2}{C_p^*(\tilde{T}_w^* - \tilde{T}_\infty)}$
$Gc = \frac{g\tilde{\beta}^* x(\tilde{C}_w - \tilde{C}_\infty)}{\tilde{U_w}^2}$	$Sc = \frac{\nu}{D_B^*}$	$Pr = \frac{\mu_\infty C_p^*}{\kappa_\infty}$	$\gamma = \frac{\Gamma_0}{d}$
$R = \frac{16\sigma^* \tilde{T}_\infty^3}{3\kappa_\infty k^*}$	$\Omega = \frac{\tilde{T}_w - \tilde{T}_\infty}{\tilde{T}_\infty}$	$K = \frac{dl_2}{\nu}$	$Re = \frac{U_w x}{\nu}$
$Br = \frac{\mu_\infty U_w^2}{\kappa_\infty \Delta \tilde{T}}$	$\nu = \frac{\mu_\infty}{\rho}$	$\Lambda = \frac{RD\tilde{C}_\infty}{\kappa}$	$\Psi = \frac{\tilde{C}_w - \tilde{C}_\infty}{\tilde{C}_\infty}$

$$\begin{cases} f = S, \;\; f' = 1 + S_f f''(0), \;\; \theta = 1 + S_t\theta'(0), \phi = 1 + S_c\phi'(0) & \text{at } \eta = 0\,, \\ f' \to 0, \;\; \theta \to 0, \phi \to 0, & \text{as } \eta \to \infty. \end{cases} \tag{15}$$

The non-dimensional thermal slip S_t, velocity slip S_f, and solutal slip S_c parameters are defined as:

$$S_t = P_0\sqrt{\frac{d}{\nu}}, \qquad S_f = H_0\rho\sqrt{d\nu}, \qquad S_c = J_0\sqrt{\frac{d}{\nu}}\,.$$

Here, prime denotes the differentiation with respect to η. In Eqs. (12)–(14), if $A = 0$, it will correspondence that the problem reduce to the steady state flow. Also, in Eq. (15), if $S > 0$, it indicates suction, $S < 0$ indicates injection. The dimensionless numbers and parameter used in Eqs. (12)–(14) are specified in Table 1.

2.1 Quantities of Physical Interest

The quantities like heat transfer rate, skin friction coefficient, and mass transfer are defined as:

$$Nu_x = \frac{xq_w}{K(\tilde{T})(\tilde{T}_w - \tilde{T}_\infty)}, \qquad C_f = \frac{\tau_w}{\frac{\rho\tilde{U_w}^2}{2}}, \qquad Sh_x = \frac{m_w x}{D_B^*\rho(\tilde{C}_w - \tilde{C}_\infty)}. \tag{16}$$

Where the surface heat flux, wall shear stress, and mass flux are given by:

$$\begin{cases} q_w = -K(\tilde{T})\left(\frac{\partial \tilde{T}}{\partial y}\right)_{y=0}, \\ \tau_w = \mu(\tilde{T})\left(\frac{\partial \tilde{u}_1}{\partial y}\right)_{y=0}, \\ m_w = -\rho D_B^*\left(\frac{\partial \tilde{C}}{\partial y}\right)_{y=0}. \end{cases} \tag{17}$$

Use Eq. (17), to rewrite quantity (16) as:

$$Nu_x = -Re_x^{\frac{1}{2}}\theta'(0), \quad C_f = 2(1+\epsilon_1(1-\theta))Re_x^{-\frac{1}{2}} f''(0), \quad Sh_x = -Re_x^{\frac{1}{2}}\phi'(0). \tag{18}$$

The quantity like Nusselt number Nu_x, skin-friction coefficientC_f, and Sherwood number Sh_x are dependent on the variation of the factors $-\theta'(0)$, $f''(0)$, and $-\phi'(0)$ respectively.

3 Entropy

Expression for entropy generation is defined as:

$$S_G''' = \frac{K(\tilde{T})}{\tilde{T}_\infty^2}\left\{\left(\frac{\partial \tilde{T}}{\partial y}\right)^2 + \frac{16\sigma^* \tilde{T}_\infty^3}{3k^* k}\left(\frac{\partial \tilde{T}}{\partial y}\right)^2\right\} + \frac{\mu(\tilde{T})}{\tilde{T}_\infty}\left(\frac{\partial \tilde{u_1}}{\partial y}\right)^2 + \frac{\sigma}{\tilde{T}_\infty}(\tilde{u_1}^2 B^2) + \frac{\mu(\tilde{T})}{\tilde{T}_\infty K_1}\tilde{u_1}^2 + \frac{RD}{\tilde{T}_\infty}\frac{\partial \tilde{T}}{\partial y}\frac{\partial \tilde{C}}{\partial y} \tag{19}$$

Characteristic entropy rate is defined as:

$$S_0''' = \frac{K(\tilde{T})(\Delta\tilde{T})^2}{x^2 \tilde{T}_\infty^2} \tag{20}$$

Dimensionless entropy generation is defined as:

$$N_G = \frac{S_G'''}{S_0'''} \tag{21}$$

Use Eqs. (20)–(21) to write the entropy generation number N_G as:

$$N_G = Re_x\left(1 + \frac{R}{1+\epsilon_2\theta}\right)(\theta')^2 + \frac{ReBr}{\Omega}\left(1+\frac{1}{K}\right)\left(\frac{1+\epsilon_1(1-\theta)}{1+\epsilon_2\theta}\right) + \frac{ReBrM}{\Omega(1+\epsilon_2\theta)}(f')^2 + Re\Psi\frac{\Lambda}{\Omega}\theta'\phi'. \tag{22}$$

Here, Re, Br, and Ω are Reynolds number, Brinkmann number, and temperature difference parameters respectively.

4 Result and Discussion

This section deals with the characteristic behaviours of different relevant parameters on velocity, temperature, entropy, and concentration profile. For numerical results, the values of pertinent parameters are given as: $M = 0.4$, $K = 0.5$, $Gr = 2$, $Pr = 0.71$, $\epsilon_1 = 0.1$, $A = 0.5$, $Gc = 2$, $\xi = \pi/4$, $\alpha = \pi/4$, $\epsilon_2 = 0.1$, $Sc = 0.6$, $Br{=}0.5$, $Re = 0.5$, $S_f = 1.5$, $S_t = 1$, $S = 0.5$, $S_p = 0.5$. Moreover, the values of specific parameters are kept constant throughout the study unless mentioned in the appropriate graphs. Figure 2 depicts the validation of the previous published work for temperature profile at $R = 1.5$ with current work. Reddy et al. [6] used the shooting technique to solve the dimensionless governing equations. And, in the current work, we used BVP4C technique that works on the collocation method.

Figure 3 shows the decreasing nature with both parameters M and ϵ_1. As the magnetic field parameter M increases from 0.5 to 3, the velocity profile $f'(\eta)$ decreases; this has happened due to an enhancement of Lorentz force that decreases the fluid velocity. It depicts the significant phenomena used in the medical field to regulate human blood by varying magnetic field. Figure 4 depicts the velocity profile with varying parameter Gr and ϵ_1. From the figure, it can be interpreted that the velocity profile $f'(\eta)$ will increase if the parameter Gr increases while the reverse trend is observed with ϵ_1. The velocity profile is shown in Fig. 5 for various values of the inclination angle α and viscosity parameter ϵ_1. The velocity profile $f'(\eta)$ decreases with an increase in both the parameters α and viscosity parameter ϵ_1. At an angle $\alpha = 0$, there is a 17.80% decrement in the velocity profile for a change in viscosity parameter ϵ_1 from 0 to 1. The decrement in velocity profile with varying α can be explained by the fact that changing the inclination angle ($\alpha = 0, \pi/4, \pi/3$) results in a reduction of the term $\cos\alpha$ in the momentum equation, which lowers the buoyancy force due to thermal diffusion. The temperature profile increases with an increase in the parameter R and thermal conductivity ϵ_2. An increment in thermal conductivity

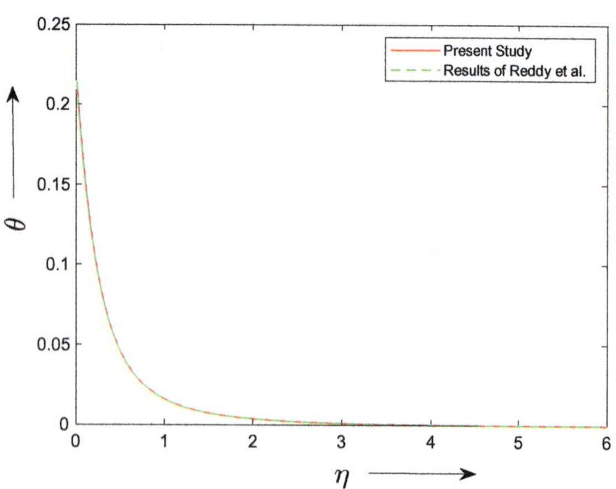

Fig. 2 Comparison on Temperature profile for $R = 1.5$

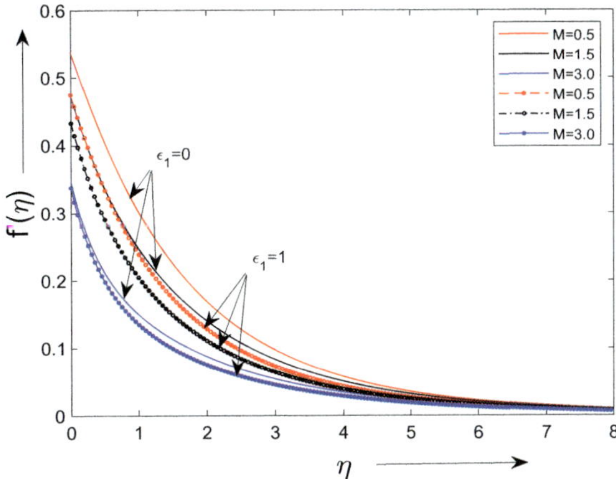

Fig. 3 Velocity profile $f'(\eta)$ by varying M

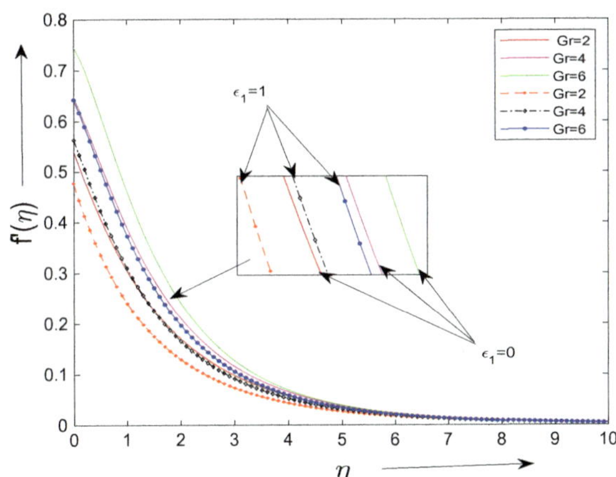

Fig. 4 Non-dimensional Velocity profile by varying Gr

signifies the faster rate at which the heat will be transferred in the fluid. It can be noted from the Fig. 6 that the temperature profile enhances as the radiation parameter varies from $R = 0$ to $R = 1.5$.

In Fig. 7, we analyzed the change in temperature profile by varying parameters ϵ_2 and Ec. Here, Ec (0.5, 1.5, 3) is positive, which signifies that the heat is transferred from the stretching sheet to the fluid. Eckert number can be characterised as the self-heating of the fluid because dissipation effect enhances the thermal boundary layer. Increasing both parameters ϵ_2 and Ec will raise the temperature profile. As Prandtl number Pr varies from 2 to 6, the fluid's thermal conductivity decreases, reducing the thermal boundary layer. Similarly, the temperature profile increases with an increase in thermal conductivity parameter, ϵ_2 from 0 to 1. Higher thermal conductivity means that the fluid will dissipate heat more rapidly. It signifies that the temperature profile decreases with an increase in the parameter Pr while the reverse trend is observed for thermal conductivity parameter ϵ_2 as depicted in Fig. 8. It is noted from Fig. 9

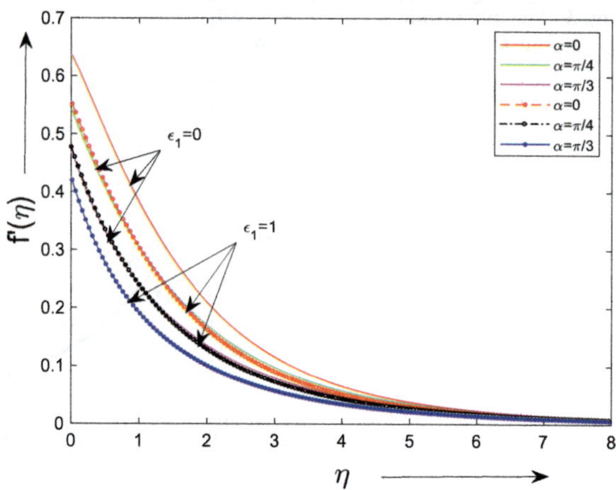

Fig. 5 Non-dimensional Velocity profile by varying α

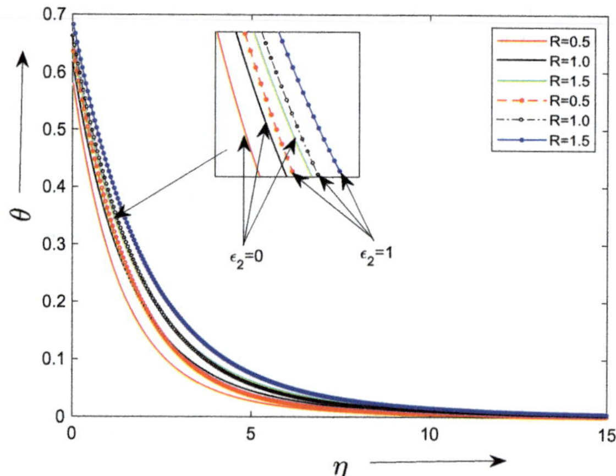

Fig. 6 Variation in temperature profile by varying R

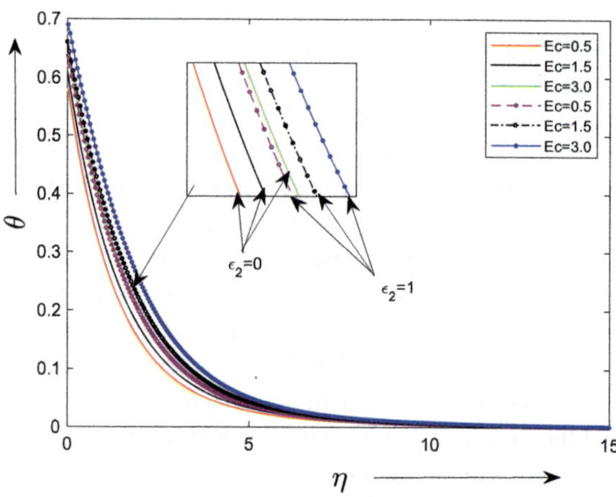

Fig. 7 Variation in temperature profile by varying Ec

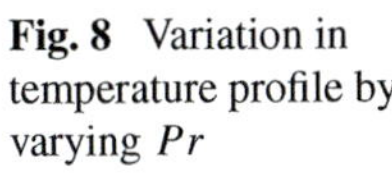

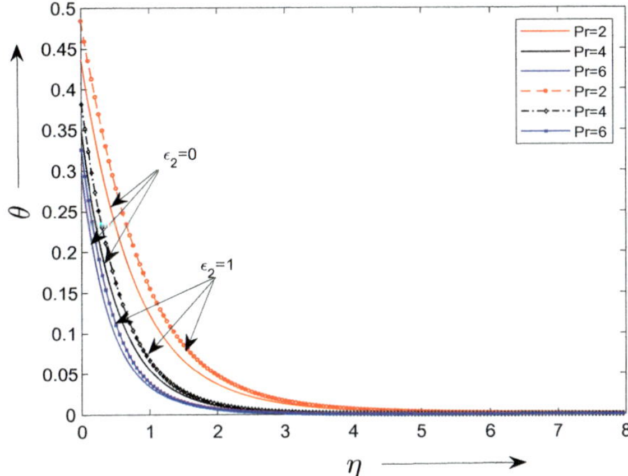

Fig. 8 Variation in temperature profile by varying Pr

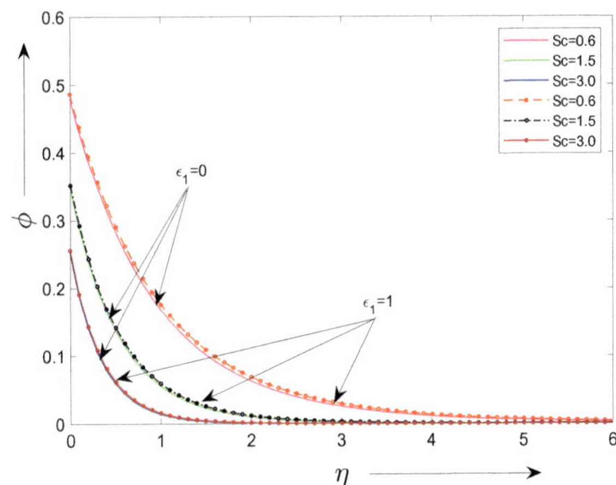

Fig. 9 Variation in concentration profile by varying Sc

that the concentration profile decreases with an increase in parameter Sc. Although, little variation is observed with a change in viscosity parameter ϵ_1 from 0 to 1. As the Schmidt number is inversely proportional to the diffusion coefficient, it declines the solutal boundary layer. Figures 10, 11, 12 and 13 illustrate the effect of different parameters on entropy number N_G.

Entropy determines the possible ways of energy distribution in a system of molecules. In other words, the measurable physical quantity that is associated with the disordered movement of particles. The knowledge of entropy optimization helps in enhancing the mechanical device's performance. Therefore, it is essential to know the effect of different parameters associated with entropy change. Figure 10 shows the variation of entropy number N_G with variation of Br and η. Entropy increases with an increase in parameter Br but decreases asymptotically for η. Brinkman number is one parameter that correlates with the system's convective heat transfer in laminar and transient flow. Its application can be seen in the microchannels flow, such as

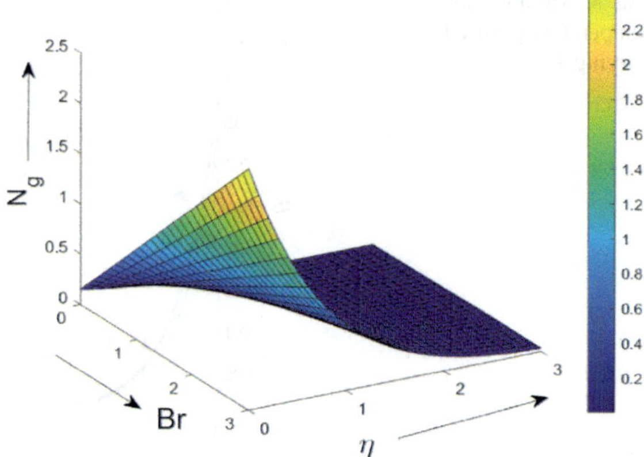

Fig. 10 Surface plot of N_G with axial direction η and Br

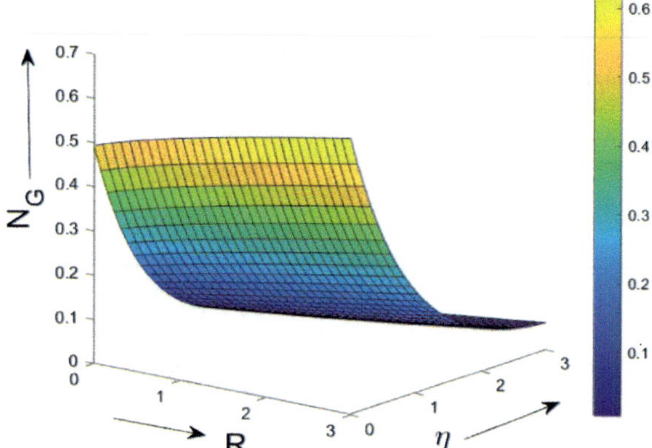

Fig. 11 Surface plot of N_G with axial direction η and R

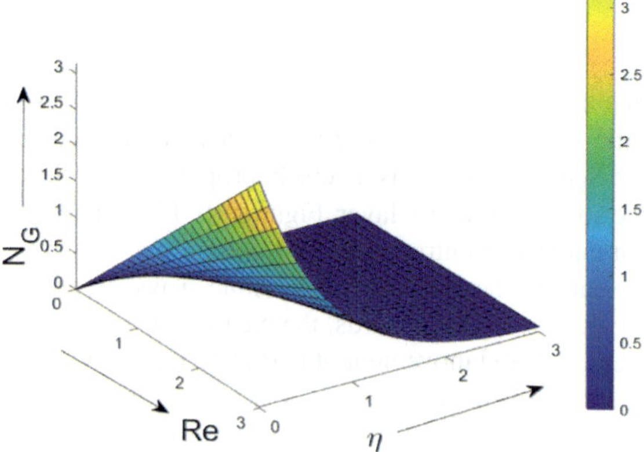

Fig. 12 Surface plot of N_G with axial direction η and Re

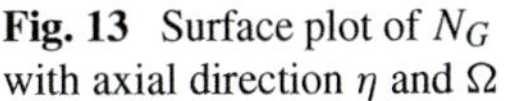

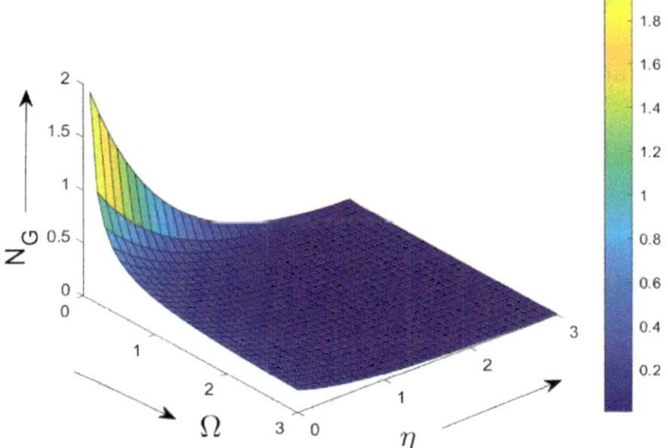

Fig. 13 Surface plot of N_G with axial direction η and Ω

cooling integrated circuit (IC) chips. The boost in entropy number due to an increase in Brinkman Br can be explained by the fact that the Brinkman number signifies the heat source in the fluid region and, together with the heat generated from the wall, enhances the entropy rate.

Increasing the radiation parameter will increase more heat in the system, resulting in the uplifting of the entropy rate. It is noted from the Fig. 12 that the entropy rate N_G increases as the Reynolds number Re increases and decreases asymptotically with η. As the value of the Reynolds number increases, the frictional resistance increases, resulting in more frictional irreversibility. This causes an increase in entropy. In contrast, the entropy decreases with an increase in the axial direction. The initial temperature difference is an essential criterion for the critical design of thermal equipment. Figure 13 display the surface plot of entropy with parameters Ω and η. It shows the decreasing trend of entropy with an increase in both the parameters Ω and η. For a small value of temperature difference, the system's entropy will be low due to the dominant effect of viscous friction. At $\eta = 0$ and with the change in magnitude of Ω from 0.103 to 3, there is a decrease of 89.22% in the entropy rate. Similarly, there is a 95.86% decrease in entropy rate with parameter η from 0 to 3.

5 Conclusion

In this study, we addressed the entropy generation on MHD fluid over a stretching sheet with variable viscosity $\mu(\tilde{T})$ and thermal conductivity $K(\tilde{T})$ by altering the parameter ϵ_1 and ϵ_2. The governing coupled PDEs that characterise the model are transformed into non-linear ODEs using similarity analysis and then solved using MATLAB software by implementing the "BVP4C" technique. The current study can aid scientists in comprehending their findings and using this approach to mechanical and industrial operations, including material processing, fuel cells, gas turbines, ventilation and renewable energy. From the above discussion, we concluded that the velocity profile declines with increased viscosity parameter ϵ_1 and magnetic field parameter M. Positive growth in an Eckert number Ec and viscosity variation parameters are responsible for enhancing the temperature profile. Increasing parameters Br and R will increase the Entropy generation N_G profile, while the reverse trend is observed with temperature difference parameter Ω.

References

1. Hayat, T., Shafiq, A., Alsaedi, A., Shahzad, S.A.: Unsteady MHD flow over exponentially stretching sheet with slip conditions. Appl. Math. Mech. **37**(2), 193–208 (2016). https://doi.org/10.1007/s10483-016-2024-8
2. Babu, M.J., Sandeep, N., Ali, M.E., Nuhait, A.O.: Magnetohydrodynamic dissipative flow across the slendering stretching sheet with temperature dependent variable viscosity. Results Phys. **7**, 1801–1807 (2017). https://doi.org/10.1016/j.rinp.2017.05.018
3. Reddy, S., Naikoti, K., Mehdi Rashidi, M.: MHD flow and heat transfer characteristics of Williamson nanofluid over a stretching sheet with variable thickness and variable thermal conductivity. Trans. A. Razmadze Math. Inst. **171**(2), 195–211 (2017). https://doi.org/10.1016/j.trmi.2017.02.004
4. Sharma, R.P., Ibrahim, S.M., Mishra, S.R., Tinker, S.: Impact of dissipative heat and radiative heat on MHD viscous flow through a slandering stretching sheet with temperature-dependent variable viscosity. Heat Transf. **50**(8), 7568–7587 (2021). https://doi.org/10.1002/htj.22243
5. Sharma, Bhupendra K., Kumawat, Chandan: Impact of temperature dependent viscosity and thermal conductivity on MHD blood flow through a stretching surface with ohmic effect and chemical reaction. Nonlinear Eng. **10**(1), 255–271 (2021). https://doi.org/10.1515/nleng-2021-0020
6. Reddy, A., Bala, P., Reddy, S.R.R., Suneetha, S.: magnetohydro dynamic flow of blood in a permeable inclined stretching surface with viscous dissipation, non-uniform heat source/sink and chemical reaction. Front. Heat Mass Transf. (FHMT) **10**, 22 (2018). https://doi.org/10.5098/hmt.10.22
7. Tripathi, B., Sharma, B.K.: Effect of variable viscosity on MHD inclined arterial blood flow with chemical reaction. Int. J. Appl. Mech. Eng. **23**(3) (2018). https://doi.org/10.2478/ijame-2018-0042
8. Tripathi, B., Kumar Sharma, B., Sharma, M.: Modeling and analysis of MHD two-phase blood flow through a stenosed artery having temperature-dependent viscosity. Eur. Phys. J. Plus **134**(9), 466 (2019) . https://doi.org/10.1140/epjp/i2019-12813-9
9. Bejan, Adrian: A study of entropy generation in fundamental convective heat transfer. J. Heat Transf. **101**(4), 718–725 (1979). https://doi.org/10.1115/1.3451063
10. Qing, J., Mubashir Bhatti, M., Ali Abbas, M., Mehdi Rashidi, M., El-Sayed Ali, M.: Entropy generation on MHD Casson nanofluid flow over a porous stretching/shrinking surface. Entropy **18**(4), 123 (2016). https://doi.org/10.3390/e18040123
11. Shit, G.C., Mandal, S.: Entropy analysis on unsteady MHD flow of Casson nanofluid over a stretching vertical plate with thermal radiation effect. Int. J. Appl. Comput. Math. **6**(1), 1–22 (2020). https://doi.org/10.1007/s40819-019-0754-4
12. Afridi, M.I., Qasim, M., Khan, I., Shafie, S., Saleh Alshomrani, A.: Entropy generation in magnetohydrodynamic mixed convection flow over an inclined stretching sheet. Entropy **19**(1), 10 (2017). https://doi.org/10.3390/e19010010
13. Shah, Zahir, Kumam, Poom, Deebani, Wejdan: Radiative MHD Casson nanofluid flow with activation energy and chemical reaction over past nonlinearly stretching surface through Entropy generation. Sci. Reports **10**(1), 1–14 (2020). https://doi.org/10.1038/s41598-020-61125-9

14. Megahed, A.M., Gnaneswara Reddy, M., Abbas, W.: Modeling of MHD fluid flow over an unsteady stretching sheet with thermal radiation, variable fluid properties and heat flux. Math. Comput. Simul. **185**, 583–593 (2021). https://doi.org/10.1016/j.matcom.2021.01.011

A Numerical Investigation on Transport Phenomena in a Nanofluid Under the Transverse Magnetic Field Over a Stretching Plate Associated with Solar Radiation

Shiva Rao and P. N. Deka

Abstract This numerical investigation considers the solar radiation effect on a nanofluid over a stretching plate acted upon by a transverse magnetic field focusing on the stagnation points. Here, linear Roseland approximation is applied for solar radiation. The physical flow problem is modeled using the sets of partial differential equations, which are then transformed into a set of non-linear ordinary differential equations by using the appropriate similarity transformation. We have a new bvp4c solver in the MATLAB platform to solve the equations numerically to investigate the solar radiation effect on various flow parameters associated with MHD nanofluids such as Brownian motion, velocity, temperature and concentration. A comparative analysis is performed for the results with previous studies in some limiting cases to prove the efficiency of the numerical approach. The results have been presented graphically as well as in tabular form to intricate the flow pattern.

Keywords MHD flow · Solar radiation · Stretching plate · Rosseland approximation · Stagnation point · Brownian motion

1 Introduction

Growing energy demand and associated energy crises coupled with environmental issues are now considered with priority across the globe. Attention towards renewable energy has increased as these can replace fossil fuels and reduce the ejection of Green House Gases. Out of different renewable energies like hydro-power from water, geothermal energy, wind energy, biomass from plants, solar energy is one of the cleanest renewables that comes directly from the sun and can be transformed into electricity directly by photoelectric effect and into heat by photo-thermal conversion. Hence, the implementation of solar energy has gained mass attention recently.

Voltaic cells and solar thermal plants are the main gateways to use solar energy and its efficiency can be increased by improving solar energy absorption. The dependence

S. Rao (✉) · P. N. Deka
Dibrugarh University, Dibrugarh, Assam 786004, India
e-mail: shivarao374@gmail.com

S. Banerjee and A. Saha (eds.), *Nonlinear Dynamics and Applications*,
Springer Proceedings in Complexity,
https://doi.org/10.1007/978-3-030-99792-2_39

on fossil fuel can be reduced by using renewable energy which mainly relies on the absorption of solar energy and conversion into thermal energy. However, there is a significant loss of energy in the absorption of sunlight by collecting panels. The weak thermo-physical properties of convectional fluids make it non-viable to construct heat exchangers with greater efficiency [1]. The water-dispersed nano-particles are found to improve the absorption of sunlight [27]. To increase the absorption efficiency different researchers have tested different nano-particles. In recent decades, nanofluids are extensively used in collectors as they elevate greater heat elimination due to their superiority in thermo-physical properties in comparison to traditional fluid [20].

Nowadays, industrial fluids are studied by researchers very intensively. Recently, there has been a great discussion about the parameters behind the heat transfer in nanofluid, despite many studies done already [33]. Nanofluids are made by the suspension of nanoparticles in the base fluid. Choi [5] was the first researcher to discover that the suspended nanoparticles in the base fluid could enhance the thermal conductivity. Lee et al. [19] measured the thermal conductivity of different metal oxides and revealed that both shape and size played an important role in enhancing thermal conductivity of the nanofluid. Nanoparticles not only increase thermal conductivity but also increase the heat transfer capacity by convection [25]. Eastman et al. [7] by their study conclude that the addition of copper nanoparticles with volume fraction less than 1% in ethylene glycol could increase the thermal conductivity up to 40%. Buonigiorno [4] attempted to explain the increase in the thermal conductivity of the nanofluid by pointing out two slip mechanism i.e., Brownian motion and thermophoresis for effective enhancement of thermal conductivity of the base fluid. MHD nanofluid has a great significance in engineering. Buongiorno's model [4] of viscous and incompressible nanofluid flow between a vertical flat plate and a porous medium was investigated by Nield and Kuznetsov [18]. Khan and Pop [16] were the first to investigated the evolution of heat transfer and nanoparticle volume fraction in a nanofluid across a stretching sheet. Rana and Bhargava [28] used the FEM approach to solve Khan and Pop's problem for the nonlinearly stretching sheet. Makinde and Aziz [23] investigated the heat transfer properties in nanofluid flow utilising convective boundary conditions. The convectional flow in a square duct in the presence of a high transverse magnetic field was investigated by Chutia and Deka [6]. Some recent work on MHD nanofluid are presented in Refs. [12, 13, 15, 30].

Thermal radiation on natural convection has become a great importance due to its wide range application is physics and engineering especially in the design of components and equipment, space technology and gas turbine, etc. Unlike conduction and convection, thermal radiation does not need any medium to transmit the heat. These properties make thermal radiation much significant in heat transfer of MHD nanofluid as it reduces the loss of heat. England and Emery [8] investigated the effect of thermal radiation on the natural convective boundary layer flow along vertical plate for absorbing and non-absorbing gases. Kumar et al. [17] presented an idea of the impact of thermal radiation on nanofluid model for flow and heat transfer over an infinite vertical plate under magnetic field and viscous dissipation.

Ali et al. [2] studied the impact on thermal radiation and non-uniform heat flux of the MHD hybrid nanofluid over the stretching cylinder. The effect of hall current which chemical reaction and thermal radiation of a nanofluid flow in a rotating channel was numerically investigated by Lv et al. [22].

Improvement in Solar collector model is one of the major priority for the use of solar energy. Nanofluid is used as a main operating fluid in most of the solar collector nowadays. Javadi ct al. [14] havc studicd thc working principlc of nanofluid on solar collector in details. Yousefi et al. [35] used Al_2O_3-water nanofluids as an operating fluid in solar collector and draw a very interesting conclusion that the nanofluid based model inceases the efficiency of solar collector by 28.3%. Faizal et al. [9] claim the possibility to make a smaller solar collector, using different nanofluid which produces the same output as the larger one. The action of CuO-water and water in a solar collector was compared by Liu et al. [21]. Sarkar and Kundu [31] studied an unsteady MHD nanofluid near a spinning sphere in the presence of solar radiation. Mushtaq et al. [24] studied the radiation effect of the MHD nanofluid flow in the two-dimensional form through the Runge–Kutta method with an appropriate shooting technique. Ghasemi et al. [10, 11] used the Keller box and Differential quadrature method (DQM) to conduct a numerical analysis of Mushtaq's work under the effects of radiation.

In this study, an investigation is done numerically with the following highlights:

- PDE's are reduced to the sets of ODE's by using similarity transformation.
- MATLAB build-in solver bvp4c is used to solve the ODE's to investigate the non-linear radiative transport phenomena in nanofluid flow under the action of transverse magnetic field under solar radiation.
- This study presents a nanofluid model for a solar collector by considering some thermal effects which can increase its efficiency to much extend.
- The study presents the velocity, temperature and concentration profiles to investigate the effect of solar radiation along with the other MHD flow parameters.
- The validity of the current results is verified by Mushtaq et al. [24]'s previous study.
- Graphical results are discussed in details with physical reasoning to clarify findings.
- The Nusselt number and Sherwood number for various parameters are thoroughly explored.

2 Mathematical Formulation

In the present study, we consider a steady two dimensional flow of a nanofluid under transverse magnetic field over a stretching sheet under the Solar radiation. As shown in Fig. 1 the stretching sheet is placed at y = 0 and the fluid start flowing towards x-axis when the sheet is stretched (force applied) along the same axis. The magnetic field B_o acts perpendicular to the direction of the flow. The stretching velocity along the x-axis is $u_w = ax$ and velocity outside the boundary layer is $u_\infty = bx$.

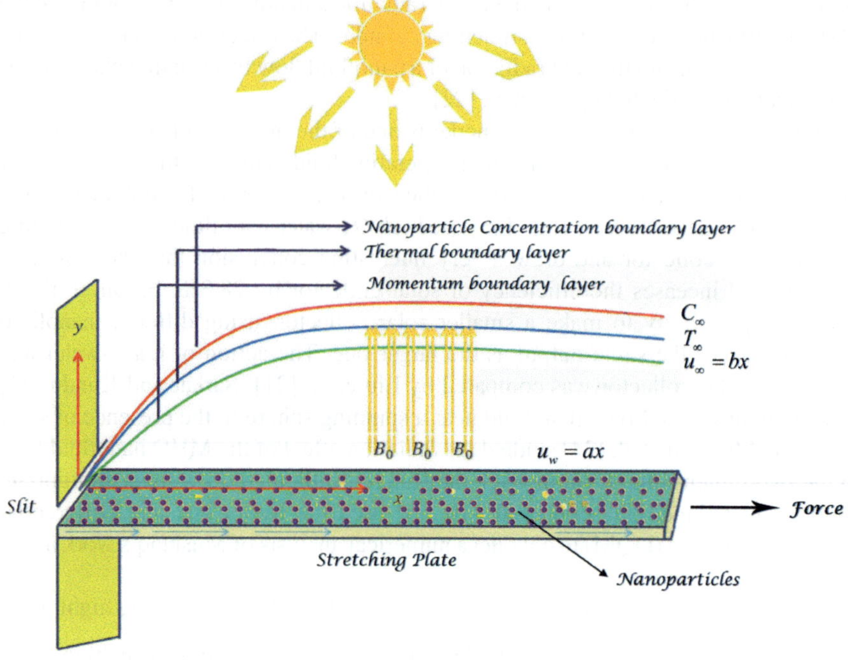

Fig. 1 Schematic diagram of the problem

The system of governing equations (see Refs. [18, 26, 32]) of the flow is given as follows:

$$\frac{\partial u}{\partial x} + \frac{\partial v}{\partial y} = 0 \tag{1}$$

$$u\frac{\partial u}{\partial x} + v\frac{\partial u}{\partial y} = u_\infty \frac{\partial u_\infty}{\partial x} + \nu_f \frac{\partial^2 u}{\partial y^2} - \frac{\sigma_e B_o^2}{\rho_f}(u - u_\infty) \tag{2}$$

$$u\frac{\partial T}{\partial x} + v\frac{\partial T}{\partial y} = \alpha \frac{\partial^2 T}{\partial y^2} + \frac{\nu_f}{C_f}\left(\frac{\partial u}{\partial y}\right)^2 - \frac{1}{(\rho C)_f}\left(\frac{\partial q_r}{\partial y}\right) + \frac{\sigma_e B_o^2}{(\rho C)_f}(u - u_\infty)^2 +$$
$$\tau\left[D_B \frac{\partial T}{\partial y}\frac{\partial C}{\partial y} + \frac{D_T}{T_\infty}\left(\frac{\partial T}{\partial y}\right)^2\right] + \frac{Q_o}{\rho_f C_p}(T - T_\infty) \tag{3}$$

$$u\frac{\partial C}{\partial x} + v\frac{\partial C}{\partial y} = D_B \frac{\partial^2 C}{\partial y^2} + \frac{D_t}{T_\infty}\frac{\partial^2 T}{\partial y^2} \tag{4}$$

where u and v are velocity component along the direction of x-axis and y-axis respectively, T is the temperature and C is the nanoparticle concentration, $\nu_f = \frac{\mu_f}{\rho}$ is the kinematic viscosity, σ_e is the electrical conductivity, $\alpha = \frac{\kappa}{(\rho C)_f}$ is the thermal diffusivity, q_r is the radiation parameter, Q_o is the internal heat generation/absorbtion coefficient, D_B and D_τ are coefficients of Brownian and thermophoretic diffusion parameter respectively. In this case, we consider the Rosseland approximation for radiation heat flux, which is mathematically expressed as (see Refs. [3, 29, 34]):

$$q_r = -\frac{-4\sigma^*}{3k^*}\frac{\partial^4 T}{\partial y} \tag{5}$$

where σ^* and k^* are Stefan-Boltzmann and mean absorption coefficient respectively.

Expanding the Taylor series and neglecting the higher order, we get

$$\mathrm{T}^4 = 4\,\mathrm{T}_\infty^3\,\mathrm{T} - 3\,\mathrm{T}_\infty^4$$

Hence Eqs. (5) becomes-

$$\frac{\partial \mathrm{q_r}}{\partial \mathrm{y}} = \frac{-16\sigma^*}{3\mathrm{k}^*}\mathrm{T}_\infty^3\frac{\partial^2 \mathrm{T}}{\partial \mathrm{y}^2} \tag{6}$$

The boundary conditions (see Ref. [16]) for the considered problem are:

$$\begin{aligned} &u = u_w(x) = ax,\, v = 0,\\ &-k\frac{\partial T}{\partial y} = h\big(T - T_f\big),\, C = C_w \ \text{ at } \ y = 0.\\ &u \to u_\infty(x) = bx,\, T \to T_\infty,\, C \to C_\infty \text{ as } y \to \infty \end{aligned} \tag{7}$$

The similarity transformation used to make Eqs. (1) to (4) dimensionless are as follows (see Ref. [18]):

$$\begin{aligned} \eta &= \sqrt{\tfrac{a}{v_f}}y,\, u = axf'(\eta),\, v = -\sqrt{av_f}\,f(\eta),\\ \theta &= \tfrac{T-T_\infty}{T_f-T_\infty},\, \phi = \tfrac{C-C_\infty}{C_w-C_\infty} \end{aligned} \tag{8}$$

The set of PDE's (2) to (4) are transformed by using the set of transformation Eq. (7) to obtain:

$$f'''(\eta) + f(\eta)f''(\eta) - (f'(\eta))^2 + \lambda^2 + M(\lambda - f'(\eta)) = 0 \tag{9}$$

$$\begin{aligned} &\frac{1}{Pr}[1 + (Rd\big(1 + (\theta_w - 1)\theta(\eta)\big)^3]\theta''(\eta) + f(\eta)\theta'(\eta) + A\theta + N_b\theta'(\eta)\phi'(\eta) +\\ &N_t(\theta'(\eta))^2 + E_c(f''(\eta))^2 + ME_c(\lambda - f'(\eta))^2 = 0 \end{aligned} \tag{10}$$

$$\phi''(\eta) + Lef(\eta)\phi(\eta) + \frac{N_t}{N_b}\theta''(\eta) = 0 \tag{11}$$

where $M = \frac{\sigma B_o^2}{\rho_f a}$ is the parameter associated with magnetic field strength, $\lambda = \frac{b}{a}$ is the ratio of rates of free stream velocity to the velocity of the stretching sheet, $Pr = \frac{\nu_f}{\alpha}$ is the Prandtl number, $R_d = \frac{16\sigma^* T_\infty^3}{3kk^*}$ is the Radiation parameter, $\theta_w = \frac{T_f}{T_\infty}$ is the Temperature parameter, $N_b = \frac{\tau D_B(C_w - C_\infty)}{\nu_f}$ is the Brownian motion parameter, $N_t = \frac{\tau D_t(T_w - T_\infty)}{T_\infty \nu_f}$ is the Theromophoresis parameter, $A = \frac{Q_o}{a\rho_f c_p}$ is the heat source and heat sink parameter for $A > 0$ and $A < 0$ respectively, $E_c = \frac{u_w}{C_p(T_w - T_\infty)}$ is the Eckert number and $Le = \frac{\nu_f}{D_B}$ is the Lewis number.

The boundary condition (7) in the dimensionless form are as follows:

$$f(0) = 0, f'(0) = 1, \theta'(0) = -\gamma[1 - \theta(0)], \phi(0) = 1$$

$$f'(+\infty) \to \lambda, \theta(+\infty) \to 0, \phi(+\infty) \to 0 \tag{12}$$

where $\gamma = \frac{h}{k\sqrt{\frac{\nu_f}{\alpha}}}$ is the Biot number.

The surface heat flux and mass flux in dimensionless form can be represented as follows:

$$\frac{Nu_x}{\sqrt{Re_x}} = -\left[1 + Rd\theta_w^2\right]\theta'(0) = Nur \tag{13}$$

$$\frac{Sh}{\sqrt{Re_x}} = -\phi'(0) = Shr \tag{14}$$

3 Method of Solutions

The dimensionless ordinary differential Eqs. (9) to (11) along with dimensionless boundary conditions (12) are solved using a MATLAB built-in solver bvp4c package. The equations are converted into the set of first order differntial equations as follows:

$$f = y_1, f' = y_1' = y_2, f'' = y_2' = y_3, \theta = y_4,$$

$$\theta' = y_4' = y_5, \phi = y_6, \phi' = y_6' = y_7$$

$$f''' = y_3' = y_2^2 - y_1 y_3 + \lambda^2 + M(\lambda - y_2) \tag{15}$$

$$\theta'' = y_5' = \frac{-\Pr[y_1 y_5 + A y_4 + N_b y_5 y_7 + N_t y_7^2 + E_c y_3^2 + M E_c(\lambda - y_2)^2]}{1 + Rd(1 - (\theta_w - 1) y_4)^3} \tag{16}$$

$$\phi'' = y_7' = -Ley_1y_7 - \frac{Nt}{Nb}y_5 \tag{17}$$

The boundary conditions are given by:

$$\begin{aligned} &y_1(0) = 0,\ y_2(0) = 1,\ y_5(0) = -\gamma[1 - y_4(0)],\ y_6(0) = 1, \\ &y_2(+\infty) = \lambda,\ y_4(+\infty) = 0,\ y_6(+\infty) = 0 \end{aligned} \tag{18}$$

4 Results and Discussions

Influence of non-dimensional parameter such as Radiation parameter R_d, along with Brownian motion parameter N_b, Thermophoresis parameter N_t, Magnetic field parameter M, Lewis number Le, Biot number γ, Prandtl number Pr and Eckert number E_c on Temperature profile $\theta(\eta)$, Nanoparticle concentration $\phi(\eta)$, reduced Nusselt number and reduced Sherwood number graphically and numerically.

The obtained numerical solutions are compared with those of prior published Mushtaq et al. [24] to justify the correctness of the employed approach. Table 1 illustrates the numerical value of heat and mass transfer, as well as the results reported by [24] in the presence and absence of radiation, which demonstrates great agreement with the results achieved in this investigation.

Table 2 shows the effect of radiation, together with other parameters, on heat and mass transfer as numerical values of $-\theta'(0)$ and $-\phi'(0)$.

Figures 2 and 3 depict the effect of N_t along with R_d on the temperature profile and Nanoparticle concentration profile respectively and it can be observed that the

Table 1 Comparison of values of $-\theta'(0)$ and $-\phi'(0)$ for the various value of N_b with Mushtaq et al. [24]

N_b	R_d	Mushtaq et al. [24]		Present study	
		$-\theta'(0)$	$-\phi'(0)$	$-\theta'(0)$	$-\phi'(0)$
0.1	0	0.078993	2.44780	0.0785	2.4478
	1	0.081387	2.40369	0.0815	2.4061
0.2	0	0.070373	2.43727	0.0704	2.4373
	1	0.078183	2.39810	0.0779	2.4001
0.3	0	0.058202	2.44012	0.0582	2.4401
	1	0.074496	2.39670	0.0735	2.3991
0.4	0	0.040852	2.44673	0.0409	2.4467
	1	0.070375	2.39623	0.0681	2.3970
0.5	0	0.018834	2.45314	0.0189	2.4531
	1	0.065911	2.39605	0.0615	2.3961

Table 2 Value of Nur and Shr for different parameters along with radiation

R_d	N_t	N_b	M	Pr	Le	γ	E_c	$-\theta'(0)$	$-\phi'(0)$
0	0.1	0.1	0.5	5.0	1.0	0.1	0.1	0.0880	0.6586
	0.2							0.0879	0.6211
	0.3							0.0878	0.5847
1	0.1	0.1	0.5	5.0	1.0	0.1	0.1	0.0871	0.6589
	0.2							0.0870	0.6213
	0.3							0.0869	0.5846
0	0.1	0.1	0.5	5.0	1.0	0.1	0.1	0.0880	0.6586
		0.2						0.0650	0.6812
		0.3						0.0847	0.6891
1	0.1	0.1	0.5	5.0	1.0	0.1	0.1	0.0871	0.6589
		0.2						0.0860	0.6801
		0.3						0.0848	0.6874
0	0.1	0.1	0.5	5.0	1.0	0.1	0.1	0.0880	0.6586
			1.0					0.0866	0.6599
			1.5					0.0853	0.6616
1	0.1	0.1	0.5	5.0	1.0	0.1	0.1	0.0871	0.6587
			1.0					0.0860	0.6580
			1.5					0.0850	0.6577
0	0.1	0.1	0.5	5.0	1.0	0.1	0.1	0.0880	0.6586
				7.0				0.0881	0.6602
				9.0				0.0880	0.6622
1	0.1	0.1	0.5	5.0	1.0	0.1	0.1	0.0871	0.6589
				7.0				0.0877	0.6581
				9.0				0.0880	0.6583
0	0.1	0.1	0.5	5.0	1.0	0.1	0.1	0.0880	0.6586
					4.0			0.0870	1.4480
					7.0			0.0865	1.9696
1	0.1	0.1	0.5	5.0	1.0	0.1	0.1	0.0871	0.6589
					4.0			0.0865	1.4408
					7.0			0.0863	1.9586
0	0.1	0.1	0.5	5.0	1.0	0.1	0.1	0.0880	0.6586
						0.2		0.1651	0.6061
						0.3		0.2329	0.5601
1	0.1	0.1	0.5	5.0	1.0	0.1	0.1	0.0871	0.6589
						0.2		0.1598	0.6147
						0.3		0.2213	0.5776

(continued)

Table 2 (continued)

R_d	N_t	N_b	M	Pr	Le	γ	E_c	$-\theta'(0)$	$-\phi'(0)$
0	0.1	0.1	0.5	5.0	1.0	0.1	0.1	0.0880	0.6586
							0.2	0.0826	0.6844
							0.3	0.0771	0.7103
1	0.1	0.1	0.5	5.0	1.0	0.1	0.1	0.0871	0.6589
							0.2	0.0830	0.6764
							0.3	0.0790	0.6940

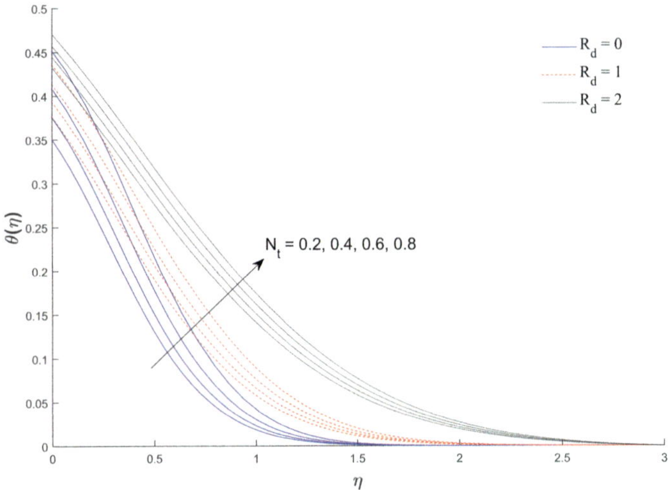

Fig. 2 Effect of $\mathrm{N_t}$ along with R_d on $\theta(\eta)$

temperature increases with the increase of both N_t and R_d whereas the nanoparticle concentration increases with N_t but decreases with an increase in radiation. Figure 4 describes the effect of N_t and R_d on the temperature derivative profile $(-\theta'(\eta))$ and it is clear that it increases with the increase of N_t and R_d. The results are important for calculating the Nusselt number since they have a direct relationship to its value when $\eta = 0$. The reason behind the fact is that the increase in N_t results the enhancement of thermophoresis forces which has the tendency to fast flow the nanoparticles from hot surface to cold surface away from stretching. This results in an increase of heat and mass transfer in the boundary layer region for nanoparticles.

Figures 5 and 6 illustrate how N_b affects the temperature and nanoparticle concentration profiles in conjunction with the specified radiation. The study elucidates that the temperature increases with the increment N_b but decreases with R_d whereas the concentration decreases with the increase in both N_b and R_d. The impact of N_b along with R_d on $-\theta'(\eta)$ is depicted in Fig. 7 and an increase is observed with the increase in N_b but an opposite trend is seen with R_d. It is well known that as Nb increases, so

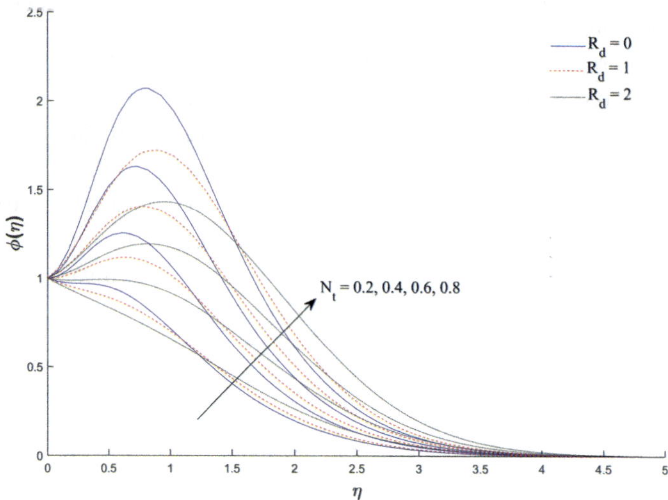

Fig. 3 Effect of N_t along with R_d on $\phi(\eta)$

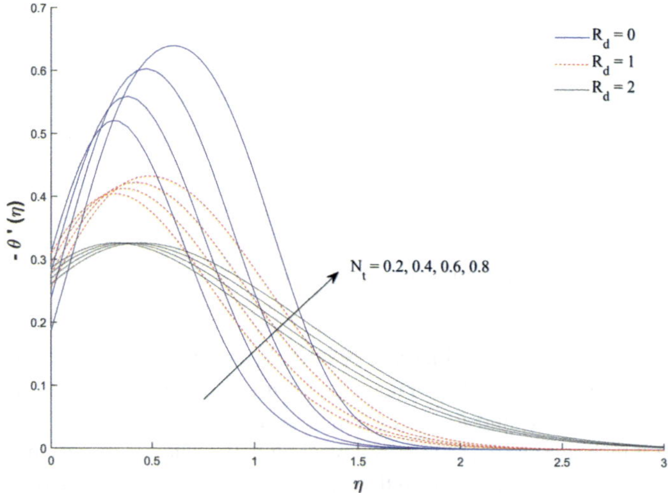

Fig. 4 Effect N_t of along with R_d on $-\theta'(\eta)$

does the random motion of nanoparticles, resulting in an increase in collisions with other nanoparticles. As a result, the kinetic energy is transformed into heat energy, and the temperature rises. But the rate of mass transfer decreases because of the tendency of the particle to get close to each other as Nb increases.

Figures 8 and 9 show the effect of Biot number γ on both temperature and nanoparticle concentration and it is observed that both increases with the increase in γ. The main reason behind the fact is that, increase in biot number means increase

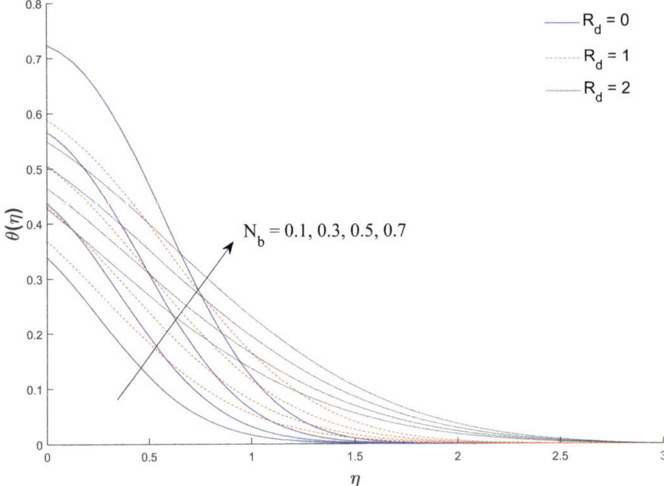

Fig. 5 Effect of N_b along with R_d on $\theta(\eta)$

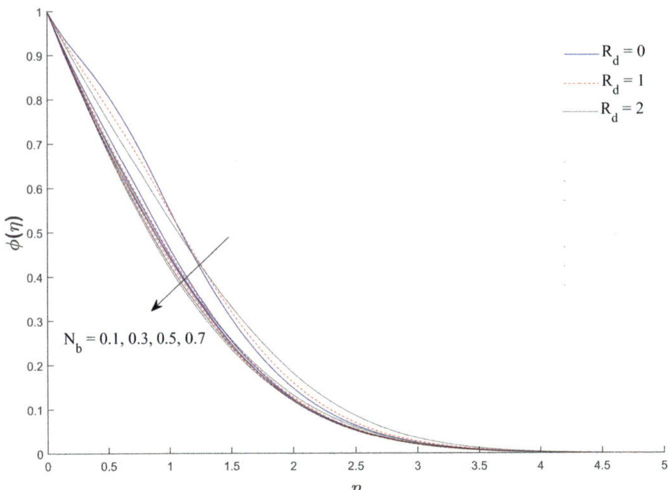

Fig. 6 Effect of N_b along with R_d on $\phi(\eta)$

in convective heat exchange at the surface which results in the increase in thermal boundary layer thickness and which in turn increases the nanoparticle concentration. Figures 10 shows that the temperature decreases with the increase of Prandtl number Pr whereas Fig. 11 shows that the nanoparticle concentration profile increases with Pr. The main reason behind the fact that with the higher value in Pr the heat diffuses

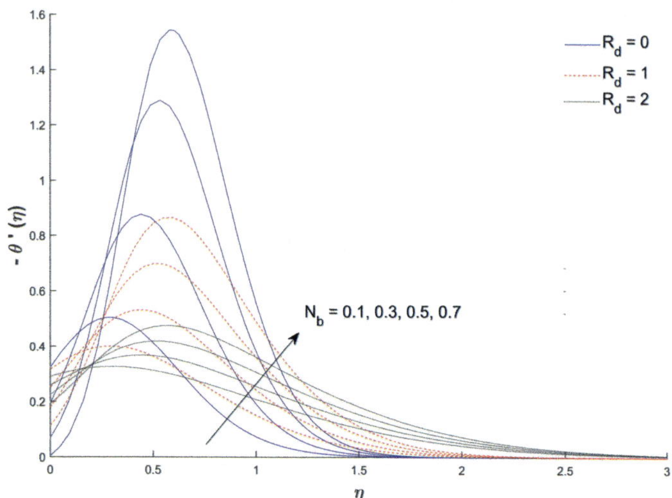

Fig. 7 Effect of N_b along with R_d on $-\theta'(\eta)$

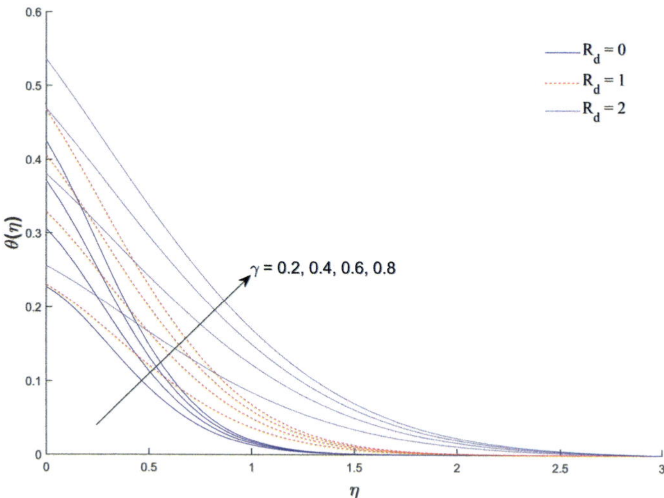

Fig. 8 Effect of γ along with R_d on $\theta(\eta)$

more rapidly than the momentum. It is also observed that at large Pr the temperature falls more drastically due to the fact that the large values of Pr leads to the low thermal conductivity.

Figure 12 depicts the impact of E_c on the temperature profile. We know that Eckert number expresses a direct relationship of flow's kinetic energy to the boundary layer enthalpy differences. This leads to the fact that the increases in E_c enhance the kinetic energy. Whereas it is well known that the temperature is an average kinetic energy.

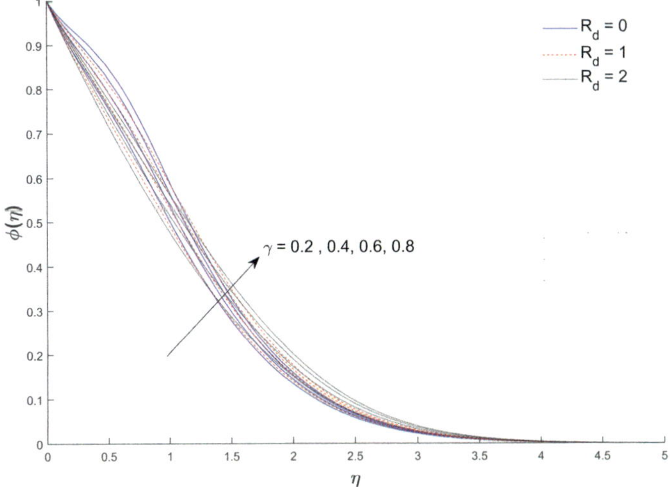

Fig. 9 Effect of γ along with R_d on $\phi(\eta)$

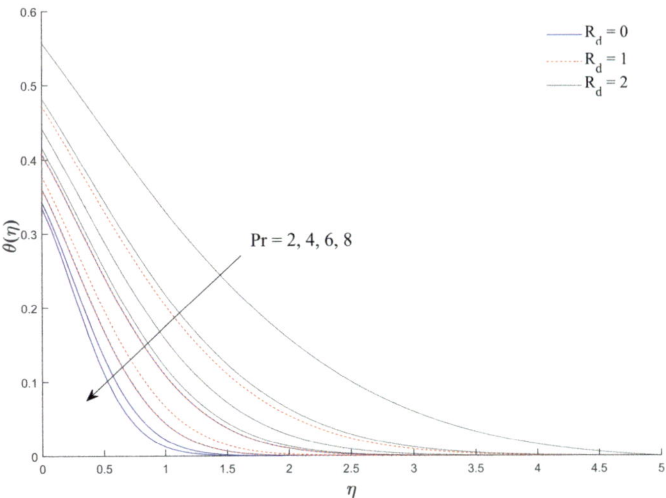

Fig. 10 Effect of Pr along with R_d on $\theta(\eta)$

Hence alternatively we can say that temperature rises with the increase in Eckert number and which can be clearly seen in the figure.

Figures 13 , 14 and 15 show the effect of magnetic parameters on velocity, temperature, and nanoparticle concentration respectively. It is noticed from the velocity profile that the velocity decreases with the increase in magnetic parameter. An exact opposite behavior is seen in the case of temperature profile where the temperature

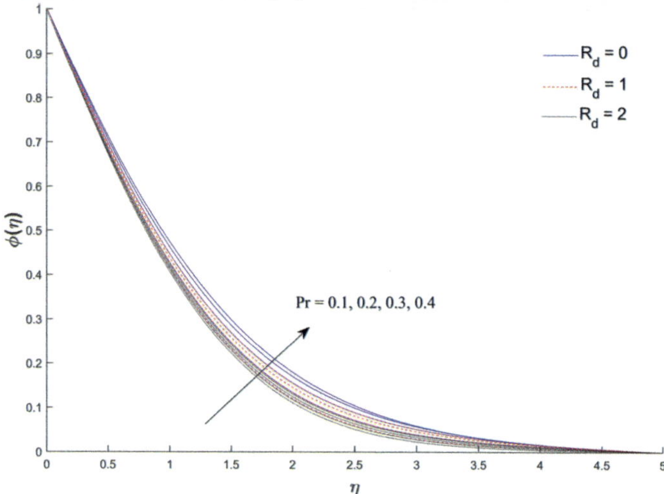

Fig. 11 Effect of Pr along with R_d on $\phi(\eta)$

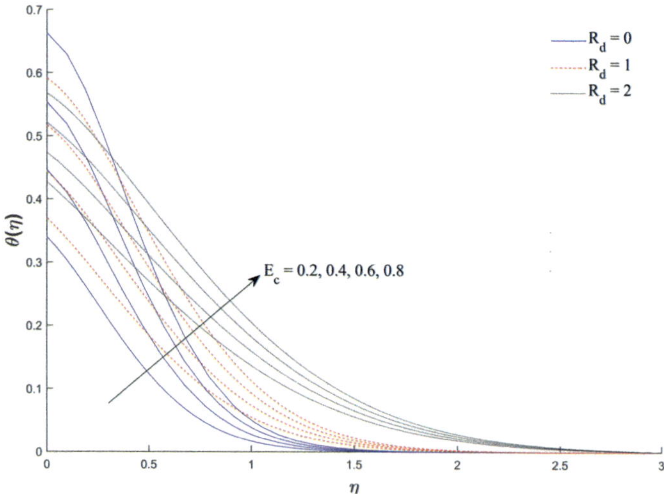

Fig. 12 Effect of E_c along with R_d on $\theta(\eta)$

increases with the increase in M and radiation R_d. In the case of nanoparticle concentration, it is clear from the graph that the concentration increases with the increasing magnetic parameter but decreases with stronger radiation. As the magnetic field parameter increases, a resistive force called a Lorentz force is produced which retards in the form magnetic pressure drop on the velocity, as a result the motion gets slowed down. Therefore the velocity decreases with the increasing value of M. Again due to the Lorentz force a resistance is offered to the flow which results in warming up

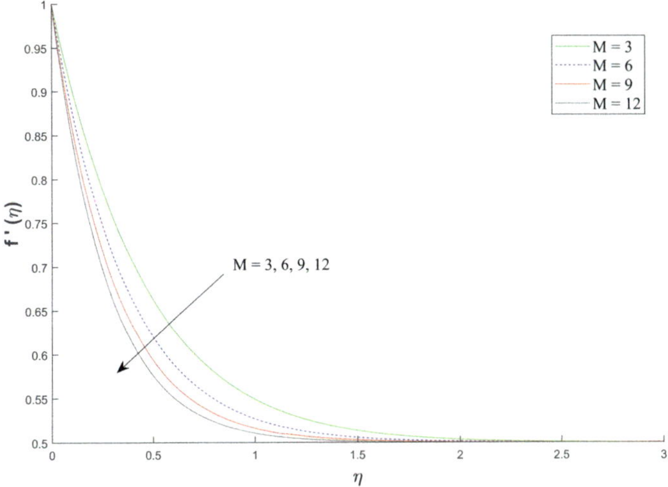

Fig. 13 Effect of M along with R_d on $f'(\eta)$

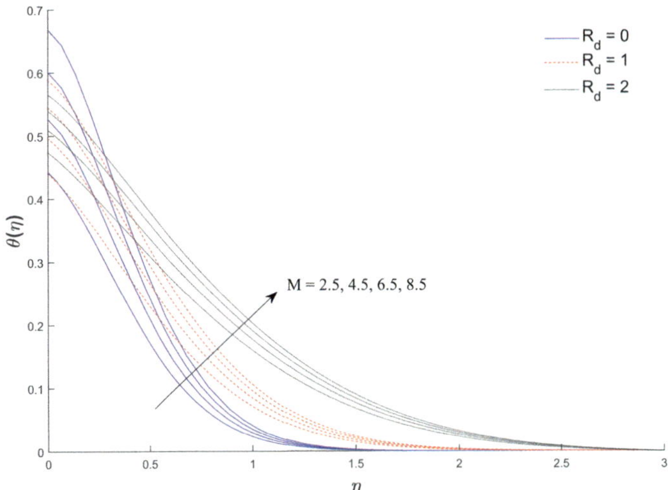

Fig. 14 Effect of M along with R_d on $\theta(\eta)$

the boundary layer region. Hence the temperature increases as the values of M gets increase.

The influence of Lewis number on temperature and nanoparticle concentration can be depicted in Figs. 16 and 17 respectively. It is noted that a growing behavior is found for temperature profile with the increment in Le and R_d whereas the concentration decreases with the increase in Le. It is observed that a smaller increase in Le results in larger differences in temperature and a thinner concentration boundary layer due to

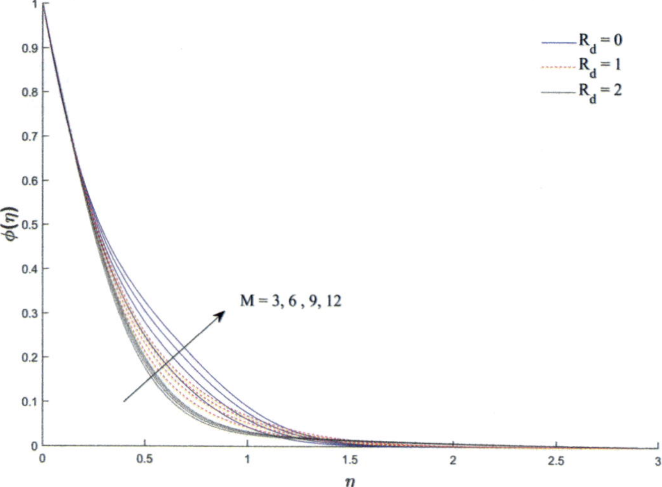

Fig. 15 Effect of M along with R_d on $\phi(\eta)$

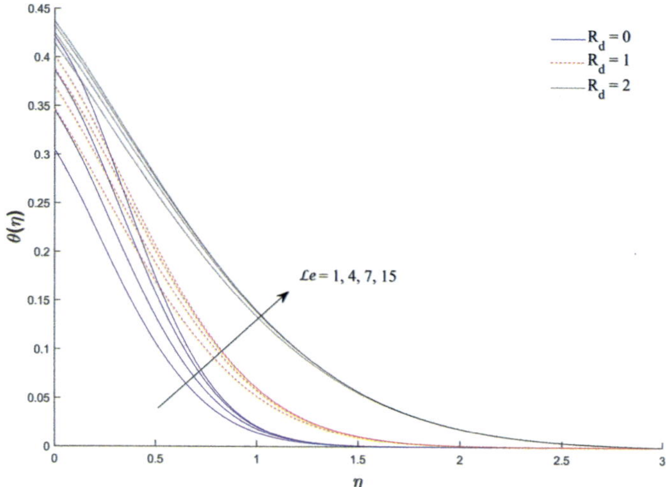

Fig. 16 Effect of Le along with R_d on $\theta(\eta)$

a weak molecular diffusivity. Figures 18 and 19 depicts the behavior of temperature profile for heat source parameter (A > 0) and heat sink parameter (A < 0) respectively. It is observed that the temperature of the thermal boundary layer increases with the increase in A (heat source parameter) and decreases with the decrease in A (heat sink parameter) under constant thermal radiation.

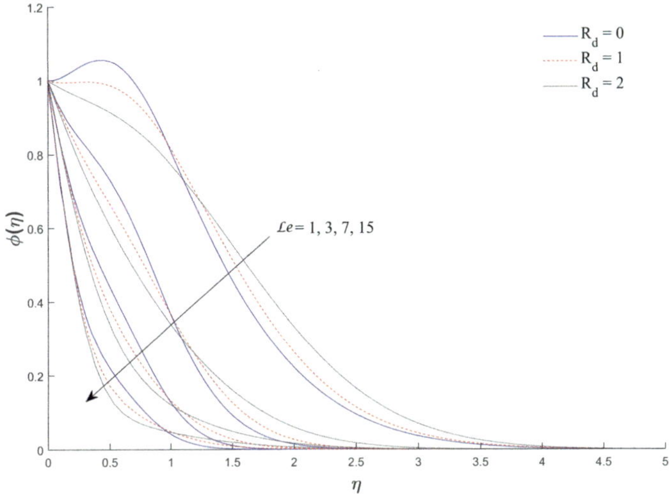

Fig. 17 Effect of Le along with R_d on $\phi(\eta)$

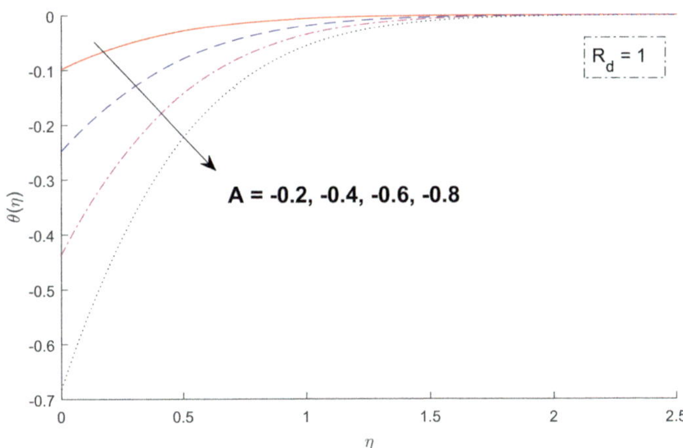

Fig. 18 Effect of $A < 0$ on $\theta(\eta)$

5 Conclusions

The influence of solar radiation on a constant two-dimensional MHD flow across a stretched plate is examined for various parameters in this study. The findings acquired in this investigation using the MATLAB programme bvp4c and the results obtained in the previous study utilizing the Runge Kutta Fourth Order Scheme showed great consistency. The impact of various parameters in our present investigation are as follows:

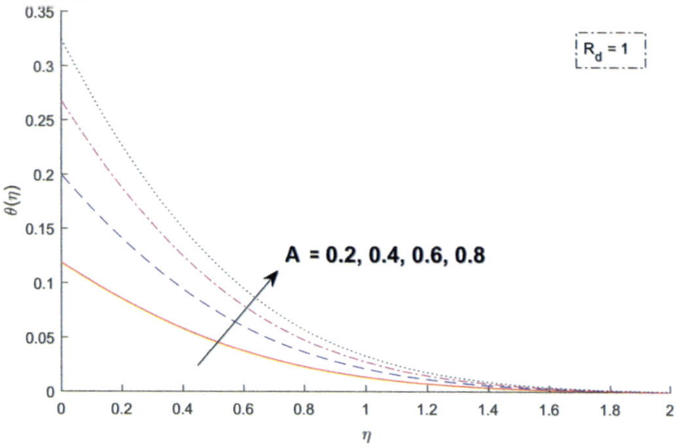

Fig. 19 Effect of $A > 0$ on $\theta(\eta)$

- The increasing value of Magnetic field parameter can decrease the nanoparticle velocity.
- The temperature distribution in the boundary layer region can be enhanced with the increment of Magnetic field parameter, Biot number Thermophoresis parameter, Brownian motion parameter, Heat source parameter, Lewis number and Eckert number.
- The temperature distribution diminishes with the increasing value of Prandtl number and Heat sink parameter.
- Nanoparticle volume fraction in the boundary layer region can be enriched by increasing the value of Thermophoresis parameter, Biot number, Magnetic field parameter, Lewis number and Prandtl number while it can decrease with the increase in Brownian motion parameter.
- The increasing value of Radiation parameter increases the Temperature profile and decreases the Nanoparticle volume fraction profile.
- The value of reduced Nusselt number is increased with the increase in Biot number and the Prandtl number and deceases with magnetic field parameter, Lewis number, Thermophoresis parameter and Brownian motion parameter.
- The value of reduced Sherwood number is increased with the increase in Brownian motion parameter, Magnetic field parameter, Prandtl number, Lewis number and Eckert number but decreases with Thermophoresis parameter and Biot number.
- It is very interesting to note that increase in Radiation parameter lead to the decrease in Reduced Nusselt number and increase in Reduced Sherwood number.
- The applicability of MATLAB's software bvp4c (Boundary layer problem of fourth-order) is ensured by verifying the findings as compared to the previously published results.
- The present study finds an application in efficient Solar collector, Cooling problems in industry, etc.

References

1. Akilu, S., Sharma, K. v., Baheta, A. T., & Mamat, R.: A review of thermophysical properties of water based composite nanofluids. In: Renewable and Sustainable Energy Reviews, vol. 66, pp. 654–678. Elsevier Ltd. (2016)
2. Bansal, J.L.: Magnetofluiddynamics of Viscous Fluids. Jaipur Publishing House, Jaipur India (1994)
3. Brewster, M.Q.: Thermal Radiative Transfer Properties. John Wiley and Sons, New York (1972)
4. Buongiorno, J.: Convective transport in nanofluids. J. Heat Transf. **128**(3), 240–250 (2006)
5. Choi, S.U.S.: Enhancing thermal conductivity of fluids with nanoparticles. ASME Int. Mech. Eng. Cong. Expo. **66**, 99–105 (1995)
6. Chutia, M., Deka, P.N.: Numerical study on MHD mixed convection flow in a vertical insulated square duct with strong transverse magnetic field. J. Appl. Fluid Mech. **8**(3), 473–481 (2015)
7. Eastman, J.A., Choi, S.U.S., Li, S., Yu, W., Thompson, L.J.: Anomalously increased effective thermal conductivities of ethylene glycol-based nanofluids containing copper nanoparticles. Appl. Phys. Lett. **78**(6), 718–720 (2001)
8. England, W.G., Emery, A.F.: Thermal radiation effects on the laminar free convection boundary layer of an absorbing gas. ASME J. Heat Transfer. **91**(1), 37–44 (1969)
9. Faizal, M., Saidur, R., Mekhilef, S., Alim, M.A.: Energy, economic and environmental analysis of metal oxides nanofluid for flat-plate solar collector. Energy Convers. Manage. **76**, 162–168 (2013)
10. Ghasemi, S.E., Hatami, M.: Solar radiation effects on MHD stagnation point flow and heat transfer of a nanofluid over a stretching sheet. Case Stud. Thermal Eng. **25** (2021)
11. Ghasemi, S.E., Hatami, M., Jing, D., Ganji, D.D.: Nanoparticles effects on MHD fluid flow over a stretching sheet with solar radiation: a numerical study. J. Mol. Liq. **219**, 890–896 (2016)
12. Gireesha, B.J., Mahanthesh, B., Shivakumara, I.S., Eshwarappa, K.M.: Melting heat transfer in boundary layer stagnation-point flow of nanofluid toward a stretching sheet with induced magnetic field. Eng. Sci. Technol. Int. J. **19**(1), 313–321 (2016)
13. Hayat, T., Imtiaz, M., Alsaedi, A., Kutbi, M.A.: MHD three-dimensional flow of nanofluid with velocity slip and nonlinear thermal radiation. J. Magn. Magn. Mater. **396**, 31–37 (2015)
14. Javadi, F. S., Saidur, R., Kamalisarvestani, M.: Investigating performance improvement of solar collectors by using nanofluids. In: Renewable and Sustainable Energy Reviews, vol. 28, pp. 232–245 (2013)
15. Khan, W.A., Makinde, O.D.: MHD nanofluid bioconvection due to gyrotactic microorganisms over a convectively heat stretching sheet. Int. J. Therm. Sci. **81**(1), 118–124 (2014)
16. Khan, W.A., Pop, I.: Boundary-layer flow of a nanofluid past a stretching sheet. Int. J. Heat Mass Transf. **53**(11–12), 2477–2483 (2010)
17. Kumar, M.A., Reddy, Y.D., Rao, V.S., Goud, B.S.: Thermal radiation impact on MHD heat transfer natural convective nano fluid flow over an impulsively started vertical plate. Case Stud. Thermal Eng. **24**, 100826 (2021)
18. Kuznetsov, A.V., Nield, D.A.: Natural convective boundary-layer flow of a nanofluid past a vertical plate. Int. J. Thermal Sci. **49**(2), 243–247 (2010)
19. Lee, S., Choi, -S, Li, S., Eastman, J.A.: Measuring Thermal Conductivity of Fluids Containing Oxide Nanoparticles (1999). http://heattransfer.asmedigitalcollection.asme.org/
20. Lenert, A., Nam, Y., Wang, N.E.: Heat transfer fluids. Ann. Rev. Heat Trans. **15**(2), 45 (2012)
21. Liu, Z.H., Hu, R.L., Lu, L., Zhao, F., Xiao, H.S.: Thermal performance of an open thermosyphon using nanofluid for evacuated tubular high temperature air solar collector. Energy Convers. Manage. **73**, 135–143 (2013)
22. Lv, Y.P., Shaheen, N., Ramzan, M., Mursaleen, M., Nisar, K.S., Malik, M.Y.: Chemical reaction and thermal radiation impact on a nanofluid flow in a rotating channel with Hall current. Sci. Rep. **11**(1) (2021)
23. Makinde, O.D., Aziz, A.: Boundary layer flow of a nanofluid past a stretching sheet with a convective boundary condition. Int. J. Therm. Sci. **50**(7), 1326–1332 (2011)

24. Mushtaq, A., Mustafa, M., Hayat, T., Alsaedi, A.: Nonlinear radiative heat transfer in the flow of nanofluid due to solar energy: a numerical study. J. Taiwan Inst. Chem. Eng. **45**(4), 1176–1183 (2014)
25. Otanicar, T.P., Phelan, P.E., Prasher, R.S., Rosengarten, G., Taylor, R.A.: Nanofluid-based direct absorption solar collector. J. Renew. Sustain. Energy **2**(3) (2010)
26. Pai, S.I.: Viscous Flow Theory: I Laminar Flow. D. VanNostrand Co., New York, USA (1956)
27. Patel, H.E., Sundararajan, T., Das, S.K.: An experimental investigation into the thermal conductivity enhancement in oxide and metallic nanofluids. J. Nanopart. Res. **12**(3), 1015–1031 (2010)
28. Rana, P., Bhargava, R.: Flow and heat transfer of a nanofluid over a nonlinearly stretching sheet: a numerical study. Commun. Nonlinear Sci. Numer. Simul. **17**(1), 212–226 (2012)
29. Raptis, A.: Radiation and free convection flow through a porous medium. Int. Commun. Heat Mass Transf. **25**, 289–295 (1998)
30. Rashidi, M.M., Vishnu Ganesh, N., Abdul Hakeem, A.K., Ganga, B.: Buoyancy effect on MHD flow of nanofluid over a stretching sheet in the presence of thermal radiation. J. Mol. Liq. **198**, 234–238 (2014)
31. Sarkar, A., Kundu, P.K.: Framing the upshot of Hall current on MHD unsteady nanofluid flow from a rotating spherical body in presence of solar radiation. Int. J. Ambient Energy (2021)
32. Schichting H.: Boundary Layer Theory, vol. 6. McGraw-Hill, New York (1964)
33. Shdaifat, M.Y.A., Zulkifli, R., Sopian, K., Salih, A. A.: Thermal and hydraulic performance of CuO/water nanofluids: a review. In: Micromachines, vol. 11, Issue 4 (2020)
34. Sparrow, E.M., Cess, R.D.: Radiation Heat Transfer. Hemisphere, Washington (1978)
35. Yousefi, T., Veysi, F., Shojaeizadeh, E., Zinadini, S.: An experimental investigation on the effect of Al_2O_3-H_2O nanofluid on the efficiency of flat-plate solar collectors. Renew. Energy **39**(1), 293–298 (2012)

Analysis of Solute Dispersion Through an Open Channel Under the Influence of Suction or Injection

Gourab Saha, Nanda Poddar, Subham Dhar, and Kajal Kumar Mondal

Abstract In presence of suction or injection, the process of dispersion of solute in an open channel flow is investigated through this work. The dispersion coefficient is analytically determined by using Mei's multi-scale analysis approach up to second order. The effect of suction or injection on dispersion coefficient are displayed graphically and discussed. The novelty of the present research is to find the effect of suction/injection Reynolds number on the dispersion coefficient and concentration distribution of the solute. It is found that the mass dispersivity decreases with the increment of suction/injection Reynolds number. The present study may play a significant role in the process of wastewater treatment. It is observed that, as dispersion time progresses, the amplitude of the concentration curves along longitudinal direction becomes flat. It is observed that the iso-concentration contour spreads along the longitudinal direction with the enhancement of injection Reynolds number.

Keywords Open channel · Multi-scale approach · Suction · Injection · Dispersivity

1 Introduction

The study of dispersion phenomena of solute through various flow situation has a great importance because the results of this study may be applied in diverse fields, namely, biology, chemical engineering, chromatography, environment fluid mechanics etc. For controlling the air or water pollution, to create a design of chemical reaction, for targeting drug delivery etc., the investigation of the mass transport of the solute is highly important. Several researchers studied the Newtonian fluid flows in different flow geometry. Suction/injection in the fluid flow can be a powerful

Supported by DST, India, UGC, India and CSIR, India.

G. Saha (✉) · N. Poddar · S. Dhar · K. K. Mondal
Cooch Behar Panchanan Barma University, Cooch Behar 736101, India
e-mail: gourabsaha2019@gmail.com

S. Banerjee and A. Saha (eds.), *Nonlinear Dynamics and Applications*,
Springer Proceedings in Complexity,
https://doi.org/10.1007/978-3-030-99792-2_40

mechanism for the flow management. This technique has an important potential in bio-medical engineering, rocket technology and food processing. In medical engineering, it plays a significant role for artificial qualitative analysis.

In an incompressible fluid flow, the longitudinal dispersion of solute through a circular tube was first studied by Taylor [1]. There were some limitations in Taylor's research work, by removing those restrictions, Aris [2] introduced a method of moment to analyze the dispersion phenomena of the tracers. Gill [3] obtained a series solution for the dispersion of solute in a laminar flow through a tube using mean concentration expansion technique. Mondal et al. [4] explored the dispersion process of settling particles in a turbulent open channel flow from an elevated continuous source. Mei [5] proposed a multiple scale analysis to investigate the dispersion of solute through various flow geometries. In this process, one can analyze the dispersion phenomena for different time scales. Later, Wu et al. [6] used the multi-scale analysis to observe the environmental dispersion in wetland flow.

Several researchers [7–10] used the boundary layer theory to analyze the impact of suction or injection in various flow field. To control fluid flow on the surface of subsonic aircraft suction/injection was imposed by Shojaefard et al. [7]. Suction/injection of a fluid has a great importance in heat transfer cooling [8]. Finite difference implicit Keller-box method applied by Ishak et al. [9] to show that how suction or injection can significantly change the flow field and it affects the rate of heat transfer from the plate. An analytical solution for energy and momentum equations in presence of suction or injection was obtained by Jha et al. [10]. Recently, Sasikumar et al. [11] investigated the effect of suction and injection on an unsteady oscillating flow of an incompressible viscous electrically conducting fluid through an asymmetric channel filled with perforated medium and oscillating wall temperature. In presence of extending surface effect with suction/injection, MHD Casson nanofluid flow over nonlinearly heated porous medium was studied by Abo-Dahab et al. [12] and they determined the wall drag, thermal and mass fluxes in the said model.

The main purpose of the present study is to investigate the effect of suction/injection Reynolds number on dispersion phenomena of solute in a laminar flow through an open channel. The analytical results are derived with the help of multi-scale homogenization process. The impact of suction/injection parameter on the dispersion coefficient, transverse concentration and longitudinal concentration distributions are discussed.

2 Mathematical Formulation

For this work, a two dimensional laminar flow of a viscous incompressible fluid through an open channel is considered. We introduce x'-axis along the direction of flow and y'-axis perpendicular to the flow. It is assumed that the depth of the channel is very small compared to the length of the channel. The velocity distribution

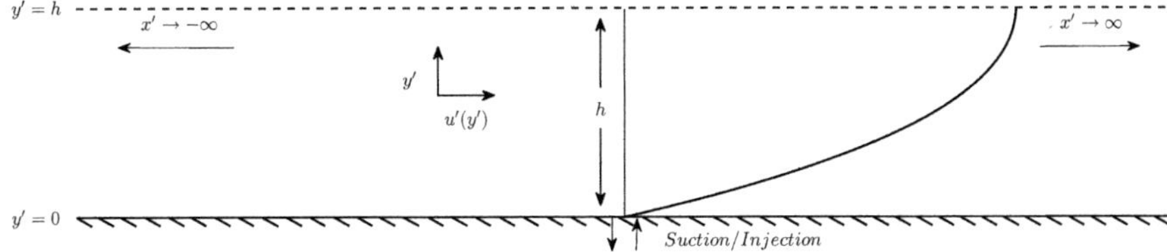

Fig. 1 Schematic diagram of the above flow

$u'(y')$ parallel to the x'-axis of the above mentioned flow satisfies the Navier-Stokes equation

$$\nu \frac{d^2u'}{dy'^2} - V_0 \frac{du'}{dy'} = -\frac{1}{\rho} \frac{\partial p}{\partial x'}, \tag{1}$$

where ν is the kinematic viscosity of the fluid. The boundary conditions of the flow field are taken as $u'(y') = 0$ at $y' = 0$ and $\frac{du'}{dy'} = 0$ at $y' = h$ (See Fig. 1).

Introducing the following dimensionless quantities

$$x = \frac{x'}{L}, y = \frac{y'}{h}, t = \frac{t'}{\frac{h^2}{D}}, u = \frac{u'}{u_c}, Sc = \frac{\nu}{D}. \tag{2}$$

The Eq. (1) takes the form

$$\frac{d^2u}{dy^2} - \gamma \frac{du}{dy} = -1, \tag{3}$$

where $u_c = -\frac{h^2}{\rho\nu}\frac{\partial p}{\partial x'}$ and $\gamma \left(= \frac{V_0 h}{\nu}\right)$ is the suction Reynolds number (-γ stands for injection Reynolds number). The respective non dimensional boundary conditions are

$$u(y) = 0 \quad at \quad y = 0, \quad \frac{du}{dy} = 0 \quad at \quad y = 1. \tag{4}$$

Solving Eq. (3) with the help of boundary condition Eq. (4), one can obtain

$$u(y) = \frac{e^{-\gamma}}{\gamma^2}\left[1 - e^{\gamma y}\right] + \frac{y}{\gamma}. \tag{5}$$

Figure 2 shows the velocity profile through the considered open channel for various values of suction/injection Reynolds number. It is observed that the velocity increases with the increment of injection parameter and if positive value of γ i.e. if suction increases, the velocity reduces in the flow field. In the limiting case of suction/injection Reynolds number i.e. when $\gamma \to 0$, the velocity of the fluid is exactly similar to that of an open channel flow and it's mathematical form is given below

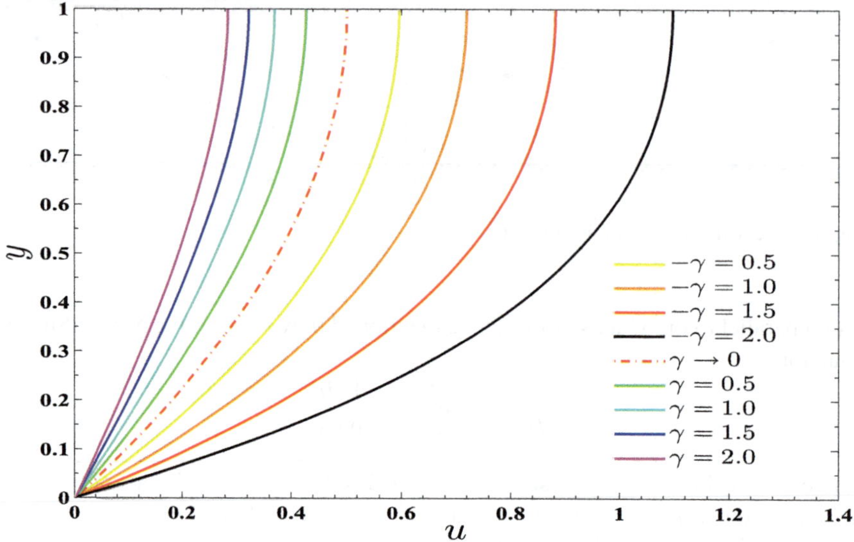

Fig. 2 Velocity profiles for different values of suction (γ) and injection ($-\gamma$) Reynolds number

$$\lim_{\gamma \to 0} u(y, \gamma) = y - \frac{y^2}{2}. \tag{6}$$

If we add the solute with constant diffusivity D in the above mentioned flow, the concentration $C(x', y', t')$ of the tracers molecules satisfies the convective-diffusion equation

$$\frac{\partial C}{\partial t'} + u' \frac{\partial C}{\partial x'} - V_0 \frac{\partial C}{\partial y'} = D \left(\frac{\partial^2 C}{\partial x'^2} + \frac{\partial^2 C}{\partial y'^2} \right), \qquad for \;\; 0 < y' < h \tag{7}$$

with the following conditions

$$C(x', y', t')\big|_{t'=0} = \delta\left(\frac{x'}{h}\right), \tag{8a}$$

$$D \frac{\partial C}{\partial y'}\bigg|_{y'=0} = 0, \tag{8b}$$

$$D \frac{\partial C}{\partial y'}\bigg|_{y'=h} = 0, \tag{8c}$$

$$C(x', y', t')|_{x' \to \pm\infty} = 0, \tag{8d}$$

where $\delta(.)$ is the Dirac delta function.

2.1 Different Time-Scales

Mei [5] first introduced multi-scale analysis to study the dispersion related problems. For the present work, three different time scales and two length scales are used. The length scales are taken as the channel depth (h) and the characteristic length of the concentration cloud (L). The timescales are defined as T_0 (i.e. the diffusion time along the depth of the channel), T_1 (i.e. the convection time along characteristic length) and T_2 (i.e. the diffusion time along the characteristic length), where

$$T_0 = h^2/D, \qquad T_1 = L/U, \qquad T_2 = L^2/D.$$

So, the ratio of the different time scales are

$$T_0 : T_1 : T_2 = 1 : \frac{1}{\varepsilon} : \frac{1}{\varepsilon^2},$$

where $\epsilon = \frac{h}{L}(\ll 1)$ is the perturbation parameter.

2.2 Dimensionless Concentration Equation

Using Eqs. (2) into(7) and (8), the dimensionless concentration equation and the respective initial and boundary conditions becomes

$$\frac{\partial C}{\partial t} + \epsilon u P_e \frac{\partial C}{\partial x} - \epsilon \gamma^* S_c \frac{\partial C}{\partial y} = \epsilon^2 \frac{\partial^2 C}{\partial x^2} + \frac{\partial^2 C}{\partial y^2}, \qquad for \;\; 0 < y < 1 \qquad (9)$$

$$C(x, y, t)|_{t=0} = \delta\left(\frac{x}{\epsilon}\right), \qquad (10a)$$

$$\left.\frac{\partial C}{\partial y}\right|_{y=0} = 0, \qquad (10b)$$

$$\left.\frac{\partial C}{\partial y}\right|_{y=1} = 0, \qquad (10c)$$

$$C(x, y, t)|_{x \to \pm\infty} = 0, \qquad (10d)$$

where $P_e = \frac{u_c h}{D}$ is the Péclet number which is the ratio of convection rate $\left(\frac{1}{h/u_c}\right)$ and diffusion rate $\left(\frac{1}{h^2/D}\right)$. Also, the suction/injection Reynolds number is defined as $\gamma = \frac{V_0 h}{\nu} = \frac{V_0 L}{\nu}\frac{h}{L} = \epsilon\gamma^*$.

2.3 Homogenization

In the present work, Mei's homogenization method is used for asymptotic analysis and we define the following time variables (fast, medium and slow) as

$$t_0 = t, \quad t_1 = \epsilon t, \quad t_2 = \epsilon^2 t. \tag{11}$$

Using chain rule, one can get

$$\frac{\partial}{\partial t} \equiv \frac{\partial}{\partial t_0} + \epsilon \frac{\partial}{\partial t_1} + \epsilon^2 \frac{\partial}{\partial t_2}. \tag{12}$$

The asymptotic expression for concentration C can be expanded into multi-scale as

$$C(x, y, t) = C_0(x, y, t_0, t_1, t_2) + \epsilon C_1(x, y, t_0, t_1, t_2) + \epsilon^2 C_2(x, y, t_0, t_1, t_2) + O(\epsilon^3). \tag{13}$$

Substituting Eqs. (12) and (13) into (9), (10b) and (10c), one can obtain

$$\left[\frac{\partial C_0}{\partial t_0} - \frac{\partial^2 C_0}{\partial y^2}\right] + \epsilon \left[\frac{\partial C_0}{\partial t_1} + \frac{\partial C_1}{\partial t_0} + u P_e \frac{\partial C_1}{\partial x} - \gamma^* S_c \frac{\partial C_0}{\partial y} - \frac{\partial^2 C_1}{\partial y^2}\right]$$

$$+\epsilon^2 \left[\frac{\partial C_0}{\partial t_2} + \frac{\partial C_1}{\partial t_1} + \frac{\partial C_2}{\partial t_0} + u P_e \frac{\partial C_0}{\partial x} - \gamma^* S_c \frac{\partial C_1}{\partial y} - \frac{\partial^2 C_0}{\partial x^2} - \frac{\partial^2 C_2}{\partial y^2}\right] + O(\epsilon^3) = 0, \quad for \ 0 < y < 1 \tag{14}$$

and

$$\frac{\partial C_0}{\partial y} + \epsilon \frac{\partial C_1}{\partial y} + \epsilon^2 \frac{\partial C_2}{\partial y} + O(\epsilon^3) = 0 \text{ at } y = 0, 1. \tag{15}$$

For the zeroth order ($O(\epsilon^0)$) perturbation, Eq. (14) and (15) gives

$$\frac{\partial C_0}{\partial t_0} = \frac{\partial^2 C_0}{\partial y^2}, \qquad for \ 0 < y < 1 \tag{16}$$

and

$$\left.\frac{\partial C_0}{\partial y}\right|_{y=0,1} = 0. \tag{17}$$

The general solution of Eq. (16) with boundary condition (17) is

$$C_0 = C_0^{(0)}(x, t_1, t_2) + \sum_{n=1}^{\infty} \text{Re}\left[C_0^{(n)}(x, t_1, t_2)e^{in\pi y}\right] e^{-n^2\pi^2 t_0}. \tag{18}$$

For large time evolution, one can consider only the first term of the expression and take the solution (18) as

$$C_0 = C_0(x, t_1, t_2). \tag{19}$$

Also, for the first order ($O(\epsilon)$) perturbation, comparing the coefficients of ϵ from Eqs. (14) to (15), we have

$$\frac{\partial C_0}{\partial t_1} + \frac{\partial C_1}{\partial t_0} + u P_e \frac{\partial C_0}{\partial x} - \gamma^* S_c \frac{\partial C_0}{\partial y} - \frac{\partial^2 C_1}{\partial y^2} = 0, \tag{20}$$

$$\left.\frac{\partial C_1}{\partial y}\right|_{y=0,1} = 0. \tag{21}$$

As C_0 is independent of y, $\frac{\partial C_0}{\partial y} = 0$. Thus, Eq. (20) reduces to

$$\frac{\partial C_0}{\partial t_1} + \frac{\partial C_1}{\partial t_0} + u P_e \frac{\partial C_0}{\partial x} - \frac{\partial^2 C_1}{\partial y^2} = 0. \tag{22}$$

Since t_0 is very large, Eq. (22) becomes

$$\frac{\partial C_0}{\partial t_1} + u P_e \frac{\partial C_0}{\partial x} - \frac{\partial^2 C_1}{\partial y^2} = 0. \tag{23}$$

Consider the depth averaged function $\langle g \rangle$ of a function g with respect to y as

$$\langle g \rangle = \int_0^1 g dy. \tag{24}$$

Now, taking depth average of Eq. (23) and applying the conditions Eqs. (21), (23) takes the form

$$\frac{\partial C_0}{\partial t_1} + \langle u \rangle P_e \frac{\partial C_0}{\partial x} = 0. \tag{25}$$

Subtracting Eq. (25) from Eq. (22), we get

$$\frac{\partial C_1}{\partial t_0} + P_e(u - \langle u \rangle)\frac{\partial C_0}{\partial x} = \frac{\partial^2 C_1}{\partial y^2}. \tag{26}$$

Considering the following substitution,

$$C_1 = P_e F(y)\frac{\partial C_0}{\partial x} \tag{27}$$

in Eq. (26) and comparing the coefficients of $\frac{\partial C_0}{\partial x}$ from both sides, one can get

$$\frac{d^2 F}{dy^2} = u - \langle u \rangle \tag{28}$$

and

$$\left.\frac{dF}{dy}\right|_{y=0,1} = 0, \tag{29a}$$

$$\langle F \rangle = 0. \tag{29b}$$

For second order $o(\epsilon^2)$ perturbation, Eqs. (14) and (15) gives

$$\frac{\partial C_0}{\partial t_2} + \frac{\partial C_1}{\partial t_1} + \frac{\partial C_2}{\partial t_0} + P_e u \frac{\partial C_1}{\partial x} - \gamma^* S_c \frac{\partial C_1}{\partial y} - \frac{\partial^2 C_0}{\partial x^2} - \frac{\partial^2 C_2}{\partial y^2} = 0, \tag{30}$$

$$\left.\frac{\partial C_2}{\partial y}\right|_{y=0,1} = 0. \tag{31}$$

Again, since t_0 is very large, Eq. (30) reduces to

$$\frac{\partial C_0}{\partial t_2} + \frac{\partial C_1}{\partial t_1} + P_e u \frac{\partial C_1}{\partial x} = \gamma^* S_c \frac{\partial C_1}{\partial y} + \frac{\partial^2 C_0}{\partial x^2} + \frac{\partial^2 C_2}{\partial y^2} \tag{32}$$

Taking depth average on Eq. (32) and applying the conditions (31) & using Eq. (27), the Eq. (32) becomes

$$\frac{\partial C_0}{\partial t_2} + \frac{\partial \langle C_1 \rangle}{\partial t_1} + P_e \langle u \frac{\partial C_1}{\partial x} \rangle = \gamma^* S_c P_e [F(1) - F(0)] \frac{\partial C_0}{\partial x} + \frac{\partial^2 C_0}{\partial x^2}$$

$$\Longrightarrow \frac{\partial C_0}{\partial t_2} + P_e \langle u \frac{\partial C_1}{\partial x} \rangle = \gamma^* S_c P_e [F(1) - F(0)] \frac{\partial C_0}{\partial x} + \frac{\partial^2 C_0}{\partial x^2}. \tag{33}$$

Subtracting Eqs.(33) from Eq. (30), on can get

$$\frac{\partial C_1}{\partial t_1} + \frac{\partial C_2}{\partial t_0} + P_e \left(u \frac{\partial C_1}{\partial x} - \langle u \frac{\partial C_1}{\partial x} \rangle \right) = \gamma^* S_c \frac{\partial C_1}{\partial y} + \frac{\partial^2 C_2}{\partial y^2} - \gamma^* S_c P_e [F(1) - F(0)] \frac{\partial C_0}{\partial x}. \tag{34}$$

Also, from Eq. (27), we have

$$\frac{\partial C_1}{\partial x} = P_e F \frac{\partial^2 C_0}{\partial x^2} \tag{35}$$

and

$$u\frac{\partial C_1}{\partial x} = P_e u F \frac{\partial^2 C_0}{\partial x^2}. \tag{36}$$

Again, taking depth average of Eq. (36), we get

$$\langle u\frac{\partial C_1}{\partial x}\rangle = P_e \langle uF\rangle \frac{\partial^2 C_0}{\partial x^2}. \tag{37}$$

Substituting Eqs. (37) in (33), it becomes

$$\frac{\partial C_0}{\partial t_2} = [1 - P_e^2\langle uF\rangle]\frac{\partial^2 C_0}{\partial x^2} + \gamma^* S_c P_e [F(1) - F(0)]\frac{\partial C_0}{\partial x}. \tag{38}$$

Multiplying Eq. (25) by ϵ and Eq. (38) by ϵ^2 and adding them, the result gives

$$\frac{\partial C_0}{\partial t_0} + \epsilon\frac{\partial C_0}{\partial t_1} + \epsilon^2\frac{\partial C_0}{\partial t_2} + \epsilon P_e[\langle u\rangle - \gamma S_c\{F(1) - F(0)\}]\frac{\partial C_0}{\partial x} = \epsilon^2[1 - P_e^2\langle uF\rangle]\frac{\partial^2 C_0}{\partial x^2}. \tag{39}$$

Since C_0 is independent of t_0, $\frac{\partial C_0}{\partial t_0} = 0$. Using the identity (12), Eq. (39) reduces to

$$\frac{\partial C_0}{\partial t} + \epsilon P_e[\langle u\rangle - \gamma S_c\{F(1) - F(0)\}]\frac{\partial C_0}{\partial x} = \epsilon^2[1 - P_e^2\langle uF\rangle]\frac{\partial^2 C_0}{\partial x^2}. \tag{40}$$

Introducing the following new dimensionless variables in the above equation

$$\tau = t, \qquad \xi = \frac{x}{\epsilon} - P_e\left[\langle u\rangle - \gamma S_c\{F(1) - F(0)\}\right]\tau, \tag{41}$$

the Eq. (40) takes the following form

$$\frac{\partial C_0}{\partial \tau} = D_T\frac{\partial^2 C_0}{\partial \xi^2}, \tag{42}$$

where

$$D_T = 1 - P_e^2\langle uF\rangle. \tag{43}$$

After solving Eq. (42) with the help of Eq. (10), one can get

$$C_0 = \frac{1}{\sqrt{4\pi D_T \tau}}\exp\left(\frac{-\xi^2}{4D_T\tau}\right). \tag{44}$$

Using Eq. (27), (36) and (37) in Eq. (34), we have

$$\frac{\partial C_2}{\partial t_0} + P_e^2\left[uF - \langle u\rangle F - \langle uF\rangle\right]\frac{\partial^2 C_0}{\partial x^2} = \gamma^* S_c P_e\left[\frac{dF}{dy} - \{F(1) - F(0)\}\right]\frac{\partial C_0}{\partial x} + \frac{\partial^2 C_2}{\partial y^2}, \quad (45)$$

where $\frac{\partial C_1}{\partial t_1} = -P_e^2\langle u\rangle F\frac{\partial^2 C_0}{\partial x^2}$. On considering the following substitution in Eq. (45)

$$C_2 = P_e^2 G(y)\frac{\partial^2 C_0}{\partial x^2} + \gamma^* P_e S_c H(y)\frac{\partial C_0}{\partial x}, \quad (46)$$

and comparing the coefficients of $\frac{\partial^2 C_0}{\partial x^2}$ and $\frac{\partial C_0}{\partial x}$ from the both sides of Eq. (45), one can get

$$\frac{d^2 G}{dy^2} = uF - \langle u\rangle F - \langle uF\rangle, \quad (47)$$

along with the conditions

$$\left.\frac{dG}{dy}\right|_{y=0,1} = 0, \quad (48a)$$

$$\langle G\rangle = 0 \quad (48b)$$

and

$$\frac{d^2 H}{dy^2} = -\left[\frac{dF}{dy} - \{F(1) - F(0)\}\right], \quad (49)$$

with

$$\left.\frac{dH}{dy}\right|_{y=0,1} = 0, \quad (50a)$$

$$\langle H\rangle = 0. \quad (50b)$$

Solving Eqs. (28), (47) and (49) with respective boundary conditions, the solutions are given below

$$F = \frac{e^{-\gamma}}{\gamma^3}\left[(e^{\gamma} - 1)\left(\frac{y^2}{2} + \frac{1}{\gamma^2} - \frac{1}{6}\right) - \frac{e^{\gamma y}}{\gamma} + y - \frac{1}{2}\right] + \frac{1}{\gamma}\left(\frac{y^3}{6} - \frac{y^2}{4} + \frac{1}{24}\right), \quad (51)$$

$$
\begin{aligned}
G &= a_1\left(\frac{y^2}{2}-\frac{1}{6}\right)+a_2\left(\frac{y^3}{6}-\frac{1}{24}\right)+a_3\left(\frac{y^4}{12}-\frac{1}{60}\right)+a_4\left(\frac{y^5}{20}-\frac{1}{120}\right)\\
&+a_5\left(\frac{y^6}{30}-\frac{1}{210}\right)+a_6\left(\frac{e^{\gamma y}}{2}-\frac{y}{\gamma}-\frac{e^{\gamma}}{\gamma^3}+\frac{1}{2\gamma}+\frac{1}{\gamma^3}+\right)\\
&+a_7\left(\frac{ye^{\gamma y}}{\gamma^2}-\frac{2e^{\gamma y}}{\gamma^3}+\frac{y}{\gamma^2}-\frac{e^{\gamma}}{\gamma^3}+\frac{3e^{\gamma}}{\gamma}-\frac{1}{2\gamma^2}-\frac{3}{\gamma^4}\right)\\
&+a_8\left(\frac{y^2e^{\gamma y}}{\gamma^2}-\frac{4ye^{\gamma y}}{\gamma^3}+\frac{6e^{\gamma y}}{\gamma^4}-\frac{2y}{\gamma^3}-\frac{e^{\gamma}}{\gamma^3}+\frac{6e^{\gamma}}{\gamma^4}-\frac{12e^{\gamma}}{\gamma^5}+\frac{1}{\gamma^3}+\frac{12}{\gamma^5}\right)\\
&+a_9\left(\frac{y^3e^{\gamma y}}{\gamma^2}-\frac{6y^2e^{\gamma y}}{\gamma^3}+\frac{18ye^{\gamma y}}{\gamma^4}-\frac{24e^{\gamma y}}{\gamma^5}+\frac{6y}{\gamma^4}-\frac{e^{\gamma}}{\gamma^3}+\frac{9e^{\gamma}}{\gamma^4}-\frac{36e^{\gamma}}{\gamma^4}-\frac{36e^{\gamma}}{\gamma^5}\right.\\
&\left.+\frac{60e^{\gamma}}{\gamma^6}-\frac{3}{\gamma^4}-\frac{60}{\gamma^6}\right)+a_{10}\left(\frac{e^{2\gamma y}}{4\gamma^2}-\frac{y}{2\gamma}-\frac{e^{2\gamma}}{8\gamma^3}+\frac{1}{4\gamma}+\frac{1}{8\gamma^3}\right),
\end{aligned}
\tag{52}
$$

where

$$
\begin{aligned}
a_1 &= \left[\frac{e^{-\gamma}}{\gamma^3}\left(e^{\gamma}-1\right)\left(\frac{1}{\gamma^2}-\frac{1}{6}\right)-\frac{e^{-\gamma}}{2\gamma^3}+\frac{1}{24\gamma}\right]\left[\frac{e^{-\gamma}\left(e^{\gamma}-1\right)}{\gamma^3}-\frac{1}{2\gamma}\right]-\langle uF\rangle,\\
a_2 &= \frac{1}{\gamma}\left[\frac{e^{-\gamma}}{\gamma^3}\left(e^{\gamma}-1\right)\left(\frac{1}{\gamma^2}-\frac{1}{6}\right)-\frac{e^{-\gamma}}{2\gamma^3}+\frac{1}{24\gamma}\right]+\frac{e^{-\gamma}}{\gamma^3}\left[\frac{e^{-\gamma}\left(e^{\gamma}-1\right)}{\gamma^3}-\frac{1}{2\gamma}\right],\\
a_3 &= \frac{e^{-\gamma}}{\gamma^4}+\left[\frac{e^{-\gamma}\left(e^{\gamma}-1\right)}{2\gamma^3}-\frac{1}{4\gamma}\right]\left[\frac{e^{-\gamma}\left(e^{\gamma}-1\right)}{\gamma^3}-\frac{1}{2\gamma}\right],\\
a_4 &= \frac{1}{\gamma}\left[\frac{e^{-\gamma}\left(e^{\gamma}-1\right)}{2\gamma^3}-\frac{1}{4\gamma}\right]+\frac{1}{6\gamma}\left[\frac{e^{-\gamma}\left(e^{\gamma}-1\right)}{\gamma^3}-\frac{1}{2\gamma}\right],\quad a_5=\frac{1}{6\gamma^2},\\
a_6 &= -\frac{e^{-\gamma}}{\gamma^2}\left[\frac{e^{-\gamma}}{\gamma^3}\left(e^{\gamma}-1\right)\left(\frac{1}{\gamma^2}-\frac{1}{6}\right)-\frac{e^{-\gamma}}{2\gamma^3}+\frac{1}{24\gamma}\right]-\frac{e^{-\gamma}}{\gamma^4}\left[\frac{e^{-\gamma}\left(e^{\gamma}-1\right)}{\gamma^3}-\frac{1}{2\gamma}\right],\\
a_7 &= -\frac{e^{-2\gamma}}{\gamma^5}-\frac{e^{-\gamma}}{\gamma^5},\quad a_8=-\frac{e^{-\gamma}}{\gamma^2}\left[\frac{e^{-\gamma}\left(e^{\gamma}-1\right)}{2\gamma^3}-\frac{1}{4\gamma}\right],\\
a_9 &= -\frac{e^{-\gamma}}{6\gamma^3},\quad a_{10}=\frac{e^{-2\gamma}}{\gamma^6},
\end{aligned}
$$

$$
\begin{aligned}
H = -\frac{e^{-\gamma}}{\gamma^3}&\left[\left(e^{\gamma}-1\right)\left(\frac{y^3}{6}-\frac{1}{24}\right)+\left(\frac{y^2}{2}-\frac{1}{6}\right)-\frac{e^{\gamma y}}{\gamma^2}+\frac{y}{\gamma}+\frac{e^{\gamma}}{\gamma^3}-\frac{1}{2\gamma}-\frac{1}{\gamma^3}\right]\\
&-\frac{1}{\gamma}\left(\frac{y^4}{24}-\frac{y^3}{12}+\frac{1}{80}\right)+[F(1)-F(0)]\left(\frac{y^2}{2}-\frac{1}{6}\right).
\end{aligned}
\tag{53}
$$

3 Result and Discussions

This section deals with a discussion of mean concentration distribution and Taylor dispersivity under the influence of suction/injection Reynolds number (γ).
The transverse concentration distribution and mean concentration distribution are given as follows

$$C = C_0 + \varepsilon C_1 + \varepsilon^2 C_2, \tag{54}$$

$$\langle C \rangle = C_0 + \varepsilon \langle C_1 \rangle + \varepsilon^2 \langle C_2 \rangle. \tag{55}$$

The Taylor dispersion coefficient is given by

$$D_T = 1 - P_e^2 \langle uF \rangle. \tag{56}$$

Figure 3 describes the dispersion coefficient of solute for various values of γ. It is seen from Fig. 3a that the dispersivity increases significantly for large values of injection parameter ($-\gamma$). On the other hand, the opposite situation arises when the suction parameter (γ) increases. In case of injection, the dispersion coefficient is large in compare to that of the case in suction parameter. Also, it is observe that, for all cases, the dispersion coefficient increases as the Péclet number enhances. Figure 3b depicts the Taylor dispersivity of the solute material for different values of injection parameter with respect to the channel height y. It is seen that as the injection Reynolds number increases, the dispersivity increases throughout the whole channel height. It is also visible from the Fig. 3b that near the bed ($y = 0$), the dispersivity is very low and as, height of the channel rises, the dispersivity also increases. It is observed that at the free surface of the open channel ($y = 1$), the Taylor dispersivity gains it's maximum value. The Taylor dispersivity is presented for various values of suction Reynolds number in Fig. 3c. The opposite phenomena on the dispersion coefficient is observed in compare to that of in Fig. 3b.

Figure 4 displays the longitudinal real concentration for different values of γ, τ and y. It is seen from Fig. 4a that with the enhancement of injection parameter, the amplitude of the concentration distribution increases. It is noted that when $-\gamma \geq 3$, the effect of injection on concentration along longitudinal direction is negligible. Influence of suction on distribution of concentration is shown in Fig. 4b and it is observed that the peak of the concentration curve reduces as suction Reynolds number increases. It is noticed from Fig. 4c that the amplitude of the longitudinal real concentration reduces significantly with the increment of dispersion time. It is interesting to note that the concentration distribution moves towards the downstream direction when transverse position raises from the bed surface (see Fig. 4d).

Iso-concentration contours are depicted in Fig. 5 for different values of injection parameter. When $-\gamma = 3$ the concentration stay at the bed surface. The concentration starts to move away longitudinally for $-\gamma = 5$, the reason behind this, the introduction of injection in the flow, the flow velocity increases, as a result convec-

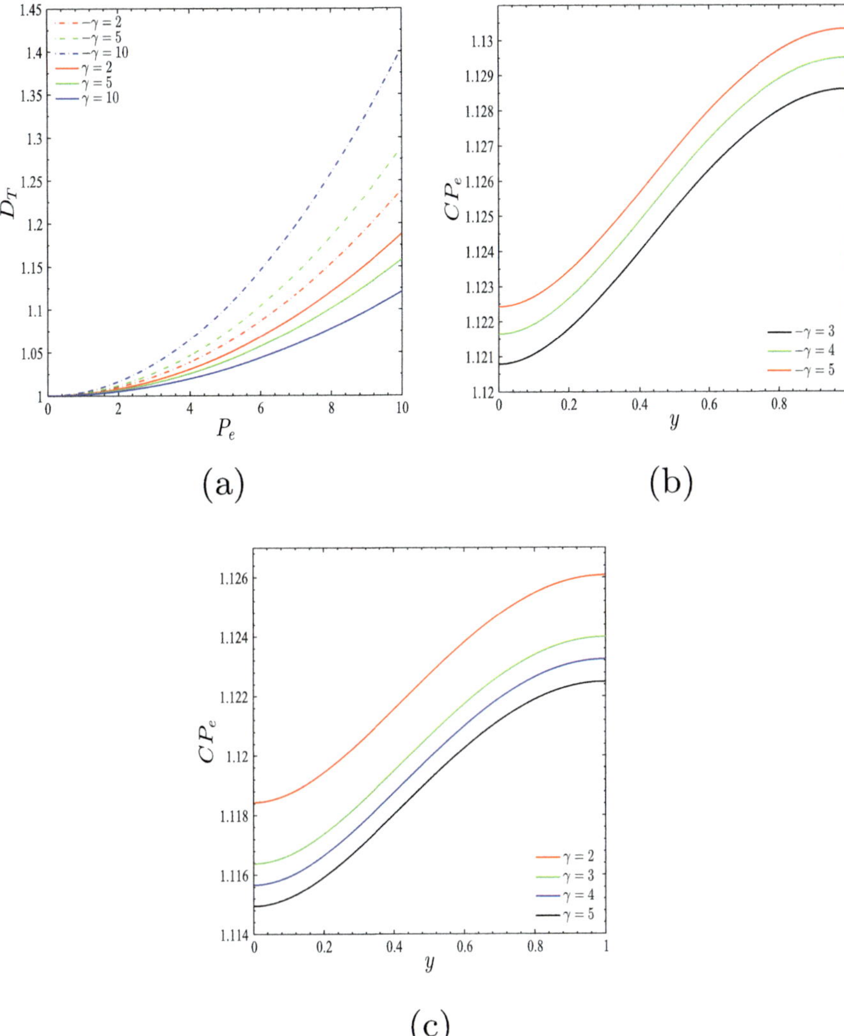

Fig. 3 **a** Dispersion coefficient for different values of injection $(-\gamma)$ and suction (γ) Reynolds number, **b** Transverse concentration for various values of injection Reynolds number $(-\gamma)$, **c** Transverse concentration for different values of suction Reynolds number (γ)

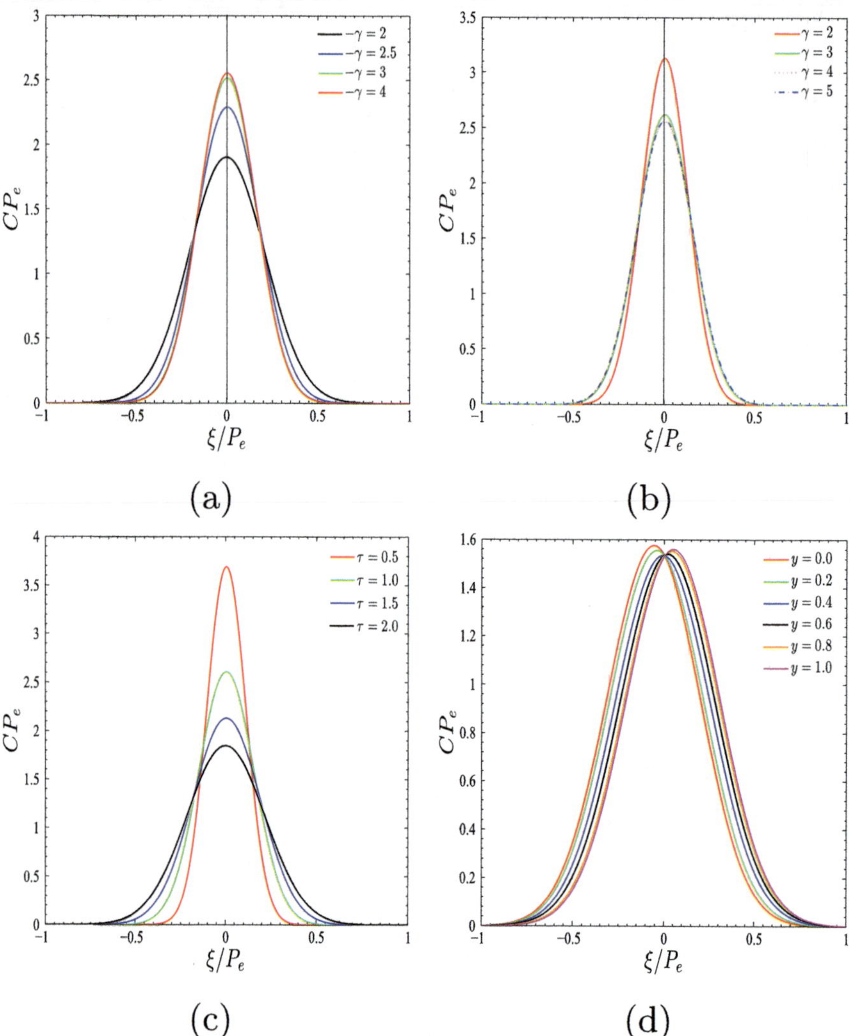

Fig. 4 Longitudinal real concentration distribution for various values of: **a** injection Reynolds number ($-\gamma$), **b** suction Reynolds number (γ), **c** dispersion time, **d** height of the channel

tion takes places in the flow. Also for $-\gamma = 10$ the rate of convection is large as compared to the previous values of injection Reynolds number.

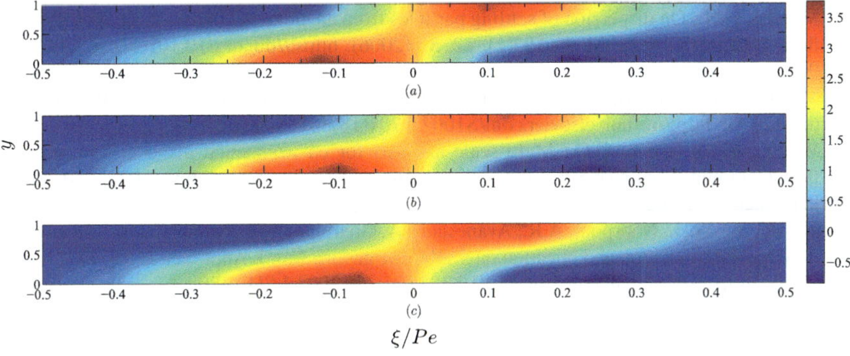

Fig. 5 Iso-concentration contour for: **a** $-\gamma = 3$, **b** $-\gamma = 5$, **c** $-\gamma = 10$

4 Conclusion

In the present study, solute transport in an open channel flow is analyzed by using multi-scale analysis technique. The real concentration distribution and Taylor dispersivity of the solute is evaluated analytically. It is observed that Taylor dispersivity decreases with the increment of suction Reynolds number and it increases with the enhancement of injection Reynolds number. The reason behind this, when γ increases, the flow decreases, as a result rate of dispersion reduces and consequently, the peak of the amplitude of the longitudinal concentration distribution decreases. The opposite scenario arises when $-\gamma$ increases, i.e. when injection Reynolds number increases. With the enhancement of $-\gamma$, the flow velocity increases which helps to increase the rate of dispersion. Therefore, one can observe the increment in the amplitude of the real concentration along the longitudinal direction. As dispersion time progresses, the amplitude of the concentration curves along longitudinal direction becomes flat. This is because, with the increment of time, the mixing of the solute increases. It is observed that the iso-concentration contour spreads along the longitudinal direction with the enhancement of injection Reynolds number ($-\gamma$), because the rate of convection becomes large for higher values of the injection Reynolds number ($-\gamma$).

References

1. Taylor, G.I.: Dispersion of soluble matter in solvent flowing slowly through a tube. Proc. R. Soc. Lond. A **219**, 186–203(1953). https://doi.org/10.1098/rspa.1953.0139
2. Aris, R.: On the dispersion of a solute in a fluid flowing through a tube. Proc. R. Soc. Lond. A **235**, 67–77 (1956). https://doi.org/10.1098/rspa.1956.0065
3. Gill, W.N.: A note on the solution of transient dispersion problems. In: Proc. R. Soc. Lond. Ser. A **298**(1454), 335–339 (1967). https://doi.org/10.1098/rspa.1967.0107

4. Mondal, K.K.: On dispersion of settling particles from an elevated sources in an open channel flow. J. Comput. Appl. Math. **193**(1), 22–37 (2006). https://doi.org/10.1016/j.cam.2005.04.068
5. Mei, C.C., Vernescu, B.: Homogeniyation methods for multiscale mechanics. In: World scientific (2010). http://refhub.elsevier.com/S1007-5704(18)30135-7/sbref0045
6. Wu, Z., Li, Y., Chen, G.Q.: Multi-scale analysis for environmental dispersion in wetland flow. Commun. Nonlinear Sci. Numer. Simul. **16**, 3168–3178 (2011). https://doi.org/10.1016/j.cnsns.2010.12.002
7. Shojaefard, M.H., Noorpoor, A.R., Avanesians, A. Ghaffapour, M.: Numerical investigation of flow control by suction and injection on a subsonic airfoil. Am. J. Appl. Sci. **2**(10), 1474–1480 (2005). https://thescipub.com/abstract/10.3844/ajassp.2005.1474.1480
8. Barnwell, R.W., Hussaini, M.Y.: Natural laminar flow and laminar flow control. Springer, New York (1992). https://doi.org/10.1007/978-1-4612-2872-1
9. Ishak, A.J., Merkin, H., Nazar, R., Pop, I.: Mixed convection boundary layer flow over a permeable vertical surface with prescribed wall heat flux. In: ZAMP: Zeitschrift *für* angewandte Mathematik und Physik, vol. 59(1), pp. 100–123 (2008). https://doi.org/10.1155/2010/659023
10. Jha, B.K., Aina, B.: Role of suction/injection on steady fully developed mixed convection flow in a vertical parallel plate microchannel. Ain Shams Eng. J. **9**(4), 747–755(2018). https://doi.org/10.1016/j.asej.2016.05.001
11. Sasikumar J, Bhati, D., Bhaskar, V.: Effect of heat and mass transfer on MHD oscillatory flow through asymmetric wavy channel in a porous medium with suction and injection. In: AIP Conference Proceedings, vol. 2277, p. 030009 (2020). https://doi.org/10.1063/5.0025530
12. Abo-Dahab, S.M., Abdelhafez, M.A., Mebarek-Oudina, F., Bilal, S.M.: MHD Casson nanofluid flow over nonlinearly heated porous medium in presence of extending surface effect with suction/injection. Indian J. Phys. **95**, 2703–2717 (2021). https://doi.org/10.1007/s12648-020-01923-z

Mathematical Modelling of Magnetized Nanofluid Flow Over an Elongating Cylinder with Erratic Thermal Conductivity

Debasish Dey, Rupjyoti Borah, and Joydeep Borah

Abstract An attempt has been made to analyze the flow behaviours of magnetized nanofluid due to an elongating cylinder with the contemporary effects of both heat and mass transference. The flow governing equations are re-modeled into a solvable form by considering a suitable similarity transformation. The MATLAB fourth-order Runge-Kutta shooting technique is implemented to work out the problem. The numerical findings are assembled on a system of tables and diagrams. These results show nanofluid properties for a wide range of circulations such as motion, thermal and mass fractions including physical dimensionless numbers. The major result of this study is that the thermal fraction of the nanofluid is an escalating function of the flow parameters namely the magnetic, Brownian motion, thermophoresis, temperature ratio and the curvature parameters.

Keywords Nanofluid · Heat transfer · Mass transfer · Extending cylinder · Variable thermal conductivity

1 Introduction

The flow of different fluid models caused due to an extending or contracting geometries have lots of applications in engineering sciences, industrial processes and medical sciences etc. Crane [1] was the first author who has discussed the fluid's flow caused due to stretching/shrinking surfaces in 1970. Again, the relevance of magnetic field on the fluid's flow caused due to different surfaces has plentiful applications in diverse fields. In the last five years, many researchers Gangadhar et al. [2], Das et al. [3], Dey et al. [4], Dey and Borah [5], Dey et al. [6, 7] etc. have

D. Dey (✉) · R. Borah
Department of Mathematics, Dibrugarh University, Dibrugarh 786004, AS, India
e-mail: debasish41092@gmail.com

J. Borah
Department of Mathematics, D. D. R. College, Chabua 786184, AS, India

S. Banerjee and A. Saha (eds.), *Nonlinear Dynamics and Applications*,
Springer Proceedings in Complexity,
https://doi.org/10.1007/978-3-030-99792-2_41

discussed the effects of magnetic field on the different fluids flow by considering stretching/shrinking surfaces.

The simultaneous effects of both thermal and mass transmission on fluid flows have multifarious applications. The effect of heat transfer has drawn many researchers due to its applications in different fields such as industrial processes, medical sciences and biological systems. The mass transfer phenomenon draws an imperative function in realizing industrial processes such as reverse osmosis, membrane separation, dissemination of chemical impurities and distillation of water etc. Manjunatha et al. [8] and Divya et al. [9] have discussed the simultaneous effects of both heat and mass transfers in the peristaltic mechanism of the Jeffrey fluid flow model. In recent time, Prasad et al. [10], Dey and Borah [11] and Vaidya et al. [12] etc. have discussed the flow behaviours with the influenced of both heat and mass transfers due to a stretching/shrinking geometries respectively. Abel et al. [13] and Jahan et al. [14] have explored the nature of flow by considering variable fluid properties.

The nanofluid signifies a crucial function in the modern time because it can enhance the thermal transmission that is important in industrial processes. It is one kind of fluid that contains solid particles with dimension less than 100 nm. Suspension of nanoparticles in fluid is one of the most appropriate techniques for enhancing heat transmission coefficients. It is a smart fluid where we can reduce or enhance the heat transfer phenomenon as requirement. The nanofluid has lots of applications such as industrial processes, nuclear reactors, transportation, electronics, extraction of geothermal power, biomedicine and food processing etc. In 1995, Choi [15] has developed the term 'Nanofluid'. Many researchers have influenced the various applications of nanofluid in modern times and put their ideas to rich the research level on nanofluids. Das et al. [16], Ghosh and Mukhopadhyay [17] and Molli and Naikoti [18] etc. have investigated the behaviour of nanofluid flow and their importance in different physical fields. Prasad et al. [10], Li et al. [19] and Narender et al. [20] have examined the flow nature of nanofluid with the effects of both heat and mass transfers. Khashi'ie et al. [21] have given the mathematical model of hybrid nanofluid flow due to a shrinking cylinder which is situated in porous medium.

The intention of this model is catalogued below:

i. Time-independent flow of magnetized nanofluid due to an extending cylinder with the effects of both thermal and concentration diffusions is considered.
ii. The governing equations viz., equation of motion, energy and species of the nanofluid are transformed into solvable form by adopting appropriate similarity transformation and hence decipher with the support of MATLAB fourth-order Runge- Kutta Shooting method.
iii. Flow behaviours of nanofluid with temperature and mass fractions are presented pictorially for assorted amount of novel flow parameters.
iv. The physical measures of curiosity such as drag force, local Nusselt number and mass diffusion rate of the fluid at the vicinity of the system are tabulated.
v. A comparison table is made to validate our results for the limiting case with the pioneer works of Jahan et al. [14].

We believed that this work carries lots of novelty such as idea of nanofluid, variable fluid properties and elongating cylinder etc. and may be applied in various industrial processes, scientific fields and medical sciences etc. In future, researchers may extend this work by considering different geometries by assisting different approaches.

2 Problem Construction

To form the mathematical model of this problem, the following suppositions are made.

(i) The steady, 2D, incompressible and viscous nanofluid flow over an extending cylinder with the synchronized influences of both heat and mass transference.
(ii) The flow is governed by (a) inertia force, (b) viscous force and (c) pressure gradient.
(iii) The cylinder of diameter $2R$ is immersed horizontally in the nanofluid with the ambient fluid's velocity is $U_e = bx$ where $b \geq 0$.
(iv) A constant magnetic field of strength (B_0) is considered in the vertical direction of the flow.
(v) The surface of the geometry is characterized by the velocity $U_W(x) = ax$, where a is the constant such that $a > 0$ signifies elongate at the surface with the prescribed wall temperature and concentration are $T_W(x) - T_\infty = x \& C_w(x) - C_\infty = x$ respectively, where T_∞ and C_∞ are the ambient temperature and concentration respectively.
(vi) $k(T)$ is supposed to fluctuate with temperature in linear manner.

The schematic illustration of this problem is drawn in Fig. 1. Following Jahan et al. [14], the leading equations are:

$$\frac{\partial}{\partial x}(ru) + \frac{\partial}{\partial r}(rv) = 0, \tag{1}$$

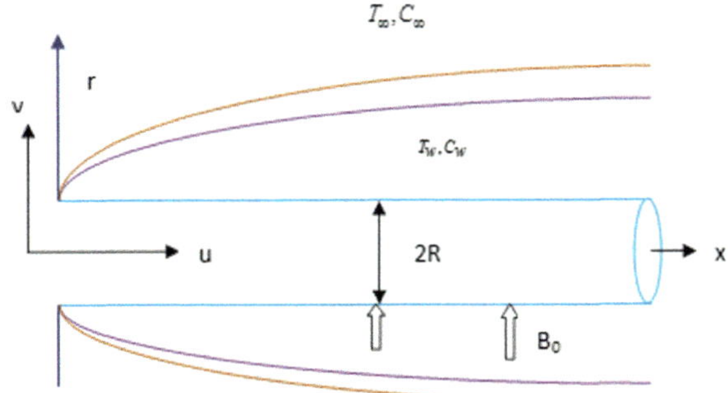

Fig. 1 Flow diagram

$$u\frac{\partial u}{\partial x}+v\frac{\partial u}{\partial r}=\upsilon\left(\frac{\partial^2 u}{\partial r^2}+\frac{1}{r}\frac{\partial u}{\partial r}\right)+\frac{\sigma B_0^2}{\rho}(u-U_e), \tag{2}$$

$$u\frac{\partial T}{\partial x}+v\frac{\partial T}{\partial r}=\frac{1}{\rho c_p}\frac{1}{r}\frac{\partial}{\partial r}\left[k(T)r\frac{\partial T}{\partial r}\right]+\tau\left[D_B\frac{\partial C}{\partial r}\frac{\partial T}{\partial r}+\frac{D_T}{T_\infty}\left(\frac{\partial T}{\partial r}\right)^2\right]+\frac{\sigma B_0^2}{\rho c_p}u^2, \tag{3}$$

$$u\frac{\partial C}{\partial x}+v\frac{\partial C}{\partial r}=D_B\left[\frac{\partial^2 C}{\partial r^2}+\frac{1}{r}\frac{\partial C}{\partial r}\right]+\frac{D_T}{T_\infty}\left[\frac{\partial^2 T}{\partial r^2}+\frac{1}{r}\frac{\partial T}{\partial r}\right]. \tag{4}$$

The relevant boundary conditions are

$$\begin{aligned}&u=U_w=ax,\, v=0,\, T=T_w,\, C=C_w \text{ at } r=R,\\ &u\to U_e=bx,\, T\to T_\infty,\, C\to C_\infty \text{ as } r\to\infty,\end{aligned} \tag{5}$$

To transform the Eqs. (1)–(4) into solvable form, the following new quantities are launched (following Jahan et al. [14]):

$$\begin{aligned}&\eta=\sqrt{\frac{a}{2\upsilon}}\left(\frac{r^2-R^2}{R}\right),\, \psi=\sqrt{\frac{\upsilon a}{2}}Rxf(\eta),\, T-T_\infty=(T_w-T_\infty)\theta(\eta),\\ &C-C_\infty=(C_w-C_\infty)\phi(\eta),\, u=\frac{1}{r}\frac{\partial\psi}{\partial r},\, v=-\frac{1}{r}\frac{\partial\psi}{\partial x}.\end{aligned} \tag{6}$$

For liquid metal, k revolutionize with temperature in linear way from $0°F-400°F$ approximately (Keys [22]). Following Abels et al. [13], we define $k(T)=k_\infty(1+\varepsilon\theta)$, where $\varepsilon=\frac{k_w-k_\infty}{k_\infty}$ is the small temperature parameter.

The Eq. (1) is satisfied by the Eq. (6) which represents the validation of the similarity transformation. Applying Eq. (6) into the Eqs. (2)–(4) and (5) we have achieved the following solvable equations along with boundary condition.

$$(2+2K\eta)f'''+ff''-(f')^2+2Kf''-M^2(f'-A)+A^2=0, \tag{7}$$

$$\begin{aligned}&(1+\varepsilon\theta)(2+2K\eta)\theta''+\varepsilon(2+2K\eta)\left(\theta'\right)^2+(1+\varepsilon\theta)K\theta'+\Pr\left(f\theta'+M^2Ecf'^2\right)\\ &+2\Pr N_b(2+2K\eta)\theta'\phi'+\Pr N_T(2+2K\eta)\theta'^2=0,\end{aligned} \tag{8}$$

$$(2+2K\eta)\phi''+K\phi'+\frac{N_t}{N_b}(2+2K\eta)\theta''+\frac{N_t}{N_b}K\theta'+Le(f\phi'-f'\phi)=0. \tag{9}$$

The surface restrictions are:

$$\begin{aligned}&f(0)=0,\, f'(0)=1,\, \theta(0)=1,\, \phi(0)=1;\\ &f'(\infty)\to A,\, \theta(\infty)\to 0,\, \phi(\infty)\to 0.\end{aligned} \tag{10}$$

where, the parameters are defined in the following way:

$$K = \frac{1}{R}\sqrt{\frac{2\upsilon}{a}},\ \Pr = \frac{\mu c_p}{k_\infty},\ M^2 = \frac{\sigma B_0^2}{\rho a},\ Ec = \frac{U_w^2}{c_p(T_w - T_\infty)},$$
$$A = \frac{b}{a},\ N_b = \frac{\tau D_B(C_w - C_\infty)}{\upsilon},\ N_t = \frac{\tau D_T(T_w - T_\infty)}{\upsilon T_\infty} \,\&\, Le = \frac{\upsilon}{D_B}.$$

The dimensionless numbers such as drag force of the fluid at surface, local Nusselt number and mass accumulation rate (determined by Sherwood number) are observed in this study. These quantities play a vital role in significant real life areas. These quantities are defined in the following way:

$$C_f = \frac{1}{\rho U_w^2}\mu\left(\frac{\partial u}{\partial r}\right)_{r=R},\ Nu_x = -\frac{x}{(T_w - T_\infty)}\left(\frac{\partial T}{\partial r}\right)_{r=R}$$
$$Sh_x = -\frac{x}{(C_w - C_\infty)}\left(\frac{\partial C}{\partial r}\right)_{r=R}. \tag{11}$$

The Eq. (6) is implemented on this Eq. (11), we have got the following form:

$$\frac{1}{\sqrt{2}}C_f \mathrm{Re}_x^{1/2} = f''(0),\ \frac{1}{\sqrt{2}}Nu_x \mathrm{Re}_x^{-1/2} = -\theta'(0),\ \frac{1}{\sqrt{2}}Sh_x \mathrm{Re}_x^{-1/2} = -\phi'(0). \tag{12}$$

3 Methodology

Following Hazarika [23] and Hazarika et al. [24], the MATLAB fourth-order Runge–Kutta shooting scheme is adopted to solve the Eqs. (7)–(9) along with the surface restriction (10). Shooting technique is a suitable scheme for solving a boundary value problem of fluid dynamics. It is a sophisticated computer oriented numerical method. This method can be used for solving both systems of linear and non linear equations, highly coupled boundary value problems of ordinary differential equations. This method does not need linearization of the equations, has been effectively applied to this class of problems.

In case of boundary value problems of second order differential equations, one condition is prescribed at either of the end points. Hence, at the initial point of integration, one condition is always missing. Shooting method estimates the missing initial condition in such a way that the estimation satisfies the condition prescribed at the boundary too, to some desired accuracy. At the beginning, the missing value is guessed and refined by using an iterative technique until the desired accuracy is obtained.

4 Results and Discussion

The above mentioned numerical technique is adopted to obtain the effects of flow parameters on the nanofluid flow in terms of velocity, temperature and mass fraction. A special emphasis is given on the effects of the magnetic (M), curvature (K), small temperature (ε), Brownian motion (N_b) and thermophoresis (N_t) parameters in the flow. We have fixed the value of the Lewis number (Le) (ratio of the Schmith number and Prandtl number) is 2 throughout the study. The greater value of Le than 1 represents that the heat will neutralize more hurriedly than species.

Influence of M on the velocity and temperature fields of the nanofluid flow are shown in Figs. 2 and 3. Application of magnetic field on the fluid decelerates the motion during the flow region which is in consistency with the reality that the Lorentz force plays as a hindering force and, subsequently, it shrinks the thickness of the boundary-layer of the motion. But, temperature of the nanofluid is an amplifying function of M. This ensues due to the strength of the applied magnetic field enhances in an electrically conducting fluid, it develops the resistive type force called 'Lorentz force'. This force decelerates the motion of the fluid. Again, the thermal force as the supplementary work done involved to defy the nanofluid against the exploit of M. It warms up the conducting nanofluid and raises the temperature field. Figures 4, 5 and 6 are depicted to demonstrate the persuade of K on the flow. To control the fluid's velocity and mass fraction, the curvature parameter is very important. That is the fluid's velocity dwindles when the curvature of the cylinder enlarges (see Fig. 4). But, the fluid's temperature is an increasing function of K (see Fig. 5) whereas, the mass fraction of the nanofluid drop downs with K (see Fig. 6). It can be accomplished that the curvature of the cylinder is a controlling parameter of fluid's velocity and mass fraction. Moreover, the larger boundary layer thickness of the momentum and concentration of the flow are obtained with lower values of K. The Brownian motion

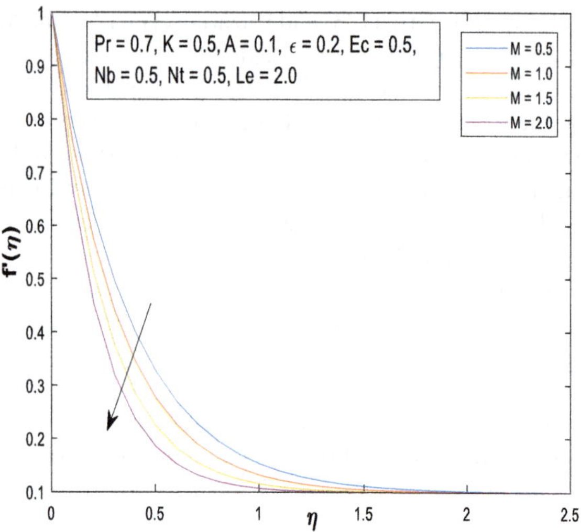

Fig. 2 Velocity outline against η for incremental amount of M

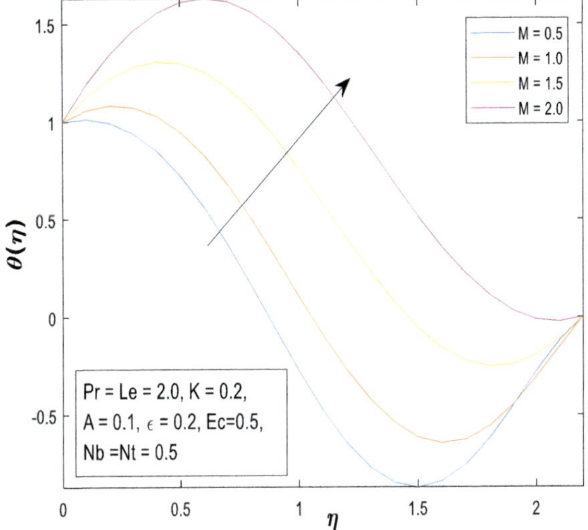

Fig. 3 Temperature field against η for incremental amount of M

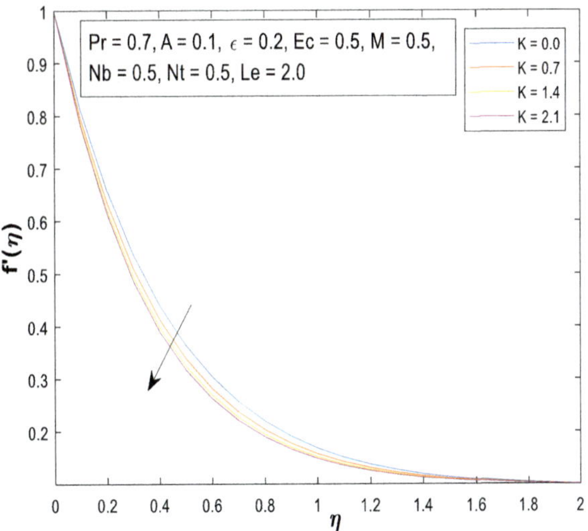

Fig. 4 Velocity outline against η for different amount of K

(represents the random motion of nano-particles) effects on the temperature and mass fraction of the nanofluid are shown pictorially in Figs. 7 and 8. From these figures, it is perceived that the fluid's temperature and mass fraction are increasing function of N_b. The reason behind this phenomenon is that the augmented values of N_b enhances the thickness of the thermal and concentration boundary layers, which eventually boost ups the fluid's temperature and concentration. Influence of Nt on the thermal and mass fraction of the fuid are elaborated in Figs. 9 and 10. The thermophoresis parameter helps to enhance the thermal transmission of the nanofluid (see Fig. 9). From this figure, it is noticed that the thermal boundary layer thickness enhances and

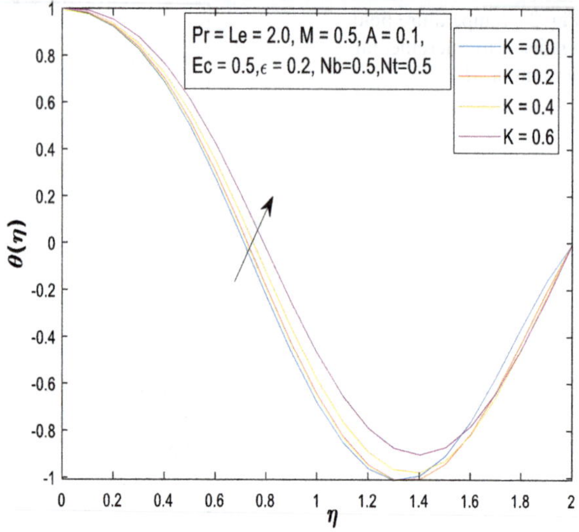

Fig. 5 Temperature field against η for different amount of K

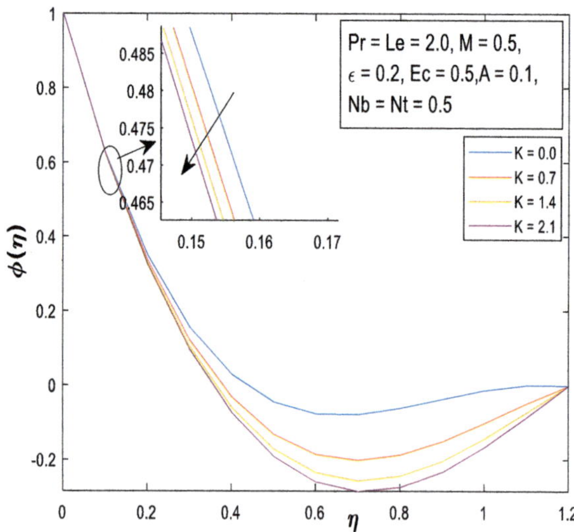

Fig. 6 Mass Fraction against η for different amount of K

tends asymptotically to zero as the distance rises from the surface of the cylinder. It can be endorsed that the escalating amount of Nt improves the liquid temperature which is happened due to development in nanoparticles percentage with Nt. But, the mass fraction of the nanoparticles lessens with the improving values of Nt (see Fig. 10). The Fig. 11 is depicted to show the influence of small temperature parameter (ε) which is responsible for variable thermal conductivity. The thermal transmission rate enhances for improving values of ε. The physics behind this characteristics is that on growing values of ε increases the thermal conductivity of the fluid which allows for a faster rate of heat transfer. The Fig. 12 shows the effect of Lewis number (Le) on the

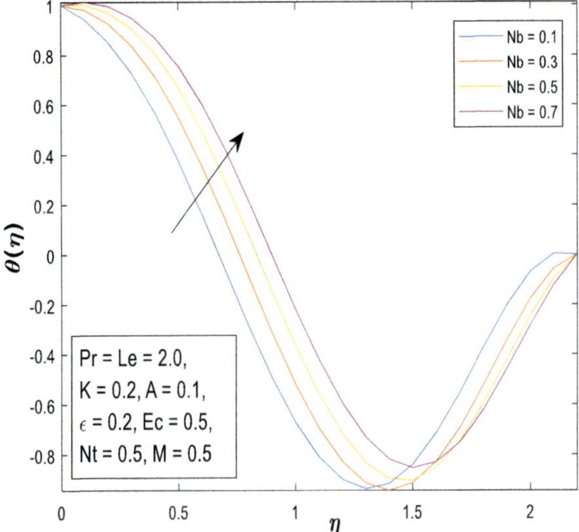

Fig. 7 Temperature field against η for different amount of Nb

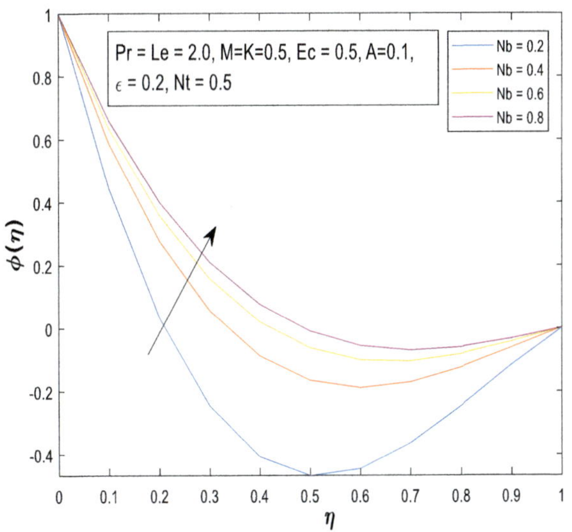

Fig. 8 Mass fraction against η for different amount of Nb

volume fraction of the nano-particles suspended in the fluid. It is discerned that the escalating values of Le significantly drops down the mass fraction of the nanofluid. Increasing values of Le signifies the larger thermal boundary film width at the outflow of dropping the width of the concentration boundary film and, consequently, the mass fraction of the nanfluid drops down.

In the nonappearance of nanofluid and concentration effects, the governing Eqs. (1)–(3) along with their boundary restrictions are fit with Jahan et al. [13] works. They have studied the boundary layer fluid's flow due to a stretching cylinder

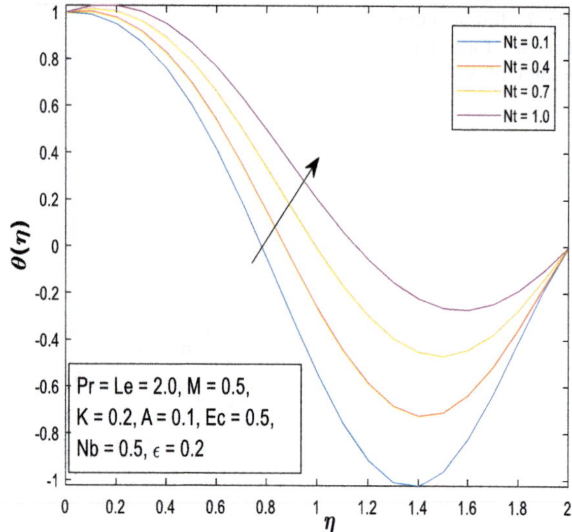

Fig. 9 Temperature field against η for different amount of Nt

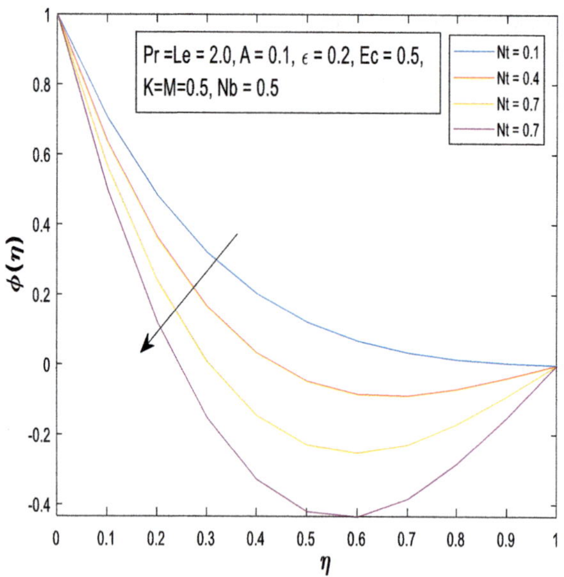

Fig. 10 Mass fraction against η for different amount of Nt

with variable thermal conductivity. We have matched up to our numerical values (bvp4c and shooting method solutions) of drag force at the surface of the cylinder with the results of Jahan et al. [13] (see Table 1).

From this Table 1, a good reasonable conformity is observed of our solutions with the work of Jahan et al. [14]. The bvp4c solution of the drag force is comparatively matched with Jahan et al. [14] works than the shooting method solution.

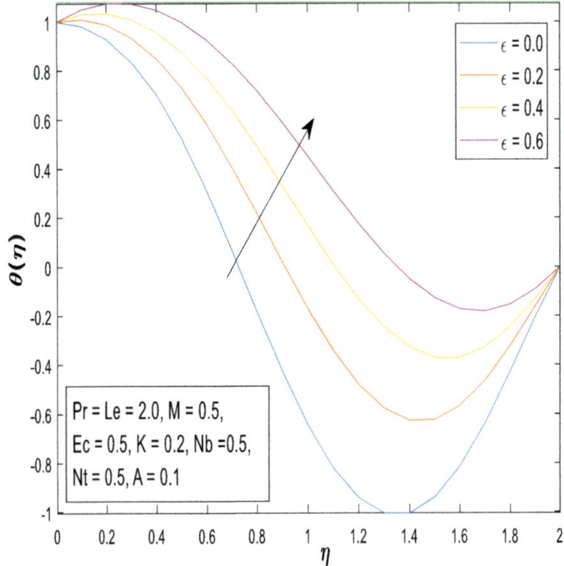

Fig. 11 Temperature field against η for different amount of ε

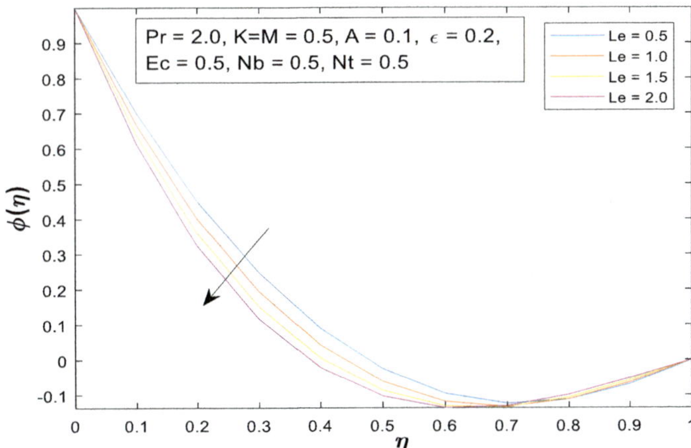

Fig. 12 Mass fraction against η for different amount of Le

The Table 2 demonstrates the outcome of the dimensionless parameter M, $K\&A$ on the skin friction coefficient of the nanofluid. From this table, it is perceived that the skin friction coefficient is a decreasing function of $M\&K$. Therefore, we can reduce the effects of drag force $(f''(0))$ of the nanofluid flow caused due to stretching cylinder by applying magnetic field. But, the drag force of the fluid increases with A. It is also seen that the numerical values of skin friction coefficient obtained by shooting method is higher than the bvp4c solver solutions. The numerical values of

Table 1 Drag force at the surface for the flow parameters $\Pr = 1, \varepsilon = 0.2, Ec = 0.5, Le = Nt = Nb = 0$

K	M	A	Skin friction coefficient		
			Jahan et al. [14] works	Present results	
				Bvp4c solution	Shooting method solution
0.0	0.2	0.1	– 0.68667	– 0.6829	– 0.7098
0.2			– 0.72698	– 0.7293	– 0.7326
0.3	0.5		– 0.81618	– 0.8122	– 0.7948
	0.2	0.2	– 0.7136	– 0.7136	– 0.7123
		0.5	– 0.5111	– 0.5011	– 0.5395

Table 2 Skin friction coefficient for the flow parameters M, K & A when $\Pr = Le = 2.0, Nt = Nb = 0.5, \varepsilon = 0.2$ & $Ec = 0.5$

M	K	A	Skin friction coefficient ($f''(0)$)	
			Bvp4c solver	Shooting method
0.2	0.2	0.1	– 0.8217	– 0.7326
0.5			– 0.8962	– 0.7873
0.5	0.2		– 0.8962	– 0.7873
	0.4		– 0.9751	– 0.8016
	0.2	0.3	– 0.7538	– 0.6711
		0.5	– 0.5756	– 0.5171

Nusselt number with different values of flow parameters are tabulated in Table 3. From this table, it is perceived that the Nusselt number reduces for the increasing values of the $M, Nt\&Nb$, whereas, the flow parameters $K\&\varepsilon$ enhance the heat transfer rate in the vicinity of the surface. Again, the Shooting method solutions of Nusselt number is smaller than the Bvp4c solver solutions. The Brownian motion (Nb), thermophoresis (Nt) and Lewis number (Le) parameters enhance the mass accumulation rate of the nanofluid at the surface of the cylinder (see Table 4). The Bvp4c solver solutions of the Sherwood number are comparatively smaller than the Shooting method solutions.

5 Conclusions

From this investigation, the following conclusions are established:

- Effects of $M\&K$ reduce the motion of the nanofluid, whereas, they enhance the temperature of the fluid.
- The temperature of the nanofluid enhances due to the presence of Nb, $Nt\&\varepsilon$.

Table 3 Nusselt number for the flow parameters M, K, Nt, Nb & ε when $\Pr = Le = 2.0$ & $Ec = 0.5$

M	K	Nt	Nb	ε	Nusselt number $(-\theta'(0))$	
					Bvp4c solver	Shooting method
0.2	0.2	0.5	0.5	0.2	0.1564	0.1057
0.5					0.1344	0.0748
0.5	0.2				0.1344	0.0748
	0.4				0.1492	0.1293
	0.2	0.4			0.1406	0.0679
		0.7			0.1224	0.0528
		0.5	0.5		0.1007	0.0748
			0.8		0.0354	0.0065
			0.5	0.2	0.1007	0.0748
				0.4	0.1095	0.1335

Table 4 Sherwood number for the flow parameters $\Pr = Le = 2.0$, $Ec = 0.5$, $M = 0.5$, $K = 0.2$ & $\varepsilon = 0.2$

Nb	Nt	Le	Sherwood number $(-\phi'(0))$	
			Bvp4c solution	Shooting method solution
0.2	0.5	2.0	1.0181	1.2550
0.5			1.1227	1.4238
0.5	0.4		1.1028	1.3345
	0.7		1.1665	1.6028
	0.5	1.0	0.7873	1.1846
		2.0	1.1227	1.4238

- The volume fraction of the nanofluid boosts up due to the effects of Nb. But, influence of Nt&Le lessen the nanofluid's volume fraction.
- By applying magnetic field, we can reduce the effects of drag force of the nanofluid at the surface of the cylinder.
- From the comparative study, we have achieved that the MATLAB built-in bvp4c solver scheme gives nearest exact solution than the Shooting technique.

Nomenclature:

ρ-density, υ-kinematic viscosity, c_p-specific heat, σ-electric charge density u-velocity along x-axis, v-velocity along r-direction, D_B-the Brownian diffusion coefficient, D_T-thermophoretic coefficient, R-radius of the cylinder, a, b-constants, ψ-dimensionless stream function, K-curvature parameter, Pr-Prandtl number, M-Magnetic field, Ec-Eckert number, A-velocity ratio parameter, N_b—Brownian motion parameter, N_t-thermophoresis parameter, ε-small temperature parameter, Le-Lewis number, $\theta(\eta)$-dimensionless temperature, $\phi(\eta)$-dimensionless concentration, Re_x-local Reynolds number, $\tau = \frac{(\rho C)_p}{(\rho C)_f}$-the ratio of effective heat capacity of the nanoparticle and base".

References

1. Crane. L.J.: Flow past a stretching plate. Zeitschrift für Angew. Math. und Phys. ZAMP **21**(4), 645–647 (1970)
2. Gangadhar, K., Ramana, K.V., Makinde, O.D., Kumar, B.R.: MHD flow of a carreau fluid past a stretching cylinder with cattaneo-christov heat flux using spectral relaxation method. Defect Diffus. Forum. **387**, 91–105 (2018)
3. Das, S., Chakraborty, S., Makinde, O.D., Jana, R.N.: Entropy analysis of MHD variable thermal conductivity fluid flow past a convectively heated stretching cylinder. Defect Diffus. Forum. **387**, 244–259 (2018)
4. Dey, D., Borah, R., Mahanta, B.: Boundary layer flow and its dual solutions over a stretching cylinder: stability analysis. In: Emerging Technologies in Data Mining and Information Security. Advances in Intelligent Systems and Computing, pp. 27–38 (2021)
5. Dey, D., Borah, R.: Stability analysis on dual solutions of second- grade fluid flow with heat and mass transfers over a stretching sheet. Int. J. Thermofluid Sci. Technol. **8**(2) (2021)
6. Dey, D., Borah, R., Khound, A.S.: Stability analysis on dual solutions of MHD Casson fluid flow with thermal and chemical reaction over a permeable elongating sheet. Heat Transf. 1–17 (2022)
7. Dey, D., Hazarika, M., Borah, R.: Entropy generation analysis of magnetized micropolar fluid streaming above an exponentially extending plane. Lat. Am. Appl. Res. **51**(4), 255–260 (2021)
8. Manjunatha, G., Rajashekhar, C., Vaidya, H., Prasad, K.V., Vajravelu, K.: Impact of heat and mass transfer on the peristaltic mechanism of Jeffery fluid in a non-uniform porous channel with variable viscosity and thermal conductivity. J. Therm. Anal. Calorim. **139**(2), 1213–1228 (2020)
9. Divya, B.B., Manjunatha, C. Rajashekhar, G., Vaidya, H., Prasad, K.V.: The hemodynamics of variable liquid properties on the MHD peristaltic mechanism of Jeffrey fluid with heat and mass transfer. Alexandria Eng. J. **59**(2), 693–706 (2020)
10. Durga Prasad, P., Kiran Kumar, R.V.M.S.S., Varma, S.V.K.: Heat and mass transfer analysis for the MHD flow of nanofluid with radiation absorption. Ain Shams Eng. J. **9**(4), 801–813 (2018)
11. Dey, D., Borah, R.: Dual solutions of boundary layer flow with heat and mass transfers over an exponentially shrinking cylinder: stability analysis. Lat. Am. Appl. Res. **50**(4), 247–253 (2020)
12. Vaidya, H., Rajashekhar, C., Manjunatha, G., Prasad, K.V., Makinde, O.D., Vajravelu, K.: Heat and mass transfer analysis of MHD peristaltic flow through a complaint porous channel with variable thermal conductivity. Phys. Scr. **95**(4) (2020)
13. Abel, M.S., Datti, P.S., Mahesha, N.: Flow and heat transfer in a power-law fluid over a stretching sheet with variable thermal conductivity and non-uniform heat source. Int. J. Heat Mass Transf. **52**(11–12), 2902–2913 (2009)
14. Jahan, S., Sakidin, H., Nazar, R.M.: MHD stagnation point flow over a stretching cylinder with variable thermal conductivity and joule heating. AIP Conf. Proc. **1787** (2016)
15. Choi, S.U.S.: Enhancing thermal conductivity of uids with nanoparticles. In: Proceedings of the 1995 ASME International Mechanical Engineering Congress and Exposition, San Francisco, USA, ASME FED 231/MD, 1995, pp. 99–105 (1995)

16. Das, K., Acharya, N., Kundu, P.K.: Influence of variable fluid properties on nanofluid flow over a wedge with surface slip. Arab. J. Sci. Eng. **43**(5), 2119–2131 (2018)
17. Ghosh, S., Mukhopadhyay, S.: Flow and heat transfer of nanofluid over an exponentially shrinking porous sheet with heat and mass fluxes. Propuls. Power Res. **7**(3), 268–275 (2018)
18. Molli, S., Naikoti, K.: MHD natural convective flow of Cu-Water nanofluid over a past infinite vertical plate with the presence of time dependent boundary condition. Int. J. Thermofluid Sci. Technol. **7**(4), 1–15 (2020)
19. Li, X., Khan, A.U., Khan, M.R., Nadeem, S., Khan, S.U.: Oblique stagnation point flow of nanofluids over stretching/shrinking sheet with Cattaneo-Christov heat flux model: existence of dual solution. Symmetry (Basel) **11**(9) (2019)
20. Narender, G., Govardhan, K., Sarma, G.S.: J. Heat Mass Trans. Res. Heat Mass Trans. Nanofluid Over Linear Stretching Surf. Viscous Dissipation Effect **6**, 117–124 (2019)
21. Khashi'ie, N.S., Arifin, N.M., Pop, I., Wahid, N.S.: Flow and heat transfer of hybrid nanofluid over a permeable shrinking cylinder with Joule heating: a comparative analysis. Alexandria Eng. J. **59**(3), 1787–1798 (2020)
22. Keys, W.M.: Convective Heat and Mass Transfer. McGraw-Hill, New York (1966)
23. Hazarika, G.C.: Shooting Method to Some Problems of Fluid Mechanics. LAP Lambert Academic Publishing (2012)
24. Hazarika, G.C., Borah, J., Konch, J.: Effects of variable viscosity and thermal conductivity on free convective MHD fluid flow over a stretching sheet. Math. Forum **27**, 2015–2019

Graphs, Networks and Communication

Structure of Protein Interaction Network Associated With Alzheimer's Disease Using Graphlet Based Techniques

Ahamed Khasim, Venkatesh Subramanian, K. M. Ajith, and T. K. Shajahan

Abstract The crucial step in analyzing a real-world network is to choose an acceptable network model. We try to select an appropriate network model for the protein-protein interaction (PPI) network of Alzheimer's disease (AD) using Graphlet-based metrics. The Relative Graphlet Frequency (RGF) count in the AD-PPI network is similar to that of the corresponding Scale-Free network. However, based on Graphlet Degree Distribution (GDD), the AD-PPI network has a good match with Geometric random graphs. The graphlet correlation statistics of the AD network show that it has a core-periphery topology.

Keywords Alzheimer's disease · Protein-protein interaction networks · Graphlets · Graphlet Correlation Matrix

1 Introduction

Alzheimer's Disease (AD) is a complex disease whose genetic architecture is believed to be polygenic [4]. Hence its genotype can be modeled with a biological network depicting the interaction between causal agents. In this paper, we model it with a Protein-Protein Interaction (PPI) network. In the case of general datasets, it is commonplace to establish their theoretical model before proceeding to analyze them. For instance, one makes statements like, *'this dataset has a normal distribution'*, before proceeding to the analysis. We would like to do the same, but for network datasets. Specifically, we study the large-scale structure of AD linked network. In the case of networks, some examples of theoretical models are Eřdos-Rényi random graph (ER) [3], Scale-Free graph (SF) [1], Geometric random graph (GEO) [9], etc. In this paper we assess which of these theoretical models are a good fit for the AD-associated PPI network.

A. Khasim · V. Subramanian · K. M. Ajith · T. K. Shajahan (✉)
National Institute of Technology Karnataka, Surathkal, Mangaluru, India
e-mail: shajahan@nitk.edu.in

S. Banerjee and A. Saha (eds.), *Nonlinear Dynamics and Applications*,
Springer Proceedings in Complexity,
https://doi.org/10.1007/978-3-030-99792-2_42

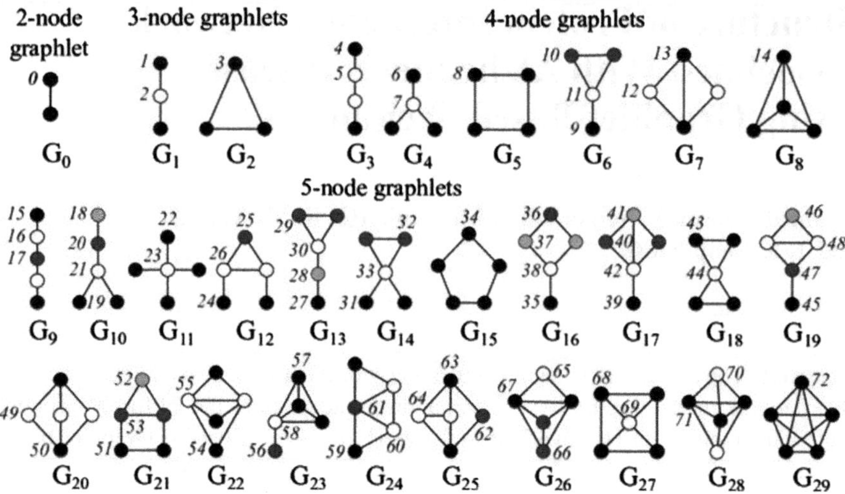

Fig. 1 G_0 to G_{29} **graphlets of 2–5 nodes.** The orbits are labeled from 0 to 72. In a particular graphlet, the similar types of nodes (automorphism orbits) are colored the same. [17]

We adopted a bottom-up method to investigate the network's local structural features and to provide a theoretical model. The degree distribution of several disease PPI networks is a power-law distribution, implying that they belong to Scale-free networks. A more general method had been implemented using graphlets by N. Prẑulj et al. [13, 14]. The graphlets are small non-isomorphic subgraphs that exist in a network (see Fig. 1). The graphlet degree distribution is one of the metrics to check the similarity between two networks. In a network, the distribution of a 2-node graphlet is the standard degree distribution, which is just one among the 30 graphlet distributions. Thus, the graphlet-based method is more general and provides a more in-depth examination of the network.

In this study, we collected AD-associated proteins from several databases and their physical interacting partners present in the human PPI network. Three graphlet-based metrics are used to analyze the network: Relative Graphlet Frequency Distance (RGFD), Graphlet Degree Distribution Agreement (GDDA), and Graphlet Correlation Matrix (GCM-11). We recognize that the constructed AD network has a core-periphery structure with features of SF, ER-DD, and GEO network models.

2 Methodology

2.1 Collection of AD-Related Genes

Genes related to Alzheimer's disease are retrieved from several databases using the methods described in the article of A Podder et al. [12]. 105 AD-associated genes are collected from NeuroDNet [16]. NeuroDNet has experimentally verified genetic information associated with neurodegenerative diseases sourced from the research literature. A set of 183 AD-associated genes extracted through text mining and genome-wide association studies (GWAS) are collected from DISEASE database [11]. AD-related genes appeared in the PubMed literature repository with more than two pieces of literature evidence (Number of PubMed IDs ≥ 2) are collected from DisGeNet and GLAD4U databases. 853 genes are collected from DisGeNet [10] and 54 genes are collected from GLAD4U [5]. AD-associated genes reported in genome-wide association studies (GWAS) are collected from two databases; GWAS Catalog [8] and GWASdb2 [7]. 98 genes are collected from GWA studies.

All of the datasets discussed above generated a total of 1293 genes associated with Alzheimer's disease. Since all databases do not use the same symbol format to represent genes, we converted all gene symbols to HUGO nomenclature committee (HGNC) approved symbol format [2].

2.2 Construction of AD-Associated Network

Direct physical interaction partners of AD-associated genes are retrieved from the Human Interactome project portal of the Center for Cancer Systems Biology (CCSB) database [15]. CCSB database has six independent datasets of binary interaction data of human proteins. Out of 1293 AD-associated genes collected from different sources, only 904 genes were listed in the CCSB database. The interacting partners of 904 genes are collected and constructed an AD-associated protein interaction network. The AD-related network has 4253 nodes and 7429 interactions. The largest connected component of the constructed network is considered for further topological analysis which comprises 4073 nodes and 7150 interactions after removing self-loops and multiple edges.

2.3 Network Topology Using Graphlet Analysis

We employed graphlet-based analysis to learn about the large-scale structure of the network. Graphlets are small connected non-isomorphic subgraphs in a large network. 30 graphlet structures of 2 to 5 nodes are shown in Fig. 1.

Relative graphlet frequency (RGF) can be used to check similarity between two networks [14]. The relative graphlet frequency of a network G is

$$F_i(G) = \frac{N_i(G)}{T(G)} \tag{1}$$

where $N_i(G)$ is the number of graphlets of type $i \in (0, 1, 2, ..29)$ and $T(G)$ is the total number of graphlets in the large network. The similarity between two networks G and H is measured using relative graphlet frequency distance and is defined as [14]

$$RGFD(G, H) = \sum_{i=0}^{29} \left| log \frac{N_i(G)}{T(G)} - log \frac{N_i(H)}{T(H)} \right| \tag{2}$$

The lower the RGFD value, the greater the similarity between two networks.

Each of the graphlets has certain symmetries and hence automorphic orbits are defined for each of them. This results in a total of 73 orbits in the 30 graphlets. In a large network, the number of nodes that touches jth orbit k times is denoted as $d_G^j(k)$. Hence the jth orbit degree distribution [13] is defined as

$$S_G^j(k) = \frac{d_G^j(k)}{k} \tag{3}$$

along with a scaling factor of $\frac{1}{k}$. We use the scaled degree distribution so that the orbits with a higher degree do not dominate the metric. We can normalize the scaled graphlet degree distribution [13] as,

$$N_G^j(k) = \frac{S_G^j(k)}{T_G^j} \tag{4}$$

where $T^j(G) = \sum_{k=1}^{\infty} S_G^j(k)$

The similarity between two networks can be measured using the graphlet degree distribution distance [13] and is defined as

$$D^j(G, H) = \left(\sum_{k=1}^{\infty} [N_G^j(k) - N_H^j(k)]^2 \right)^{\frac{1}{2}} \tag{5}$$

The GDD agreement of jth orbit between two networks [13]

$$A^j(G, H) = 1 - D^j(G, H) \tag{6}$$

Considering all orbits $j \in \{0, 1, 2, \ldots 72\}$ from 30 graphlets; the GDD agreement [13]

$$GDDA(G, H) = \frac{1}{73}\sum_{j=0}^{72} A^j(G, H) \tag{7}$$

We compared the RGFD and GDDA values of the created AD network to those of several model networks. Model networks include (1) Erdös-Rényi random graphs (ER) with the same number of nodes and edges as AD networks, (2) Erdös-Rényi random graphs with the same degree distribution as AD networks (we denote as ER-DD), (3) Scale-Free networks with the same number of nodes (SF), (4) 2-dimensional geometric random graph (GEO 2D), and (5) 3-dimensional geometric random graph (GEO 3D). From each type, 20 model networks have been constructed. To assess the similarity of model networks, the RGFD and GDDA of the AD network and 100 model networks are computed using the Graphcrunch software [6].

In a large network, each node may be represented as a 73-dimensional vector, with each dimension corresponding to the number of connected orbits in contact with the node. This vector is termed as Graphlet Degree Vector (GDV). GDV of ith node is $GDV_i = (C_0, C_1, C_2, ...C_{72})$ where C_0 is the number of orbit-0 in contact with node i and similarly for other orbits also. By considering only 11 non-redundant orbits of 2- to 4-node graphlets, the graphlet degree vector may be reduced to 11 dimensions [18]. A metric is defined to characterize a network based on the correlations between the 11 non-redundant orbits (j={0,1,2,4,5,6,7,8,9,10,11}) of 2- to 4-node graphlets [18]. 11 dimensional GDV of each node is calculated and are arranged in an $n \times 11$ matrix with each row corresponds to a node. Now, Spearman correlation coefficient (r_s) is calculated for every possible pair of columns. Then a 11×11 matrix is made, called the Graphlet Correlation Matrix (GCM), with the r_s values [18]. That is, one computes r_s for columns 1 and 2 of the $n \times 11$ matrix and sets it as GCM_{12}, then computes r_s for columns 1 and 3 and sets it as GCM_{13}, and so on. Thus, regardless of the number of nodes in the network, one ends up with an 11×11 matrix.

3 Results and Discussion

3.1 RGFD Between AD-Network and Model Networks

We calculated the relative graphlet frequency of the AD-associated network and compared it with the model networks. The similarity between the AD network and model network is quantified using RGFD value (see Table 1). Based on the RGFD value, the AD network is far from ER but close to ER-DD and SF model networks. A similar kind of analysis using graphlet frequency count (of 3–5 nodes graphlets) had been done by N. Prẑulj et al. [14] on 4 PPI network data sets of fruit fly and yeast. All those PPI had a better fit with GEO, except for one network data set of fruit fly, which had more noise. But in our network, the edge density is 0.086%, hence the chance of noise is low. In our analysis, in addition to 3–5 node graphlets, we accounted for

Table 1 The RGFD and GDDA value between AD-PPI network and different model networks. The RGFD and GDDA between AD-PPI and 20 model networks from each type were calculated, and their average is tabulated here

Model Networks	RGFD	GDDA
ER	8.6027	0.6823
ER-DD	3.6045	0.7925
SF	3.6971	0.7668
GEO 2D	6.1738	0.8100
GEO 3D	5.7149	0.8214

the frequency of 2 nodes graphlet (G_0) also. Based on the RGFD metric we say that AD-associated PPI has more fit with SF and ER-DD networks.

3.2 Graphlet degree distribution agreement

The graphlet degree distribution of 73 orbits present in the AD network and the corresponding 100 model networks are calculated. GDDA between AD network and 20 model networks of each type (ER, ER-DD, SF, GEO 2D, and GEO 3D) are calculated and their average is given in Table 1. The AD network has more than 60% agreement with all model networks. The agreement is highest for GEO 3D graphs (GDDA = 0.8214) compared to ER, ER-DD, SF, and GEO 2D. The agreement of GEO 3D with PPI networks had been reported in [13]. Since all model networks show an agreement of more than 60%, we cannot claim that the AD-associated PPI is strictly geometric random graphs.

3.3 Graphlet correlation matrix of AD-related PPI network

The graphlet correlation matrix (Fig. 2) of the AD network is constructed for 11 non-redundant orbits. The GCM-11 have two separated correlated clusters; correlation cluster of orbits $\{1, 6, 9\}$ and orbits $\{0, 2, 4, 5, 7, 8, 10, 11\}$. Orbits 1 and 6 are single degree nodes in graphlets G_1 and G_4 respectively. Hence, the correlation of orbits 1 and 6 can be characterized as the existence of degree-1 nodes in the networks.

Orbits 7 and 8 exhibit a significant correlation in GCM-11's second correlation cluster (nearly equal to 1). Orbit 7 and 8 are mediator nodes in G_4 and G_5. This correlation indicates the existence of mediator nodes in the network. Similarly, the high correlations of orbits 7 with 2, 5 with 11, and 8 with 2 are observed in GCM-11. Orbits $\{11\}$ corresponds to higher degree mediator nodes while {2,5,7,8} corresponds to just broker type nodes (act as a link between two nodes) in the networks. In short,

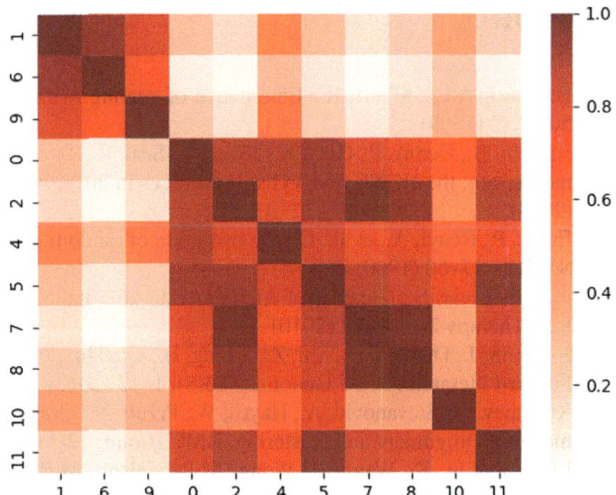

Fig. 2 GCM-11 of the AD-related PPI network. Non-redundant orbits from 0 to 11 are labeled on the horizontal and vertical axes. Each cell's color gradient represents Spearman's correlation value between corresponding orbits. The strongly correlated orbits are brought together in the illustration

the correlation between the orbits in the second cluster indicates that the network comprises mediator nodes that are either highly connected or just brokers.

A similar correlation pattern was identified in GCM-11 of world trade network (WTN) [18]. Countries that participate in global markets are either on the periphery or at the center of WTN. Similarly, in the AD-associated PPI network, the proteins are either sparsely distributed on the network's outer shell or located in the network's inner core as mediators or hubs.

4 Conclusion

Graphlet-based metrics were used in this study to assess the resemblance of the protein protein interaction network (PPI) associated with Alzheimer's Disease (AD) to several theoretical network models. The Relative Graphlet Frequency Distance (RGFD) and Graphlet Degree Distribution Agreement (GDDA) are used to calculate the resemblance of created AD-network to theoretical model networks. According to the RGFD value, the network is more similar to the Erdös-Rényi random graph with degree distribution same as that of AD network (ER-DD) and Scale-Free (SF) networks. Our results contradict the prior conclusion, in which PPI was near to 3 Dimensional Geometric random graph (GEO 3D) [14]. Using GDDA data, our AD-associated PPI outperforms other model networks in terms of agreement with GEO 3D. The resemblance with other model networks is likewise noteworthy (greater than 60% agreement). Based on the RGFD and GDDA metrics, AD network shows features of all three models: i.e., Geometric random, Scale-Free, or Erdös-Rényi random. We can also see from the graphlet correlation matrix of the AD-related network that the proteins are either on the network's periphery or at its center. That is, the network has a core-periphery structure.

References

1. Barabási, A.L., Albert, R.: Emergence of scaling in random networks. Science **286**(5439), 509–512 (1999)
2. Braschi, B., Denny, P., Gray, K., Jones, T., Seal, R., Tweedie, S., Yates, B., Bruford, E.: Genenames. org: the HGNC and VGNC resources in 2019. Nucl. Acids Res. **47**(D1), D786–D792 (2019)
3. Erdős, P., Rényi, A., et al.: On the evolution of random graphs. Publ. Math. Inst. Hung. Acad. Sci **5**(1), 17–60 (1960)
4. Ertekin-Taner, N.: Genetics of Alzheimer disease in the pre-and post-GWAS era. Alzheimer's Res. Therapy **2**(1), 1–12 (2010)
5. Jourquin, J., Duncan, D., Shi, Z., Zhang, B.: Glad4u: deriving and prioritizing gene lists from Pubmed literature. BMC Genomics **13**(8), 1–12 (2012)
6. Kuchaiev, O., Stevanović, A., Hayes, W., Pržulj, N.: Graphcrunch 2: software tool for network modeling, alignment and clustering. BMC Bioinf. **12**(1), 1–13 (2011)
7. Li, M.J., Liu, Z., Wang, P., Wong, M.P., Nelson, M.R., Kocher, J.P.A., Yeager, M., Sham, P.C., Chanock, S.J., Xia, Z., et al.: Gwasdb v2: an update database for human genetic variants identified by genome-wide association studies. Nucl. Acids Res. **44**(D1), D869–D876 (2016)
8. Morales, J., Welter, D., Bowler, E.H., Cerezo, M., Harris, L.W., McMahon, A.C., Hall, P., Junkins, H.A., Milano, A., Hastings, E., et al.: A standardized framework for representation of ancestry data in genomics studies, with application to the nhgri-ebi gwas catalog. Genome Biol. **19**(1), 1–10 (2018)
9. Penrose, M.: Random Geometric Graphs, vol. 5. OUP Oxford (2003)
10. Piñero, J., Queralt-Rosinach, N., Bravo, A., Deu-Pons, J., Bauer-Mehren, A., Baron, M., Sanz, F., Furlong, L.I.: Disgenet: a discovery platform for the dynamical exploration of human diseases and their genes. Database **2015** (2015)
11. Pletscher-Frankild, S., Pallejà, A., Tsafou, K., Binder, J.X., Jensen, L.J.: Diseases: text mining and data integration of disease-gene associations. Methods **74**, 83–89 (2015)
12. Podder, A., Pandit, M., Narayanan, L.: Drug target prioritization for alzheimer's disease using protein interaction network analysis. OMICS: A J. Integr. Biol. **22**(10), 665–677 (2018)
13. Pržulj, N.: Biological network comparison using graphlet degree distribution. Bioinformatics **23**(2), e177–e183 (2007)
14. Pržulj, N., Corneil, D.G., Jurisica, I.: Modeling interactome: scale-free or geometric? Bioinformatics **20**(18), 3508–3515 (2004)
15. Rolland, T., Taşan, M., Charloteaux, B., Pevzner, S.J., Zhong, Q., Sahni, N., Yi, S., Lemmens, I., Fontanillo, C., Mosca, R., et al.: A proteome-scale map of the human interactome network. Cell **159**(5), 1212–1226 (2014)
16. Vasaikar, S.V., Padhi, A.K., Jayaram, B., Gomes, J.: Neurodnet-an open source platform for constructing and analyzing neurodegenerative disease networks. BMC Neurosci. **14**(1), 1–13 (2013)
17. Yaveroglu, O.N.: Graphlet correlations for network comparison and modelling: World Trade Network example. Ph.D. thesis, Imperial College London, UK (2013)
18. Yaveroğlu, Ö.N., Malod-Dognin, N., Davis, D., Levnajic, Z., Janjic, V., Karapandza, R., Stojmirovic, A., Pržulj, N.: Revealing the hidden language of complex networks. Sci. Rep. **4**(1), 1–9 (2014)

On Divisor Function Even(Odd) Sum Graphs

S. Shanmugavelan and C. Natarajan

Abstract The theory of numbers has numerous applications in cryptography and information security. Number theoretic graphs are a novel and hybrid branch of number theory and graph theory. In this paper, we introduce a new structure of graphs called divisor function even sum graphs and divisor function odd sum graphs, which are analogous to divisor function graphs, and investigate some of their properties. In addition, we present a MATLAB code for generating the graph in $O(n)$ time.

Keywords Number theoretic graphs · Divisor function even sum graph · Divisor function odd sum graph

1 Introduction

For each natural number n, Number Theory is compounded with a variety of functions such as the Euler function $\phi(n)$, sigma function $\sigma(n)$, divisor function $D(n)$, Möbius function $\mu(n)$ and so on. These functions are used to construct number theoretic graphs, a developing field that has captivated researcher's interest ([9, 13]). In the last decade, various research studies about such graphs have been conducted ([2, 7, 14, 15, 17]). In spectral graph theory, a number theoretic graph known as Ramanujan networks is employed to solve optimization problems in communication network theory ([4, 18]) and is also used in cryptography [1]. Engineering and communication networks make advantage of the cycle structure of arithmetic Cayley graphs ([10, 11]), which are number theoretic graphs. For notations and terminologies in graph theory and number theory that are not defined here, we generally follow ([3, 5]).

S. Shanmugavelan (✉) · C. Natarajan
Department of Mathematics, Srinivasa Ramanujan Centre, SASTRA Deemed University, Kumbakonam 612001, India
e-mail: shanmugavelan@src.sastra.edu

C. Natarajan
e-mail: natarajan_c@maths.sastra.edu

S. Banerjee and A. Saha (eds.), *Nonlinear Dynamics and Applications*, Springer Proceedings in Complexity,
https://doi.org/10.1007/978-3-030-99792-2_43

Throughout our discussion, the set of all factors for each natural number n is denoted as Divisor function $D(n)$, and whose cardinality is known as the tau function $\tau(n)$, which will be computed from the prime factorization of $n = p_1^{r_1} \times p_2^{r_2} \times \ldots \times p_n^{r_n}$, where p_i's are distinct primes and r_i's are positive integers.

In 2015, Kannan et al. [8] introduced the divisor function graph $G_{D(n)}$. In addition, they proved that $G_{D(n)}$ is always connected and complete graph if and only if no two proper divisors in $D(n)$ are relatively prime. Also, for perfect square numbers, the chromatic number for $G_{D(n)}$ is at least 3 and it is Eulerian. For further studies on divisor function graph, we may refer ([12, 16]). Chalapathi and Kiran Kumar [6] presented Euler even graphs ε_n and Euler odd graphs O_n for $2n$ and $2n+1$ in 2016, examined connectedness and completeness, and established that the graph ε_{2n+1} is isomorphic to disjoint union of two complete graphs. They also proved that the graph O_{2n+1} is Eulerian if and only if $\phi(2n+1)$ is a multiple of 4. Motivated by this, we introduce divisor function even (odd) sum graphs and study their properties. We also write a MATLAB code to get graphical representation, order and size, which will be used to study numerous graph theory concepts.

2 Divisor Function Even Sum Graph

In this section, we introduce the Divisor function even sum graph and study its basic properties.

Definition 1 For any natural number n, the Divisor function even sum graph is a simple graph (V, E) where $V = \{d_i : d_i \mid n, n \text{ is any positive integer}\}$ is the vertex set and the edge set E= $\{d_i d_j : d_i + d_j = \text{an even number}\}$. We denote it by $G_{eD(n)}$.

Example 1 The Divisor function even sum graph of Ramanujan number 1729 which has all its divisors as odd is shown in Fig. 1.

Theorem 1 *$G_{eD(n)}$ is complete iff it has only odd divisors.*

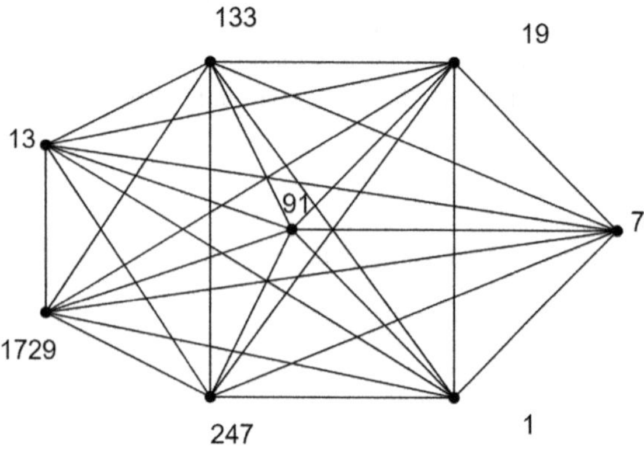

Fig. 1 Divisor function even sum graph $G_{eD(1729)}$

Proof **Necessity:** Assume that $G_{eD(n)}$ is complete. Clearly, the sum of an edge's endpoints is even. Suppose that one of the divisors is an even number. Because 1 is a divisor for any integer, the sum of such end vertices with 1 leads to an odd sum, implying that there are no edges in this case, which is impossible.

Sufficiency: Assume that n has only odd divisors for every number n. The sum of two odd divisors, on the other hand, is always even. As a result, every distinct pair of vertices has an edge, and the graph is complete.

Corollary 1 *$G_{eD(2n+1)}$ is complete.*

Corollary 2 *$G_{eD(p^n)}$ is complete, $p \neq 2$.*

Proof Since all the divisors of prime powers are $\{1, p, p^2, \ldots, p^{n-1}, p^n\}$ which are clearly odd. By Theorem 1, the proof follows.

Theorem 2 *$G_{eD(n)}$ is not a bipartite graph.*

Proof Suppose that $G_{eD(n)}$ is bipartite.

Case 1: n is an Odd Integer

In this case, one of the partitions, say X, only has one vertex 1. The second partition, Y, is comprised of all the remaining vertices, and because odd numbers only have odd divisors, the sum of 1 and the other odd divisor is even. The sum of two odd divisors in Y, on the other hand, is even, which is not possible.

Case 2: n is an Even Number

One of the partitions say X, consists of improper divisors of n only. Note that even numbers may have both odd and even divisors.

Case 2a: Odd Proper Divisors

It is not possible to have odd proper divisors in X, since the sum of 1 with such odd proper divisors would result in an even sum. Then, by Case 1, all such divisors must be placed in Y, which is impossible.

Case 2b: Even Proper Divisors

Here, it is not possible to have even proper divisors in X, since the sum of n with such even proper divisors will results in an even sum. Then all such divisors must be placed in Y, which is impossible in this instance as well because the sum of any two even proper divisors is also even.

Theorem 3 *For an odd composite number n, $G_{eD(n)}$ is an Euler graph iff and it has odd number of divisors.*

Proof **Necessity:** Assume that $G_{eD(n)}$ is an Euler graph. We claim that n has an odd number of divisors. If n has an even number of odd divisors, then all of the vertices will have an odd degree, since the sum of any two odd divisors in $G_{eD(n)}$ is even, contradicting our hypothesis.

Sufficiency: Assume that n is odd composite and has odd number of divisors. Clearly, the graph's vertices should then be adjacent to each other. Because the number of divisors is odd, every vertex should receive an even degree, and so $G_{eD(n)}$ is Eulerian.

Theorem 4 *Any regular non-complete $G_{eD(2n)}$ can be decomposed into disjoint union of $2K_{\frac{\tau(2n)}{2}}$, $n \geq 2$.*

Proof Let the prime factorisation of $2n = 2^a \times p_1^{r_1} \times p_2^{r_2} \times \ldots \times p_k^{r_k}$, where p_i's are odd. Hence, $\tau(2n) = |V(G_{eD(n)}| = (a+1)(r_1+1)\ldots(r_k+1)$. Let the total number of odd divisors of $2n = (r_1+1) \times \ldots \times (r_k+1) = \mathcal{O}$ (say).

Claim: $\mathcal{O} = \frac{\tau(2n)}{2}$.

Clearly, $\mathcal{O} \leq \frac{\tau(2n)}{2}$.

Suppose that $\tau(2n) < 2\mathcal{O}$.

(i.e.) $(a+1) \times (r_1+1) \times \ldots \times (r_k+1) < 2\mathcal{O}$.

Then, $(a+1) \times (r_1+1) \times \ldots (r_k+1) < 2(r_1+1) \times \ldots (r_k+1)$ which makes $a < 1$ and so a is either 0 or negative integer. Since $n \geq 2$, a cannot be negative. Therefore, the only probability is $a = 0$ (i.e.) $\tau(2n) = p_1^{r_1} \times \ldots p_k^{r_k}$, which is a factorization as product of odd primes only, which is a contradiction and hence, $\frac{\tau(2n)}{2} \geq \mathcal{O}$.

Let us label the vertices of odd factors of $2n$ as $o_1, o_2 \ldots o_{\tau(2n)/2}$ whose induced subgraph is isomorphic to $K_{\tau(2n)/2}$. Now, the total number of even divisors (say) ($\mathcal{E}$). Now,

$$
\begin{aligned}
\mathcal{E} &= \tau(2n) - \mathcal{O} \\
&= (a+1) \times (r_1+1) \times \ldots (r_k+1) - (r_1+1) \times \ldots (r_k+1) \\
&= [a+1-1](r_1+1)\ldots(r_k+1) \\
&= a \times [(r_1+1)\ldots(r_k+1)].
\end{aligned}
$$

Let us label the vertices of even divisors of $2n$ as $e_1, e_2 \ldots e_\xi$ whose induced sub graph is a complete of order ξ. Also note that there is no edge connecting the vertices of $\mathcal{O}_i$'s to $\mathcal{E}_j$'s, $i = 1, 2 \ldots \frac{\tau(2n)}{2}$, $j = 1, 2 \ldots \xi$. Now it is enough to prove $\xi = \tau(2n)/2$.

Suppose that $2\xi > \tau(2n)$. This implies, $2 \times a \times (r_1+1) \times \ldots \times (r_k+1) < (a+1) \times (r_1+1) \times \ldots \times (r_k+1)$, a similar contradiction is arrived. Hence the proof.

The following MATLAB Code is used to get a graph of divisor even sum graph with its order and size.

Input: Any natural number $n > 1$.

Output: A graph of Divisor even sum graph with its order and size.

```
function Divisorfunction_evensum_graph(n)
n=input('Enter any natural number>1');
sum=0;
j=0;
for i=1:n
if mod(n,i)==0
j=j+1;
a(j)=i;
end
end
disp('The  Vertices of  Divisor Even sum ...
    graph are');
disp(a);
disp(' ');
x=1;y=1;
disp('The Divisor Even sum graph Edges are');
for i=1:j-1
for k=i+1:j
sum=a(i)+a(k);
if mod(sum,2)==0
fprintf('{%d,%d}',a(i),a(k));
    disp(' ');
    s(x)=a(i);
    t(y)=a(k);
    x=x+1;y=y+1;
    end
    end
    end
    G=graph(s,t);
    o=0;
    for q=1:n
    if(q≠a)
    o=o+1;
    m(o)=q;
    end
    end
    GE=rmnode(G,[m]);
    PO=plot(GE);
    PO.NodeColor='k';
    PO.EdgeColor='k';
    PO.NodeLabel=[a];
    title('Divisor Even Sum  Graph of ',n);
    m=numnodes(GE);
    disp('The Order of Divisor Even Sum  ...
        Graph is');
    disp(m);
    ne=numedges(GE);
    disp('The size of Divisor Even Sum  ...
        Graph is');
    disp(ne);
    end
```

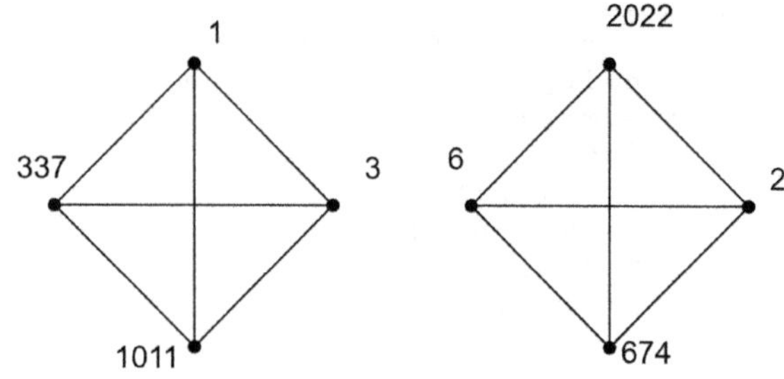

Fig. 2 Divisor Even sum graph of 2022

Example 2 The sample output of divisor even sum graph of $n = 2022$ (satisfies the hypothesis of Theorem 4 which is isomorphic to union of two complete graphs K_4 is shown in Fig. 2.

3 Divisor Function Odd Sum Graph

Let us define a Divisor function odd sum graph:

Definition 2 For any natural number n, Divisor function odd sum graph is a simple (V, E) graph such that $V = \{d_i : d_i \mid n, n$ is any positive integer$\}$ and the edge set $E = \{d_i d_j : d_i + d_j =$ an odd number$\}$. We denote it by $G_{oD(n)}$.

Example 3 The Divisor function odd sum graph for $n = 1234$ is shown in Fig. 3.

Theorem 5 *$G_{oD(2^n)}$ is always a star for any positive integer n.*

Proof Clearly, there is no edge between 2^i and 2^j, $1 \leq i, j \leq n$. But, $1 \in V[G_{oD}(2^n)]$, the sum of 1 with remaining n vertices leads to odd sum only. As a result, a star graph is formed by keeping all even powers in one partition and 1 in another.

Theorem 6 *$LG_{oD(2^n)}$ is complete for any positive integer n.*

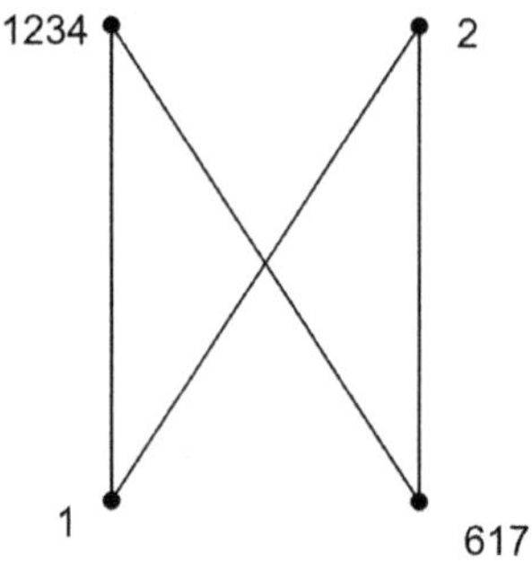

Fig. 3 Divisor function odd sum graph $G_{oD(1234)}$

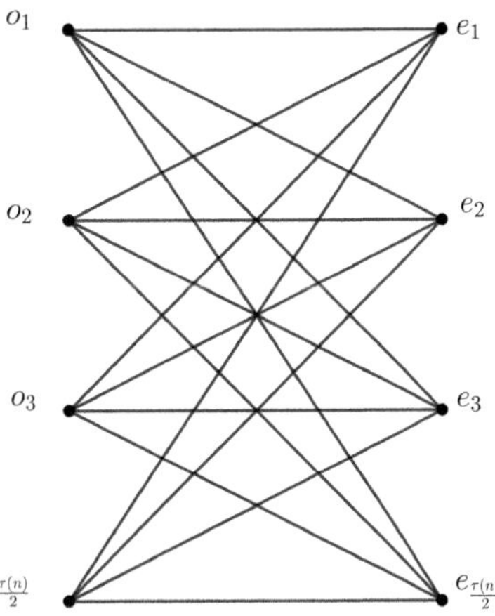

Fig. 4 k-Regular even $G_{oD(n)}$

Proof Clearly, $V(G_{oD(2^n)}) = \{2^0, 2^1 \dots 2^k\}$ and $G_{oD(2^k)} \simeq K_{1,k}$ because 1 is the only odd divisor of 2^n and remaining n divisors are even, all of those can be kept in same partition. Moreover, it is obvious that the line graph of a star graph is always complete, the proof follows.

Theorem 7 *For a k−Regular even* $G_{oD(n)}, \alpha = \beta = \frac{\tau(n)}{2}$.

Proof Assume that $G_{oD(n)}$ be a k-Regular graph of even order p. Clearly, $D(n)$ contains a set of all even(odd) divisors (say) $E(O)$ respectively. Since $G_{oD(n)}$ is k-regular, it follows that $|O| = |E| = \frac{\tau(n)}{2}$. Let the odd divisors be $o_1, o_2, \dots o_{\frac{\tau(n)}{2}}$ and the even divisors be $e_1, e_2, \dots e_{\frac{\tau(n)}{2}}$. Note that there exists an edge from o_i to e_i, $i = 1, 2 \dots \tau(n)$ and there is no edge from o_i to o_j and e_i to $e_j, i \neq j$.

Clearly, $G_{oD(n)} \simeq K_{\frac{\tau(n)}{2}, \frac{\tau(n)}{2}}$ (refer Fig. 4) and hence any minimum vertex cover must include vertices only from O or only from E. Henceforth,

$$\begin{aligned} |O| &= \beta = \frac{\tau(n)}{2} \\ Now, \alpha &= p - \beta \\ &= p - \frac{\tau(n)}{2} = \frac{\tau(n)}{2}. \end{aligned}$$

Theorem 8 *If n is factorised as an equal number of odd and even proper divisors, then* $G_{oD(n)}$ *is Hamiltonian.*

Proof Without loss of generality, let us assume n be an even integer, which may have both odd and even divisors. Let O denotes a set of all odd proper divisors and label the vertices of O as $\{o_1, o_2 \dots o_s\}$ and E denotes a set of all even proper divisors which are labelled as $\{e_1, e_2 \dots e_s\}$.

Let us prove the theorem by induction on s. For $s = 1$, the resulting graph is C_4 which has a spanning cycle with vertices $\{1, o_1, e_1, n\}$. Assume that the result is true

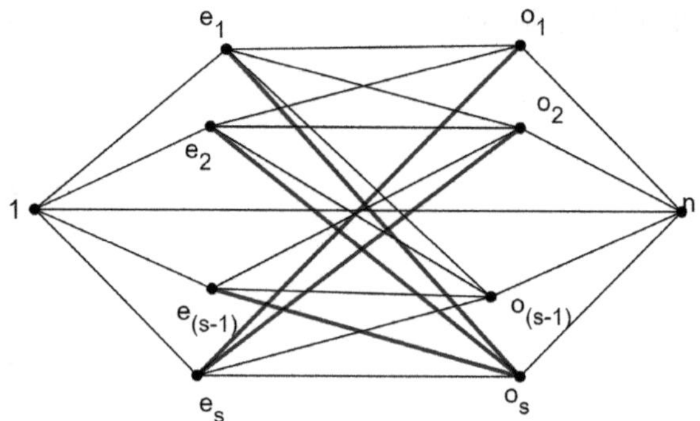

Fig. 5 Hamiltonian $G_{oD(n)}$

for all integer n which has at most $(s-1)$ odd proper divisors. Consider the case in which n has s odd proper divisors. Note that the induced subgraphs $G[\{1\} \cup V(E)]$ and $G[\{n\} \cup V(O)]$ forms a star.

The vertex 1 is not adjacent any odd proper divisor o_i's , the vertex n is not adjacent to any even proper e_j's and there exists an edge between every o_i's and e_j's, $1 \leq i \leq (s-1)$, $1 \leq j \leq (s-1)$ (Refer Fig. 5). By hypothesis, we have a spanning cycle C for any arbitrary vertex (say) e_i, $1 \leq i \leq (s-1)$ which comprises of a walk $W : \{e_i o_i, o_i e_{i+1}, e_{i+1} o_{i+1} \ldots e_{(s-1)} o_{(s-1)}\}$ followed by edges $\{o_{(s-1)} n, n1, 1e_i\}$ such that the length of C is $2(s-1)+3 = 2s-1$. Now, consider a subgraph $C' = C - \{e_{(s-1)} o_{(s-1)}, o_{(s-1)} n\}$ adding the edges $\{e_{s-1} o_s, o_s e_s, e_s o_{s-1}\}$ to C' again forms spanning cycle which includes all $2s+2$ vertices.

4 Conclusion

In this article, we introduced a new class of number theoretic graph structures termed divisor even sum (odd sum) graphs. We also studied several properties of these graphs, such as covering number, completeness, etc. In addition, we created a MATLAB method to draw these graphs, which will allow us to explore various other properties of these graphs. This research work has possible applications in communication networks and will be studied further in the future.

Acknowledgements The authors would like to thank the referees for their valuable comments. Also, the authors thank the Department of Science and Technology, Government of India for the financial support to the Department of Mathematics, SASTRA Deemed to be University under FIST Programme -Grant No. : SR/FST/MSI-107/2015(c).

References

1. Costache, A., Feigon, B., Lauter, K., Massierer, M., Puskas, A.: Ramanujan graphs in cryptography. J. Cryptol. Res. 1–33 (2018). https://eprint.iacr.org/2018/593.pdf
2. Aravinth, R.H., Vignesh, R.: Mobius Function Graph $M_n(G)$. International Journal of Innovative Technology and Exploring Engineering. **8**(10), 1481–1484 (2019)
3. Bondy, J.A.: USR Murty: Graph Theory with Application. Springer, New York (2017)
4. Bien, F.: Construction of telephone networks by group representations. Not. Am. Math. Soc. **36**(1), 5–22 (1989)
5. Burton, D.M.: Elementary Number Theory. 7th edn. The McGraw-Hill Companies (2010)
6. Chalapathi, T., Kumar, K.: Graph structures of Euler totient numbers. DIU J. Sci. Technol. **11**(2), 19–29 (2016)
7. Saliha, H.F.M., Badal Ibrahim, N.: Generalized the divisor sum T_k- function of graph. General Lett. Math. **8**(2), 67–74 (2020)
8. Kannan, K., Narasimhan, D., Shanmugavelan, S.: The graph of divisor function $D(n)$. Int. J. Pure Appl. Math. **102**(3), 483–494 (2015)
9. Somer, L., Krizek, M.: On a connection of number theory with graph theory. Czechoslovak Math. J.**54**(129), 465–485 (2004)
10. Madhavi, L., Maheswari, B.: Enumeration of traingles and Hamilton cycles in quadratic residue Cayley graphs. Chamchuri J. Math. **1**, 95–103 (2009)
11. Madhavi, L., Maheswari, B.: Enumeration of Hamilton cycles and triangles in Euler totient Cayley graphs. Graph Theory Notes of New York **LIX**, 28–31 (2010)
12. Narasimhan, D., Elamparithi, A., Vignesh, R.: Connectivity, independency and colorability of divisor function graph $G_{D(n)}$. Int. J. Eng. Adv. Technol. **8**(2S), 209–213 (2018)
13. Alan, N., Erdos, P.: An application of graph theory to additive number theory. Eur. J. Combon.**6**(3), 201–203 (1985)
14. Jose, R., Susha, D.: $\mu-$ graph of a finite group. Int. J. Math. Trends Technol. **67**(1), 129–135 (2021)
15. Shanmugavelan, S.: The Euler function graph $G(\phi(n))$. Int. J. Pure Appl. Math. **116**(1), 45–48 (2018)
16. Shanmugavelan, S., Thanga Rajeswari, K., Natarajan, C.: A note on indices of primepower and semiprime divisor function graph. TWMS J. Appl. Eng. Math. **11**(special issue), 51–62 (2021)
17. V M S S Kiran Kumar, R., Chalapathi, T.: Difference divisor graph of the finite group. Int. J. Res. Ind. Eng. **7**(2), 235–242 (2018)
18. Chow, Y.-T., Shi, W., Wu, T., Yin, W.: Expander Graph and Communication-Efficient Decentralized Optimization (2016). https://arxiv.org/abs/1810.01053v1

A Visible Watermarking Approach Likely to Steganography Using Nonlinear Approach

Sabyasachi Samanta

Abstract Digital watermarking is the course of action of set in information into digital content. Authenticity or integrity is one of the key features of watermarking. Perceptibility or existence of information on cover signal depends on nature of watermarking process. In this effort, a massive amount of message may embed in cooperation with nonlinear pixel and bit positions of image. Without embedding any specific position, data embedded to entire image stating from an arbitrary position chosen by the secret key. The number of effected pixels of carrier image is proportionally got higher with the level of message. Choice of pixel position entirely depends on the size of message. At decryption era, without inspect all of the pixels of cover image, data bits are composed from that special pixel and bit positions. With embedding of large amount of information, sometimes it produces some visible scrap to the carrier. Though it's visible to us but it's still secure to us, as it's embedded through some key to some nonlinear pixel positions. This approach is highly convinced the visible watermarking approach as well as the steganography approach also.

Keywords Information Security · Payload · Nonlinear Pixel Position (NPP) · Steganography · Visible Watermarking

1 Introduction

Digital Watermarking is the development of embedding a message on a host signal. A watermark can either be visible or invisible [16]. In visible watermarking the information comes to be visible on the image or video or picture [2, 4]. That is typically used for logos or text. Invisible digital watermarking is a type of steganography that aims to secrete information through medium to substantiate ownership, truthfulness or deliver additional information [9–12]. Steganography hides the visibility through the carrier like invisible watermarking [13–15]. Here I have proposed the Nonlinear Pixel Position-4 bit (NPP-4 bit) method, which is based on nonlinear pixel position

S. Samanta (✉)
Department of Information Technology, Haldia Institute of Technology, Haldia, WB, India
e-mail: sabyasachi.smnt@gmail.com

S. Banerjee and A. Saha (eds.), *Nonlinear Dynamics and Applications*,
Springer Proceedings in Complexity,
https://doi.org/10.1007/978-3-030-99792-2_44

selection through the image. The pixel positions are selected depending on the size of the message. Six digits key value is taken as a private key. Using the key value the initial and total pixel positions are selected. Then the pixel positions are calculated by doing the exponential function with the height and width of the image. In this proposed technique, the embedding process is made by the exponential values started from nonlinear pixel position of the carrier image. The bit position is calculated from the exponential values. Depending on the magnitude of the message the next pixel positions are selected by the combination of key. In this process the primary pixel position is not unique to all, i.e. it's varied from process to process and entirely depends on key. Also depending on payload and size of the image, the pixel positions varied from one to other. All over the data bits are being embedded to the entire carrier image. Figure 1 describes the 4 bits embedding process to the different bit positions for color images.

Section 2 represents the related works related with the watermarking and steganography technique. Section 3 represents an implementation of the technique of visible watermarking approach. Section 4 demonstrates the evidence of experimental result. Section 5 is with the performance analysis of NPP-4 bit methodology. Section 6 draws the conclusion of the work.

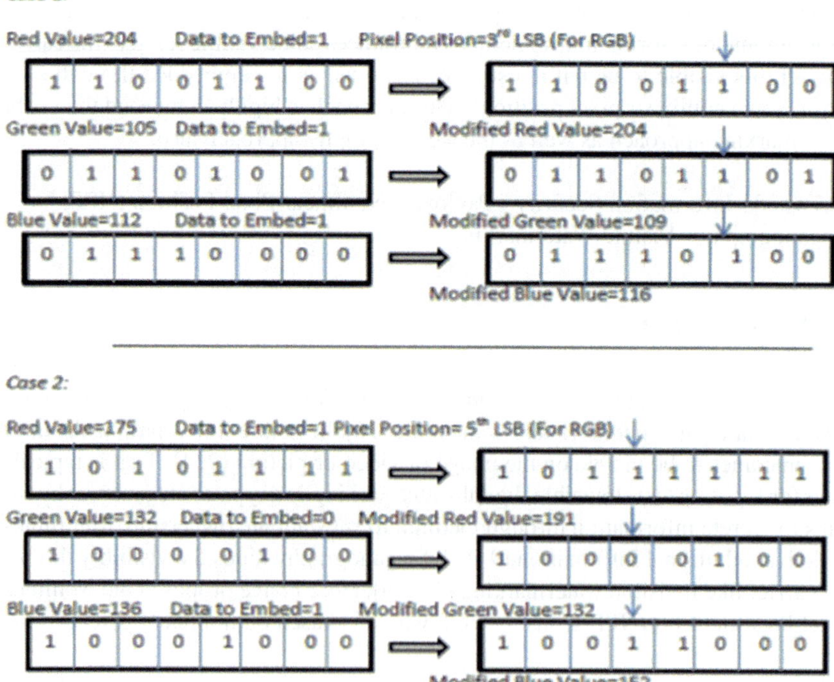

Fig. 1 Working principle of NPP-4 bit for color image

2 Related Works

Jiasheng Qu et al. [1] proposed a visible image watermarking scheme based on gradient-weighted class activation mapping. They have proposed the region of interest selection strategy to locate the secluded body of images for watermark embedding. They have used the similar technique for gray scale and each component of colour image.

Kamaldeep Joshi et al. [3] planned an image steganography using 7th bit of a pixel as indicator by introducing the successive temporary pixel in the gray scale image. One bit is hidden at the selected pixel position and the second bit is hidden to the next pixel position. On the basis of the seventh bit of the pixels of an image, a mathematical function is applied to that bit position of the pixels.

Hide in Picture (HIP) [5] was created by Davi Tassinari de Figueiredo. HIP technique uses only for bitmap images. If payload is high, it's necessary to modify more than one bit from each byte of the image. This process uses password protection to hide files and only those who know the password, are able to retrieve the data.

QuickStego [6]: QuickStego modify the pixels of the image. It encodes the secret text by adding small deviation in color to the image. In practice these small differences does not appear to the human eye. This technique does not encrypt the secret text message by any encoding scheme.

Mansoor Fateh et al. [7] proposed an improved version of the LSB matching. Thry proposed the scheme which contains two phases including embedding and extracting the message. Here a pair of bits of the secret message is hidden in a pair of pixels with only one change.

Wenfa Qi et al. [8] proposed an adaptive embedding method for visible watermarking. Here they have embedded watermark image the host image Just Noticeable Difference coefficient. Mukherjee et al. [17] proposed the mid position value technique to embed data bits within the tangled cover image. Also they applied the inverse Arnold transformation on stego image.

3 The Scheme

This section represents the algorithm for visible watermarking approach. Section A explains the total encryption process and section B depicts the decryption of plain text from the stego image.

A. Algorithm for data embedding in cover image.

Step I: Take message input, compute the length and create an array.

Step II: Select the initial pixel position using the key.

(i) Calculate the number of required pixel positions.
(ii) Calculate the exponential of function using key (or key set).

(iii) Take the digits up to "e". Accumulate first three digits to A[x], next three digits to A[y] and last digit to A[z].

Step III: Replace of data bits with R, G & B values of pixels.

(i) Calculate the width and height of the image as (M x N).
(ii) Select the pixel position and set the value of x and y with the value of M and N ((0, 0) to (M-1, N-1)).

If (x > (M-1)) or (y > (N-1)) then Set P (x, y) = P (0 + (x %(M-1)), (0 + (y %(N)) else set P (x, y) = (x, y).
(iii) To select the bit position, set z = A[z] of a pixel.
(iv) The four data bits are replaced in analogous bit position of each RGB elements and pixels are reformed.

Step IV: Repeat Step II to Step IV.

Step V: Stop.

B. Algorithm for data extraction from stego image.

Step I: To search out the pixel and bit position in R, G & B of selected pixels follow Step II and step III of Algorithm A.

Step II: Retrieve the embedded data bits from the selected bit positions of selected pixels and accumulate it to data array.

Step III: Initially find out the length of the message from the encrypted array.

Step IV: Find out the embedded data bits from the stego image and bring together the delivered message.

Step V: Stop.

4 Implementation of Process and Experimental Result

Figure 2 stands for the cover and stego image of LENA and MONALISHA after embedding 1000 characters. Figure 3 shows the histogram of LENA image (512 × 512) as cover and stego image. Figure 4 shows the histogram comparison using PoV for Red, Green, for Blue using LENA Image.

5 Performance Analysis of NPP-4 Methodology

Invisible digital watermarking is called a kind of steganography. Embedding of 500 more characters using NPP-4 bit method creates visible dissimilarities at the output image. Sometimes NPP-1 bits may also create tiny spots in stego image. As a result

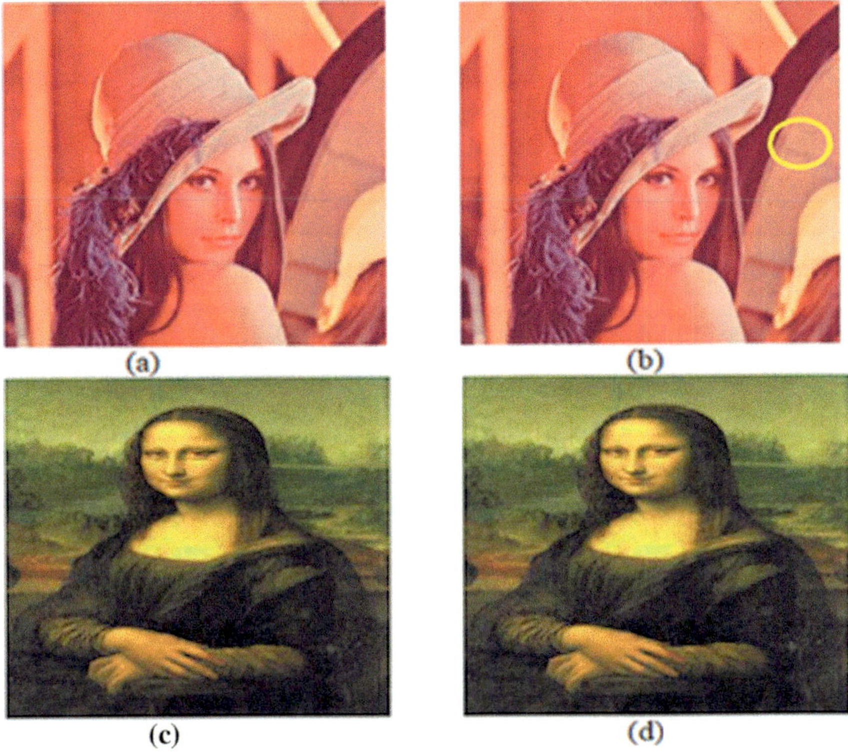

Fig. 2 **(a, c)** Cover and **(b, d)** Stego image of LENA and MONALISHA after embedding 1000 characters using NPP-4 bit

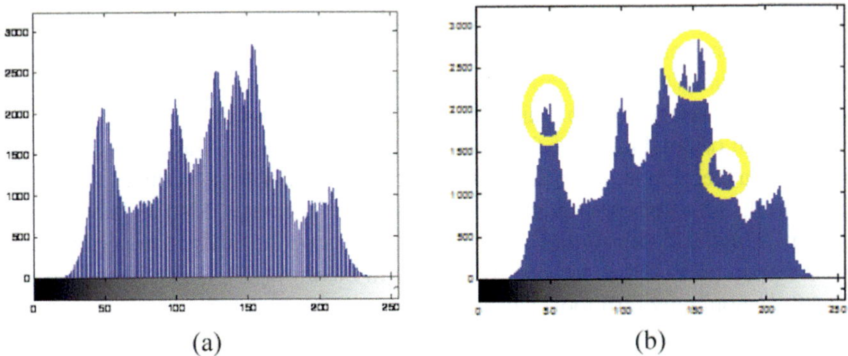

Fig. 3 Histogram of LENA as (**a**) Cover and (**b**) Stego image after embedding 1000 characters using NPP-4 bit

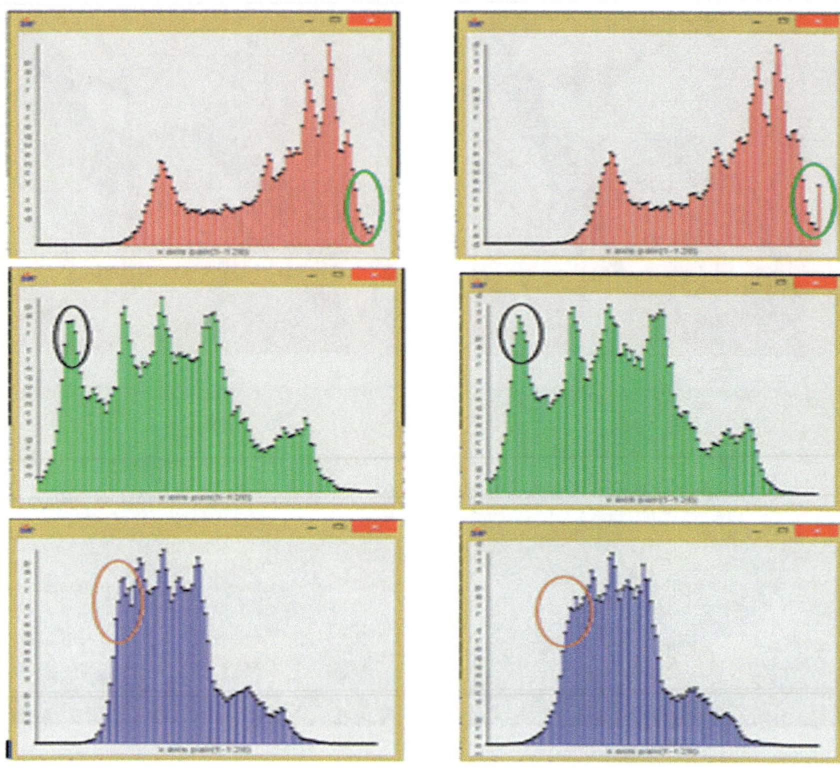

Fig. 4 Histogram comparison using PoV for Red, Green, and Blue using LENA image

its do not support the benchmarks of invisible digital watermarking. The visible variations or patches on cover image do not break any principles of information security or confidentiality. As the data bits have embedded to arbitrary pixel and bit positions of entire image. Without applying the embedding algorithm or key value, the extraction of original message from noticeable zones of stego image is quite impossible to attackers. In steganography, the embedding algorithm produces the image output as the attacker does not minds the presence of message. Also invisible digital watermarking produces similar like output. To measure the qualitative performance of developed NPP-4bit approache, HIP (Hide In Picture) [18], QuickStego [19] and Xiao Steganohraphy [20] tools have choosen.

The x–y direction of the following figures represents the methodologies and different performance metrics respectively. The x-direction of following igures represent the NPP-4 bit and other well known existing methodologies like HIP, QuickStego and Xiao Steganohraphy. Figures 5 and 6 represent the comparison for PSNR between NPP-4 bit and other well known approaches for color and gray scale image using LENA (128 × 128) as a cover respectively. Figures 7 and 8 subsequently signify the superiority of SSIM between NPP 4bit and other well known approaches for color and gray scale image as same cover image. As I have explained the embedding of

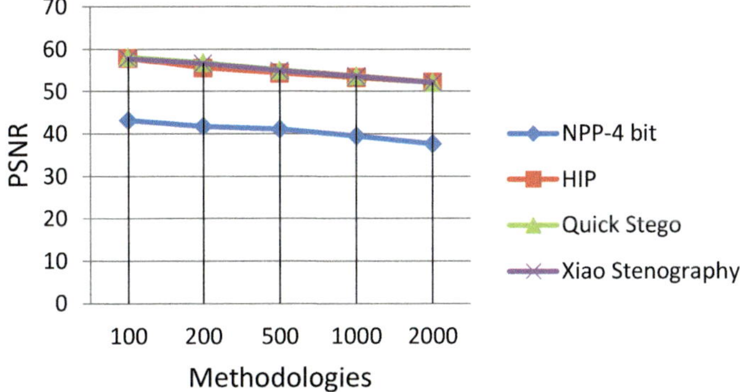

Fig. 5 PSNR comparison of different NPP approaches with other methodologies using LENA/(Color, 128 × 128) Images

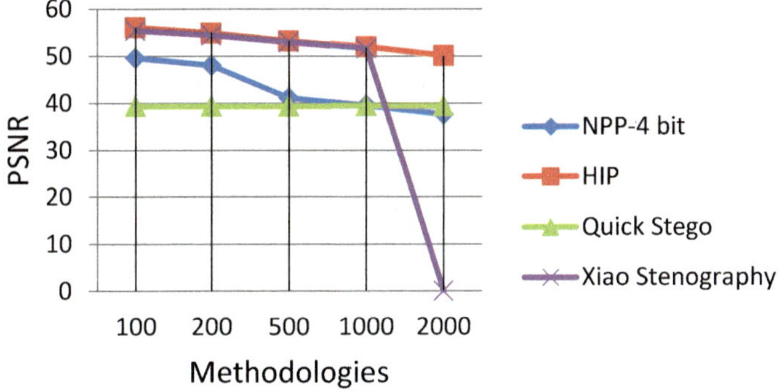

Fig. 6 PSNR comparison of different NPP approaches with other methodologies using LENA/(Gray Scale, 128 × 128) Images

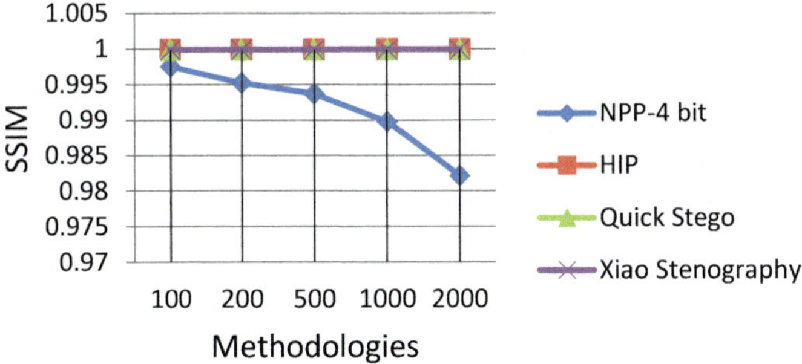

Fig. 7 SSIM comparison of different NPP approaches with other methodologies using LENA/(Color, 128 × 128) Images

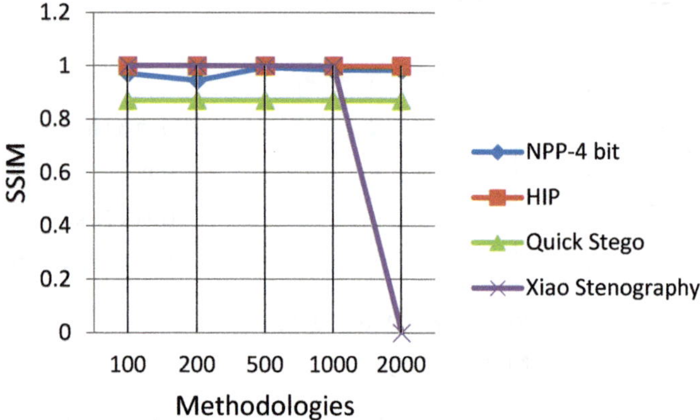

Fig. 8 SSIM comparison of different NPP approaches with other methodologies using LENA/(Gray Scale, 128 × 128) Images

four bits i.e. NPP-4 bit generate the tiny spot on image. So for the PSNR for both color and gray scale stego image sometimes generates lower result for the newly developed methodology. The SSIM measure for color images produces the likewise result for NPP-1 bit and NPP-2 bit techniques. The SSIM measure for both the color and gray scalge images generates similar graph for newly developed methodology.

6 Conclusion

Here I have proposed a visible watermarking approach likely to steganography using nonlinear approach and private key cryptographic technique. In this approach, four data bits have placed in every selected pixel element with random pixel positions about the image. Moreover for fewer number of content, sometimes it produces the similar images in naked eye. But for more it produces nonlinear tiny spots throughout the image. So for this approach I have named as a visible watermarking approach likely to steganography. Though the tiny spots are visible to us as effected location, but it's quite impossible to retrieve the embedded message from stego image. Only using the key and proper set of pixel and bit positions embedded message can retrieve from the watermarked image. So it's the visible watermarking approach through NPP but the methodology hold up the unadulterated steganography approaches.

References

1. Jiasheng, Q., Wei, S., Xiangchun, L., Lizhi Zhao, and Xiaobing Zhao: A Novel Improved Reversible Visible Image Watermarking Algorithm Based on Grad-CAM and JND. Secur.

Commun. Netw. Article ID 6652897, 1–17 (2021). https://doi.org/10.1155/2021/6652897
2. Mahbuba, B., Mohammad, S.U.: Analysis of digital image watermarking techniques through hybrid methods. Adv. Multimed. Article ID 7912690, 1–12 (2020). https://doi.org/10.1155/2020/7912690
3. Kamaldeep, J., Swati, G., Rajkumar, Y.: A new method of image steganography using 7th bit of a pixel as indicator by introducing the successive temporary pixel in the gray scale image. J. Comput. Netw. Commun. Article ID 9475142, 1–11 (2018). https://doi.org/10.1155/2018/9475142
4. Nandhini, S., Somaya, A.-M., Ahmed, B.: Image steganography. A Review of the Recent Advances. IEEE Access, vol. 9, pp. 1–15. Digital Object Identifier (2021). https://doi.org/10.1109/ACCESS.2021.3053998
5. Pedram, H., Vidyasagar, P., Elizabeth, C.: A Survey of Steganographic and Steganalytic Tools for the Digital Forensic Investigator, pp. 1–12. http://www.pedramhayati.com/images/docs/survey_of_steganography_and_steganalytic_tools.pdf
6. Naveen Brahma Teja, K., Madhumati, G.L., Rama Koteswara Rao, K.: Data hiding using EDGE based steganography. Int. J. Emerging Technol. Adv. Eng. **2**(11), 354–360(2012). ISSN 2250–2459
7. Mansoor, F., Mohsen, R., Yasser, I.: A new method of coding for steganography based on LSB matching revisited. Secur. Commun. Netw. Article ID 6610678, pp 1–15(2021). https://doi.org/10.1155/2021/6610678
8. Wenfa, Q., Yuxin, L.,Sirui, G., Xiang, W., Zongming, G.: An adaptive visible watermark embedding method based on region selection. Secur. Commun. Netw. Article ID 6693343, pp 1–15 (2021). https://doi.org/10.1155/2021/6693343
9. Robert, L., Shanmugapriya, T.: A study on digital watermarking techniques. Int. J. Recent Trends Eng. **1**(2), 223–225 (2009)
10. Janu, N., Kumar, A., Dadheech, P., Sharma, G., Kumar, A., Raja, L.: Multiple watermarking scheme for video & image for authentication & copyright protection, IOP conference series. Mater. Sci. Eng. (2021). https://doi.org/10.1088/1757-899X/1131/1/012020,pp.1-12,ICETCE 2021
11. Shweta, W., Deepa, K., Ankit, R., Aruna, J., Vishal, J.: A comprehensive review on digital image watermarking. Workshop on Computer Networks & Communications, Chennai, India, May 01, 2021, pp. 126–143 (2021)
12. Fu, Z., Li, E., Cheng, X., Huang, Y., Hu, Y.: Recent advances in image steganography based on deep learning. J. Comput. Res. Dev. **58**(3), 548–568 (2021)
13. Dinu, C.: Low distortion transform for reversible watermarking. IEEE Transactions on Image Processing, Jan. 2012, vol. 21, no. 1, pp. 412–417 (2012).
14. Caldelli, R., Macaluso, G., Bartolini, F., Barni, M.: Near Lossless Image authentication Transparent to near Lossless Coding. Security, Steganography, and Watermarking of Multimedia Contents, VI, SPIE, vol. 5306, pp. 737–747(2004).
15. Afroja, A., Muhammad A.U.: Digital watermarking with a new algorithm. IJRET: Int. J. Res. Eng. Technol. **03**(03), 212–217(2014). eISSN: 2319–1163, pISSN: 2321–7308
16. Supriadi, R., De Rosal, I.M.S., Abdul, S., Pulung, N.A.: Inverted LSB image steganography using adaptive pattern to improve imperceptibility. J. King Saud Univ.—Comput. Inf. Sci. 1–10 (2021). https://doi.org/10.1016/j.jksuci.2020.12.017
17. Mukherjee, S., Roy, S., Sanyal, G.: Image Steganography using mid position value Technique. In: International Conference on Computational Intelligence and Data Science (ICCIDS 2018), Procedia Computer Science, vol. 132, pp. 461–46 (2018)
18. Image Steganography tool: http://www.softpedia.com/get/Security/Security-Related/Hide-In-Picture.shtml
19. Image Steganography tool: http://download.cnet.com/QuickStego/3000-2092_4-75593140.html
20. Image Steganography tool: http://download.cnet.com/Xiao-Steganography/3000–2092_4–10541494.html

A New Public Key Encryption Using Dickson Polynomials Over Finite Field with 2^m

Kamakhya Paul, Madan Mohan Singh, and Pinkimani Goswami

Abstract The Dickson polynomials are widely used in different context. Because of the permutation behaviors and semi-group property under composition, its application can also be noticed in cryptography. In this paper, we propose a new public key encryption scheme based on Dickson polynomial of first kind over a finite field with 2^m, where m is the product of two large primes p and q, with primes $2^p - 1$ and $2^q - 1$. The security of the proposed scheme is equally dependent on the Integer Factorization Problem (IFP) and the Discrete Dickson Problem (DDP). We also compared the encryption-decryption performance of the proposed scheme with some other existing schemes and found that the performance of the proposed scheme is better compared to them.

Keywords Dickson polynomial · Integer Factorization Problem · Discrete Dickson Problem · Discrete Logarithm Problem · Encryption scheme

1 Introduction

Public key cryptography, also known as asymmetric cryptography, where transmission of messages takes place over an open network, was first introduced by Diffie and Hellman [1] in 1976. Here separate keys are used for encryption and decryption,

K. Paul (✉)
North Eastern Hill University, Shillong 793022, ML, India
e-mail: kamakhyapaul4@gmail.com

M. M. Singh
Department of Basic Sciences & Social Sciences, North Eastern Hill University, Shillong 793022, ML, India
e-mail: mmsingh2004@gmail.com

P. Goswami
Department of Mathematics, University of Science and Technology Meghalaya, Ri-Bhoi 793101, ML, India
e-mail: pinkimanigoswami@yahoo.com

S. Banerjee and A. Saha (eds.), *Nonlinear Dynamics and Applications*,
Springer Proceedings in Complexity,
https://doi.org/10.1007/978-3-030-99792-2_45

and hence the insecurity of transmitting the secret key over an insecure channel was overcome. This is observed in symmetric cryptosystems.

Dickson [2] first introduced a type of polynomial, which was later labeled by Schur [3] as Dickson polynomial. With the proposition by Dickson [2], extensive research and looking into the properties of the polynomial had started and hence followed by application of Dickson polynomial in cryptography [4–7]. Lidl [7], also surveyed the algebraic properties of the Dickson polynomial over $\mathbb{F}_q$ and over the integers $\mathbb{Z}_n$.

Most of the cryptographic schemes are being developed based on the hard problems that can be solved both ways only if prior knowledge of the related problem is known. Initially, the hard problems that were being used includes discrete logarithm and factoring of a large composite number in terms of primes, taken only one hard problem at a time. McCurley [8] in 1988 was the first to use two different number theoretic assumptions in the development of a single key distribution protocol. Many more cryptosystems were later proposed [9–12, 14–16, 24, 25], which were based on the merging of two hard problems such as Discrete Logarithm and Factoring of a large composite number, Elliptic Curve Discrete Logarithm, Knapsack problem, and many more. Which prompted us to use two hard problem in Dickson polynomial to propose a public key encryption scheme.

In this paper, we have operated Discrete Logarithm and Integer Factorization in the Discrete Dickson Problem (DDP) over the finite field with cardinality 2^m and proposed a cryptosystem whose security is based on the hardness of solving IFP and DDP.

2 Dickson Polynomial

A type of polynomial was introduced by Dickson [2] in 1896 in the form of

$$x^k + k \sum_{i=1}^{(k-1)/2} \frac{(k-i-1)...(k-2i+1)}{2.3...i} a^i x^{k-2i}, \; k \; is \; odd, \tag{1}$$

over the finite field F_q, which later came to be known as the Dickson polynomial.

2.1 Definition (Dickson Polynomial of First Kind)

Assume n be a positive integer and $a \in \mathbb{F}_q$, then the Dickson polynomial $D_n(x, a)$ of the first kind over any finite field $\mathbb{F}_q$ is defined by

$$D_n(x, a) = \sum_{i=0}^{\lfloor \frac{n}{2} \rfloor} \frac{n}{n-i} \binom{n-i}{i} (-a)^i x^{n-2i} \tag{2}$$

where $\lfloor \frac{n}{2} \rfloor$ is the largest integer less than or equal to $\frac{n}{2}$.

The Dickson polynomial satisfy the recurrence relation: $D_n(x, a) = \mathrm{x}D_{n-1}(x, a) - \mathrm{a}D_{n-2}(x, a)$, $n \geq 2$, under the initial condition $D_0(x, a) = 2$ and $D_1(x, a) = x$ and a few initial polynomials are given below:

$D_2(x, a) = x^2 - 2a$
$D_3(x, a) = x^3 - 3ax$
$D_4(x, a) = x^4 - 4ax^2 + 2a^2$
$D_5(x, a) = x^5 - 5ax^3 + 5a^2x$

Commutativity under composition is of considerable importance and is satisfied by the Dickson polynomial for a = 0 or 1 [18], and hence it satisfies the semi-group property under composition:

$D_{mn}(x, 1) = D_m(D_n(x, 1), 1) = D_m(x, 1) \circ D_n(x, 1) = D_n(x, 1) \circ D_m(x, 1) = D_n(D_m(x, 1), 1)) = D_{nm}(x, 1)$.

2.2 Definition (Modified Dickson Polynomial)

Let us define a map, $D_p : \mathbb{Z}_n \rightarrow \mathbb{Z}_n$, which is defined as $y = D_p(x) \pmod n$, with x and n being integers. Here, we call $y = D_p(x) \pmod n$ as the modified Dickson polynomial. Below are a few properties satisfied by modified Dickson polynomial.

1. The modified Dickson polynomial is commutative under composition, that is $D_p(D_q(x) \pmod n) = D_{pq}(x) \pmod n = D_q(D_p(x) \pmod n)$
2. Let q be an odd prime and let $x \in \mathbb{Z}$ such that $0 \leq x < q$. Then the period of the sequence $D_n(x) (\mathrm{mod}\ q)$ for $n = 0, 1, 2, 3, 4, \ldots\ldots.$ is a divisor of $q^2 - 1$.

The first key exchange cryptosystem which was based on Dickson polynomial was introduced by Müller and Nöbauer [18] in 1981, where the power functions of the RSA system, introduced by Rivest et al. [17] in 1978, was replaced by Dickson polynomials $D_n(x, a)$ with parameter $a = -1, 0, 1$. It was also observed that the RSA cryptosystem was equivalent to the Dickson system for parameter $a = 0$ [18]. In 2011, Wei [13] introduced in his paper that the Dickson polynomial $D_n(x, 1)$ over a finite field 2^m is a permutation polynomial if and only if n is odd and proved that solving a Discrete Dickson Problem (DDP) is as difficult as solving Discrete Logarithmic Problem (DLP). Note that the hardness of DLP was also observed by McCurley [20]. It is also observed that computable groups where DLP is hard to solve [21–23] are of great importance in cryptography.

2.3 Definition (Discrete Dickson Problem)

Let R be a commutative ring with unity, for any $n \in \mathbb{Z}^+$, and given y and x, the problem of calculating the value of n such that $y = D_n(x, 1)$ is called the Discrete Dickson Problem(DDP).

It is observed throughout the paper that we have used for $a = 1$, $D_n(x, 1) = D_n(x)$.

3 Our Proposed Public Key Encryption Scheme

Here we propose our scheme, which consists of three parts, that includes, Key generation, encryption, and decryption.

Key Generation

The generation of the key includes the below given steps:

1. Choose two random large primes, p and q, of the same size, such that $2^p - 1$ and $2^q - 1$ are prime.
2. Using the above p and q, compute $n = 2^m$, where $m = p \times q$.
3. For the value of n, find $\phi(n)$, where $\phi(n) = (2^{2p} - 2^{p+1})\times(2^{2q} - 2^{q+1})$.
4. Choose e, such that $1 < e < \phi(n)$ and $gcd(e, \phi(n)) = 1$.
5. Find d, such that $ed \equiv 1 \pmod{\phi(n)}$, where d is the modular inverse of e.
6. Choose a, such that $0 \leq a \leq \phi(n)$ - 1.
7. Choose a random $\alpha \in \mathbb{Z}_n^*$ and compute $x = \frac{1}{2} D_a(2\alpha) \pmod{n}$.

- ***PUBLIC KEY***: (n, e, x, α)
- ***PRIVATE KEY***: (p, q, a, d)

Encryption

Here the process of encrypting the simple plain text into cipher text is performed, so that an intruder doesn't get to read the message. For the message $M \in \mathbb{Z}_n$,

1. Select a random $r \in \mathbb{Z}_n^*$ and for the selected r, compute $k_1 = \frac{1}{2} D_e(2r) \pmod{n}$.
2. Similarly select $c \in \mathbb{Z}_n^*$ and for the selected c, compute $k_2 = \frac{1}{2} D_c(2\alpha) \pmod{n}$.
3. Now finally compute k_3 using selected r and the given x, where $k_3 = \frac{M}{4} D_c(2x)$ $D_e(2r) \pmod{n}$.

For the plain text message "M", the encrypted ciphertext is (k_1, k_2, k_3), which will be received by the decoder to generate the message.

Decryption

On receiving the encrypted message (k_1, k_2, k_3), the receiver performs the below given steps:

1. Firstly, he/she deals with obtaining the value of r, by computing $\frac{1}{2}D_d(2k_1) \pmod{n}$.
2. Followed by computing Y, where $Y = k_1^{-1} \pmod{n}$.
3. Compute W, where $W = k_3 Y \pmod{n}$.
4. Then compute Z, where $Z = \frac{1}{2}D_{a^{\phi(n)+1}}(2k_2) \pmod{n} = \frac{1}{2}D_c(2x) \pmod{n}$.
5. Finally obtain the plain-text message $M = WZ^{-1} \pmod{n}$.

Verification

First verification is dealing with obtaining the value r in the decryption.

$$
\begin{aligned}
&\tfrac{1}{2}D_d(2k_1) \pmod{n}\\
&= \tfrac{1}{2}D_d(2\tfrac{1}{2}D_e(2r)) \pmod{n}\\
&= \tfrac{1}{2}D_d(D_e(2r)) \pmod{n}\\
&= \tfrac{1}{2}D_{de}(2r) \pmod{n}\\
&= \tfrac{1}{2}2r\\
&= r
\end{aligned}
$$

As Dickson polynomial is commutative under composition, we have $D_d(D_e(\alpha)) \pmod{n} = D_{(de)}(\alpha) \pmod{n} = D_e(D_d(\alpha) \pmod{n}) = D_1(\alpha) \pmod{n} = \alpha$, also

$$
\begin{aligned}
&Z\\
&= \tfrac{1}{2}D_{a^{\phi(n)+1}}(2k_2) \pmod{n}\\
&= \tfrac{1}{2}D_a(2\tfrac{1}{2}D_c(2\alpha)) \pmod{n}\\
&= \tfrac{1}{2}D_c(D_a(2\alpha)) \pmod{n}\\
&= \tfrac{1}{2}D_c(2x) \pmod{n}
\end{aligned}
$$

And the verification of the message to be generated is

$$
\begin{aligned}
&WZ^{-1} \pmod{n}\\
&= k_3 Y Z^{-1} \pmod{n}\\
&= k_3 k_1^{-1}(\tfrac{1}{2}D_c(2x))^{-1} \pmod{n}\\
&= \tfrac{M}{4}D_c(2x)D_e(2r)(\tfrac{1}{2}D_e(2r))^{-1}(\tfrac{1}{2}D_c(2x))^{-1} \pmod{n}\\
&= \tfrac{M}{4}D_c(2x)D_e(2r)2D_e^{-1}(2r)2D_c^{-1}(2r) \pmod{n}\\
&= M
\end{aligned}
$$

4 Explanation with a Simple Example Using Wolfram Mathematica

Key Generation

1. Let $p = 3$ and $q = 5$, so $2^3 - 1$ and $2^5 - 1$ is prime.
2. So $n = 2^{15}$, where $m = 3 * 5$.
3. For $n = 32768$, $\phi(n) = 46080$.
4. $e = 11$ and $gcd(e, \phi(n)) = 1$.
5. $d = 41891$, for $ed \equiv 1 \pmod{n}$.
6. Choose $a = 122$.
7. For $\alpha = 177$, $x = \frac{1}{2}D_{122}(2 \times 177) \pmod{32768} = 15041$.

- ***PUBLIC KEY***: (32768, 11, 15041, 177)
- ***PRIVATE KEY***: (3, 5, 122, 41891)

Encryption

For message $M = 4433$.

1. Taking $r = 223$, $k_1 = \frac{1}{2}D_{11}(2 \times 223) \pmod{32768} = 6623$.
2. Taking $c = 19$, $k_2 = \frac{1}{2}D_{19}(2 \times 177) \pmod{32768} = 29745$.
3. $k_3 = \frac{4433}{4}D_{19}(2 \times 15041)D_{11}(2 \times 223) \pmod{32768} = 20943$.

Encrypted message (6623, 29745, 20943).

Decryption

1. $\frac{1}{2}D_{41891}(2 \times 6623) \pmod{32768} = 223$.
2. $Y = 6623^{-1} \pmod{32768} = 25119$.
3. $W = 20943 \times 25119 \pmod{32768} = 9745$.
4. $Z = \frac{1}{2}D_{122\phi(32768)+1}(2 \times 29745) \pmod{32768} = \frac{1}{2}D_{19}(2 \times 15041) \pmod{32768}$ $= 6337$.
5. $6337^{-1} \pmod{32768} = 30529$.
6. Finally $9745 \times 30529 \pmod{32768} = 4433 = M$.

   And hence the original message is generated.

5 Security

In the proposed cryptosystem, the security is found to be completely built upon the Integer Factorization Problem (IFP) and the Discrete Dickson Problem (DDP). Here we have observed a few cases of common attacks where the proposed cryptosystem was well secured.

As the encrypted message can be assessed by an intruder, he/she can have an assess to (k_1, k_2, k_3). Now, for him/her to generate the message M, he/she has to obtain the value of p and q of m and so the value of d, followed by finding a from $\frac{1}{2}D_a(2\alpha) \pmod{n}$. And this can only be achieved if both the Integer Factorization Problem and Discrete Dickson Problem can be solved. The value of p and q is chosen in such a way that the size of m is 1024-bit or above, so no known algorithm can be used to factor m. And also to find a from $\frac{1}{2}D_a(2\alpha) \pmod{n}$, the intruder have to solve DDP. Also, the value of α and r should be large enough to prevent exhaustive search attack. It should be kept in mind that to encrypt different messages different values of r and c should be used. Because if a sender uses same parameters for the encryption of two different messages M_1 and M_2, then the intruder can obtain $k_3 = \frac{M_1}{4}D_c(2x)D_e(2r) \pmod{n}$ and $k_3' = \frac{M_2}{4}D_c(2x)D_e(2r) \pmod{n}$. And hence from the relation $M_2 = k_3' k_3^{-1} M_1$, the intruder can have the message M_2 on knowing M_1. So on choosing different values of r and c, the message M_2 cannot be known even on knowing M_1.

Suppose the intruder somehow manages to find the values of p and q and then computes $r = \frac{1}{2}D_d(2k_1) \pmod{n}$ and $W = k_3 Y \pmod{n} = k_3 k_1^{-1} \pmod{n} = \frac{M}{2}D_c(2x) \pmod{n}$. To find the message M from above, one has to know c, which is a computationally impossible assumption of the Discrete Dickson Problem which is equivalent to solving Discrete Logarithm Problem (DLP).

6 Performance Analysis

Here we have observed the performance of our proposed cryptosystem.

Let T_{ddp}, T_{mul}, T_{exp}, T_{ch}, and T_{inv} represents the time for execution of Discrete Dickson Problem, time for modular multiplication, time for modular exponential operation, time for Chebyshev map and time for modular inverse computation respectively. And the time for computation is $1T_{ddp} = 0.172$ s, $1T_{mul} = 0.00207$ s, $1T_{exp} = 5.37$ s, $1T_{ch} = 0.172$ s and $1T_{inv} = 0.0207$ s. Hence in our cryptosystem, the total computational complexity required is $6T_{ddp} + 8T_{mul} + 2T_{inv}$ and the total time is equivalent to 1.09012 s, which is comparatively a smaller amount of time taken for computation.

Below we have compared the computational complexity of the proposed scheme with few other existing schemes (Table 1).

Table 1 Comparison Table

Scheme	Encryption	Decryption	Time encryption (s)	Time decryption(s)	Total time(s)
Gowsami et al. [9]	$6T_{exp} + 3T_{mul}$	$4T_{exp} + 3T_{mul} + 3T_{inv}$	32.22621	21.54831	53.77452
Poulakis [10]	$6T_{exp} + 4T_{mul}$	$3T_{exp} + 2T_{mul} + 2T_{inv}$	32.22828	16.15554	48.38302
Goswami et al. [25]	$4T_{exp} + 2T_{mul}$	$3T_{exp} + 2T_{mul} + 2T_{inv}$	21.48414	16.15554	37.63968
Tahat et al. [24]	$6T_{ch} + 3T_{mul}$	$4T_{ch} + 3T_{mul} + 3T_{inv}$	1.03821	0.75631	1.79452
Proposed Scheme	$4T_{ddp} + 4T_{mul}$	$2T_{ddp} + 4T_{mul} + 2T_{inv}$	0.69628	0.39384	1.09012

7 Conclusion

In this paper, we have proposed a public key cryptosystem based on IFP and DDP. The use of IF and DDP has highly enhanced the security of the proposed cryptosystem and it is also observed that the computational complexity is quite low compared to that of other schemes.

References

1. Diffie, W., Hellman, M.: New directions in cryptography. IEEE Trans. Inf. Theory **22**(6), 644–654 (1976)
2. Dickson, L.E.: The analytic representation of substitutions on a power of a prime number of letters with a discussion of the linear group. Ann. Math. **11**(1), 65–120 (1896/1897)
3. Schur, I.: Arithmetisches über die Tschcbyscheffschen Polynome. Gesammelte Abhandlungen **3**, 422–453 (1973)
4. Lidl, R., Müller, W.B.: Permutation polynomials in RSA-cryptosystems, In: Chaum, D. (eds.), Advances in Cryptology, pp. 293–301. Springer, Boston, MA (1984). https://doi.org/10.1007/978-1-4684-4730-923
5. Lidl, R., Müller, W.B.: A note on polynomials and functions in algebraic cryptography. Ars Combinatoria **17**, 223–229 (1984)
6. Lidl, R., Müller, W.B.: On commutative semigroups of polynomials with respect to composition. Monatshefte für Mathematik **102**(2), 139–153 (1986)
7. Lidl, R.: Theory and applications of Dickson polynomials. Top. Polynomials One Sev. Var. Their Appl. Vol. Dedic. Mem. PL Chebyshev **1821–1894**, 371–395 (1993)
8. McCurley, K.S.: A key distribution system equivalent to factoring. J. Cryptol. **1**(2), 95–105 (1988)
9. Goswami, P., Singh, M.M., Bhuyan, B.: A new public key scheme based on integer factorization and discrete logarithm. Palest. J. Math. **6**(2), 580–584 (2017)
10. Poulakis, D.: A public key encryption scheme based on factoring and discrete logarithm. J. Disc. Math. Sci. Cryptograp. **12**(6), 745–752 (2009)

11. Guo, R., Wen, Q., Jin, Z., Zhang, H.: Pairing based elliptic curve encryption scheme with hybrid problems in smart house. In: 2013 Fourth International Conference on Intelligent Control and Information Processing (ICICIP), pp. 64–68. Institute of Electrical and Electronics Engineers (IEEE), Beijing, China (2013)
12. Shao, Z.: Signature schemes based on factoring and discrete logarithms. IEE Proc.-Comput. Digit. Tech. **145**(1), 33–36 (1998)
13. Wei, P., Liao, X., Wong, K.W.: Key exchange based on Dickson Polynomials over finite field with 2^m. J. Comput. **6**(12), 2546–2551 (2011)
14. Ismail, E.S., Hijazi, M.S.N.: A new cryptosystem based on factoring and discrete logarithm problems. J. Math. Stat. **7**(3), 165–168 (2011)
15. Mohamad, M.S.A., Ismail, E.S.: Threshold cryptosystem based on factoring and discrete logarithm problems. In: AIP Conference Proceedings, pp. 1020–1023. American Institute of Physics, Selangor, Malaysia (2013)
16. Baocang, W., Yupu, H.: Public key cryptosystem based on two cryptographic assumptions. In: IEE Proceedings—Communications, vol. 152(6), pp. 861–865 (2005). https://doi.org/10.1049/ip-com:20045278
17. Rivest, R., Shamir, A., Adleman, L.: A method for obtaining digital signatures and public key cryptosystems. Commun. ACM **21**(2), 120–126 (1978)
18. Müller, W.B., Nöbauer, R.: Some remarks on public key cryptography. Studia Scientiarum Mathematicarum, Hungarica **16**, 71–76 (1981)
19. Mullen, G.L., Panario, D.: Handbook of Finite Fields (Discrete Mathematics and Its Applications), 1st edn. CRC Press, Boca Raton (2013)
20. McCurley, K.S.: The discrete logarithm problem. In: Proceedings of Symposia in Applied Mathematics, Cryptology and Computational Number Theory, vol. 42, pp. 49–74. American Mathematical Society, Boulder Colorado (1990)
21. Menezes, A.J., Van Oorschot, P.C., Vanstone, S.A.: Hand Book of Applied Cryptographhy. CRC Press, Bacon Raton (2018)
22. Álvarez, R., Tortosa, L., Vicent, J.F., Zamora, A.: Analysis and design of a secure key exchange scheme. Inf. Sci. **179**(12), 2014–2021 (2009)
23. Coppersmith, D., Odlyzko, A.M., Schroeppel, R.: Discrete logarithms in GF(p). Algorithmica **1**(1), 1–15 (1986)
24. Tahat, N., Tahat, A.A., Abu-Dalu, M., Albadarneh, R.B., Abdallah, A.E., Al-Hazaimeh, O.M.: A new public key encryption scheme with chaotic maps. Int. J. Electr. Comput. Eng. **10**(2), 1430–1437 (2020)
25. Goswami, P., Singh, M.M., Bhuyan, B.: A new public key scheme based on DRSA and generalized GDLP. Disc. Math. Algorithms Appl. **8**(4), 1650057 (2016)

Strongly k-Regular Dominating Graphs

Anjan Gautam and Biswajit Deb

Abstract A dominating set D of $V(G)$ is defined to be a *k-regular dominating set* if the subgraph of G induced by D is k-regular. We define a graph to be a *strongly k-regular dominating graph* if each dominating set of it is k-regular. Some classes of graphs that are strongly *0-regular and 1-regular* are characterized. In particular, all trees T with $diam(T) \leq 7$ for which trees are strongly 0-regular dominating are characterized.

Keywords Tree · Domination number · k-regular dominating set · Triangular Snake graph

1 Introduction

A graph G is a pair $(V(G), E(G))$, where $V(G)$ is a finite nonempty set and $E(G)$ is a set of unordered pairs of distinct elements of $V(G)$. The degree of a vertex $v \in V(G)$ is the number of edges incident to it in G and it is denoted by $deg(v)$. If all the vertices of G have the same degree k, then G is k-regular, or simply regular.

The eccentricity $\epsilon(v)$ of a vertex v is the greatest distance between v and any other vertex, that is

$$\epsilon(v) = \max_{u \in V} d(v, u).$$

The center of a graph is the set of all vertices with minimum eccentricity. The diameter of a graph is the maximum eccentricity of any vertex in the graph. A graph H is said to be a subgraph of G if $V(H) \subseteq V(G)$ and $E(H) \subseteq E(G)$. The induced subgraph $\langle S \rangle$ is the graph whose vertex set is $S \subseteq V$ and whose edge set consists of all of the edges in E that have both endpoints in S.

The open neighborhood $N(v)$ of $v \in V(G)$ is defined as

A. Gautam (✉) · B. Deb
Department of Mathematics, Sikkim Manipal Institute of Technology, Sikkim Manipal University, Majitar, Rangpo, East-Sikkim 737136, India
e-mail: anjangautam12@gmail.com

S. Banerjee and A. Saha (eds.), *Nonlinear Dynamics and Applications*, Springer Proceedings in Complexity,
https://doi.org/10.1007/978-3-030-99792-2_46

$$N(v) = \{u \in V : uv \in E\}$$

and the closed neighborhood $N[v]$ of v is defined as

$$N[v] = \{u \in V : uv \in E\} \cup \{v\}.$$

A vertex of degree one is called a pendant vertex, its neighbor a support vertex, and its incident edge a pendant edge or leaves. If a vertex v is adjacent to two or more leaves, v is said to be a strong support vertex. For standard terminology in graphs, we refer the book by Harary [2].

The set S is called a dominating set of G if every vertex in $V \setminus S$ has a neighbor in S. The domination number $\gamma(G)$ is the cardinality of a minimum dominating set of G. A set $D \subseteq V(G)$ is a totally dominating set of a graph G if every vertex v in $V(G)$ is adjacent to some vertex in D. The total domination number $\gamma_t(G)$ is the cardinality of the minimum totally dominating set. For concepts related to domination we refer [7]. Recently in 2019, a new lower bound on the domination number of a graph is observed [10]. A dominating set D of $V(G)$ is defined to be a k-regular dominating set if $\langle D \rangle$ is k-regular. We define a graph to be a strongly k-regular dominating graph if each minimum dominating set of it is k-regular (For example see Fig. 1). The concept of the domination number in a graph was defined by Berge [1] as, "coefficient of external stability". Later in 1962, Ore [3] used the name "domination number" for the same. The connected domination number was studied by Sampathkumar and Walikar [4]. In 1994 Sampathkumar and Latha studied set domination in graphs [5]. The concept of regular set domination number was introduced by Kulli and Janakiram [6] in 2009. For a survey on domination in graphs, we refer book "Topics in Domination in Graphs" [8].

Consider a region that is suffering from a pandemic situation like COVID-19. Suppose that regular testing has to be done of the entire population and to achieve this new test centers are to be established. With limited resources, test center with medical facilities are to be placed with back up in strategic locations. The problem is to decide about the locations where such testing center needs back up. Let vertices represent different localities of the region and an edge connects two localities that

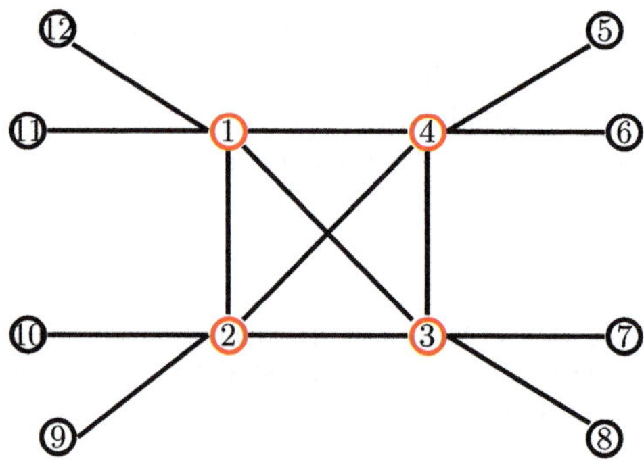

Fig. 1 *Strongly 3-regular dominating graph* with dominating set $D = \{1, 2, 3, 4\}$

share boundary. A locality (or vertex) and all of the regions that are adjacent to it we can refer as the neighborhood of that locality. A testing center in a location can serve the people in the localities in its neighborhood. Here the domination problem is to choose localities at which place these service facility to be located such that each neighborhood contains at least one and if required their must be a backup testing center in its neighborhood. An optimal solution for this problem can be achieved by obtaining a k-regular dominating set for $k \geq 1$ with minimum number of vertices. It is to be noted that a connected dominating set may not be a good idea to solve the above problem due to the scarcity of the resources and high demand of the test centers.

2 Results

2.1 Strongly 0-Regular Dominating Graph

Theorem 1 *Let P_n be a path with n vertices. Then P_n is strongly 0-regular dominating graph if and only if $n \equiv 0$ or 2 (mod 3).*

Proof The domination number of a path with n vertices is $\gamma(P_n) = \lceil \frac{n}{3} \rceil$. Let $P_n = v_1 v_2 \ldots v_n$ be strongly 0-regular dominating graph. If possible let $n \equiv 1 \ (mod\ 3)$. Consider the partition $A_0, A_1, \ldots, A_{m-1}, m = \lceil \frac{n}{3} \rceil$, where $A_i = \{v_{3i+1}, v_{3i+2}, v_{3i+3}\}$ for $i = 0, 1, \ldots, m-2$ and $A_{m-1} = \{v_{3(m-1)+1}\}$. The vertex v_{3i+2}, covers A_i, for $i = 0, 1, \ldots, m-2$, and vertex $v_{3(m-1)+1}$ is covered by $v_{3(m-2)+3}$ or itself. If we consider a dominating set $D = \{v_2, v_5, \ldots, v_{3(m-2)+2}, v_{3(m-2)+3}\}$ of size $m = \lceil \frac{n}{3} \rceil$, then $v_{3(m-1)+2}$ and $v_{3(m-1)+3}$ are adjacent as shown in Fig. 2 so, $\langle D \rangle$ is not 0-regular, a contradiction. Thus $n \equiv 0$ or 2 $(mod\ 3)$.

Conversely, let $n \equiv 0$ or 2 $(mod\ 3)$.

Case 1: $n \equiv 0 (mod\ 3)$.
The set $S = \{v_2, v_5, \ldots, v_{3(m-1)+2}\}$ is the only dominating set of size $m = \lceil \frac{n}{3} \rceil$. If possible let, there exist a dominating set $T = \{u_1, u_2, \ldots, u_m\}$ of size m other than S. As $n \equiv 0(\text{mod } 3)$ each vertex $u_i, i = 1, 2, \ldots, m$ must cover 2 other vertices. If $u_1 = v_1$ then u_1 will cover just one vertex v_2 and if $u_1 = v_3$ then u_1 will cover v_2 and v_4, which makes v_1 uncovered. Thus, we can say $u_1 = v_2$. Similarly $u_2 = v_5, u_3 = v_8, \ldots, u_m = v_{3(m-1)+2}$. As $\langle S \rangle$ is 0-regular which implies P_n is strongly 0-regular dominating graph.

Case 2: $n \equiv 2(\text{mod } 3)$.
In this case there are only two dominating set of size m.
Let $D = \{v_2, v_5, \ldots, v_{3(m-2)+2}, v_{3(m-1)+2}\}$
and $S = \{v_2, v_5, \ldots, v_{3(m-2)+2}, v_{3(m-1)+1}\}$ be dominating set of size m.
One can similarly show as Case 1, that any other dominating set T of size m is either D or S. As $\langle D \rangle$ and $\langle S \rangle$ are both 0-regular, thus P_n is strongly 0-regular dominating graph.

Fig. 2 A Path P_7 with dominating set $D = \{2, 5, 6\}$

Theorem 2 *Let C_n be a cycle with n vertices. Then C_n is strongly 0-regular dominating graph if and only if $n \equiv 0$ or 2 $(mod\ 3)$.*

We can write the proof of Theorem 2 on the same line as Theorem 1.

Lemma 1 *A strongly 0-regular dominating graph G cannot have a pair of pendant vertices at distance 3 apart.*

Proof Let G be a graph with a path of length 3 between any two pendent vertices u, v. Let the path be $\{u, u_1, v_1, v\}$ and D be a minimum dominating set of G. As u and v are pendent vertices, we may consider u and v to be in D but alternately, we can also consider $u_1, v_1 \in D$ to cover u and v. As $\{u_1, v_1\}$ is an edge so G is not strongly 0-regular dominating graph.

Remark 1 If a tree has no pendant vertices at distance 3, then no two support vertices in T are adjacent.

Theorem 3 *A tree T with $diam(T) \leq 5$ is strongly 0-regular dominating graph if and only if no two pendant vertices are at distance 3.*

Proof Let a tree T with $diam(T) \leq 5$ is 0-regular dominating graph. If possible let two pendant vertices are at distance 3 then by Lemma 1 graph T is not strongly 0-regular dominating graph, a contradiction.

Conversely, let T has no pendant vertices at distance 3. Let D be a minimum dominating set. For a tree T with $diam(T) = 1, 2$, the result holds as dominating set consists of one vertex. Tree which holds the condition with $diam(T) = 3$ does not exist.

Consider a tree T with $diam(T) = 4$ and $u_1u_2u_3u_4u_5$ be its diametric path. Any minimum dominating set D consists of either support vertex or pendant vertex. By Remark 1 D is 0-regular, thus T with $diam(T) = 4$ is strongly 0-regular dominating graph.

Consider $diam(T) = 5$ and $u_1u_2 \ldots u_6$ be a diametric path. Diametric path and pendant vertices of u_2, u_5 will be covered by u_2 and u_5. If u_3 has a tail $u_3v_1v_2$, then v_1 or v_2 will cover its tail. Thus any minimum dominating set D comprises of strong support vertices, and support vertices or pendent vertices. As two support vertices are not adjacent thus D is 0-regular dominating set.

Theorem 4 *A tree T with $diam(T) = 6$ is strongly 0-regular dominating graph if and only if*

(i) no two pendant vertices are at distance 3, and
(ii) center of T contains a pendant vertex or a tail of length 2.

Proof Let T be a strongly 0-regular dominating tree with $diam(T) = 6$ and $u_1u_2 \dots u_7$ be a diametric path in it. By Lemma 1 G cannot have a pendant vertices at distance 3 apart. If possible let $center(T)$ does not contains a pendant vertex or a tail of length 2. Let D be a minimum dominating set containing all the support vertices. A set aggregating only of all support vertices will not dominates $center(T) = u_4$, as v_3 or v_5 has no pendant vertices. Thus D consisting of support vertices along with v_3 or v_5 is not 0-regular dominating set, a contradiction. Hence, center of T contains a pendant vertex or a tail of length 2.

Conversely, assume that (i) and (ii) holds for a tree T. Let D be any minimum dominating set of T.

Case 1: Center of T contains a pendant vertex.
As (i) implies u_3 and u_5 are not support vertices, so $u_3, u_5 \notin D$. For, if $u_3 \in D$, then u_3 covers u_2, u_4 and the support vertex w_1 of any tail $u_3w_1w_2$ at u_3, if there is any. To cover the pendant vertices at u_2, u_4 and w_1 we need 3 more vertices other than u_3 in D. But all these vertices may be covered by $\{u_2, u_4, w_1\}$, contradicting our assumption that D is minimum. Similarly we can argue that $u_5 \notin D$. Therefore, $u_4 \in D$ and $u_3, u_5 \notin D$. Any other vertex in D will be a support vertex or a pendant vertex. By (i) no two support vertices are adjacent, so $< D >$ is 0-regular.

Case 2: Center of T contains a tail of length 2.
Let $u_4v_1v_2$ be a tail of length 2 at u_4. We claim that, $u_3, u_4, u_5 \notin D$. For if $u_4 \in D$, then either v_1 or $v_2 \in D$.
If $v_1 \in D$, then u_4 must cover either u_3 or u_5. Suppose u_4 covers u_3. Then either $deg(u_3) = 2$ or there is a tail at u_3 other than $u_3u_2u_1$, whose pendant vertex is in D. Also either u_1 or $u_2 \in D$. Contradicting the minimality of D.
If $v_2 \in D$, then u_4 is either covering itself or u_3 or u_5. If u_4 covering u_3 or u_5, then arguing in the same way as above, we can show that D is not minimum. If u_4 covering itself, than dropping u_4 and v_2 from D and including v_1 we can get a dominating set smaller than D, contradiction to the minimality of D. Therefore, $u_4 \notin D$ and $v_1 \in D$.
For if $u_3 \in D$, then u_3 is either covering itself or u_2 or w_1 of any tail $u_3w_1w_2$ at u_3. But u_3 does not cover u_2 or w_1 as it is covered by its pendant vertex. If u_3 covers itself then removing u_3 and u_1 from D and adding u_2 we can get dominating set smaller than D, a contradiction. Therefore, $u_3 \notin D$, similarly $u_5 \notin D$. Thus, any vertex in D is either support vertex or a pendant vertex, so $\langle D \rangle$ is 0-regular.

As every minimum dominating set D of T is 0-regular, so T is a strongly 0-regular dominating graph.

Definition: For any two vertices $u_1, u_2 \in V(G)$, by saying that (u_1, u_2) has a tail of length (a_1, a_2) we mean that there is a tail at u_i of length a_i, for $i = 1, 2$.

Theorem 5 *Let T be a tree with $diam(T) = 7$ and center $\{u_4, u_5\}$. Then T is strongly 0-regular dominating graph if and only if*

(i) no two pendant vertices in T are at distance 3, and
(ii) (u_4, u_5) does not have a tail of length $(2, 0)$ and $(2, 3)$.

Proof Consider a tree T with $diam(T) = 7$ and center $\{u_4, u_5\}$ that is strongly 0-regular dominating graph, and $u_1u_2 \ldots u_8$ be a diametric path. Let D be a minimum dominating set of T. By Lemma 1 T cannot have pendant vertices at distance 3 apart. If possible let there exist a tail at (u_4, u_5) of length $(2, 0)$ or $(2, 3)$. Let $u_4v_1v_2$ be a tail at u_4 and $u_5w_1w_2w_3$ at u_5.

Case 1: (u_4, u_5) has a tail of length $(2, 0)$.
As u_6, u_4 has no pendant vertex, D consisting of u_2, u_7, v_1, support vertices of tails at u_3 or u_6 if any, along with u_4 or u_6 or u_5 will cover T. Considering $u_4 \in D$, u_4v_1 will be an edge in $\langle D\rangle$, therefore it is not 0-regular, a contradiction.

Case 2: (u_4, u_5) has a tail of length $(2, 3)$.
Similarly as above, D consisting of u_2, u_7, v_1, support vertices of tails at u_3 or u_6 along with u_4 or u_6 or u_5 or w_1 will cover T. Considering u_4 or u_6 or $w_1 \in D$, $\langle D\rangle$ will not be 0-regular, a contradiction.

Thus, if T is strongly 0-regular dominating graph then (i), (ii) holds.

Conversely, let (i), (ii) holds and D be any minimum dominating set of T. We consider the following cases.

Case 1: $u_4, u_5 \notin$ closed neighbor of any support vertex.
In this case (u_4, u_5) may have tail of length $(3, 3)$ or $(0, 3)$ or $(3, 0)$ or they may have no tail. In any case D consist of support vertex or pendant vertex together with u_4 or u_5, so $\langle D\rangle$ is 0-regular. Hence T is strongly 0-regular dominating graph.

Case 2: Either u_4 or $u_5 \in$ closed neighbor of any support vertex.
If $u_4 \in$ closed neighbor of any support vertex, then (u_4, u_5) may have tail of length $(1, 0)$ or $(1, 2)$ or $(1, 3)$ or $(2, 2)$. If (u_4, u_5) have tail of length $(1, 0)$ or $(1, 3)$, then u_4 must be in D to cover u_5 and pendant vertex of u_4. If $u_4 \notin D$ then we need two more vertex in D, contradicting the minimality of D. Therefore, D comprise of support vertex or pendant vertex together with u_4, thus $\langle D\rangle$ is 0-regular.
If (u_4, u_5) have tail of length $(1, 2)$, then $u_5 \notin D$. For if $u_5 \in D$ we will need two more vertex in D to cover pendant vertex at u_4 and pendant vertex of a tail of length 2 at u_5. But if we just consider u_4 and support vertex of a tail at u_5, then they will cover same number of vertices as, u_6 can be covered by u_7. In this case any D will consist of support vertices or pendant vertices, Thus $\langle D\rangle$ is 0-regular.
If (u_4, u_5) have tail of length $(2, 2)$, then support vertices of tails from u_4 and u_5 will cover them respectively. Arguing as above $\langle D\rangle$ will be 0-regular. For if u_4 and u_5 have more than one tail of different lengths then we can argue similarly as above that, any D will consist of support vertices or pendant vertices. Thus, in any case $\langle D\rangle$ is 0-regular. Hence T is strongly 0-regular dominating graph.

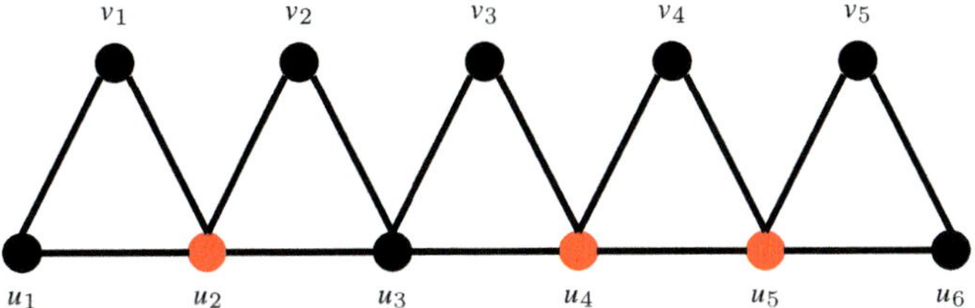

Fig. 3 A Triangular Snake T_6 with dominating set $D = \{u_2, u_4, u_5\}$

Definition: A Triangular Snake graph T_n is obtained from a path $u_1, \ldots, u_n$ by joining u_i and u_{i+1} to a new vertex v_i for $1 \leq i \leq n-1$. That is, every edge of a path is replaced by a triangle.

Theorem 6 *A Triangular Snake T_n is strongly 0-regular dominating graph if and only if $n \equiv 1$ $(mod\ 2)$.*

Proof Let a Triangular Snake T_n be strongly 0-regular dominating graph. Domination number of Triangular Snake $T_n = \lfloor \frac{n}{2} \rfloor$. If possible let $n \equiv 0$ $(mod\ 2)$.

Consider the partition $A_0, A_1, \ldots, A_{m-1}$, $m = \lfloor \frac{n}{2} \rfloor$, where $A_0 = \{u_1, u_2, u_3, v_1, v_2\}$, $A_i = \{u_{2(i+1)}, u_{2(i+1)+1}, v_{2i+1}, v_{2(i+1)}\}$, for $1 \leq i \leq m-2$ and $A_{m-1} = \{u_n, v_{n-1}\}$. The vertex u_2 dominates A_0 and $u_{2(i+1)}$ dominates A_i for $1 \leq i \leq m-2$. A_{m-1} is dominated by u_n or by u_{n-1}. Let us consider the dominating set of size $\lfloor \frac{n}{2} \rfloor$, $D = \{u_2, u_4, \ldots, u_{n-2}, u_{n-1}\}$, as shown in Fig. 3 (for $n = 6$). As $\langle D \rangle$ contains an edge $u_{n-2}u_{n-1}$ so T_n is not strongly 0-regular dominating graph, a contradiction.

Conversely, let $n \equiv 1$ $(mod\ 2)$. Then the set $D = \{u_2, u_4, \ldots, u_{n-1}\}$ is the only dominating set of size $\lfloor \frac{n}{2} \rfloor$. If possible let $S = \{w_1, w_2, \ldots, w_m\}$, be a another dominating set of size $m = \lfloor \frac{n}{2} \rfloor$. If $w_1 = u_1$ then u_3 and v_2 is not covered and if $w_1 = u_3$ then u_1 and v_1 is not covered. Thus $w_1 = u_2$ will cover all the vertices of partition A_0. Similarly we can show $w_2 = u_4, \ldots, w_m = u_{n-1}$. As $\langle D \rangle$ is 0-regular, therefore T_n is strongly 0-regular dominating graph.

Remark 2 For $k \geq 2$ a tree T is not strongly k-regular dominating graph.

2.2 *Strongly 1-Regular Dominating Graphs*

Double star: Double star $S_{m,n}$ is the graph obtained by joining the center of two stars $K_{1,n}$ and $K_{1,m}$ with an edge.

Remark 3 Double star $S_{m,n}$ is strongly 1-regular dominating graph for $m, n \geq 2$.

Book Graph: The m-book graph is defined as the graph Cartesian product $B_m = K_{m,1} \Box P_2$, where $K_{m,1}$ is a star graph and P_2 is the path graph on two vertices.

Fig. 4 A Path P_4 with dominating set $D = \{2, 4\}$ or $\{2, 3\}$ or $\{1, 4\}$ or $\{1, 3\}$ and total dominating set $S = \{2, 3\}$

Theorem 7 *Book graph B_m is strongly 1-regular dominating graph for $m \geq 2$.*

Proof Consider a book graph $B_m = K_{m,1} \Box P_2$, where $K_{m,1}$ is a star graph and P_2 is the path graph on two vertices. Let $V(K_{m,1}) = 1, 2, \ldots, m+1$, where the center is 1, and $V(P_2) = a, b$. Now, $V(B_m) = (1, a), (1, b), (2, a), (2, b), \ldots, (m+1, a), (m+1, b)$. From the definition of Cartesian product the set $D = \{(1, a), (1, b)\}$ is sufficient to cover all the vertices of B_m, for all m. For $m \geq 2$, every dominating set $D = \{(1, a), (1, b)\}$, thus $\langle D \rangle$ is 0-regular, and B_m is strongly 1-regular dominating graph.

Lemma 2 *A strongly 1-regular dominating graph cannot have a path of length 3, with degree sequence of its vertices as $1, a, 2, 1$, for $a \geq 2$.*

Proof Let G be a graph with a path of length 3 and degree sequence of vertices of path $\{u_1, u_2, u_3, u_4\}$ is $1, a, 2, 1$, $a \geq 2$. Let D be a minimum dominating set of G. We can consider $u_2, u_3 \in D$ or $u_2, u_4 \in D$ to dominate vertices of path. As we have $u_2, u_4 \in D$ so, G is not strongly 1-regular dominating graph.

Theorem 8 *If a graph is strongly 1-regular dominating graph, then every minimum dominating set of G is equal to its minimum total dominating set.*

Proof Let a graph G be strongly 1-regular dominating graph and D be any minimum dominating set. As every vertex of $V(G)$ has a neighbor in D, thus D is a total dominating set.

Remark 4 The converse of Theorem 8 is not true. For example, see Fig. 4, here domination number is equal to total domination number, but induced subgraph of every dominating set is not 1-regular.

2.3 Construction of Strongly k-Regular Dominating Graph

Given a graph G we can construct strongly k-regular dominating graph.

- Take a complete graph K_{k+1}.
- Put at least two pendant vertices to each vertex of K_{k+1}.
- Now select two vertices u, v of K_{k+1} and add one more vertex to both of them (those vertices has degree at least $k+3$).
- Take an edge $xy \in E(G)$ and replace x and y with any one of the pendant vertices of u and v.

- Take $|E(G)|$ copies of K_{k+1} with pendant vertices.
- Repeat the above step with each edge of G.
- Dominating set $\gamma(G)$ contains all the vertices of K_{k+1} which is k-regular.

3 Conclusion

In this study, the concept of a strongly k-regular dominating graph is introduced and some classes of strongly 0-regular and 1-regular dominating graphs are characterized. Specifically, all 0-regular dominating trees T with $diam(T) \leq 7$ are characterized. Application of domination number is discussed in various field such as Computer Communication Networks, Radio Stations, Facility Location Problems, Coding Theory etc. [9].

References

1. Berge, C.: Theory of Graphs and Its Applications. Dunod Paris (1958)
2. Harary, F.: Graph Theory. Wiley Online Library (1969)
3. Ore, O.: Theory of Graphs. American Mathematical Society Colloquium Publications, vol. 38, pp. 206–212 (1962)
4. Sampathkumar, E., Walikar, H.B.: The connected domination number of a graph. J. Math. Phys **13**, 607–613 (1979)
5. Sampathkumar, E., Latha, L.P.: Set domination in graphs. J. Graph Theory **18**, 489–495 (1994)
6. Kulli, V.R., Janakiram, B.: Regular set domination in graphs. Natl. Acad. Sci. Lett. (India) **32**, 351–355 (1981)
7. Haynes, T.W., Hedetniemi, S., Slater, P.: Fundamentals of Domination in Graphs. CRC press 1 (2013)
8. Haynes, T.W., Hedetniemi, S.T., Henning, M.A.: Topics in Domination in Graphs. Springer, 64 (2020)
9. Gupta, Preeti: Domination in graph with application. Indian J. Res. **2**, 115–117 (2013)
10. Hajian, M., Henning, M.A., Rad, N.J.: A new lower bound on the domination number of a graph. J. Comb. Optim. **38**, 721–738 (2019)

Chaotic Based Image Steganography Using Polygonal Method

Dipankar Dey, Solanki Pattanayak, and Sabyasachi Samanta

Abstract Information security is the most concerning factor in the time of tremendous innovative headway. Immense specialized information cleared a way for the development of information taking methods. Steganographic methods help to veil the mysterious data with some other media that goes about as a spread to the data. This strategy makes it hard for programmers to see the data since it shows up as a media instead of as data itself. Image steganography assists with concealing mystery data in any dimension and worth of images. These pictures can be traded without looking for any consideration of programmers to the mystery data inside it. Here the information is encoded into the image through an arbitrary rectangular region, to make it more secure. Using chaotic map we have generated random number series through which data embedded in rectangular region with uneven interval. This can be unscrambled at the beneficiary side with the best possible calculation alongside the secret key. We have scrambling the data and concealing it in the image utilizing the chaotic map. Removing the data and unscrambling it is done at the beneficiary. Embedding capacity of the proposed methodology is also comparable with other well known existing methodologies. Performance analysis is also measured and the proposed methodology generates enhanced result than the others.

Keywords Steganography · Chaotic map · Rectangular region · Encryption · Decryption · Cover image

D. Dey
Depatment of Computer Science and Technology, Global Institute of Science and Technology, Haldia, WB, India
e-mail: dey_dipankar@rediffmail.com

S. Pattanayak
Department of Computer Science, Haldia Institute of Management, Haldia, WB, India
e-mail: solankipattanayak16@gmail.com

S. Samanta (✉)
Department of Information Technology, Haldia Institute of Technology, Haldia, WB, India
e-mail: sabyasachi.smnt@gmail.com

S. Banerjee and A. Saha (eds.), *Nonlinear Dynamics and Applications*,
Springer Proceedings in Complexity,
https://doi.org/10.1007/978-3-030-99792-2_47

1 Introduction

Various strategies such as encryption and steganography are used to protect the data on the Internet. Encryption is the study of transforming a mysterious message into another structure, and the ultimate goal is not universally understandable. Steganography is a strategy for making sure about data by concealing it in some other medium, with the end goal that the presence of data is disguised to everybody aside from the planned sender and collector. Steganography suggest to the craftsmanship and study of concealing mystery data from other media [7, 8]. The data that is hidden is called a mystery message, and the medium in which the data is hidden is called a distribution report. A distributed dataset containing implied messages is called a stego report. A calculation used to obfuscate a message on the sender's propagating media and remove the obfuscated message from the saver's stego archive called stego framework [9–12]. Here in this study introduces image steganography technology the contributions of this work is very inventive. Initially we have inserted a gray scale image with size of M*N and chosen a rectangular region to hide the text inside the image using chaotic map to generate the random number. We have used the mathematical functions which give the better output collect to other existing techniques and unauthorized users can not access our information.

Section 2 gives an overview of some existing related systems. The proposed system is presented and the analysis of the proposed system is evaluated in Sect. 3. Section 4 explains the performance of the proposed system with the relevant existing system. Finally, the conclusion of this work is in Sect. 5.

2 Literature Review

Information covering by using LSB [1] is one of the most basic and habitual strategies so far. This methodology works by concealing information at all noteworthy bits of the pixels (LSB). Huge however it's yet at the same time a few changes in the picture may disable the implanted information. The Pixel Value Differencing (PVD) [2], recommended by Wu and Tsai is one of the convincing cryptographic approach. It structures pixel hinders from the spread and adjusts the pixel contrast in each pair of squares for information inclusion. More prominent the distinction more is the adjustment made. In light of PVD, another technique for tri method of pixel esteem differencing is proposed by Chang et al. [3]. This new strategy has shown better performance in terms of payload [4, 18] and PSNR. Gray Level Correction (GLM) [5] proposed by Potder Et al. is another strategy for mapping information by reforming the dark level estimates of the pixels present in the image. The gray level estimates for the selected pixels are distinguished based on the secret information bits. Several strategies have also been proposed to combine PVD and GLM techniques with the goal of increasing payload. One such technology is Safarpour Et al. is taught by [6]. When inspected, the image is screened and split at the correct pixels using bit substitution. These pixels are randomly selected rather than in order. Chen Et al. adopted the strategy used when using the side match technique [13]. Using particle swarm optimization and chaos theory, authors propose a steganography technique that finds the optimal pixel position in the cover image and hides confidential data while preserving the quality of the resulting stego image. To increase the embedded capacity, the host and secret images are divided into blocks, each block containing the appropriate amount of secret bits. In [14], Hsiao Et al. suggested a method of data hiding based on a 16 × 16 Sudoku matrix by taking a 16 × 16 Sudoku matrix and extending it to a two-layer magic square. Low-cost data embedding methods are also being considered to improve PSNR and maintain good image quality with the same embedding capacity. Here Manikandan et al. [15] proposes a method to validate the receiver by incorporating the E-mail and OTP verification to get the encrypted image. Mansoor Fateh et al. [16] proposed an improved version of the LSB matching, which works with two phases including embedding and extracting the message. In this method two bits of the secret message are hidden in two pixels with only one change. Wenfa Qi et al. [17] proposed an adaptive embedding method for visible watermarking where they embedded watermark image in the host image.

3 Proposed Work and Algorithms

3.1 Process of the Proposed Work

The following is an overview of a new plan using the chaos system with the goal of being able to change the intensity values for all pixels. To hide the secret data in the proposed plot, (a) we make a square shape district of dimension 512 × 512 of a unique picture (dimension of the gray scale picture is 512 × 512); (b) input different payload type text as 1024 bytes text and store in s1 variable (c) Chaotic maps are used to generate private keys. (d) A bitwise XOR operation is applied between the target pixel and the private key to get the final encrypted image. Figure 1 shows a block diagram of the proposed method.

3.2 Algorithm for Encoding Process

Step 1: Input gray scale image of size 512 × 512 and a text (size of text vary with respect to programming.

Step 2: Initialize the variable $q = 0.9898$ (the value of q is determined from chaotic map) and$s = 5, xc = \frac{row}{2}, yc = \frac{col}{2}$, where row and col is the size of the image. Choose the value of n (secret key) as in following order of varying text file$(n, s1)$. Here $s1$ is the size of text file. The text file is stored in p_k in one dimensional array.

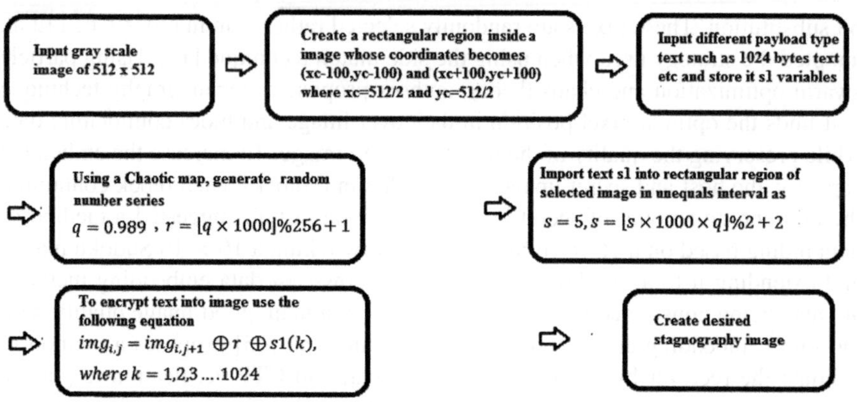

Fig. 1 Block diagram of the proposed method

$\{(4, 100), (8, 500), (12, 1000), (17, 2000), (22, 3000), (26, 5000), (30, 10000), (33, 20000)$ *and* $(36, X)\}$, here X is the remaining text file size. Store the value of n into image location which will be used in decryption process for extract the original text file. The value of n is stored in image location as $img\left(\frac{row}{2} + 10, \frac{col}{2} + 10\right) = n$. Here image location is also a secret key.

Step 3: Initialize $i = xc - n,\ j = yc - n$.

Step 4:$If(i \leq xc + nandj \leq yc + n)$

Step 5: Then

$$r = \lfloor q \times 1000 \rfloor \%125, \quad q = 3.5 \times q \times (1 - q),$$
$$img_{i,j} = img_{i,j+1} \oplus r \oplus p_k, where\, k\, is\, the\, size\, of\, text\, file,$$
$$s = \lfloor s \times 1000 \times q \rfloor \%2 + 2,$$

Iterate uneven gap which is s times.

Step 6: Repeat Step 4 until condition is false.

Step 7: Create the desired steganography image.

Step 8: Stop.

3.3 Flow Chart of Encoding Process

3.4 Algorithm for Decoding Process

Step 1: Input gray scale image (steganography img) of size 512 × 512 and a text (cover text may differ from process to process).

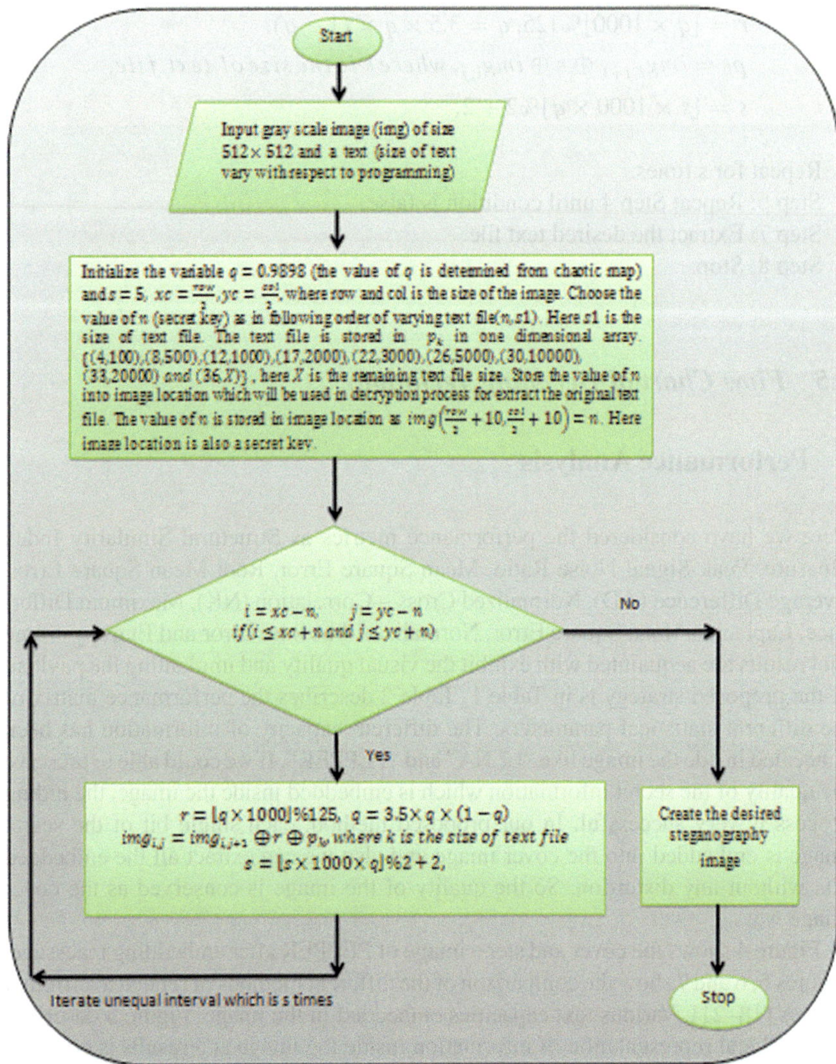

Fig. 2 Encoding flow chart

Step 2: Initialize the variable $q = 0.9898$ (the value of q is determined from chaotic map) and $s = 5, xc = \frac{row}{2}, yc = \frac{col}{2}$, where row and column is the measurement of the picture.

The merit of n is accepted via image location as $n = img\left(\frac{row}{2} + 10, \frac{col}{2} + 10\right)$. Here image location is defined during encryption process.

Step 3: Initialize as $i = xc - n, j = yc - n$.

Step 4: $If(i \leq xc + nandj \leq yc + n)$

Step5: Then

$$r = \lfloor q \times 1000 \rfloor \%125, q = 3.5 \times q \times (1 - q),$$
$$p_k = img_{i,j+1} \oplus r \oplus img_{i,j}, where\, k\, is\, the\, size\, of\, text\, file,$$
$$s = \lfloor s \times 1000 \times q \rfloor \%2 + 2,$$

Repeat for s times.
Step 6: Repeat Step 4 until condition is false.
Step 7: Extract the desired text file.
Step 8: Stop.

3.5 Flow Chart of Decoding Technique

4 Performance Analysis

Here we have considered the performance metrics as Structural Similarity Index Measure, Peak Signal Noise Ratio, Mean Square Error, Root Mean Square Error, Average Difference (AD), Normalized Cross—Correlation (NK), Maximum Difference, Laplacian Mean Square Error, Normalized Absolute Error and Entropy. Some trial results are acquainted with exhibit the visual quality and implanting the payload of the proposed strategy is in Table 1. Table 2 describes the performance matrix of the different statistical parameters. The different capacity of information has been concealed inside the image like "LENA" and "PEPPER". If we could able to preserve the quality of the secret information which is embedded inside the image, the hiding process is very successful. In our proposed method, each single bit of the secret image is embedded into the cover image and also we can extract all the embedded bits without any distortion. So the quality of the image is conserved as the cover image was.

Figure 4 shows the cover and stego image of PEPPER after embedding maessage. Figures 5, 6 and 7 show the comparison of the different methods of related to different images [18–21]. Various text capacities embedded in the image. Figure 5 describes the graphical representation of information inside the image and results is compare with the other existing scheme with our proposed scheme. Here, the comparision result shows that our proposed scheme is extremely secure in the scheme of information hiding through image. Figure 6 shows the comparison of PSNR value with the other existing scheme with our proposed scheme. Our chaotic method's based algorithm has been found the optimized value of PSNR and the value of this statisticale parameter is better which has been compare with other scheme.

Figure 7 illustrates the Similarity Measure for Chaotic Method over other Existing Technique. The SSIM result of our proposed scheme describe that the quality of information has been preserved 100% that is there are no difference between information (before embedding and after extraction) which has been hidden inside the images.

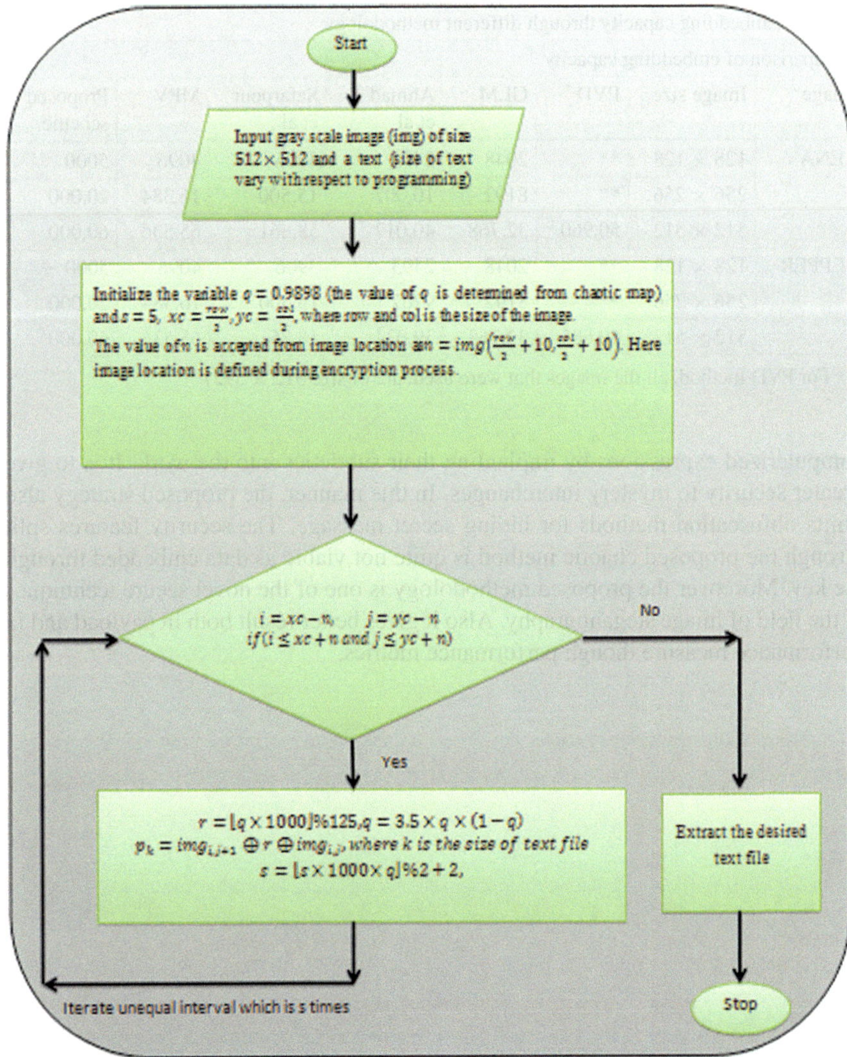

Fig. 3 Decoding flow chart

5 Conclusion

Steganography is incredible and powerful for the correspondence of mystery information. For image steganography, different techniques have been utilized. Here, we have proposed an approach that hides the mystery messages in the image utilizing a chaotic map. Image steganography strategies can be utilized to spare significant passwords and keys without recognizable to the outside world. It can likewise be utilized by the advanced craftsmanship creators to secure the copyrights of their

Table 1 Embedding capacity through different methodology

Comparison of embedding capacity							
Image	Image size	PVD	GLM	Ahmad T et al	Safarpour et al	MPV	Proposed scheme
LENA	128 × 128	**	2048	2493	3906	4096	5000
	256 × 256	**	8192	10,007	15,500	16,384	20,000
	512 × 512	50,960	32,768	40,017	58,861	65,536	60,000
PEPPER	128 × 128	**	2048	2493	3906	4096	5000
	256 × 256	**	8192	9767	15,500	16,384	20,000
	512 × 512	50,685	32,768	39,034	58,861	65,536	60,000

(** For PVD method, all the images that were used, are of size 512 × 512)

computerized expressions by implanting their subtleties into the work. It is to give greater security to mystery interchanges. In this manner, the proposed strategy also limits obfuscation methods for hiding secret message. The security features split through the proposed chaotic method is quite not viable as data embedded through the key. Moreover the proposed methodology is one of the novel secure techniques in the field of image steganography. Also it gives better result both in payload and in performance measure though performance metrics.

Table 2 Performance metrics for LENA image

Images	Similarity parameters	Length of the embedding character							
		500	1000	2000	3000	5000	10000	20000	60000
Lena 512 × 512	SSIM	1.000000	1.000000	0.999984	0.999980	0.999990	1.000000	0.999990	1.000000
	PSNR	65.065000	63.906000	63.524000	59.291000	58.616000	57.596000	57.253000	57.151000
	MSE	0.020256	0.026451	0.028889	0.076553	0.089428	0.113110	0.122390	0.125310
	RMSE	0.142320	0.162640	0.169970	0.276680	0.299050	0.336320	0.349840	0.353990
	AD	0.000282	0.000061	0.000729	0.000557	0.000103	0.000214	0.000214	0.000629
	NK	1.000000	1.000000	1.000000	1.000000	1.000000	1.000000	1.000000	0.999990
	MD	59.000000	62.000000	49.000000	80.000000	79.000000	121.000000	123.000000	134.000000
	LMSE	0.001032	0.001348	0.001472	0.003900	0.004556	0.005763	0.006235	0.006384
	NAE	0.000004	0.000005	0.000007	0.000011	0.000013	0.000015	0.000016	0.000015
	ENTROPY	7.445600	7.445600	7.445600	7.445600	7.445600	7.445600	7.445600	7.445600

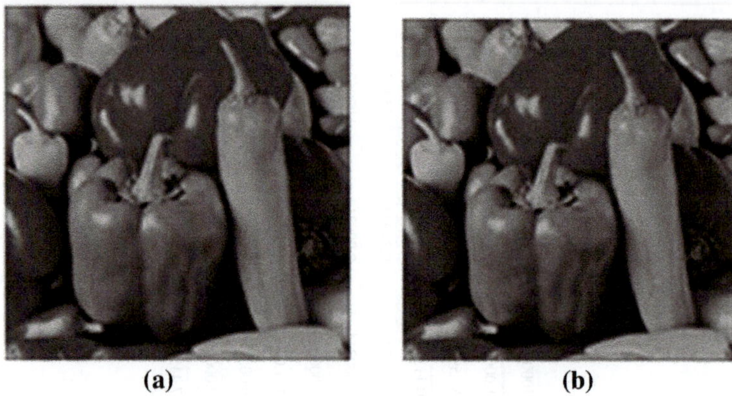

Fig. 4 (**a**) Cover and (**b**) Stego image of PEPPER after embedding 20,000 characters

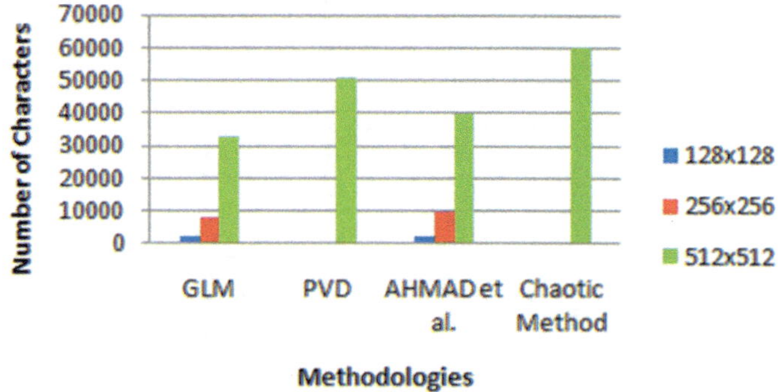

Fig. 5 Graphical representation of embedding capacity for chaotic method over other existing

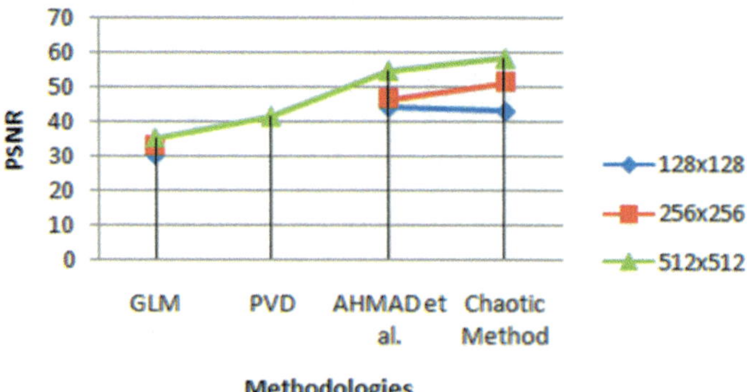

Fig. 6 Comparisn of PSNR for chaotic method over other existing techniques

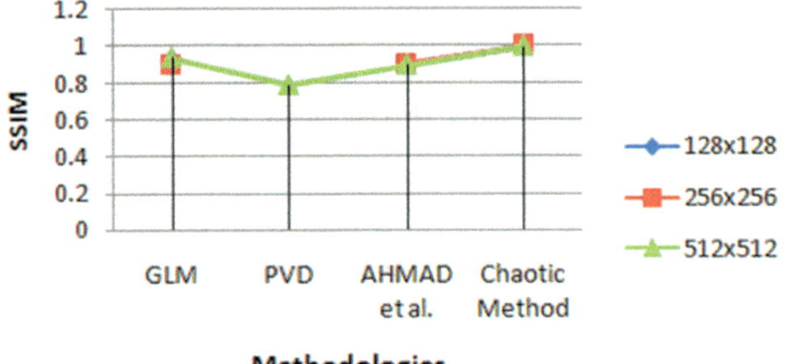

Fig. 7 Comparison of SSIM for chaotic method over other existing techniques

References

1. Rustad, S., Syukur, A., Andono, P.N.: Inverted LSB image steganography using adaptive pattern to improve imperceptibility. J. King Saud Univ. – Comput. Inf. Sci. 1–10 (2021). https://doi.org/10.1016/j.jksuci.2020.12.017
2. Sancheti, A.: Pixel value differencing image steganography using secret key. Int. J. Innov. Technol. Explor. Eng. **2**(1), 68–72 (2012). ISSN: 2278-3075
3. Huang, P., Chang, K.C., Chang, C.P., Tu, T.M.: A novel image steganography method using tri-way pixel value differencing. J. Multimed. (2008). https://doi.org/10.4304/jmm.3.2.37-44,pp. 37-44
4. Lan, T., Mansour, M., Tewfik, A.: Robust high capacity data embedding. In: International Conference on Image Processing—Vancouver, BC, Canada, Proceedings 2000, pp. 581–584 (2000). https://doi.org/10.1109/icip.2000.901025
5. Potdar, V., Chang, E.: Gray level modification steganography for secret communication. In: IEEE International Conference on Industrial Informatics, Berlin, Germany, pp. 355–368 (2004)
6. Safarpour, M., Charmi, M.: Capacity enlargement of the PVD steganography method using the GLM technique, pp. 1–6 (2016). https://www.researchgate.net/publication/289406922
7. Al-Taani, A.T., AL-Issa, A.M.: A novel steganographic method for gray-level images. Int. J. Comput., Inf., Syst. Sci., Eng. **3**(3), 574–579 (2009). https://www.researchgate.net/publication/238749025
8. Kanan, H., Nazeri, B.: A novel image steganography scheme with high embedding capacity and tunable visual image quality based on a genetic algorithm. Expert Syst. Appl. **41**(14), 6123–6130 (2014)
9. Mukherjee, S., Sanyal, G.: Enhanced position power first mapping (PPFM) based Image Steganography. Int. J. Comput. Appl. (IJCA), Taylor and Francis **39**, 59–68 (2017)
10. Almohammad, A., Ghinea, G.: Stego-image quality and the reliability of PSNR. In: Image Processing Theory, Tools and Applications. IEEE, Paris, France (2010). https://doi.org/10.1109/IPTA.2010.5586786
11. Ash, S., Mukherjee, S., Sanyal, G.: A DWT based steganographic method using prime first mapping (PFM). In: Advances in Computing and Communicational Engineering, ICACCE, pp. 471–476 (2015)
12. Mukherjee, S., Roy, S., Sanyal, G.: Image steganography using mid position value technique. In: International Conference on Computational Intelligence and Data Science (ICCIDS 2018), Procedia Computer Science vol. 132, pp. 461–46 (2018)
13. B¨ohme, R., Westfeld, A.: Breaking cauchy model-based JPEG steganography with first order statistics. In: 9th European Symposium on Research Computer Security, vol. 3193, pp. 125–140 (2009)

14. Hsiao, T., Liu, D., Chen, T., Chen, C.: Research on image steganography based on sudoku matrix. symmetry **13**(3), 387, 1–11 (2021. https://doi.org/10.3390/sym13030387
15. Manikandan, T., Muruganandham, A., Babuji, R., Nandalal, V., Mazher, Iqbal, J.l.: Secure E-health using images steganography. J. Phys., Conf. Ser. **1917**(012016), 1–7 (2021). https://doi.org/10.1088/1742-6596/1917/1/012016
16. Fateh, M., Rezvani, M., Irani, Y.: A new method of coding for steganography based on LSB matching revisited. Secur. Commun. Netw. **2021**(6610678), 1–15 (2021) https://doi.org/10.1155/2021/6610678
17. Qi, W., Liu, Y., Guo, S., Wang, X., Guo, Z.: An adaptive visible watermark embedding method based on region selection. Secur. Commun. Netw. **2021**(6693343), 1–11 (2021). https://doi.org/10.1155/2021/6693343
18. Sun, W., Zhou, J., Li, Y., Cheung, M., She, J.: Robust high-capacity watermarking over online social network shared images. IEEE Trans. Circuits Syst. Video Technol. 1–14 (2020). https://doi.org/10.1109/TCSVT.2020.2998476
19. Image Steganography tool.: http://www.softpedia.com/get/Security/Security-Related/Hide-In-Picture.shtml
20. Image Steganography tool.: http://download.cnet.com/QuickStego/3000-2092_4-75593140.html
21. Image Steganography tool.: http://download.cnet.com/Xiao-Steganography/3000-2092_4-10541494.html

On the Construction Structures of 3 × 3 Involutory MDS Matrices over $\mathbb{F}_{2^m}$

Meltem Kurt Pehlivanoğlu, Mehmet Ali Demir, Fatma Büyüksaraçoğlu Sakallı, Sedat Akleylek, and Muharrem Tolga Sakallı

Abstract In this paper, we propose new construction structures, in other words, transposition-permutation path patterns for 3×3 involutory and MDS permutation-equivalent matrices over $\mathbb{F}_{2^3}$ and $\mathbb{F}_{2^4}$. We generate 3×3 involutory and MDS matrices over $\mathbb{F}_{2^3}$ and $\mathbb{F}_{2^4}$ by using the matrix form given in [1], and then all these matrices are analyzed by finding all their permutation-equivalent matrices. After that, we extract whether there are any special permutation patterns, especially for this size of the matrix. As a result, we find new 28,088 different transposition-permutation path patterns to directly construct 3×3 involutory and MDS matrices from any 3×3 involutory and MDS representative matrix over $\mathbb{F}_{2^3}$ and $\mathbb{F}_{2^4}$. The 35 patterns are in common with these finite fields. By using these new transposition-permutation path patterns, new 3×3 involutory and MDS matrices can be generated especially for different finite fields such as $\mathbb{F}_{2^8}$ (is still an open problem because of the large search space). Additionally, the idea of finding the transposition-permutation path patterns can be applicable to larger dimensions such as 8×8, 16×16, and 32×32. To the best of our knowledge, the idea given in this paper to find the common and unique transposition-permutation path patterns over different finite fields is the first work in the literature.

M. Kurt Pehlivanoğlu (✉) · M. Ali Demir
Kocaeli University, Kocaeli 41001, Turkey
e-mail: meltem.kurt@kocaeli.edu.tr

M. Ali Demir
e-mail: 215112045@kocaeli.edu.tr

F. Büyüksaraçoğlu Sakallı · M. Tolga Sakallı
Trakya University, Edirne 22030, Turkey
e-mail: fbuyuksaracoglu@trakya.edu.tr

M. Tolga Sakallı
e-mail: tolga@trakya.edu.tr

S. Akleylek
Ondokuz Mayıs University, Samsun 55200, Turkey
e-mail: sedat.akleylek@omu.edu.tr

S. Banerjee and A. Saha (eds.), *Nonlinear Dynamics and Applications*,
Springer Proceedings in Complexity,
https://doi.org/10.1007/978-3-030-99792-2_48

Keywords MDS matrix · Lightweight cryptography · Diffusion matrices · Permutation-equivalent matrices

1 Introduction

Substitution-Permutation Networks (SPNs) are one of the main structures used in designing a block cipher. SPNs include three components: key scheduling, non-linear layer (e.g., Substitution boxes (S-boxes)), and linear layer (or diffusion layer).

Maximum Distance Separable (MDS) matrices have the maximum branch number that ensures the maximum diffusion [2]. Recently, studies based on designing efficient MDS diffusion layers aim to minimize the circuit depth and reduce the required number of logical gates (especially Exclusive-OR (XOR) gates). XOR count metric defines the hardware implementation cost of a diffusion layer. To design circuit implementation of a diffusion layer with the lowest circuit depth ensures low latency.

Basically, the construction methods for MDS matrices are divided into three methods: direct construction methods, search-based methods, and hybrid methods. Cauchy matrices [4], companion matrices [5], Vandermonde matrices [6], and the 3×3 matrix form given in [1] are direct construction methods. Because of their algebraic structures, an MDS matrix can be generated directly (without searching) by using these structures. In search-based methods, the special matrix forms such as circulant matrices, Hadamard matrices, Toeplitz matrices are used to generate MDS matrices. Hybrid construction methods combine direct construction method and search-based method. Generalized Hadamard (GHadamard) [7] matrix form is a hybrid construction method and generates directly new (involutory) MDS matrices by using an (involutory and MDS) Hadamard matrix of which MDS property can be confirmed by search. But, all these construction methods are evaluated as local optimization technique that focuses on the diffusion matrix coefficients with minimum XOR count. They do not guarantee to find the best circuits. Recent studies have been focused on optimizing the diffusion matrix circuits globally [3, 11–13].

In this paper, efficient construction structures of involutory and MDS matrices are evaluated. In [8], the authors proved the existence of special matrix types (patterns) for constructing involutory Hadamard MDS matrices. In [9], the authors generated lightweight involutory and MDS matrices by using elementary block matrices. They also focused on the properties of block matrices and gave some construction methods. In [10], the authors defined new constructions based on the idea of the subfield construction method to identify involutory MDS matrices. In [11], the authors extracted special construction methods. They used permutation-equivalent classes to search for involutory MDS matrices.

In this paper, we focus on using the permutation-equivalent classes for finding whether there are any special transposition-permutation path patterns, especially for 3×3 involutory and MDS matrices over $\mathbb{F}_{2^3}$ and $\mathbb{F}_{2^4}$ defined by the irreducible polynomials $x^3 + x + 1$ (the hexadecimal notation `0xb` corresponds to this polynomial notation), $x^3 + x^2 + 1$ (`0xd`), $x^4 + x + 1$ (`0x13`) and $x^4 + x^3 + 1$ (`0x19`), respectively. The 3 × 3 matrix form given in [1] is used to generate all involutory and MDS matrices over these finite fields. Then, we find the representative matrices which can represent a class of all permutation-equivalent matrices. After that, we extract all the transposition-permutation paths where P_c, P_r and T express column permutation, row permutation, and matrix transposition, respectively. To find new transposition-permutation path patterns, column permutation (P_c), row permutation (P_r) and matrix transposition (T) steps are applied repetitively on each representative matrix. There are 1,176 and 37,800 3 × 3 involutory and MDS matrices over $\mathbb{F}_{2^3}$ and $\mathbb{F}_{2^4}$ (See Proposition 1. [1]). So, we analyze 2,352 and 75,600 3 × 3 involutory and MDS matrices in total over $\mathbb{F}_{2^3}$ and $\mathbb{F}_{2^4}$, respectively.

1.1 Motivation and Contribution

To design an efficient involutory and MDS matrix is a challenging open problem [2]. Recent studies have focused on the global optimization of the lightweight involutory and MDS matrices to extract the lightest circuits. In [3], it has also remained as future work to focus on the integration of local and global optimization techniques to increase the potential of all these techniques. Especially, it is important to construct involutory and MDS matrices with lightweight coefficients in the local optimization. But first, it is important to know that how one can directly construct an involutory and MDS matrix in an efficient way. Then the local optimization and global optimization techniques are used to reduce the hardware cost. The contributions of this paper can be summarized as follows:

- To the best of our knowledge, the idea based on finding the common and unique transposition-permutation path patterns over different finite fields is the first work in the literature.
- We give all found transposition-permutation path patterns results publicly available at https://github.com/mkurtpehlivanoglu/transposition-permutation-path-patterns.git.
- We give new transposition-permutation path patterns especially for 3 × 3 involutory and MDS matrices over $\mathbb{F}_{2^3}$ and $\mathbb{F}_{2^4}$. For all generated 3 × 3 involutory and MDS matrices over these finite fields, permutation-equivalent matrices are grouped together, in other words, permutation-equivalent classes are found. Then, we extract all the transposition-permutation paths. Finally, we find 28,088 different transposition-permutation path patterns that construct all the permutation-equivalent classes. The 35 patterns are in common with these finite fields.
- These patterns could be applied to any involutory and MDS matrix, then its permutation-equivalent matrices could be constructed directly without any searching cost. Additionally, by using these new transposition-permutation path patterns, new 3 × 3 involutory and MDS matrices can be generated especially for different finite fields which have large search space such as $\mathbb{F}_{2^8}$. Moreover, the idea of

finding the transposition-permutation path patterns can be applied not only 3×3 dimension but also to different larger dimensions.

1.2 Organization

This paper is organized as follows, the preliminaries on the involutory and MDS matrices and the special matrix form used in generating all 3×3 involutory and MDS matrices are given in Sect. 2. In Sect. 3, we show that how to find new transposition-permutation path patterns and we give these structures. The conclusion is given in the last section.

2 Preliminaries

In this paper, we use the $\mathbb{F}_{2^m}/p(x)$ notation to express the finite field $\mathbb{F}_{2^m}$ defined by the irreducible polynomial $p(x)$. The hexadecimal notation of the irreducible polynomial $p(x)$ can be used instead of the polynomial notation, e.g. the hexadecimal notation 0xb corresponds to polynomial notation of the irreducible polynomial $x^3 + x + 1$ in $\mathbb{F}_{2^3}$, and it can be represented by $\mathbb{F}_{2^3}/\texttt{0xb}$. In Definition 1, we recall the MDS matrix definition.

Definition 1 Let L be an $n \times n$ matrix in $M_n(\mathbb{F}_{2^m})$, L is an MDS matrix if and only if it's all square sub-matrices are non-singular [11].

Theorem 1 *Let A and B be two $n \times n$ matrices in $M_n(\mathbb{F}_{2^m})$. If A is an MDS matrix, and A and B are permutation-equivalent matrices, if and only if B is an MDS matrix [11].*

From Theorem 1, if we permute the rows and columns of an MDS matrix, the MDS property will be preserved.

Corollary 1 *Let A and B be two $n \times n$ matrices in $M_n(\mathbb{F}_{2^m})$. If A and B are permutation-equivalent matrices, then A and B have the same implementation cost (XOR count) [11].*

Definition 2 Let $A = \begin{bmatrix} a_{11} & a_{12} & a_{13} \\ a_{21} & a_{22} & a_{23} \\ a_{31} & a_{32} & a_{33} \end{bmatrix}$ be a 3×3 over $\mathbb{F}_{2^m}$ for $a_{12} = (a_{11} + 1)b_0, a_{13} = (a_{11} + 1)b_1, a_{21} = (a_{22} + 1)b_0{}^{-1}, a_{23} = (a_{22} + 1)b_0{}^{-1}b_1, a_{31} = (a_{11} + a_{22})b_1{}^{-1}, a_{32} = (a_{11} + a_{22})b_1{}^{-1}b_0, a_{33} = a_{11} + a_{22} + 1$, where $a_{11} \neq a_{22} \neq a_{33}$ and $b_0, b_1 \in \mathbb{F}_{2^m} - \{0\}$ [1].

The 3 × 3 matrix form IM can be represented as follows:

$$\boldsymbol{IM} = \begin{bmatrix} a_{11} & (a_{11}+1)b_0 & (a_{11}+1)b_1 \\ (a_{22}+1)b_0{}^{-1} & a_{22} & (a_{22}+1)b_0{}^{-1}b_1 \\ (a_{11}+a_{22})b_1{}^{-1} & (a_{11}+a_{22})b_1{}^{-1}b_0 & a_{11}+a_{22}+1 \end{bmatrix} \quad (1)$$

By using Eq. 1, all 3 × 3 involutory and MDS matrices over the $\mathbb{F}_{2^m}$ (See proof [1]) can be generated. Moreover, the number of all 3 × 3 involutory and MDS matrices over $\mathbb{F}_{2^m}$ is $(2^m - 1)^2 \cdot (2^m - 2) \cdot (2^m - 4)$, where $m > 2$ [1].

In this paper, we use the matrix form given in Eq. 1, then we generate all 3 × 3 involutory and MDS matrices over the $\mathbb{F}_{2^3}$/`0xb`, $\mathbb{F}_{2^3}$/`0xd`, $\mathbb{F}_{2^4}$/`0x13` and $\mathbb{F}_{2^4}$/`0x19`.

The number of 3 × 3 involutory and MDS matrices over $\mathbb{F}_{2^3}$ is $(2^3 - 1)^2 \cdot (2^3 - 2) \cdot (2^3 - 4) = 1,176$, and the number of 3 × 3 involutory and MDS matrices over $\mathbb{F}_{2^4}$ is $(2^4 - 1)^2 \cdot (2^4 - 2) \cdot (2^4 - 4) = 37,800$ for each irreducible polynomial.

3 Findings on New Transposition-Permutation Path Patterns

Totally 77,952 3 × 3 involutory and MDS matrices are generated over $\mathbb{F}_{2^3}$ and $\mathbb{F}_{2^4}$. We search all the permutation-equivalent matrices and grouped them together, i.e. permutation-equivalent classes are found. Then, one representative matrix is picked from each class. As a result, we find 100 different 3 × 3 involutory and MDS representative matrices over $\mathbb{F}_{2^3}$/`0xb` and $\mathbb{F}_{2^3}$/`0xd` finite fields, and also find 3,132 and 3,119 different 3 × 3 involutory and MDS representative matrices over $\mathbb{F}_{2^4}$/`0x13` and $\mathbb{F}_{2^4}$/`0x19`, respectively. After that, P_c, P_r and T steps are applied repetitively on each representative matrix to find its own permutation-equivalent matrices, and all these transposition-permutation paths are recorded. We find a total of 28,088 different transposition-permutation path patterns which give the possible transposition-permutation steps to construct whole permutation-equivalent matrices from one representative matrix. In Table 1, we give 35 common path patterns which are found over $\mathbb{F}_{2^3}$ and $\mathbb{F}_{2^4}$. In Table 2, we give most commonly used 50 path patterns over $\mathbb{F}_2^4$. The $P_{c_{i,j}}$, $P_{r_{i,j}}$ and T, notations given in the tables represent changing (permutation) the ith column with the jth column, changing the ith row with the jth row, and transposition of the given matrix, respectively, where $0 \le i, j \le 2$. The path numbers (the first column) which are given in bold in the tables indicate the common paths over two finite fields. All the common transposition-permutation path patterns for $\mathbb{F}_{2^3}$ and $\mathbb{F}_{2^4}$ finite fields are given in bold in Tables 1 and 2.

In Table 1, we give a total of 35 common transposition-permutation path patterns. Normally, we find totally 537 different path patterns over these fields. The second column of the table represents which steps have to be applied to the representative matrix, and the third column represents how many 3 × 3 involutory and MDS representative matrix uses the related pattern to construct its permutation-equivalent matrices. For example, for the third path pattern "$P_{c_{1,0}}$, $P_{r_{1,0}}$" indicates two steps:

Table 1 Generated common transposition-permutation path patterns for $\mathbb{F}_{2^3}$ and $\mathbb{F}_{2^4}$.

No	Transposition-permutation path patterns	Number of matrices			
		$\mathbb{F}_{2^3}$/0xb	$\mathbb{F}_{2^3}$/0xd	$\mathbb{F}_{2^4}$/0x13	$\mathbb{F}_{2^4}$/0x19
1	$P_{c_{1,2}}, P_{r_{1,2}}$, T	13	12	415	399
2	$P_{c_{0,1}}, P_{r_{0,1}}$, T	12	24	424	412
3	$P_{c_{1,0}}, P_{r_{0,1}}$	6	13	422	411
4	$P_{c_{2,0}}, P_{r_{0,2}}$	35	15	434	398
5	$P_{c_{2,1}}, P_{r_{1,2}}$, T	4	17	446	402
6	$P_{c_{1,2}}, P_{r_{2,1}}$	4	14	415	415
7	$P_{c_{1,0}}, P_{r_{1,0}}$, T	19	13	374	421
8	$P_{c_{0,1}}, P_{r_{0,1}}$	9	29	411	418
9	$P_{c_{0,2}}, P_{r_{0,2}}$	12	12	381	404
10	$P_{c_{1,2}}, P_{r_{1,2}}$	11	15	418	399
11	$P_{c_{2,0}}, P_{r_{0,2}}$, T	28	14	373	418
12	$P_{c_{2,1}}, P_{r_{2,1}}$	19	2	421	380
13	$P_{c_{2,0}}, P_{r_{2,0}}$	9	6	411	416
14	$P_{c_{0,1}}, P_{r_{1,0}}$, T	18	11	414	398
15	$P_{c_{2,0}}, P_{r_{2,0}}$, T	10	5	427	444
16	$P_{c_{1,2}}, P_{r_{2,1}}$, T	8	14	385	379
17	$P_{c_{1,0}}, P_{r_{0,1}}$, T	2	5	425	370
18	$P_{c_{2,1}}, P_{r_{1,2}}$	12	18	383	430
19	$P_{c_{1,0}}, P_{r_{1,0}}$	27	6	438	399
20	$P_{c_{1,2}}, P_{r_{2,1}}$, T-$P_{c_{0,2}}, P_{r_{2,0}}$	1	1	9	16
21	$P_{c_{2,1}}, P_{r_{2,1}}$, T	24	2	408	392
22	$P_{c_{0,2}}, P_{r_{2,0}}$	14	15	404	455
23	$P_{c_{0,2}}, P_{r_{2,0}}$, T	17	19	408	395
24	$P_{c_{0,2}}, P_{r_{0,2}}$, T	8	9	405	413
25	$P_{c_{1,2}}, P_{r_{2,1}}$, T-$P_{c_{0,2}}, P_{r_{2,0}}$, T	1	4	9	11
26	$P_{c_{0,1}}, P_{r_{1,0}}$	14	16	415	406
27	$P_{c_{0,2}}, P_{r_{2,1}}$, T-$P_{c_{1,0}}, P_{r_{2,1}}$, T	6	1	12	11
28	$P_{c_{0,1}}, P_{r_{2,0}}$, T-$P_{c_{1,2}}, P_{r_{0,2}}$	5	2	12	10
29	$P_{c_{1,0}}, P_{r_{1,2}}$, T-$P_{c_{2,0}}, P_{r_{1,2}}$	7	5	6	16
30	$P_{c_{2,0}}, P_{r_{0,1}}$, T-$P_{c_{0,1}}, P_{r_{2,0}}$	6	8	9	15
31	$P_{c_{2,0}}, P_{r_{1,0}}$, T-$P_{c_{0,2}}, P_{r_{2,1}}$, T	7	2	19	19
32	$P_{c_{1,0}}, P_{r_{1,2}}$, T-$P_{c_{2,0}}, P_{r_{1,2}}$, T	4	4	12	11
33	$P_{c_{2,1}}, P_{r_{2,0}}$, T-$P_{c_{2,1}}, P_{r_{1,0}}$, T	2	1	10	14
34	$P_{c_{1,0}}, P_{r_{1,2}}$, T-$P_{c_{2,0}}, P_{r_{2,1}}$	2	4	10	15
35	$P_{c_{2,0}}, P_{r_{2,0}}$, T-$P_{c_{0,1}}, P_{r_{0,1}}$	1	3	9	14

firstly replace the second column of this matrix with first column, and then replace the second row of the matrix with first row. In total 25 and 814 representative matrices use this patterns over $\mathbb{F}_{2^3}$ and $\mathbb{F}_{2^4}$, respectively.

Table 2 Most commonly used 50 transposition-permutation path patterns for $\mathbb{F}_{2^4}$.

No	Transposition-permutation path patterns	Number of matrices	
		$\mathbb{F}_{2^4}$/0x13	$\mathbb{F}_{2^4}$/0x19
1	$P_{c_{2,1}}, P_{r_{1,2}}$, T	446	402
2	$P_{c_{1,0}}, P_{r_{1,0}}$	438	399
3	$P_{c_{2,0}}, P_{r_{0,2}}$	434	398
4	$P_{c_{2,0}}, P_{r_{2,0}}$, T	427	444
5	$P_{c_{1,0}}, P_{r_{0,1}}$, T	425	370
6	$P_{c_{0,1}}, P_{r_{0,1}}$, T	424	412
7	$P_{c_{1,0}}, P_{r_{0,1}}$	422	411
8	$P_{c_{2,1}}, P_{r_{2,1}}$	421	380
9	$P_{c_{1,2}}, P_{r_{1,2}}$	418	399
10	$P_{c_{1,2}}, P_{r_{2,1}}$	415	415
11	$P_{c_{0,1}}, P_{r_{1,0}}$	415	406
12	$P_{c_{1,2}}, P_{r_{1,2}}$, T	415	399
13	$P_{c_{0,1}}, P_{r_{1,0}}$, T	414	398
14	$P_{c_{0,1}}, P_{r_{0,1}}$	411	418
15	$P_{c_{2,0}}, P_{r_{2,0}}$	411	416
16	$P_{c_{0,2}}, P_{r_{2,0}}$, T	408	395
17	$P_{c_{2,1}}, P_{r_{2,1}}$, T	408	392
18	$P_{c_{0,2}}, P_{r_{0,2}}$, T	405	413
19	$P_{c_{0,2}}, P_{r_{2,0}}$	404	455
20	$P_{c_{1,2}}, P_{r_{2,1}}$, T	385	379
21	$P_{c_{2,1}}, P_{r_{1,2}}$	383	430
22	$P_{c_{0,2}}, P_{r_{0,2}}$	381	404
23	$P_{c_{1,0}}, P_{r_{1,0}}$, T	374	421
24	$P_{c_{2,0}}, P_{r_{0,2}}$, T	373	418
25	$P_{c_{2,1}}, P_{r_{0,1}}$, T $P_{c_{2,1}}, P_{r_{2,0}}$, T	27	14
26	$P_{c_{2,0}}, P_{r_{1,2}}$, T $P_{c_{0,2}}, P_{r_{1,0}}$	24	15
27	$P_{c_{1,0}}, P_{r_{0,1}}$, T $P_{c_{0,1}}, P_{r_{1,0}}$	24	15
28	$P_{c_{0,2}}, P_{r_{2,1}}$, T $P_{c_{0,1}}, P_{r_{2,1}}$, T	24	13
29	$P_{c_{2,0}}, P_{r_{1,2}}$, T $P_{c_{2,1}}, P_{r_{0,2}}$	22	11
30	$P_{c_{2,1}}, P_{r_{1,0}}$, T $P_{c_{2,1}}, P_{r_{2,0}}$	21	15
31	$P_{c_{0,2}}, P_{r_{2,0}}$, T $P_{c_{1,0}}, P_{r_{1,0}}$	21	8
32	$P_{c_{1,2}}, P_{r_{1,2}}$, T $P_{c_{0,1}}, P_{r_{1,0}}$	20	17
33	$P_{c_{2,1}}, P_{r_{0,1}}$, T $P_{c_{2,0}}, P_{r_{0,1}}$, T	20	12
34	$P_{c_{0,2}}, P_{r_{0,2}}$, T $P_{c_{0,1}}, P_{r_{1,0}}$, T	20	11
35	$P_{c_{0,2}}, P_{r_{1,0}}$, T $P_{c_{2,0}}, P_{r_{1,2}}$	20	7
36	$P_{c_{2,0}}, P_{r_{1,0}}$, T $P_{c_{0,2}}, P_{r_{2,1}}$, T	19	19

(continued)

Table 2 (continued)

No	Transposition-permutation path patterns	Number of matrices	
		$\mathbb{F}_{2^4}/\texttt{0x13}$	$\mathbb{F}_{2^4}/\texttt{0x19}$
37	$P_{c0,2}, P_{r2,1}, \text{T} P_{c1,2}, P_{r2,0}$	19	14
38	$P_{c2,1}, P_{r0,1}, \text{T} P_{c2,1}, P_{r2,0}$	19	12
39	$P_{c2,1}, P_{r1,2}, \text{T} P_{c1,2}, P_{r1,2}$	19	12
40	$P_{c1,0}, P_{r2,0}, \text{T} P_{c0,1}, P_{r1,2}, \text{T}$	19	10
41	$P_{c2,0}, P_{r1,2}, \text{T} P_{c0,1}, P_{r1,2}$	19	9
42	$P_{c1,2}, P_{r1,0}, \text{T} P_{c0,2}, P_{r1,0}$	18	16
43	$P_{c0,2}, P_{r1,2}, \text{T} P_{c2,0}, P_{r1,0}, \text{T}$	18	16
44	$P_{c1,0}, P_{r1,2}, \text{T} P_{c1,2}, P_{r0,1}$	18	16
45	$P_{c0,2}, P_{r2,1}, \text{T} P_{c1,0}, P_{r2,1}$	18	16
46	$P_{c1,0}, P_{r1,2}, \text{T} P_{c1,2}, P_{r1,0}$	18	15
47	$P_{c0,1}, P_{r0,1}, \text{T} P_{c1,2}, P_{r2,1}, \text{T}$	18	14
48	$P_{c0,1}, P_{r0,2}, \text{T} P_{c1,0}, P_{r1,2}, \text{T}$	18	14
49	$P_{c2,1}, P_{r2,1}, \text{T} P_{c1,0}, P_{r1,0}, \text{T}$	18	13
50	$P_{c2,0}, P_{r1,0}, \text{T} P_{c2,1}, P_{r0,1}$	18	13

In Table 2, we give the most commonly used 50 transposition-permutation path patterns over $\mathbb{F}_{2^4}$. Normally, we find 27,551 different path patterns over $\mathbb{F}_{2^4}/\texttt{0x13}$ and $\mathbb{F}_{2^4}/\texttt{0x19}$ and 2,722 of these paths are common.

By using these new transposition-permutation path patterns, new 3×3 involutory and MDS matrices can be generated especially for different finite fields such as $\mathbb{F}_{2^8}$. Finding lightweight involutory MDS matrices over $\mathbb{F}_{2^8}$ is a challenging problem because there are $\approx 2^{31.95}$ 3×3 involutory and MDS matrices over these fields. Instead of searching involutory and MDS matrices, finding direct construction structures is crucial. Finding new transposition-permutation path patterns for different finite fields will ensure the extraction of many common paths. Moreover, the idea given in this paper can be extended to larger dimensions. Firstly, some representative matrices for these large dimensions can be found, then new transposition-permutation path patterns given in this paper can be used to generate new involutory MDS matrices. That will be more efficient because it does not require any search cost.

4 Conclusion and Future Works

Designing of involutory and MDS matrices is a challenging problem. In this paper, we focus on finding new transposition-permutation path patterns from 3×3 representative matrices over the finite fields $\mathbb{F}_{2^3}$ and $\mathbb{F}_{2^4}$. We generate 28,088 different and 35 common transposition-permutation path patterns for these finite fields. It is remarkable that all the common path patterns over $\mathbb{F}_{2^3}/\texttt{0xb}$ and $\mathbb{F}_{2^3}/\texttt{0xd}$ are also take place over $\mathbb{F}_{2^4}/\texttt{0x13}$ and $\mathbb{F}_{2^4}/\texttt{0x19}$. It is clear that all the patterns could be applied to any involutory and MDS matrix, then permutation-equivalent matrices could be computed directly without any search cost.

As a future work, it would be interesting to focus on finding new transposition-permutation path patterns for 3 × 3 involutory and MDS matrices over different finite fields. Then, all these path patterns would be compared to extract the common paths. Thus, these common paths would ensure a direct construction method for finding permutation-equivalent matrices of any representative involutory and MDS matrix. Moreover, the idea given in this paper for 3 × 3 involutory and MDS matrices could be applied to larger dimensions such as 8 × 8, 16 × 16, and 32 × 32.

References

1. Guzel, G.G., Sakallı, M.T., Akleylek, S., Rijmen, V., Cengellenmis, Y.: A new matrix form to generate all 3 × 3 involutory MDS matrices over $\mathbb{F}_{2^m}$. Inf. Process. Lett. **147**, 61–68 (2019)
2. Duval, S., Leurent, G.: MDS matrices with lightweight circuits. IACR Trans. Sym. Crypt. **2**, 48–78 (2018)
3. Tan, Q.Q., Peyrin, T.: Improved heuristics for short linear programs. IACR Trans. Cryptogr. Hardw. Embed. Syst. **2020**(1), 203–230 (2020)
4. Youssef A.M., Mister S., Tavares S.E. On the design of linear transformation for substitution permutation encryption networks. In: Selected Areas in Cryptography (SAC), pp. 40–48, Ottawa Ontario, Canada, (1997)
5. Guo, J., Peyrin, T., Poschmann, A.: The PHOTON family of lightweight hash functions. In: Rogaway, P. (eds.) Advances in Cryptology - CRYPTO 2011. CRYPTO 2011. LNCS, vol 6841. pp. 222–239. Springer, Heidelberg (2011). https://doi.org/10.1007/978-3-642-22792-9_13
6. Gupta, K.C., Ray, I.G.: On Constructions of circulant MDS matrices for lightweight cryptography. In: Huang X., Zhou J. (eds) Information Security Practice and Experience. ISPEC 2014. LNCS, vol. 8128, pp. 29–43. Springer, Cham (2013). https://doi.org/10.1007/978-3-319-06320-1_41
7. Pehlivanoglu, M.K., Sakallı, M.T., Akleylek, S., Duru, N., Rijmen, V.: Generalisation of Hadamard matrix to generate involutory MDS matrices for lightweight cryptography. IET Inf. Secur. **12**(4), 348–355 (2018)
8. Zhou, L., Wang, L., Sun, Y.: On efficient constructions of lightweight MDS matrices. IACR Trans. Symmetric Cryptol. **2018**(1), 180–200 (2018)
9. Wang, S., Li, Y., Tian, S., Zeng, X.: Four by four MDS matrices with the fewest XOR gates based on words. Adv. Math. Commun. 1–28 (2021)
10. Li, S., Sun, S., Li, C., Wei, Z., Hu, L.: Constructing low-latency involutory MDS matrices with lightweight circuits. IACR Trans. Symmetric Cryptol. **2019**(1), 84–117 (2019)
11. Yang, Y., Zeng, X., Wang, S.: Construction of lightweight involutory MDS matrices. Des. Codes Cryptogr. **89**, 1453–1483 (2021)
12. Maximov, A., Ekdahl, P.: New circuit minimization techniques for smaller and faster AES Sboxes. IACR Trans. Cryptogr. Hardw. Embed. Syst. **2019**(4), 91–125 (2019)
13. Xiang, Z., Zeng, X., Lin, D., Bao, Z., Zhang, S.: Optimizing implementations of linear layers. IACR Trans. Symmetric Cryptol. **2020**(2), 120–145 (2020)

Fractional System and Applications

A Novel Generalized Method for Evolution Equation and its Application in Plasma

Santanu Raut, Subrata Roy, and Ashim Roy

Abstract This article presents a new class of the kink soliton, anti-kink soliton solution for the Zakharov-Kuznetsov-Burgers (ZKB) equation. To establish the existence of such type of model in a real physical situation, an unmagnetized viscous plasma containing cold ions and the electrons obeying Cairns-Tallis distribution is considered, and employing reductive perturbation method (RPM) classical ZKB equation is derived. The Generalised Kudryashov method (GKM) is employed to explore the solution of the aforesaid equation and the symbolic software package Maple is adopted in carrying out the complicated algebraic computation. Finally, the physical significance of different parameters on wave propagation is demonstrated through numerical understanding.

Keywords Zakharov-Kuznetsov-Burgers equation · Cairns-Tallis distribution · Reductive perturbation method · Generalised Kudryashov method

1 Introduction

During the last few decades, nonlinear evolution equations (NLEEs) have gained a lot of attention from the authors, due to their vast applications in different branches of nonlinear sciences. For example, NLEEs have been utilized to formulate various problems associated with protein chemistry, chemical kinetics, quantum mechanics, plasma physics, the propagation of shallow-water waves, etc. The classical KdV equation is an example of NLEE which is extensively utilized to model weakly non-

S. Raut (✉)
Department of Mathematics, Mathabhanga College, Coochbehar 736146, India
e-mail: raut_santanu@yahoo.com

S. Roy
Department of Mathematics, Cooch Behar Panchanan Barma University, Cooch Behar 736101, India

A. Roy
Department of Mathematics, Alipurduar Univeristy, Alipurduar 736121, India

S. Banerjee and A. Saha (eds.), *Nonlinear Dynamics and Applications*,
Springer Proceedings in Complexity,
https://doi.org/10.1007/978-3-030-99792-2_49

linear long waves. In many works [1, 2], the investigations done were restricted to the only one-dimensional flow of the ions and the electrons. However, in many situations, the classical KdV equation becomes inappropriate when one encounters a situation where the higher dimensional coordinate system is included. The Kadomtsev-Petviashili (KP) equation and the Zakharov-Kuznetsov (ZK) equation [3] are adopted to investigate the properties of solitary waves in various nonlinear systems with higher dimensional systems. To determine characteristics of IAW wave, as well as DAW, is determined in different plasma mediums the ZK model is utilized in many situations [4]. For the first time, Munro and Parkes [5] found that the governing equation ZK turns into a modified form (mentioned as modified ZK equation) in a plasma environment when the non-isothermal type electrons are considered. Further, it was also found that choosing a suitable form of electron number density in a particular form of plasma environment suggested by Schamel [6], causes to form modified form of ZK equation through RPT [7]. It is found that in most of the observations for finding propagating characteristics of IAW in a plasma environment the impact of viscosity, collisions, ion streaming velocity is neglected. But, a large number of experimental studies has been carried out in different plasma states and it has been observed that the evolution of solitary wave in a dissipative system significantly depends on inter-particle collisions and viscosity, etc. [8–10]. Recently, some authors observed IAW in the ZK model along with the consideration of Burgers term. For instance, Moselem et al. [11] report some analytical solutions of the ZKB equation to investigate IAW propagating in a magnetized dusty plasma containing isothermal electron and cold positive ions. Bedwehy and Moslem [12] again applied the ZK model to observe the characteristic of the shock waves in three-component plasma. Yin et al. [13] find some solitary wave solution of $(2+1)$-dimensional ZKB equation in order to find the generalized beta effect in the system. Seadawy in the year 2015, find some nonlinear wave solutions for the $(3+1)$-dimensional ZKB equation in [14] and discuss the stability of the solutions briefly. In the year 2016, Yang et al. [15] reports some three-dimensional Rossby solitary waves solutions for the ZKB model. Abdullah et al. [16] employed a modified extended mapping method in order to obtain a solitary wave solution for $(3+1)$-dimensional ZKB equation. To solve different kinds of NLEEs, several researchers have proposed and applied various analytical, as well as numerical techniques, such as the modified trial equation method (MTEM), the modified Kudryashov method, (G'/G)-expansion method, the functional variable method, and many other symbolic techniques involving tedious computations, [17–19]. Recently, The generalized Kudryashov method (GKM) is employed to construct traveling wave solutions of different evolution equations. Compared with other nonlinear techniques, the GKM is more efficient to construct directly the exact solutions of high order nonlinear partial differential equations [19]. To acquire the traveling wave solutions of the ZKB equation the GKM is applied in the present investigation.

We observe the weakly nonlinear IAW in collisionless, unmagnetized, plasma system containing nonextensive electrons and cold ions. To aim this, we first derive the (2+1)-dimensional ZKB equation and apply GMK various types of solution such as kink and anti-kink soliton have been derived. In the present investigation, we consider Cairns-Tallis distribution because such a two-parameter mixed distribution function may describe effectively various nonlinear phenomena in a nonthermal plasma environment. For the first time, Tribeche et al. [20] proposed this distribution and addressed the effectivity of the distribution when nonthermal and non-extensivity may act simultaneously. Model equation and Definition of Corresponding Parameters and derivation of the ZK-Burgers equation are describe in Sect. 2. An outline for the proposed nonlinear method and a set of the new solution are derived using GKM in Sect. 3. Section 4 briefly discusses the numerical structure of the solutions. Finally, concluding remarks are given in Sect. 5.

2 Model Equation and Definition of Corresponding Parameters

To verify the reliability of our results, a magnetized three-component plasma system with cold ions, electrons obeying q nonextensive nonthermal distribution and immobile negatively charged dust grains is considered. The external magnetic field is directed along $x - axis$, i.e. $\mathbf{B} = B_0\widehat{x}$. At equilibrium, the charge neutrality condition can be written as $n_{i0} = n_{eo} + Z_d n_{d0}$ where n_{j0} represent the equilibrium densities of jth species ($j = e, i, d$ stands for electron, ion and dust grains respectively) and Z_d denotes the charge number of dust grains. Assuming the existence of such a plasma environment, the dynamics of IAWs is governed by

$$\frac{\partial n_i}{\partial t} + \nabla.(n_i u_i) = 0 \tag{1a}$$

$$\frac{\partial u_i}{\partial t} + (u_i.\nabla)u_i = -\nabla\phi + \Omega(u_i \times \hat{x}) + \nu_i \nabla^2 u_i \tag{1b}$$

$$\nabla^2\phi = n_e - n_i + \delta \tag{1c}$$

where $\delta = \frac{Z_d n_{d0}}{n_{i0}}$ and n_j is the density of the jth species ($j = e, i, d$ stands for electron, ion and dust grains respectively). $\mathbf{u}_i$ is the ion fluid velocity and ϕ is chosen an electrostatic potential. Here n_j is normalized by the unperturbed ion density n_{i0}. u_i is normalized by the ion-acoustic speed $C_{si} = \sqrt{\frac{T_e}{m_i}}$. ϕ, the electrostatic wave potential is normalized by $\frac{T_e}{e}$. The space and time variables are in units of the ion Debye length $\lambda_{di} = \sqrt{\frac{T_e}{4\pi n_{i0} e^2}}$ and the ion plasma period $\omega_{pi}^{-1} = \sqrt{\frac{m_i}{4\pi n_{i0} e^2}}$. Also $\Omega = \frac{\omega_{ci}}{\omega_{pi}}$ where the ion gyrofrequency ω_{ci} is defined as $\omega_{ci} = \frac{eB_0}{m_i c}$ and c is the velocity of light. $\nu_i = \frac{\nu_0}{\omega_{pi}\lambda_{di}^2}$ represent the normalized ion kinematic viscosity where ν_0 is the unnormalized kinematic viscosity. e is the magnitude of electron charge, m_i is mass of ion.

The electron density is given by

$$n_e(\phi) = n_{e0}\left\{1 + (q-1)\frac{e\phi}{T_e}\right\}^{\frac{q+1}{2(q-1)}}\left\{1 + L_1\left(\frac{e\phi}{T_e}\right) + L_2\left(\frac{e\phi}{T_e}\right)^2\right\} \tag{2}$$

where

$$L_1 = -\frac{16q\alpha}{3 - 14q + 15q^2 + 12\alpha}$$
$$L_2 = \frac{16q\alpha(2q-1)}{3 - 14q + 15q^2 + 12\alpha}$$

Here, α stands to present the spectral index and the density (n_e) of nonextensive nonisothermal electrons can be written as

$$n_e = \mu\{1 + (q-1)\phi\}^{\frac{q+1}{2(q-1)}}\{1 + L_1\phi + L_2\phi_2\} \tag{3}$$

which implies

$$\begin{aligned} n_e &= \mu\left\{1 + \frac{1+q}{2}\phi + \frac{(1+q)(3-q)}{8}\phi^2 + \frac{(1+q)(3-q)(5-3q)}{48}\phi^3\right\} \times \\ &\quad \{1 + L_1\phi + L_2\phi_2\} \\ &= \mu\{1 + \beta_1\phi + \beta_2\phi^2 + \beta_3\phi^3 + \cdots\} \end{aligned} \tag{4}$$

where

$$\mu = \frac{n_{e0}}{n_{i0}}$$
$$\beta_1 = L_1 + \frac{1+q}{2}$$
$$\beta_2 = L_2 + L_1\frac{1+q}{2} + \frac{(1+q)(3-q)}{8}$$
$$\beta_3 = L_2\frac{1+q}{2} + L_1\frac{(1+q)(3-q)}{8} + \frac{(1+q)(3-q)(5-3q)}{48}$$

Here μ denotes the unpertubated density ratios of electrons to ions ($= \frac{n_{e0}}{n_{i0}}$). We express Eqs. (1a)–(1c) as follows

$$\frac{\partial n_i}{\partial t} + \frac{\partial(n_i u_{ix})}{\partial x} + \frac{\partial(n_i u_{iy})}{\partial y} = 0 \tag{5a}$$

$$\frac{\partial u_{ix}}{\partial t} + \left(u_{ix}\frac{\partial}{\partial x} + u_{iy}\frac{\partial}{\partial y}\right)u_{ix} = -\frac{\partial\phi}{\partial x} - \nu_i\left(\frac{\partial^2}{\partial x^2} + \frac{\partial^2}{\partial y^2}\right)u_{ix} \tag{5b}$$

$$\frac{\partial u_{iy}}{\partial t} + \left(u_{ix}\frac{\partial}{\partial x} + u_{iy}\frac{\partial}{\partial y}\right)u_{iy} = \frac{\partial\phi}{\partial y} - \nu_i\left(\frac{\partial^2}{\partial x^2} + \frac{\partial^2}{\partial y^2}\right)u_{iy} + \Omega u_{iz} \tag{5c}$$

$$\frac{\partial u_{iz}}{\partial t} + (u_{ix}\frac{\partial}{\partial x} + u_{iy}\frac{\partial}{\partial y})u_{iz} = \frac{\partial \phi}{\partial y} - \nu_i\left(\frac{\partial^2}{\partial x^2} + \frac{\partial^2}{\partial y^2}\right)u_{iz} + \Omega u_{iy} \tag{5d}$$

$$\left(\frac{\partial^2}{\partial x^2} + \frac{\partial^2}{\partial y^2}\right)\phi = n_e - n_i + \delta \tag{5e}$$

Independent variables are stretched as

$$\xi = \epsilon^{\frac{1}{2}}(x - v_p t),\ \eta = \epsilon^{\frac{1}{2}} y,\ \tau = \epsilon^{\frac{3}{2}} t \tag{6}$$

Actually to derive ZKB equation, the dependent variables are expanded as [21]

$$n_i = 1 + \epsilon n_{i1} + \epsilon^2 n_{i2} + \cdots \tag{7a}$$

$$u_{ix} = \epsilon u_{ix1} + \epsilon^2 u_{ix2} + \cdots \tag{7b}$$

$$u_{iy} = \epsilon^2 u_{iy1} + \epsilon^3 u_{iy2} + \cdots \tag{7c}$$

$$u_{iz} = \epsilon^{3/2} u_{iz1} + \epsilon^{5/2} u_{iz2} + \cdots \tag{7d}$$

$$\phi = \epsilon \phi_1 + \epsilon^2 \phi_2 + \cdots \tag{7e}$$

$$\nu_i \approx \epsilon^{1/2} \nu_{i0} \tag{7f}$$

Here $v_p = \sqrt{\frac{1}{\mu\beta_1}}$ is phase velocity. Using RPT method [7] and after setting $\phi_1 = \Phi$, we finally get a relation that can be claimed as the (2+1)-dimensional ZKB equation

$$\frac{\partial \Phi}{\partial \tau} + A\Phi\frac{\partial \Phi}{\partial \xi} + B\frac{\partial^3 \Phi}{\partial \xi^3} + C\frac{\partial^3 \Phi}{\partial \xi \partial \eta^2} + D\left(\frac{\partial^2}{\partial \xi^2} + \frac{\partial^2}{\partial \eta^2}\right)\Phi = 0 \tag{8}$$

where

$$A = \frac{3}{2v_p} - \mu\beta_2 v_p^3,\quad B = \frac{v_p^3}{2},\quad C = \frac{v_p^3}{2}(1 + \Omega^{-2}),\quad D = -\frac{\nu_{i0}}{2}.$$

3 Solution of (2+1)-Dimensional ZKB Equation

3.1 Description of the Generalized Kudryashov Method

In the present section, a generalized structure of the GKM is described in brief. We consider the fractional differential equation as,

$$u_\tau = \mathcal{N}(u_\xi, u_{\xi\xi}, u_{\tau\xi}, u_{\xi\zeta\zeta}, u_{\xi\xi\xi}...) \tag{9}$$

where $u = u(\xi, \eta, \tau)$. The main steps of GKM are stated below as,
Step 1. We introduce a new variable θ to substitute the variables ξ, η and τ. Now, we write

$$u(\xi, \eta, \tau) = V(\theta), \quad \theta = k\xi + n\eta - \omega\tau \tag{10}$$

where k, n and ω, are the constants to be determined later. Then, the Eq. (9) is converted into a nonlinear ordinary differential equation (NODE) as,

$$\omega V' = \mathcal{Q}(V, V', V'', V''', \cdots) \tag{11}$$

for $V = V(\theta)$ where $\mathcal{Q}$ presents a polynomial of V along with its derivatives (the superscripts stands for presenting the differential w.r.t. θ).
Step 2. Now, we choose the solution of Eq. (11) in the particular form presented below as,

$$V(z) = \frac{\sum_{i=0}^{r} a_i Y^i(\theta)}{\sum_{j=0}^{p} a_j Y^j(\theta)} = \frac{a_0 + a_1 Y(\theta) + a_2 Y^2(\theta) + \cdots + a_r Y^r(\theta)}{b_0 + b_1 Y(\theta) + b_2 Y^2(\theta) + \cdots + b_p Y^p(\theta)}. \tag{12}$$

where $Y = Y(\theta)$ obeys the NODE,

$$Y' = Y^2 - Y, \tag{13}$$

Combining Eqs. (12) and (11) we find the polynomial equation Y. Using the homogeneous balance principle the relation between p and r is determined.
Step 3. Equating the coefficients of the term Y^m a system of algebraic equation is determined. Utilizing the symbolic system package Maple this system is solved. Finally, the values of a_i, $i = 0, 1, 2, 3, \ldots, r$ and b_j, $j = 0, 1, 2, \ldots, p$ are determined.
Step 4. Substituting the parametric values a_i and b_j, and setting the general solution of Eq. (13) $Y = 1/(1 \pm e^{\theta})$ in Eq. (12), the solutions of Eq. (9) are obtained.

3.2 *Application of the Generalized Kudryashov Method on ZKB Equation*

In this sub-section generalized kudryashov method is employed to construct a set of exact analytic solution of ZK-Burgers equation. To get traveling wave solution of Eq. (8), we consider the transformation

$$\Phi(\xi, \eta, \tau) = V(\theta), \quad \theta = k\xi + n\eta - \omega\tau. \tag{14}$$

Substituting Eq. (14) into Eq. (8), we have obtained the following nonlinear differential equation:

$$-\omega V' + AkVV' + Bk^3V''' + Ckn^2V''' + Dk^2V'' + Dn^2V'' = 0 \tag{15}$$

Integrating we get

$$-\omega V + \frac{AkV^2}{2} + Bk^3V'' + Ckn^2V'' + Dk^2V' + Dn^2V' = 0 \tag{16}$$

Applying homogeneous balance principle between V'' and V^2, we find a relationship for r, p as,

$$r = p + 2$$

Taking $p = 1$, gives $r = 3$. We choose a trial solution of Eq. (16) as

$$V = \frac{a_0 + a_1Y + a_2Y^2 + a_3Y^3}{b_0 + b_1Y} \tag{17}$$

Putting Eq. (17) into Eq. (16) a system of algebraic equations is obtained and by solving this system with the help of symbolic computation software Maple, we find the values of the involved coefficients. These are determined as,

Set 1.

$$a_2 = -\frac{12Db_0(k^2+n^2)}{5Ak},\ B = -\frac{5Cn^2 - Dk^2 - Dn^2}{5k^2},\ \omega = -\frac{6D(k^2+n^2)}{5},$$
$$a_0 = 0,\ a_1 = 0,\ a_2 = a_2,\ a_3 = 0,\ b_0 = b_0,\ b_1 = 0 \tag{18}$$

By using Eq. (18), the solution can be written as

$$\Phi_{2a} = -\frac{3D(k^2+n^2)}{5Ak}\left(1 - tanh\left(\frac{\theta}{2}\right)\right)^2 \tag{19}$$

where $\theta = k\xi + n\eta + \frac{6D(k^2+n^2)\tau}{5}$.

Set 2.

$$B = -\frac{Cn^2}{k^2},\ \omega = -D(k^2+n^2),\ a_0 = 0,\ a_1 = -\frac{2Db_0(k^2+n^2)}{kA},\ a_2 = 0,$$
$$a_3 = 0,\ b_0 = b_0,\ b_1 = 0. \tag{20}$$

By using Eq. (20), the soliton solution can be written as

$$\Phi_{1a} = -\frac{D(k^2+n^2)}{kA}\left(1 - tanh\left(\frac{\theta}{2}\right)\right) \tag{21}$$

where $\theta = k\xi + n\eta + D(k^2+n^2)\tau$.

Set 3.

$$B = -\frac{5Cn^2 - Dk^2 - Dn^2}{5k^2},\ \omega = \frac{6D(k^2+n^2)}{5},\ a_0 = -a_2,\ a_1 = 0,\ a_2 = a_2,$$
$$a_2 = \frac{-12Db_0(k^2+n^2)}{5kA},\ a_3 = 0,\ b_0 = b_0,\ b_1 = 0 \tag{22}$$

By using Eq. (22), the soliton solution can be termed as

$$\Phi_{3a} = \frac{3D(k^2+n^2)}{5kA}\left(1 - \left(1 - tanh\left(\frac{\theta}{2}\right)\right)^2\right) \tag{23}$$

where $\theta = k\xi + n\eta - \frac{6D(k^2+n^2)\tau}{5}$.

4 Results and Discussion

This work presents a new class of effective solutions for ZK-Burgers model. In fact, we choose some particular numerical values for the coefficients (depending on various parameters) of the equation and plot some 2D and 3D graphs accordingly. Figure 1a is drawn to exhibits the effect of Burgers term in shock profile. Figure 1a clearly shows that the amplitude of the wave substantially depends on dust kinematic viscosity. Increasing ν_{i0} leads for rising dissipation in a system and naturally the amplitudes of the shock rises. The effect of the variation of unperturbed density ratio of electrons to that of ions (μ) in wave profile is demonstrated in Fig. 1(b) where enhancing μ causes for rising of the amplitude of the shocks. In order to show the effect of variation of the parameters spectral index (α) and nonextensive parameter q Fig. 1c, d are plotted, where enhancing of both the parameters causes for diminishing of the amplitude of the shocks. Utilizing the symbolic computation system Mathematica, two three dimensional graphs of the obtained solutions of ZKB medium are drawn in Fig. 2. Figure 2a clearly shows the propagation of shock in space zone whereas the significant effect of the viscosity parameter ν_{i0} is shown in Fig. 2b.

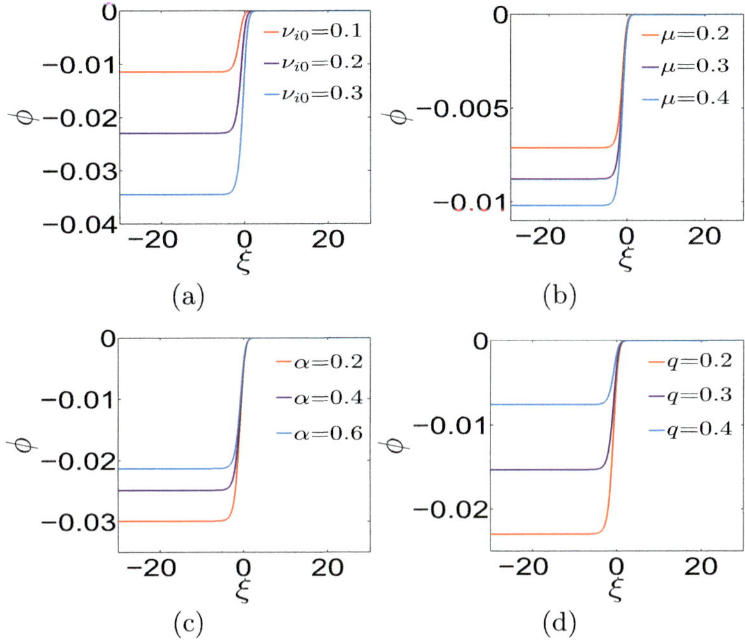

Fig. 1 **a** The 2D Profiles of Solution (19) for $\alpha = 0.5,\ q = 0.2,\ \tau = 1,\ n = \frac{1}{\sqrt{2}},\ k = \frac{1}{\sqrt{2}},\ \eta = 1,\ \mu = 0.5,\ \omega = 0.5,$ **b** when $\alpha = 0.5,\ q = 0.2,\ \tau = 1,\ n = \frac{1}{\sqrt{2}},\ k = \frac{1}{\sqrt{2}},\ \eta = 1,\ \nu_{i0} = 0.1,\ \omega = 0.5$ **c** when $\nu_{i0} = 0.1,\ q = 0.2,\ \tau = 1,\ n = \frac{1}{\sqrt{2}},\ k = \frac{1}{\sqrt{2}},\ \eta = 1,\ \mu = 0.5,\ \omega = 0.5,$ **d** when $\alpha = 0.5,\ \nu_{i0} = 0.1,\ \tau = 1,\ n = \frac{1}{\sqrt{2}},\ k = \frac{1}{\sqrt{2}},\ \eta = 1,\ \mu = 0.5,\ \omega = 0.5$

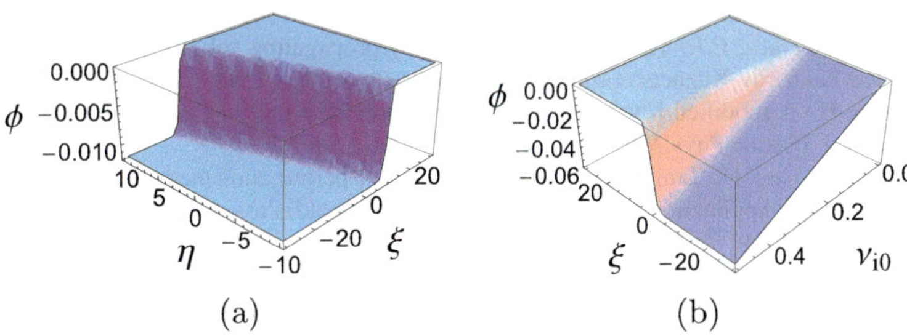

Fig. 2 **a** The 3D Profiles of Solution (19) for $\alpha = 0.5,\ q = 0.2,\ \tau = 1,\ n = \frac{1}{\sqrt{2}},\ k = \frac{1}{\sqrt{2}},\ \mu = 0.5,\ \omega = 0.5,\ \nu_{i0} = 0.1,$ **b** for $\alpha = 0.5,\ q = 0.2,\ \tau = 1,\ n = \frac{1}{\sqrt{2}},\ k = \frac{1}{\sqrt{2}},\ \eta = 1,\ \mu = 0.5,\ \omega = 0.5$

5 Conclusion

We have utilized the ZKB model to observe IAW propagating in a non-extensive plasma medium, this example has been provided basically to establish the existence of such types of models in different physical situations. Actually, in the light of our observation, emphasis is imposed on deriving the model solutions. The GKM method, a newly developed novel nonlinear technique, is employed to ZKB equation and a new class of effective solutions such as kink soliton, anti-kink soliton, are derived which may be useful in the various dynamic systems of physical situations like the water wave phenomena, the elastic media, the quantum mechanics, the dynamics of adiabatic parameters, problems on the industrial phenomena, etc. Some complicated algebraic calculations have been solved with the help of the symbolic computations package Maple. Finally, some 2D and 3D figures are also drawn to show the impact of different parameters on wave structures.

References

1. Das, G.C., Paul, S.N.: Ionacoustic solitary waves in relativistic plasmas. Phys. fluids **28**(3), 823–825 (1985)
2. Roychoudhury, R.K., Bhattacharyya, S.: Effect of ion temperature on ion-acoustic solitary waves: a pseudopotential approach. Can. J. Phys. **65**(7), 699–702 (1987)
3. Zakharov, V.E., Kuznetsov, E.A.: On threedimensional solitons. Zhurnal Eksp. Teoret. Fiz **66**, 594–597 (1974)
4. Mondal, K.K., Roy, A., Chatterjee, P., Raut, S.: Propagation of ion-acoustic solitary waves for damped forced Zakharov Kuznetsov equation in a relativistic rotating magnetized electron-positron-ion plasma. Int. J. Appl. Comput. Math. **6**(3), 1–17 (2020)
5. Munro, S., Parkes, E.J.: The stability of obliquely-propagating solitary-wave solutions to a modified Zakharov-Kuznetsov equation. J. Plasma Phys. **70**(5), 543–552 (2004)
6. Schamel, H.: A modified Korteweg-de Vries equation for ion acoustic waves due to resonant electrons. J. Plasma Phys. **9**(3), 377–387 (1973)
7. Kakutani, T., Ono, H., Taniuti, T., Wei, C.C.: Reductive perturbation method in nonlinear wave propagation II. Application to hydromagnetic waves in cold plasma. J. Phys. Soc. Jpn. **24**(5), 1159–1166 (1968)
8. El-Bedwehy, N.A., Moslem, W.M.: Zakharov-Kuznetsov-Burgers equation in super thermal electron-positron-ion plasma. Astrophys. Space Sci. **335**(2), 435–442 (2011)
9. Raut, S., Mondal, K.K., Chatterjee, P., Roy, A.: Two-dimensional ion-acoustic solitary waves obliquely propagating in a relativistic rotating magnetised electron-positron-ion plasma in the presence of external periodic force. Pramana **95**(2), 1–13 (2021)
10. Raut, S., Mondal, K.K., Chatterjee, P., Roy, A.: Propagation of dust-ion-acoustic solitary waves for damped modified Kadomtsev-Petviashvili-Burgers equation in dusty plasma with a q-non extensive nonthermal electron velocity distribution. SeMA J. 1–23 (2021)
11. Moslem, W.M., Sabry, R.: Zakharov-Kuznetsov-Burgers equation for dust ion acoustic waves. Chaos Solitons Fractals **36**(3), 628–634 (2008)
12. El-Bedwehy, N.A., Moslem, W.M.: Zakharov-Kuznetsov-Burgers equation in super thermal electron-positron-ion plasma. Astrophys. Space Sci. **335**(2), 435–442 (2011)
13. Yin, X., Yang, L., Liu, Q., Wu, G.: (2+ 1)-dimensional ZK-Burgers equation with the generalized beta effect and its exact solitary solution. Comput. Math. Appl. **77**(1), 302–310 (2019)

14. Seadawy, A.R.: Nonlinear wave solutions of the three-dimensional Zakharov-Kuznetsov-Burgers equation in dusty plasma. Physica A: Stat. Mech. Appl. **439**, 124–131 (2015)
15. Yang, H.W., Xu, Z.H., Feng, X.R., Yin, B.S., Dong, H.H.: ZK-Burgers equation for three-dimensional Rossby solitary waves and its solutions as well as chirp effect. Adv. Differ. Equ. **2016**(1), 1–22 (2016)
16. Seadawy, A.R., Jun, W.: Mathematical methods and solitary wave solutions of three-dimensional Zakharov-Kuznetsov-Burgers equation in dusty plasma and its applications. Results in Phys. **7**, 4269–4277 (2017)
17. Yel, G., Sulaiman, T.A., Baskonus, H.M.: On the complex solutions to the (3+ 1)-dimensional conformable fractional modified KdV-Zakharov-Kuznetsov equation. Mod. Phys. Lett. B **34**(05), 2050069 (2020)
18. Islam, M.H., Khan, K., Akbar, M.A., Salam, M.A.: Exact traveling wave solutions of modified KdV-Zakharov-Kuznetsov equation and viscous Burgers equation. SpringerPlus **3**(1), 1–9 (2014)
19. Ryabov, P.N., Sinelshchikov, D.I., Kochanov, M.B.: Application of the Kudryashov method for finding exact solutions of the high order nonlinear evolution equations. Appl. Math. Comput. **218**(7), 3965–3972 (2011)
20. Tribeche, M., Amour, R., Shukla, P.K.: Ion acoustic solitary waves in a plasma with nonthermal electrons featuring Tsallis distribution. Phys. Rev. E **85**, 037401 (2012)
21. Mushtaq, A., Shah, H.A.: Nonlinear Zakharov-Kuznetsov equation for obliquely propagating two-dimensional ion-acoustic solitary waves in a relativistic, rotating magnetized electron-positron-ion plasma. Phys. Plasmas **12**(7), 072306 (2005)

Impact of Fear and Strong Allee Effects on the Dynamics of a Fractional-Order Rosenzweig-MacArthur Model

Hasan S. Panigoro and Emli Rahmi

Abstract This paper discusses the impact of fear and strong Allee on the dynamical behaviors of the prey and predator relationship following the Rosenzweig-MacArthur model using fractional-order derivative as the operator. As results, four equilibrium points are identified namely the origin point, a pair of axial points, and the interior point. The origin is always locally asymptotically stable while others are conditionally asymptotically stable. The occurrence of transcritical bifurcation around the axial and Hopf bifurcation in the interior are also successfully investigated. The numerical simulations are conducted to support analytical findings. Some interesting dynamics such as forward bifurcation and bistability condition are also provided numerically.

Keywords Fractional-order · Rosenzweig-MacArthur · Allee effect · Fear effect

1 Introduction

Food chain schemes are always found in nature. Every organism may become a predator to others due to its need for food. As a result, each organism has a chance to go extinct as an impact of this ecological mechanism. Therefore, studying the existence of organisms that have prey and predator relationship always be a crucial issue for researchers. One of the much-publicized ways is using mathematical modeling.

In 1963, a mathematical model is developed by Rosenzweig and MacArthur based on the Lotka-Volterra predator-prey model which assumes that the population of prey grows logistically and its hunting by the predator for foods following Holling type-II as the predator functional response [1]. Nowadays, the Rosenzweig-MacArthur model becomes an attractive reference to establish a novel predator-prey model by

H. S. Panigoro (✉) · E. Rahmi
Department of Mathematics, State University of Gorontalo, Bone Bolango 96119, Indonesia
e-mail: hspanigoro@ung.ac.id

E. Rahmi
e-mail: emlirahmi@ung.ac.id

S. Banerjee and A. Saha (eds.), *Nonlinear Dynamics and Applications*,
Springer Proceedings in Complexity,
https://doi.org/10.1007/978-3-030-99792-2_50

Table 1 The biological interpretation of variables and parameters

Variables and parameters	Biological interpretation
x	The density of prey
y	Density of predator
r	Intrinsic growth rate of prey
k	Level of fear
K	Environmental carrying capacity of prey
b	Allee threshold
m	Predation rate
a	Half saturation constant of predation
n	Predator growth rate which converted from the predation process
d	Predator death rate

involving some ecological components associated with real phenomena in nature. For example, see [2, 3] and references therein.

In this paper, we assume that the growth rate of prey is influenced by the indirect impact of the predator through the fear effect [4]. We also assume that this intrinsic growth rate could also decrease by the intraspecific competition and difficulty in finding mates is known as the Allee effect [5]. Thus, we have the following model.

$$\begin{aligned} \frac{dx}{dt} &= \frac{rx}{1+ky}\left(1-\frac{x}{K}\right)(x-b) - \frac{mxy}{a+x}, \\ \frac{dy}{dt} &= \frac{nxy}{a+x} - dy. \end{aligned} \tag{1}$$

See Table 1 for the biological interpretation of variables and parameters. The term $(x-b)$ represents the Allee effect where for $b \leq 0$ called weak Allee effect and $b > 0$ called strong Allee effect. In our work, we assume that the intrinsic growth rate of prey affected by strong Allee effect. Due to biological purpose, other parameters also positive constant and both $x(t)$ and $y(t)$ satisfy $(x, y) \in \mathbb{R}_+^2$ where $\mathbb{R}_+^2 := \{(x, y) \mid x \geq 0, y \geq 0, \ x \in \mathbb{R}, \ y \in \mathbb{R}\}$.

Since the current state of both prey and predator depends on all of their previous conditions, using fractional-order derivative is considered more appropriate in expressing the model better than classical integer-order derivative [3, 6, 7]. Following a similar way with [3, 7] such as replacing the first-order with fractional-order derivative and scaling the time dimension, we obtain the new model as follows.

$$\begin{aligned} {}^{C}\mathcal{D}_t^{\alpha} x &= \frac{rx}{1+ky}\left(1-\frac{x}{K}\right)(x-b) - \frac{mxy}{a+x}, \\ {}^{C}\mathcal{D}_t^{\alpha} y &= \frac{nxy}{a+x} - dy, \end{aligned} \tag{2}$$

where ${}^{C}\mathcal{D}_t^{\alpha}$ is Caputo fractional-order derivative defined by

$$ {}^{C}\mathcal{D}_t^{\alpha} f(t) = \frac{1}{\Gamma(1-\alpha)} \int_a^t \frac{f'(\tau)}{(t-\tau)^{\alpha}}\, d\tau, $$

$\alpha \in (0, 1]$ is the order of the derivative and $\Gamma(\cdot)$ is Euler Gamma function [8].

In Sasmal [9], the predator-prey model involving fear and Allee effects has been studied. Sasmal's model is quite similar to ours both in assumptions and the deterministic model. The big difference which becomes the novelty of our works lies in the predator functional response and the operator of the model. In our works, the Michaelis-Menten type is used as the predator functional response which is considered more realistic than bilinear ones. The fractional-order derivative is also used to replace the first-order derivative as the operator to cover the memory effect.

The rest of the paper is arranged as follows. In Sect. 2, the feasibility and local stability of equilibrium points are verified. Furthermore, the existence of transcritical and Hopf bifurcations are examined in Sect. 3. Several numerical simulations are explored in Sect. 4 not only to support the analytical findings but also to show other dynamical behaviors such as the occurrence of forward bifurcation and bistability conditions. We finally end our work by giving a conclusion in Sect. 5.

2 Feasibility and Stability of Equilibrium Points

The feasible equilibrium points of model (2) are acquired by finding the the positive solution of the following equations.

$$ \left[\frac{r(x-b)}{1+ky}\left(1-\frac{x}{K}\right) - \frac{my}{a+x}\right] x = 0, $$
$$ \left[\frac{nx}{a+x} - d\right] y = 0. $$

Therefore, four equilibrium points are identified as follows.

(i) The origin $E_0 = (0, 0)$ which represents the extinction of both populations.
(ii) A pair of axial points $E_1 = (b, 0)$ and $E_2 = (K, 0)$ which represent the existence of prey and the extinction of predator.
(iii) The interior point $E_3 = (\hat{x}, \hat{y})$ which represents the existence of both populations where $\hat{x} = \frac{ad}{n-d}$ and $\hat{y}$ is the positive solution respect to y of the following equation.

$$ y^2 + \frac{y}{k} + \frac{\hat{m}}{4k^2m} = 0, \tag{3} $$

where $3\hat{m} = \frac{4(\hat{x}-K)(\hat{x}-b)(\hat{x}+a)kr}{K}$. Since $E_i \in \mathbb{R}_+^2$ $\forall i = 0, 1, 2$, then they always exist. Furthermore, the existence condition of E_3 is given by the following theorem.

Theorem 1 *If $n > d$ and (i) $m \leq \hat{m}$ then the interior point does not exist; (ii) $m > \hat{m}$ then there exists an interior point.*

Proof Since $n > d$ then $\hat{x}$ is always positive. Thus, the existence of E_3 depends on the positive solution of quadratic equation (3). If $m < \hat{m}$ then the solution of equation (3) is a pair of complex conjugate numbers and hence the interior point does not exist. When $m = \hat{m}$, we have $\hat{y} = -\frac{1}{2k} < 0$, and hence E_3 also does not exist. For $m > \hat{m}$, the only positive solution of equation (3) is given by $\hat{y} = -\frac{1}{2k}\left(1 - \sqrt{1 - \frac{\hat{m}}{m}}\right)$. This completes the proof. □

Now, we discuss the local stability for each equilibrium point. The following theorems are presented.

Theorem 2 *The origin $E_0 = (0, 0)$ is always locally asymptotically stable.*

Proof The linearization around E_0 gives the Jacobian matrix as follows.

$$\mathcal{J}(x, y)|_{E_0} = \begin{bmatrix} -br & 0 \\ 0 & -d \end{bmatrix}.$$

The eigenvalues of $\mathcal{J}(x, y)|_{E_0}$ are $\lambda_1 = -br$ and $\lambda_2 = -d$ which give $|\arg(\lambda_i)| = \pi > \alpha\pi/2 \, \forall i = 1, 2$. According to the Matignon condition [10], E_0 is always locally asymptotically stable. □

Theorem 3 *The axial point $E_1 = (b, 0)$ is locally asymptotically stable if $b > K$ and $n < \frac{(a+b)d}{b}$.*

Proof For the axial point E_1, we have the Jacobian matrix

$$\mathcal{J}(x, y)|_{E_1} = \begin{bmatrix} -\frac{(b-K)br}{K} & -\frac{bm}{a+b} \\ 0 & \frac{bn}{a+b} - d \end{bmatrix}, \tag{4}$$

which give eigenvalues $\lambda_1 = -\frac{(b-K)br}{K}$ and $\lambda_2 = \frac{bn}{a+b} - d$. Based on Matignon condition [10], the local asymptotic stability condition are satisfied when $\lambda_i < 0, \ i = 1, 2$ which are given by $b > K$ and $n < \frac{(a+b)d}{b}$. □

Theorem 4 *The axial point $E_2 = (K, 0)$ is locally asymptotically stable if $b < K$ and $n < \frac{(a+K)d}{K}$*

Proof The Jacobian matrix evaluated at E_2 is given by

$$\mathcal{J}(x, y)|_{E_2} = \begin{bmatrix} (b-K)r & -\frac{mK}{a+K} \\ 0 & \frac{nK}{a+K} - d \end{bmatrix}, \tag{5}$$

where the eigenvalues are $\lambda_1 = (b - K)r$ and $\lambda_2 = \frac{nK}{a+K} - d$. If $b < K$ and $n < \frac{(a+K)d}{K}$ then $|\arg(\lambda_i)| = \pi > \alpha\pi/2, \ i = 1, 2$ that obeys the Matignon condition [10]. □

Theorem 5 *The interior point $E_3 = (\hat{x}, \hat{y})$ is locally asymptotically stable if (i) $\xi_1 < 0$, or (ii) $\xi_1 > 0$, $\xi_1^2 < 4\xi_2$, and $\alpha < \hat{\alpha}$, where $\xi_1 = -\frac{(3\hat{x}^2 - 2(b+K)\hat{x} + bK)r}{(1+k\hat{y})K} - \frac{am\hat{y}}{(a+\hat{x})^2}$, $\xi_2 = \frac{(1+2k\hat{y})ad^3m\hat{y}}{(1+k\hat{y})n^2\hat{x}^2}$, and $\hat{\alpha} = \frac{2}{\pi}\tan^{-1}\left(\frac{\sqrt{4\xi_2 - \xi_1^2}}{\xi_1}\right)$.*

Proof At E_3, we have

$$\mathcal{J}(x, y)|_{E_3} = \begin{bmatrix} \xi_1 & -\frac{n\xi_2\hat{x}^2}{ad^2\hat{y}} \\ \frac{ad^2\hat{y}}{n\hat{x}^2} & 0 \end{bmatrix}. \tag{6}$$

Therefore, the polynomial characteristic is obtained as follows.

$$\lambda^2 - \xi_1\lambda + \xi_2 = 0. \tag{7}$$

Since $\xi_2 > 0$, by obeying Proposition 1 in [11], the stability conditions given in Theorem 5 are proven. □

3 Bifurcation Analysis

In this section, we give two types of bifurcations phenomena namely transcritical and Hopf bifurcations by following theorems.

Theorem 6 *Suppose that $n < \min\left\{\frac{(a+b)d}{b}, \frac{(a+K)d}{K}\right\}$. Two axial points E_1 and E_2 exchange their stability via transcritical bifurcation when b crosses K.*

Proof Since $n < \min\left\{\frac{(a+b)d}{b}, \frac{(a+K)d}{K}\right\}$, we have $|\arg(\lambda_2)| = \pi > \alpha\pi/2$ for each Jacobian matrix (4) and (5). Therefore, the stability of E_1 and E_2 depend on the sign of λ_1. When $b < K$, $|\arg(\lambda_1)| = \pi > \alpha\pi/2$ for Jacobian matrix (5) and $|\arg(\lambda_1)| = 0 < \alpha\pi/2$ for Jacobian matrix (4). Hence, E_1 is a saddle point while E_2 is locally asymptotically stable. When $b = K$, $E_1 = E_2$ and $|\arg(\lambda_1)| = \alpha\pi/2$ which represents a non-hyperbolic equilibrium point. For $b > K$ the sign of $|\arg(\lambda_1)|$ for Jacobian matrices (4) and (5) are switched which indicates the stability of E_1 and E_2 changes. According to those circumstances, the transcritical bifurcation occurs driven by the Allee threshold (b). □

Theorem 7 *Let $\xi_1 > 0$ and $\xi_1^2 < 4\xi_2$. A Hopf bifurcation occurs around the interior point $E_3 = (\hat{x}, \hat{y})$ when α passes through $\hat{\alpha}$.*

Proof From (7), the appropriated eigenvalues are given by

$$\lambda_{1,2} = \frac{1}{2}\left(\xi_1 \pm \sqrt{\xi_1^2 - 4\xi_2}\right). \tag{8}$$

Since $\xi_1 > 0$ and $\xi_1^2 < 4\xi_2$, the eigenvalues (8) are a pair of complex conjugate numbers with positive real parts. It is also valid that $m(\hat{\alpha}) = \hat{\alpha}\pi/2 - \min_{1\leq i\leq 2} |\arg(\lambda_i)| = 0$ and $\left.\frac{dm(\alpha)}{d\alpha}\right|_{\alpha=\hat{\alpha}} \neq 0$. According to Theorem 4.6 in [12], Hopf bifurcation occurs around E_3 driven by α with $\hat{\alpha}$ is the critical point. □

4 Numerical Simulation

Some numerical simulations are demonstrated using a generalized predictor-corrector scheme given by Diethelm et al. [13]. This scheme is applied to numerical software called *Python-3* to produce some figures such as bifurcation diagrams and time series. In this paper, we study numerically the influence of the Allee threshold (b) and the order of the derivative (α) to the dynamical behaviors of model (2). Since the model does not discuss a specific case, all parameter values are chosen hypothetically by considering the previous analytical results. We first set the parameter as in Table 2 and varying the Allee threshold (b) in interval $[0.4, 2.4]$, see Fig. 1.

From the bifurcation diagram given by Fig. 1a, when b is varied in the interval $[0.4, 2.4]$, the dynamical behaviors change two times. For $0.4 \leq b < 1$, we have a locally asymptotically stable equilibrium point E_2 and an unstable point E_1. The stability of both E_1 and E_2 change sign when b crosses $\hat{b}_1 = 1$ which confirm the existence of transcritical bifurcation given by Theorem 6. This dynamical behaviors are maintained for $\hat{b}_1 < b < \hat{b}_2 = 1.8$. Denote that the interior point E_3 does not exist for interval $0.4 \leq b < \hat{b}_2$. When b passes $\hat{b}_2$, the axial point E_1 again losses its stability, and a locally asymptotically stable point E_3 emerges which indicates the existence of forward bifurcation. This conditions are preserved for $\hat{b}_2 < b \leq 2.4$. Remember that E_0 is always locally asymptotically stable and hence the bistability condition always occurs for each case when the dynamical behaviors change. We perform the phase portraits by picking the values of $b = 0.5, 1.5, 2.3$, which presents the dynamical behavior for each interval. See Fig. 1b, c, d. The stability shifts from E_2 to E_1 and finally to E_3 while E_0 always locally asymptotically stable. This means, the bistability condition always exists for $[0.4, 2.4]$ except in every bifurcation point. This means that the existence of populations depends on the initial values. From those phase portraits, we show that for the given two close initial values, the solutions tend to distinct equilibrium points. Both populations could be extinct or only the existence of prey is preserved.

The next simulation aims to show the influence of the order of the derivative (α) to the dynamical behaviors of model (2). The parameter values are chosen as

Table 2 Parameter values for numerical simulations given in Fig. 2

Parameters	r	k	K	m	a	n	d	α
Values	0.4	0.8	1	0.3	0.9	0.15	0.1	0.9

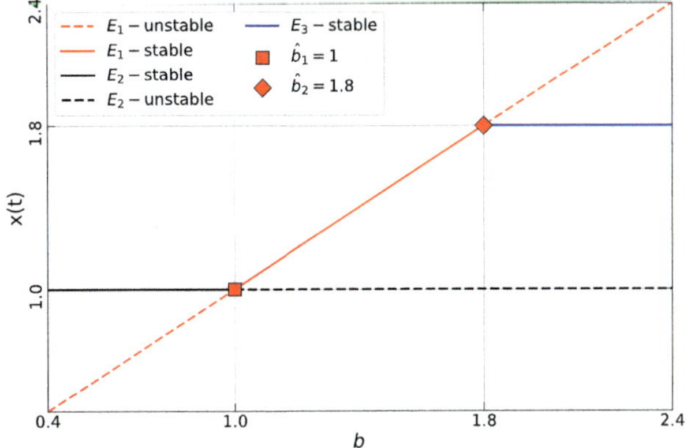

(a) Bifurcation diagram driven by the Allee threshold b

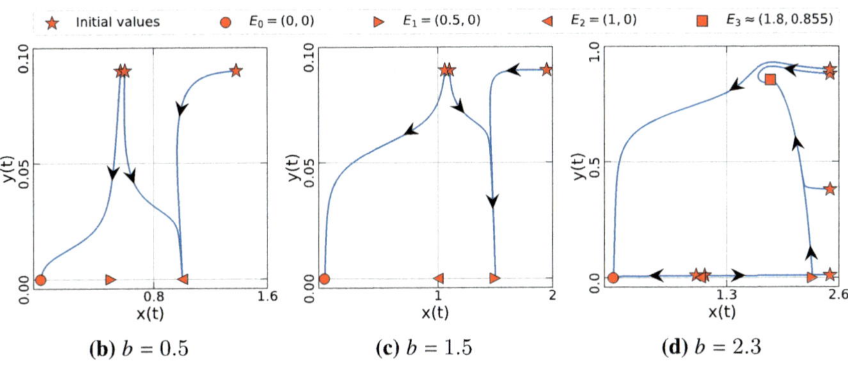

(b) $b = 0.5$ **(c)** $b = 1.5$ **(d)** $b = 2.3$

Fig. 1 Bifurcation diagram and phase portraits of model (2) with parameter values as in Table 2

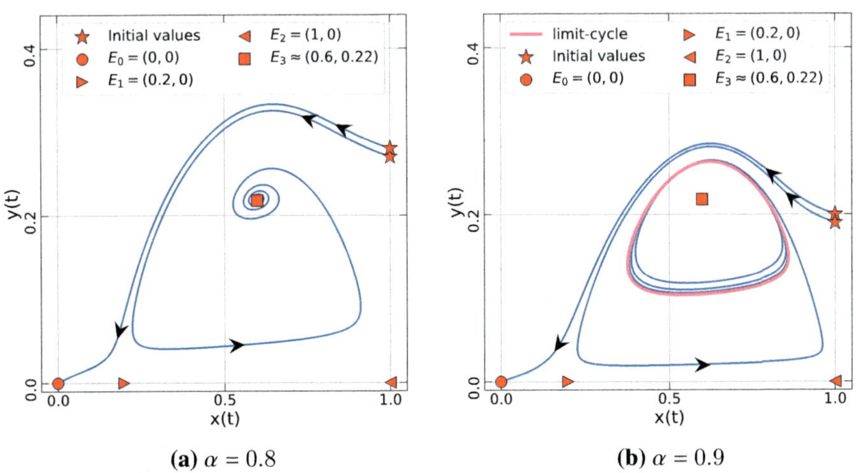

(a) $\alpha = 0.8$ **(b)** $\alpha = 0.9$

Fig. 2 Phase portraits of model (2) with parameter values as in Table 3

Table 3 Parameter values for numerical simulations given in Fig. 2

Parameters	r	k	K	b	m	a	n	d
Values	0.4	0.8	1	0.2	0.3	0.6	0.2	0.1

in Table 2. Based on Theorem 5, The Jacobian matrix (6) has a pair of complex conjugate eigenvalues with positive real parts. Thus, from Theorem 7, the interior point E_3 undergoes a Hopf bifurcation when α passes through the critical point $\hat{b}$. By using these parameter values, we confirm that the critical point is $\hat{\alpha} \approxeq 0.84304$. To show this condition, we pick $\alpha = 0.8$ and $\alpha = 0.9$ and the numerical results given by the phase portraits in Fig. 2. When $\alpha = 0.8$, two locally asymptotically stable equilibrium points occur i.e. $E_0 = (0, 0)$ and $E_3 \approx (0.6, 0.22)$. As the impact, the model (2) leads to bistability condition. For two close initial values, the solutions convergent to different equilibrium points namely E_0 and E_3. When α is increased to 0.9, E_3 losses its stability and nearby solution convergent to a periodic signal namely limit-cycle. Although the interior point is unstable, both populations are still preserved periodically around the interior point. This ends our numerical simulations.

5 Conclusion

The dynamical behaviors of a fractional-order Rosenzweig-MacArthur model involving fear and strong Allee effects have been studied. The model has four equilibrium points namely the origin, a pair in axial, and a unique interior point. Those two equilibrium points in the axial may exchange their stability via transcritical bifurcation. For the interior point, the stability may change via Hopf bifurcation driven by the order of the derivative. To support the analytical findings, numerical simulations are provided including a bifurcation diagram and phase portraits. We have found numerically that the model undergoes transcritical bifurcation, forward bifurcation, Hopf bifurcation, and bistability conditions. From the biological viewpoint, these circumstances mean that the existence of both prey and predator are threatened due to predation mechanism, fear, and allee effects.

References

1. Rosenzweig, M.L., MacArthur, R.H.: Graphical representation and stability conditions of predator-prey interactions. Am. Nat. **97**, 209–223 (1963). https://doi.org/10.1086/282272
2. Moustafa, M., Mohd, M.H. , Ismail, A.I., Abdullah, F.A.: Stage structure and refuge effects in the dynamical analysis of a fractional order Rosenzweig-MacArthur prey-predator model. Prog. Fract. Differ. Appl. **5**, 49–64 (2019). https://doi.org/10.18576/pfda/050106
3. Panigoro, H.S., Suryanto, A., Kusumawinahyu, W.M., Darti, I.: A Rosenzweig-MacArthur model with continuous threshold harvesting in predator involving fractional derivatives

with power law and mittag-leffler kernel. Axioms **9**, 122 (2020). https://doi.org/10.3390/axioms9040122

4. Barman, D., Roy, J., Alrabaiah, H., Panja, P., Mondal, S.P., Alam, S.: Impact of predator incited fear and prey refuge in a fractional order prey predator model. Chaos Solitons Fractals **142**, 110420 (2021). https://doi.org/10.1016/j.chaos.2020.110420
5. Allee, W.C.: Animal Aggregations, A Study in General Sociology. The University of Chicago Press, Chicago (1931)
6. Panigoro, H.S., Suryanto, A., Kusumawinahyu, W.M., Darti, I.: Global stability of a fractional-order Gause-type predator-prey model with threshold harvesting policy in predator. Commun. Math. Biol. Neurosci. **2021**, 63 (2021). https://doi.org/10.28919/cmbn/6118
7. Rahmi, E., Darti, I., Suryanto, A.: Trisilowati: a modified Leslie-Gower model incorporating Beddington-DeAngelis functional response, double Allee effect and memory effect. Fractal Fract. **5**, 84 (2021). https://doi.org/10.3390/fractalfract5030084
8. Petras, I.: Fractional-Order Nonlinear Systems: Modeling, Analysis and Simulation. Springer, London, Beijing (2011)
9. Sasmal, S.K.: Population dynamics with multiple Allee effects induced by fear factors - A mathematical study on prey-predator interactions. Appl. Math. Model. **64**, 1–14 (2018). https://doi.org/10.1016/j.apm.2018.07.021
10. Matignon, D.: Stability results for fractional differential equations with applications to control processing. Comput. Eng. Syst. Appl. , 963–968 (1996). https://doi.org/10.1.1.40.4859
11. Ahmed, E., El-Sayed, A.M.A., El-Saka, H.A.A.: On some Routh-Hurwitz conditions for fractional order differential equations and their applications in Lorenz, Rössler, Chua and Chen systems. Phys. Lett. A. **358**, 1–4 (2006). https://doi.org/10.1016/j.physleta.2006.04.087
12. Baisad, K., Moonchai, S.: Analysis of stability and Hopf bifurcation in a fractional Gauss-type predator-prey model with Allee effect and Holling type-III functional response. Adv. Differ. Equ. **2018**, 82 (2018). https://doi.org/10.1186/s13662-018-1535-9
13. Diethelm, K., Ford, N.J., Freed, A.D.: A predictor-corrector approach for the numerical solution of fractional differential equations. Nonlinear Dyn. **29**, 3–22 (2002). https://doi.org/10.1023/A:1016592219341

Stabilization of Fractional Order Uncertain Lü System

Manoj Kumar Shukla

Abstract This paper presents a stabilization strategy for fractional order chaotic systems (FOCS) with unknown parameters. Control of FOCS is a less explored area and very limited approaches have been put forward in the literature. Here, an effort has been made to propose an improved stabilizing controller for a particular class of systems. The control structure is obtained by using a systematic adaptive backstepping procedure which is based on fractional order extension of Lyapunov stability results. The designed controller also avoids the singularity problem common in the traditional backstepping procedure. Parameter update laws achieved while applying adaptive backstepping strategy give estimates of the uncertain parameters of the system. The numerical simulation results given at the end validate the control strategy proposed for the stabilization of uncertain Lü system.

Keywords Fractional calculus · Chaotic system · Adaptive backstepping · Lü system

1 Introduction

Fractional derivative and integration have found wide applications in the past two decades. Control system has become one of the areas of application of fractional calculus. Nonlinear systems especially chaotic systems find applicability in almost every area of engineering and science. The fractional order version of the chaotic systems also called as fractional order chaotic systems (FOCS) have become the point of discussion in the past 15 years and various researchers have put forward the analysis and control of different FOCS [1, 2].

Several techniques that have been employed for control of integer order chaotic systems (IOCS) [3, 4], have been extended for FOCS also [5, 6].

One of the main advantages of going for fractional order chaotic systems is that such systems display chaos for a range of values of fractional order and hence can

M. K. Shukla (✉)
Lovely Professional University, Phagwara, Punjab, India
e-mail: manoj.22223@lpu.co.in

S. Banerjee and A. Saha (eds.), *Nonlinear Dynamics and Applications*,
Springer Proceedings in Complexity,
https://doi.org/10.1007/978-3-030-99792-2_51

be used for different applications. Fractional order version of various IOCS has been studied in the literature. Some prominent contributions can be found in [7, 8], etc. The backstepping control technique developed by Kristic et al. [9] is widely used for the control of IOCS with known or unknown parameters. It is based on the Lyapunov stability technique and ensures the global stability of systems. Podlubny et al. [10] extended Lyapunov theory to fractional order nonlinear systems which is further based on Mittag- Leffler stability concept and both of these combined with backstepping approach have been utilized in the present manuscript.

The main contribution of the work is to propose a stabilizing controller for a class of FOCS. The technique is further implemented for control of fractional order Lü System. Various approaches have been proposed in the literature for designing feedback controller for this system but these techniques have some drawbacks and also most of these cannot handle the case of unknown parameters. In [11, 12], the system stability is analyzed via traditional Lyapunov stability, whereas in the present work, Lyapunov method is extended for FOCS on the basis of Mittag–Leffler stability which proves to give better performance in comparison to traditional methods. Multiple controllers are to be designed in the techniques presented in [13, 14]. On the other hand, only one controller needs to be designed in the strategy presented in the present manuscript.

Further, Sect. 2 gives the basic idea of fractional calculus and the stability of FOCS. The controller design approach is given in Sect. 3. Section 4 gives numerical simulations results. Section 5 concludes the contributions made here.

2 Basic Preliminaries of Fractional Calculus

The fractional order derivative and integral can be defined as

$$ {}_aD_t^q = \begin{cases} \frac{d^q}{dt^q} & q > 0 \\ 1 & q = 0 \\ \int_a^t (d\tau)^{-q} & q < 0 \end{cases} \tag{1}$$

The major definitions are expressed as

Grunwald–Letnikov Definition

$$ D_t^q f(t) = \lim_{h \to 0} \frac{1}{h^q} \sum_{j=0}^{\infty} (-1)^j \binom{q}{j} f(t - jh) \tag{2}$$

Riemann–Liouville Definition

$$\mathcal{I}^q f(t) \triangleq \frac{1}{\Gamma(q)} \int_0^t (t-\tau)^{q-1} f(\tau) d\tau \tag{3}$$

Caputo Definition (Derivative)

$$\mathcal{D}^q f(t) \triangleq \mathcal{I}^{m-q} \mathcal{D}^m f(t) = \frac{1}{\Gamma(m-q)} \int_0^t \frac{f(\tau)^m}{(t-\tau)^{q-m+1}} f(\tau) d\tau \tag{4}$$

The solution of nonlinear fractional order differential equation (FODEs) can be derived from Grunwald–Letnikov definition [1, 15, 16]. The following expression gives the numerical solution of the nonlinear FODE of form ${}_aD_t^q y(t) = f(y(t), t)$

$$y(t_k) = f(y(t_k), t_k) h^q - \sum_{j=1}^{k} c_j^{(q)} y(t_{k-j}) \tag{5}$$

3 Stabilization Strategy

The approach for obtaining the stabilizing controller by using adaptive version of backstepping control for the systems of a particular class. The system parameters are taken to be unknown. The general form of the class of the systems is given below:

$$D_t^{q_1} x_1(t) = \theta_1 (x_2(t) - x_1(t))$$

$$D_t^{q_2} x_2(t) = -x_1(t) x_3(t) + \theta_2 x_1(t) + \theta_3 x_2(t) \tag{6}$$

$$D_t^{q_3} x_3(t) = x_1(t) x_2(t) - \theta_4 x_3(t) + u$$

where, $\theta_1, \theta_2, \theta_3$ and θ_4 are unknown constant parameters u is the controller. Table 1 gives a list of these types of systems.

The backstepping technique is modified to tackle the singularity problem. The result can be written in form of the following theorem:

Theorem 1 *The system in* (6), *can be controlled by the controller which is expressed as*

$$u = -k z_1^2 + \hat{\theta}_4 \alpha_2 + D^q \alpha_2 \tag{7}$$

Table 1 FOCS represented by the class of systems given in (6)

Name of System	Dynamics
Liu system	$D_t^{q_1} x_1(t) = a(x_2(t) - x_1(t))$
	$D_t^{q_2} x_2(t) = -kx_1(t)x_3(t) + bx_1(t)$
	$D_t^{q_3} x_3(t) = x_1(t)x_2(t) - bx_3(t)$
Lü system	$D_t^{q_1} x_1(t) = a(x_2(t) - x_1(t))$
	$D_t^{q_2} x_2(t) = -x_1(t)x_3(t) + cx_2(t)$
	$D_t^{q_3} x_3(t) = x_1(t)x_2(t) - bx_3(t)$
Chen system	$D_t^{q_1} x_1(t) = a(x_2(t) - x_1(t))$
	$D_t^{q_2} x_2(t) = -x_1(t)x_3(t) + dx_1(t) + cx_2(t)$
	$D_t^{q_3} x_3(t) = x_1(t)x_2(t) - bx_3(t)$

The controller guarantees the asymptotic stabilization of the system, provided, $\frac{\theta_3}{\theta_1} < k < 1$ *with an assumption* $\theta_1 > 0$, *where k is the design parameter. The variable* $z_1 = x_1$, α_2 *is the virtual controller given as* $\alpha_2 = \widehat{\theta}_1 + \widehat{\theta}_2 + k\widehat{\theta}_3 - k^2\widehat{\theta}_1 + k\widehat{\theta}_1$, *where,* $\widehat{\theta}_i$; $i = 1, 2, 3, 4$; *are the estimates of unknow system parameters which can be expressed as:*

$$
\begin{aligned}
D^q\widehat{\theta}_1 &= \gamma z_1 z_2\big(1 - k^2 + k\big);\\
D^q\widehat{\theta}_2 &= \gamma z_1 z_2;\\
D^q\widehat{\theta}_3 &= \gamma k z_1 z_2;\\
D^q\widehat{\theta}_4 &= -\gamma z_3 \alpha_2
\end{aligned}
$$

Proof By choosing $q_1 = q_2 = q_3 = q$, let $z_1 = x_1$ and $z_2 = x_2 - \alpha_1$, where α_1 is the virtual controller. It gives,

$$
D^q z_1 = \theta_1(x_2 - x_1) = \theta_1 z_2 + \theta_1\alpha_1 - \theta_1 x_1 \tag{8}
$$

One can chose the as $V_1 = \frac{1}{2}z_1^2$. One can differentiate V_1 w.r.t. time while taking a fractional order q,

$$
D^q V_1 \leq z_1 D^q z_1 \Rightarrow D^q V_1 \leq z_1(\theta_1 z_2 + \theta_1\alpha_1 - \theta_1 x_1)
$$

The dynamics gets modified to,

$$
D^q z_1 = \theta_1 z_2 + \theta_1(kz_1 - z_1) = \theta_1 z_2 - \theta_1(1 - k)z_1 \tag{9}
$$

where, $\alpha_1 = kz_1$. Also, the fractional derivative of V_1 now satisfies the following condition:

$$D^q V_1 \leq -\theta_1(1-k)z_1^2 + \theta_1 z_1 z_2$$

In a similar fashion, with the help of equations in (6) the second transformation variable will be represented as:

$$D^q z_2 = -z_1 x_3 + \theta_2 x_1 + \theta_3 x_2 - k(D^q z_1)$$

$$\Rightarrow D^q z_2 = -z_1(z_3 + \alpha_2) + \theta_2 z_1 + \theta_3 z_2 + \theta_3 k z_1 - k\theta_1 z_2 + k(1-k)\theta_1 z_1 \quad (10)$$

where, $z_3 = x_3 - \alpha_2$. The Lyapunov function for the overall system up to this stage shall be expressed as:

$$V_2 = V_1 + \frac{1}{2}z_2^2 + \frac{1}{2}\gamma^{-1}\left(\theta_1 - \hat{\theta}_1\right)^2 + \frac{1}{2}\gamma^{-1}(\theta_2 - \hat{\theta}_2)^2 + \frac{1}{2}\gamma^{-1}(\theta_3 - \hat{\theta}_3)^2$$

which further can be written as,

$$D^q V_2 \leq -\theta_1(1-k)z_1^2 + \theta_1 z_1 z_2 + z_2 D^q z_2 - \gamma^{-1}\left(\theta_1 - \hat{\theta}_1\right)\left(D^q\hat{\theta}_1\right)$$

$$-\gamma^{-1}\left(\theta_2 - \hat{\theta}_2\right)\left(D^q\hat{\theta}_2\right) - \gamma^{-1}\left(\theta_3 - \hat{\theta}_3\right)\left(D^q\hat{\theta}_3\right)$$

$$\begin{aligned}&\Rightarrow D^q V_2 \leq -\theta_1(1-k)z_1^2 - (\theta_1 k - \theta_3)z_2^2\\&- z_1 z_2 z_3 + z_1 z_2\left\{\theta_1 + \theta_2 + \theta_3 k + \theta_1 k - \theta_1 k^2 - \alpha_2\right\}\\&- \gamma^{-1}\left(\theta_1 - \hat{\theta}_1\right)\left(D^q\hat{\theta}_1\right) - \gamma^{-1}\left(\theta_2 - \hat{\theta}_2\right)\left(D^q\hat{\theta}_2\right)\\&- \gamma^{-1}\left(\theta_3 - \hat{\theta}_3\right)\left(D^q\hat{\theta}_3\right)\end{aligned}$$

Here, $\hat{\theta}_1, \hat{\theta}_2$ and $\hat{\theta}_3$ estimates of system parameters θ_1, θ_2 and θ_3, respectively. Also, γ is a parameter that controls the convergence of the estimates of the parameters. For stabilizing the system and to satisfy the stability criterion α_2 may be selected as:

$$\alpha_2 = \hat{\theta}_1 + \hat{\theta}_2 + k\hat{\theta}_3 - k^2\hat{\theta}_1 + k\hat{\theta}_1 \quad (11)$$

The following expressions give the parameter update laws as,

$$D^q\hat{\theta}_1 = \gamma z_1 z_2\left(1 - k^2 + k\right);\quad D^q\hat{\theta}_2 = \gamma z_1 z_2;\quad D^q\hat{\theta}_3 = \gamma k z_1 z_2 \quad (12)$$

The update laws and the controller led to the following expression:

$$D^q V_2 \leq -\theta_1(1-k)z_1^2 - (\theta_1 k - \theta_3)z_2^2 - z_1 z_2 z_3$$

With the assumption, $\theta_1 > 0$, one has to ensure that, $\frac{\theta_3}{\theta_1} < k < 1$, so that stability can be ensured. Further, from the 3rd expression in (6), one can have,

$$D^q z_3 = x_1 x_2 - \theta_4 x_3 + u - D^q \alpha_2 = z_1 z_2 + k z_1^2 - \theta_4 z_3 - \theta_4 \alpha_2 + u - D^q \alpha_2 \quad (13)$$

Finally, the overall Lyapunov function for the transformed system dynamics in (8), (10) and (13), can be chosen as:

$$V_3 = V_2 + \frac{1}{2} z_3^2 + \frac{1}{2} \gamma^{-1} (\theta_4 - \widehat{\theta}_4)^2$$

which further leads to the following expression:

$$\begin{aligned} D^q V_3 \leq & -\theta_1 (1-k) z_1^2 - (\theta_1 k - \theta_3) z_2^2 - z_1 z_2 z_3 \\ & + z_3 (z_1 z_2 + k z_1^2 - \theta_4 z_3 - \theta_4 \alpha_2 + u - D^q \alpha_2) \end{aligned}$$

The final control law and the expressions for parameter update for stabilization of the whole system are given below:

$$u = -k z_1^2 + \alpha_2 \widehat{\theta}_4 + z_3 \widehat{\theta}_4 + D^q \alpha_2 \quad (14)$$

$$\& \quad D^q \hat{\theta}_4 = -\gamma z_3 \alpha_2 \quad (15)$$

The above selection of the controller and update law, leads to the following stability condition:

$$D^q V_3 \leq -\theta_1 (1-k) z_1^2 - (\theta_1 k - \theta_3) z_2^2 - \theta_4 z_3^2 \quad (16)$$

The results in (14), (15) and (16) confirm the finite time convergence of transformation variables z_1, z_2 and z_3 to zero, which further ensures stabilization of the system states x_1, x_2 and x_3. The controller obtained here avoids the singularity problem.

4 Simulation Results for Fractional Order Lü System

The controller design approach is illustrated in this section with the help of an example system. Fractional order Lü system which belongs to the category of systems given in Table 1, can be described as:

$$D_t^q x_1(t) = \theta_1 (x_2(t) - x_1(t))$$

$$D_t^q x_2(t) = -x_1(t) x_3(t) + \theta_2 x_2(t) \quad (17)$$

$$D_t^q x_3(t) = x_1(t)x_2(t) - \theta_3 x_3(t) + u$$

Here, parameters θ_1, θ_2 and θ_3 are considered to be uncertain. For $q_1 = q_2 = q_3 = q$, the final controller structure using the approach mentioned in Sect. 3, is described as follows:

$$u = \widehat{\theta}_3\alpha_2 + D^q\alpha_2 - kz_1^2 \tag{18}$$

and the updating laws for parameter estimates are given as,

$$D^q\widehat{\theta}_1 = \gamma x_1(x_2 - kx_1)\left(1 - k^2 + k\right)$$

$$D^q\widehat{\theta}_2 = \gamma x_1(x_2 - kx_1)(1 + k) \tag{19}$$

$$D^q\widehat{\theta}_3 = \gamma(x_3 - \alpha_2)\alpha_2$$

where, $z_1 = x_1$, $z_2 = x_2 - \alpha_1$ and $z_3 = x_3 - \alpha_2$. Here, $\theta_1 > 0$, and also, $\frac{\theta_2}{\theta_1} < k < 1$. With controller u in action, on the basis of the same arguments given in the previous section, the stability of the system can be ensured.

The system parameters are chosen as $(\theta_1, \theta_2, \theta_3) = (36, 3, 20)$ and the design constant k is taken as: $k = 0.7$. The initial values of the system states have been chosen as: $(x_1(0), x_2(0), x_3(0)) = (-0.1, 0.5, 0.2)$ and the order of derivative is taken as, $q = 0.95$. The convergence parameter is selected as $\gamma = 1.5$. The simulation time is taken as 10 s with a step size of $h = 0.005$ s. Figure 1 depicts the convergence of the x_1 and x_2 converge to zero in finite time, whereas, state x_3 can be seen to be bounded as time tends towards infinity. Therefore, one can conclude that the proposed control strategy leads the systems towards stability in finite time, even when the system parameters are uncertain.

The controller for the particular system has been derived by using the backstepping strategy which is further based on the Lyapunov stability criterion. While applying the backstepping technique, in each step Lyapunov function has to be framed and its derivative is calculated which should be proven to be negative definite for a particular choice of controller. After the subsequent steps, the final controller is obtained while ensuring that the time derivative of the overall Lyapunov function is negative definite. this ensures the stability of the whole system, whatever be the dynamics.

With the right choice of the controller and ensuring the stability of the whole system, it can be concluded that the states of the systems are converging which is evident from the simulation results.

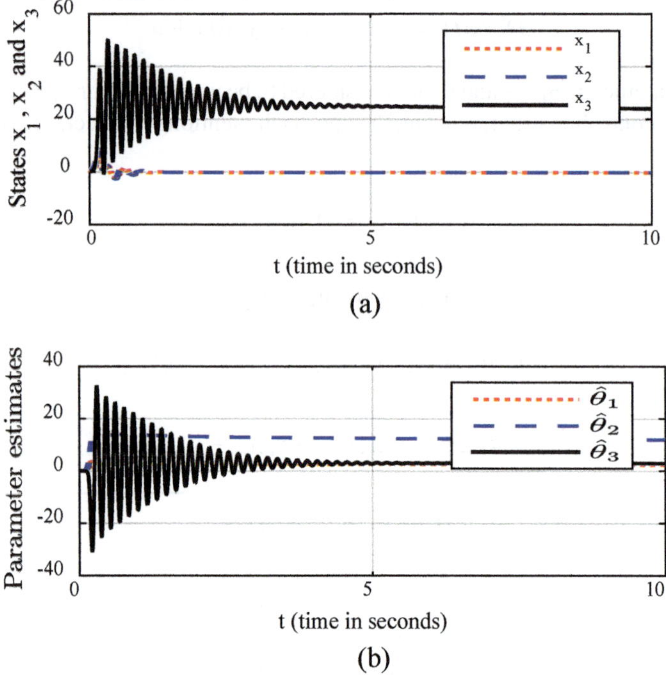

Fig. 1 **a** Convergence of system states **b** Estimates of the unknown parameters

5 Conclusion

The paper presents a novel method for the stabilization of uncertain FOCS of a particular class. The backstepping method used here is systematic and ensures asymptotic stability. The proposed controller has been employed for the stabilization of uncertain fractional order Lü system. The simulation results validate the efficacy of the same. The controllers presented here can be extended for synchronization of the chaotic systems which further can be used for different practical applications like secure communication etc.

References

1. Petras, I.: Fractional-Order Nonlinear Systems: Modeling, Analysis, and Simulation. Springer Science & Business Media (2008)
2. Shukla, M.K., Sharma, B.B.: Control and synchronization of a class of uncertain fractional order chaotic systems via adaptive Backstepping control. Asian J Control **20**, 707–720 (2018). https://doi.org/10.1002/asjc.1593
3. Yassen, M.T.: Chaos synchronization between two different chaotic systems using active control. Chaos Solitons Fractals **23**, 131–140 (2005)

4. Astakhov, V.V., Anishchenko, V.S., Kapitaniak, T., Shabunin, A.V.: Synchronization of chaotic oscillators by periodic parametric perturbations. Phys. D Nonlinear Phenom. **109**, 11–16 (1997)
5. Sharma, B.B., Kar, I.N.: Contraction theory based adaptive synchronization of chaotic systems. Chaos Solitons Fractals **41**, 2437–2447 (2009)
6. Sharma, B.B., Kar, I.N.: Stabilization and tracking controller for a class of nonlinear discrete-time systems. Chaos Solitons Fractals **44**, 902–913 (2011). https://doi.org/10.1016/j.chaos.2011.07.009
7. Hartley, T.T., Lorenzo, C.F., Qammer, H.K.: Chaos in a fractional order Chua's system. IEEE Trans. Circuits Syst. I Fundam. Theory Appl. **42**, 485–490 (1995)
8. Lu, J.G.: Chaotic dynamics of the fractional-order Lü system and its synchronization. Phys. Lett. A **354**, 305–311 (2006)
9. Krstic, M., Kanellakopoulos, I., Kokotovic, P.V.: Nonlinear and Adaptive Control Design. Wiley (1995)
10. Li, Y., Chen, Y., Podlubny, I.: Stability of fractional-order nonlinear dynamic systems: Lyapunov direct method and generalized Mittag—Leffler stability. Comput. Math. with Appl. **59**, 1810–1821 (2010). https://doi.org/10.1016/j.camwa.2009.08.019
11. Chen, D., Liu, Y., Ma, X., Zhang, R.: Control of a class of fractional-order chaotic systems via sliding mode. Nonlinear Dyn. **67**, 893–901 (2012)
12. Aghababa, M.P.: Robust stabilization and synchronization of a class of fractional-order chaotic systems via a novel fractional sliding mode controller. Commun. Nonlinear Sci. Numer. Simul. **17**, 2670–2681 (2012). https://doi.org/10.1016/j.cnsns.2011.10.028
13. Odibat, Z.M.: Adaptive feedback control and synchronization of non-identical chaotic fractional order systems. Nonlinear Dyn. **60**, 479–487 (2010). https://doi.org/10.1007/s11071-009-9609-6
14. Radwan, G., Moaddy, K., Salama, K.N., et al.: Control and switching synchronization of fractional order chaotic systems using active control technique. J. Adv. Res. **5**, 125–132 (2014). https://doi.org/10.1016/j.jare.2013.01.003
15. Vinagre, B.M., Chen, Y.Q., Petráš, I.: Two direct Tustin discretization methods for fractional-order differentiator/integrator. J. Franklin Inst. **340**, 349–362 (2003)
16. Dorcak, L.: Numerical models for the simulation of the fractional-order control systems. Slovak Acad. Sci. (1994)

Artificial Intelligence, Internet of Things and Smart Learning

The Transfer Trajectory onto the Asteroid for Mining Purposes Using LPG-Algorithm

Vijil Kumar and Badam Singh Kushvah

Abstract In this research, a new methodology named as LPG-Algorithm is designed to determine the transfer trajectory between two celestial bodies or any two locales in space. It is constructed by combining Lambert's problem with the genetic algorithm (GA). In this algorithm, the initial state of the transfer trajectory has been optimized with the help of GA, which is already attained by solving Lambert's problem. We have successfully applied this algorithm to obtain the transfer trajectory of a spacecraft from the Low Earth orbit (LEO) to some desired asteroid within a time frame. For this purpose, 8 Near-Earth Asteroids (NEAs) have been selected. These are likely assets to help space industrialization, as they have the earmarks of being the least affordable source of certain required crude materials like valuable metals and semiconducting elements. The convergence of the genetic algorithm to the optimal initial state of the transfer trajectory is also shown in this research.

Keywords Asteroid mining · Genetic algorithm · Lambert's problem · Orbital mechanics · Transfer trajectory trajectory

1 Introduction

The asteroids are also a part of our solar system like all planets. All asteroids were formed at the beginning of the solar system. On January 1, 1801, the first asteroid $1Ceras$ was discovered by G. Piazzi from the Palermo astronomical observatory. About 100 years later the first Near Earth Asteroid $433Eros$ was discovered by G. Witt. The astronomers are constantly searching for asteroids in our solar system and have so far identified about 600,000 asteroids. Nearly 10,000 asteroids of total asteroids, passes close to Earth's orbit, and orbiting around the Sun, are called

V. Kumar (✉) · B. S. Kushvah
Department of Mathematics and Computing, Indian Institute of Technology (ISM), Dhanbad 826004, Jharkhand, India
e-mail: vijilchoudhary@gmail.com

S. Banerjee and A. Saha (eds.), *Nonlinear Dynamics and Applications*, Springer Proceedings in Complexity,
https://doi.org/10.1007/978-3-030-99792-2_52

Near-Earth Asteroids (NEAs) [22]. The NEAs are some of the most accessible real estate in the solar system. Onto the 17% NEAs, is much easier to reach than a soft landing on the Earth's moon.

Only in one Platinum Group Metals (PGMs) rich asteroid of diameter 50-m, may contain 174 times the monthly world output of PGMs [7]. The most useful PGMs are platinum, iridium, osmium, palladium, rhodium and ruthenium. Some asteroids may found metallic elements like iron, nickel and copper sometimes in incredible quantities [17, 26]. In addition, asteroids may contain water and other gases, like CO_2, CO, nitrogen and methane. Some semiconductors non-metallic & metallic also found in asteroids. The semiconductors like tellurium, antimony, indium, cadmium, selenium, arsenic, germanium, gallium and phosphorus are used in micro electronic manufacturing. These are the key to supporting life in the present timespan [2, 6].

On the other hand, the two-point boundary value problem (TPBVP) in the two-body dynamical environment is known as Lambert's problem. To solve this problem, we required the positions of any two celestial bodies respectively and also the transfer time between them. The solution of the Lambert's problem gives the initial and final velocity of the transfer trajectory. A brief knowledge about Lambert's problem is given by Blanchard [5]. They discuss all the various cases of Lambert's theorem in to a single form, which especially suitable for numerical work. The determination of an orbit having a specified flight time and connecting two position vectors, frequently referred to as Lambert's problem. A variety of methods of dealing with this problem has been discussed over the years by many writers. There are many solutions of this problem that may be found in the literature. Most of the earlier methods have been characterized by a particular formulation of the time of flight equation and a particular independent variable to be used in a Newton-Rephson style of iteration [21]. Finding low energy transfer of satellite, minimum transfer time, optimal burnout angle of transfer and so on, they are just a few examples of the trajectory design problem in astrophysics [20]. Levine showed that the true anomaly of the point in an orbit where the velocity vector is parallel to the line of sight from an initial point to the terminal point is independent of the orbit [15]. In his research, they developed a new corollary to the famous Lambert's problem and apply a new property of two body boundary value problem [24]. The elementary form of Kepler's equation provides the analytic description of the time of flight [3, 13].

Thenceforth, the convergence is remarkable rapid and almost uniform as well as being essentially independent of the initial guess [4]. In a research, the author finds the new conic direction that interfaces two points of a gravity field in a given time, which is represented by a set of transcendental equations due to Lagrange. The Lagrange equation for the orbital transfer time can be expressed as a series expansion for all cases. Lambert's theorem is one of the most useful tool for interplanetary transfer trajectory design. The interplanetary transfer trajectory has three phases. The first phase is the powered phase. In which spacecraft gets an impulse to jump from parking orbit of departure planet to transfer orbit. The second phase is called the free-flight phase, in which the spacecraft freely move towards the destination point with the help of center gravitational force. The final phase is re-entry phase. In this phase the spacecraft enter the parking orbit of destination planet with the help of backward impulse [14, 28]. The backward impulse is used to decrease the velocity of a spacecraft. Besides, the accuracy of interplanetary transfer trajectory depends significantly on the free flight phase.

However, The purpose of issue here is that the free flight phase is the biggest dynamic factor in the transfer trajectory. It will take 90% to 98% time of whole transfer time and also has the least control. This implies that the accuracy of the free flight phase depends on the accuracy of the powered and re-entry phase. So, these must be selected with as greater precision as possible. We calculate the synodic period of every asteroid to know the next encounter with the Earth. The synodic period is calculated to dividing the product of two orbital periods by their differences, i.e. when the two orbital periods are nearly equal then the synodic period is quite long. We are getting the close encounter epoch of the asteroid and the Earth from the JPL Small-Body Database Browser. The collision point of an asteroids with the Earth is determined. The important figuring is done to counteract the collision, for example, the diversion in the trajectory of the asteroid by hitting the Earth. In this way, with the assistance of spacecraft, the asteroid could be pushed far away from the Earth's impact point.

On the other hand, the genetic algorithm has been included in LPG-Algorithm for optimizing the investigation of the interplanetary trajectory. The genetic algorithm has been effectively applied to a few streamlining undertakings, going in size from 2-variable to the 7-variable problem [8, 31]. In every case, its performs significantly better then the grid search technique that is commonly used [9, 19]. Most of the basic and advanced properties of the genetic algorithm are given by the author Mitsuo gen in his research. They described detail of the use of the genetic algorithm in the area of optimization research like advanced planning and scheduling model, real time task scheduling models, reliability optimization models, communication network model, interplanetary transfer trajectory design, multi-objective rendezvous model and many more [11]. A classical multi-objective technique requires a prior problem information. Since the GA uses population points, they might have the option to numerous Pareto-optimal arrangements at the same time [18]. The results of recent research suggests that the non-dominated sorting GA can be effectively used to discover various Pareto-optimal solution, the information on which could be helpful to the best optimal launch date and optimal initial velocity for trajectory design [27, 29]. Many of the methods like differential correction methods have been used to improve the initial state. Because the solution of the transfer trajectory is very sensitive for the initial state of the transfer trajectory [16, 30].

We introduced a new algorithm to obtain a suitable initial state of the transfer trajectory. i.e., LPG-Algorithm. The results indicate the promising performance of the proposed method in providing an appropriate initial state for reaching the spacecraft to the asteroid. LPG-Algorithm is specially made for interplanetary transfer trajectory. In this research, this technique have used to transfer a spacecraft on the near Earth asteroids. It is guaranteed to impact the spacecraft to the asteroid even if the solution of Lambert's problem is not optimal. The solution of the Lambert's problem gives us the initial velocity vector. If we have a non-optimal solution then it is required more fuel and also longer maneuver time. Then the genetic algorithm is applied on it. Thus, the fittest part of the last generation of the LPG-Algorithm will be the optimal initial state of the transfer trajectory. And the spacecraft will precisely land on the pre-selected asteroid with the least error in position.

2 LPG-Algorithm

A new algorithm has been developed for transfer trajectory from one celestial body onto another celestial body. It's named **LPG-Algorithm**. We combine Lambert's problem with the Genetic Algorithm (GA) in this methodology. And it has used to solve the most famous astrophysics problem with high accuracy. The Lambert's problem is an approach to fathom for the direction interfacing two position vectors with a given time of transfer. Figure 1 demonstrates that $\mathbf{r_0}$ is the initial position vector at the time t_0 and $\mathbf{r_f}$ is the final position vector at the time t_f. $\Delta t = t_f - t_0$ is the transfer time of the spacecraft between the two positions and $\Delta \nu$ is the transfer angle between the two positions. The transfer trajectory is depicted in Fig. 1 by the red color arc. The Earth's orbit and the asteroid's orbit is portrayed by blue and green color, respectively. The departure and arrival position is portrayed by the red dot. Lambert's problem is the well known problem in astrophysics. So, we are not going in details to the solution of the Lambert's problem. Any interested reader may go through the references papers [5, 24] for more explanation. Further, the genetic algorithm is a population based stochastic process. A single variable, which the genetic algorithm will optimize, is called genes. The collocation of all genes is called the chromosomes. And the number of chromosomes used in a single iteration is denoted as population, which is depicted in Fig. 2. It generates randomly and then reaches the best optimal value by the genetic algorithm process. The three leading operators are used in it, namely; Selection, Crossover and Mutation. There are many

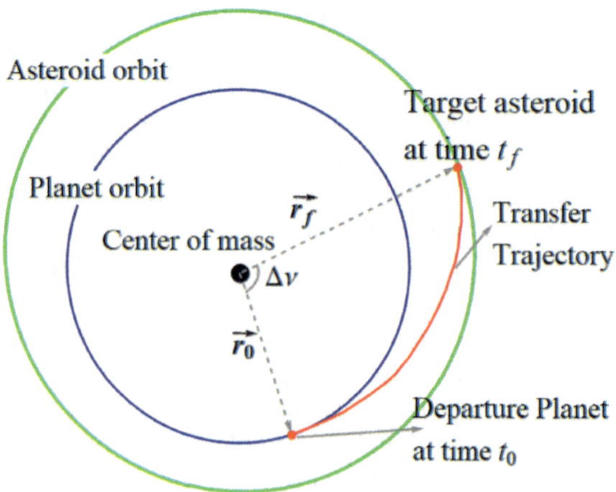

Fig. 1 Transfer trajectory using solution of Lambert problem without scale

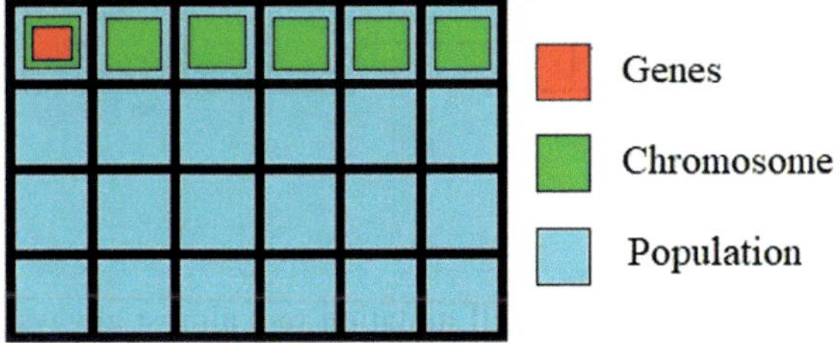

Fig. 2 Distribution of population for the genetic algorithm

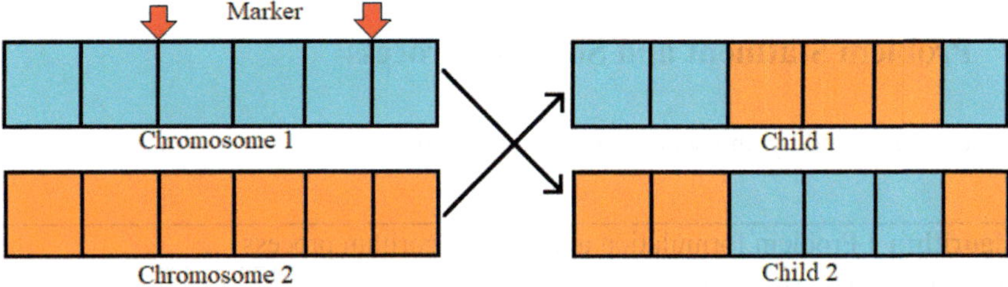

Fig. 3 Crossover process

methods that were found in literature for the selection process. We are using one of the most common method that are the roulette wheel selection process [23].

In the crossover process, we select two parents chromosomes and create randomly two markers on the chromosomes. Then interchange the genes between the markers and generate two new child chromosomes. it is also showing in Fig. 3. We can see that the three genes of parent 1 (orange) are replaced by parent 2 (cyan) and generate two new child chromosomes. After creating the child chromosomes, we select the better one between the parent and the child chromosomes.

The mutation operators are for the most part used to give investigation. The crossover operators is generally used to lead the population to converge on one good solution found up until now. Thus, while crossover tries to converge to a particular point in the scene, the mutation forth a valiant effort to stay away from assembly and investigate more regions. We like to investigate substantially more toward the start of the pursuit procedure. Then again, we lean toward more exploitation toward the finish of the inquiry procedure to guarantee the convergence of the population to the global optimum [10]. There is only an exemption; when population joins to a nearby optimum, we should expand the population assorted variety to investigate different regions. As indicated by the above actualities, too high mutation rate builds the likelihood of looking through more regions in search space, however, it prevents the population to converge to an optimum solution. On the other hand, too small mutation rates may result in falling to local optima instead of the global optimum i.e., too high mutation rate reduces the search ability of the genetic algorithm to a

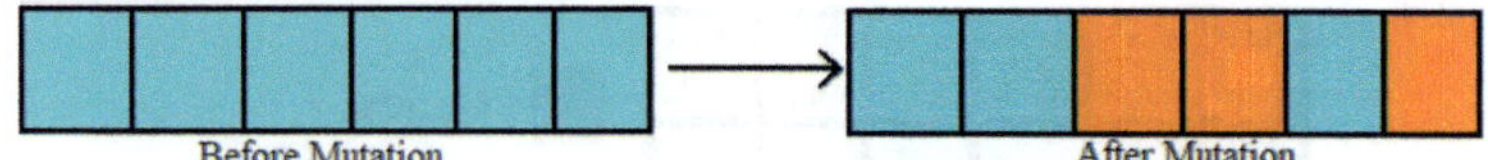

Fig. 4 Mutation process

simple random walk while a too small mutation rate almost always falls to a local optimum. So, in our computation we are using 35% mutation factor (Fig. 4).

3 Problem Statment and Solution Process

It is demonstrated by an pseudo code and given below:

Algorithm 1 Problem formulation and LPG-Algorithm process

procedure LPG- ALGORITHMM PROCEDURE

Step 1: We assume the equation of motion of two body problem [20] under the central gravitational force as the Sun.

Step 2: To solve the Lambert problem for the velocity of spacecraft $[v_a, v_b]$ at the low Earth orbit and the pre-selected asteroid respectively.

Step 3 (goal): Next, we formulate an objective function for genetic algorithm, that minimize the distance between the spacecraft position $[X, Y, Z]$ and the target position $[R_x, R_y, R_z]$.

Step 4: Now, we update the initial state $[r_0, v_a]$ of the transfer trajectory using genetic algorithm.

Step 5 (output): Finally, we get the optimized initial state $[r_0, v_a]$ and described the needed data in the tables and figures.

3.1 Advantage and Limitations

The merits and demerits of the proposed methodology are as follows: This concept is easy to understand. It searches from a population of points, not a single point, which means it calculates the value of the objective function and chooses the minimum of them. LPG-Algorithm is robust concerning the optimum local value of the objective function. It is stochastic and can be operated on many representations.

However, this requires less information about the problem, but designing an objective function and the representation and operators can be difficult. It is a computationally expensive algorithm i.e., it is time-consuming.

4 Apply on the Equation of Motion

Let us suppose two celestial bodies that are rotating about their center of mass. r_0 & v_0 are the position & velocity vectors of the first celestial body at the departure time t_0. r_f & v_f are the position & velocity vectors of the second celestial body at the arrival time t_f and is defined as:

$$r_0 = [r_{0x},\ r_{0y},\ r_{0z}], \tag{1}$$

$$v_0 = [v_{0x},\ v_{0y},\ v_{0z}], \tag{2}$$

$$r_f = [r_{fx},\ r_{fy},\ r_{fz}], \tag{3}$$

$$v_f = [v_{fx},\ v_{fy},\ v_{fz}] \tag{4}$$

where the subscript "f" and "0" denote the arrival and departure states respectively. The subscript "x", "y" and "z" denote the unit vectors along with the x-axis, y-axis & z-axis respectively. Now find the solution of Lambert's problem using r_0, r_f and the transfer time $\Delta t = t_f - t_0$. The solution comes in the form of velocity vectors of the transfer trajectory at the departure and arrival position. Suppose that velocity vectors are:

$$v_a = [v_{ax}, v_{ay}, v_{az}], \tag{5}$$

$$v_b = [v_{bx}, v_{by}, v_{bz}] \tag{6}$$

The initial state for the transfer trajectory from r_0 onto r_f is given:

$$\text{initial state} = [r_{0x}, r_{0y}, r_{0z}, v_{ax}, v_{ay}, v_{az}] \tag{7}$$

where, the first three quantity $[r_{0x}, r_{0y}, r_{0z}]$ are the departure position of spacecraft in 3-D coordinate respectively. And last three quantity $[v_{ax}, v_{ay}, v_{az}]$ are the velocity of the spacecraft in x, y, z direction respectively. The velocity vectors $v_a = [v_{ax}, v_{ay}, v_{az}]$ are obtained from the solution of Lambert's problem. This initial state is non optimal because, the spacecraft fails to reach the desired position using this initial state. We apply the Genetic algorithm to optimize it. For the Genetic algorithm, we generate the random chromosomes as follows: We generate the number of chromosomes is equal to the number of population size. Here, the position vectors is fixed, because the departure position of spacecraft does not change. Further, calculate the fitness value of every chromosomes using two-body simulation as the solution of Keplerian orbit [12]. The central gravitational force in two-body simulation and Lambert's solution are the same. The gravitational force of the Sun is $1.3271 \times e^{11}$ $km^3 s^{-2}$. A new optimization function is introduced which is defined below (Fig. 5):

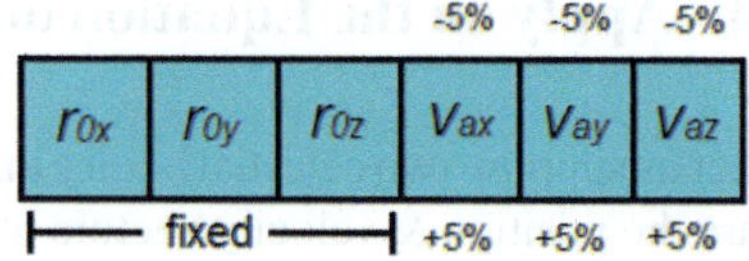

Fig. 5 Non-optimal initial state of transfer trajectory

$$F_{opt} = \{\|X - r_{fx}\| + \|Y - r_{fy}\| + \|Z - r_{fz}\|\} \tag{8}$$

subject to:

$$\sqrt{(R_x - r_{fx})^2 + (R_y - r_{fy})^2 + (R_z - r_{fz})^2} = R_{park} \tag{9}$$

where, $\|\cdot\|$ is denotes the modulas value, $[X,\ Y,\ Z]$ is the position vector of the spacecraft. We obtain this from the solution of two-body problem. The position vector $[r_{fx},\ r_{fy},\ r_{fz}]$ of target celestial body and is defined in Eq. 3 and R_{park} is a user-defined quantity, which may be varies for the different planets or asteroids. It denotes the radius of parking orbit around the target celestial body. The Eq. 9 gives the constrained condition of the objective function (8). We can understand the LPG-Algorithm in a better way from the short python program, which is given in Fig. 6. Where the python function *TwoBody()* contains the equations of motion of the two-body problem. The python function *RKF45()* is defined for integration, which is based on Runge-Kutta-Fehlberg Method. It has a methodology to decide whether the proper step size h is being utilized. At each progression, two unique approximations for the solution are made and compared. In the event that the two answers are in close understanding, the estimate is accepted. On the off chance that the two answers don't consent to a predefined tolerance, the step size is diminished. And also if the answer consent to more significant digits than required, the step size is expanded. The input required for the function is a mathematical model (here *TwoBody()*), the initial state of the mathematical model (given in equation-7), the initial and final time for integration time (here $[t_0,\ t_f]$) and a initial step size (here h). We set the relative tolerance as 1.0 e^{-10}. The python function *CalculateF()* is made to calculate the fitness value for every generated chromosomes. The inputs required for this function are the position vectors of the target celestial body (here r_f), the spacecraft (here $[X,\ Y,\ Z]$) and the radius of parking orbit around the target celestial body (here R_{park}).

The function *GeneticAlgorithm()* is the main optimization process function. It takes input a mathematical model function *TwoBody()*, a integration method *RKF45()*, a function to calculate the fitness value *CalculateF()* and the initial state of the model to integrate it. With in this function, we take the population size is equal to 100. It generates randomly chromosomes equal to the population size as the initial state of the mathematical model. Further, integrates the mathematical model for every chromosomes using given integrating function and also calculate the fitness value using the fitness function (8). Then the function *GeneticAlgorithm()*, picks up a most optimal chromosome refers to the optimal initial state. For future investigation,

```
i, MaxIteration, ClosePoint = 0, 10000, 1.0e-3
while(F_opt <= ClosePoint):
        v_a^new = GeneticAlgorithm(TwoBody,RKF45,CalculateF,[r0, v_a^old])
        '''
        this function takes 100 chromosomes at once and select
        most appropriate velocity vector & returns it
        '''
        solution = RKF45(TwoBody, [r0, v_a^new], [t0, tf], h, tol=e^-10)
        '''
        this returns float array of 6 element in which first
        3 is position and last 3 is velocity of the spacecraft
        '''
        X, Y, Z = solution[0], solution[1], solution[2]
        F_opt = CalculateF(r_f, [X, Y, Z], R_park)
        '''
        CalculateF function is calculate the value of F_opt using
        optimize function (defined in equation -7) & returns it
        '''
        # store the nukmerical data of every iteration in a file
        file = open('TransferData.txt','a')
        file.write(i, X, Y, Z, F_opt)
        file.close()
        i=i+1 # update number of iteration
        v_a^old = v_a^new # change velocity vector for next iteration
        if (i == MaxIteration):
                break
```

Fig. 6 A python program to demonstrate the basic of LPG-Algorithm

we make a separate data text file to store the numerical data of every iteration. We repeat this process either till they complete the *MaxIteration* or satisfies the optimal condition $F_{opt} <= 1 \cdot e^{-9}$.

5 Numerical Simulation

We are introducing here a new methodology for the interplanetary transfer trajectory design. A genuine interplanetary transfer trajectory experiences different gravitational forces consistently such as radiation pressure, magnetic forces, atmospheric drag, sun oriented radiation, etc. The interplanetary transfer trajectory design problem can be expressed as an optimization issue. Where one of the essential objectives is to minimize fuel requirements. Some other optimization problems are intermediate planetary flybys, type of arrival, mission duration and velocity constraints, etc. The primary thought of the proposed methodology comprises of focusing on an asteroid that might be a good source of extraterrestrial Platinum Group Metals (PGMs). The asteroids are chosen from the close Earth class with the fly-by distance from the Earth smaller then $2.647e^{-2}\ AU$. From the past few decades, these types of asteroids have attracted the attention of researchers for their spacious storage of precious metals. An asteroid must be selected before starting the asteroid mining mission. The selection of asteroids for mining purposes, and mission design follows the following these steps.

Table 1 Name of selected asteroid with their properties and the close encounter epoch to the Earth

Asteroid name	Diameter ($m.$)	Mass($kg.$)	Close encounter epoch (MJD)	Close encounter distance (AU)
1996 GT	670	$5.44 \times e^{11}$	59856.40833	$7.123 \times e^{-2}$
1999 RA32	170	$9.37 \times e^{9}$	60756.32569	$9.184 \times e^{-2}$
1998 SF36	380	$1.04 \times e^{11}$	63679.66319	$8.529 \times e^{-2}$
1999 JU3	380	$1.01 \times e^{11}$	63952.24513	$4.744 \times e^{-2}$
1986 EB	1800	$1.11 \times e^{13}$	64408.32569	$2.647 \times e^{-1}$
2011 UW158	280	$3.94 \times e^{10}$	69293.18402	$4.315 \times e^{-2}$
1996 FG3	550	$3.13 \times e^{11}$	70134.60347	$5.646 \times e^{-2}$
1999 RQ36	180	$9.91 \times e^{9}$	73725.02500	$5.007 \times e^{-3}$

- In the first step, the required minerals for mining must be determined.
- Then, the asteroids with the probable of being endowed in the target minerals must be found.
- Before examining the subtleties of these two stages, it is important to talk about the conceivable objective bodies and minerals that can be found in the core of the Near-Earth Asteroid (NEAs).

In Table 1, we listed some selected asteroids who will pass near the Earth in the upcoming four decades. The name of the selected asteroids are given in the first column. The diameter of the asteroids is shown in the second column. The mass of the asteroids is shows in the next column. The fourth column shows the close encounter epoch of the asteroid in Modified Julian Date(MJD) format, which referred from JPL HORIZONS Web-Interface. And the last column denote the close encounter distance between the Earth and the asteroid. The spacecraft follows the following steps to reach on the surface of the asteroid.

1. Start of spacecraft from Low Earth Orbit (LEO) to asteroid along transfer trajectory using a solution of Lambert's problem.
2. The Lambert solution fails to get optimal initial state to reach onto the desired asteroid.
3. Next, we update the initial state of the spacecraft using the LPG-Algorithm.
4. Finally, we get an optimal initial state of the spacecraft with good accuracy.

The significant errand of the mission design everywhere comprises of the picked of all factor free parameters. The first step is to choose an launch date few hours before the close encounter epoch of the asteroid. Now we execute the mission from solving Lambert's problem, which allows transferring the zero mass object from one space point to another along Keplerian orbit in given transfer time. The coordinates of the Earth and the asteroid are taken from JPL small body database. The gravitational field is supposed to the center of the Sun. It should be mentioned that on the phase of calculation of asteroid motion. Then find the transfer time between these two position

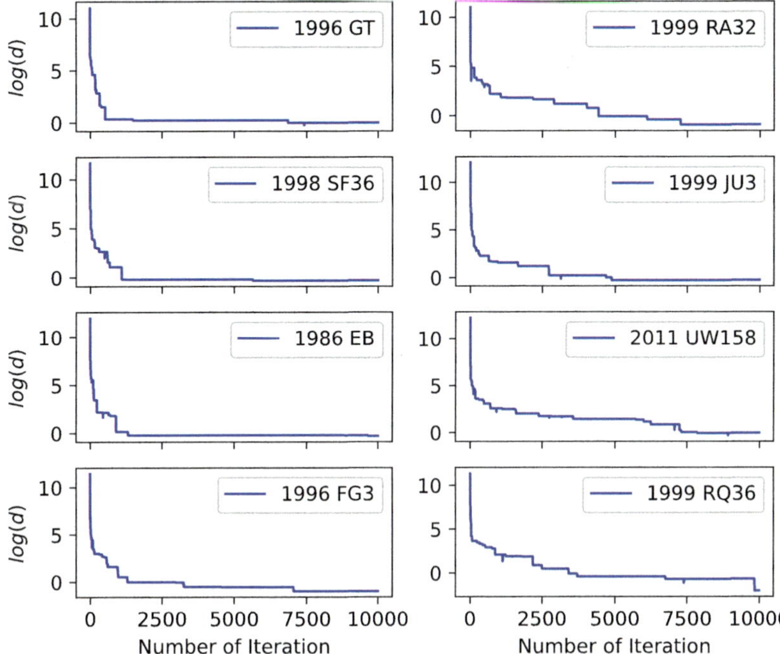

Fig. 7 Iteration of all asteroids versus logarithmic function of distance

vectors $\Delta t = t_f - t_0$. We have three parameters here i.e., departure position vector, arrival position vector and the transfer time. Now we apply the LPG-Algorithm successfully. It gives the optimal initial state, allowing the spacecraft to reach as close to the prearranged target asteroid as possible. Here the value of R_{park} is set equal to radius of the target asteroid, because our goal is to reach onto the asteroid.

We use python interpreted language for the computational purpose. For the numerical calculation of the LPG-Algorithm, we run it up to 10 thousand iterations and save the data of each iteration in a separate text file. Figure 7 portrays a graphs between the number of iterations and the distance between the spacecraft & the asteroids. We see that when the iteration of the genetic algorithm is increased up to 10000, the distance is going to decrease continuously up to $1e^{-3} Km$ from the asteroid. In first, 10% iterations, the distance is decreased rapidly and in the remaining iterations it is approximately constant (decreasing slowly). In the first 10% iterations, the distance reduces from 0.15 million kilometers to 10 km, which is hard to show in a single plot because of a huge difference along with the iterations. So, here we have changed the scale of the distance to the logarithmic function with base "e". i.e., 0.15 million kilometers on y-axis denoted by 11.91839057 and 10 km on y-axis denoted by 2.30258509. It is an easy way to show a large distance and a small distance in a graph. All the sub-plot of Fig. 7 are shown the number of iteration of the genetic algorithm and logarithmic function of distance(d) with base "e" of all pre-selected asteroids.

Table 2 All necessary calculated data of transfer trajectory to the asteroids

Asteroid Name	Departure $(Earth)$	Arrival $(Asteroid)$	Transfer time $(Hours)$	ΔV_1 $(\frac{Km.}{sec})$	ΔV_2 $(\frac{Km.}{sec})$	Computational time $(Minutes)$
1996 GT	59839.20000	59856.40833	413	3.49857	4.07962	10.2356
1999 RA32	60732.15902	60756.32569	579	4.16823	5.39587	12.8542
1998 SF36	63659.99652	63679.66319	482	4.57381	3.76428	13.5168
1999 JU3	63932.99513	63952.24513	473	5.07286	3.10864	9.4687
1986 EB	64371.40208	64408.32569	571	5.78610	6.37561	10.2584
2011 UW158	69273.64236	69293.18402	469	2.50942	3.82617	8.0962
1996 FG3	70114.06180	70134.60347	493	5.68423	4.11682	10.5962
1999 RQ36	73709.23333	73725.02500	379	3.18354	6.74251	9.5316

Our computer has intel $i7$, 7th generation, 16 GB RAM, 3.5 GHz processor and Windows 10. And we use this computer for all numerical simulations. The necessary computation of the transfer trajectory is shown in Table 2. All the pre-selected asteroids names are given in the first column of the Table 2, which is also given in Table 1. The second and third column describes departure and arrival epoch of the spacecraft in MJD format respectively. In the fourth column, the transfer time is given in hours. We chose the arrival time few hours before the close encounter epoch. The fifth column represents the minimum required impulse (ΔV_1), when the spacecraft is going to leave Low Earth Orbit (LEO). The required second impulse (ΔV_2) is given in the next column. The spacecraft enter in the parking orbit of the asteroid after applying the second impulse. The total computational time of the respective asteroid is given in the last column.

Figure 8 has portrayed the graph of the transfer trajectory from Low Earth Orbit to the asteroids. These asteroids are selected randomly from the list of pre-selected asteroids. In left side sub-figure Fig. 8a, c, e are the asteroid **1996 GT**, **1999 RA32** and **1999 RQ36** respectively. These sub-figures demonstrates:

1. Earth orbit with a blue line and Earth position on departure time with a blue dot.
2. Asteroid orbit with a green line and the asteroid position on an arrival time with a green dot.
3. The yellow dotted line shows the keplerian orbit of the transfer trajectory by using Lambert's problem.
4. The red line shows the keplerian orbit of the transfer trajectory by using LPG-Algorithm.

We portray the transfer trajectory in these left side graph but it is not visible because of too close. It can not be seen without zoom in. So, we have added extra sub-figure in the right side. These all sub-figures Fig. 8b, d, f show a particular part (vicinity of the asteroid) of the left side sub-figures Fig. 8a, c, e respectively. Here, in the right side sub-figures shows:

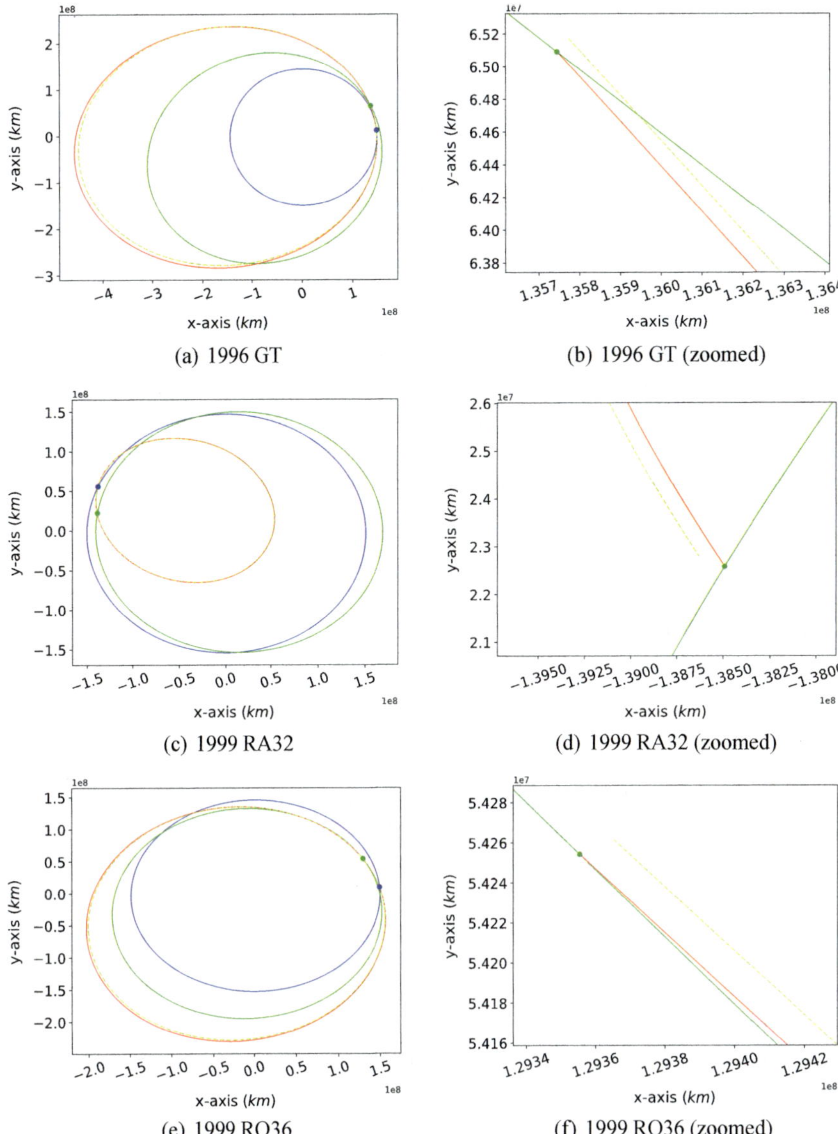

Fig. 8 The difference in the transfer trajectory using Lambert's problem and the LPG-Algorithm

1. The green dot shows the arrival position of the asteroid.
2. The blue and green lines shows the Earth's orbit and the asteroid's orbit respectively.
3. The yellow dotted line shows the transfer trajectory by using Lambert's problem.
4. The red line shows the transfer trajectory by applying LPG-Algorithm.

We observe that from the figures and calculations, the transfer trajectory by using Lambert's problem and by using LPG-Algorithm are approximately same in periods. Only difference is in accuracy of results. We find that the error in target position is approximately $1e^{-3}$ by the Lambert solution. And by the LPG-Algorithm this error can be reduced to less than $1e^{-9}$. The LPG-Algorithm gives more accurate optimal initial state of the transfer trajectory. And LPG-Algorithm is effective and transfers the spacecraft exactly to the desired asteroid even if the solution of the Lambert's problem is far from the global optimum. Only three variable i.e., velocity component of the initial state is needed for a interplanetary transfer trajectory from Low Earth Orbit to the asteroid [1, 25]. It must renovates the possibility to reach the desired asteroid.

6 Conclusions

In this article, we present the LPG-Algorithm, a new substantial methodology for solving the Lambert problem using a genetic algorithm. It has also been demonstrated that, because of its low complexity, it has numerous advantages in terms of generality and performance. It has been implemented to find the transfer trajectory from the Low Earth Orbit to the 8 selected asteroids. The asteroids, which may have perilous metals. All these asteroids are in Near-Earth Asteroid (NEA) group and they also close fly-by to the Earth. The results show that two small impulses on the spacecraft can transfer it to the asteroid. The numerical simulation data and figure show the clear difference between the solution of Lambert's problem and the LPG-Algorithm. In both cases, the transfer time is the same. The LPG-Algorithm is very useful for interplanetary transfer trajectory design and gives more accurate results. This algorithm can be implemented in an interplanetary transfer trajectory design and a ballistic missile trajectory.

Acknowledgements All the required data has downloaded through the following:
1. https://ssd.jpl.nasa.gov/sbdb.cgi
2. https://ssd.jpl.nasa.gov/horizons.cgi
3. http://www.asterank.com/.

References

1. Abdelkhalik, O., Mortari, D.: N-impulse orbit transfer using genetic algorithms. J. Spacecraft Rockets **44**(2), 456–460 (2007)
2. Andrews, D.G., Bonner, K., Butterworth, A., Calvert, H., Dagang, B., Dimond, K., Eckenroth, L., Erickson, J., Gilbertson, B., Gompertz, N., et al.: Defining a successful commercial asteroid mining program. Acta Astronautica **108**, 106–118 (2015)
3. Battin, R.H., Fill, T.J., Shepperd, S.W.: A new transformation invariant in the orbital boundary-value problem. J. Guidance Control **1**(1), 50–55 (1978)
4. Battin, R.H., Vaughan, R.M.: An elegant lambert algorithm. J. Guidance Control Dyn. **7**(6), 662–670 (1984)
5. Blanchard, R., Devaney, R., Lancaster, E.: A note on lambert's theorem. J. Spacecraft Rockets **3**(9), 1436–1438 (1966)
6. Brophy, J.R., Friedman, L., Culick, F.: Asteroid retrieval feasibility. In: 2012 IEEE Aerospace Conference, pp. 1–16. IEEE (2012)
7. Busch, M.: Profitable asteroid mining. JBIS **57**, 301–305 (2004)
8. Cage, P., Kroo, I., Braun, R.: Interplanetary trajectory optimization using a genetic algorithm. In: Astrodynamics Conference, p. 3773 (1994)
9. Coit, D.W., Smith, A.E.: Reliability optimization of series-parallel systems using a genetic algorithm. IEEE Trans. Reliab. **45**(2), 254–260 (1996)
10. Dos Santos, D.P., Prado, A.F.: Minimum fuel multi-impulsive orbital maneuvers using genetic algorithms. Adv. Astron. Sci. **145**, 1137–1150 (2012)
11. Gen, M., Lin, L.: Genetic algorithms. Wiley Encyclopedia of Computer Science and Engineering, pp. 1–15 (2007)
12. Hinckley, D.W., Hitt, D.L.: Evolutionary approach to lambert's problem for non-keplerian spacecraft trajectories. Aerospace **4**(3), 47 (2017)
13. Ivashkin, V., Lan, A.: Construction of the optimal trajectories for the earth-asteroid-earth mission under high-thrust flight. Cosmic Res. **58**, 111–121 (2020)
14. Ivashkin, V., Lang, A.: Optimum trajectories for an earth–asteroid–earth mission with a high thrust flight. In: Reports of the Academy of Sciences, vol. 484, pp. 161–166 (2019)
15. Izzo, D.: Revisiting lambert's problem. Celestial Mech. Dyn. Astron. **121**(1), 1–15 (2015)
16. Jiang, R., Chao, T., Wang, S., Yang, M.: Adaptive genetic algorithm in rendezvous orbit design. In: 2016 35th Chinese Control Conference (CCC), pp. 5677–5682. IEEE (2016)
17. Kargel, J.S.: Metalliferous asteroids as potential sources of precious metals. J. Geophys. Res. Planets **99**(E10), 21129–21141 (1994)
18. Kim, Y.H., Spencer, D.B.: Optimal spacecraft rendezvous using genetic algorithms. J. Spacecraft Rockets **39**(6), 859–865 (2002)
19. Kumar, V., Kushvah, B.: Computation of periodic orbits around l 1 and l 2 using pso technique. Astron. Reports **64**(1), 82–93 (2020)
20. Lawden, D.: Minimal rocket trajectories. J. Amer. Rocket Soc. **23**(6), 360–367 (1953)
21. Lei, H., Xu, B.: Families of impulsive transfers between libration points in the restricted three-body problem. Monthly Notices Royal Astron. Soc. **461**(2), 1786–1803 (2016)
22. Lewicki, C., Diamandis, P., Anderson, E., Voorhees, C., Mycroft, F.: Planetary resources-the asteroid mining company. New Space **1**(2), 105–108 (2013)
23. Lipowski, A., Lipowska, D.: Roulette-wheel selection via stochastic acceptance. Phys. A: Stat. Mech. Appl. **391**(6), 2193–2196 (2012)
24. Nelson, S.L., Zarchan, P.: Alternative approach to the solution of lambert's problem. J. Guidance Control Dyn. **15**(4), 1003–1009 (1992)
25. Rauwolf, G.A., Coverstone-Carroll, V.L.: Near-optimal low-thrust orbit transfers generated by a genetic algorithm. J. Spacecraft Rockets **33**(6), 859–862 (1996)
26. Sonter, M.J.: The technical and economic feasibility of mining the near-earth asteroids. Acta Astronautica **41**(4–10), 637–647 (1997)
27. Srinivas, N., Deb, K.: Muiltiobjective optimization using nondominated sorting in genetic algorithms. Evolut. Comput. **2**(3), 221–248 (1994)

28. Wheelon, A.D.: Free flight of a ballistic missile. ARS J. **29**(12), 915–926 (1959)
29. Wu, G.q., Tan, L.G., Li, X., Song, S.M.: Multi-objective optimization for time-open lambert rendezvous between non-coplanar orbits. Int. J. Aeronaut. Space Sci., 1–16 (2019)
30. Yokoyama, N., Suzuki, S.: Modified genetic algorithm for constrained trajectory optimization. J. Guidance Control Dyn. **28**(1), 139–144 (2005)
31. Zhang, D., Song, S., Duan, G.: Fuel and time optimal transfer of spacecrafts rendezvous using lambert's theorem and improved genetic algorithm. In: 2008 2nd International Symposium on Systems and Control in Aerospace and Astronautics, pp. 1–6. IEEE (2008)

Prediction of Chaotic Attractors in Quasiperiodically Forced Logistic Map Using Deep Learning

J. Meiyazhagan and M. Senthilvelan

Abstract We forecast two different chaotic dynamics of the quasiperiodically forced logistic map using the well-known deep learning framework Long Short-Term Memory. We generate two data sets and use one in the training process and the other in the testing process. The predicted values are evaluated using the metric called Root Mean Square Error and visualized using the scatter plots. The robustness of the Long Short-Term Memory model is evaluated using the number of units in the layers of the model. We also make multi-step forecasting of the considered system. We show that the considered Long Short-Term Memory model performs well in predicting chaotic attractors upto three steps.

Keywords Logistic map · Chaos · Prediction · Deep learning · Long short-term memory

1 Introduction

Recently, Machine Learning (ML) and Deep Learning (DL) models have been used in various fields of physics [1–3]. In the study of dynamics of nonlinear systems, ML and DL algorithms are extensively used for the prediction and discovery of the behaviour of the chaotic and complex systems. For example, they have been used to identify chimera states [4, 5], in the replication of chaotic attractors [6], using symbolic time series for network classification [7], separating chaotic signals [8], learning dynamical systems in noise [9] and in the prediction of extreme events [10–15]. Very recently, the authors of Ref. [16] have considered Hénon map and used a ML algorithm, namely Artificial Neural Network (ANN), to study the extreme events in it. The authors have focussed on binary classification and classified the data points as extreme and non-extreme [16].

J. Meiyazhagan · M. Senthilvelan (✉)
Department of Nonlinear Dynamics, Bharathidasan University,
Tiruchirappalli 620024, Tamil Nadu, India
e-mail: senthilvelan.m@bdu.ac.in

S. Banerjee and A. Saha (eds.), *Nonlinear Dynamics and Applications*,
Springer Proceedings in Complexity,
https://doi.org/10.1007/978-3-030-99792-2_53

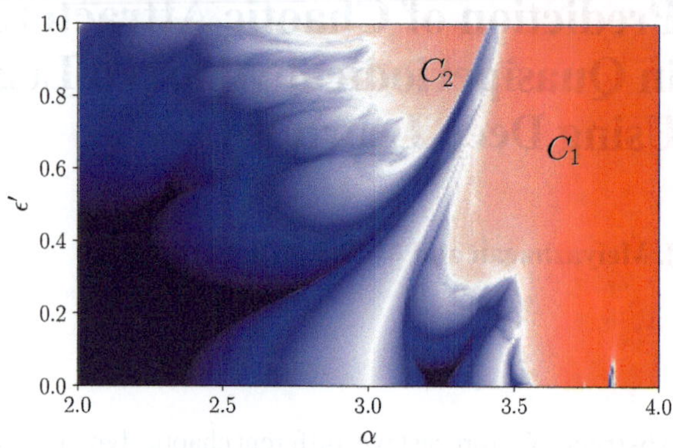

Fig. 1 Schematic phase diagram of the quasiperiodically forced logistic map. C_1 and C_2 are two different chaotic regimes

In our studies, we consider logistic map with quasiperiodic forcing and predict the time series of the system which is not continuous. The system exhibits chaos in two different regimes. We predict both the chaotic attractors of this system with the help of the DL framework, namely Long Short-Term Memory (LSTM). The logistic map with quasiperiodic forcing is described by the following equations, namely [17, 18]

$$x_{n+1} = \alpha[1 + \epsilon \cos(2\pi\phi_n)]x_n(1 - x_n), \tag{1a}$$

$$\phi_{n+1} = \phi_n + \omega \ (\text{mod } 1), \tag{1b}$$

where ϵ and $\omega = (\sqrt{5} - 1)/2$ are the forcing amplitude and irrational driving frequency respectively. The authors in Ref. [17] redefined the driving parameter as $\epsilon' = \epsilon/(4/\alpha - 1)$ to study the dynamics of the system in the regimes of $0 \leq x \leq 1$, $0 \leq \phi \leq 1$ and $0 \leq \epsilon \leq 1$. The schematic phase diagram [17] of the system is given in Fig. 1. The system shows various dynamic behaviours, namely periodic, strange non-chaotic and chaotic attractors which can be characterized by the nonzero Lyapunov exponent Λ [17], where

$$\Lambda = \lim_{N\to\infty} \frac{1}{N} \sum_{i=1}^{N} \ln |\alpha[1 + \epsilon \cos(2\pi\phi_i)](1 - 2x_i)|. \tag{2}$$

From Fig. 1, we can notice the interesting behaviour of the considered system which has two chaotic regimes, namely C_1 and C_2. The C_1 regime is the continuation of the chaotic regime in the logistic map for $\epsilon = 0$ at the end of the period-doubling cascade, at $\alpha = 3.5699...$. The chaos in C_2 regime is due to low nonlinearity and large amplitude forcing [17]. Our aim is to predict chaotic attractors in both the regime using LSTM model since it is capable of forecasting the data which is in the form of a sequence.

We organize our work as follows. In Sect. 2, we discuss the generation of training and testing data. In Sect. 3, we consider a DL framework called LSTM and train it using training set data and predict the test set data. The performance of the LSTM model is discussed in Sect. 4. We present the conclusion in Sect. 5.

2 Data Preparation

Generating data is the foremost task in prediction because prediction is done only by learning the relationship between the given data. We calculate the value of x for 10^5 iterations using Eqs. (1a, 1b) in both the regimes C_1 and C_2. This discrete space data is then converted into supervised learning data by taking x_n as input and x_{n+1} as output. The chaotic attractors in the both regimes C_1 and C_2 are shown in Fig. 2. The Fig. 2a, b corresponding to the regime C_1 and Fig. 2c, d correspond to the C_2 regime.

The values of the parameters are taken as (a) $\alpha = 3.6$, $\epsilon' = 0.5$, (b) $\alpha = 3.9$, $\epsilon' = 1.0$, (c) $\alpha = 3.0$, $\epsilon' = 1.0$ and (d) $\alpha = 3.1$, $\epsilon' = 0.8$. We divide the data into two parts: (i) training set and (ii) test set. Training set data are used during the training process of the DL model and test set data are used for the evaluation of the ability of the DL model. In Fig. 2, the blue dots are the data used for training purpose and the

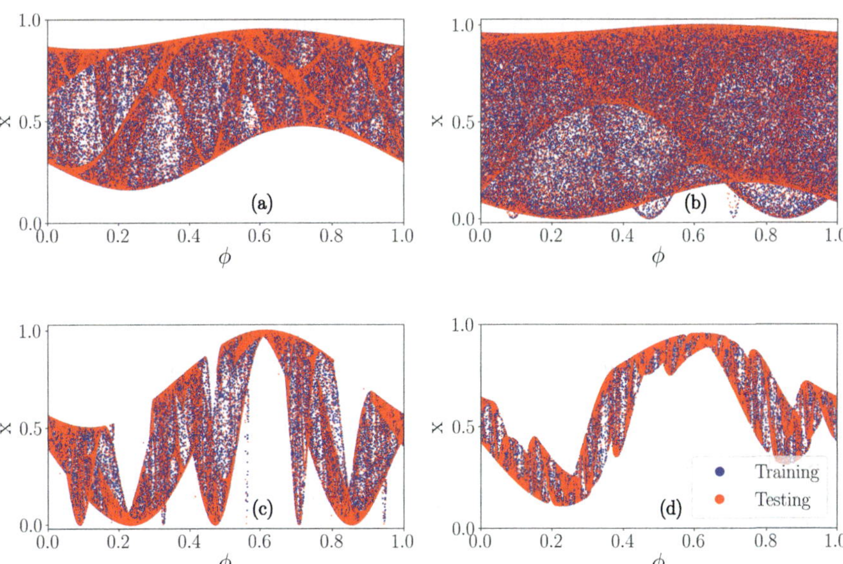

Fig. 2 Chaotic attractors in two different regimes C_1 and C_2. **a** $\alpha = 3.6$, $\epsilon' = 0.5$ and **b** $\alpha = 3.9$, $\epsilon' = 1.0$ correspond to C_1. **c** $\alpha = 3.0$, $\epsilon' = 1.0$ and **d** $\alpha = 3.1$, $\epsilon' = 0.8$ correspond to C_2. The points in blue (colour online) denoting the training set data and red (colour online) denoting the test set data

red coloured data are used for testing. We use 6×10^4 data as training set data and 4×10^4 data as test set data.

These two sets of data are rescaled using min-max normalization which is given by the formula [19],

$$x_i^{rescaled} = a + \frac{(x_i - x_{min})(b - a)}{x_{max} - x_{min}}, \qquad i = 1, 2, 3, \ldots, n, \tag{3}$$

where x_{min} and x_{max} are the minimum and maximum value of the data set respectively. We fix $a = -1$ and $b = +1$ in order to scale the data between -1 and $+1$. During the testing phase, this preprocessing scaling step is reversed after obtaining the output from the DL model in order to compare the results with the actual data.

3 Deep Learning Framework: Long Short-Term Memory

When the data is in a sequential form one can make use of the Recurrent Neural Networks (RNN) [20] which is a type of ANN. For the present study we consider a DL framework known as LSTM [21] which is a special kind of RNNs. In recent years, LSTM framework has proven to be capable of forecasting time series of the chaotic systems even when there are extreme events in the time series [10, 13, 14]. The main feature that differentiates LSTM from the other RNNs is that the latter has only one activation function for the neurons that is tanh but in the case of the former, a sigmoid function is used for recurrent activations and tanh is used for the activation of neurons. The sigmoid activation function is defined by [22],

$$\sigma(z) = \frac{1}{1 + e^{-z}}. \tag{4}$$

We construct the LSTM model in the following way. We consider two LSTM layers each having 16 units in it and followed by a layer which has one neuron for output. During the training, we give both the input and the corresponding output to the model, that is we give x_n as the input and x_{n+1} as the output. By doing this, the model will learn the nonlinear relations between the given data. After training, the learned model is used to forecast the data steps. During the testing phase, we feed only the input data and ask the model for the corresponding output. The predicted values at the output given by the LSTM model are compared with actual values to determine the efficiency of the model in forecasting the chaotic attractors of the considered system.

4 Results and Discussion

To visualize the performance of the considered DL model, predicted data are plotted over the actual data in Fig. 3. Black dots (colour online) denote the actual value and green dots (colour online) denote the predicted value. The data in Fig. 3a, b correspond to C_1 regime and Fig. 3c, d correspond to C_2 regime. From the plots, we can see that relatively all predicted data coincide with the actual data. To have a clear understanding of the efficiency of the model we calculate the Root Mean Square Error (RMSE) value using the formula,

$$RMSE = \sqrt{\sum_{i=1}^{N_{Test}} \frac{(\hat{Y}_i^{Test} - Y_i^{Test})^2}{N_{Test}}}, \tag{5}$$

where $\hat{Y}_i^{Test}$, Y_i^{Test} and N_{Test} denote the predicted values, actual values and total number of data in the test set respectively. We make use of the scatter plots which are plotted by taking actual values in the x-axis and predicted values in the y-axis (see Fig. 4). From Fig. 4a, b we can see that the RMSE values for the regime C_1 are 0.015 and 0.012 respectively for the parameter values $\alpha = 3.6$, $\epsilon' = 0.5$ and $\alpha = 3.9$, $\epsilon' = 1.0$. The outcome of the scatter plots almost fit in straight line, thereby indicating that the difference between predicted and actual values are very low. From Fig. 4c, d

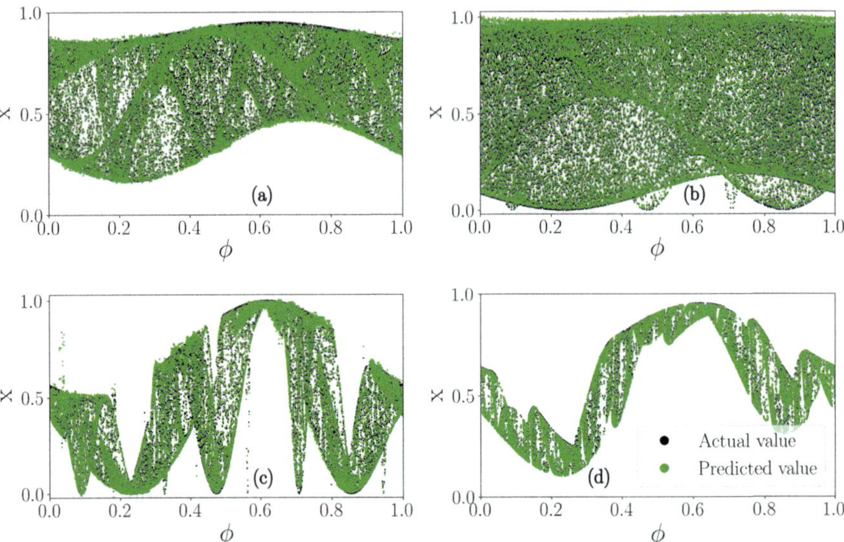

Fig. 3 Plots of forecasted values over the actual values for four different sets of α and ϵ values as mentioned in Fig. 2. The Figures **a**, **b** correspond to C_1 regime and **c**, **d** correspond to C_2 regime. Black dots denote the actual value and green dots denote the predicted value

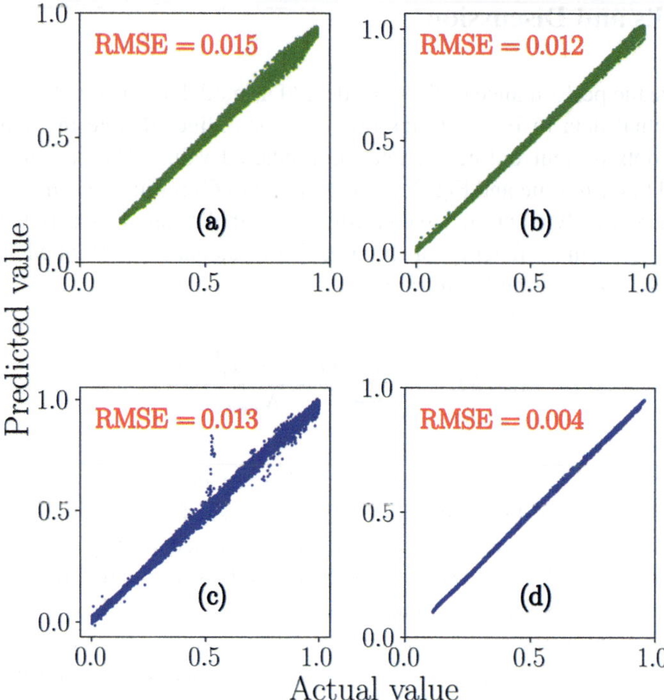

Fig. 4 Scatter plots and RMSE values of the four different cases. The green colour plots correspond to the regime C_1 and blue colour corresponds to C_2

we can see that the results of the second regime C_2 are calculated as 0.013 and 0.004 respectively for the parameter values $\alpha = 3.0$, $\epsilon' = 1.0$ and $\alpha = 3.1$, $\epsilon' = 0.8$. The scatter plots for the test set data of regime C_2 also show very little scatter points, thereby indicating the best fit of predicted data with the actual data.

4.1 *Effect of Model Architecture*

To study the effect of model architecture on the performance of the considered model we vary the number of units and analyse the performance based on the RMSE values. For this purpose, we change the units in both LSTM layers and train the model. Then each trained model is evaluated using the test set data. The outcome is shown in Fig. 5. For the C_1 regime, we evaluate the model with the data corresponding to $\alpha = 3.6$, $\epsilon' = 0.5$ and plot the results in Fig. 5a. For the C_2 regime, we evaluate the model with the data corresponding to $\alpha = 3.0$, $\epsilon' = 1.0$ and plot the results in Fig. 5b. The RMSE value changes while varying the number of units in the LSTM layers.

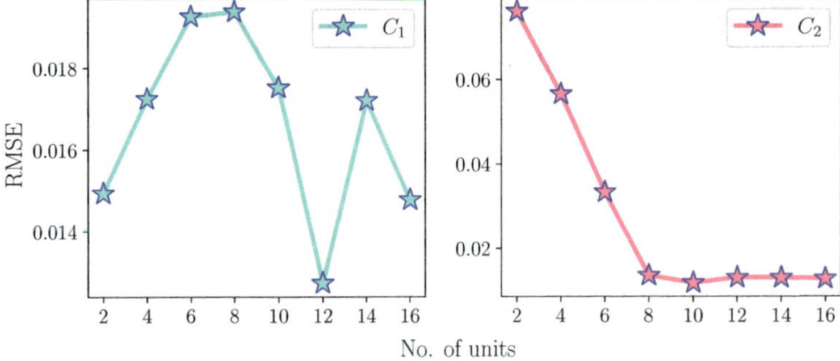

Fig. 5 RMSE values for various number of units in LSTM layers. **a**, **b** corresponds to the regimes C_1, $\alpha = 3.6$, $\epsilon' = 0.5$ and C_2, $\alpha = 3.0$, $\epsilon' = 1.0$ respectively

4.2 *Multi-step Forecasting*

Now, we consider the task of multi-step forecasting. To do this, while preparing the supervised learning data, instead of having only one future step value, we take more than one value at the output. For this we consider the data in both the regimes C_1 ($\alpha = 3.6$, $\epsilon' = 0.5$) and C_2 ($\alpha = 3.0$, $\epsilon' = 1.0$). The results of multi-step forecasting are shown in Fig. 6.

From this figure we can infer that in the forecasting of multi-steps two and three, the considered model outperformed our expectations in the prediction task, the plots have fewer scatter points and the RMSE values are in admissible range. But for steps

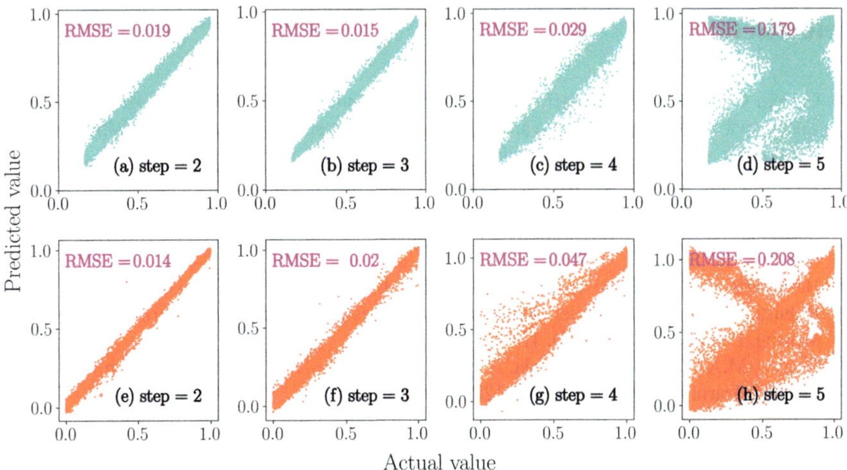

Fig. 6 Scatter plots with RMSE values for the multi-step forecasting. **a–d** correspond to C_1 regime $\alpha = 3.6$, $\epsilon' = 0.5$ and **e–f** correspond to C_2 regime $\alpha = 3.0$, $\epsilon' = 1.0$

four and five, the model failed to give accurate values in both the regimes. It can be seen from Fig. 6c, d, g, h, the points are scattered much when making forecasting with fourth and fifth steps.

5 Conclusion

In this work, we have considered the logistic map with quasiperiodic forcing. The system exhibits chaos in two different regimes. We employed a DL framework LSTM, for the prediction of two different chaos. For this, we have generated 10^5 data totally and used 6×10^4 data for training and the remaining 4×10^4 data for the purpose of testing. We forecast the chaos corresponding to the two regimes C_1 and C_2. The outcome of the experiments are evaluated using the performance metric RMSE value and they are analyzed through the scatter plots which have been plotted between the predicted value and actual value. Further, we have checked the effect of the number of units of the LSTM layers on the performance of the model. In this connection, we have done multi-step forecasting in order to predict more than one future value of the considered map. From the obtained results, we conclude that the developed LSTM framework can be used for forecasting the chaotic dynamics of the discrete system, namely quasiperiodically forced logistic map described by Eqs. (1a, 1b).

Acknowledgements JM thanks RUSA 2.0 project for providing a fellowship to carry out this work. MS acknowledges RUSA 2.0 project for providing financial support in procuring a high-performance GPU server which highly assisted this work.

References

1. Carleo, G., Cirac, I., Cranmer, K., Daudet, L., Schuld, M., Tishby, N., Vogt-Maranto, L., Zdeborová, L.: Machine learning and the physical sciences. Rev. Mod. Phys. **91**, 045002 (2019)
2. Choudhary, A., Lindner, J.F., Holliday, E.G., Miller, S.T., Sinha, S., Ditto, W.L.: Physics-enhanced neural networks learn order and chaos. Phys. Rev. E **101**, 062207 (2020)
3. Miller, S.T., Lindner, J.F., Choudhary, A., Sinha, S., Ditto, W.L.: The scaling of physics-informed machine learning with data and dimensions. Chaos Solitons Fractals: X **5**, 100046 (2020)
4. Barmparis, G.D., Neofotistos, G., Mattheakis, M., Hizanidis, J., Tsironis, G.P., Kaxiras, E.: Robust prediction of complex spatiotemporal states through machine learning with sparse sensing. Phys. Lett. A **384**(15), 126300 (2020)
5. Ganaie, M.A., Ghosh, S., Mendola, N., Tanveer, M., Jalan, S.: Identification of chimera using machine learning. Chaos **30**(6), 063128 (2020)
6. Pathak, J., Zhixin, L., Hunt, B.R., Girvan, M., Ott, E.: Using machine learning to replicate chaotic attractors and calculate lyapunov exponents from data. Chaos **27**(12), 121102 (2017)
7. Panday, A., Lee, W.S., Dutta, S., Jalan, S.: Machine learning assisted network classification from symbolic time-series. Chaos **31**(3), 031106 (2021)
8. Krishnagopal, S., Girvan, M., Ott, E., Hunt, B.R.: Separation of chaotic signals by reservoir computing. Chaos **30**(2), 023123 (2020)

9. Mukhopadhyay, S., Banerjee, S.: Learning dynamical systems in noise using convolutional neural networks. Chaos **30**(10), 103125 (2020)
10. Meiyazhagan, J., Sudharsan, S., Senthilvelan, M.: Model-free prediction of emergence of extreme events in a parametrically driven nonlinear dynamical system by deep learning. Europ. Phys. J. B **94**(8), 1–13, 100046 (2021)
11. Meiyazhagan, J., Sudharsan, S., Venkatasen, A., Senthilvelan, M.: Prediction of occurrence of extreme events using machine learning. Eur. Phys. J. Plus **137**,1–20 (2022)
12. Pyragas, V., Pyragas, K.: Using reservoir computer to predict and prevent extreme events. Phys. Lett. A **384**(24), 126591 (2020)
13. Ray, A., Chakraborty, T., Ghosh, D.: Optimized ensemble deep learning framework for scalable forecasting of dynamics containing extreme events. Chaos **31**(11), 111105 (2021)
14. Nag Chowdhury, S., Ray, A., Mishra, A., Ghosh, D.: Extreme events in globally coupled chaotic maps. J. Phys. Complex. **2**(3), 035021 (2021)
15. Asch, A., Brady, E., Gallardo, H., Hood, J., Chu, B., Farazmand, M.: Model-assisted deep learning of rare extreme events from partial observations. arXiv:2111.04857 (2021)
16. Lellep, M., Prexl, J., Linkmann, M., Eckhardt, B.: Using machine learning to predict extreme events in the hénon map. Chaos **30**(1), 013113 (2020)
17. Prasad, A., Mehra, V., Ramaswamy, R.: Strange nonchaotic attractors in the quasiperiodically forced logistic map. Phy. Rev. E **57**(2), 1576 (1998)
18. Heagy, J.F., Hammel, S.M.: The birth of strange nonchaotic attractors. Physica D: Nonl. Phenomena **70**(1–2), 140–153, 100046 (1994)
19. Al Shalabi, L., Shaaban, Z., Kasasbeh, B.: Data mining: a preprocessing engine. J. Comput. Sci. **2**(9), 735–739 (2006)
20. Rumelhart, D.E., Hinton, G.E., Williams, R.J.: Learning representations by back-propagating errors. Nature **323**(6088), 533–536 (1986)
21. Hochreiter, S., Schmidhuber, J.: Long short-term memory. Neural Comput. **9**(8), 1735–1780, 100046 (1997)
22. Goodfellow, I., Bengio, Y., Courville, A.: Deep Learning. MIT Press (2016)

Dynamic Calibration of a Stress-Wave Force Balance Using Hybrid Soft Computing Approach

Sima Nayak and Niranjan Sahoo

Abstract Aerodynamic vehicles come across the influence of impulsive forces and these are the major concerns associated with high-speed atmospheric vehicles. These shock wave induced impulsive forces impart hazardous effects on the surface of the vehicle. So, the magnitude of these forces is required for the design and modification of aerospace vehicles. Due to practical constraints, the real-time experiment is very difficult. Therefore, the ground-based test facilities are carried out using an aerodynamic model in shock tubes and shock tunnels. These models are required to be calibrated properly before carrying out the actual experiments. In the present study, a bi-cone model with a stress-wave force balance is used to perform the calibration task. The balance is mounted inside the model with strain gauge which records strain signal related to the applied force acting on the nose of bi-cone model. The strain signals of impulsive forces are captured for different magnitude and these signals are used for training and recovery of forces. Two different methods have been adopted for the recovery of the forces; one through classical de-convolution technique and another using the hybrid soft-computing approach, Adaptive neuro-fuzzy inference system (ANFIS). The forces recovered through both the techniques are compared with the known forces and also with each other. This provided an insight about the feasibility and applicability of the soft computing approach towards the inverse recovery of unknown forces for short duration experiments.

Keywords ANFIS · De-convolution · Short duration force recovery · Soft computing · Strain gauge

1 Introduction

In aerodynamic industry impulsive forces are the key concerns allied with hypersonic space vehicles like missiles, aircrafts, space shuttle, etc. The impulsive forces are resulted due to the shock waves generated in high-speed flow conditions. The impact

S. Nayak (✉) · N. Sahoo
Department of Mechanical Engineering, Indian Institute of Technology Guwahati, Guwahati, India
e-mail: n.sima@iitg.ac.in

S. Banerjee and A. Saha (eds.), *Nonlinear Dynamics and Applications*,
Springer Proceedings in Complexity,
https://doi.org/10.1007/978-3-030-99792-2_54

of such forces on the aerodynamic bodies causes serious damages to the structures causing huge losses to the aerospace programs. Pre-determination of the magnitude of such forces helps in the modification of the existing aero structural designs and also in the development of new designs. Nevertheless, such forces are highly transient and impulsive in nature that exists for a very small duration of time. Owing to the constraints associated with actual aerodynamic experiments, ground-based test facilities like shock tubes and shock tunnels are often employed to create such an impulsive environment where similar aero models are tested for impulsive forces [1, 2]. As a common practice, the aero models are tested in the test section of the shock tubes or shock tunnels where the high-speed flows having Mach numbers in the range of supersonic flow are imparted on the model surface. For the measurement purpose, the sensors are placed on any attachment mounted on the aero models where the effect of the forces can be felt by the attached structure [3]. These mounting structures are popularly known as stress-wave force balance which is judiciously designed to capture the appropriate response from the sensors [4, 5]. Two types of force balances are mostly used for calibration purposes; one is inertia-dominated and another is stiffness-based. The shock wave imparts an impulsive force on the body of the model where the magnitude in general is higher at the nose of the model. It may be noted here that, due to the short duration of action, direct measurement of the steady-state value of the force is difficult. Therefore, it is inversely predicted from strain gauge or accelerometer responses.

The response from the strain gauge or accelerometer is required to be post-processed to get the imparted force through inverse calculation. The convolution-deconvolution approach is popularly used for the recovery of such impulsive forces. For a single-component, linear system, with an applied load u(t) and a single-output signal y(t), a convolution integral relates between input forces and output responses through an impulse response function g(t) as presented by Eq. (1).

$$y(t) = \int_0^t g(t - \tau)\, u(\tau)\, d\tau \tag{1}$$

The solution of this equation is obtained in the time domain by discretizing the signals with a time step Δt and Eq. (1) is represented in the following form in which g(t) is obtained via calibration tests.

The researchers make use of different calibration strategies in order to evaluate g(t) for which the mathematical formulation relies on a linear relationship between the input and output responses. Therefore, this process is mathematically complex and cumbersome. However, with the advancement of the soft-computing approach, a simplified model can be adopted for the recovery of such forces. In fact, the soft computing approach is mathematically less intensive and can also provide accurate prediction results [6]. The soft computing approach is very popular among researchers nowadays due to its simplicity and versatility. Therefore, its use is found in various fields of engineering such as image recognition, control theory, power sectors, etc. [7, 8]. The most popular soft computing approaches are the neural network, fuzzy logic, genetic algorithms, etc. However, the hybrid method proved to be better in recent times. One such hybrid approach is the "adaptive neuro-fuzzy inference system" (ANFIS) which is a combination of fuzzy logic and artificial neural network (ANN). It is an assembly of Takagi–Sugeno type fuzzy inference system embedded in the network of ANN. Moreover, it is a data-driven approach driven by the ANN with a fuzzy system acting as a front-end pre-processor for the input–output task [9, 10].

Most of the researchers have adopted the de-convolution approach for the inverse prediction of impulsive forces whereas, a very limited use of the ANFIS technique is observed in the open literature. Therefore, the objective of the present work is to implement and test the usage of the ANFIS technique for the recovery of impulsive forces in the case of short-duration aerodynamic experiments. To accomplish the objective one aero model (blunt bi-cone model) has been fabricated (DASA CTV) in-house and a calibration task has been performed using an impulse hammer and strain gauges. A suitable stiffness-based force balance has been fabricated and the location of maximum strain has been identified through numerical simulation using the finite element method (ANSYS Workbench 18.0). The strain gauge is mounted at the location of maximum strain corresponding to the axial force. The strain data is recorded for known impulsive forces (applied using the impulse hammer) of different magnitude. These "time-strain-force" signals are used to train the ANFIS system and the force is predicted for time-strain input using the trained ANFIS system and compared with the known forces as well as with the forces recovered through deconvolution technique. The detailed working procedure has been elaborated in the Fig. 1.

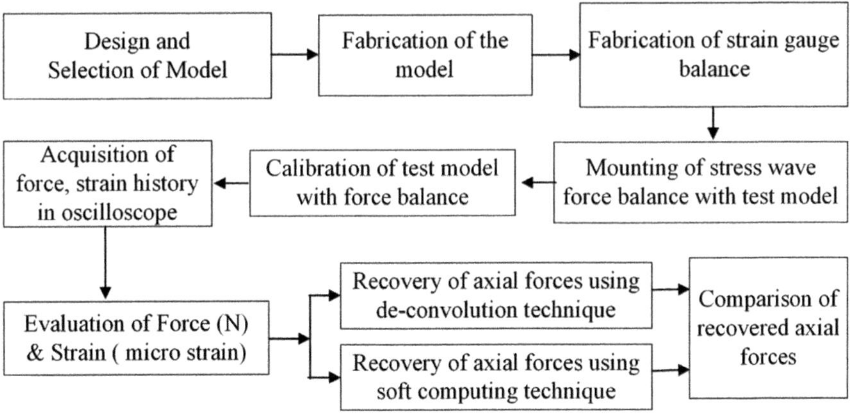

Fig. 1 Layout of the proposed work

2 Calibration Experiment

To check the effectiveness of the ANFIS system for the prediction of impulsive forces during short duration experiments is the aim of the present work. To train the ANFIS system there is a requirement of strain and force signals. Therefore, a calibration task has been planned by mounting stiffness-based force balance on the fabricated model [11]. Typically, force measurement using any force balance (Fig. 2) includes various considerations such as; the design of force balance for specific sensors, their mounting structure, choice of sensing location, balance calibration and force recovery [12–15]. Stiffness dominated force balance is generally used for measurement of forces through the measurement of strain signal using semiconductor strain gauges. In this category of force balance, model is assumed to behave as a rigid body. When the model is subjected to an impulsive force (using impulse hammer), it imparts a strain signal corresponding to the applied force where, the response is captured using a strain gauge module and data acquisition system. The strain gauge is mounted at the specified axial location (Fig. 2) on the force balance and connected to a Wheatstone bridge circuit. The balance is mounted on a fixed support using a bench vice. The strain signal is captured by an Oscilloscope (Model: MDO3024, Tektronix, USA, 2.5GS/s) corresponding to the impulsive force applied by the impulse hammer. These known forces along with their corresponding strain signals are used to establish a relation between the strain and force so that the unknown forces can be inversely estimated using the strain signal. In the present studies, the calibration has been done at the nose of the test model with the help of the impulse hammer.

The direction of the impulse hit should be normal to the hitting surface of the test model. Strain signals are recorded in the oscilloscope during calibration experiments for different impulsive forces. Typical signals of the impulsive force and strain responses for the axial direction are shown in Fig. 3. The initial strain values show the values in the negative direction indicating a compression strain. This is expected

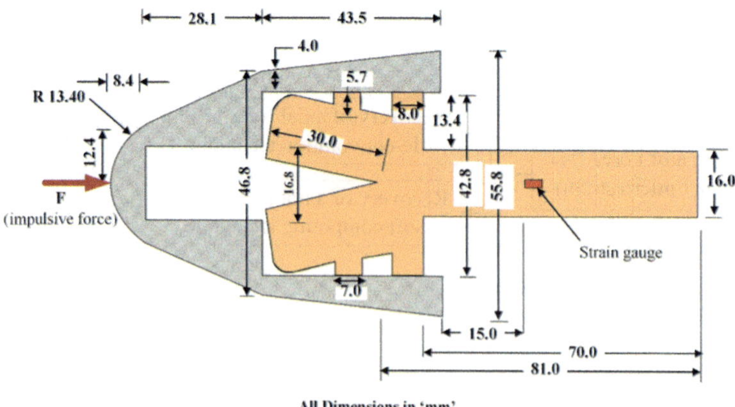

Fig. 2 Sectional view of the test model integrated with force balance

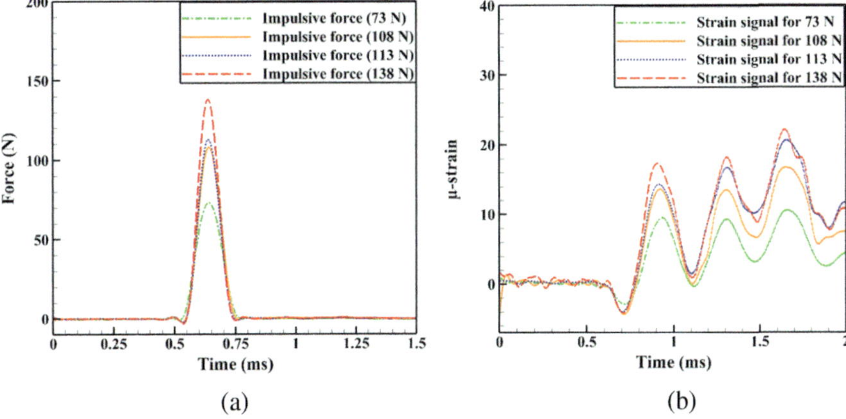

Fig. 3 Force and strain signals from the calibration **a** Impulsive force from the impulse hammer **b** Strain responses from the strain gauge

as the balance is mounted on support. The known set (4 sets) of "time-strain-force" signals are used to train the ANFIS system and the other forces are recovered and compared with the known data. Similarly, the strain signals are used to predict the forces through de-convolution technique.

3 Recovery of Forces Using ANFIS Method

The recovery of impulsive forces using the soft computing-based ANFIS approach is the central theme of the present work. The strain signals obtained from the calibration experiment corresponding to four different (73 N, 108 N, 113 N, and 138 N) impulsive loads are used for the training and recovery of force through the ANFIS method. As inferred from the literature, the prediction through ANFIS technique is better when the recovered datasets are within the training range. Therefore, for the present case, 108 N and 113 N forces are recovered individually, after training the other three data sets. As mentioned previously, the ANFIS is a data-driven method dealing with input–output mapping. It is a multi-input single-output system (MISO) consisting of one input layer, four hidden layers and one output layer [16]. The ANFIS architecture is shown in Fig. 4. One of the standard modules of MATLAB is equipped with ANFIS architecture is used for the training and testing purpose. It works with hybrid and steepest descent algorithm providing a better output. As mentioned previously ANFIS is a combination of neural network and fuzzy logic; their combined effect help in better training and recovery of the impulsive force [17]. Therefore, the "time-strain-force" combinations for all four sets are used for this purpose. The system is trained with three sets of data excluding the set of data to be recovered. Successful implementation of ANFIS for the present exercise needs a proper understanding of

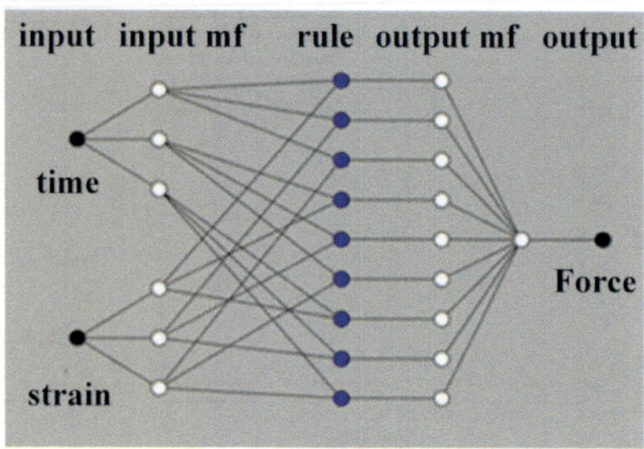

Fig. 4 ANFIS architecture for optimized combinations

the process and algorithm to be followed. All such steps require different user inputs to achieve proper training and recovery [18].

The ANFIS training system starts with the fuzzification of the training data sets. The fuzzy logic part of the hybrid approach helps in the fuzzification of the data sets. The parametric variations for fuzzification depend on the type of membership functions (MFs) to normalize the input training data set [19]. The available options for these functions are from the broad categories as piecewise linear functions, Gaussian functions, bell-shaped functions, and sigmoid functions. Therefore, the standard input MFs used are triangular MFs, trapezoidal MFs, bell-shaped MFs, sigmoid MFs, Gauss MFs, singleton MF, pi MF, etc. Similarly, there are two types of output MFs viz. "linear" and "constant". Out of the different MFs, the selection of a suitable one depends on the type of data and the user experience. With the help of MFs, the fuzzy system can be tuned by adopting an optimization technique out of the available ones as "backpropagation method" and "hybrid method" such that the input–output modelling can be carried out. The backpropagation method adopts the steepest descent algorithm for modifying all parameters during training whereas, the hybrid model uses the least square method for parameters linked to the output MFs and backpropagation for the parameters linked to the input MFs [20]. Since various options are available and the best one needs to be selected, parametric studies are performed to find out the input requirements of ANFIS, and its capability is judged for the force recovery. According to the evidence from literature, the grid partition type of clustering technique is suitable for problem definitions having a smaller number of input variables (less than 6). After few iterations with the available parameters, it is observed that a combination of "three linguistics variables", "gauss-gauss2 IMF combinations", constant output MF, and hybrid optimization technique yields a better recovery of the data set due to less root mean square error (RMSE). Therefore, all the results are obtained using these optimized parameters which are discussed in the next section.

4 Recovery of Forces Using De-convolution Technique

As mentioned previously, the classical way of inverse prediction of the forces is using the convolution-deconvolution approach. Here, the time domain is converted to the frequency domain and the response function is generated which acts as the linkage parameter and remains nearly constant. For the present case, two intermediate forces are recovered to compare with the ANFIS. The 108 N and 113 N forces are recovered by using the time strain signal corresponding to the other three forces individually. A Mat Lab based programming is used to recover the unknown forces and compared with the known value. All these results are compared with the ANFIS results.

5 Results and Discussion

The accuracy of the force prediction on applying impulse loads in axial directions on nose of model is analyzed using ANFIS method as well as Deconvolution method. The impulsive forces in axial direction of the model along with the corresponding strain signals are filtered using a low pass filter of 12.5 kHz to remove the noise. The system is trained with three sets of data leaving the one to be recovered. Out of four different magnitudes of forces (73 N, 108 N, 113 N, and 138 N), 108 N force is recovered after training the ANFIS with 73 N, 113 N and 138 N force. Similarly, 113 N force is recovered after training the ANFIS with 73 N, 108 N and 138 N force.

The peak magnitudes of these forces have been compared with the original signals and the percentage of deviation is calculated for all the cases. The ANFIS is able to predict the unknown forces with a certain degree of accuracy for the different data sets. An error of 2.87% and 3.54% are obtained for 108 N and 113 N force, respectively when compared with the actual signal (Table.1). Similarly, the intermediate forces (108 N and 113 N) are recovered individually, through deconvolution technique by

Table 1 Error calculation between the recovered and the input value

Categories of forces: Actual Value = 108 N	Force (N)	Error (%) $\left\lvert \frac{F_{actual} - F_{recovered}}{F_{actual}} \right\rvert \times 100$	Categories of forces: Actual value = 113 N	Force (N)	Error (%) $\left\lvert \frac{F_{actual} - F_{recovered}}{F_{actual}} \right\rvert \times 100$
Deconvolution (DC-1)	111.9	3.6	Deconvolution (DC-4)	108.4	4.77
Deconvolution (DC-2)	112.6	4.25	Deconvolution (DC-5)	120.7	6.81
Deconvolution (DC-3)	114.0	5.55	Deconvolution (DC-6)	122.6	8.49
ANFIS recovered	104.9	2.87	ANFIS recovered	109.0	3.54

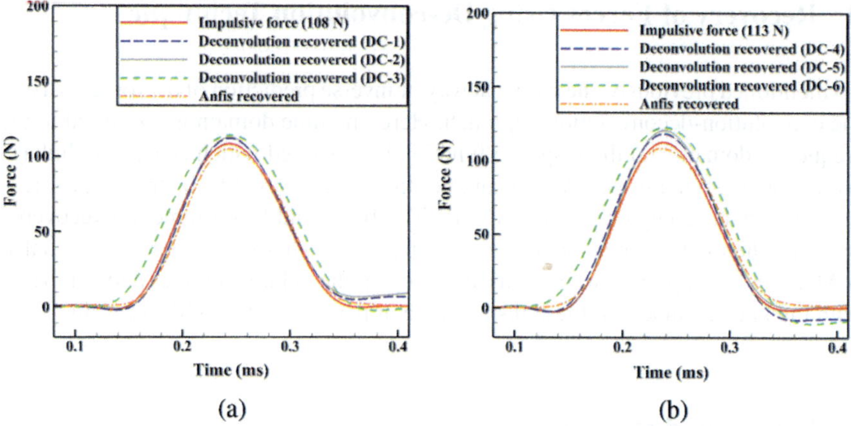

Fig. 5 Recovered signals using ANFIS and Deconvolution technique; (a) 108 N; (b)113 N

obtaining the response function from other three forces. The 108 N is recovered with the help of 113 N (DC-1), 138 N (DC-2) and 73 N (DC-3). Similarly, 113 N is recovered with the help of 108 N (DC-4), 138 N (DC-5) and 73 N (DC-6).

All the results obtained for 108 N and 113 N are compared with the original signal as well as with the ANFIS and deconvolution recovered results (Fig. 5). For 108 N recovery, a deviation of 2.87% is observed amongst ANFIS predicted results as compared with the original results and a maximum deviation of 5.55% with the results obtained through deconvolution technique. Similarly, for 113 N recovery, a deviation of 3.54% is observed amongst ANFIS predicted results as compared with the original result and a maximum deviation of 8.49% with the results obtained through deconvolution technique. The recovery signal trend is matching well with the original results. Therefore, the results obtained through ANFIS recovery matches well within the uncertainty band.

6 Conclusion

The stress wave force balance and bi-cone model have been successfully fabricated and by using a semiconductor strain gauge module, the calibration experiment has been performed. The strain signals corresponding to four sets of impulsive forces (applied with the help of an impulse hammer) is recorded with the help of an oscilloscope and the results are filtered using a low pass filter at a frequency of 12.5 kHz. These "time-strain-force" signals are used to train and recover the forces through an ANFIS based soft computing approach. The system is trained with three sets of data leaving the data to be recovered. All the known data sets are recovered and their peak magnitudes are compared with the actual ones. It is observed that the ANFIS is able to predict the forces within a certain range of accuracy. Similarly, the forces are also recovered using deconvolution technique and the results are compared with the ANFIS recovered results as well with the original ones. The ANFIS results are found obeying the signal trend pretty well. Compared to the original value, the least error of peak magnitude value is less in case of ANFIS recovered results (2.87%) as compared to the de-convolution results (3.6%). However, the maximum error of 3.54% is observed in case of ANFIS as compared to 8.49% in deconvolution results. Therefore, the ANFIS method can be used for short duration force recovery well within the training range where the prediction can fall in the acceptable error range.

References

1. Mee, D.J.: Dynamic calibration of force balances for impulse hypersonic facilities. Shock Waves **12**(6), 443–455 (2003)
2. Sahoo, N., Reddy, K.P.J.: Force measurement techniques for hypersonic flows in shock tunnels. Int. J. Hypersonics **1**(1), 31–58 (2010)
3. Naumann, K.W., Ende, H., Mathieu, G.: Technique for aerodynamic force measurement within milliseconds in shock tunnel. Shock Waves **1**(3), 223–232 (1991)
4. Abdel-Jawad, M.M., Mee, D.J., Morgan, R.G.: New calibration technique for multiple-component stress wave force balances. Rev. Sci. Instrum. **78**(6) (2007)
5. Tuttle, S.L., Mee, D.J., Simmons, J.M.: Drag measurements at Mach 5 using a stress wave force balance. Exp. Fluids **19**(5), 336–341 (1995)
6. Nanda, S.R., Kulkarni, V., Sahoo, N., Menezes, V.: An innovative approach for prediction of aerodynamic coefficients in shock tunnel testing with soft computing techniques. Meas. J. Int. Meas. Confed. **134**, 773–780 (2019)
7. Rout, A.K., Nanda, S.R., Sahoo, N., Kalita, P., Kulkarni, V.: Soft computing—a way ahead to recover heat flux for short duration experiments. J. Therm. Sci. Eng. Appl. **14**(3), 1–11 (2022)
8. Zhu, F., Wu, Y.: A rapid structural damage detection method using integrated ANFIS and interval modeling technique. Appl. Soft Comput. **25**, 473–484 (2014)
9. Azari, A., Poursina, M., Poursina, D.: Radial forging force prediction through MR, ANN, and ANFIS models. Neural Comput. Appl. **25**(3–4), 849–858 (2014)
10. Kumar Rout, A., Ranjan Nanda, S., Sahoo, N., Kalita, P., Kulkarni, V.: Implementation of soft computing technique for recovery of impulsive heat loads. J. Thermophys. Heat Transf. 1, 1–10 (2021)
11. Abdel-Jawad, M.M., Mee, D.J., Morgan, R.G.: New calibration technique for multiple-component stress wave force balances. Rev. Sci. Instrum. **78**(6), 1–7 (2007)
12. Deka, S., Kamal, A., Pallekonda, R.B., Rahang, M., Kulkarni, V.: Measurement technique for ideal selection of sensors and accurate force recovery on aerodynamic models. Exp. Tech. (2021)
13. Nanda, S.R., Kulkarni, V., Sahoo, N., Menezes, V.: A comparison of accelerometer and piezofilm-based force balances for hypersonic shock tunnels. Proc. Inst. Mech. Eng. Part G J. Aerosp. Eng. **233**(14), 5310–5320 (2019)
14. Wang, Y., Liu, Y., Luo, C., Jiang, Z.: Force measurement using strain-gauge balance in a shock tunnel with long test duration. Rev. Sci. Instrum. **87**(5) (2016)
15. Deka, S., Pallekonda, R.B., Rahang, M.: Comparative assessment of modified deconvolution and neuro-fuzzy technique for force prediction using an accelerometer balance system. Meas. J. Int. Meas. Confed. **171** (2019)
16. Jang, J.S.R.: ANFIS: adaptive-network-based fuzzy inference system. IEEE Trans. Syst. Man Cybern. **23**(3), 665–685 (1993)

17. Ramesh, P., Nanda, S.R., Kulkarni, V., Dwivedy, S.K.: Application of neural-networks and neuro-fuzzy systems for the prediction of short-duration forces acting on the blunt bodies. Soft Comput. **23**(14), 5725–5738 (2019)
18. Nanda, S.R., Kulkarni, V., Sahoo, N., Menezes, V.: Sensitivity studies of ANFIS based force recovery technique towards prediction of aerodynamic load. Flow Meas. Instrum. **80**, 101969 (2021)
19. Nanda, S.R., Kulkarni, V., Sahoo, N.: Design of artificial neuro-fuzzy based methodology for six component force balance. Procedia Eng. **144**, 528–536 (2016)
20. Pratihar, D.K.: Soft computing: fundamentals and applications. Alpha Sci. Int. Ltd (2013)

Environment-Friendly Smart City Solution with IoT Application

Ayush Kumar, Saket Kumar Jha, and Jitendra Singh Tamang

Abstract In the proposed system, a whole city will be spectated and observed. Since a Smart City is considered, it is likely to be a metropolitan city so that basic necessities like transportation, internet, telecommunication, etc. are presumed to be available. The proposed system will be equally expensive for both brownfield and Greenfield projects of Smart Cities. With the advancing times, IoT technology is getting familiar to every section of society and by the virtue of advent of the 5G system, the IoT appears to be the primus inter pares for its application in the futuristic scenario. The sensor arrays and input/output devices are being used in different sections of the devices to perform the required operation. The introduction of solar powered independent systems proved to be an autonomous robust technology. The low latency communication between node clusters can be a significant factor for the development and transformation of smart cities in a modern fashion that is the root basis for the introduction of futuristic models.

Keywords IoT · 5G · Sensor array · Solar powered · Low latency · Node clusters · Smart city · Futuristic model

1 Introduction

In 1998 the term smart city erupted [1] and has gained its popularity in recent years. This new generation holds youth of technical era; every sector of society is shifting from the product centric pattern to the service-oriented pattern to follow up with latest trends of Internet of Things. Now, Societies ponder Smart cities as an ecosystem of infrastructure and services due to its all-rounder essence with context of its reasonable facilities [2]. In attempt to provide these services globally, one must have a holistic view of every smart city problem and its comprehensive solution. The IoT and 5G combination enable a unified ecosystem for wide range of

A. Kumar · S. Kumar Jha (✉) · J. Singh Tamang
Department of Electronics and Communication Engineering, Sikkim Manipal Institute of Technology, Sikkim Manipal University, Majitar 737136, Sikkim, India
e-mail: saketjha55288@gmail.com

S. Banerjee and A. Saha (eds.), *Nonlinear Dynamics and Applications*,
Springer Proceedings in Complexity,
https://doi.org/10.1007/978-3-030-99792-2_56

applications in metropolitan vicinity, with the benefit that authorities can operate the e-administration and services for the means to produce inexpensive and resilient smart city infrastructure. In addition, for a holistic smart city development, a centralized approach for the processing of large amounts of data flow in the same network infrastructure is needed for data processing that can be further used for providing the instruction to the actuators using edge computing paradigm, [3] which pushes computation and data processing away from the core datacenters to the outer edge of networks that are closer to the data sources. The main benefit of this work is it makes the process faster in computation, processing of data and increases the ability as well as scalability. The IoT era means a whole new world of applications and services. This includes the Smart City surveillance and evaluation where a set of smart sensors and IoT devices monitors everyday smart city activities and helps in forecasting, reducing energy consumption, keeping the administration updated about big and small incidents of vicinity with centralized management server. It is a robust technology for the futuristic management which just not only insures real-time service delivery potential but simultaneously prevents the global warming and facilitates the residents with latest infrastructure. Perfect mixtures of small and big technologies are preferable so that innovation possibilities could cope up with future advancements. Currently, widely used platforms that are addressing up with our needs are Radiofrequency identification (RFID), Bluetooth Low-Energy (BLE), Near Field Communication (NFC) and Fourth Generation of cellular systems (4G) and for the long range of applications, the Low Power Wide Area (LPWA) has its superset LoRa WAN protocol and the future cellular IoT [4]. Smart City applications potentially benefit from IoT technologies; particularly they take advantage of the diversified and reactive character of IoT infrastructure. Others have addressed smart city applications like typical grid computing, using cloud services as a measure to work around the scope and complexity. This has led to extreme utilization of multiple parallel processing methodologies such as general data transmission or data sharing that have been long used to develop Distributed Systems applications, [5] where it was observed as a primitive coordination model and illustrates a more advanced version of the dataflow model. Data management is a component in the IoT for enabling the smart cities infrastructure which consists of data acquisition, processing and dissemination [6]. Data standards, quality, and utilization are the different aspects of data acquisition. When 5G is used in conjunction with its advanced characteristics, it enables an increase in the number of peripherals, acquires optimum bandwidth for uplink and downlink, ultra—low latency and supports energy-efficient data transmission [7]. Smart cities are built on the experiences and success of decades [8] which resembles a nation's innovation capabilities and success approaching determination [9]. There is no doubt that future of administration lies in the lap of IoT and corresponding steps by the administration is necessary to deal the future [10].

Conflict arises when existing IoTs are being compared with futuristic 5G IoT system. However, 5G beats the existing technology in all manners except for capital. These 5G antennas have very narrow coverage length and hence require huge physical infrastructure investment for its successful implementation. Moreover, it is found out to be a one-time investment as it balances the chart with its efficient and worthwhile returns in account of futuristic demands.

In this paper we will talk about our approach of designing a networking system in smart city and will be talking about different layers of our model in detail. The model will explain about methodology of fetching the data from physical layer to processing and managing the data in the servers and its utilization. This system will be managing environmental parameters primarily, fetched from different parts of region along with some other data sets and will transmit details to the centers. This will not only help in getting to know about the real-time local conditions but will also help administration to identify the relatively polluted regions and work as needed. The data acquitted can be used for various statistical purposes in future.

2 Smart City Solution (Model Approach)

The concept entails splitting the city into 3–4 specified regions, each with its own Base Station. A Central Hub must be created in the heart of the city for cohesive strategic planning and as a location for preserving maintenance data. The base must be separated into distinct network stations in order to have practical non-erroneous data analysis capabilities that can handle ground reports and maintain the databases up to date with real-time environment parameter reports. Such characteristics are to be harvested from node clusters generated by taking into account certain nodes from specified locations.

The nodes will be having 5G connectivity from which the communication with peer nodes and network stations are to be established. From Fig. 1, the sensed data are being recorded by nodal database and then transmitted using 5G protocol.

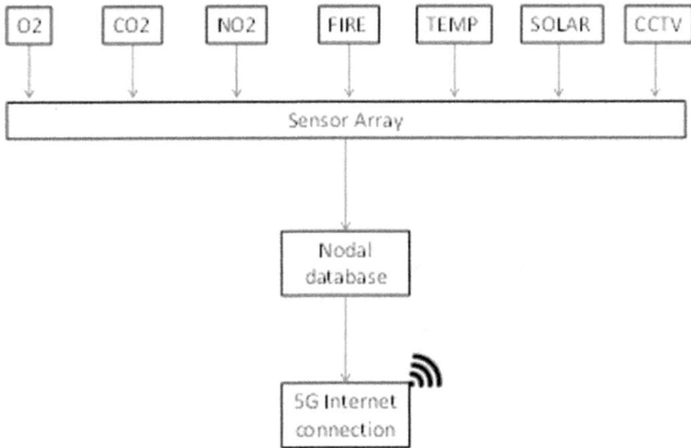

Fig. 1 Basic structure of a node

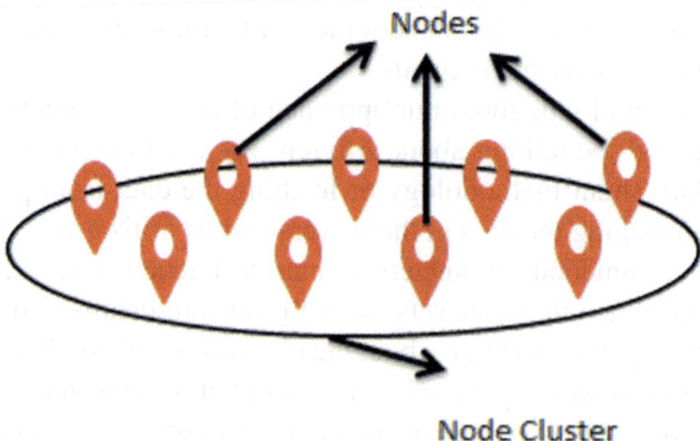

Fig. 2 A node cluster

A cluster of nodes will consist of 8–10 nodes in a region and will be required for certain locations in order to centralize infrastructure that can be used as unified eco-space for better and more efficient human surveillance in a confined region. This is necessary to avoid machine malfunctions and to ensure that the installed infrastructure is properly maintained. The node cluster is seen in Fig. 2. Clusters are described as communication between peer nodes in defined areas, despite the fact that no specific node is necessary to bind nodes to create a cluster. The importance of constructing a cluster is that network stations can observe an average and accurate information about a region's climatic component. However for very local and root level details of small area data obtained from nodes are best suited.

The Network Stations will have access to the data collected by the nodes. The Network Stations will keep an eye on various nodes and may be regarded a place where actual human beings are sitting and observing data. Any unexpected findings from the node clusters must first be validated by a person before being sent to the Base Station who oversees the numerous network stations. The Base Station will continue to supervise the Network Stations' appropriate operation and administration. Network Stations authority would be able to make critical decisions at the local level to avert public disruption and improve service performance. If Network Stations detect a fire or a water leak through sensors, police may check the situation using a CCTV camera and then quickly make remedial arrangements for the risky situation. Furthermore, Network Stations may connect with local police stations to take required steps in emergency situations, such as evacuation of regions, etc.

The Node Clusters, Network Stations, Base Stations, and Central Hub are aligned hieratically in Fig. 3 to build a systematic smart city model, with the Central Hub having to serve as the top authority with administration, execution, and crucial decision-making powers. Multiple base stations will be linked to the central hub. Base Stations are subservient to the Central Hub, which monitors Network Station activities, supervises their coordination, and double-checks the smooth functioning and decision-making. It also maintains the link between the Central Hub and the Network Stations. Network stations on the other hand are to be seen as local

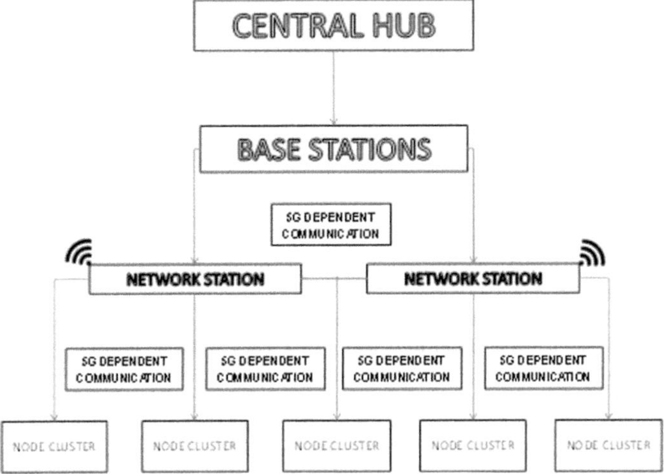

Fig. 3 5G Structured communication model

level smart city authority which will be monitoring all the Node Clusters under its boundaries and will be keeping a watch on instances being observed by the nodes. In case of non-urgent but important decision needed, it should be relayed to the Central Hub via Base Stations so that it can be carried out correctly and methodically.

3 Future Scope and Conclusion

The optimum use of 5G technology occurs when it is combined with Internet of Things. It is very remarkable to witness data being delivered in real-time. Environmental characteristics from a confined area under observation are being acquired using sensors placed at each node having 5G connectivity. The benefit of adopting 5G communications technology is the ability to communicate with peer nodes and Network Stations with minimal latency. One may speculate that the sensors employed can only perceive a restricted area, but this goes in hand with 5G technology, which offers more bandwidth while communicating in a limited area since bandwidth and distance of a channel are inversely related to each other. As a result, bringing nodes closer together would result in perfect synchronization with 5G low-latency communication and viable surveillance. Each node will have its own local data acquisition system, allowing data to be retained locally for future reference.

Henceforth 5G technology when combined with IoT for noble purpose like Environment monitoring as a pollution control measure can prove to be a boon for society. This will not just keep us updated about our environment parameter real-time but can also help government to take steps in maintaining pollution levels on different parts of city. Also, this will enable us to get information of our very local environment parameters as well important and urgent activities on our fingertips through mobile phone. Thus, this paper gives a structural overview to a method for working and maintenance of a noble purpose like Smart city environment monitoring.

References

1. Van Bastelaer, B.: Digital cities and transferability of results. In: 4th EDC Conference on Digital Cities, Salzburg, pp. 61–70 (1998)
2. Bruneo, D., Distefano, S., Giacobbe, M., Minnolo, A.L., Longo, F., Merlino, G., Mulfari, D., et al.: An IoT service ecosystem for smart cities: The# smartme project. Internet Things **5**, 12–33 (2019)
3. Cicirelli, F., Guerrieri, A., Spezzano, G., Vinci, A.: An edge-based platform for dynamic smart city applications. Future Gener. Comput. Syst. **76**, 106–118 (2017)
4. Goudos, S.K., Dallas, P.I., Chatziefthymiou, S., Kyriazakos, S.: A survey of IoT key enabling and future technologies: 5G, mobile IoT, sematic web and applications. Wirel. Pers. Commun. **97**(2), 1645–1675 (2017)
5. Giang, N.K., Lea, R., Blackstock, M., Leung, V.C.: On building smart city IoT applications: a coordination-based perspective. In: Proceedings of the 2nd International Workshop on Smart, pp. 1–6 (2016)
6. Gharaibeh, A., Salahuddin, M.A., Hussini, S.J., Khreishah, A., Khalil, I., Guizani, M., Al-Fuqaha, A.: Smart cities: a survey on data management, security, and enabling technologies. IEEE Commun. Surv. Tutor. **19**(4), 2456–2501 (2017)
7. Rao, S.K., Prasad, R.: Impact of 5G technologies on smart city implementation. Wirel. Pers. Commun. **100**(1), 161–176 (2018)
8. Repko, J., DeBroux, S.: Smart cities literature review and analysis. IMT 598 Spring 2012 Emerg. Trends Inf. Technol. (2012)
9. Nam, T., Pardo, T.A.: Smart city as urban innovation: focusing on management, policy, and context. In: Proceedings of the 5th International Conference on Theory and Practice of Electronic Governance, pp. 185–194 (2011)
10. Kumar, S., Tiwari, P., Zymbler, M.: Internet of Things is a revolutionary approach for future technology enhancement: a review. J. Big Data **6**, 111 (2019)

Parametric Optimization of WEDM Process on Nanostructured Hard Facing Alloy Applying Metaheuristic Algorithm

Abhijit Saha, Pritam Pain, and Goutam Kumar Bose

Abstract Wire Electric Discharge Machining (WEDM) is broadly utilized for manufacturing geometrically intricate and hard material parts. Taguchi orthogonal array using L25 is applied as design of experiment for the five input process parameters. Artificial Neural Network is utilized for testing and validation of the experimental data. This is trailed by employing a multi-objective optimization through Genetic Algorithm (GA) approach. As the outcomes got through GA infer a set of possible solutions. Grey Relation Analysis (GRA) is applied theory to find out the best parametric amalgamation among the arrangement of practical other options. In this case the weights are considered through Fuzzy set. Lastly the optimized results are obtained with regards to discharge pulse time (0.5 μs), discharge stop time (12 μs), servo voltage (36 V), wire tension (501gm), wire feed rate (8 m/min) correspondingly.

Keywords Artificial Neural Network (ANN) · Multi objective Genetic Algorithm (M-GA) · Grey relational analysis · Wire Electric Discharge Machining (WEDM) · Orthogonal array

1 Introduction

The capability to fabricate coatings is significant to manufacturing blueprint. Numerous segments need utilization of extra layers to enhance the mechanical properties and secure against hazardous environments. Nanostructured hard facing alloy are extensively used in various die and tool making industries. Machining of these materials is rather challenging if higher machined quality is warranted. Wire electric discharge machining (WEDM) process has evolved as a promising cutting strategy for cutting such hard to cut conductive materials.

A. Saha (✉) · P. Pain · G. Kumar Bose
Department of Mechanical Engineering, Haldia Institute of Technology, Haldia 721657, West Bengal, India
e-mail: saha.abhijit1798@gmail.com

S. Banerjee and A. Saha (eds.), *Nonlinear Dynamics and Applications*, Springer Proceedings in Complexity,
https://doi.org/10.1007/978-3-030-99792-2_57

In WEDM process, the workpiece material is taken out by a progression of series of distinct electrical spark between the anode workpiece and cathode wire. Due to the presence of dielectric fluid those sparks are extremely focused and due to this the temperature of that local area increases rapidly. That temperature is high enough to melt and instantly vaporize the workpiece material.

Based on the past research work by several researchers a brief literature review is presented here. Somashekhar et al. [1] employed ANN for modeling and optimize the input process parameters in micro-electric discharge machining. Ming et al. [2] also used ANN to predict cutting parameters in WEDM.

Genetic algorithms depend on procedure of natural selection and natural heredity qualities, which are supplementary strong and bound to find the universal best possible combinations. A few uses of GA-based method in nontraditional process optimization issues have been accounted in the literature. Kuriakose and Shunmugam [3] developed multiple regression models to represent association among WEDM input parameters and output responses. GA was used to obtain a most favorable amalgamation of process parameters. Mahapatra and Patnaik [4] employed genetic algorithm technique to optimize the WEDM process parameters. Prasad and Krishna [5] used GA to optimize the cutting speed and surface roughness in WEDM process. Pasam et al. [6] also evaluated the control parameters of WEDM titanium alloy (Ti6Al4V) using GA.

Moreover, Kuruvila and Ravindra [7] applied Taguchi's technique and a genetic algorithm to resolve parametric influence and optimal process parameters in WEDM process. Kumar and Agarwal [8] also used a GA to optimize the machining setting in WEDM. Zhang et al. [9] proposed hybrid method of RSM and GA for multi-criteria optimization of WEDM process parameters. Padhi et al. [10] also employed non dominating sorting GA technique to optimize the multiple responses of WEDM process. Sharma et al. [11] developed mathematical model using RSM and optimization has been carried using GA. Varun and Venkaiah [12] applied grey relational analysis coupled with GA based hybrid technique to simultaneously optimize the process responses in WEDM. Nair et al. [13] used GRA and GA to optimize the multi-performance characteristics in WEDM of Inconel 617. Shandilya et al. [14] applied various techniques viz. teaching–learning-based optimization, grey relational analysis and genetic algorithm for parametric optimization of WEDM process parameters. Various specialists have concentrated on the impacts of ideal machining boundaries on exhibitions in WEDM process utilizing a variety of optimization techniques [15–17].

Additionally, it is found however, from the machining literature that a not many researchers [18–20] assessed the best amalgamation of different input process parameters on finishing attributes in WEDM for machining nanostructured hardfacing material. Very little investigation has been accessible on optimization of WEDM condition for nanostructured hard facing alloy using multi-objective GA. It is essential to explore the quality characteristics of any finished component in order to illustrate its performance. In order to find out the best parametric blend of GA results a Multi-Criteria Decision Making (MCDM) technique which is commonly known as GRA method has been employed.

2 Experimental Setup

The current experiments were done on 5 axes CNC Wire Electric Discharge Machining. The figure of the machining system is shown in the Fig. 1.

To reduce the experiments machining were carried out following L_{25} Orthogonal Array taking five important input parameters like discharge pulse time (T_{on}), discharge stop time (T_{off}), servo voltage (SV), wire tension (WT) and wire feed rate (WF) [16]. The factors and their levels as considered for testing is shown in the Table 1.

After the experiments average surface roughness (Ra) of all machined surface were measured by using Taylor Hobson's Talysurf. The MRR is calculated by measuring the area of the work surface removed by time. The machining time is calculated by using stopwatch.

Fig. 1 WEDM machine

Table 1 Machining inputs with their ranges [16]

Machining parameter	Unit	Range
T_{on}	μs	0.3–0.5
T_{off}	μs	8–12
SV	V	35–47
WT	gm	500–900
WF	m/min	5–9

3 Results and Discussions

Artificial Neural Network (ANN) has been applied for identifying and learning the link between the input statistics and corresponding responses. The outcomes can be predicted once the training, validation and testing of the trial data is over by applying ANN. Figure 2 shows the simulink model for ANN.

3.1 ANN Analysis

For training 80% of the data have been considered, 10% of the data have been considered for validation and for testing 10% of data are considered during ANN analysis of MRR. MSE achieved after 6 iterations and it was ended. Figures 3 and 4 show regression and performance plot of the MRR.

Here in Fig. 2 shows, training R = 0.99854, validation R = 0.90286 and testing R = 0.96343. Thus overall value of the R is 0.91931. Hence, the training data suggests a noble fit because the validation and test results showing the values of R is greater than 0.9. Similarly Fig. 5a and 5b illustrate the final performance plot for machining time and Ra respectively.

3.2 Multi-objective Optimization of Responses

In this current examination, an effort has been made to choose the ideal estimations of input process parameters to get the main quality outputs inside the boundaries ranges. A practical advancement system, GA is created to deal with the streamlining issue for this assessment. In GA the numerical condition is outlined to give as the input parameter. As the outcomes acquired through GA suggest a domain of solutions, for that reason where the weight of the control parameters are calculated via Fuzzy

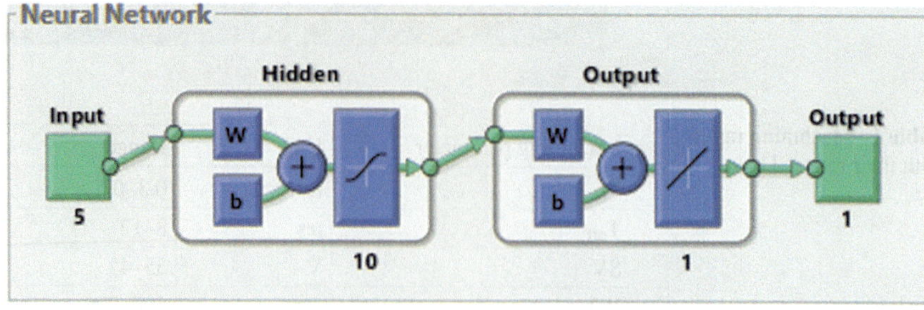

Fig. 2 Simulink diagram

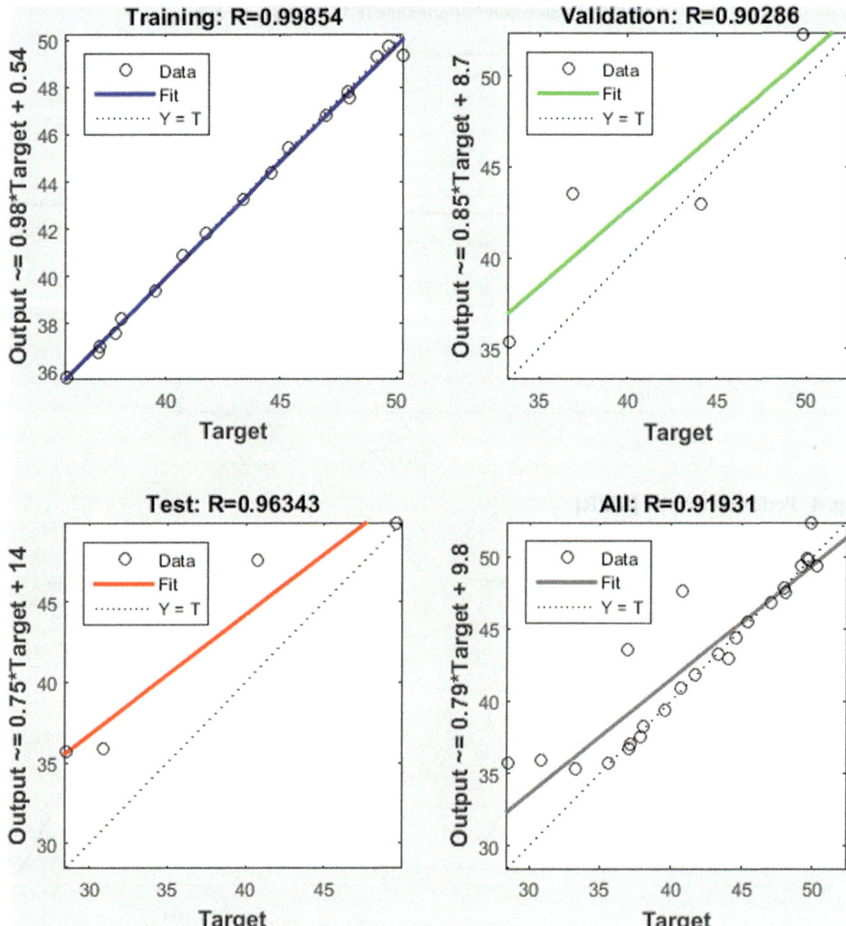

Fig. 3 Regression plot [MRR]

set hypothesis, there GRA can be applied. This can calculate the best parametric alternatives from the solution region (Fig. 6).

The specified weights of MRR, machining time and Ra are 33%, 26% and 41% respectively as determined by using Fuzzy set theory. Table 2 exhibits the results of grade and their ranks. It has been found that experiment number 3 has the maximum grey relational grade point. Consequently, the experimental run of 3 which has parametric combination Ton 0.5 µSec, Toff 12 µSec, SV 36 V, WT 501 g and WF 8 m/min is the finest amongst other experimental trail for having high MRR and lower machining time and Ra.

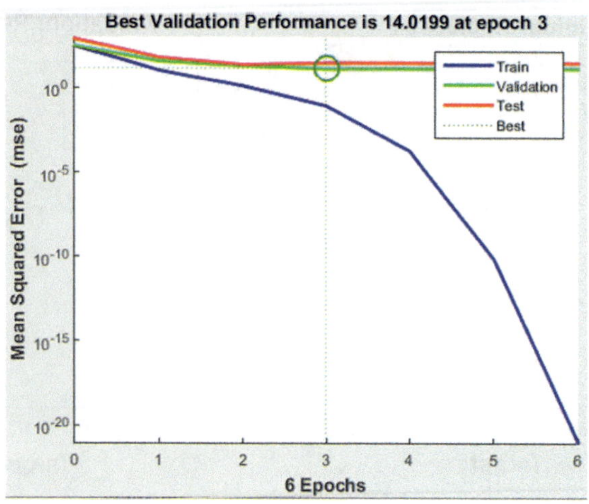

Fig. 4 Performance plot [MRR]

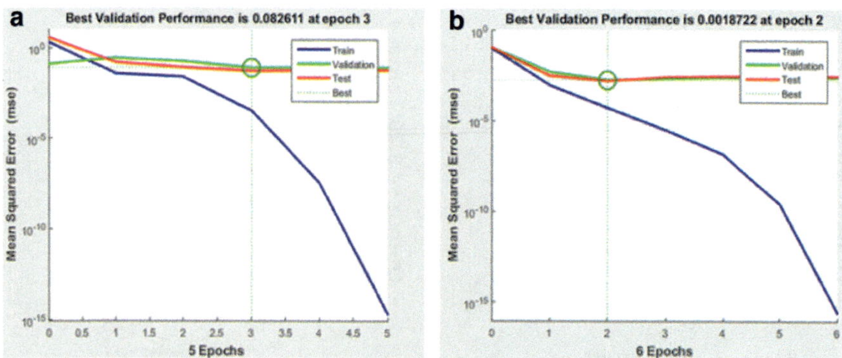

Fig. 5 Performance plot **a** Machining time **b** Surface roughness

3.3 Confirmation Experiments

Confirmation tests were performed in order to validate the optimization results by setting input parameters ($T_{on} = 0.5$, $T_{off} = 12$, SV = 36, WT = 501 and WF = 8). The corresponding responses (MRR, machining time and R_a) were measured. Predicted error lies within the limit of 5% and hence, can be considered as significant for acceptance (Table 3).

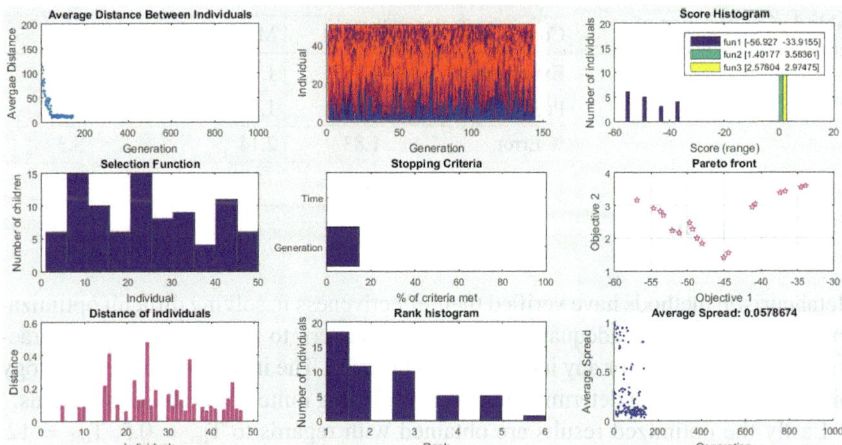

Fig. 6 Plot functions for GA

Table 2 Optimization results

Exp. no	Control parameters					Responses			Grey grade	Rank
	T_{on}	T_{off}	SV	WT	WF	*MRR*	Machining	R_a		
1	0.3	12	44	523	5	33.915	3.58	2.58	0.485	7
2	0.4	12	42	510	5	41.236	2.96	2.61	0.481	10
3	**0.5**	**12**	**36**	**501**	**8**	**44.987**	**1.40**	**2.72**	**0.640**	**1**
4	0.3	12	44	523	5	34.527	3.54	2.58	0.484	8
5	0.3	12	44	522	5	36.658	3.42	2.58	0.492	6
6	0.5	8	36	508	7	52.081	2.24	2.97	0.436	15
7	0.5	9	44	506	7	54.722	2.92	2.91	0.460	13
8	0.5	8	46	506	7	56.927	3.16	2.93	0.519	3
9	0.5	9	42	507	8	53.387	2.70	2.88	0.445	14
10	0.4	12	44	522	5	37.349	3.37	2.58	0.492	5
11	0.5	10	41	506	7	49.769	2.47	2.81	0.423	18
12	0.4	12	43	522	5	40.734	3.05	2.60	0.483	9
13	0.5	10	36	506	8	48.690	2.00	2.89	0.435	16
14	0.5	12	36	510	7	44.412	1.55	2.69	0.588	2
15	0.5	8	36	512	8	51.147	2.16	2.97	0.429	17
16	0.5	10	36	506	8	47.943	1.82	2.83	0.474	11
17	0.5	12	41	502	7	49.284	2.27	2.67	0.512	4
18	0.5	10	44	506	7	53.768	2.80	2.84	0.460	12

Table 3 Confirmation of optimization results

Comparison	MRR	Machining time	R_a
Experimental	44.987	1.40	2.72
Predicted	45.811	1.37	2.63
% Error	1.83	2.14	3.3

4 Conclusions

Metaheuristic methods have verified their effectiveness in solving difficult optimization problems. It can adequately uphold the leaders to decide the ideal interaction boundary setting for any industrialized practice. The incorporated methodology might be utilized for a determination issue including quite a few choice standards.

Lastly the optimized results are obtained with regards to $T_{on} = 0.5$, $T_{off} = 12$, $SV = 36$, $WT = 501$ and $WF = 8$ respectively. This examination work might be broadened further by considering some other exhibition qualities, for example, kerf, form factor as well as relevant input process parameters such as peak current, wire diameter in WEDM.

References

1. Somashekhar, K.P., Ramachandran, N., Mathew, J.: Optimization of material removal rate in micro-EDM using Artificial Neural Network and Genetic Algorithms. Mater. Manuf. Process. **25**(6), 467–475 (2010). https://doi.org/10.1080/10426910903365760
2. Ming, W., Hou, J., Zhang, Z., Huang, H., Xu, Z., Zhang, G., Huang, Y.: Integrated ANN- LWPA for cutting parameter optimization in WEDM. Int. J. Adv. Manuf. Technol. 1–18 (2015)
3. Kuriakose, S., Shunmugam, M.S.: Multi-objective optimization of wire-electro discharge machining process by non-dominated Sorting Genetic Algorithm. J. Mater. Process. Technol. **170**, 133–141 (2005)
4. Mahapatra, S.S., Patnaik, A.: Optimization of wire electrical discharge machining (WEDM) process parameters using Taguchi method. Int. J. Adv. Manuf. Technol. **34**(9–10), 911–925 (2007)
5. Prasad, D.V.S.S.S.V., Krishna, A.G.: Empirical modeling and optimization of wire electrical discharge machining. Int. J. Adv. Manuf. Technol. **43**, 914–925 (2009)
6. Pasam, V.K., Battula, S.B., Valli, P.M., Swapna, M.: Optimizing surface finish in WEDM using the taguchi parameter design method. J. Braz. Soc. Mech. Sci. Eng. **32**(2), 107–113 (2010)
7. Kuruvila, N., Ravindra, H.V.: Parametric influence and optimization of wire EDM of hot die steel. Mach. Sci. Technol. **15**(1), 47–75 (2011)
8. Kumar, K., Agarwal, S.: Multi-objective parametric optimization on machining with wire electric discharge machining. Int. J. Adv. Manuf. Technol. **62**, 617–633 (2012)
9. Zhang, G., Zhang, Z., Ming, W., Guo, J., Huang, Y., Shao, X.: The multi-objective optimization of medium-speed WEDM process parameters for machining SKD11 steel by the hybrid method of RSM and NSGA-II. Int. J. Adv. Manuf. Technol. **70**, 2097–2109 (2014)
10. Padhi, P.C., Mahapatra, S.S., Yadav, S.N., Tripathy, D.K.: Multi-objective optimisation of machining parameters in wire electrical discharge machine using non-dominating sorting genetic algorithm. Int. J. Prod. Qual. Manag. **14**, 107–129 (2014)
11. Sharma, N., Khanna, R., Gupta, R.D.: WEDM process variables investigation for HSLA by response surface methodology and genetic algorithm. Eng. Sci. Technol. **18**, 171–177 (2015)

12. Varun, A., Venkaiah, N.: Simultaneous optimization of WEDM responses using grey relational analysis coupled with genetic algorithm while machining EN 353. Int. J. Adv. Manuf. Technol. **76**, 675–690 (2015)
13. Nair, A., Kumanan, S., Shanavas, K.P.: Multi-performance optimization in wire EDM of Inconel 617 using GRA and genetic algorithm. Mater. Today: Proc. (2021). ISSN: 2214-7853. https://doi.org/10.1016/j.matpr.2021.08.279
14. Shandilya, P., Rouniyar, A.K., Saikiran, D.: Multi-objective parametric optimization on machining of Inconel-825 using wire electrical discharge machining. Proc. Inst. Mech. Eng. C J. Mech. Eng. Sci. **234**(20), 4056–4068 (2020). https://doi.org/10.1177/0954406220917706
15. Majumder, H., Paul, T.R., Dey, V., Dutta, P., Saha, A.: Use of PCA-grey analysis and RSM to model cutting time and surface finish of Inconel 800 during wire electro discharge cutting. Measurement **107**, 19–30 (2017). ISSN: 0263-2241 (Elsevier)
16. Paul, T.R., Saha, A., Majumder, H., Dey, V., Dutta, P.: Multi-objective optimization of some correlated process parameters in EDM of inconel 800 using a hybrid approach. Int. J. Braz. Soc. Mech. Sci. Eng. (2019). https://doi.org/10.1007/s40430-019-1805-9. (Springer (Accepted, SCI))
17. Saha, A., Mondal, S.C.: Optimization of wire electric discharge machining process: a review and reflection. Int. J. Prod. Qual. Manag. **22**(3), 340–362 (2017)
18. Saha, A., Mondal, S.C.: Statistical analysis and optimization of process parameters in wire cut machining of welded nano-structured hardfacing material. Silicon (2018). https://doi.org/10.1007/s12633-018-9924-y. (Springer (Accepted, SCI))
19. Saha, A., Mondal, S.C.: Experimental investigation and modelling of WEDM process for machining nano-structured hardfacing material. J Braz. Soc. Mech. Sci. Eng. **39**, 3439–3455 (2016)
20. Saha, A., Mondal, S.C.: Multi-objective optimization in WEDM process of nanostructured hardfacing materials through hybrid techniques. Measurement **94**, 46–59 (2017)

Object Detection: A Comparative Study to Find Suitable Sensor in Smart Farming

Mohit Kumar Mishra and Deepa Sonal

Abstract Crop-loss is one of the major factors that led the farmers to ruin their lives and everything they possess. Internet of Things (IoT) is an advanced technology that is effectively benefitting various fields nowadays. We have tried to use this technology to save the lives of farmers by saving their crops from being ruined. For this, we have proposed a model that is going to be used in the agricultural fields to prevent wildlife attacks that is one of the major factors of crop loss. As we know that Animal hunting is one of the punishable offences in many countries. We want the animals to run away from the field without any physical harm. In this paper, we have tried to do a comparative study of sensors that would be most preferable for our proposed model. We want to find out which sensor is most suitable for the detection of wild animals in the agriculture field. For finding that we have done some experiments and observed the results to justify our sensor selection. As we want to choose the most suitable sensors for our crop protection model.

Keywords IoT · Sensors · Crop protection · IR sensor · Ultrasonic sensor · Attack detection · Comparative study · Another keyword

1 Introduction

There is no doubt that a significant amount of research has been done in the realm of software and hardware applications in Internet of Things (IoT). One of them is the suggested research project. In most cases, a typical pest repeller will emit a consistent frequency. This consistent frequency will only deter the pest for a short time and is ineffective [1]. It is because strong and unusual sounds, such as ultrasonic sounds that rats can hear, frequently scare them and cause transient aversion lasting anywhere

M. Kumar Mishra
Department of Electronics and Communication Engineering, Manipal University, Jaipur, India

D. Sonal (✉)
Department of Computer Science, V.K.S. University, Arrah 802301, India
e-mail: deepsonapwc@gmail.com

S. Banerjee and A. Saha (eds.), *Nonlinear Dynamics and Applications*,
Springer Proceedings in Complexity,
https://doi.org/10.1007/978-3-030-99792-2_58

from a few minutes to a few weeks. In this research, we have tried to find out the most suitable sensors used for detecting the animals or pests in this animal repelling system [2]. One important commitment, we are going to make here is we don't want to physically harm any animal or pest for repulsion [3]. We want only to scare them and ran them away from the agricultural field as they enter the field [4]. An IoT based system is implemented to detect the Animal and produce repelling sound and repelling frequency to scare them away from the crop field [5].

We have two options for animal detection sensors. First one is ultrasonic sensor and second one is IR sensor. Both the sensors can be used to detect any object comes in the way. Let us first know about both the sensors individually. Some of the recent studies done in this field is given in [6] where the author has described that ultrasonic sensor can work well with sponge, wood, Tile and plastic whereas IR sensor can work only with Paper sheet typen of products.

1.1 IR Sensor

Infrared Sensor or IR Sensor is an equipment used to detect the objects present in surrounding by sending infrared waves which is usually has frequency in range of 300GHz to 400 THz. This sensor works on the theory of reflected light waves. Reflected infrared light or light emitted from an infrared remote or beacon. Distance and vicinity are also determined via infrared sensors [7]. After detecting the reflected light, a range calculation between the sensor and the target is computed [8]. The Fig. 1 given below shows the diagram of IR sensor.

1.2 Ultrasonic Sensor

An ultrasonic sensor is also used to detect the obstacle in the way and determine the distance of the object. It works on the principle of reflecting sound waves [9, 10]. The ultrasonic sensor emits sound waves, which are reflected if an obstacle is before it. Ultrasound is a type of sound that is inaudible to the individual ear since its lowest working frequency is still higher than the human hearing threshold of roughly 20 kHz. Ultrasound spans a wide range of frequencies, from 40 kHz for

Fig. 1 IR sensor

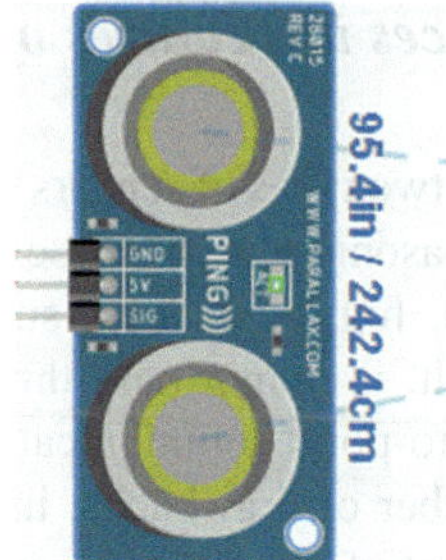

Fig. 2 Ultrasonic sensor

proximity sensors to 1 MHz and above for diagnostic imaging. The range finding by an ultrasonic sensor can be seen in the Fig. 2 given.

So, objective of this research paper is to find out the most suitable sensor for animal detection so that Animal repeller system can work efficiently. In first section of this paper, we have given the introduction about the system and sensors. In second section, we have described the methodology for showing difference in the sensors theoretically and experimentally. In third section, we have done analysis on data observed in experiment and in fourth section we have given the conclusion of this research.

2 Methodology

Choosing a sensor for particular application might be difficult in either project. The dependability of the sensor and other application equipment has a big impact on the effectiveness of the machine. There are a few factors to contemplate when choosing a sensor for your applications. Accuracy refers to how accurate the measurement is in relation to the genuine distance. The minimal reading or variation in readings that can be recorded is called resolution.

Precision refers to the tiniest measurement that can be taken frequently and dependably.

To compare and study the behavior of both the above sensors, it is very important to list out the features of IR and ultrasonic sensor theoretically. Then we have done some experiments on both the sensors by attaching it to Arduino board one by one and checking out its performance through observations. And finally checking out the observation results through data analysis whether the experiment result is significant or not.

2.1 Theoretical Differences in Features of Both the Sensors

The main distinction among the two types of sensors is that IR sensors sense electromagnetic radiation, whereas ultrasonic sensors sense mechanical or auditory energy. Hard things (e.g. walls, doors), fog, dust, mist, sunlight, and other factors alter infrared frequencies. As a result, it cannot pass through walls or doors. It works on line of sight (LoS), so point-to-point communication is done.

Infrared sensors have a number of drawbacks, including the inability to utilize them in direct sunlight due to interference. It can make it difficult to use outdoor applications or dismal indoor applications. Ultrasonic sensors identify obstructions through sound waves and are unaffected by a variety of circumstances. Ultrasonic sensors are more efficient than infrared sensors if reliability is a priority in the sensor choices.

2.2 Experimental Observations for Both the Sensors

We have done experiments also to take observations for checking efficiency of both the sensors one by one. We have collected data readings through serial monitor, some during day time and some at night time. Various parameters for both the sensors can be seen in the given dataset table Table 1.

First we have taken the readings for ultrasonic sensor in day time and night time for object detection. Figure 3 here shows the reading taken on experimental setup of

Table 1 Various parameters for experimental setup

Parameters	IR Sensor	Ultrasonic sensor
Range	10–80 cm	2–10 cm
Beam-width	75°	30°
Beam Pattern	Narrow (Line)	Canonical
Frequency	353 THz	40 kHz
Unit cost	~750 INR	~130 INR

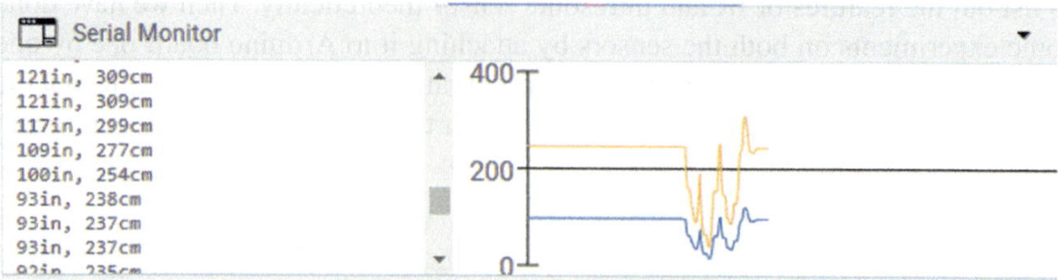

Fig. 3 Ultrasonic sensor data displayed on serial monitor of arduino IDE

ultrasonic sensor with Arduino UNO microcontroller. Figure 4 is showing the circuit diagram of experimental setup.

The circuit diagram in Fig. 4 has been designed using tinkercad.com online app. Experimental setup of given circuit design has been done which is shown in Fig. 5. The same experiment we have done with IR sensor and taken the observations to check its efficiency in day time and night time.

The Fig. 6 shows the observation taken during the daytime in sunlight using both the sensors one-by-one. It shows the efficiency of both the sensors of sensing the

Fig. 4 Circuit design of ultrasonic sensor with arduino UNO

Fig. 5 Experimental setup with LCD display to show distance

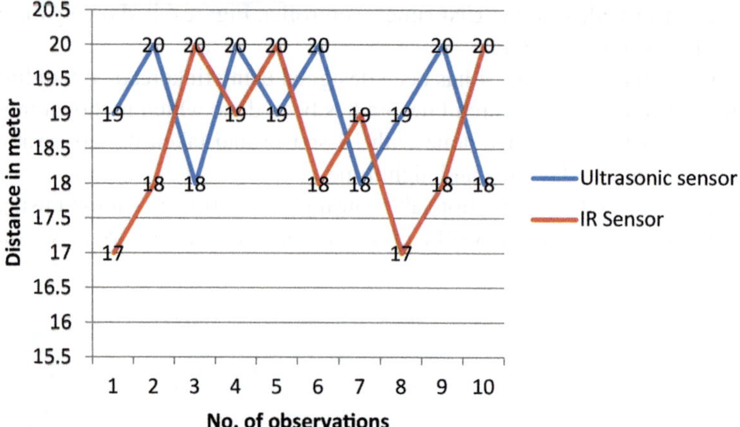

Fig. 6 Detection of an object coming toward both sensors in the daytime

object coming towards them. It is visible from the above chart that the IR sensor is also working efficiently during day-time in Sunlight. IR sensor is able to detect the object as efficiently as ultrasonic sensor can detect.

(a) Then we perform the same experiment at night and dark environment to check the efficiency of both the sensors using the same experimental setup. And based on the readings of the experiment for both the sensors, Fig. 7 has been plotted as given.

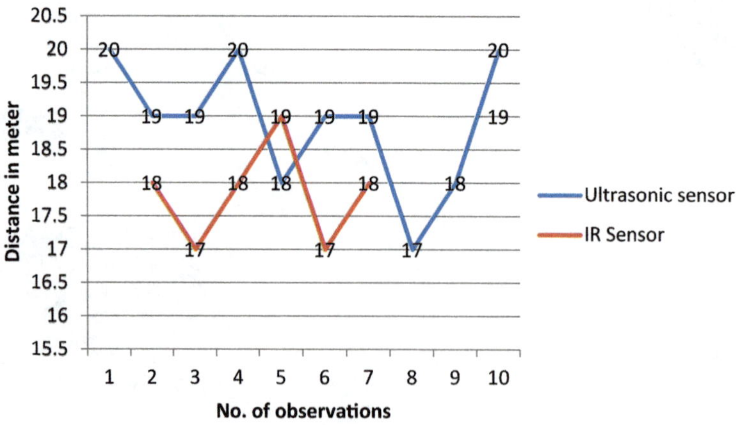

Fig. 7 Detection of an object coming toward both sensors in the night time

We can see the difference in working efficiency of both the sensors at night/dark. It is very much clear from the above plotted graph that IR sensor was unable to detect the coming object three times. It shows that IR sensor is not that much reliable during night.

3 Results and Discussion

From these experimental observations, we can discuss that IR sensor can be used efficiently in day time and for indoor experiments, while for outdoor experiments that needs to be done at night or in darkness; ultrasonic sensor is much more efficient than the IR sensor. For strongly supporting our experiment, we have done data analysis also that can show the significance of our experiment being carried out.

Both the experiments are analyzed and we have tried to find out the significance of experiment conducted during day and night time respectively (as shown in Tables 2 and 3). In Table 3, total no. of observations is 7 whereas actually total 10 observations

Table 2 Data analysis showing significance of experiment during daytime

SUMMARY						
Groups	Count	Sum	Average	Variance		
Column 1	10	191	19.1	0.766667		
Column 2	10	186	18.6	1.377778		
ANOVA						
Source of variation	SS	df	MS	F	P-value	F crit
Between groups	1.25	1	4.25	4.165803	0.294521	4.413873
Within groups	19.3	18	1.072222			
Total	20.55	19				

Table 3 Data analysis at night/dark

SUMMARY						
Groups	Count	Sum	Average	Variance		
Column 1	10	189	18.9	0.988889		
Column 2	7	126	18	0.666667		
ANOVA						
Source of variation	SS	df	MS	F	P-value	F crit
Between groups	3.335294	1	3.335294	3.878249	0.067672	4.543077
Within groups	12.9	15	0.86			
Total	16.23529	16				

were taken. IR sensor was not able to sense object 3 times in night/dark. The F-tabulated and F-calculated values are also very closed in observational analysis during day time. It shows the significance of observation and experiment. It implies that IR sensor is not reliable for object detection in fields. As we have to make guarding system especially for nights when Farmers are either sleeping or not in the agricultural fields. In fact in day time if there is a cloudy day, then also it will not work effectively because of lack of light. So it is most efficient to work with ultrasonic sensor than IR sensor where 24×7 guarding is required.

4 Conclusion

We can conclude that we should go for ultrasonic sensor instead of infrared sensor in our Animal repeller proposed model. We have to design and develop a final cropland guarding system that can help out farmers not only in daytime but especially during night [11]. So we should use the best components in our animal repeller system. Before repelling the wild animals, it is very much essential to detect the animals at right time and quickly [3]. So that crop harm can be minimized or removed. Thus we have concluded to use ultrasonic sensor for our animal detection system.

References

1. Louis, L.: Working principle of arduino and using it as a tool for study and research. Int. J. Control. Autom. Commun. Syst. **1**(2), 21–29 (2016). https://doi.org/10.5121/ijcacs.2016.1203
2. Ragavi, B., Pavithra, L., Sandhiyadevi, P., Mohanapriya, G.K., Harikirubha, S.: Smart sgriculture with AI sensor by using agrobot. In: 2020 Fourth International Conference on Computing Methodologies and Communication (ICCMC), pp. 1–4 (2020)
3. Haque, M.A., Sonal, D., Haque, S., Nezami, M.M., Kumar, K: An IoT-based model for defending against the novel coronavirus (COVID-19) outbreak. Solid State Technol. 592–600 (2020)
4. Kumar, S.: Artificial Intelligence in Indian Irrigation (2019)
5. Samarasinghe, M.G.P.M.: Use of IoT for Smart Security Management in Agriculture, vol. 978, pp. 65–73 (2019)
6. Adarsh, S., Kaleemuddin, S.M., Bose, D., Ramachandran, K.I.: Performance comparison of Infrared and Ultrasonic sensors for obstacles of different materials in vehicle/ robot navigation applications. IOP Conf. Ser. Mater. Sci. Eng., **149**(1), (2016). https://doi.org/10.1088/1757-899X/149/1/012141
7. David, S., Anand, R.S., Sagayam, M.: Enhancing AI based evaluation for smart cultivation and crop testing using agro-datasets. J. Artif. Intell. Syst. **2**(1), 149–167 (2020)
8. Divya, R., Chinnaiyan, R.: Reliable AI-based smart sensors for managing irrigation resources in agriculture—a review. In: International Conference on Computer Networks and Communication Technologies, pp. 263–274 (2019)
9. Smart Agriculture Sensors|Mouser. https://www.mouser.in/applications/smart-agriculture-sensors/. Last accessed 30 Apr 2021

10. Baranwal, T.: "Nitika and PK Pateriya" Development of IoT based smart security and monitoring devices for agriculture. In: Proceeding of the 2016 6th International Conference-Cloud Sysem Big Data Engineering Confluence 2016, no. November 2020, pp. 597–602 (2016). https://doi.org/10.1109/CONFLUENCE.2016.7508189
11. Parashar, M.: Candidate Declaration "IoT Based Smart Agriculture Monitoring System," vol. 151042

Robust Adaptive Controller for a Class of Uncertain Nonlinear Systems with Disturbances

Ngo Tri Nam Cuong, Le Van Chuong, and Mai The Anh

Abstract This paper presents a method to synthesize the controller for uncertain nonlinear systems based on a combination of sliding mode control, adaptive control, and radial basis function (RBF) neural network. We propose an adaptive control law based on the RBF neural network to identify and compensate for variable parameter components, nonlinear function vectors, and external disturbance. The main linear component is built based on a sliding control. The designed controller has the advantage of being resistant to the elements of uncertainty and has a high control quality.

Keywords Nonlinear systems · Adaptive control · System identification

1 Introduction

In practice, the uncertain nonlinear systems are affected by external disturbances which are very common. The existence of uncertain parameters adversely affects the performance of the system. Control design for such a class of objects has attracted the attention of many researchers in past decades. A combination of the adaptive control method and the neural network has been shown in the researches [1–4], in which nonlinear components and external disturbance are identified using the neural network to generate a compensation control signal for the uncertain components. Some researches on adaptive control have been implemented for such variable nonlinear systems, where variable parameters are identified and adjusted by adaptive control law [5–7]. In [8], a control law is built based on sliding mode control in which uncertain components are considered for control design, and thus the designed system is stability. The control design using backstepping and fuzzy techniques is

N. T. N. Cuong
Systemtec JSC, Hanoi, Vietnam

L. V. Chuong (✉) · M. T. Anh
Vinh University, Vinh city, Nghean, Vietnam
e-mail: lvchuong85@gmail.com; chuonglv@vinhuni.edu.vn

S. Banerjee and A. Saha (eds.), *Nonlinear Dynamics and Applications*,
Springer Proceedings in Complexity,
https://doi.org/10.1007/978-3-030-99792-2_59

implemented in [9], where the external disturbance is identified and compensated by fuzzy logic, the variable parameters are determined based on the backstepping technique. In the papers [10, 11], identifying and correcting uncertain parameters are also implemented based on the backstepping control. Thus, there are many different methods to synthesize control systems for the class of nonlinear objects under the affection of variable parameters and external disturbance. Many results have been shown advantages of diffirent methods, however enhancing performmance of the designed system is still problems of interest to many researches. This paper presents a method of synthesizing a stable adaptive controller based on combining adaptive control, sliding control, and RBF neural network.

2 Problem Formulation

A multi-input multi-output (MIMO) nonlinear system will be considered in the paper:

$$\dot{\mathbf{x}} = \mathbf{A}\mathbf{x} + [\mathbf{B} + \mathbf{\Delta B}]\mathbf{u} + \mathbf{f}(\mathbf{x}) + \mathbf{d}(t), \tag{1}$$

where $\mathbf{x} = [x_1, x_2 ... x_n]^T$ is state vector; $\mathbf{u} = [u_1, u_2 ... u_m]^T$ is control vector; $\mathbf{A} \in \mathbb{R}^{n \times n}$ is Hurwitz matrix with fixed elements; $\mathbf{B} \in \mathbb{R}^{n \times m}$ is matrix with fixed elements; $\mathbf{\Delta B} \in \mathbb{R}^{n \times m}$ matrix matched uncertainty; $\mathbf{f}(\mathbf{x}) = [f_1(\mathbf{x}), f_2(\mathbf{x}), ..., f_n(\mathbf{x})]^T$ is smooth nonlinear vetor, matched uncertainty; $\mathbf{d}(t) = [d_1(t), d_2(t), ..., d_n(t)]^T$ is external disturbance vector with slow varible elements, matched uncertainty $|d_i(t)| \leq d_M$.

The block diagram of the designed system using identification structure with compensation of uncertain component and external disturbance is shown in Fig. 1. MODEL is the identification model; IDENT is the identification block; COMP is the compensation block of uncertain components and external disturbance; SMC is the sliding mode controller.

The control signal can be considered as follow:

$$\mathbf{u} = \mathbf{u}_{smc} + \mathbf{u}_c, \tag{2}$$

where $\mathbf{u}_{smc}$ is control signal vector of SMC; $\mathbf{u}_c$ is control signal vector for compensation of uncertain component and external disturbance.

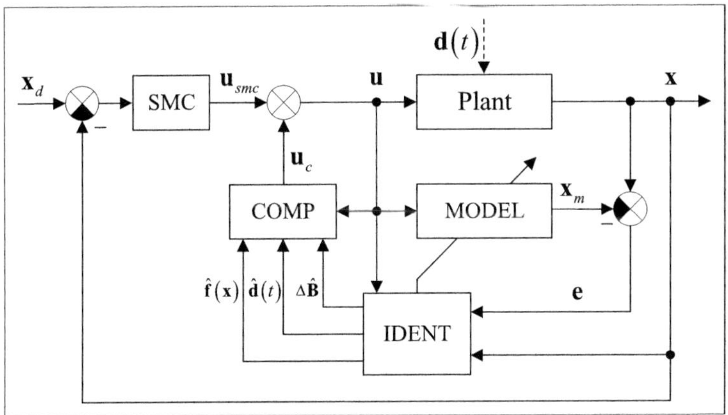

Fig. 1 Block diagram of the designed system

3 Algorithm for Identification and Compensation of Uncertain Parameters

Uncertain components $\mathbf{\Delta B}, \mathbf{f}(\mathbf{x}), \mathbf{d}(t)$ in (1) need to be identified and adjusted for compensation. The identification model for uncertain parameters in (1) can be written:

$$\dot{\mathbf{x}}_m = \mathbf{A}\mathbf{x}_m + [\mathbf{B} + \mathbf{\Delta}\hat{\mathbf{B}}]\mathbf{u} + \hat{\mathbf{f}}(\mathbf{x}) + \hat{\mathbf{d}}(t), \tag{3}$$

where $\mathbf{x}_m = [x_{m1}, x_{m2} \ldots x_{mn}]^T$ is state vector of the model; $\mathbf{\Delta}\hat{\mathbf{B}}$ is the estimated matrix of $\mathbf{\Delta B}$ which is defined by elements Δb_{ij}; $\hat{\mathbf{f}}(\mathbf{x}) = [\hat{f}_1(\mathbf{x}), \hat{f}_2(\mathbf{x}), ..., \hat{f}_n(\mathbf{x})]^T$ is the estimated vector of $\mathbf{f}(\mathbf{x})$; $\hat{\mathbf{d}}(t) = [\hat{d}_1(t), \hat{d}_2(t), ..., \hat{d}_n(t)]^T$ is the estimated vector of $\mathbf{d}(t)$.

From (1) and (3), we have:

$$\dot{\mathbf{e}} = \mathbf{A}\mathbf{e} + \mathbf{\Delta}\tilde{\mathbf{B}}\mathbf{u} + \tilde{\mathbf{f}}(\mathbf{x}) + \tilde{\mathbf{d}}(t), \tag{4}$$

where $\mathbf{e} = \mathbf{x} - \mathbf{x}_m$; $\mathbf{\Delta}\tilde{\mathbf{B}} = \mathbf{\Delta B} - \mathbf{\Delta}\hat{\mathbf{B}}$; $\tilde{\mathbf{f}}(\mathbf{x}) = \mathbf{f}(\mathbf{x}) - \hat{\mathbf{f}}(\mathbf{x})$; $\tilde{\mathbf{d}}(t) = \mathbf{d}(t) - \hat{\mathbf{d}}(t)$. Identification progress will be converging when $\mathbf{\Delta}\tilde{\mathbf{B}} \to 0, \tilde{\mathbf{f}}(\mathbf{x}) \to 0, \tilde{\mathbf{d}}(t) \to 0$. Because $\mathbf{A}$ is defined by a Hurwitz matrix, so $\mathbf{e} \to 0$, and (4) is stability.

With $\mathbf{f}(\mathbf{x})$ is a smooth function vector, by using a RBF neural network for the approximation. The elements of $\mathbf{f}(\mathbf{x})$ can be written:

$$f_i(\mathbf{x}) = \sum_{j=1}^{L} w_{ij}^* \phi_{ij}(\mathbf{x}) + \varepsilon_i, \tag{5}$$

$\forall i = \overline{1, n}; j = \overline{1, L}$, where L is number of basis function with a large enough number to guarantee the error $|\varepsilon_i| < \varepsilon_i^m$, $w_{ij}^* = const$ is the ideal weights. The basis functions are selected by the following form:

$$\phi_{ij}(\mathbf{x}) = \exp\left(\frac{\|\mathbf{x} - \mathbf{c}_{ij}\|^2}{2\sigma_{ij}^2}\right), \tag{6}$$

where $\mathbf{c}_{ij}$ are the position of the center of the basis functions $\phi_{ij}(\mathbf{x})$, and σ_{ij} are the standard deviation of the basis functions. The evaluation vector $\hat{\mathbf{f}}(\mathbf{x})$ is defined by (6) with adjusted weights $\hat{w}_{ij}$:

$$\hat{f}_i(\mathbf{x}) = \sum_{j=1}^{L} \hat{w}_{ij}\phi_{ij}(\mathbf{x}), i = \overline{1, n}. \tag{7}$$

Training of the RBF neural network is implemented by adjustment of the weights $\hat{w}_{ij}$ in comparison with the ideal weights w_{ij}^*:

$$\tilde{w}_{ij} = w_{ij}^* - \hat{w}_{ij}, \tag{8}$$

from (5), (7) and (8), we have:

$$f_i(\mathbf{x}) = \hat{f}_i(\mathbf{x}) + \varepsilon_i \rightarrow \tilde{f}(\mathbf{x}) = \sum_{j=1}^{L} \tilde{w}_{ij}\phi_{ij}(\mathbf{x}) + \varepsilon_i, \tag{9}$$

ε_i is the approximate error with a sufficiently small value.

Theorem 1 *Equation (4) are stable when the following conditions are satisfied:*

$$\|\mathbf{e}\| > \frac{2\sum_{i=1}^{n} \varepsilon_i \|\bar{\mathbf{P}}_i\|}{r_{\min}(\mathbf{Q})}; \tag{10}$$

$$\mathbf{u}^T \Delta\tilde{\mathbf{B}}^T \mathbf{P}\mathbf{e} + \sum_{i=1}^{n}\sum_{j=1}^{m} \Delta\dot{\tilde{b}}_{ij}\Delta\tilde{b}_{ij} = 0; \tag{11}$$

$$\mathbf{e}^T\mathbf{P}\begin{bmatrix} \sum_{j=1}^{L} \tilde{w}_{1j}\phi_{ij}(\mathbf{x}) \\ \vdots \\ \sum_{j=1}^{L} \tilde{w}_{nj}\phi_{ij}(\mathbf{x}) \end{bmatrix} + \sum_{i=1}^{n}\sum_{j=1}^{L} \dot{\tilde{w}}_{ij}\tilde{w}_{ij} = 0; \tag{12}$$

$$\mathbf{e}^T \mathbf{P}\tilde{\mathbf{d}}(t) + \sum_{i=1}^{n} \dot{\tilde{d}}_i \tilde{d}_i = 0. \tag{13}$$

P *is a positive definite symmetric matrix* □

Proof. For Eq. (4), the Lyapunov function is selected as follows::

$$V = \mathbf{e}^T \mathbf{P}\mathbf{e} + \sum_{i=1}^{n}\sum_{j=1}^{m} \Delta\tilde{b}_{ij}^2 + \sum_{i=1}^{n}\sum_{j=1}^{L} \tilde{w}_{ij}^2 + \sum_{i=1}^{n} \tilde{d}_i^2. \tag{14}$$

The Eq. (4) will be stable if the derivative (14) $\dot{V} < 0$. From (14), we have:

$$\dot{V} = \mathbf{e}\dot{\mathbf{P}}\mathbf{e} + \mathbf{e}^T P\dot{\mathbf{e}} + 2\sum_{i=1}^{n}\sum_{j=1}^{m} \Delta\dot{\tilde{b}}_{ij}\Delta\tilde{b}_{ij} + 2\sum_{i=1}^{n}\sum_{j=1}^{L} \dot{\tilde{w}}_{ij}\tilde{w}_{ij} + 2\sum_{i=1}^{n} \dot{\tilde{d}}_i\tilde{d}_i. \tag{15}$$

Substitute (4) into (15):

$$\begin{aligned}\dot{V} &= \mathbf{e}^T \left(\mathbf{A}^T\mathbf{P} + \mathbf{P}\mathbf{A}\right)\mathbf{e} + 2\mathbf{u}^T \mathbf{\Delta}\tilde{\mathbf{B}}^T \mathbf{P}\mathbf{e} + 2\mathbf{e}^T \mathbf{P}\tilde{\mathbf{f}}(\mathbf{x}) + 2\mathbf{e}^T \mathbf{P}\tilde{\mathbf{d}}(t) + \\ &+2\sum_{i=1}^{n}\sum_{j=1}^{m} \Delta\dot{\tilde{b}}_{ij}\Delta\tilde{b}_{ij} + 2\sum_{i=1}^{n}\sum_{j=1}^{L} \dot{\tilde{w}}_{ij}\tilde{w}_{ij} + 2\sum_{i=1}^{n} \dot{\tilde{d}}_i\tilde{d}_i.\end{aligned} \tag{16}$$

From (16) and (9), we have:

$$\begin{aligned}\dot{V} &= \mathbf{e}^T \left(\mathbf{A}^T\mathbf{P} + \mathbf{P}\mathbf{A}\right)\mathbf{e} + 2\mathbf{e}^T\mathbf{P}\boldsymbol{\varepsilon} + 2(\mathbf{u}^T \mathbf{\Delta}\tilde{\mathbf{B}}^T \mathbf{P}\mathbf{e} + \sum_{i=1}^{n}\sum_{j=1}^{m} \Delta\dot{\tilde{b}}_{ij}\Delta\tilde{b}_{ij}) + \\ &+2(\mathbf{e}^T\mathbf{P}\begin{bmatrix}\sum_{j=1}^{L} \tilde{w}_{1j}\phi_{ij}(\mathbf{x}) \\ \vdots \\ \sum_{j=1}^{L} \tilde{w}_{nj}\phi_{ij}(\mathbf{x})\end{bmatrix} + \sum_{i=1}^{n}\sum_{j=1}^{L} \dot{\tilde{w}}_{ij}\tilde{w}_{ij}) + 2(\mathbf{e}^T \mathbf{P}\tilde{\mathbf{d}}(t) + 2\sum_{i=1}^{n} \dot{\tilde{d}}_i\tilde{d}_i).\end{aligned} \tag{17}$$

Substitute (11), (12), and (13) into (17)):

$$\dot{V} = \mathbf{e}^T \left(\mathbf{A}^T\mathbf{P} + \mathbf{P}\mathbf{A}\right)\mathbf{e} + 2\mathbf{e}^T\mathbf{P}\varepsilon. \tag{18}$$

The Eq. (18) can be written:

$$\dot{V} = -\mathbf{e}^T\mathbf{Q}\mathbf{e} + 2\sum_{i=1}^{n} \varepsilon_i\bar{\mathbf{P}}_i\mathbf{e}, \tag{19}$$

$\mathbf{Q} = -(\mathbf{A}^T\mathbf{P} + \mathbf{P}\mathbf{A})$, $\bar{\mathbf{P}}_i$ is the i-th row of the matrix **P**.

Using inequality transformations [12], the Eq. (19) can be written:

$$\dot{V} = -\mathbf{e}^T\mathbf{Q}\mathbf{e} + 2\sum_{i=1}^{n}\varepsilon_i\bar{\mathbf{P}}_i\mathbf{e} < -r_{\min}(Q)\|\mathbf{e}\|^2 + 2\sum_{i=1}^{n}\varepsilon_i\left\|\bar{\mathbf{P}}_i\right\|\|\mathbf{e}\|. \quad (20)$$

Substitute (10) into (20), we have $\dot{V} < 0$, the Eq. (4) is stable. □

The expressions (11), (12), and (13) of the Theorem 1 contain identification algorithms $\Delta\mathbf{B}$, $\mathbf{f}(\mathbf{x})$, and $\mathbf{d}(\mathbf{t})$.

The Eq. (11) contains slowly variable elements, i.e. $\Delta\dot{b}_{ij} \approx 0$. The matrix $\mathbf{\Delta B}$ with uncertain parameters is identified by the matrix $\mathbf{\Delta\hat{B}}$ using the update law:

$$\Delta\dot{\hat{b}}_{ij} = u_j\bar{\mathbf{P}}_i\mathbf{e}. \quad (21)$$

From (7) and (12), because of $w_{ij}^* = const$, we have $\dot{w}_{ij}^* = 0$. The vector $\hat{\mathbf{f}}(\mathbf{x})$ for identification of the nonlinear function $\mathbf{f}(\mathbf{x})$ can be written:

$$\hat{f}_i(\mathbf{x}) = \sum_{j=1}^{L}\hat{w}_{ij}\phi_{ij}(\mathbf{x}), i = \overline{1, n}. \quad (22)$$

The update weights can be defined:

$$\dot{\hat{w}}_{ij} = \bar{\mathbf{P}}_i\mathbf{e}\phi_{ij}(\mathbf{x}). \quad (23)$$

From (13), because of slow-varying external disturbance $\dot{d}(t) \approx 0$. The vector $\hat{\mathbf{d}}(t)$ for identification of $\mathbf{d}(t)$ can be written:

$$\dot{\hat{d}}_i(t) = \bar{\mathbf{P}}_i\mathbf{e}. \quad (24)$$

The received results from (21), (22), (23), and (24) are used to synthesis the compensation control law $\mathbf{u}_c$.

The Eq. (1) can be again written as follows:

$$\dot{\mathbf{x}} = \mathbf{A}\mathbf{x} + \mathbf{B}\mathbf{u} + \mathbf{I}\mathbf{f}_\Sigma(t), \quad (25)$$

where $\mathbf{f}_\Sigma(t) = \mathbf{\Delta B}\mathbf{u} + \mathbf{f}(\mathbf{x}) + \mathbf{d}(t), \mathbf{f}_\Sigma(t) = [f_1^\Sigma, f_2^\Sigma, ..., f_n^\Sigma]^T$; $\mathbf{I}^{n\times n}$ has main diagonal elements $I_{ij} = 1,\ i = j = \overline{1, n}$ are rows which corresponds to the vector $\mathbf{f}_\Sigma(t)$ in the case $|f_i^\Sigma| \neq 0$; other elements $I_{ij} = 0$ in the case $i \neq j$ and $|f_i^\Sigma| = 0$.

Substitute (2) into (25):

$$\dot{\mathbf{x}} = \mathbf{A}\mathbf{x} + \mathbf{B}\mathbf{u}_{smc} + \mathbf{B}\mathbf{u}_c + \mathbf{I}\mathbf{f}_\Sigma(t). \quad (26)$$

The vector $\mathbf{u}_c$ can be selected:

$$\mathbf{u}_c = -\mathbf{H}\hat{\mathbf{f}}_\Sigma(t), \quad (27)$$

$$\hat{\mathbf{f}}_{\Sigma}(t) = \Delta\hat{\mathbf{B}}\mathbf{u} + \hat{\mathbf{f}}(\mathbf{x}) + \hat{\mathbf{d}}(t); \tag{28}$$

$\Delta\hat{\mathbf{B}}$, $\hat{\mathbf{f}}(\mathbf{x})$, and $\hat{\mathbf{d}}(t)$ are presented in (21), (22), (23), and (24).

Substitute (27) into (26):

$$\dot{\mathbf{x}} = \mathbf{A}\mathbf{x} + \mathbf{B}\mathbf{u}_{smc} - \mathbf{B}\mathbf{H}\hat{\mathbf{f}}_{\Sigma}(t) + \mathbf{I}\mathbf{f}_{\Sigma}(t). \tag{29}$$

From (29) we can see that uncertain elements will be compensated with the condition:

$$-\mathbf{B}\mathbf{H}\hat{\mathbf{f}}_{\Sigma}(t) + \mathbf{I}\mathbf{f}_{\Sigma}(t) = 0. \tag{30}$$

The Eq. (30) will be satisfied with the following condition:

$$\mathbf{B}\mathbf{H} = \mathbf{I}. \tag{31}$$

The Eq. (31) will be satisfied with:

$$\mathbf{H} = \mathbf{B}^{+}. \tag{32}$$

where $\mathbf{B}^{+}$ is the pseudo-inverse matrix of $\mathbf{B}$.

Thus, the article has synthesized the compensation control law $\mathbf{u_c}$ (27) with identification vectors $\hat{\mathbf{f}}_{\Sigma}(t)$ (28), $\mathbf{H}$ (32).

Using the compensation control law (27), the Eq. (29) can be written:

$$\dot{\mathbf{x}} = \mathbf{A}\mathbf{x} + \mathbf{B}\mathbf{u}_{smc}. \tag{33}$$

Thus, in this section, the identification and compensation control law $\mathbf{u_c}$ (27) for the uncertain components of (1) have been presented, and then (1) is rewritten to (33). For (33), the control law is synthesized based on the sliding mode control.

4 Synthesis of the Sliding Mode Control Law

The error vector between the state vector $\mathbf{x}$ and the desired state vector $\mathbf{x}_d$:

$$\tilde{\mathbf{x}} = \mathbf{x} - \mathbf{x}_d \rightarrow \mathbf{x} = \tilde{\mathbf{x}} - \mathbf{x}_d. \tag{34}$$

Substitute (34) into (33), we have:

$$\dot{\tilde{\mathbf{x}}} = \mathbf{A}\tilde{\mathbf{x}} + \mathbf{B}\mathbf{u}_{smc} + \mathbf{A}\mathbf{x}_d - \dot{\mathbf{x}}_d. \tag{35}$$

For (35), the hyper sliding surface is chosen as follows [13]:

$$\mathbf{s} = \mathbf{C}\tilde{\mathbf{x}}, \tag{36}$$

where $\mathbf{C}$ is the parameter matrix of hyper sliding surface, $\mathbf{s} = [s_1, s_2, ..., s_n]^T$.

The next problem is to define the control law $\mathbf{u}_{smc}$ which ensures movement of the system (35) towards the hyper sliding surface (36) and keep it there.

The control signal $\mathbf{u}_{smc}$ can be written by:

$$\mathbf{u}_{smc} = \begin{cases} \mathbf{u}_s & \text{if} \quad \mathbf{s} \neq 0 \\ \mathbf{u}_{eq} & \text{if} \quad \mathbf{s} = 0 \end{cases}, \tag{37}$$

$\mathbf{u}_s$ is the control signal that moves the system (35) towards the hyper sliding surface (36); $\mathbf{u}_{eq}$ is the equivalent control signal that keeps the system (35) on the hyper sliding surface (36).

The Eq. (37) can be rewritten as:

$$\mathbf{u}_{smc} = \mathbf{u}_{eq} + \mathbf{u}_s, \tag{38}$$

$\mathbf{u}_{eq}$ is defined in [13]:

$$\dot{\mathbf{s}} = \mathbf{C}\dot{\tilde{\mathbf{x}}} = 0. \tag{39}$$

From (35) and (39), we have:

$$\mathbf{C}\left(\mathbf{A}\tilde{\mathbf{x}} + \mathbf{B}\mathbf{u}_{eq} + \mathbf{A}\mathbf{x}_d - \dot{\mathbf{x}}_d\right) = 0. \tag{40}$$

From (40), the equivalent control signal can be defined as follows:

$$\mathbf{u}_{eq} = -[\mathbf{C}\mathbf{B}]^{-1}\left[\mathbf{C}\mathbf{A}\tilde{x} + \mathbf{C}\mathbf{A}\mathbf{x}_d - \mathbf{C}\dot{\mathbf{x}}_d\right]. \tag{41}$$

Next, we define the control signal $\mathbf{u}_s$ that moves the system (35) towards the hyper sliding surface (36).

For the hyper sliding surface (36), the Lyapunov function can be selected by:

$$V = \frac{1}{2}\mathbf{s}^T\mathbf{s}. \tag{42}$$

Condition for the existence of slip mode can be written:

$$\dot{V} = \mathbf{s}^T\dot{\mathbf{s}} < 0. \tag{43}$$

Substitute (35) and (38) into (43), with attention to (39), (40) we have:

$$\dot{V} = \mathbf{s}^T\left[\mathbf{C}\left(\mathbf{A}\tilde{\mathbf{x}} + \mathbf{B}\mathbf{u}_{eq} + \mathbf{A}\mathbf{x}_d - \dot{\mathbf{x}}_d\right) + \mathbf{C}\mathbf{B}\mathbf{u}_s\right] < 0. \tag{44}$$

Inequality (43) can be written as:

$$\mathbf{s}^T\left[\mathbf{C}\mathbf{B}\mathbf{u}_s\right] < 0. \tag{45}$$

So to satisfy the condition (43), the control signal from (45) can be defined as follows:

$$\mathbf{u}_s = -[\mathbf{CB}]^{-1}\left[\delta \text{sgn}\left(s_1\right), \delta \text{sgn}\left(s_2\right), ..., \delta \text{sgn}\left(s_n\right)\right]^T, \tag{46}$$

δ is a positive coefficient. Substituting (41) and (46) into (37), the control signal can be defined by $\mathbf{u}_{smc}$ as follows:

$$\mathbf{u}_{smc} = \begin{cases} -[\mathbf{CB}]^{-1}\left[\delta \text{sgn}\left(s_1\right), \delta \text{sgn}\left(s_2\right), ..., \delta \text{sgn}\left(s_n\right)\right]^T & if \quad \mathbf{s} \neq 0 \\ -[\mathbf{CB}]^{-1}\left[\mathbf{CA}\tilde{\mathbf{x}} + \mathbf{CA}\mathbf{x}_d - \mathbf{C}\dot{\mathbf{x}}_d\right] & if \quad \mathbf{s} = 0 \end{cases}. \tag{47}$$

Finally, the control signals (27) and (47) are used for (2), and the control laws of (1) have been synthesized successfully.

5 Results and Discussion

Simulations are implemented on the Matlab environment for the controller (2) where parameter matrix, nonlinear function vectors, disturbance vectors of the system (1) are defined as follows:

$$\mathbf{A} = \begin{bmatrix} -3.7376 & 0.0779 \\ 2.3515 & -4.1702 \end{bmatrix}; \mathbf{B} = \begin{bmatrix} 0.7014 & 0.5629 \\ 0.7248 & 0.4541 \end{bmatrix}; \mathbf{\Delta B} = \begin{bmatrix} 0.2104 & 0.1689 \\ 0.2174 & 0.1362 \end{bmatrix};$$
$$\mathbf{f}\left(\mathbf{x}\right) = \begin{bmatrix} 0.02\sin\left(x_1\right)\sin\left(x_2\right) \\ 0.02x_1^2 \end{bmatrix}; \mathbf{d}\left(t\right) = \begin{bmatrix} 0.2\sin\left(0.5t\right) \\ 0.2\cos\left(0.7t + 2\right) \end{bmatrix}. \tag{48}$$

With the desired signal $\mathbf{x_d} = [1.5, 1.0]^T$. The simulation results are shown in Figs. 2, 3 and 4.

The results of the identification of variable parameter components, nonlinear function vectors, and external disturbance are shown in Fig. 2. The results after using the compensation signal from the identification rule for uncertain components are presented in Fig. 3. From Figs. 2 and 3, we can see that the uncertain components are identified and compensated with an asymptotic error of zero. Figure 4 depicts responses of the system which present the result of tracking the state vectors of the system with the desired signal vector. These simulation results once again prove the correctness and effectiveness of the proposed control law.

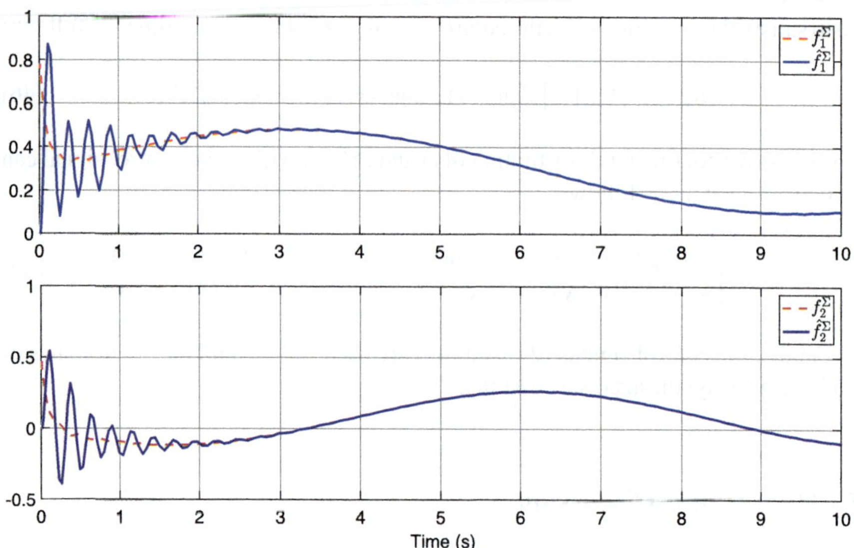

Fig. 2 The identification vectors $\hat{\mathbf{f}}_\Sigma$ (28)

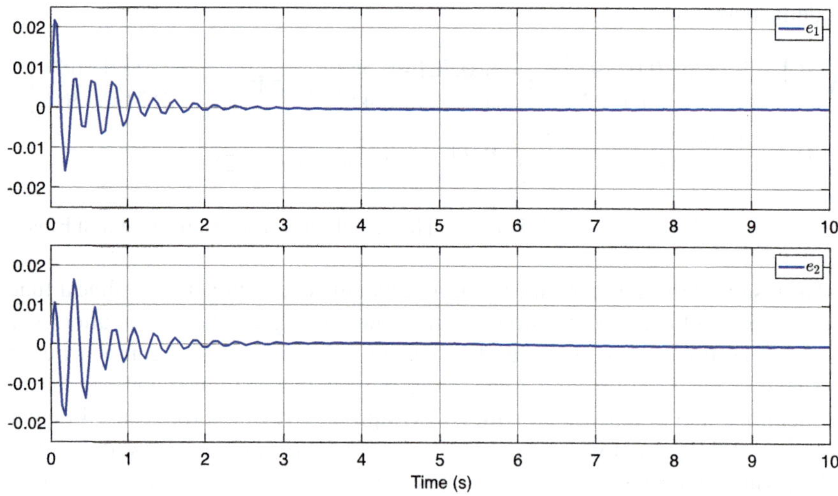

Fig. 3 The error between (1) and linear model (33) with compensation for uncertain components

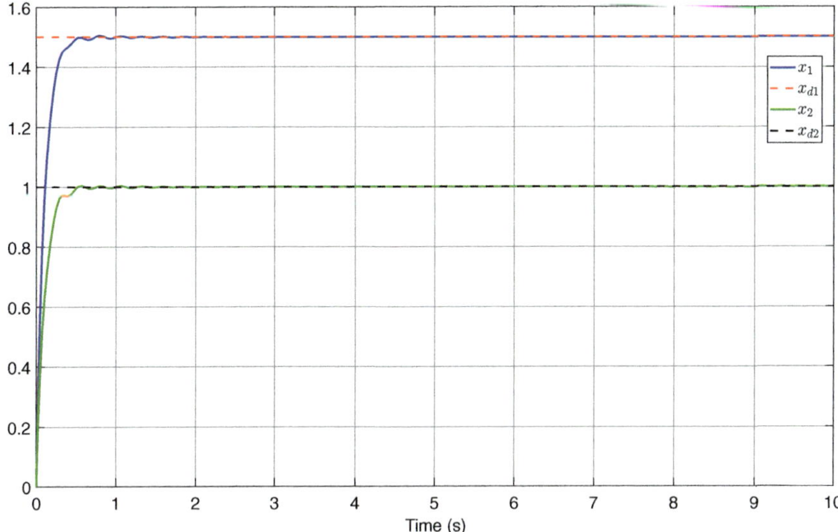

Fig. 4 Responses of the system for the desired signals $\mathbf{x}_d$

6 Conclusion

The article has synthesized the controller for a class of nonlinear objects. Lyapunov stability theory is used to design the adaptive update law which allows identifying uncertain parameter components, nonlinear function vectors, and external disturbance. The identification results are used to generate adaptive control rules that compensate for the uncertain components. And then, the linear part of the control law is synthesized based on sliding mode control. The simulation results show that the variable parameter components, nonlinear function vectors, and external disturbance are identified and compensated according to the algorithm proposed by the article; the output vector of the system tracks to the desired set signal vector with high controllability.

References

1. Huang, S.N., Tan, K.K., Lee, T.H.: A combined PID/adaptive controller for a class of nonlinear systems. Automatica **37**(4), 611–618 (2001). https://doi.org/10.1016/S0005-1098(00)00195-3
2. Arefi, M., Jahed-Motlagh, M., Karimi, H.: Adaptive neural stabilizing controller for a class of mismatched uncertain nonlinear systems by state and output feedback. IEEE Trans. Cybern. **45**(8), 1587–1596 (2015). https://doi.org/10.1109/TCYB.2014.2356414
3. Oveisi, A., Jeronimo, M., Nestorović, T.: Nonlinear observer-based recurrent wavelet neuro-controller in disturbance rejection control of flexible structures. Eng. Appl. Artif. Intell. **69**,

50–64 (2018). https://doi.org/10.1016/j.engappai.2017.12.009
4. Zhang, Y., Qin, S.: Adaptive actuator fault compensation for linear systems with matching and unmatching uncertainties. J. Process. Control. **19**(6), 985–990 (2009). https://doi.org/10.1016/j.jprocont.2008.12.008
5. Tao, G., Chen, S., Joshi, S.: An adaptive actuator failure compensation controller using output feedback. IEEE Trans. Autom. Control. **47**(3), 506–511 (2002). https://doi.org/10.1109/9.989150
6. Tang, X., Tao, G., Wang, L., Stankovic, J.: Robust and adaptive actuator failure compensation designs for a rocket fairing structural-acoustic model. IEEE Trans. Aerosp. Electron. Syst. **40**(4), 1359–1366 (2004). https://doi.org/10.1109/TAES.2004.1386887
7. Maity, A., Höcht, L., Holzapfel, F.: Time-varying parameter model reference adaptive control and its application to aircraft. Eur. J. Control. **50**, 161–175 (2019). https://doi.org/10.1016/j.ejcon.2019.04.007
8. Zhang B., Liu L., Ding K.: Optimal integral sliding mode control with feedforward compensation for nonlinear systems and its applications. In: 34th Chinese Control Conference (CCC), 2015-September, pp. 3367–3372 (2015). https://doi.org/10.1109/ChiCC.2015.7260159
9. Huang, Y., Wang, T., Wang, J., Ma, K., Zhang, C., Huang, X.: Extended fuzzy adaptive event-triggered compensation control for uncertain nonlinear systems with input hysteresis. IEEE Access **7**, 89658–8966 (2019). https://doi.org/10.1109/ACCESS.2019.2926280
10. Quang L., Putov V., Sheludko V., Kuznetsov A., Chernyshev M.: Adaptive robust control of an uncertain multi-degree-of-freedom elastically deformable electromechanical plant with adaptive compensation for an unknown disturbance. In: 10th Mediterranean Conference on Embedded Computing (MECO), pp 1–6 (2021). https://doi.org/10.1109/MECO52532.2021.9460305
11. Xu, Z., Li, L., Yao, J., Hu, X., Liu, Q., Xie, N.: State constraint control for uncertain nonlinear systems with disturbance compensation. IEEE Access **7**, 155251–155261 (2019). https://doi.org/10.1109/ACCESS.2019.2947629
12. Ortega J.: Matrix Theory. Springer US (1987). https://doi.org/10.1007/978-1-4899-0471-3
13. Utkin, V.: Sliding Modes in Control and Optimization. Springer, Berlin (1992). https://doi.org/10.1007/978-3-642-84379-2